ACS SYMPOSIUM SERIES 475

Radiation Effects on Polymers

Roger L. Clough, EDITOR
Sandia National Laboratory

Shalaby W. Shalaby, EDITOR
Clemson University

Developed from a symposium sponsored
by the Division of Polymer Chemistry, Inc.,
at the 200th National Meeting
of the American Chemical Society,
Washington, D.C.,
August 26–31, 1990

American Chemical Society, Washington, DC 1991

Library of Congress Cataloging-in-Publication Data

Radiation effects on polymers / Roger L. Clough, Shalaby W. Shalaby, editors.

p. cm.—(ACS symposium series; 475)

"Developed from a symposium sponsored by the Division of Polymer Chemistry, Inc., at the 200th National Meeting of the American Chemical Society, Washington, D.C., August 26–31, 1990."

Includes bibliographical references and index.

ISBN 0–8412–2165–0

1. Polymers—Effect of radiation on—Congresses.

I. Clough, Roger L. (Roger Lee), 1949– . II. Shalaby, Shalaby W. III. American Chemical Society. Division of Polymer Chemistry. IV. American Chemical Society. Meeting (200th: 1990: Washington, D.C.) V. Series.

QD381.9.R3R32 1991
620.1'9204228—dc20 91–31557
CIP

The paper used in this publication meets the minimum requirements of American National Standard for Information Sciences—Permanence of Paper for Printed Library Materials, ANSI Z39.48–1984. ♾

PRINTED IN THE UNITED STATES OF AMERICA

ACS Symposium Series

M. Joan Comstock, *Series Editor*

Foreword

THE ACS SYMPOSIUM SERIES was founded in 1974 to provide a medium for publishing symposia quickly in book form. The format of the Series parallels that of the continuing ADVANCES IN CHEMISTRY SERIES except that, in order to save time, the papers are not typeset, but are reproduced as they are submitted by the authors in camera-ready form. Papers are reviewed under the supervision of the editors with the assistance of the Advisory Board and are selected to maintain the integrity of the symposia. Both reviews and reports of research are acceptable, because symposia may embrace both types of presentation. However, verbatim reproductions of previously published papers are not accepted.

Contents

TECHNOLOGICAL APPLICATIONS OF RADIATION TO POLYMERS

DEGRADATION AND STABILIZATION OF IRRADIATED POLYMERS

REFERENCES AND INDEXES

Preface

INTEREST IN THE EFFECTS OF HIGH-ENERGY RADIATION on macromolecular materials is presently enjoying a strong resurgence. In recent years, significant progress has been made both in the fundamental understanding of radiation chemistry and in application-oriented work on polymer–radiation interactions. Increasing interest in this field is driven by growth in applications that can be grouped into two categories.

The first category comprises applications in which radiation is employed as a critical part of the manufacturing or processing of a material or component. Examples include the curing and cross-linking of coatings, rubbers, composites, hydrogels, and a growing number of other material types. Polymerization of monomers under specialized conditions; generation of graft copolymers; and immobilization of drugs, enzymes, and catalysts onto polymer surfaces are procedures that are also being extensively researched. Because of concerns about the environmental and health effects of ethylene oxide as a sterilant of biomedical products (such as plastic syringes and vials), interest is growing in using radiation sterilization. A related application of radiation is in the sterilization of food products. X-ray and electron-beam polymeric-resist materials are now being developed for use in the production of the next generation of microelectronic circuits. The primary motivation behind this research is to achieve smaller feature sizes, which because of refraction effects are unavailable with conventional photoresists.

The second category comprises applications in which macromolecular materials are necessarily exposed to radiation throughout their useful lifetimes. Here, reliability, lifetime prediction, and radiation resistance of materials are the critical issues. Examples include materials used in containment buildings in nuclear power plants, in large-scale high-energy physics installations, in small-scale radiation-producing equipment and radioisotope laboratories, in containment and encapsulation schemes for storage and disposal of radioactive waste, in space vehicles, and in magnets for future fusion reactors. In addition to the long-standing interest in radiation effects on mechanical properties, a growing effort to understand and control radiation-induced changes in optical properties of polymers exists today. Current research includes efforts to produce radiation-hard polymer-based scintillators, optical fibers, and radioluminescent lights.

Progress in the fundamental understanding of polymer radiation chemistry includes advances in identifying and determining the kinetics of reactive species formed during the early stages of radiation interactions with organic materials. Pulse radiolysis and electron spin resonance techniques have been quite productive in this area. Advances have been made in our understanding of the detailed chemistry underlying changes in molecular structure and bonding in irradiated polymers and also in our understanding of the basis of crystallinity changes. The nature of species involved in radiation oxidation chemistry and the mechanisms of energy transfer in macromolecular systems are now better defined. Several advanced analytical techniques, including nuclear magnetic resonance spectroscopy and detailed microprofiling methods, are now providing a clearer picture of the basis of macroscopically observed changes in irradiated materials.

The symposium on which this book is based brought together research communities studying radiation effects on polymers in a wide variety of applications, including integrated fundamental and applied studies. In addition to the chapters provided by the symposium speakers, a number of chapters were also solicited from distinguished researchers in the field to further enhance the comprehensiveness of this volume. Many chapters provide a brief review of major progress in particular aspects of the field in recent years together with current research results. As such, this book should provide a valuable reference for scientists who are concerned with polymer radiation effects.

Many of the chapters provide an important, fundamental understanding of radiation-induced changes in macromolecules and relate that understanding to a major application or category of applications. Furthermore, the chemical basis of radiation effects is nearly always the same, so a given study may provide insights that are useful for many different applications. Thus, a number of the chapters could arguably have been placed equally well in the fundamental studies section or in one of the later sections.

Given the broad coverage of the field that is provided in this text, including progress in most of the major areas of polymer radiation chemistry, and to make this book useful to the reader as a general tool, we have provided a reference list of earlier work in polymer radiation effects. This list may be found on pages 610–611.

We particularly thank the authors represented in this book for the many valuable manuscripts we received. Thanks are also owed to colleagues who provided careful reviews of these manuscripts. We are grateful to those who served as session chairpersons at the symposium, including Jim O'Donnell of the University of Queensland; A. Charlesby, editor of the Journal *Radiation Physics and Chemistry*; D. Wiles and D. Carlsson of NRC Canada; Ken Gillen of Sandia National Laboratories; Elsa Reichmanis of AT&T Bell Laboratories; and (the late) Malcolm Dole. We are also indebted to Maureen Rouhi and the production staff of the Books Department of the American Chemical Society.

ROGER L. CLOUGH
Sandia National Laboratories
Albuquerque, NM 87185

SHALABY W. SHALABY
Clemson University
Clemson, SC 29634

August 1, 1991

Fundamental Studies of Polymer Radiation Effects

THE CHANGES INDUCED IN MACROMOLECULAR MATERIALS under the influence of ionizing radiation are complex. This section provides an up-to-date description of our understanding of radiation-induced phenomena in organic materials and the techniques used to study these effects.

The early events in the interaction of radiation with organic matter are particularly complicated and involve a wide variety of short-lived, reactive species. Pulse radiolysis techniques have been very useful in sorting these complicated events, and several chapters describe the understanding gained with this technique. Electron spin resonance (ESR) studies have provided an understanding of the nature and fate of radical species created in solid polymers; several chapters provide important ESR results. Two chapters discuss the influence of differences in linear energy transfer, resulting from different types of radiation involving charged particles, on polymer radiation chemistry. Two other chapters focus on radiation-induced changes in crystallinity in polymers and the impact of these changes on material properties. The final two chapters in this section describe the application of nuclear magnetic resonance spectroscopy to analyzing the detailed nature of molecular structural changes in irradiated polymers, including changes in macromolecular stereochemistry and the formation of cross-links.

Chapter 1

Pulse Radiolysis Studies of Polymers

Seiichi Tagawa

Research Center for Nuclear Science and Technology, University of Tokyo, 2–22 Shirakata-Shirane, Tokai-mura, Ibaraki, 319–11 Japan

Recent or picosecond (PS) and nanosecond (NS) electron beam (EB), ion beam (IB), and synchrotron radiation (SR) pulse radiolysis studies on polymers are reviewed. Transient absorption and emission spectra of short-lived reactive intermediates such as electrons, radical anions and cations, excited states and neutral radicals have been observed in irradiated solid polymers and polymers solutions, such as polyethylene model compounds and ethylene-propylene copolymers, polystyrene and related compounds including resists, and sigma-conjugated polymers such as polysilanes and polygermanes. Very recent progress in our picosecond EB pulse radiolysis systems, especially new LL (linac-laser) twin system, is also shown. IB and SR pulse radiolysis systems (1 ns and subnanosecond time resolution for emission spectroscopy, respectively) are developed very recently and applied to polymers. The history of pulse radiolysis studies on polymers is also reviewed shortly.

A large number of papers on radiation effects on polymers for various fields have been published over the last 40 years [1,2,3,4]. However, the complexity of reactions of reactive intermediates is so large that studies on the detailed reaction mechanisms are very difficult. In principle, various kinds of ions and excited states are initially formed by ionization and excitation of polymers. Beginning with these species, further new reactive intermediates such as ions and radicals are often formed. Final products are produced through a lot of reactions of various reactive intermediates. Direct measurements of short-lived reactive intermediates are essential to make clear the detailed reaction mechansims of irradiated polymers. A pulse radiolysis technique, which can detect short-lived reactive intermediates directly, is very powerful method to study detailed reaction mechanisms of reactive intermediates. Although several research groups published papers on pulse radiolysis studies of polymer solutions, few groups studied pulse radiolysis of solid polymers and no clear transient absorption due to polymer ions and excited states was observed in irradiated solid polymers before 1983 [5-9]. Very recently clear transient

0097–6156/91/0475–0002$08.25/0

absorption spectra of polymer ions and excited states have been observed in irradiated solid polymers [10-13]. New types of pulse radiolysis techniques such as ion beam [14-16] and synchrotron radiation [17,18] pulse radiolysis have also been applied to polymers, especially to LET effect studies on polymers.

The present paper mainly describes recent progress in electron beam, ion beam and synchrotron radiation pulse radiolysis studies on polymers.

2. OUR PULSE RADIOLYSIS SYSTEMS AND THEIR TYPICAL APPLICATIONS TO POLYMERS

2.1 Electron Beam Pulse Radiolysis

A 35 MeV S-band (2856MHZ) electron linear accelerator (linac) with capability of producing a picosecond single electron pulse (pulse width; 10 ps) has been operates since 1977 [19].

The picosecond pulse radiolysis with optical emission spectroscopy [20] has been applied to polymers in solutions [21]. The time resolution is about 10 ps for the emission spectroscopy.

Picosecond and nanosecond electron beam pulse radiolysis with optical absorption spectroscopy has been applied to polymers such as resists [10,22], polystyrene and related polymers [12,23,24], ethylene-propylenediene terpolymer [25], ethylene-proyrene copolymers [13], epoxy resin [11], polysilanes [26-30] and polygermanes [28].

The standard absorption spectroscopy system is composed of a Xe light source, a monochromator, a fast response photodiode, a transient digitizer or sampling oscilloscope with a fast pulse amplifier and a computer system [32]. Fig. 1 shows the block diagram of the system and the formation process of polymethylpropylsilane radical anion through the reaction of polysilane with the solvated electron in tetrahydrofuran (THF) monitored by the system. The time-resolution is about 700 ps for the transient digitizer system and 60 ps for the sampling oscilloscope system. An absorption spectroscopy system with a time response of about 50 ps has also developed by using a streak camera with a gate option [33].

Recently a new absorption spectroscopy system which has a very high time resolution of 20 ps, has been developed by using two parallel linacs [34]. The system, which is called "twin linac picosecond pulse radiolysis system," has been applied to polyethylene model compounds [13] and polysilanes [27]. Fig. 2 shows the layout of that twin linac system [34]. Fig. 3 shows the formation process of the polymethylpropylsilane through the reaction of polysilane with the solvated electron in THF monitored by that twin linac system [27].

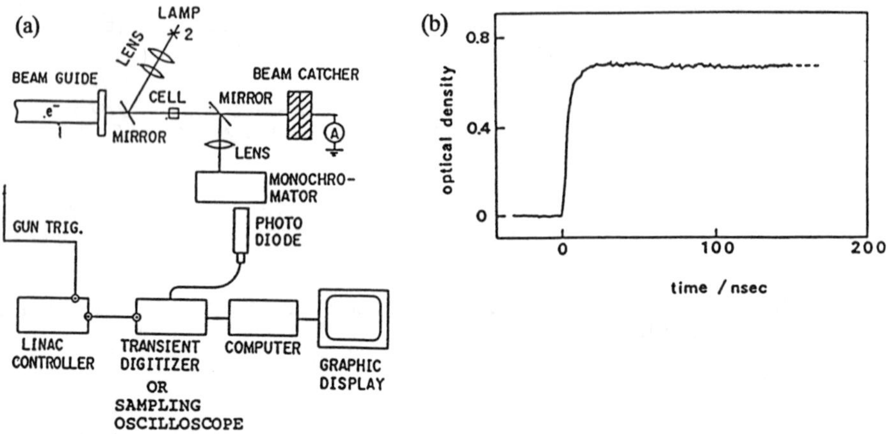

Fig. 1. a, The block diagram of the pulse radiolysis system with absorption spectroscopy. (Reproduced with permission from reference 32. Copyright 1983 Pergamon.) b, Optical density change of 93 mM polymethylpropylsilane solution in tetrahydrofuran after the pulse, which was monitored at 365 nm. (Reproduced from reference 26. Copyright 1987 American Chemical Society.)

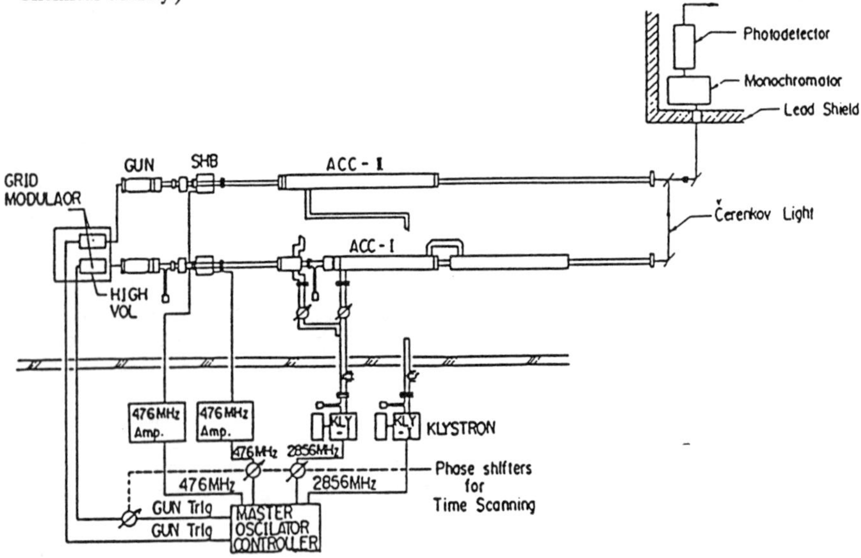

Fig. 2. Layout of the new twin linac system. (Reproduced with permission from reference 34. Copyright 1986 Elsevier Sequoia.)

2.2 New picosecond Pulse Radiolysis (LL Twin System)

Very recently LL (Linac-Laser) twin picosecond pulse radiolysis system has been developed [35.36]. Conventional twin linac system uses Cerenkov light as a monitor light, which is not intense enough in wavelength region higher than 500 nm. But many important reactive intermediates have absorption at higher than 500 nm. New LL twin picosecond pulse radiolysis system uses pulsed laser diode synchronized

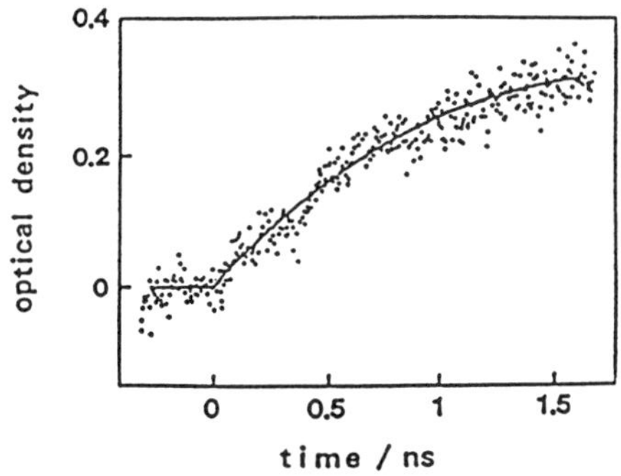

Fig. 3. Primary process of the formation of polymethylpropylsilane radical anion observed with the twin linac system. (Reproduced from reference 27. Copyright 1988 American Chemical Society.)

with linac microwave, which is intense in near infra-red and infra-red region. Both conventional twin linac system and new LL twin system are the stroboscopic type methods. The principle of both systems is shown as shown in Fig. 4. Both systems use 10 picosecond electron pulses from the linac as irradiation sources. As analyzing light, the twin linac system and the LL twin system use Cerenkov light and laser pulses, respectively. The system is shown in Fig. 5. The setup of new LL twin system is much easier than the conventional twin linac system. Fig. 6 shows the decay of the radical cation of n-dodecane as the model compound of polyethylene in irradiated neat dodecane monitored by the new LL twin system.

2.3 Ion Beam Pulse Radiolysis

Our ion beam pulse radiolysis system, which is mainly composed of about 1 ns ion beam pulses and optical detection system, has been applied to polymer systems [14-16]. Ion beams of H^+, He^+ and N^+ are most often used. The system is shown in Fig. 7. The beams were obtained from Van de Graaff ion accelerator (the terminal voltage is from 0.4 through 3.75 MV). The emission produced by the pulsed ion beam impact is detected through a monochromater by two optical detection systems: a fast photomultiplier tube operated in a counting mode and a optical multi-channel analyzer with 10 ns gate pulse system. The time profile of the emission is obtained by a coincidence measurement between a photon and a pulsed beam, and finally accumulated in a pulse height analyzer controlled by a personal computer. The time resolution of our ion beam pulse radiolysis system is about 1 ns. This value was mainly dominated by the pulse width of the ion beam. Figs. 8 and 9 shows the emission spectra and the time profile of emission observed in thin films of polystyrene spin coated on silicon wafer and irradiated by ion beam pulses.

2.4 Picosecond Synchrotron Radiation (SR) Pulse Radiolysis

The first picosecond SR pulse radiolysis experiments using one of the Accumulation Ring (AR) beam lines of the TRISTAN

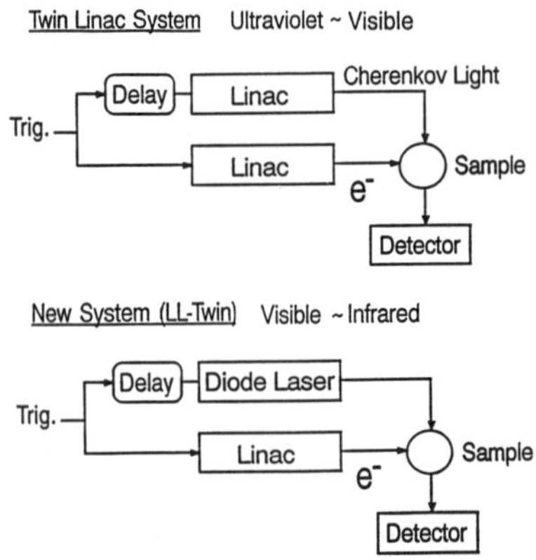

Figure 4. Principle of the twin linac and LL (linac–laser) systems.

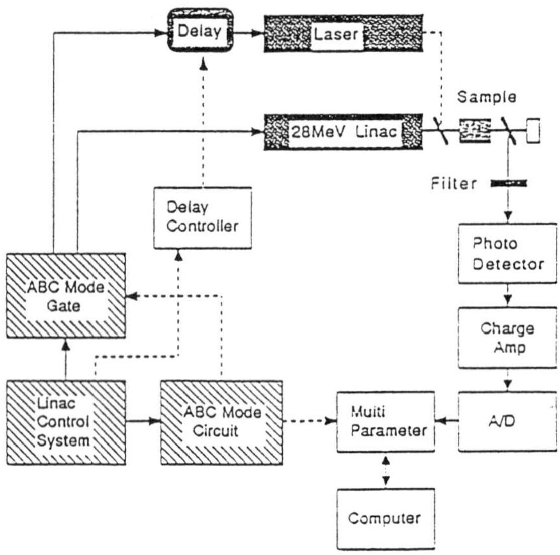

Figure 5. Block diagram of new LL–twin systems.

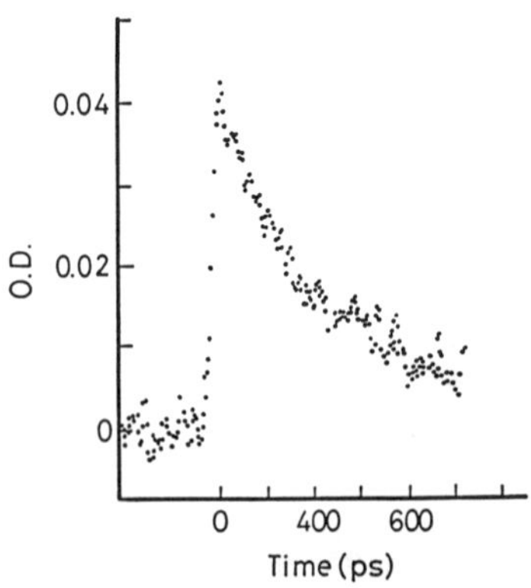

Figure 6. The time profile of the radical cation of n-dodecane monitored at 806 nm by using the LL twin system.

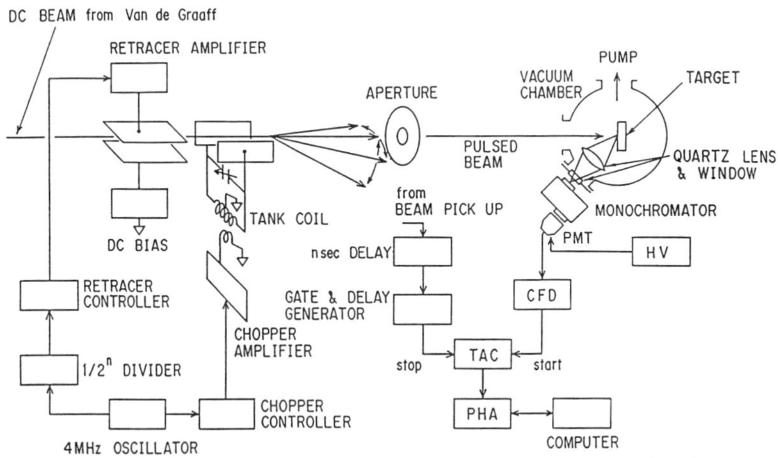

Fig. 7. The schematic diagram of the pulsing system and the ion beam radiolysis system with an optical emission spectroscopy. PMT denotes photomultiplier tube; HV, high-voltage power supply; CFD, constant fraction discriminator; TAC, time-to-amplitude converter; and PHA, pulse-height analyzer. (Reproduced with permission from reference 14. Copyright 1989 Pergamon.)

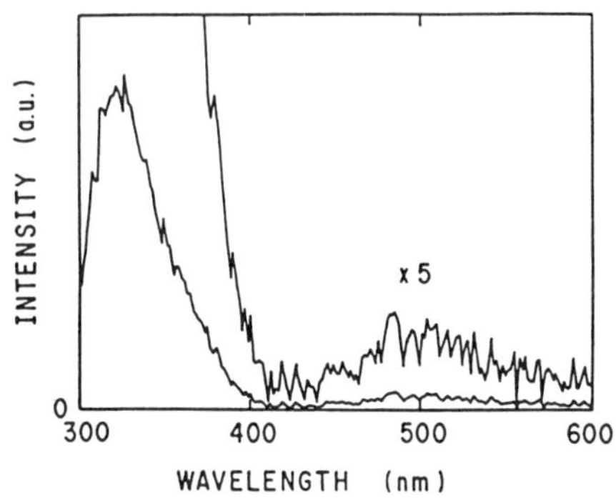

Fig. 8. The emission spectrum from a polystyrene resist film (0.5 μmm thickness) excited by 2.5 MeV He^{+}. The vertical axis was not calibrated for the efficiency of the optical system and the detector. (Reproduced with permission from reference 15. Copyright 1989 Pergamon.)

accelerators (a specific accelerator complex) of National Laboratory for High Energy Physics (KEK) is now starting [17,18]. The TRISTAN consist of the AR and the main ring (MR). The AR is primarily a beam accumulator and every booster for the MR, a positron-electron collider. SR of the AR has been utilized since 1987, but that of the MR is now only planned.

The picosecond SR pulse radiolysis system at TRISTAN has several excellent characteristics for time resolved measurements of short-lived reactive intermediates in radiation and photoinduced reactions such as high time resolution, the long time interval between neighboring bunches and the wide energy range. The pulse width of SR pulse from TRISTAN is very short. The SR pulse from the AR

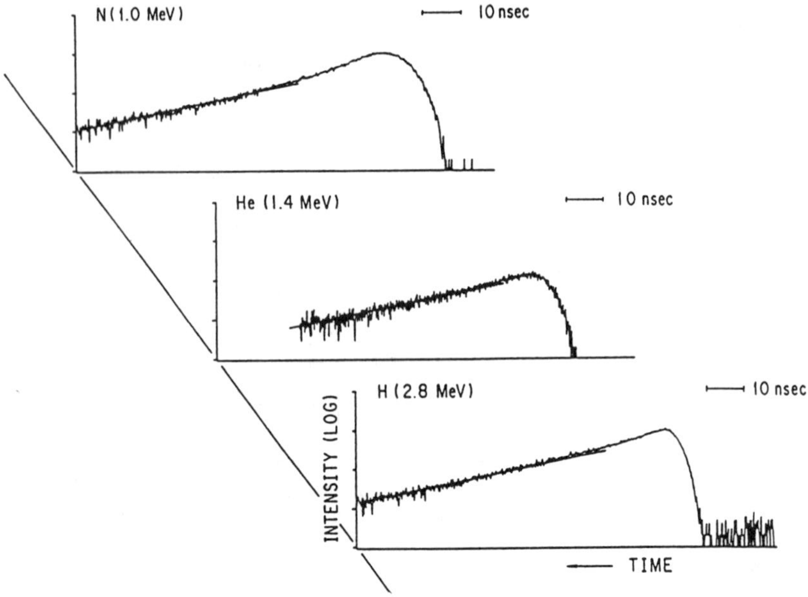

Fig. 9. Time profiles of the excimer fluorescence from polystyrene resist films (0.5 μmm thickness) excited by ion beam impacts. These time profiles were not influenced by the quenching seen in Fig. 6. The straight lines correspond to the fluorescence lifetime obtained by the electron pulse radiolysis study of polystyrene (Itagaki et al., 1983; Tagawa, 1986). Reproduced with permission from reference 15. Copyright 1989 Pergamon.)

is now about 60 ps, which could be very short. The SR pulse from the AR is now about 60 ps, which could be below 10 ps by using the MR of TRISTAN. The time intervals between neighboring bunches are very long, compared with other SR facilities: 1.25 us at the AR and 5 us at the MR. The long intervals between neighboring bunches can afford powerful means to detect the accurate formation and decay times of reactive intermediates and to analyze the complicated reactions of reactive intermediates. The energy range of SR from TRISTAN is very wide from VUV to MeV region as shown in Fig. 10. The SR pulse radiolysis experiments have carried out from 1988 and have also been applied to solid polymers such as plastic scintillators and polystyrene films as shown in Figs. 11 and 12.

3. RECENT PROGRESS IN PICOSECOND AND NANOSECOND PULSE RADIOLYSIS STUDIES ON POLYMERS.

3.1 Polystyrene and Related Polymers

Recently electron beam pulse radiolysis studies on polystyrene films and related polymers have been done. Very recently both ion beam and synchrotron radiation pulse radiolysis studies on polystyrene films have been carried out. Recently both absorption and emission spectra of main reactive intermediates such as sandwich type dimer radical cations, excimer type excited singlet states, triplet states

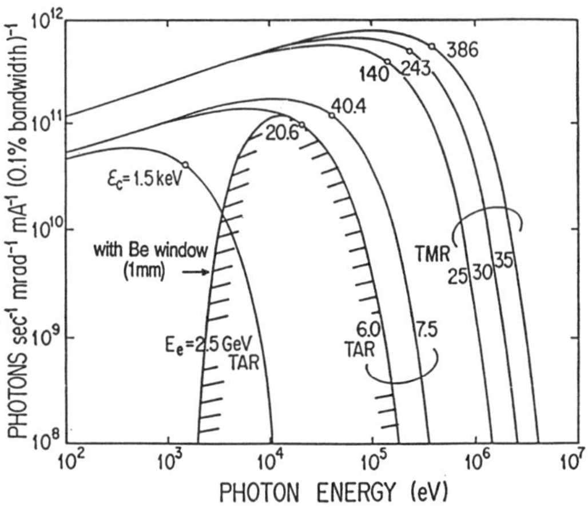

Fig.10. Spectra and critical energies of radiation from the TRISTAN accelerators at various beam energies.

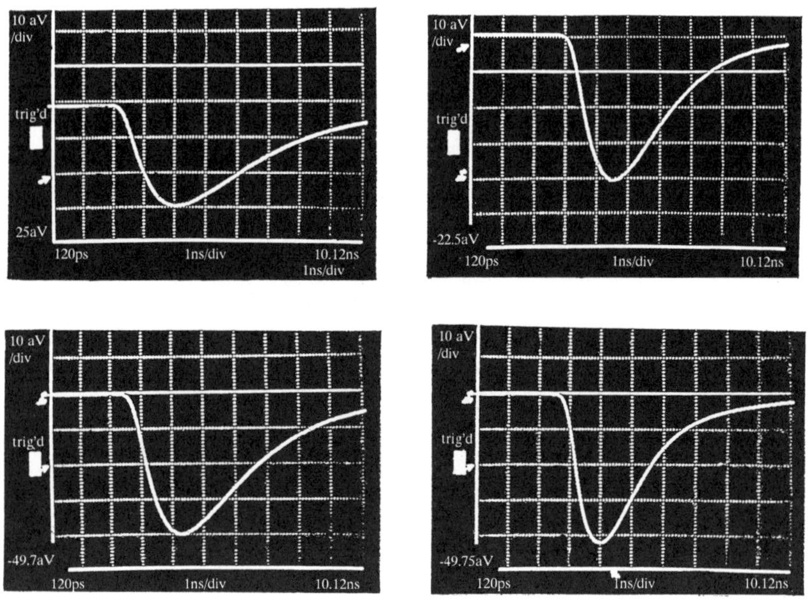

Fig.11. Time profiles of plastic scintillators observed by SR pulse radiolysis (1ns/div).

and anions of polystyrene have been observed very clearly for irradiated polymer films [10,12] as shown in Figs. 13 and 14, although pioneering work of pulse radiolysis studies on solid polystyrene showed energy and charge transfer processes from polymer matrix to solute molecules and did not detect clearly polymer ions and excited states [6-8]. The polystyrene base resists such as partially

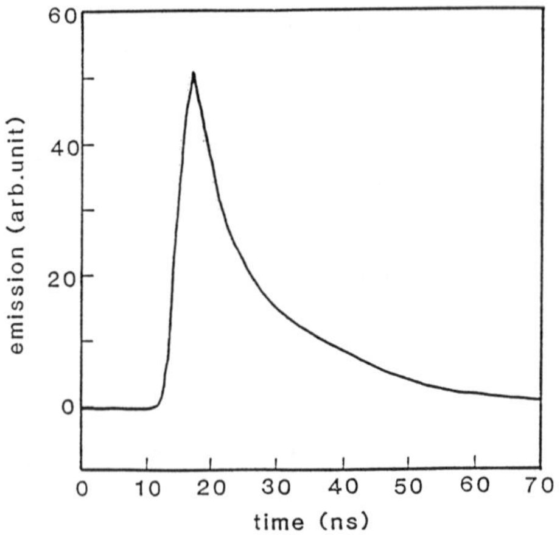

Fig.12. Time profile of excimer fluorescence of polystyrene film observed by SR pulse radiolysis.

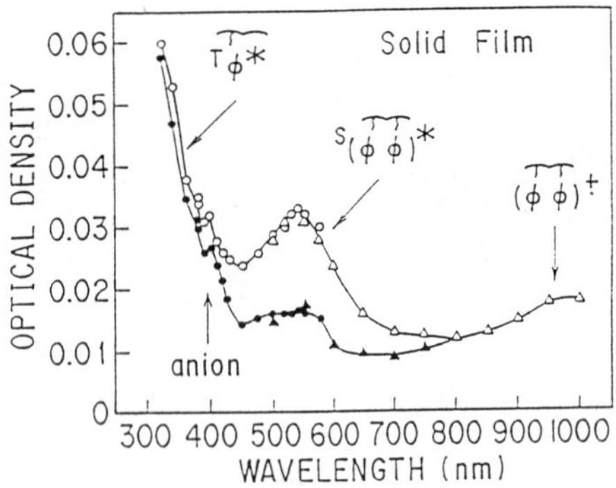

Fig.13. Transient absorption spectra observed in the pulse radiolysis of polystyrene solid films at 10 ns (○,△) and 30 ns (●,▲) after 2 ns electron pulses. (Reproduced with permission from reference 12. Copyright 1986 Pergamon.)

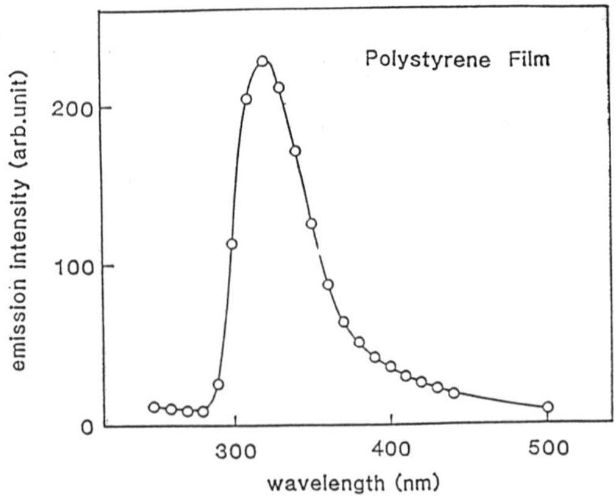

Fig.14. Emission spectrum at 1.5 ns after e−irradiation.

chloromethylated polystyrene (CMS) have been studied by pulse radiolysis [10,12]. Figs. 15, 16 show transient absorption spectra observed in pulse radiolysis of CMS films and CMS in benzene. Figs. 17 and 18 show transient absorption spectra observed for irradiated partially chloromethylated polymethylstyryne in benzene and THF, which is sensitive electron beam resist with high dry etching durability.

Pulse radiolysis studies on similar types of silicon containing resists such as partly chloromethylated polydiphenylsiloxane (SNR) [22,29], chloromethylated polysilanes [26.29], and copolymers have been done. Figs. 19 and 20 show transient absorption spectra observed for irradiated polydiphenylsiloxane (base polymer for SNR) in benzene and $CHCl_3$. Figs. 21 and 22 show transient absorption spectra observed for irradiated SNR in Benzene and $CHCl_3$. Fig. 23 shows transient absorption spectra observed for irradiated polysilane base polymer and chloromethylated polysilanes in THF. Figs. 24 and 25 show transient absorption spectra observed for irradiated trimethylsilylstyrene and trimethylsilylstyrene-chloromethylated styrene copolymer in THF.

3.2 Saturated Hydrocarbon Polymers

Very recently, absorption due to main reactive intermediates such as polymer ions, excited states and alkyl radicals have been observed clearly in pulse radiolysis of saturated hydrocarbon polymers and their model compounds [13], although pioneering work [5] and the following work [9] did not detect polymer ions, excited states, and alkyl radicals.

The broad absorption bands observed in visible and near infrared region on subnanosecond and nanosecond time scale at room temperature for irradiated ethylene-propylene copolymer (EPR) are mainly due to EPR excited states and partly due to tail parts of the absorption bands of both EPR

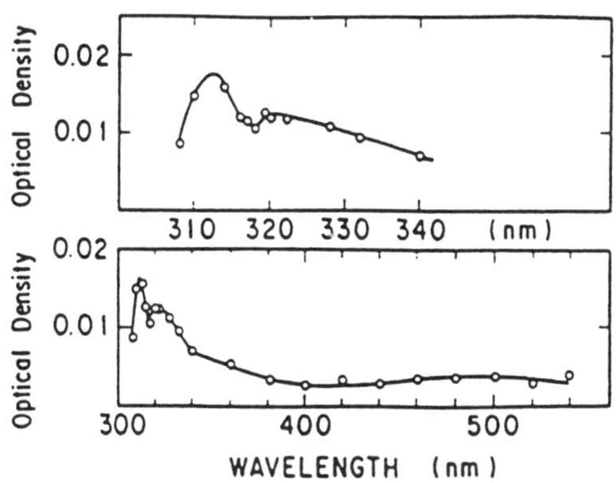

Fig. 15. Transient absorption spectra observed in the pulse radiolysis of CMS solid films. (Reproduced with permission from reference 12. Copyright 1986 Pergamon.)

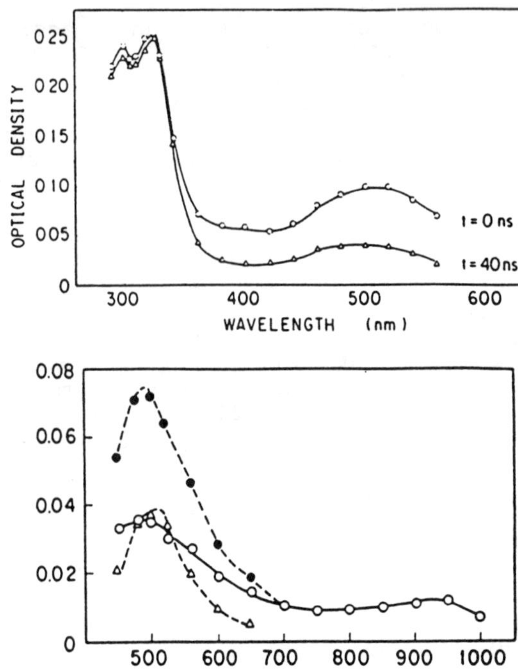

Fig.16. Transient absorption spectra observed in the pulse radiolysis of 200 mM (A) CMS solution in benzene immediately (○) and at 40 ns (△) after 2 ns pulses and 80 mM (B) CMS solutions in benzene immediately (●) and 15 ns (○) after 2 ns pulses, and the difference between these two points (△). (Reproduced with permission from reference 12. Copyright 1986 Pergamon.)

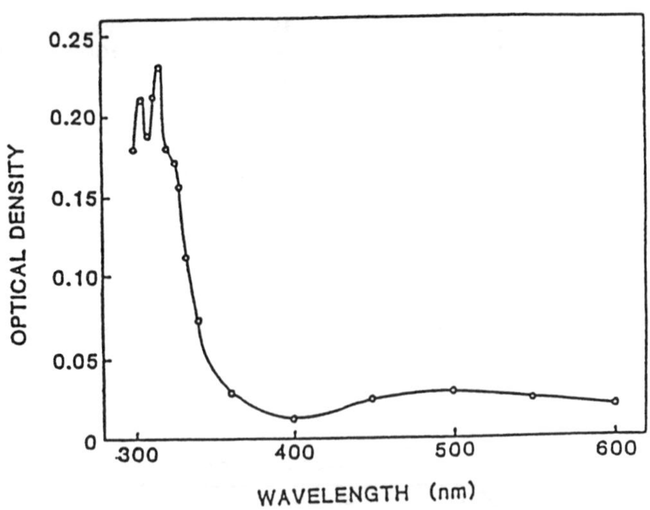

Fig. 17. Transient absorption observed in pulse radiolysis of partially chloromethylated polymethylstyrene in benzene. (Reproduced with permission from reference 23. Copyright 1985 Pergamon.)

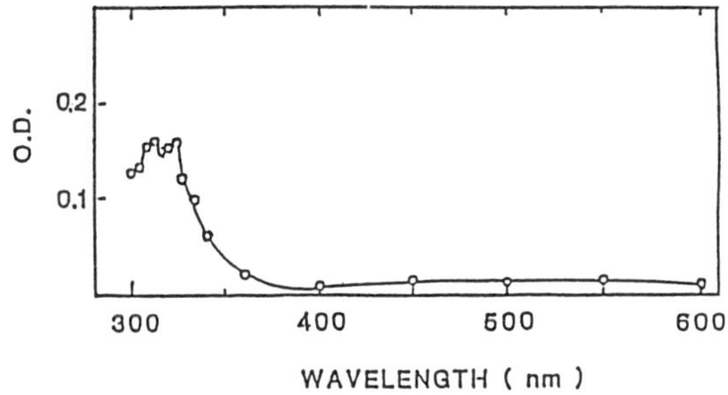

Fig. 18. Transient absorption observed in pulse radiolysis of partially chloromethylated polymethylstyrene in tetrahydrofuran. (Reproduced with permission from reference 23. Copyright 1985 Pergamon.)

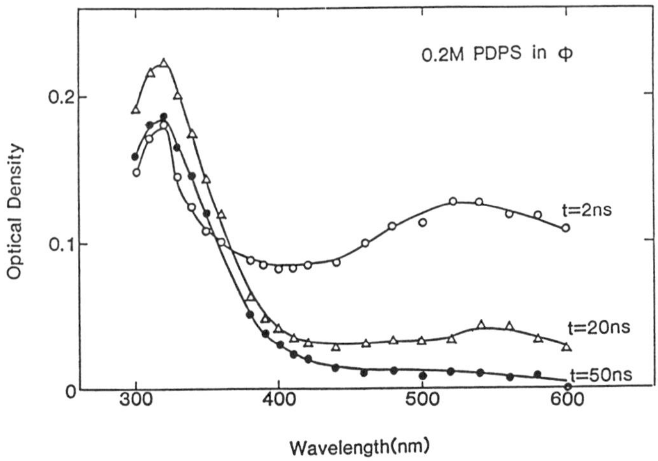

Fig.19. Transient absorption spectra observed in pulse radiolysis of polydiphenylsiloxane in benzene.

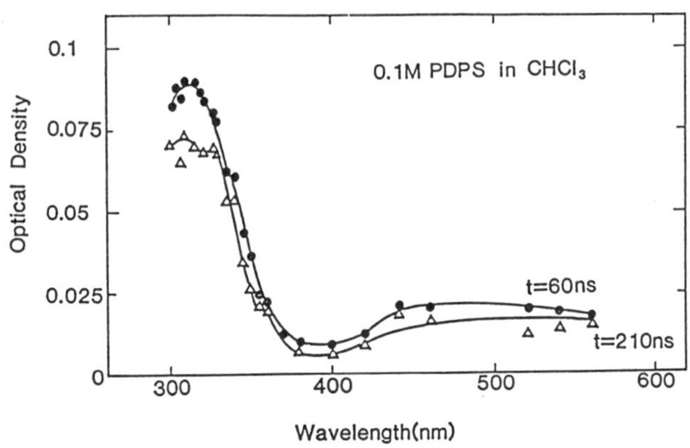

Fig.20. Transient absorption spectra observed in pulse radiolysis of polydiphenylsiloxane in $CHCl_3$.

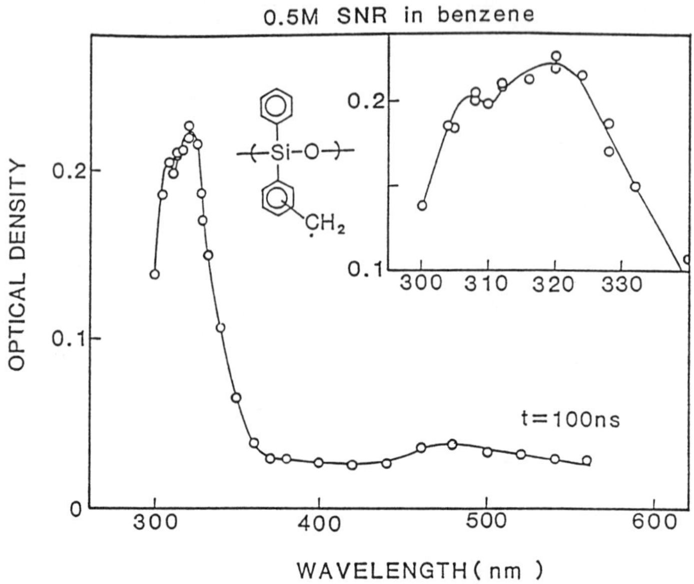

Fig.21. Transient absorption spectra observed in pulse radiolysis of SNR in benzene.

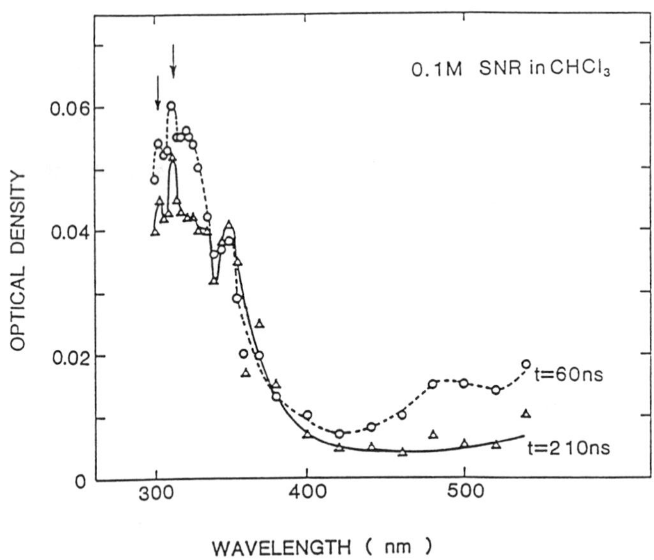

Fig.22. Transient absorption spectra observed in pulse radiolysis of SNR in $CHCl_3$.

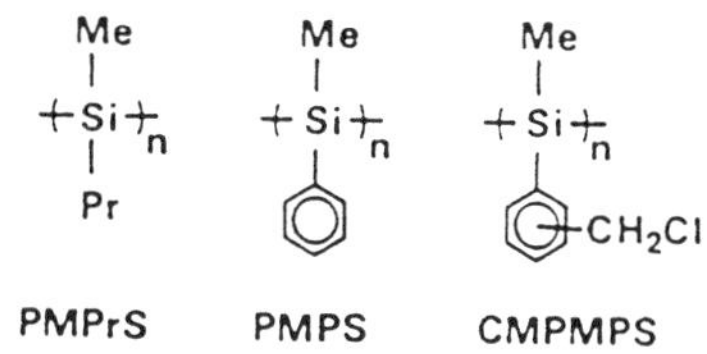

Molecular structures of poly(methylpropylsilane) (PMPrS), poly(methylphenylsilane) (PMPS), and chloromethylated PMPS (CMPMPS).

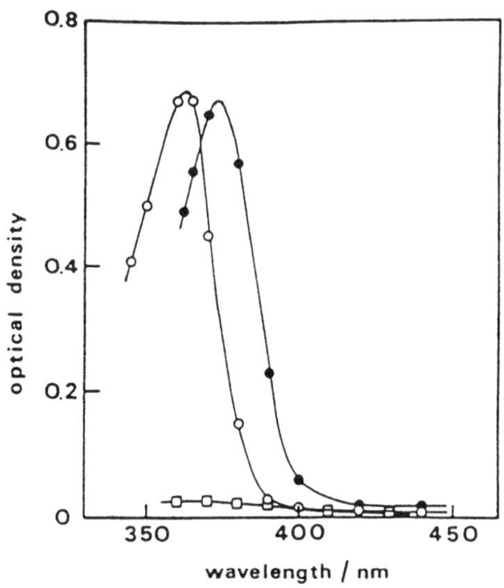

Fig.23. Transient UV absorption spectra of PMPrS, PMPS, and CMPMPS in THF solutions at 40 ns after an electron pulse: (○) 93 mM PMPrS, (●) 123 mM PMPS, (□) 2 mM CMPMPS. (Reproduced from reference 26. Copyright 1987 American Chemical Society.)

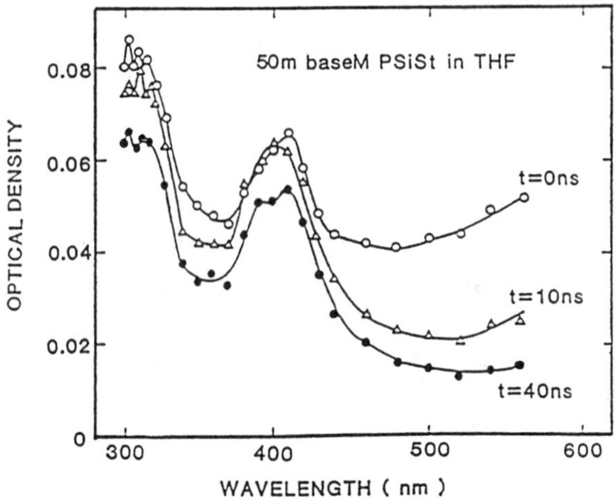

Fig.24. Transient absorption spectra observed in pulse radiolysis of trimethylsilystyrene in tetrahydrofuran.

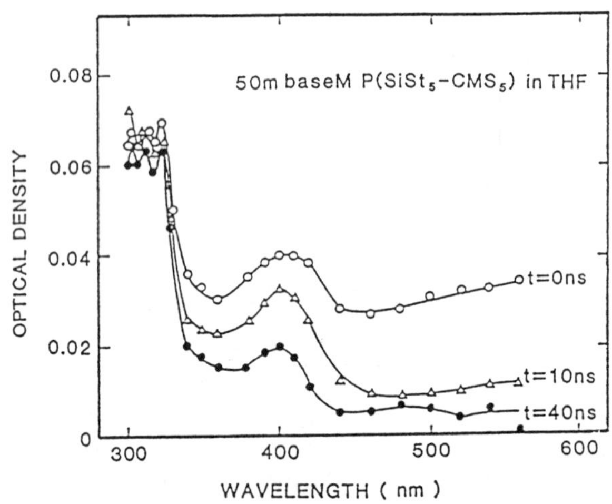

Fig.25. Transient absorption spectra observed in pulse radiolysis of trimethylsilylstyrene–chloromethylated styrene copolymer in tetrahydrofuran.

radical cations and trapped electrons as shown in Fig. 26. The radical cation of EPR were clearly observed in irradiated EPR containing CC14 as shown in Fig. 27. The lifetime of the radical cation was very short probably due to the reaction of the radical cation such as proton transfer reaction even if in solid state alkane polymers. The wavelengths of absorption peaks of alkane radical cations and excited states become longer with increasing the number of carbon atoms of low molecular n-alkanes as shown in Fig. 28. Absorption spectra of alkyl radicals were observed for irradiated linear, cyclic, and branched alkanes [13]. The UV absorption due to alkyl radicals was not confirmed in irradiated EPR lack of transparency of polymer films.

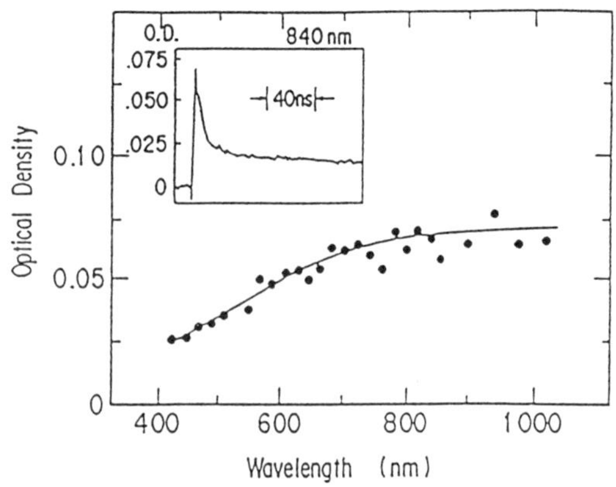

Fig. 26. Transient optical absorption spectrum observed at room temperature in pulse radiolysis of ethylene–propylene copolymer (EPR) immediately after a 2 ns electron pulse irradiation. Insert: the time profile of the absorption monitored at 840 nm. (Reproduced with permission from reference 13. Copyright 1989 Pergamon.)

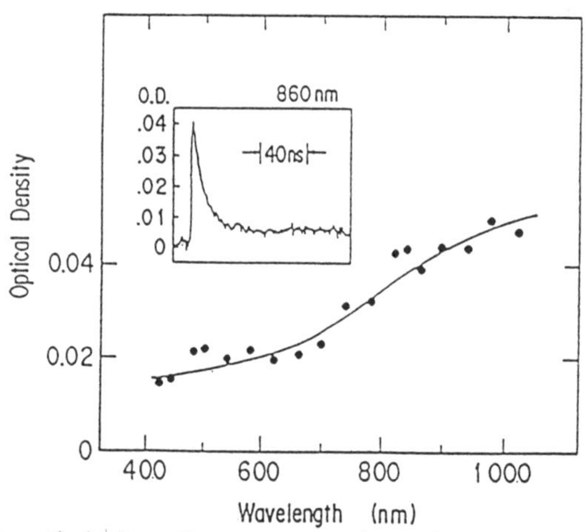

Fig. 27. Transient optical absorption spectrum observed at room temperature in pulse radiolysis of ethylene–propylene copolymer (EPR9) containing CCl_4 immediately after a 2 ns pulse. Insert: the time profile of the absorption monitored at 860 nm. (Reproduced with permission from reference 13. Copyright 1989 Pergamon.)

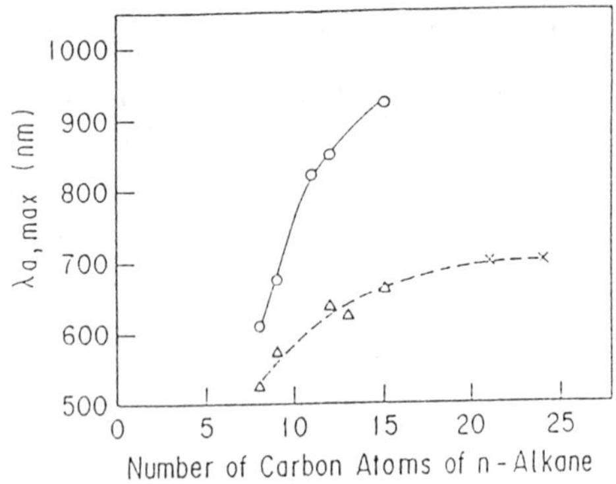

Fig.28. The wavelengths of absorption peaks of alkane radical cations (○) and alkane excited states (△,x) observed in pulse radiolysis of liquid n–alkanes. (○,△) and (x) were measured at about 18 and 78 C, respectively. **(Reproduced with permission from reference 13. Copyright 1989 Pergamon.)**

3.3 Polysilanes and Polygermanes

Very intense transient absorption due to radical anions [26] and cations [30] of polysilane high polymers was reported in 1987. Then further studies on polysilane radical anions [27] and polysilane radical cations [29,31] have been reported. Figs. 23 and 29 show transient absorption of radical anions of polysilanes. Fig. 30 shows transient absorption spectrum of the radical cation of polymethylphenylsilane [29]. The absorption spectra of polysilane radical anions and cations are very similar as shown in Fig. 31 (pairing properties). Polysilane radical anions and cations have observed in irradiated solid polymethylpropylsilane and cations have observed in irradiated solid polymethylpropylsilane shown in Figs. 32.

Polygermane radical anions [28] and cations have also been observed as shown in Fig. 33.

4. SHORT REVIEW OF PULSE RADIOLYSIS ON POLYMERS

Pulse radiolysis studies on polymers in liquid solutions and solid (molten) polymers are summarized in Table 1 and 2, respectively. The kinds of pulse radiolysis methods applied to polymers are also summarized in Table 3.

5. CONCLUSION

Pulse radiolysis is very powerful to detect reactive intermediates in irradiated polymer systems. Electron beam pulse radiolysis studies on polymer solutions and polymer films have been carried out very successfully. Ion beam and synchrotron radiation pulse radiolysis studies on polymers have been done very recently and will be very important to make clear detailed mechanisms of ion beam and x-ray induced characteristic reactions of polymers.

Molecular Weights and Compositions of Polysilanes

polymer	composition[a]	$M_w \times 10^{-3}$	M_w/M_n	R^b
PPnMS	$(MePhenSi)_n$	702	1.9	
DMS-MHxS	$(Me_2Si)_{0.4}(Me\text{-}c\text{-}HexSi)_{0.6}$	665, 13	2.5, 2.3	0.62
DMS-MPrS	$(Me_2Si)_{0.3}(MePrSi)_{0.7}$	985, 11	2.2, 1.8	0.19
DMS-PnMS	$(Me_2Si)_{0.4}(MePhenSi)_{0.6}$	279, 10	1.7, 1.9	0.27
DMS-MPS	$(Me_2Si)_{0.4}(MePhSi)_{0.6}$	1080, 8	2.9, 2.1	0.77
DPS-DMS	$(Ph_2Si)_{0.5}(Me_2Si)_{0.5}$	471, 3	1.9, 1.8	0.19
DPS-MPrS	$(Ph_2Si)_{0.5}(MePrSi)_{0.5}$	1010, 4	3.2, 2.1	0.71
DPS-PnMS	$(Ph_2Si)_{0.5}(MePhenSi)_{0.5}$	719	4.2	

[a]Phen = phenethyl; Me = methyl; c-Hex = cyclohexyl; Pr = propyl; Ph = phenyl. [b]R is the ratio o molecular weight fraction of the bimodal distribution to the low molecular weight fraction from elution profile.

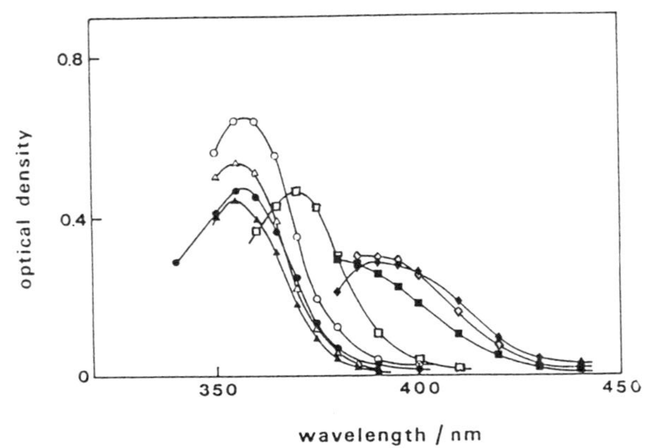

Fig.29. Transient UV absorption spectra of organopolysilanes at 40 ns after an electron pulse; cell length = 1 cm: (○) PPnMS; (●) DMS–MHxS;(△) DMS–MPrS; (▲) DMS–PnMS; (□) DMS–MPS; (■) DPS=DMS; (◇) DPS–MPrS; (◆) DPS–PnMS. (Reproduced from reference 27. Copyright 1988 American Chemical Society.)

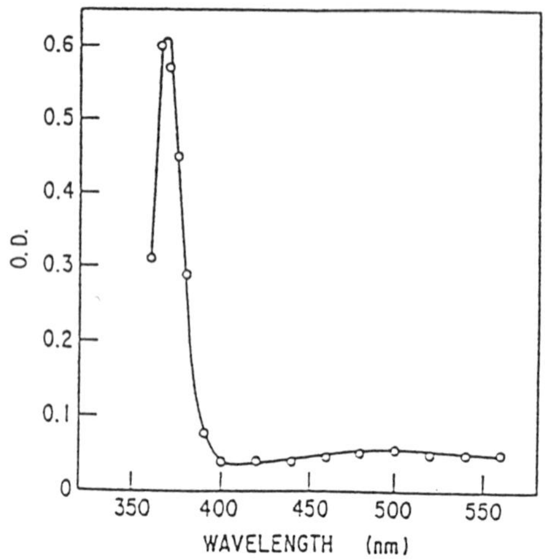

Fig. 30. Transient absorption spectrum of 100 mM polymethylphenylsilane (PMPS) solutions in CH_2Cl_2. (Reproduced with permission from reference 29. Copyright 1988 Fotoporima Konwakaishi Henshuiinkai.)

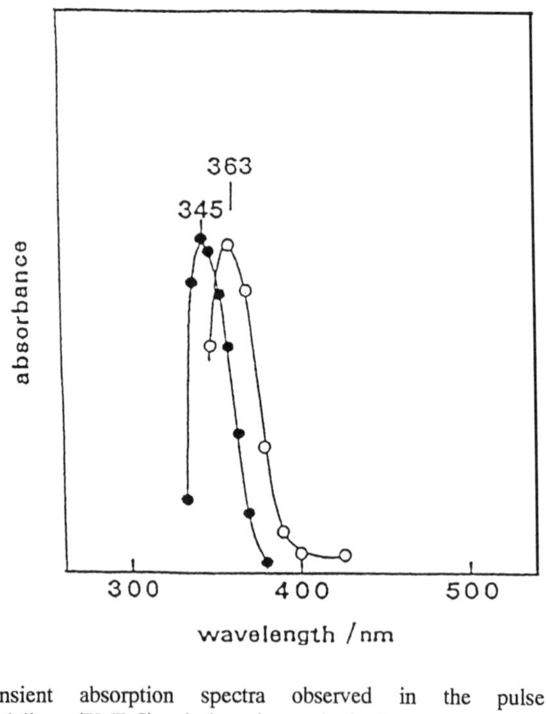

Fig.31. Transient absorption spectra observed in the pulse radiolysis of polymethylpropylsilane (PMPrS) solutions in tetrahydrofuran (the radical anion of PMPrS) (o) and in methylene chloride (the radical cation of PMPrS) (●) immediately after 2 ns electron pulse.

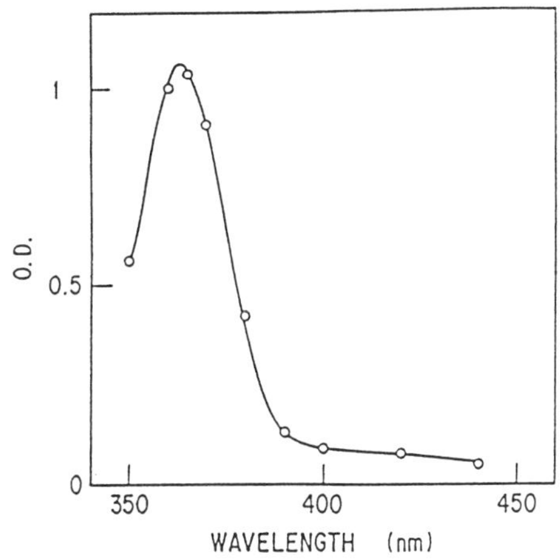

Fig.32. Transient absorption spectrum observed in pulse radiolysis of molten polymethylpropylsilane at 67 C.

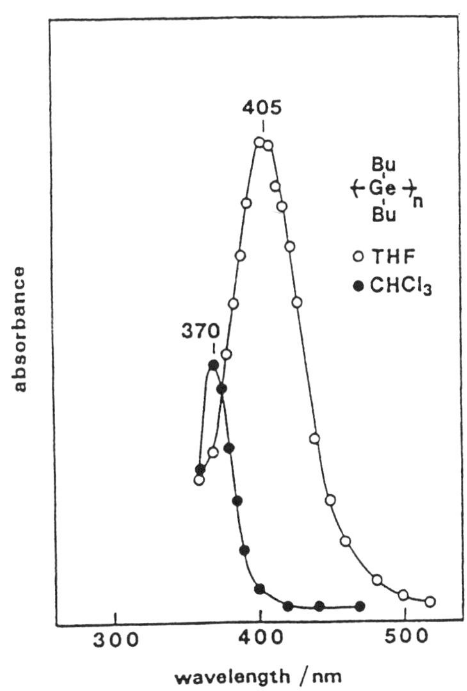

Fig.33. Transient absorption spectra of radical anions and cations observed in pulse radiolysis of polygermane in tetrahydrofuran and $CHCl_3$, respectively.

Table 1. Pulse Radiolysis Studies on Polymers in Liquid Solutions

Polymers	Transient Species	Methods	References
Polystyrene	P(E)*, CT Complex P(D)+, P(T)+	OA	1-6
Polystyrene	P(S)*, P(E)	OE	1-4
Polystyrene	P(S)*, P(E)	OE	8
Oligostyrene	P(S)*, P(E)	OE	8-11
Poly-α-methylstyrene	P(E)*, CT Complex	OA	1,4
Polystyrene-type Resists(CMS,MCMS)	CT Complex, R	OA	3,4,5,7
Polystyrene-type Resists (CMS, MCMS)	P(E)*	OE	3,4
Polysiloxane-type Resists (SNR)	CT Complex, R	OA	7
Polyvinylcarbazole	P+	OA	12,13
Polyvinylcarbazole	P(S)*,P(E)*	OE	14-16
Polyisobutene	Fragments	LS	17
PMMA	Fragments	LS	18,19
PMMA and Related Polymers	P-	OA	20
Polyethylene Oxide	R	OA	21
Polyethylene Oxide	OH	OA	29,30
Polyethylene Oxide	Fragments	LS	19,22
Polysilane	P-,P+	OA	23-27
Polygermane	P-,P+	OA	28
Polyvinylpyrrolidone	OH	OA	29
Sodium Polyacrylate	OH	OA	29,31
Dextran	OH	OA	29
Polymethylvinylketone	Fragments	LS	32
Polyvinylnaphthalene	P(D)+	OA	33
Polyvinylbiphenyl	P-,P+	OA	34,35
Polyvinylbiphenyl	P-	OA	36

Where P(S)* and P(T)* denote singlet and triplet excited states of polymers, respectively. P- and P+ denote radical anions and cations of polymers, respectively. P(E)* and P(d)+ denote excimer type polymer excited states and dimer cation type polymer cations, respectively. R denotes polymer radicals. OH denotes OH radical from radiolysis of water. OA, OE, and LS denote optical absorption, optical emission, and light scattering methods, respectively.

Table 2 Pulse Radiolysis Studies on Solid (Molten) Polymers

Polymers	Transient Species	Methods	References
Polystyrene	P-O	A	37,38
Polystyrene	P-O	A	39
Polystyrene	P(E)*P(T)*,P-,P(D)+	OA	3,5
Polystyrene	P(E)	OE	3,40
Polystyrene	P(E)*	OE	41-44
Resist(CMS)	CT Complex, R	OA	3-7
Resist(CMS)	P(E)*	OE	3
Molten Polyethylene	---	OA	45
Polyethylene	e-	OA	46
Molten n-Alkane	R(M)	OA	47,48
Neat n-Alkane	M+,M(S)*,M(=), R(M),e-	OA	48
EPR(Copolymer	P*,P+,e-	OA	40,49
EPDR(Terpolymer)	energy transfer	OE	50
PMMA	energy transfer	OA	37,38
PMMA and Related Polymers	P+,P-	OA	51
Epoxy Resin	P*,Exciplex	OE	52
Epoxy Resin	P+	OA	52

P(S)* and P(T)* denote singlet and triplet excited states of polymers, respectively. P- and P+ denote radical anions and cations, respectively. P(E)* and P(D)+ denote excimer type polymer excited states and dimer cation type polymer cations, respectively. R denotes polymer radicals. CT complex denotes charger-transfer complex between phenyl rings and chlorine atoms. R(M) denotes alkyl radicals M(S) and M+ denote the excited states and the radical cations of alkanes, respectively. M(=)+ denotes olfin radical cations. OA and OE denote optical absorption and emission methods, respectively.

Table 3 Pulse Radiolysis Methods Applied to Polymer Science

Beams	Methods	Time Resolution	References
Electron Beam	OA	s and ps	1-7,12-14,16,20,21,23-31,33-40,45-49,51,52
Electron Beam	OE	s and PS	1-4,8-11,14-16, 40-44, 50,52
Electron Beam	LS	s	17-19,22,32
Ion Beam	OE	ns	41-44
Neutron	OA	s	53
SR	OE	ns and ps	54-56

Where SR denotes synchrotron radiation. OA, OE and LS denote optical absorption, optical emission, and light scattering measurements, respectively. S, ns, and ps denote microsecond, nanosecond, and picosecond, respectively.

6. ACKNOWLEDGEMENTS

The author thanks his co-workers whose names appear in the cited references for their contributions. Especially the author thanks Dr. Y. Yoshida and Dr. H Shibata for collaboration about recent work.

REFERENCES

1. M. Dole, The Radiation Chemistry of Polymers, Academic, New York, 1972.

2. J,H, O'Donnell and D.F. Sangster, Principles of Radiation Chemistry, American Elsevier, New York, 1970.

3. H. Schoenbacher, Proc. Intern. Radiat. Degradation of Polymers and Radiat. Resistant Materials, Takasaki, 1989, p.1.

4. R.L. Clough and K.T. Gillen, ibid., p. 13

5. J. Silverman and S.O. Nielsen, Polym. Preprints, Amer. Chem. Soc., 1968, 9, 296.

6 S.K. Ho, S. Siegel and H.J. Schwartz, J. Phys. Chem., 1967, 71, 4527.

7. S.K.Ho and S. Siegel, J. Chem. Phys., 1969, 50, 1142.

8. J.K. Thomas, J. Chem. Phys., 1969, 51, 550 and 1969, 51, 770.

9. G.R.A. Johnson and A. Willson, Radiat. Phys. Chem., 1977, 10, 89.

10. Y. Tabata, S. Tagawa and M. Washio, "Materials for Micro-lithography", L.F. Thompson, C.G. Willson and J.M.J.Frechet, Eds., ACS Symposium Series, American Chemical Society, Washington, D.C., 1984, 266, 151.

11. S. Tagawa, M. Washio, N. Hayashi and Y. Tabata, J. Nuclear Material, 1985, 133, 785.

12. S. Tagawa, Radiat. Phys. Chem., 1986, 27, 455.

13. S. Tagawa, N. Hayasi, Y. Yoshida, M. Washio and Y. Tabata, Radiat. Phys. Chem., 1989, 34, 503.

14. N. Kouchi, Y. Aoki, H. Shibata, S. Tagawa, H. Kobayashi and Y. Tabata, Radiat. Phys. Chem., 1989, 34, 379.

15. N. Kouchi, S. Tagawa, H. Kobayashi and Y. Tabata, Radiat. Phys. Chem., 1989, 34, 453.

16. S. Tagawa, N. Kouchi, H. Shibata and Y. Tabata, "Advanced Resist Technology and Processing VI," SPIE 1989, 1086, 65.

17. A. Ogata and S. Tagawa, Review Sci. Instr., 1989, 60, 2197.

18. Y. Yoshida, H. Shibata, S. Tagawa, M Washio, Y. Tabata, N. Kouchi and A. Ogata, "Advances Resist Technology and Processing VI," SPIE, 1989, 1086, 274.

19. Y. Tabata, J. Tanaka, S. Tagawa, Y. Katumura, T. Ueda and K. Hasegawa, J. Fac. Eng. Univ. Tokyo, 1978, 34,12.

20. S. Tagawa, Y. Katumura and Y. Tabata, Chem. Phys. Lett., 1979, 64, 258.

21. S. Tagawa, M. Washio and Y. Tabata, Chem. Phys. Lett., 1979, 68, 276.

22. S. Tagawa, "Polymers for High Technology," M.J. Bowden and S.R. Turner, Eds., ACS Symposium Series, American Chemical Society, Washington, D.C., 1987, 346, 37.

23. Y. Tabata, S. Tagawa, M. Washio and N. Hayashi, Radiat. Phys. Chem., 1985, 25, 305.

24. H. Itagaki, K Horie I Mita, M. Washio, S. Tagawa, Y. Tabata, H. Sato and Y. Tanaka, Macromolecules, 1987, 20, 2774.

25. S. Kawanishi, M. Hagiwara, Y.Katsumura, Y. Tabata and S. Tagawa, Radiat. Phys. Chem., 1985, 26, 705.

26. H. Ban, K. Sukegawa and S. Tagawa, Macromolecules, 1987, 20, 177.

27. H. Ban, K Sukegawa and S. Tagawa, Macromolecules, 1988, 21, 45.

28. H. Ban, A. Tanaka, N. Hayashi, S. Tagawa and Y. Tabata, Radiat, Phys. Chem., 1989, 34, 587.

29. S. Tagawa, M. Washio, Y. Tabata, H. Ban and S. Imamura, J. Photopolym. Sci. Tech., 1988, 1, 323.

30. S. Tagawa, Abstract of IBM Polymer Colloquium '87, Japan, 1987.

31. S. Irie, K. Oka and M. Irie, Macromolecules, 1988, 21, 110.

32. H. Kobayashi, T. Ueda, T. Kobayashi, M. Washio, Y. Tabata and S. Tagawa, Radiat. Phys. Chem., 1983, 21, 13.

33. H. Kobayashi, T. Ueda, T. Kobayashi, S. Tagawa, Y. Yoshida and Y. Tabata, Radiat. Phys. Chem., 1984, 23, 393.

34. Y. Tabata, H. Kobayashi, M. Washio, Y Yoshida, N. Hayashi and S. Tagawa, J. Radioanal. Nucl. Chem. 1986, 101, 163.

35. Y. Yoshida, T. Ueda, T. Kobayashi, H. Shibata and S. Tagawa, Proceedings of Symp. Radiat. Chem. of Japan, October, 1990 (in Japanese).

36. Y. Yoshida, T. Ueda, T. Kobayashi, K. Miya and S. Tagawa, Proceedings of Fall Meeting of the Atomic Energy Society of Japan, C8, p122, 1990.

REFERENCES IN TABLES

1. Tagawa, S., Schnabel, W. Washio, M., and Tabata, Y., Radiat, Phys. Chem. 18, 1087, 1981.

2. Washio, M., Tagawa, S., and Tabata, Y., Radiat, Phy. 1983.

3. Tabata, Y., Tagawa, S., and Washio, M., "Materials for Microlithography" Thompson, L.F., Willson, C.G., Frechet, J.M., eds. ACS Symposium Series 266, 151, 1984.

4. Tabata, Y., Tagawa, S., Washio, M., Chem. 25, 305, 1985.

5. Tagawa, S., Radiat. Phys. Chem. 27, 455, 1986.

6. Tagawa, S., Proc. Int. Ion Eng. Congr., Takagi, eds., Kyoto, Japan, 3, 1681, 1983.

7. Tagawa, S., "Polymers for High Technology," Bowden, M.J. and S.R., eds., ACS Symposium Series 346, 37, 1987.

8. Itagaki, H., Horie, K., Y., J. Chem. Phys. 79, 3996, 1983.

9. Itagaki, H., Horie, K., Mita, I., Washio Sato, H. and Tanaka, Y., Chem. Phys. Lett. 120, 547, 1985.

10. Itagaki, H., Horie, K., Mita, I., Washio, M., Tagawa, S., Tabata, Y., Sato, H., and Tanaka, Y., Macromolecules, 20, 2774, 1987.

11. Itagaki, H., Horie, K., Mita, I., Washio, M., Tagawa, S., and Tabata. Y., Radiat. Phys. Chem. 34, 597, 1989.

12. Tagawa, S., Tabata, Y., Arai, S., and Imamura, M., J. Polym. Sci., Polym. Lett. Ed. 12, 545, 1974.

13. Washio, M., Tagawa, S., and Tabata, Y., Polym, J. 13, 935, 1981.

14. Tagawa, S., and Tabata, Y., Polym. Prep., Am. Chem, Soc. Div. Polym, Chem. 20, 411, 1979.

15. Tayawa, S., Washio, M., and Tabata, Y., Chem. Phys. Lett. 68, 276, 1979.

16. Tabata, Y., Katsumura, Y., Kobayashi, H., Washio, M., and Tagawa, S., "Picosecond Phenomena II," Hochstrasser, R.M., Kaiser, W., and Shank, C.V., eds., Springer-Verlag, New York, N.Y. Springer Series in Chemical Physics, 14, 266, 1980.

17. Beck, G. Lindenau, D., and Schnabel, W., Eur. Polym. J. 11, 761, 1975.

18. Beck, G., Lindenau, D., and Schnabel, W., Macromolecules, 10, 135, 1977.

19. Schnabel, W., Radiat, Phys. Chem. 28, 303, 1986.

20. Ogasawara, M. Tanaka, M., and Yoshida, J. Phys. Chem. 91, 937, 1987.

21. Borgwardt, U., Schnabel. W., and Henglein, A., Makromol. Chem. 127, 176, 1969.

22. Grollmann, U., and Schnabel, W., Makromol. Chem. 181, 1215, 1980.

23. Ban, H., Sukegawa, K., and Tagawa, S., Macromolecules, 20, 1775, 1987.

24. Ban, H., Sukegawa, K., and Tagawa, S., Macromolecules, 21, 45, 1988.

25. Ban, H., Tanaka, A., Hayashi, N., Tagawa, S., Tabata, Y., Radiat. Phys. Chem. 34, 587, 1989.

26. Tagawa, S., IBM Polymer Colloquium '87, Japan, 1987.

27. Tagawa, S., Washio, M., Tabata, Y., Ban, H., and Imamura, S., J. Photopolym. Sci, Tech. 1, 323, 1988.

28. Irie, S., Oka, K., and Irie, M., Macromolecules, 21, 110, 1988.

29. Behzadi, A., Borgwardt, U., Henglein, A., Schamberg, E., and Schnabel, W., Ber. Bunsenges. Phys. Chem. 74, 649, 1970.

30. Matheson, M., Mamou, A., and Silverman, J., and Rabani, J., J. Phys. Chem. 77, 2420, 1973.

31. Behzadi, A., and Schnabel, W., Macromolecules 6, 824, 1973.

32. Lindenau, D., Beaven, S.W., Beck, G., and Schnabel, W., Eur. Polym. J. 13, 819, 1977.

33. Irie, S., and Irie, M., Macromolecules 19, 2182, 1986.

34. Ogasawara, M., Tanaka, M., and Yoshida, H., J. Phys. Chem. 91, 937, 1987.

35. Tanaka, M., Ogasawara, M., and Yoshida, H., Radiat. Phys. Chem. 34, 591, 1989.

36. Miyazaki, T., Fueki, K., and Kato, N., Abstracts of Intern. Symp. of Fast Excitation Processess, March 1988, Tokyo.

37. Ho, S.K., Siegel, S., and Schwartz, H.J., J. Phys. Chem. 71, 4527, 1967.

38. Ho, S.K., and Siegel, S., J. Chem. Phys. 50, 1142, 1969.

39. Thomas, J.K., J. Chem. Phys. 51, 550, 1969 and 51, 770, 1969.

40. Washio, M., Tabata, Y., and Tagawa, S., "Pulse Radiolysis of Irradiated Systems," Edited by Tabata, Y., CRC Press, Florida, in press.

41. Tagawa, S., Kouchi, N., Aoki, Y., and Shibata, H., SPIE Advances in Resist Technology and Processing III, Vol. 631, 22, 1986.

42. Kouchi, N., Tagawa, S., Kobayashi, H., and Tabata, Y., Radiat. Phys. Chem. 34, 453, 1989.

43. Kouchi, N., Aoki, Y., Shibata, H., Tagawa, S., Kobayashi, H., and Tabata, Y., Radiat. Phys. Chem. 34, 759, 1989.

44. Kouchi, N., and Tagawa, S., "Pulse Radiolysis of Irradiated Systems," edited by Tabata, Y., CRC Press, Florida, in press.

45. Silverman, J., and Nielsen S.O., Polym. Preprints, Amer. Chem. Soc. 9, 296, 1968.

46. Johnson, G.R.A., and Willson, A., Radiat. Phys. Chem. 10, 89, 1977.

47. Tabata, Y., Kobayashi, H., Washio, M., Yoshida, Y., Hayashi, N., and Tagawa, S., J. Radioanal. Nuclear Chem. 101, 163, 1986.

48. Tagawa, S., Hayashi, N., Yoshida, Y., Washio, M., and Tabata, Y., Radiat. Phys. Chem. 34, 503, 1989.

49. Tagawa, S., Washio, M., and Tabata, Y., Polym. Prep., Japan, Vol. 37, 1N15, 1988.

50. Kawanishi, S., Hagiwara, M., Katsumura, Y., Tabata, Y., and Tagawa, S., Radiat. Phys. Chem. 26, 705, 1985.

51. Tabata, M., Nillsson, G., Lund, A., and Shoma, J., J. Polym. Sci., Polym. Chem. Ed. 21, 3257, 1983.

52. Tagawa, S., Washio, M., Hayashi, N., and Tabata, Y., J. Nuclear Materials. 133/134, 785, 1985.

53. Tabata, Y. and Tagawa, S., Proc. US/Japan Seminar on Fast Pulse Reactor, Long, R.L., An, S., and Wakabayashi, H., Tokai-mura, Japan, p. 463, 1976.

54. Ogata, A., and Tagawa, S., Review Sci. Instr. 60, 2197, 1989.

55. Tagawa, S., Yoshida, Y. Washio, M., Kouchi, N., Shibata, H., Tabata, Y., and Ogata, A., "Advanced Resist Technology and Proecssing VI," SPIE, 1989.

56. Tagawa, S., Proc. Conf. Radiat. Curing Asia, 49, 1988.

RECEIVED August 30, 1991

Chapter 2

Cross-Linking of Hydrocarbon Polymers and Their Model Compounds

Linear-Energy-Transfer Effects

Y. Tabata

Department of Nuclear Engineering, School of Engineering, Tokai University, 1117 Kitakaname, Hiratsuka, Kanagawa, Japan

Radiation effects on a series of n-paraffins and hydrocarbon polymers such as polyethylene and ethylene–propylene rubber have been investigated. A clear linear energy transfer (LET) effect on those materials has been demonstrated by various experimental methods. It has been found that crosslinking is not distributed homogeneously. So-called microheterogeneous spatial distributions by γ and e-beams and macroheterogeneous distributions by high LET ion beams have been shown to be formed.

Linear energy transfer (LET) effects on hydrocarbon compounds, including a series of n-alkanes, squalane, polyethylene, and ethylene–propylene rubber, have been studied. A heterogeneous crosslinking formation in the hydrocarbon polymers by γ and electron irradiation has been extensively investigated. Measurements of transient species by pulse radiolysis and product analysis by various methods have been carried out. γ-rays from ^{60}Co, electrons from accelerators, fast neutrons from a nuclear reactor, and ions from various accelerators including a cyclotron and Van de Graaff were used for the irradiation.

Experimental Methods

The pulse radiolysis were carried out by using a 35-MeV linear accelerator including, the Twin Linac Pulse Radiolysis System at the Nuclear Engineering Research Laboratory, The University of Tokyo (1). Measurements of gel formation were made by direct product analysis (2). Gas

0097–6156/91/0475–0031$06.00/0

chromatography of evolved gases and mass spectrometry of the irradiated products were performed (3). Electron spin resonance measurements were also carried out at low temperatures (4).

Results and Discussion

Results of the pulse radiolysis both in bulk n-dodecane (n-$C_{12}H_{26}$) and in the presence of carbon tetrachloride (100 mM) are shown in Figure 1. As was already reported (5,6), most transient species can be observed and have been assigned:

n-dodecane	RH
alkyl radical	R·
olefinic cation-radical	$R(-H)^{+\cdot}$
singlet excited state	RH*
alkane cation radical	$RH^{+\cdot}$
trapped electron	e_t

The measurements were extended to higher number of paraffins up to C_{30} and ethylene–propylene rubber.

Pulse radiolysis of thin films of ethylene–propylene rubber in the absence and presence of CCl_4 was carried out, and the results are summarized at 2 ns after the pulse in Figure 2.

Time profiles at 860 nm are shown in the same figure. Three transient species, singlet polymer excited state P*, polymer cation radical $P^{+\cdot}$, and olefinic cation radical $P(-H)^{+\cdot}$ are formed in a shorter wave length region of 240–350 nm. These have been confirmed by electron spin resonance measurements at very low temperatures of various hydrocarbons including polymers (6) as well as the pulse radiolysis experiments (7, 8). It was readily observed by our Twin Linac Pulse Radiolysis System with time resolution of 20 ps for the absorption experiment that alkyl radicals R· formed very fast within the time resolution of 20 ps for various saturated hydrocarbons (1,8).

>From those experimental results, the mechanism of alkyl radical formation in hydrocarbons including polymers has been proposed. The scheme is presented in Figure 3.

Highly excited cation radical is most important for the formation of alkyl radical. About 10% contribution for the radical formation come from the superexcited state of hydrocarbons, RH**. The crosslinking mechanism could be described as follows:

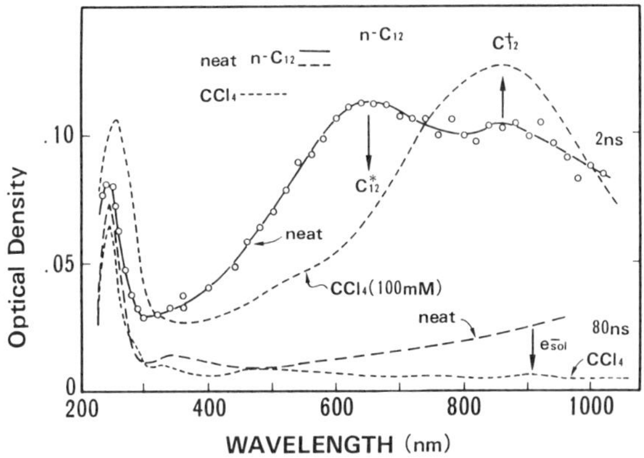

Figure 1. Pulse radiolysis of n-$C_{12}H_{26}$ at 2 ns and 80 ns after the pulse.

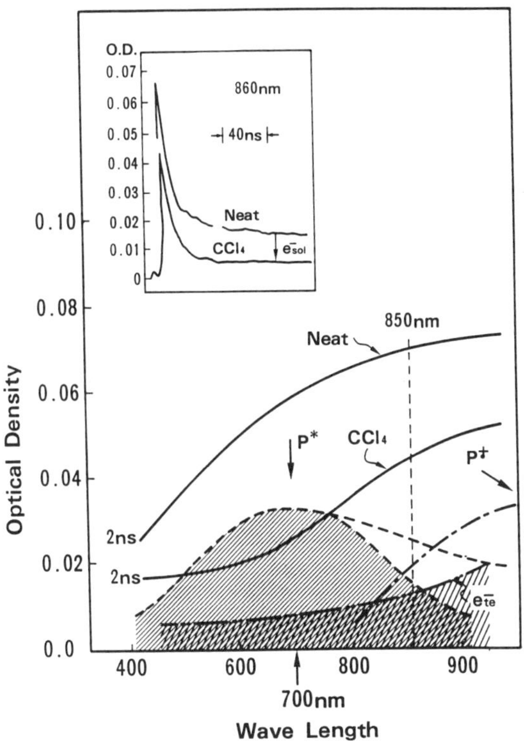

Figure 2. Pulse radiolysis of ethylene–propylene rubber films in the absence and presence of CCl_4.

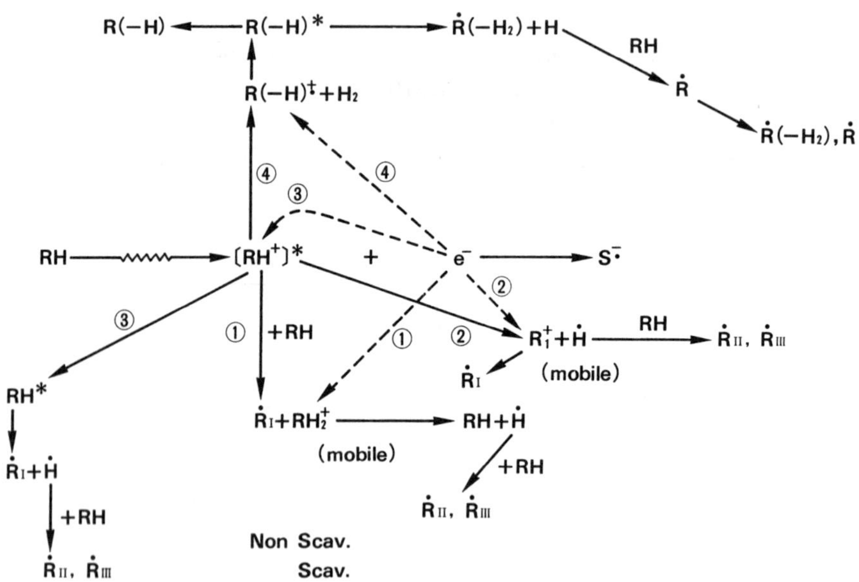

$$RH^{**} \longrightarrow R\cdot + H^{*}_{\cdot} \qquad RH^{**} \longrightarrow R(-H) + H_2$$

$$RH + H^{*}_{\cdot} = R\cdot + H_2$$

Figure 3. Mechanism of radical formation in saturated hydrocarbons.

A. Deprotonation

$$\mathrm{RH} \rightsquigarrow \mathrm{RH}^{*+}_{\cdot} + e^-$$
$$\mathrm{RH}^{*+}_{\cdot} + \mathrm{RH} \longrightarrow \mathrm{R_1}\cdot + \mathrm{RH_2}^{+}_{\cdot}$$
$$\mathrm{RH_2}^{+}_{\cdot} + e^- \longrightarrow \mathrm{RH} + \mathrm{H}\cdot$$
$$\mathrm{RH} + \mathrm{H}\cdot \longrightarrow \mathrm{R_{II,III}}\cdot + \mathrm{H_2}$$

Pair-wise Formation

B. Dissociation of Excited Cation Radical

$$\mathrm{RH}^{*+}_{\cdot} \longrightarrow \mathrm{R_I}^{+} + \mathrm{H}$$
$$\mathrm{R_1}^{+} + e^- \longrightarrow \mathrm{R_I}\cdot$$
$$\mathrm{RH} + \mathrm{H}\cdot \longrightarrow \mathrm{R_{II,III}}\cdot + \mathrm{H_2}$$

Pair-wise Formation

It is known that in crystalline alkanes, more than 90% of alkyl radicals are formed as pair radicals at very low temperatures (6). About 40% of pair radicals are located at the end of chains in the crystal. It has been demonstrated by gel permeation chromatography that 50% of dimers are almost linear due to the crosslinking at the ends of chains.

C. Olefinic Cation Radical

$$\mathrm{RH} \rightsquigarrow \mathrm{RH}^{**},\ \mathrm{RH}^{+*}_{\cdot} + e^-$$
$$\mathrm{RH}^{**} \longrightarrow \mathrm{R(-H)} + \mathrm{H_2}$$
$$\mathrm{RH}^{*+}_{\cdot} \longrightarrow \mathrm{R(-H)}^{+}_{\cdot} + \mathrm{H_2}$$
$$\mathrm{R(-H)}^{+}_{\cdot} + e^- \longrightarrow \mathrm{R(-H)}^{*} \longrightarrow \mathrm{R(-H)}$$
$$\mathrm{R(-H)}^{*} \longrightarrow \mathrm{R(-H_2)}\cdot + \mathrm{H}\cdot$$
$$\mathrm{RH} + \mathrm{H}\cdot \longrightarrow \mathrm{R}\cdot + \mathrm{H_2}$$

Pair-wise Formation

D. Charge Transfer

$$\mathrm{R(-H)} + \mathrm{RH}^{+}_{\cdot} \longrightarrow \mathrm{R(-H)}^{+}_{\cdot} + \mathrm{RH}$$
$$\mathrm{R(-H)}^{+}_{\cdot} + e^- \longrightarrow \mathrm{R(-H)}^{*}$$
$$\mathrm{R(-H)}^{*} \longrightarrow \mathrm{R(-H_2)}\cdot + \mathrm{H}\cdot$$
$$\mathrm{RH} + \mathrm{H}\cdot \longrightarrow \mathrm{R}\cdot + \mathrm{H_2}$$

Pair-wise Formation

This process can be scavenged by cation scavengers.

E. Thermal Process of RH*

$$\mathrm{RH}^{**} \longrightarrow \mathrm{RH}^{*}\ \text{relaxation}$$
$$\mathrm{RH}^{+*}_{\cdot} \longrightarrow \mathrm{RH}^{+}$$
$$\mathrm{RH}^{*} \longrightarrow \mathrm{R}\cdot + \mathrm{H}\cdot$$
$$\mathrm{RH} + \mathrm{H}\cdot \longrightarrow \mathrm{R}\cdot + \mathrm{H_2}$$
$$\mathrm{RH}^{+}_{\cdot} + e^- \longrightarrow \mathrm{RH}^{*}$$

Pair-wise Formation

F. Radical Migration (intra- and inter-)

$$R(-H) + R\cdot \longrightarrow R(-H_2)\cdot + RH$$

$$\underset{\text{(mobile)}}{RH + R\cdot} \longrightarrow R\cdot + RH$$

Processes A and B are mainly not scavengeable ones, and process C is partially scavengeable. Processes D, E, and F are almost scavengeable.

Usually radicals are formed pairwise. Therefore, there is a tendency that crosslinking occurs between the radical pairs. The majority of the crosslinking takes place through the pairs inhomogeneously, not through a homogeneous diffusion process. Of course, the process depends largely upon the condition of the molecular aggregation states, either in the crystal, the amorphous, or the molten state. If double bonds are formed, selective charge transfer occurs from cation radical to the site, then radicals are formed in the vicinity of the double bond pairwise. On the other hand, it is known that radicals can migrate through either intramolecular or intermolecular processes. Then the radical is trapped stably as an allyl radical. Another alkyl radical can reach the allyl radical through migration in the medium to give crosslinking.

In addition, once a crosslink is formed, both olefinic cation radical and tert-alkyl radical are easily formed in that location. This makes further crosslinking in the same site easier. The easier formation of tertiary alkyl radical has been demonstrated by electron spin resonance (spin trapping technique; 9,10), and olefinic cation radical is shown to be formed easier in branched hydrocarbons.

It is strongly suggested that through those crosslinking processes mentioned above, crosslinking in hydrocarbon polymers, particularly linear hydrocarbons, occurs rather heterogeneously. This is an intrinsic tendency for the crosslinking process of hydrocarbon polymers.

Formation of dimer, trimer, tetramer and higher oligomers can be detected and analyzed by both gel permeation chromatography and mass spectroscopy. The quantity of unsaturated bonds can be also analyzed. These experimental results indicate clearly that inhomogeneous crosslinking occurs. Through the processes of C, D, and F, an enrichment of unsaturation in higher oligomers goes on. From those experimental results mentioned above and analysis of the results, heterogeneous crosslinking has been concluded; the schematic representations are shown in Figures 4 and 5, which represent crystalline and molten states, respectively. These can be used to simulate crosslinking of polyethylene or hydrocarbon polymers, that is, the former would be able to simulate crosslinking in the crystalline phase and the latter in the amorphous phase of polyethylene.

LET Effects on Hydrocarbon Polymer

Almost 15 year ago, LET effect on n-eicosane was studied by means of electron spin resonance at 77K (4). The sample was irradiated with γ-rays

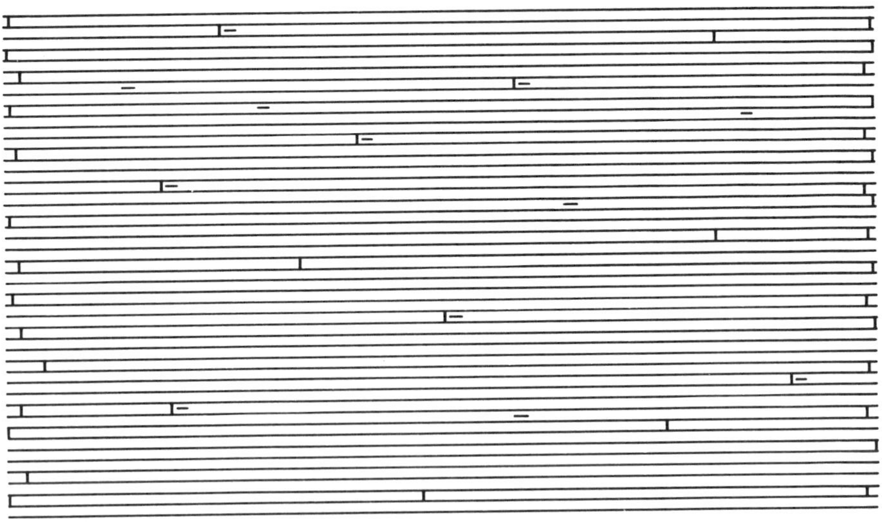

ı crosslinking, -double bond

Figure 4. Crosslinking and double formation in crystalline paraffin.

Figure 5. Crosslinking of n-alkane in the molten state.

from a ^{60}Co or H^+ (60 MeV) from a synchro-cyclotron at the University of Maryland, He^{2+} (23 MeV), C^{4+} (85 MeV), and N^{4+}(82 MeV) from a cyclotron at the Institute of Physical and Chemical Research. After irradiation with a dose of 4.3 × 10^{21} ev by C^{4+} ion, the ESR was measured at 77K. A very broad spectrum due to strong spin–spin interaction among radicals formed in the track, which is quite different from that of isolated radicals formed by γ-irradiation, was obtained. The decay of the broad component is faster than that of γ-irradiated isolated radicals by a factor of 5.5. About 30–40% of the radicals disappear below 150K. This is indicating that one-third of the radicals which survived at 77K combine with each other within the track of C^{4+} ion to make crosslinks. This must give rise to heterogeneous crosslinking in n-eicosane crystals. (See Figure 6.)

According to our other experiments on proton (60 MeV) (13) irradiation of n-eicosane at 77K, only 5% of isolated radicals decay up to 200K; on the other hand, about 80% of pair radicals decay up to the same temperature. Pair radicals disappear completely at 230K, and only 20% of isolated radicals decay at the same temperature. As the experimental conditions are different to some extent between H^+ and C^{4+} irradiations, it is difficult to conclude accurately the difference. However, one can say that three different decay processes occur in different temperature regions, that is, first there is decay of high-density radicals in the track of C^{4+} up to ca 130K, second is decay of pair radicals up to 210K, and finally there is decay of isolated (single) radicals at higher temperatures up to room temperature.

Fast heterogeneous crosslinking in the track of ion C^{4+}, then crosslinking through recombination of pair radicals, and finally slow crosslinking through encounter of isolated radicals occur successively. As certain amounts of the high-density radicals and the pair radicals are already reacted at 77K, these must be taken into consideration for estimation of the overall crosslinking yield.

Experimental results on irradiation by fast neutrons from a nuclear reactor (a fast neutron source reactor, Yayoi, at Nuclear Engineering Research Laboratory at the University of Tokyo) are shown for polyethylene (PE) and ethylene–propylene rubber (ET–PP), together with experimental results using ^{60}Co γ-rays. Gel fractions as a function of irradiation doses are shown in Figure 7.

It is quite clear from the figure that no difference between fast neutron and γ-irradiations has been found in a wide range of irradiation doses, as far as the gel content measurements are concerned. These experiments have been extended to rather heavier ions, like He^{2+} and N^{4+}, by Sasuga et.al. One of their experimental results is shown in Figure 8.

A big difference among different charged particles has been found with respect to mechanical properties. Within 1 MGy irradiation, residual elongation of the irradiated samples depends significantly on the kind of charged particles, that is, on the LET of the radiation. Above 1 ev/A, the

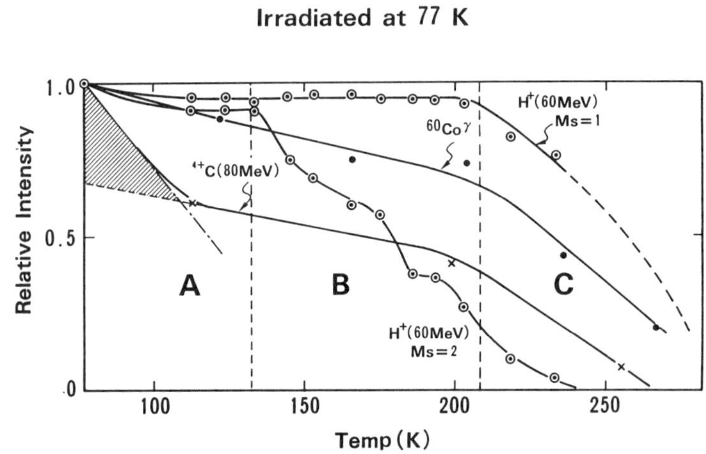

Figure 6. Decay of alkyl radical in n-eicosane for γ, H^+, and C^{4+} irradiations.

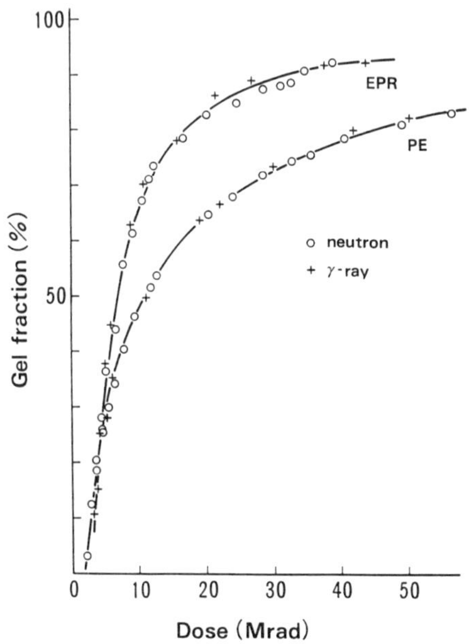

Figure 7. Gel fraction as a function of dose in γ and fast-neutron irradiation for polyethylene and ethylene–propylene rubber. (Reproduced with permission from *Radiat. Phys. Chem.* Copyright 1991 Pergamon.)

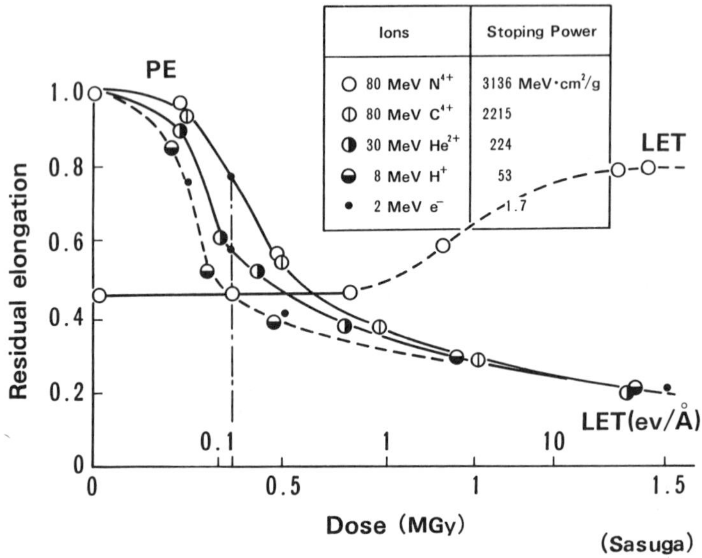

Figure 8. Residual elongations of polyethylene as a function of irradiation dose for 5 different LET. (Reproduced with permission from *Radiat. Phys. Chem.* **1991,** *37,* 135–140. Copyright 1991 Pergamon.)

effect becomes significant for polyethylene. This phenomena could be explained by *macro*heterogeneous crosslinking in the track due to high-density radical formation by high-LET charged particles, compared with low-LET radiation (and high-energy electrons) which produce rather *micro*heterogeneous crosslinking. Concerning the residual elongation, there is no difference between γ-rays or electron beam and proton H^+.

There exists, however, a big difference among H^+, He^{2+}, C^{4+}, and N^{4+}, as shown in the figure under the experimental conditions. For the irradiated polyethylene sheets, heterogeneous crosslinking under a fixed number of crosslinks below 1 MGy irradiation makes easier the stretching and is more favorable for stretching than homogeneous crosslinking. Figure 9 shows the scheme qualitatively.

For gel formation, no differences among different radiation sources have been observed. This is suggesting that the G-value of crosslinking is almost the same for all kinds of radiations examined. In a dose range up to 2 MGy, a significant difference in tensile strength of the irradiated samples between e^- and H^+ irradiations has been found. This is suggesting that a suitable condition for higher tensile strength does exist for the proton irradiation in a dose range of 1–2 MGy, compared with that of electron irradiation. Therefore, it may be possible to distinguish the difference of radiation effect between e^- and H^+ by this method, in spite of the fact that no differences can be found by measurements of the gel fraction and the stretching. These results are summarized in Table I.

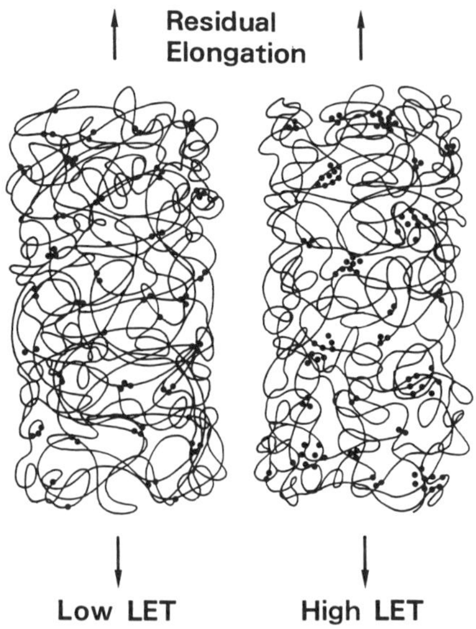

Figure 9. Schematic representation of heterogeneous crosslinking for low and high LET radiations

Table I. LET Effect on Polyethylene

Test Method \ Radiation Source	$\gamma(Co)^{60}$	e^-	n	H^+	He^{2+}	C^{4+}	N^{4+}
Residual Elraction a)	S	S	S	S	M	L	L
Gel Fraction b)	E	E	E	E			
Swelling c)	S	S		L			L
Tensile Strength d)	L′	L′		H			

a) 0.1～1MGy, b) 0.02～2MGy, c) ～2MGy, d) 1～2MGy

S : small, M : medium, L : large, E : equivalent, L′ : low, H : high

Conclusion

Radiation effects on a series of n-paraffins have been studied as model compounds of hydrocarbon polymers. It has been pointed out that heterogeneous crosslinking in both crystalline and molten (amorphous) states occur in different ways, even if radiation sources are γ-rays from ^{60}Co or high-energy electrons with a low LET. An LET effect has been clearly observed among different radiations of e^-, H^+, He^{2+}, C^{4+}, and N^{4+}. The effect was detected and confirmed by various methods. Radiation effects on polymers could be well simulated by that of their model compounds. In saturated linear hydrocarbons, no chain scission occurs above a certain number of hydrocarbons, probably ca 20 carbons. The crosslinking yield is not much affected by LET under our experimental conditions, because there is no LET dependence on the gel formation. The main LET effect comes from difference in spatial distribution of active species (including alkyl radicals and unsaturated bonds) and crosslinks.

Literature Cited

1. Kobayashi, H.; Tabata, Y. *Radiat. Phys. Chem.* **1989,** *34,* 447–451.
2. Seguchi, T.; Hayakawa, N.; Tamura, N.; Hayashi, N.; Katsumura, Y.; Tabata, Y. *Radiat. Phys. Chem.* **1989,** *33,* 119–128
3. Seguchi, T.; Hayakawa, N.; Tamura, N.; Hayashi, N.; Katsumura, Y.; Tabata, Y. *Radiat. Phys. Chem.* **1988,** *32,* 753–760.
   Seguchi, T.; Arakawa, K.; Tamura, N.; Katsumura, Y.; Hayashi, N.; Tabata, Y. *Radiat. Phys. Chem.* **1990,** *36,* 259–266.
4. Hamanoue, K.; Kamantauskas, V.; Tabata, Y.; Silverman, J. *J. Chem. Phys.* **1974,** *61,* 3439–3443.
   Kimura, K.; Matsui, M.; Karasawa, T.; Imamura, M. Ogawa, M.; Tabata, Y.; Oshima, K. *J. Chem. Phys.* **1975,** *63,* 1797–1802.
5. Tagawa, S.; Hayashi, N.; Yoshida, Y.; Washio, M.; Tabata, Y. *Radiat. Phys. Chem.* **1989,** *34,* 503–511.
6. Iwasaki, M.; Toriyama, K.; Fukaya, M.; Muto, H,; Nunome, K. *J. Phys. Chem.* **1985,** *89,* 5278.
   Miyazaki, T. *Radiat. Phys. Chem.* **1991,** *37,* 11–14.
   Toriyama, K.; Nunome, K.; Iwasaki, M. *J. Chem. Phys.* **1986,** *90,* 6836.
   Trifunac, A. D.; Werst, D. W.; Percy L. T. *Radiat. Phys. Chem.* **1989,** *34,* 547.
7. Brede, O.; Bös, J.; Naumann, W.; Mehnert, R. *Radiochem. Radioanal. Lett.* **1978,** *35,* 85.
   Jonah, C. D. *Radiat. Phys. Chem.* **1983,** *21,* 53.
   Klassen, N. V.; Teather, G. G. *J. Phys. Chem.* **1985,** *89,* 2048.
   Mehnert, R.; Brede, O.; Cserep, G. *Radiat. Phys. Chem.* **1985,** *26,* 353.

8. Tabata, Y.; Kobayashi, H.; Washio, M.; Yoshida, Y.; Hayashi, N.; Tagawa, S. *J. Radioanal. Nucl. Chem.* **1986,** *101,* 163.

9. Tabata, M.; Sohma, J.; Yamaoka, H.; Matsuyama *Chem. Phys. Lett.* **1985,** *119,* 256.

10. Katsumura, Y., The University of Tokyo, unpublished data.

11. Katsumura, Y.; Tabata, Y.; Seguchi, T.; Hayakawa, N.; Yoshida, K.; Tamura, N. *Radiat. Phys. Chem.* **1985,** *26,* 211.
Seguchi, T.; Hayakawa, N.; Yoshida, K.; Tamura, N.: Katsumura, Y.; Tabata, Y. *Radiat. Phys. Chem.* **1985,** *26,* 221.

12. Seguchi, T.; Sasuga, T.; Kawakami, W.; Hagiwara, M.; Kohno, I.; Kamitubo, H. *Proc. 11th Int. Conf. Cyclotronon and Their Application,* **1987,** 667.
Sasuga, T.; Kawanishi, S.; Kohno, I. *Proc. Int. Conf. Radiation Damage to Organic Materials in Nuclear Reactors and Radiation Environments,* Takasaki, Japan, July 17–20, 1989.
Sasuga, T.; Kawanishi, S.; Seguchi, T.; Kohno, I. *Polymer* **1989,** *30,* 2054–2059.

RECEIVED May 21, 1991

Chapter 3

Linear-Energy-Transfer Effects on Polymers

W. Schnabel, Q. Q. Zhu, and S. Klaumünzer

Hahn Meitner Institut, Berlin GmbH, D–1000, Berlin 39, Germany

Various polymers (polymethacrylonitrile, PMCN, polyvinyl–acetate, PVAc, poly(ethylene oxide), PEO, poly(ethylene–co–propylene), CP–EP, and polystyrene, PSt) have been irradiated at room temperature and in the absence of oxygen with γ–rays, and various ions of different kinetic energy E. Thus, a wide range of electronic stopping power, dE/dx, from $2x10^{-4}$ to 2.5 MeV/μm was accessible.

PMCN predominantly undergoes main–chain scission and, in analogy to the previously examined polymethylmethacrylate, PMMA, the corresponding radiation chemical yield G(S), decreases with increasing stopping power. PSt predominantly crosslinks and the corresponding yield, G(X), is small and independent of dE/dx. PVAc, PEO and CP–EP also crosslink predominantly under the action of ionizing radiation and the gel doses increase with increasing stopping power. This finding indicates that, for these polymers, both radiation chemical yields, G(S) and G(X) decrease with increasing dE/dx. Moreover, for PEO and PVAc, irradiated with ^{40}Ar ions, the gel dose increases by a factor of 4 upon increasing E from 6.5 MeV to 150 MeV although dE/dx remains unaltered at 2.2. MeV/μm.

These effects are discussed within the framework of the track structure model of Chatterjee and Magee, and some difficulties in applying this model to polymers are pointed out.

General considerations. Radiation chemical research on polymers concentrated for several decades on the effects of radiations of low linear energy transfer (LET) such as γ–rays and electron beam radiation. At present, however, there is growing interest in effects caused by high LET radiation such as ion–beam radiation (1–17). This is due to the fact that ion–beam radiation can be applied in surface modification and techniques used in the production of wave guides and microelectronic devices. The interest in practical applications also initiated various activities concerning the fundamentals of radiation effects caused by high LET radiation. The essential

0097–6156/91/0475–0044$06.00/0

questions arising in comparing chemical effects on polymers induced by different kinds of radiation concern the dependence of chemical yields, G, on stopping power, dE/dx, initial particle energy, E, and absorbed dose rate, $\dot{D}$:

$$G = f(dE/dx, E, \dot{D}) \tag{1}$$

Research work concerning the dependence of G–values of crosslinking and main–chain scission, G(X) and G(S), respectively, on dE/dx and E was carried out with several polymers at the Hahn–Meitner–Institute in Berlin using ion–beam radiations generated by the accelerator VICKSI (van de Graaff Isochron Cyclotron Kombination für schwere Ionen). Prior to the presentation and discussion of the experimental data a brief review concerning the dissipation of the energy of heavy ions in polymers appears to be suitable. As has been shown previously by other authors (18) not only the stopping power but also the size of the volume around the trajectory of the particle in which its energy is initially dissipated (in the following denoted as track) depend significantly on the particle energy. Therefore, the structure of the tracks varies upon changing the particle energy. This can give rise to differences in the distribution of chemically reactive intermediates (radicals, ions, excited species) and thus influence the radiation chemical yields.

Dependence of the stopping power on the particle energy. Principally, heavy particles lose energy by various modes of interaction with the absorbing material. With respect to particles dealt with in this article the important modes are: (a) interaction with shell electrons referred to as "electronic" enery loss and (b) interaction via the Coulomb field of the target nuclei, i.e. elastic collisions referred to as "nuclear" energy loss:

$$(dE/dx)_{total} = (dE/dx)_{el} + (dE/dx)_{nucl} \tag{2}$$

The loss fractions applying to nuclear and electronic interaction can be calculated. Typical results of such calculations obtained with the aid of the computer code TRIM (19) for argon ions and polystyrene as stopping material are presented in Fig.1. It can be seen that, at relatively low particle energies, energy loss by nuclear collisions is predominant. At high particle energies, on the other hand, the nuclear energy loss is relatively small and can

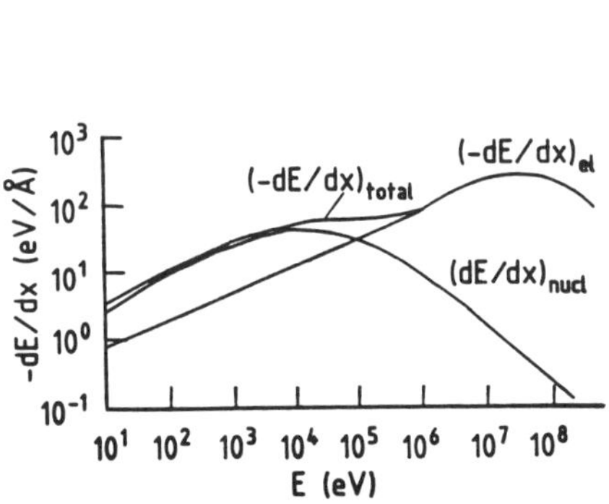

Fig.1. Dependence of the stopping power of polystyrene against ^{40}Ar ions on the particle energy. (Calculated with the aid of the computer code TRIM (19).

be neglected in discussions concerning possible differences in the mode of action of ion–beam radiations. Additional results are presented in Table I and Fig.2. Table I contains data relevant to ion–beams of relatively low energy that were utilized for surface modifications of polymers in several laboratories. Obviously, the nuclear energy loss fraction is rather high in the

Table I. Total stopping power of polystyrene and fractions of electronic and nuclear energy loss as calculated with the aid of the computer code TRIM (19) for He, Ne and Ar ions of relatively low energy

Particle Radiation	$(dE/dx)_{total}$ (MeV/μm)	f_{el} *) (%)	f_{nucl} **) (%)
He (E_0= 100 keV)	0.142	98.6	1.4
Ne (E_0= 200 keV)	0.37	78	22
Ar (E_0= 400 keV)	0.70	71	29

*) $(dE/dx)_{el}/(dE/dx)_{total}$ **) $(dE/dx)_{nucl}/(dE/dx)_{total}$

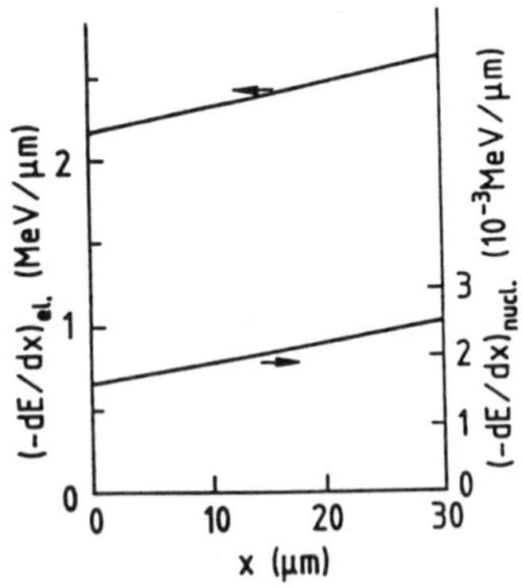

Fig.2. Stopping power of methacrylonitrile against $^{20}Ne^{7+}$ ions (E_0 = 275 MeV) vs. the particle penetration depth x. The total particle range is 425 μm. The range on the abscissa applies approximately to the thickness d of irradiated polymer films. Left ordinate: electronic stopping power. Right ordinate: nuclear stopping power.

cases of 0.2 MeV Ne and 0.4 MeV Ar ions. Fig.2 presents data concerning the energy absorption of 275 MeV Ne ions in polymethacrylonitrile. Here, $(dE/dx)_{el}$ and $(dE/dx)_{nucl}$ are plotted vs. the penentration depth of the projectiles in the target. In this case, up to a total target thickness of 30 μm, the fraction of energy loss due to nuclear collisions contributes only very little to the total energy loss.

Collisions of heavy particles with nuclei might lead to chemical alterations in the absorbing material differing qualitatively and/or quantitatively from those resulting from interactions with shell electrons. This problem has not been addressed in detail in previous publications and deserves certainly future attention. In this work, however, the ion energy and the target thickness have been chosen so that contributions of $(dE/dx)_{nucl}$ to radiation chemical yields can be safely neglected.

Dependence of the track radius on the particle energy. Regarding possible track structure effects the reader's attention is to be drawn again to Fig.1. For $E > 1$ MeV, dE/dx passes through a maximum, and, notably, argon ions of quite different energy are dissipating energy in the polymer with the same dE/dx. For example, about equal stopping power values apply for the particle energies $E = 6.5$ MeV and $E = 150$ MeV. Since the ion energy determines the velocity distribution of the secondary electrons the initial energy distribution around the ions' path is quite different for these two ion energies. This difference can cause differences in G values. In other words, the influence of track structure effects on radiation chemical yields could be principally investigated by carefully measuring the dependence of G(S) and G(X) on the particle energy.

The problem of the local energy distribution induced by the passage of fast heavy ions along their trajectories has been dealt with generally by Chatterjee and Magee (18). According to their model, which summarizes both experimental and theoretical work, the track structure is grossly characterized by two regions: (1) a track core region of radius r_c, in which more than 50 % of the energy is deposited, resulting in an extremely high local concentration of chemically active intermediates, and (2) a penumbra region of radius r_p, in which the remaining fraction of the energy is deposited in ionization and excitation events by energetic secondary electrons generated by the primary particle in the center of the core. As can be seen from Fig.3 the two radii depend quite differently on the particle energy E, which is here expressed in units of MeV per nucleon. The drastic increase of r_p relative to r_c implies that chemical effects induced by secondary electrons should become more and more important as the particle energy increases. In Table II, core and penumbra radii being of interest with respect to the ions examined in this work are presented. Obviously, r_c becomes unrealistically small for low kinetic energies per nucleon. In this case, the energy lost by the ion is concentrated in the small volume corresponding to r_p which envelopes the ion trajectory.

Provided the radiation chemical yields would directly reflect the energy distribution given immediately after the passage of an ion, significant differences in G–values are expected regarding irradiations by 6.5 MeV and 180 MeV Ar ions. On the other hand, G–values should differ less significantly if polymers are irradiated by 150 MeV Ar ions and 275 MeV Ne ions. In this connection, it should be pointed out that both r_c nd r_p are based on purely physical considerations. Both radii characterize the energy distribution immediately, i.e. about 10^{-15}s after the passage of the ion. Subsequent processes like conversion and dissipation of electronic energy, generation and

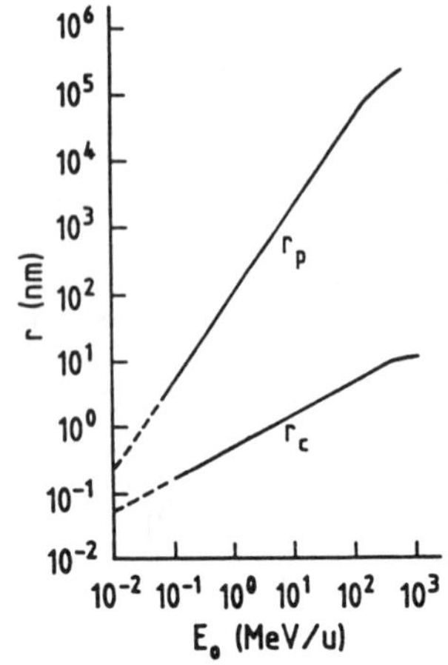

Fig.3. Penumbra radius r_p and core radius r_c vs. the particle energy. (After Chatterjee and Magee (18))

Table II. Penumbra and core radii calculated on the basis of the track structure theory (18) for polystyrene

Particle Radiation	E_0 (MeV/u)	r_p (nm)	r_c (nm)
0.1 MeV $^{4}He^{2+}$	0.025	0.8	—
6.5 MeV $^{40}Ar^{2+}$	0.16	10	0.2
180 MeV $^{40}Ar^{8+}$	4.5	700	1.1
275 MeV $^{20}Ne^{7+}$	13.75	2700	2.0

migration of free radicals as well as their reactions are rather complex and, at present, essentially unknown for most polymers. Thus, a comprehensive interpretation of radiation chemical yields is impossible. Nevertheless, a search for track structure effects in polymers may provide valuable information being of general importance for the field of radiation chemistry.

RESULTS

Polymers undergoing predominantly crosslinking.

Polystyrene, PSt. Gel doses, D_{gel}, and 100 eV–yields for crosslinking, G(X), calculated according to the relationship $G(X) = 50\ N_a/M_w\ D_{gel}$ are shown in Table III (N_a: Avogadro's number, M_w: initial weight average molar mass). Obviously, neither a track structure nor a stopping power effect was detected in this case: G(X) is independent of dE/dx and E within the error limit of our measurements. Notably, this result is at variance with results of other authors who found G(X) increasing with increasing stopping power (7,8,13–17). Moreover, G(X) was reported to depend on E (6). Possible reasons for these discrepancies have been discussed elsewhere (1).

Other polymers. Gel doses obtained by irradiating poly(vinyl acetate), PVAc, poly(ethylene oxide), PEO, and an ethylene–propylene copolymer, CP–EP with ^{60}Co–γ–rays or Ar ions of different energy are shown in Table IV. In all cases, gel doses are significantly larger for Ar ion–beam irradiation than for γ–irradiation. It is known from former work that PVAc and PEO undergo, apart from crosslinking, also main–chain scission to an appreciable extent if irradiated with radiation of low stopping power. According to eq.(3)

$$D_{gel} = 100\ N_a/M_w[2G(X) - G(S)/2] \tag{3}$$

gel formation is possible only if $G(S)/G(X) < 4$. Principally, the higher gel doses found for Ar ion–beam irradiations could be due either to a decrease in G(X) or to an increase in G(S) with increasing stopping power. For γ–irradiation $G(S)/G(X) \approx 1$ was found earlier in the case of PEO (21). Therefore, the increase in D_{gel} from 0.2×10^{21} (γ–rays) to 2.1×10^{21} eV/g (150 MeV Ar ions) could be due to an increase in G(S). However, an increase in D_{gel} by a factor of 10, at constant G(X), would result in $G(S)/G(X) > 4$ with the consequence of predominant main–chain scission. From the fact that gel formation was observed under all circumstances it is inferred that, also at high stopping power, $G(S)/G(X) < 4$ indicating a decrease in both G(S) and G(X). The most interesting result of these experiments concerns the finding that, at almost constant stopping power, the gel doses for PVAc and PEO depend on E: the gel doses at E = 150 MeV are about 4 times larger than those at E = 6.5 MeV. This result seems to indicate the existence of a track effect. Notably, a track effect was not found in the case of CP–EP, where the gel dose does not depend on E, within the error limit of our measurements.

Polymers undergoing predominantly main–chain scission

Polymethacrylonitrile (PMCN). Both ion–beam and ^{60}Co–γ–irradiation lead to main–chain cleavage. Results are presented in Table V. G(S) decreases with increasing stopping power. Obviously, PMCN exhibits the same behavior as polymethylmethacrylate PMMA (3). The dependence of G(S) on E has not yet been investigated in this case.

Table III. Crosslinking of polystyrene induced by radiation of different stopping power in the absence of oxygen (1)

Radiation	E (MeV)	dE/dx (MeV/μm)	M_w	D_{gel} (10^{21}eV/g)	G(X)
^{60}Co–γ–rays	ca.1$^{*)}$	$2x10^{-4}$			0.05
$^{20}_{10}Ne^{7+}$	275	0.37	$1.6x10^5$	3.6	0.05
			$5.3x10^5$	1.1	0.05
$^{40}_{18}Ar^{8+}$	180	2.3	$4.8x10^5$	2.0	0.03
$^{40}_{18}Ar^{2+}$	6.5	2.2	$2.7x10^5$	2.3	0.05
			$1.8x10^6$	0.4	0.04

$^{*)}$ energy of γ–photon

Table IV. Gel doses for polyvinylacetate, poly(ethylene oxide) and poly(ethylene–co–propylene) in the absence of O_2

Radiation	E (MeV)	dE/dx (MeV/μm)	D_{gel}(10^{21}eV/g)		
			PVAc a)	PEO b)	CP–EP c)
^{60}Co–γ–rays	ca.1	$2x10^{-4}$	0.9	0.2	0.07
$^{40}_{18}Ar^{2+}$	6.5	2.3	2.4	0.5	0.62
$^{40}_{18}Ar^{8+}$	150	2.2	8.5	2.1	0.52

$^{a)}M_w = 2x10^6$; $^{b)}M_w = 8.7x10^5$; $^{c)}M_w = 3.58x10^5$

Table V. 100 eV–yields of main–chain scission of polymethacrylonitrile in the absence of O_2

Radiation	E (MeV)	dE/dx (MeV/μm)	G(S)	References
^{60}Co–γ–rays	ca.1$^{*)}$	$2x10^{-4}$	3.3	(20)
$^{20}_{10}Ne^{7+}$	275	0.43	0.5	(2)
$^{40}_{18}Ar^{8+}$	180	2.5	0.4	(2)

$^{*)}$ energy of photon

DISCUSSION

As already pointed out above general conclusions regarding LET and track structure effects in polymers cannot yet be arrived at as far as chemical mechanisms are concerned. From a theoretical point of view there is a considerable lack of data input concerning the formation, the spatial distribution, and the chemical reactions of free radicals in the tracks of fast ions. Theoretically, the gap ranging from the very short period of energy deposition ($t \approx 10^{-15}$s) to the period of the formation of the final products ($t > 10^{-6}$s) has to be bridged. Experimentically, the latter are far from being completely characterized. In particular, the available results usually do not include radiation chemical yields of products of low molar mass. It appears that the important question whether densely ionizing radiation produces bond scissions in neighboring repeat units of the polymer chain cannot be answered until a material balance comprising all radiolysis products is available. However, what can be concluded on the basis of the results obtained until now applies to the efficiency of heavy ions with respect to crosslinking and main–chain scission. It seems that there is a general trend regarding G(X) and G(S) values. Both yields were found to decrease with increasing LET for all polymers investigated apart from polystyrene. On the basis of the assumption that densely ionizing particles are as effective as low LET radiations in producing primary products (ions, radicals, electronically excited species) the inefficiency of heavy ions to reduce the molar mass of polymers such as PMMA and PMCN or to increase the gel doses of polymers such PEO, PVAc and CP–EP might be related to the very high local concentration of reactive species. This leads to early deactivation reactions, especially to the combination of radicals (cage reactions). Simultaneous bond scissions in neighboring repeat units should contribute to the origin of inefficiency of heavy ions. This appears to be quite feasible as far as the high fraction (> 50 %) of energy absorbed in the track core is concerned.

Notably, a theoretical approach, on the basis of the Chatterjee–Magee track model, to track effects concerning the Fricke dosimeter system resulted in a rather satisfying agreement with the experimental data (18). In particular, this agreement refers to the general trend of decreasing G–values with increasing stopping power. Although the mechanisms concerning the chemical reactions occurring in the Fricke dosimeter system and in polymers are quite different, there might be common physical features that could explain the similar trend in the decrease of G–values. For example, in the case of the fluid Fricke dosimeter system, which consists of an aqueous solution, the magnitude of the radiation chemical yields is strongly dependent on the diffusion rate of free radicals. In the case of rigid polymers, free radical diffusion is certainly not an important factor. However, in this case energy dissipation from the core to rather remote regions in the form of transfer of excitation energy might lead to similar results as the diffusion of free radicals in fluid systems. This applies of course, to the case that energy dissipaion ultimately results in the formation of reactive free radicals or radical ions. It is also feasible that radical migration mechanisms involving, for example, hydrogen abstraction become operative in rigid polymers.

Finally it should be pointed out that the independence of G(X) on dE/dx observed in our laboratory in the case of polystyrene is still a matter of controversy with researchers of other laboratories. For the sake of clarity it appears necessary to carry out again additional crucial experiments in this case.

REFERENCES

[1] Schnabel, W., Klaumünzer, S., Int.J.Radiat.Appl.Instrum., Part C, Radiat.Phys.Chem. **1991**, 37, 131

[2] Schnabel, W., Klaumünzer, S., Int.J.Radiat.Appl.Instrum., Part C, Radiat.Phys.Chem. **1989**, 33, 23

[3] Schnabel, W., Klaumünzer, Sotobayashi, H., Asmussen, F., Tabata, Y., Macromolecules, **1984**, 17, 2108

[4] Brown, W.L., Nucl.Instr.Meth.Phys.Res. B, **1989**, 37/38, 270

[5] Brown, W.L., Rad. Effects, **1986**, 98, 281

[6] Aoki, Y., Kouchi, N., Shibata, H., Tagawa, S., Tabata, Y., Immamura, S., Nucl.Instr.Meth.Phys.Res. B, **1988**, 33, 799

[7] Egusa, S., Ishigure, K., Tabata, Y., Macromolecules, **1979**, 12, 939

[8] Zverev, A.B., Lavrentovitch, Ya.I., Kabakchi, A.M., Khim.Vys. Energ., **1969**, 3, 453

[9] Rudolph W.,Grambole, D., Groetzschel, R., Heiser, C., Herrmann, F., Knothe, P., Neelmejer, C., Nucl.Instr.Meth.Phys. Res. B, **1988**, 33 ,803

[10] Adesida, I., Nucl.Instr.Meth.Phys.Res. B, **1983**, 209/210, 79

[11] Adesida, I., Nucl.Instr.Meth.Phys.Res. B, **1985**, 7/8, 923

[12] Mladenov, G.M., Emmoth, B., Appl.Phys. Lett., **1981**, 38, 1000

[13] Puglisi, O, Licciardello, A. Pignataro, S., Calcagno, L.,Foti, G., Rad. Effects, **1986**, 98, 327

[14] Calcagno, L., Foti, G., Licciardello, A., Puglisi, O., Appl.Phys. Lett. **1987**, 51, 907

[15] Puglisi, O., Licciardello, A., Calcagno, L.,Foti, G.,Nucl.Instr. Meth.Phys. Res. B, **1987**, 19/20, 865

[16] Puglisi, O., Licciardello, A., Calcagno, L.,Foti, G., J.Mater. Res., **1988**, 3, 1247

[17] Licciardello, A. Puglisi, O., Calcagno, L.,Foti, G., Nucl.Instr. Meth.Phys.Res. B, **1990**, 46, 338

[18] Chatterjee, A. and Magee, J.L., in Radiation Chemistry, Principles and Applications, Farhataziz and Rodgers, M.A.J., Eds., VHC, Weinheim, **1987**, Chapters 1, 5 and 6

[19] Ziegler,J.F, Biersack, J.P., Littmark, U., "The Stopping and Range of Ions in Solids". In The Stopping and Ranges of Ions in Matter, Ziegler, J.F., Ed., Pergamon Press, New York, **1985**, Vol.1., p. 109

[20] Helbert, J.N., Poindexter, E.H., Stahl, G.A., Chen, C.Y., Pittman, C.U., Lai, J.H., J.Polym.Sci.,Polym.Chem Ed., **1979**, 17, 49

[21] Schnabel, W., Makromol.Chem., **1970**, 131, 287

RECEIVED April 2, 1991

Chapter 4

Pulse Radiolysis of Polystyrene Films

S. P. Mezyk, S. Yamamura, and J. K. Thomas

Department of Chemistry and Biochemistry, University of Notre Dame, Notre Dame, IN 46556

Pulse radiolysis studies, using 3 ns, 0.40 MeV electrons from a Febetron 706, have been used to investigate the early phenomena in the irradiation of thin (~ 100 μm) polystyrene (PS) films. In pure films, the radiolysis produces polystyrene excimers, which are identified by their luminescence and absorption spectra. The doping of these films with molecules such as pyrene and anthracene, gives excited singlet and triplet states, cations and anions of these arenes, concomitant with a decrease in the polystyrene excimer yield. These species are seen to form rapidly; well within the 3 ns electron pulse. The events leading to the formation of these solute intermediates are discussed.

The mode of formation of excited states by low-energy radiation, i.e., photochemistry, is well established. Excitation into the first excited state leads to long lived species which can fluoresce, intersystem cross, or produce chemistry. Excitation into states above the first excited state can sometimes lead to unique chemistry, which competes with rapid internal conversion leading to the first excited state, and the chemistry that results is that of the first excited state. However, the mode of formation of excited states by high-energy radiation, i.e., radiolysis, is still debated. For a number of years the formation of excited states on radiolysis with γ rays, fast electrons or α particles has been used as a method of measurement of the energy of the radiation and is the basis of the scintillation technique (*1*). Various theories are used to account for the formation of these excited states. The most popular theory is that the radiolysis initially produces ions whose rapid recombination

0097–6156/91/0475–0053$06.00/0

results in the energy appearing as excited states (*2*). However, an alternative theory, although by no means as well established, is that of direct irradiation. This is called the optical theory of the energy loss (*3*). A simple calculation shows that both processes could occur, the formation of excited states by direct excitation and also via subsequent ion recombination (*4*). In practice there is a rule of thumb that states that excited states are formed predominantly on radiolysis of liquids or materials of low dielectric constant or low polarity such as benzene or cyclohexane, while ionic products are formed on radiolysis of more polar liquids such as water or alcohols (see Figure 1). In photochemistry, excitation of a material inevitably leads to the singlet excited state, and any triplet states are formed by intersystem crossing from this state.

Intersystem crossing from singlet excited states also occurs in radiation chemistry; however, triplet states are also formed independently of excited singlet states (*5*). Nanosecond studies have established that some of the triplet excited states are certainly formed by ion recombination, which is also a possible source of singlet excited states (*6*). Picosecond studies have shown that the mode of the formation of the excited state is by no means simple and that some of the singlet excited states and ions are formed rapidly (*7*). Such behaviour could be caused by an initial action of the radiation to form ions, electrons, and positive holes. The fast recombination then results from electrons, which are partially trapped or solvated in the liquid, reacting with the positive holes. Such a process is expected to be very rapid, as the electrons have high mobilities in these systems. The slow process could be due to positive holes recombining with electrons which have become trapped on solute molecules. This latter process is slower than the former as anions move at a much slower rate than electrons in these liquids. These events are depicted pictorially in Figure 1.

As shown in Table I, an increase in the polarity of the liquid causes a diminution in the predominance of excited state production in the radiolytic process, and an increase in long lived (≥ μsec) ion chemistry is also evident. In very polar liquids such as water and alcohols, the cation is a solvated proton and the chemistry of interest is that of the solvated electron.

At first sight it would appear that similar events play a major role in the radiolysis of polymer films, but due to their rigidity polymer films freeze out much of the chemistry that occurs in local regions of high energy release. The diffusion of reactive species, which can mask important events in liquid systems, is minimized, and rapid electronic energy and charge transfer (*10*) emphasized.

To date several studies are available which emphasize radiolytic events in polymer films (*11-16*). Earlier work (*11-13*) was much in the vein of radiation chemistry; while later studies drew attention to its

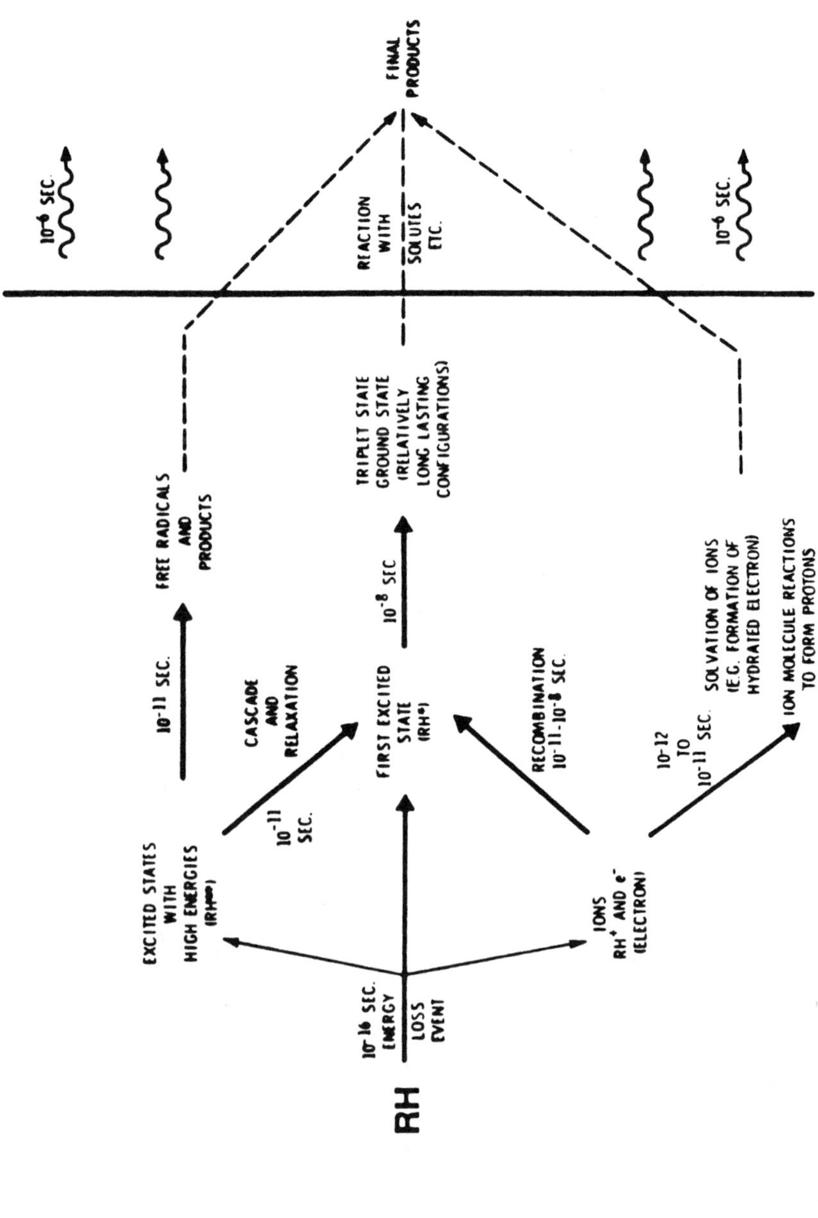

Figure 1. The temporal behaviour of species evolution in the high energy irradiation of a molecular system. Data taken from references 8 and 9.

Table I. Yields of Singlet and Triplet Excited States, Ions and Total Yield in Radiolysis of Solvents

Solvent	G(T)	G(S)	G(ions) (a)	G(total) (b)
O-xylene	1.7	2.5	~0.1	4.3
M-xylene	1.8	2.7	~0.1	4.5
P-xylene	2.4	2.0	~0.1	4.5
Pseudo cumene	1.8	1.6	...~0.1	3.5
Mesitylene	1.8	1.6	~0.1	3.5
Toluene	2.4	2.1	~0.1	4.6
Benzene	3.8	1.6	~0.1	5.5
Benzyl alcohol	1.1	0.7	2.1	5.9
Benzonitrile	1.4	1.2	1.4	4.0
Dimethylaniline	3.1	0.9	<0.2	4.2
Phenol	0	0	3.0	3.0
Cyclohexane	0.7	0.7	1.6	3.0
Tetrahydrofuran	0.1	0.04	0.66	0.8
Dioxane	-	1.03	0.12	1.15
Methanol	0	0	2.0	2.0
Water	0	0	3.5	3.5

NOTE: More recent studies on short time scales may give larger yields. However, the overall concept of the data remains unchanged.

SOURCE: Data taken from reference 6.

(a) Yield of anion or cation

(b) G(total) = G(ions) + G(excited states).

importance in the newly emerging fields of imaging and polymer packaging. The present study looks in depth at the nature and the origin of the early processes leading to excited state production.

Experimental

The laser flash photolysis equipment has been described in earlier publications (*17*). The pulse radiolysis equipment was similar to that used in laser photolysis, with the laser being replaced by a Field Emission Corporation 706 Febetron, which generated 3 ns (FWHM) pulses of 0.4 MeV electrons. The detection equipment had a rise time of less than 1.0 ns. Radiation dosimetry was carried out by placing the samples in front of a suitable dosimeter that absorbed the entire beam, e.g., a water sample or a microscope slide. The relative amount of energy lost to the film was measured by the loss in energy to the dosimeter with and without the film in place. Absolute dosimetry was carried out by observing the hydrated electron in water, $e^-_{(aq)}$, at 600 nm (pH 12.0, 0.10 M ethanol), where the extinction coefficient was 12,000 $M^{-1}cm^{-1}$ (*18*).

Polymer films of ~ 100 μm thickness were made by drying concentrated solutions for 24 hours. The concentration of solutes in the films was measured, utilizing established spectra, on a Perkin Elmer 55B spectrophotometer.

Results

Excited Singlet States. Figure 2 shows typical data for fluorescence decay following irradiation of 0.10 M pyrene in polystyrene (PS) films by either 3 ns pulse of 0.4 MeV electrons, or by 10 or 0.2 ns pulse of light from an N_2 laser at λ = 337.1 nm. Oxygen was removed from the films by pumping at 10^{-3} torr for 1 hour prior to study. This figure shows the fitting of the data by a single exponential of the form

$$I = I_0 e^{-kt}$$

where I_0, I, k, and t refer to the fluorescence intensity at t = 0, the intensity at time t, the rate constant and the time respectively. Some data are also fitted by the expression

$$\frac{I(t)}{I_0} = \frac{\int_{-\infty}^{\infty} \exp(-X^2) \exp(-\bar{k}t \exp(\gamma x)] \, dx}{\int_{-\infty}^{\infty} \exp(-X^2) \, dx}$$

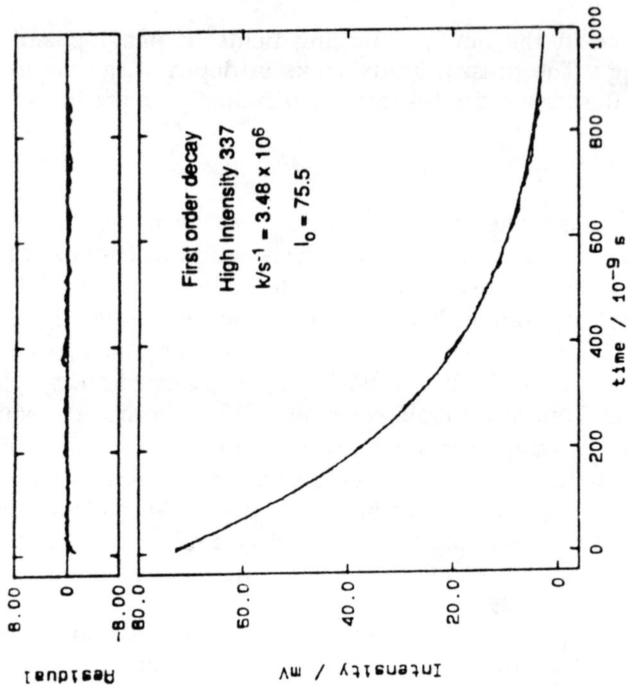
First order decay
High Intensity 337
k/s-1 = 3.48 x 10^6
I0 = 75.5
Residual
Intensity / mV
time / 10-9 s

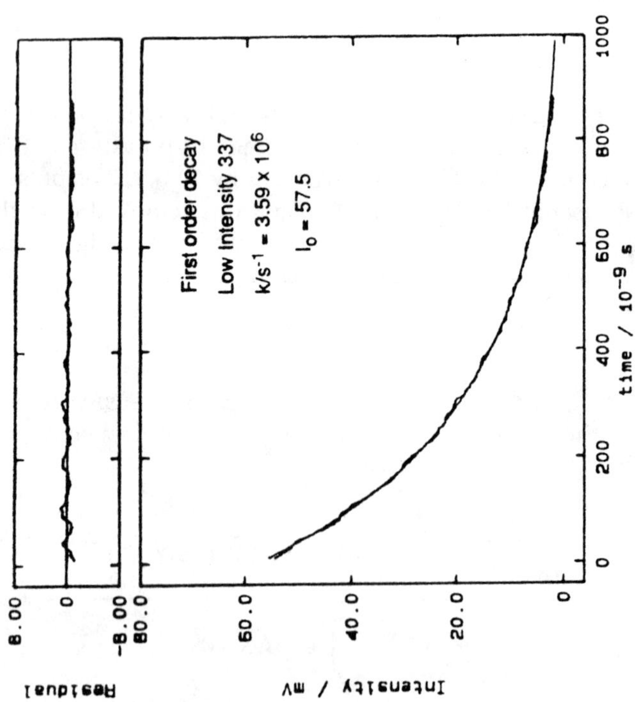
First order decay
Low Intensity 337
k/s-1 = 3.59 x 10^6
I0 = 57.5
Residual
Intensity / mV
time / 10-9 s

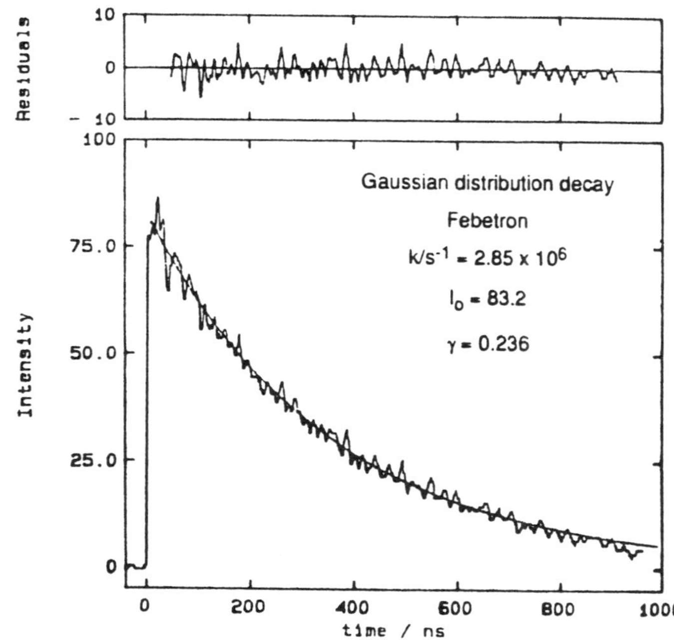

Figure 2a. Typical data from pulsed electron (Febetron) and laser (low and high intensity) irradiation of pyrene in polystyrene films for 10^{-3} M solute. The mode of fitting and the obtained kinetic parameters are given with the data.

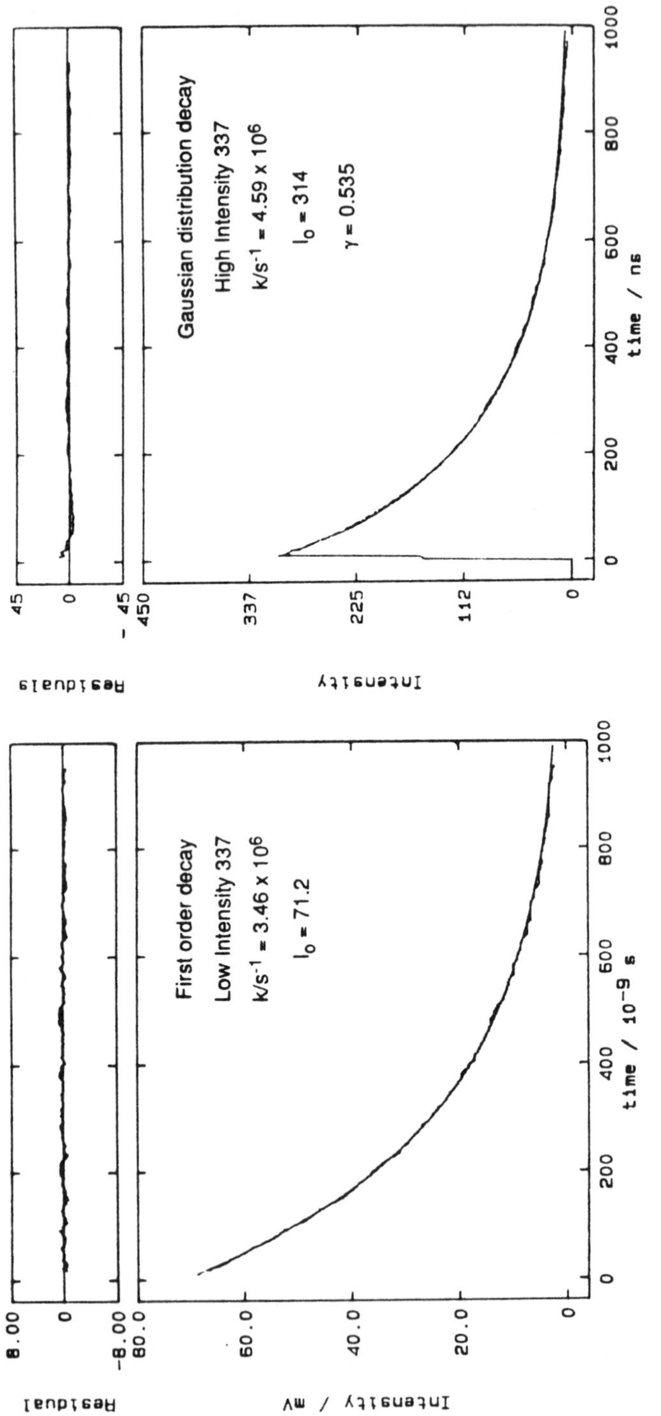
First order decay
Low Intensity 337
k/s^{-1} = 3.46 x 10^6
I$_o$ = 71.2
Residual
Intensity / mV
time / 10^{-9} s
Gaussian distribution decay
High Intensity 337
k/s^{-1} = 4.59 x 10^6
I$_o$ = 314
γ = 0.535
Residuals
Intensity
time / ns

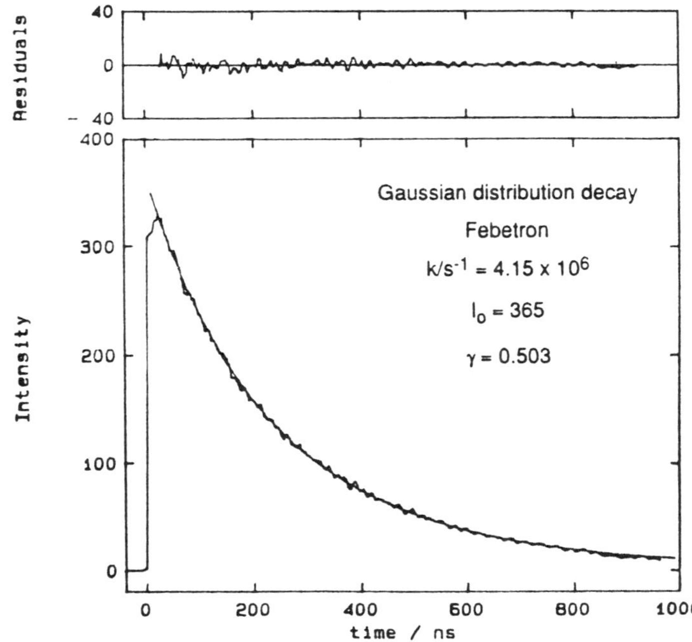

Figure 2b. Typical data from pulsed electron (Febetron) and laser (low and high intensity) irradiation of pyrene in polystyrene films for 10^{-2} M solute. The mode of fitting and the obtained kinetic parameters are given with the data.

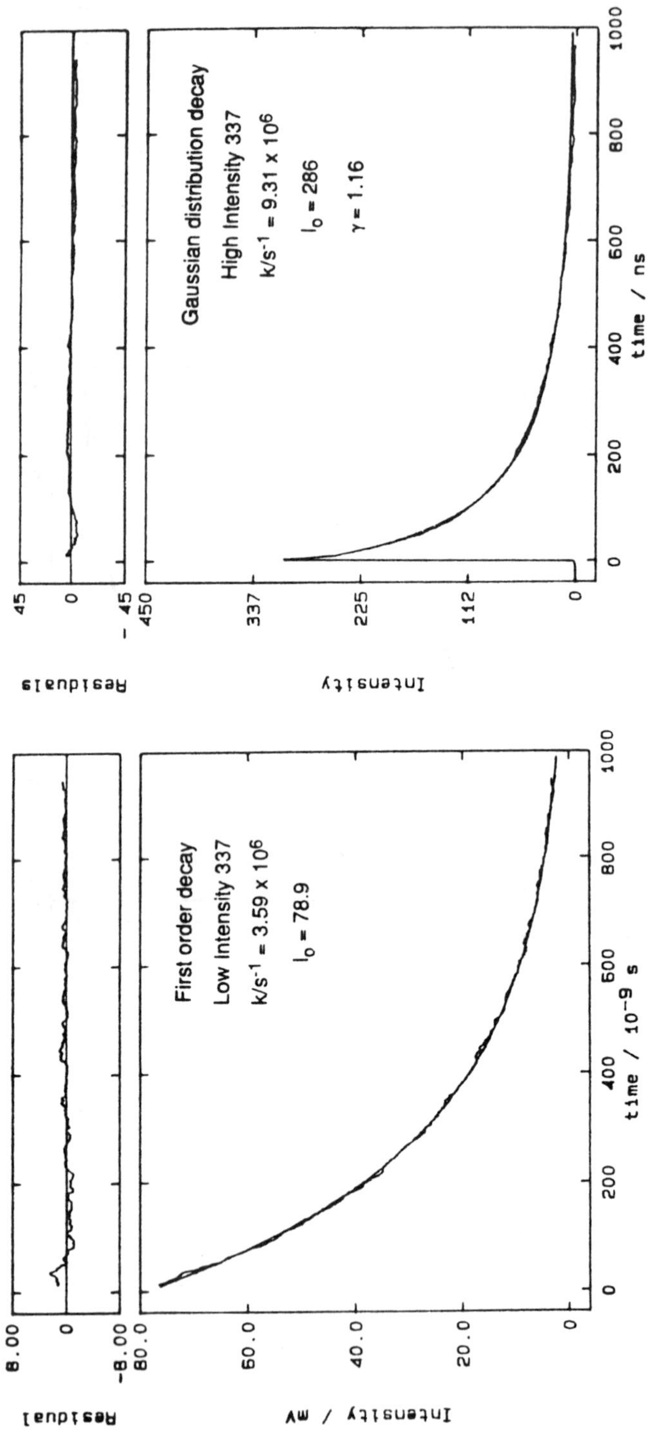
Gaussian distribution decay
High Intensity 337
k/s-1 = 9.31 x 10^6
I0 = 286
γ = 1.16
Residuals
Intensity
time / ns
First order decay
Low Intensity 337
k/s-1 = 3.59 x 10^6
I0 = 78.9
Residual
Intensity / mV
time / 10^-9 s

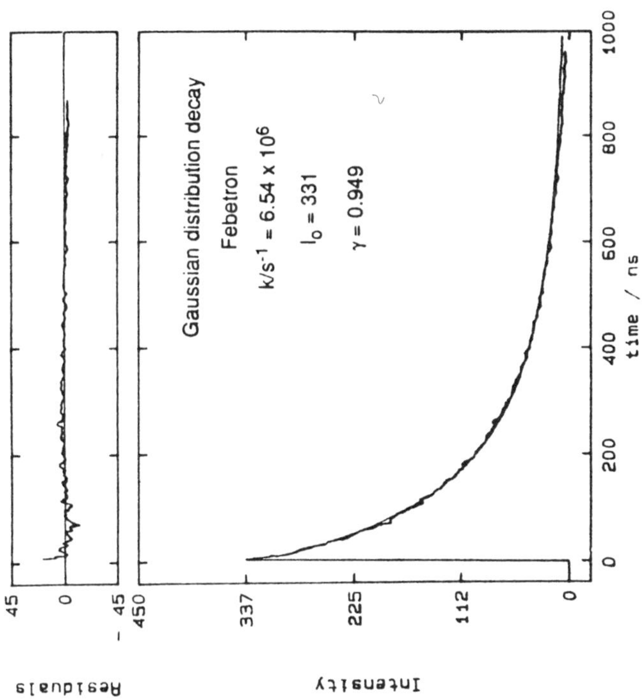

Figure 2c. Typical data from pulsed electron (Febetron) and laser (low and high intensity) irradiation of pyrene in polystyrene films for 10^{-1} M solute. The mode of fitting and the obtained kinetic parameters are given with the data.

which corresponds to a Gaussian distribution in ln(k) (*19*),

$$\ln(k) = \ln(\bar{k}) \pm \gamma x$$

where ($\bar{k}$) is the average and γ is the width of the distribution.

The comparison of the obtained data shows that the decay of excited pyrene is mono exponential when excited with low intensity laser pulses, with a rate constant of $k = 3.46 \times 10^6\ s^{-1}$, and quite non-exponential on excitation with electrons. The best fit to the data in the latter case is by the Gaussian distribution in ln(k). This effect is observed over the concentration range 10^{-3} to 10^{-1} M. Furthermore, excitation at high laser intensity or at a concentration of 0.20 M pyrene also produces non-single exponential decay kinetics of the pyrene excited state. The pertinent data are contained within the figure and the legend.

These data are interpreted in terms of local regions of high energy released by the 0.4 MeV electrons. The irradiation leads to local high concentrations of excited pyrene triplet and excited singlet states. This effect has been called a "spur" in radiation chemistry, and typically 40 eV of energy is lost in a volume of radius ~ 20 $Å^3$ (*20*). Hence with a G value of 3.0, (typical for most liquid systems), the local excitation state-free radical population is ~ 0.1 M. Similar effects are simulated at high pyrene concentrations and at high laser intensity.

Excited Singlet States and Ions. Figure 3 shows the transitory spectra observed in the pulse radiolysis of evacuated 0.10 M pyrene in polystyrene. Three major peaks are observed at 415, 455 and 493 nm. These correspond to the excited triplet state (*21*), the cation and the anion (*22*)of pyrene respectively. These species are produced within the 3 ns electron pulse, even at the lowest pyrene concentration (10^{-3} M), while the decays occur over the μsec to msec region. Similar behaviour is observed on irradiation of 0.10 M anthracene in polystyrene where an intermediate spectrum centered at 435 nm, is obtained, which corresponds to the anthracene triplet (*21*).

The addition of oxygen to these systems causes the pyrene absorptions at 415 and 493 nm, and the anthracene absorption at 435 nm to decay much faster. This is illustrated in Figure 4 for anthracene. The measured rate constant for this quenching process is $k_q = 2.6 \times 10^7$ $M^{-1}s^{-1}$, in agreement with photochemical measurements (*23*). In contrast, the pyrene 455 nm species is not significantly affected by the presence of oxygen; this is consistent with its assignment as a cation.

Figure 5 shows the yields of all arene species as a function of pyrene and anthracene concentrations in irradiated polystyrene. The concentration effect of these probes on the precursor polystyrene excimer species (monitored at $\lambda = 550$ nm) is shown for anthracene in Figure 6.

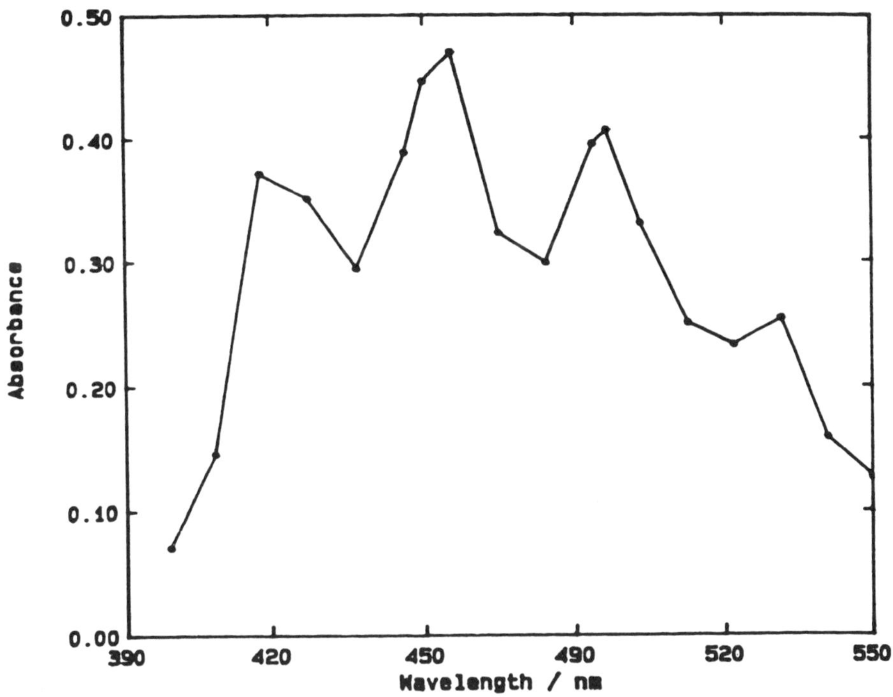

Figure 3. The transient absorption spectrum obtained from the pulsed electron irradiation of polystyrene film containing 10^{-1} M pyrene.

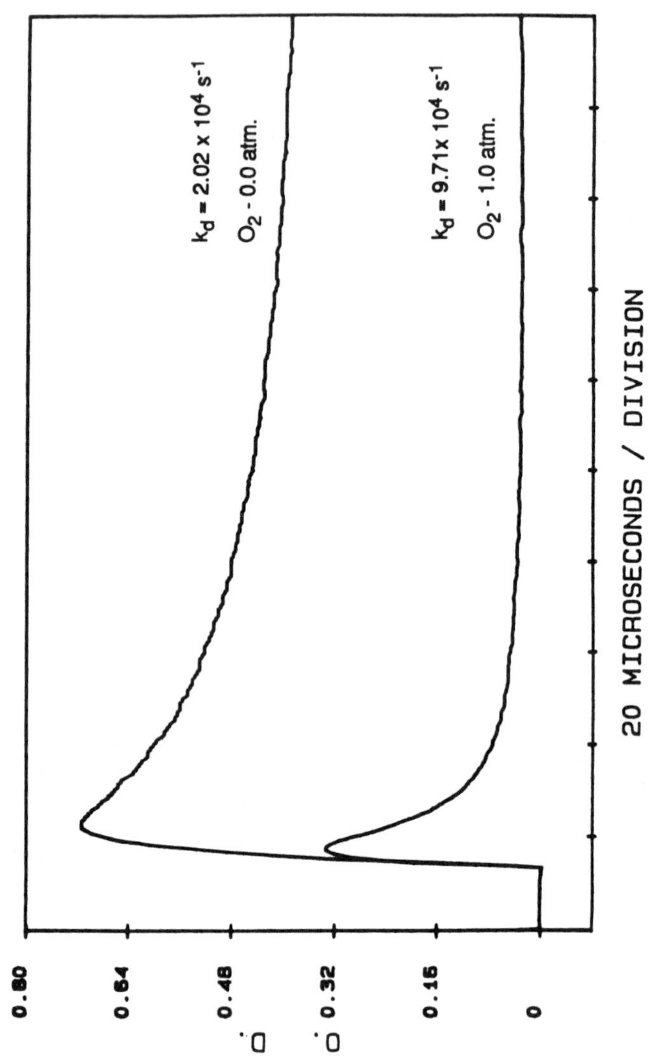

Figure 4. The oxygen quenching of the anthracene triplet decay in polystyrene film.

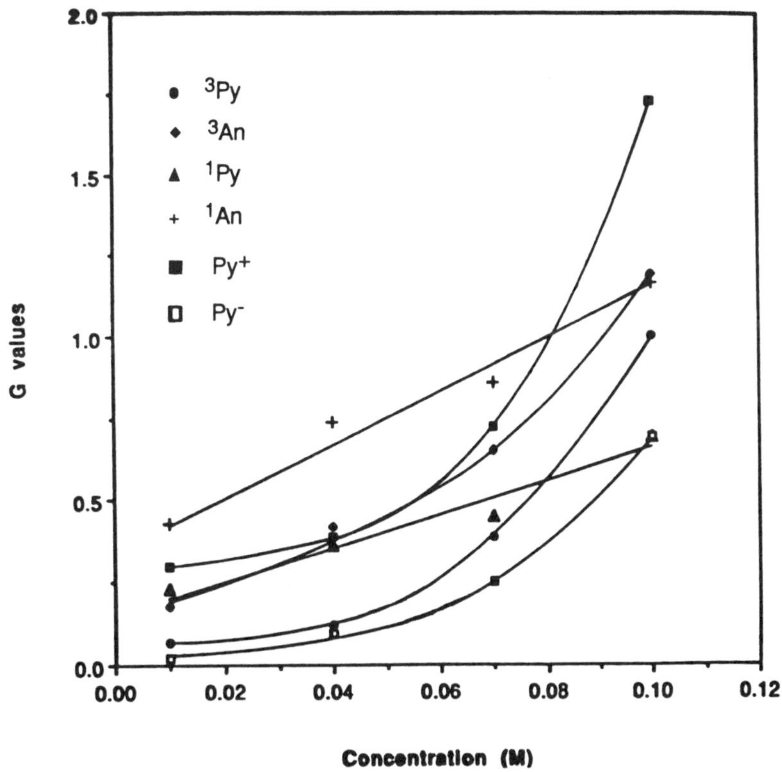

Figure 5. The yields for the transient species obtained upon the pulsed electron irradiation of polystyrene films containing pyrene and anthracene. The absorption extinction coefficients used were ^{3}An - 56,000 $M^{-1}cm^{-1}$, ^{3}Py - 34,000 $M^{-1}cm^{-1}$, Py^+ - 26,000 $M^{-1}cm^{-1}$ and Py^- - 45,000 $M^{-1}cm^{-1}$.

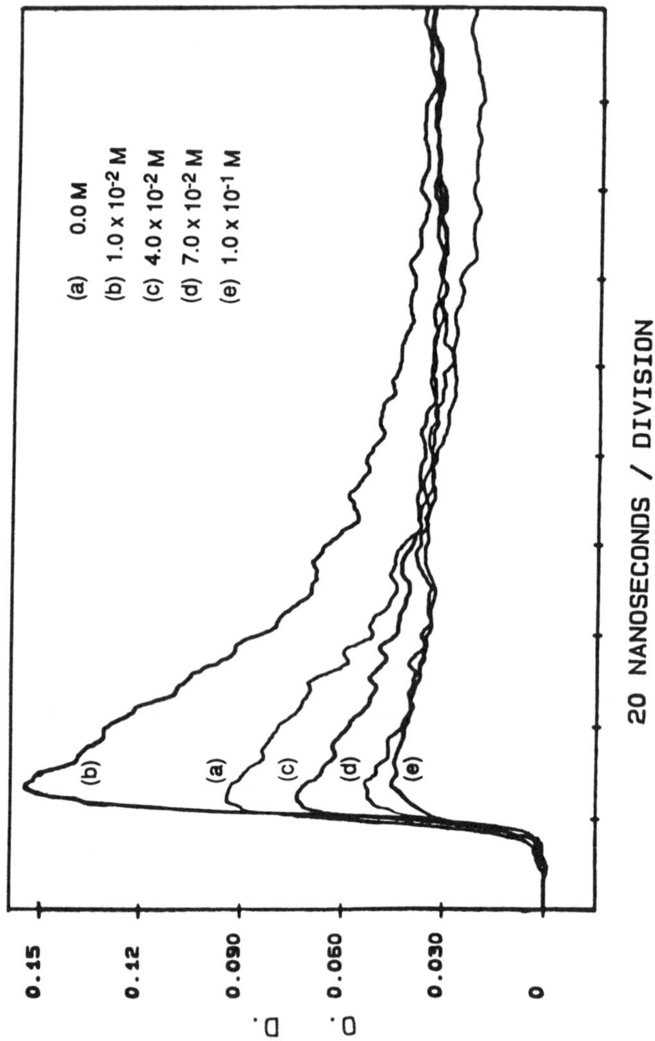

Figure 6. The effect of anthracene concentration on the polystyrene excimer yield observed at 550 nm.

Discussion

In the present studies it has been shown that the pulse radiolysis of pure polystyrene films gives rise to several different short lived species. In particular, the species absorbing at λ = 550 nm has been identified as the polystyrene excimer, in agreement with previous studies in polystyrene (*16*) and in liquid benzene (*10*). This species was also found to emit at 330 nm. Additional absorption maxima at 400 and ~ 900 nm are identified as being due to the polystyrene cation and anion respectively (*16*). No evidence has been obtained for a triplet state in this work, although derivatives of the triplet were found earlier in the work on liquid benzene (*10*).

The addition of solutes such as pyrene and anthracene to the polystyrene films give rise to fluorescence characteristic of these arenes, and the triplet, cation and anion absorptions for pyrene, and the triplet absorption for anthracene, were also observed. These species are formed by electronic and charge transfer from the polystyrene film to the added solute. The data also confirm these assignments by their spectral position (Figure 3) and by their reactivity with species such as oxygen (Figure 4).

The yields of the different solute ions and excited states are shown in Figure 5, and indicate a continuous rise with increasing solute concentration. However, the shapes of these curves are quite different for the singlet excited states and for the triplets, cation and anion. The latter show an accelerated increase in yields with increasing arene concentration while the singlet excited states is fairly linear over the concentration range 10^{-2} to 10^{-1} M. The absolute yields of these species are of the order of unity, in agreement with the singlet yield of G = 1.6 for the benzene excimer. This value was obtained from the established extinction coefficient in liquid benzene, and used directly for the polystyrene film.

The exact nature of the transfer of the electronic and excitation energy from the film to the solutes is a matter of some debate at this stage. However, Figure 6 indicates that addition of anthracene to the polystyrene gives rise to a decrease in the yield of the polystyrene excimer, without noticeable effect on its lifetime. Similar behaviour is obtained for the addition of pyrene. This effect is very abrupt, increasing sharply around 4 x 10^{-2} M An. This is <u>not</u> characteristic of energy transfer processes such as radiative or Förster transfer, as these exhibit a much smaller dependence of transfer rate on concentration.

The data is indicative, however, of the interference of the added solute with a precursor of the polystyrene excimer. This could be the monomer polystyrene, which is initially excited by the radiolysis event and then rapidly travels to a low energy trap; the excimer. As the concentration of solute increases there is a progressive interference with this process to capture the monomer energy directly. This process which is extremely rapid, would be too fast to observe on the present nanosecond timescale.

An alternative process is that the excited monomer and excimer polystyrene species have ionic precursors, and that the solutes interfere with the recombination of the ions that give rise to these states. At this stage of the investigation it is difficult to distinguish exactly between these two processes.

The formation of the solute ions is not as difficult to visualize. As previously indicated, both cations and anions of polystyrene are formed, and their charge can be transferred to the solute, either before these species have fully settled into their traps or by direct reaction.

The triplet yield is not as significant as would appear from the measurements. In the case of anthracene the majority of the triplet comes via intersystem crossing from the excited singlet. For pyrene, however, the singlet yield is lower and thus a larger proportion of the triplet appears to come from another source. This again could be transfer of triplet energy through the film to the solute, although the yield of this process is considerably smaller than any other processes that we have observed, or alternatively, it could be due to an ionic recombination which could be competing to give either singlet of triplet. At this stage we cannot distinguish between these two processes but we can indicate is that the yield of triplets from this process is small.

A clear feature in the present studies, however, is the non-homogeneous loss of energy to the film in the radiolytic excitation. This feature of "spur-like" events is analogous to that discussed for aqueous systems, but the present data give a clear snapshot of these events "frozen out" in the polymer film. By comparison to the high intensity laser studies, the local concentration of species can be estimated to be as high as 0.10 M, in agreement with the various estimates of spur size and degree of energy loss in the irradiated zones *(20)*.

Conclusion

The present work has shown the advantages of using a medium such as a polymer film to study early events in radiolysis which eliminate the diffusion processes that occur in liquids. These initial studies indicate the power of this method, and combined with data from photochemical techniques, which are presently underway, will enable us to pinpoint the mechanisms of electronic and ionic transfer occurring in these systems.

Acknowledgments

The authors would like to thank NSF for support of this work and Argonne National Laboratory for the loan of the Febetron.

Literature Cited

(1) Birks, J. "The Theory and Practice of Scintillation Counting"; Pergamon Press: Oxford, 1964.
(2) Magee J. L.; Huang, J. K. *J. Phys. Chem.* **1972**, *76*, 3801.
(3) Platzman, R. L. In "Radiation Research", Silini, G., Ed.; North Holland: Amsterdam, 1967, p 20.
(4) Thomas, J. K."The Chemistry of Excitation at Interfaces"; American Chemical Society: Washington, DC, 1984; ACS Monograph No. 181, p 17.
(5) Thomas, J. K. *Int. J. Radiat. Phys. Chem.* **1976**, *8* ,13.
(6) Beck, G.; Thomas, J. K. In "Liquid Scintillation Counting", Peng, Alphen, Horrocks, Eds.; Academic Press: New York, 1980; Vol. 5, p 17.
(7) Beck, G.; Thomas, J. K. *J. Phys. Chem.* **1985**, *89* , 4062.
(8) Thomas, J. K. *Record of Chem. Prog.* **1971**, *32*, 145.
(9) Thomas, J. K. *Int. J. Radiat. Phys. Chem.* **1976**, *8*, 1.
(10) Thomas, J. K.; Mani, I. *J. Chem. Phys.* **1969**, *51*, 1834.
(11) Thomas, J. K. *J. Chem. Phys.* **1969**, *51*, 770.
(12) Ho, S. K.; Siegel, S. *J. Chem. Phys.*, **1969**, *50*, 1142.
(13) Wilske, J.; Heusinger, H. *Int. J. Radiat. Phys. Chem.* **1970**, *2*, 131.
(14) Kira, A; Sumamura, M. *J Phys. Chem.* **1984**, *88*, 1865.
(15) Tagawa, S; Washio, M; Tabata, Y. *Chem. Phys. Lett.* **1979**, *68*, 276.
(16) Tagawa, S. *Radiat. Phys. Chem.* **1986**, *27*, 455.
(17) Milosavljevic, B.; Thomas, J. K. *J. Phys. Chem.* **1985**, *89*, 1830.
(18) Anka, M.; Hart, E. J. "The Hydrated Electron", Wiley: New York, 1969.
(19) Krasnansky, R.; Koike, K.; Thomas, J. K. *J. Phys. Chem.* **1990**, *94*, 4521.
(20) Pimblott, S. M.; Mozumder, A.; Green, N. J. B. *J. Chem. Phys.*, **1989**, *90*, 6595.
(21) Kikuchi, K. "Triplet-Triplet Absorption Spectra", Bunshin: Tokyo, 1989.
(22) Wallace, S.; Grätzel, M.; Thomas, J. K. *Chem. Phys. Lett.* **1973**, *23*, 359.
(23) Chu, D.-Y.; Kucynski, J.; Thomas, J. K. *Macromolecules* **1988**, *21*, 2094.

RECEIVED February 21, 1991

Chapter 5

Pulse Radiolysis of Doped Polyethylene in Molten State

Ortwin Brede

Central Institute of Isotope and Radiation Research, Permoserstrasse 15, O–7050, Leipzig, Germany

Pulse radiolysis experiments at 393 K with transparent, molten, but form-stable polyethylene samples enable the study of transient processes in PE under kinetically homogeneous conditions. As PE species cations of the olefin type and radicals were observed. Scavenger radicals were formed by PE exciton trapping and reactions of molecular alkyl radicals. It is reported on some applied PE pulse radiolysis studies which gave information for the antioxidants chemistry and some PE modification processes.

The time-resolved study of transient processes in polymers is of interest in radiation chemistry but also more generally in physical and polymer chemistry. Such prosesses driven mainly by radical species are responsible, e.g., for the radiation-induced crosslinking and grafting, for the polymer ageing during manufacture and in the subsequent use, for the selection of stabilizers etc.

Pulse radiolysis as the time-resolved technique of radiation chemistry enables the direct observation of reactive species as solvated or trapped electrons, ions, electronically excited states and radicals. In comparison to laser flash photolysis, pulse radiolysis has the advantage of the energy absorption proportionally to the electron increments of the sample, i.e., of transient generation within the matrix. Hence, reactions of transients within the polymer and with added scanvengers can be analyzed.

For the study of polyethylene (PE) two main difficulties exist that are caused by the semicristallinity of the polymer at room temperature: PE is non-transparent and represents a non- homogeneous system from the kinetical point of view. Therefore, until now PE as original material has been studied in pulse radiolysis with optical detection only in case of thin foils *(1)*. To overcome the mentioned limiting factors we started with PE pulse radiolysis in molten state which will be briefly reported in this paper.

0097–6156/91/0475–0072$06.00/0

Experimental

The experiments were performed at 393 K with transparent, molten, but higly viscous and form-stable polyethylene samples. Additives were admixed by heat-rolling at about 410 K or diffusion into the sample in the case of liquids. The samples were prepared by cutting of 4 mm thick PE plates to pieces of the dimensions of $4 \times 10 \times 20\ mm^3$. As polymer matrices different PE types were used. Most of the experiments reported here were made with the Leuna LDPE Al21FA having a cristallinity of 45 per cent *(4)*.

The pulse radiolysis experiments were performed mainly with 40 ns pulses of 1 MeV electrons of an ELIT type accelerator (dose per pulse between 100 and 200 Gy) *(2)*. In the μs-time scale some experiments were undertaken with the 3 MeV LINAC in Budapest (2.5 μs pulses with a similar dose) *(3)*.The detection system consisted of a boosted xenon lamp, a 1 P 28 photomultiplier, a grating monochromator and a 500 MHz real time oscilloscope. Details of the equipment are given elsewhere *(2)*.

Species in Polyethylene

In the PE radiolysis as reactive transients hydrogen atoms, alkyl radicals, cations, trapped electrons and vibronically excited molecules have to be taken into account. But only a part of these species is expected to be observable under the conditions of pulse radiolysis of the molten polymer.

Figure 1 shows optical transient absorption spectra taken in pulse radiolysis of molten pure PE. The spectra were measured point by point in different runs and a relatively good reproducibility was reached. There is a broad spectral tail coming from the UV range decaying to the VIS and continued in the NIR (not demonstrated here). From the kinetics analyzed in representative spectral points (insets in Figure 1) two general transient types could be distinguised - a short-living one ($\tau_{1/2} = 500$ ns) absorbing mainly in the VIS part and a long-living species ($\tau_{1/2} \geq 10\ \mu$s) absorbing mainly in the UV.

Experiments with PE samples containing typical cation scavengers as aliphatic amines and ethers gave arguments for the interpretation of the short-living species as PE cations and the long-living transients as PE radicals of different structures *(4)*. Figure 2 shows a typical result of such a cation scavenger action: in the presence of an oligomeric glycol ether the cation lifetime was reduced, the radical kinetics was not influenced and the radical yield dropped down.

It is reasonable that because of the non-uniformity of the polymer we can define only structure types of the observed species. In analogy to low-molecular olefin radical cations that have absorption bands at the UV border and in the red spectral range *(5)* the PE cations are interpreted to be of the olefin type, i.e., localized at a double bond.

As probable generation channels the charge transfer from the PE parent cations to alkene groups contained in the polymer (1) and the fragmentation of the molecular cations yielding olefin cations and molecular by-products (2) have to be taken into account.

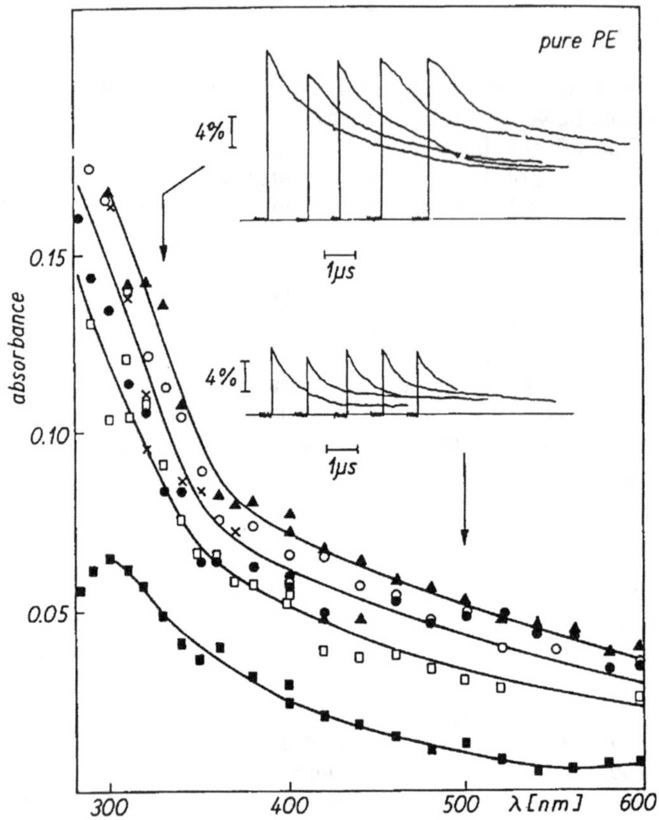

Figure 1. Transient spectra in pure PE taken immediately after a 40 ns pulse (various points and the curve set are due to different runs) and after 3 μs (■). Insets show time profiles at λ = 330 and 500 nm for the different runs. (Adapted from ref. *4*)

$$C_nH_{2n+2}^{\cdot +} + \overset{|}{\underset{|}{C}} = \overset{|}{\underset{|}{C}} \longrightarrow C_nH_{2n+2} + \overset{|}{\underset{|}{C}}{}^{\cdot} - \overset{|}{\underset{|}{C}}{}^{+} \qquad (1)$$

$$C_nH_{2n+2}^{\cdot +} \longrightarrow \text{PE-}\underset{|}{C}{}^{\cdot} - \overset{|}{\underset{|}{C}}{}^{+} + H_2 \ ; \ CH_4 \ ; \ C_2H_6 \quad \text{etc.} \qquad (2)$$

Then, the mentioned scavenger reaction with an ether or an amine should be a deprotonation of the PE cation.

$$\overset{|}{\underset{|}{C}}{}^{\cdot} - \overset{|}{\underset{|}{C}}{}^{+} + R_2O \longrightarrow R_2OH^{+} + \overset{|}{\underset{|}{C}} = \underset{|}{C} - \underset{|}{C}{}^{\cdot} - \qquad (3)$$

Radical Formation by Exciton Trapping

The depletion of PE radical yield in presence of added substances (cf. Figure 2) is accompanied by the formation of scavenger radicals taking place in times much shorter than our best time resolution of about 10 ns. This is demonstrated on an

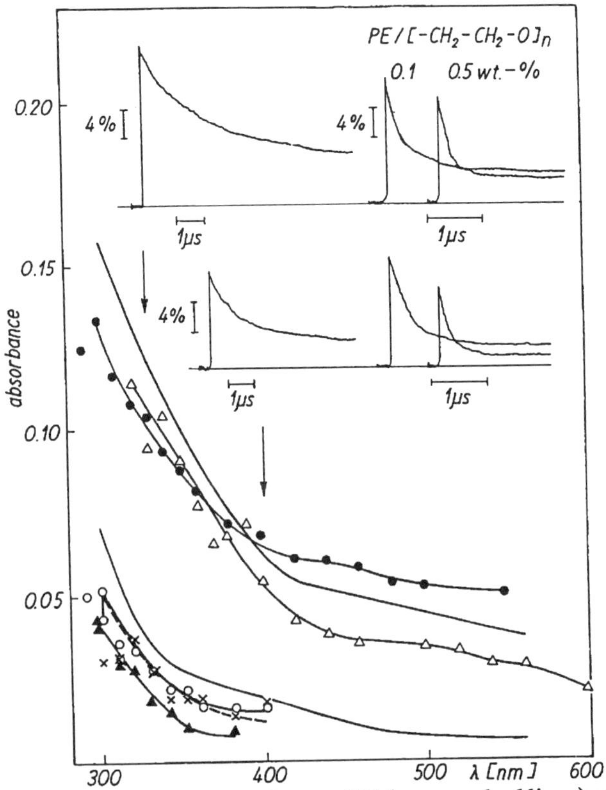

Figure 2. Optical absorption spectra in pure PE (non-marked lines) and samples containing 0.1 (● , o ,x) and 0.5 (△ , ▲) per cent oligomer glycol ether taken immediately after the pulse (upper set) and after 3 μs (lower curves). Insets show time profiles at 330 and 400 nm (PE left-hand side). (Adapted from ref. *4*)

example in that the optical absorption spectrum of the scavenger radicals lyes within our spectral observation range. In samples doped with benzophenone the well-known and very pronounced benzophenone ketyl radical spectrum is observed (Figure 3) also at very low additive concentrations($\geq 10^{-3}$ mol dm^{-3}). This fast formation of radicals within the PE matrix cannot be explained by a molecular process. In accordance with a hypothesis developed by Partridge *(6)* for explaining results of matrix isolation studies the described phenomenon will be interpreted by trapping of strongly-coupled singlet excitons. These excitons should be very mobile energy packets moving through the polymer chain with a frequency higher than that of the nuclear vibration rate. The decay by trapping either within the macromolecule (at internal irregularities) or by impurities or additives. Some considered conditions of exciton trapping according to the Partridge model are skeched in Figure 4.

As alrealy mentioned, the exciton trapping by scavengers takes place in competition to the internal trapping inside of the macromolecule as formulated in (4).

$$PE^* \longrightarrow PE^{\cdot} + H \qquad (4a)$$

$$PE^* \xrightarrow{+\ SH} S^{\cdot} + H + PE \qquad (4b)$$

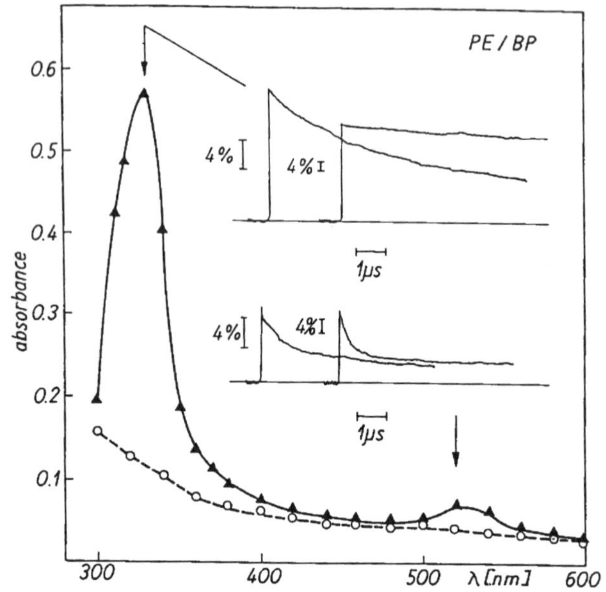

Figure 3. Transient spectrum in PE doped with 0.2 wt.-% benzophenone taken immediately after the 40 ns pulse (▲). (o) = PE background absorption. Insets show time profiles at 330 and 520 nm (PE pure left-hand curves). (Adapted from ref. *4*)

The external exciton scavenging seems to be a general phenomenon in all polyolefins and has been studied in a lot of cases *(4,7-9)*. But in contrast to the interpretation of Partridge *(6)* as products only radicals were observed.

The ratio between internal and external trapping can be changed by introducing additional internal traps in the PE erther by use of different PE types or by variation of the quality of the polymer, e.g., by oxidation of the PE. Hence, in respect to its number of internal exciton traps it is posible to characterize the quality of a PE by the determination of the yield of the scavenger radicals using, e.g., benzophenone in form of its ketyl radical as monitor *(10)*.

Reactions of Massive PE Radicals

The PE radicals represented in Figure 1 by the long-living spectral tail include very different structures as, e.g., alkyl radicals in different positions, hetero-atom centered radicals etc.. Nevertheless, taking them formally as a unity it is possible to describe their kinetic fate.

In pure PE the radicals decay with a rate of $3 \times 10^4\ s^{-1}$ yielding a very long-lasting product observable by a residual optical absorption. This can be seen by the left-hand time profile given in Figure 5. This effect is interpreted as a conversion of PE alkyl radicals into allylic ones (5) known to proceed at very low temperatures in the time scale of seconds. Because of our pulse radiolyses results obtained in molten PE the process has been formulated in one step, but may happen also in a more-step sequence for that arguments were given by Dole *(11)*.

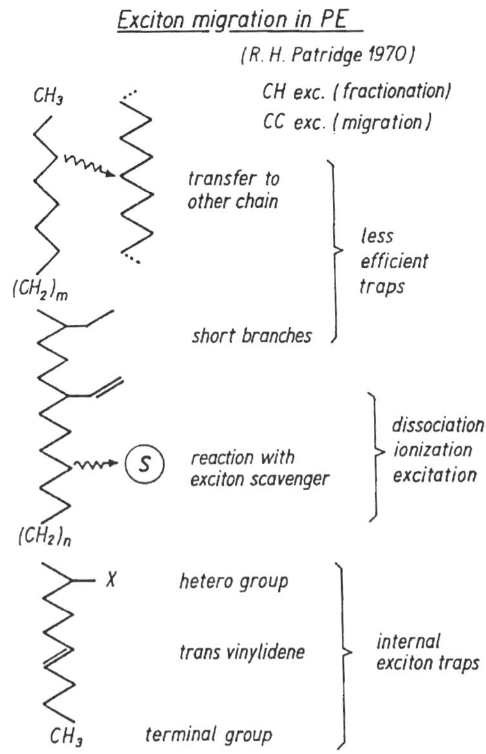

Figure 4. Qualitative considerations on the PE consistency connected with the exitation transfer efficiency (modes of exciton trapping).

$$PE^{\cdot} \equiv -\underset{|}{\overset{|}{C}}-\underset{|}{\overset{|}{C}}-\underset{|}{\overset{|}{C}}^{\cdot} \dashrightarrow H_2 + -\underset{|}{C}=\underset{|}{C}-\underset{|}{C}^{\cdot}- \quad (5)$$

In the presence of a scavenger as, e.g., diphenylamine the PE alkyl radicals (perhaps also the other primary radicals) react under generation of diphenylaminyl radicals. This happens in the millisecond time scale and can be derived from the time-resolved part of the right-hand time profile of Figure 5 taken near the aminyl absorption maximum *(4)*.

$$PE^{\cdot} + \Phi_2NH \longrightarrow \Phi_2N^{\cdot} + PE \quad (6)$$

Similar effects are observable also using other types of scavengers *(9,12)*.

Sterically Hindered Phenols as Scavengers

The observation of transients in molten PE can serve as a basis for analyzing the elementary reactions of antioxidant action within the polymer matrix. This was performed with a series of sterically hindered phenols *(9)*. As an example the case of bis-(2-hydoxy-3-tert.-butyl-5-methylpenyl)sulfide will be elucidated.

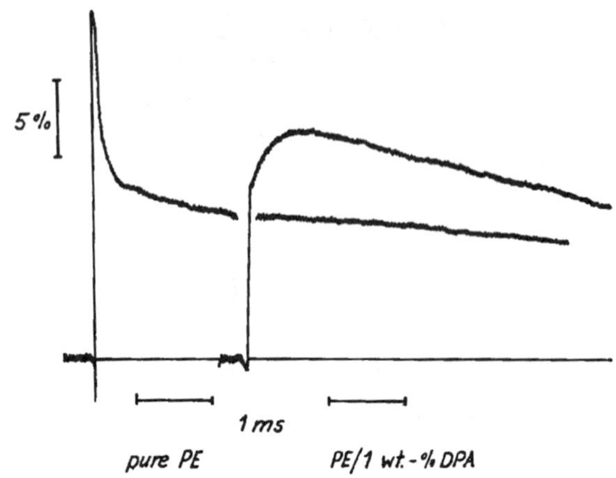

Figure 5. Time profiles taken in pure PE at $\lambda = 320$nm (on the left) and PE containing 1 wt-% diphenylamine at $\lambda = 350$ nm (on the right hand) with μs-pulse radiolysis.

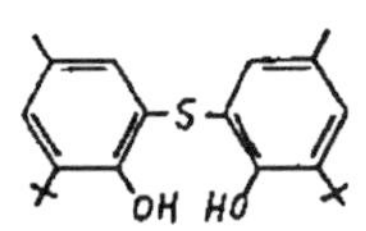

Ionizing irradiation generates in this sulfur-bridged phenol three different species which could be identified by comparative experiments in liquid alkane solutions *(13)*. Figure 6 gives the transient absorption spectum of such a phenol / PE sample taken immediately after a 40 ns electron pulse. Three absorption maxima can be distinguished identified as follows:
- λ_{max} = 340 nm, phenoxyl type radical (Φ-O $^{\cdot}$),
- λ_{max} = 390 nm, phenolate anion (Φ-O $^{\cdot -}$) and
- λ_{max} = 480 nm, sulfidyl radical (-S $^{\cdot}$-).

The phenoxyls are formed by different reaction channels as, e.g., PE exciton trapping (7) and reaction of massive PE radicals (8) that are analogeous to reactions (4b) and (6).

$$\Phi\text{-OH} \;+\; \text{PE}^{*} \longrightarrow \Phi\text{-O}^{\cdot} \;+\; \text{H} \;+\; \text{PE} \qquad (7)$$
$$\Phi\text{-OH} \;+\; \text{PE}^{\cdot} \longrightarrow \Phi\text{-O}^{\cdot} \;+\; \text{PE} \qquad (8)$$

As additional species phenolate anions were found in consequence of the dissociative electron attachment reaction (9).

$$\Phi\text{-OH} \;+\; e_{tr}^{-} \longrightarrow \Phi\text{-O}^{\cdot -} \;+\; \text{H} \qquad (9)$$

Electrons are not directly observed in our experiments with molten PE. But they exist, certainly, in the picosecond time range and can be scavenged by additives forming anions. The neutralization of Φ-O$^{\cdot -}$ (see time profiles in Figure 6) delivers a further part of phenoxyls and also S-centered radicals (formed only by this path)

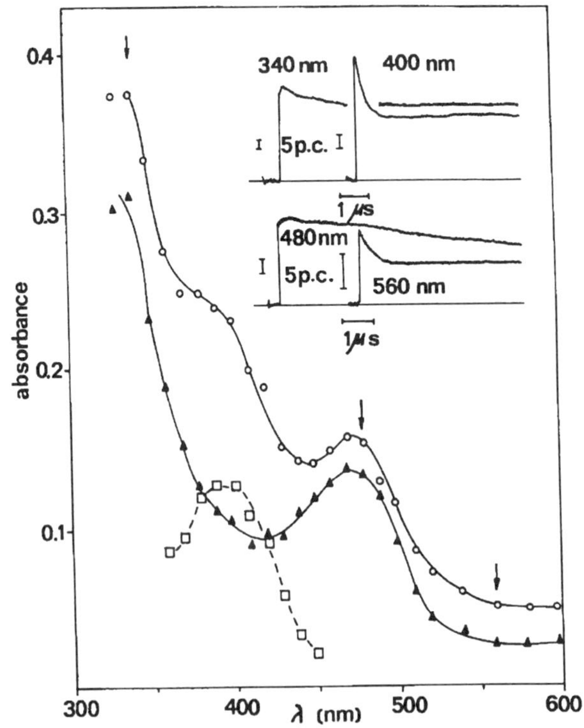

Figure 6. Transient absorption spectra of PE samples containing 0.2 (o - immediately after pulse) and 0.5 wt-% of the S-phenol (o - immediately, ▲ -after 3 μs, □ -difference between o and ▲). Insets show time profiles for the o.5 wt-% containing sample). (Adapted from ref. *9*)

which can be explained by the mesomery of the phenolate anion as formulated in (10).

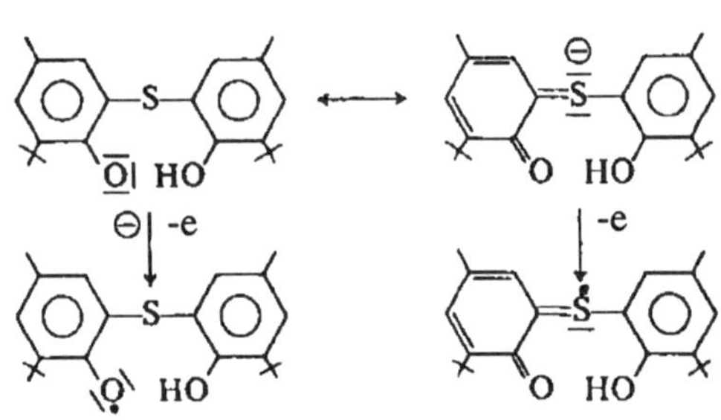

These sketched experiments should demonstrate the power of pulse radiolysis in kinetic studies in the antioxidant chemistry. Further details are given in *(14)*.

PE Doped with Aromatic Olefins

Connected with considerations on the radiation-induced grafting of monomers onto PE *(15)* a pulse radiolysis study on PE samples doped with aromatic olefins was made *(16)*. Figure 7 shows transient spectra of PE samples containing styrene (ST). The

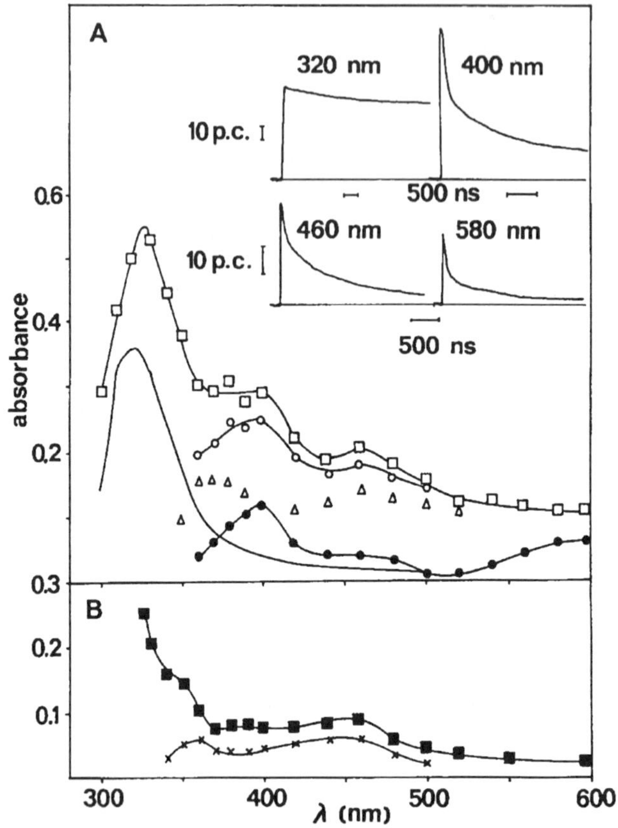

Figure 7. Transient absorption spectra taken from PE samples after lying 31 h in ST (part A) and 26 h in a 1:10 (v:v) mixture of ST and CCl_4 (part B), resp. (□ ,■) immediately after the 40 ns pulse, (-) non-marked curve after 3 μs, (o) difference spectrum between (□) and (-), (●) anion spike decaying within 200 ns, (△) cationic part, (x) cation spike taken up to 1 μs. Insets show time profiles in PE / ST. (Adapted from ref. *16*)

manifold of species could be cleared up by using carbon tetrachloride as additional scavenger and by comparing the results with those of liquid state experiments *(17)*:

- λ_{max} = 320 nm, benzyl type radicals (ST ·),
- λ_{max} = 350 and 460 nm, dimer styrene radical cations and
- λ_{max} = 400 nm styrene radical anions.

It was found that carbon tetrachloride scavenges the electrons as the precursors of styrene anions and reduces the yield of benzyl radicals. In Figure 7, looking on the difference spectra taken from transient absorptions at different time delay and analyzing the transient kinetics the mentioned identification of the species could be further verified.

Also at a very low styrene concentration of about 0.05 mol dm^{-3} all transients were found to be generated within the electron pulse and no time-resolved formation could be observed. At a styrene concentration $< 10^{-2}$ mol dm^{-3} only the benzyl type radicals could be found. This suggests that also these radicals are generated by the exciton trapping process as elucidated above.

$$PE^{*} \longrightarrow PE^{\cdot} + H \qquad (11a)$$

$$PE^{*} \xrightarrow{+ST} [PE...ST]^{*} \longrightarrow [PE^{\cdot} + ST^{\cdot}] \qquad (11b)$$

The radical pair formed in reaction (11b) can recombine efficiently (12a) or can be separated by diffusion (12b) where the latter one is very restricted within the polymer. Therefore, only a small amount of free PE$^{\cdot}$ can survive being able to initiate a grafting in form of longer styrene chains (13).

$$[PE^{\cdot} + ST^{\cdot}] \longrightarrow PE\text{-}ST \qquad (12a)$$

$$[PE^{\cdot} + ST^{\cdot}] \longrightarrow PE^{\cdot} + ST^{\cdot} \qquad (12b)$$

$$PE^{\cdot} + ST \longrightarrow PE\text{-}ST^{\cdot} \xrightarrow{+ST} PE\text{-}ST\text{-}ST^{\cdot} \qquad (13)$$

$$ST^{\cdot} + ST \longrightarrow ST\text{-}ST^{\cdot} \quad \text{ect.} \qquad (14)$$

The main part of all the radicals underlyes deactivation according to (12a) or can initiate styrene homopolymerization (14) which was also found to dominate by steady-state experiments *(15)*.

The experimentally supported considerations show that exciton scavenging in PE is a very efficient process. But because of the competition of internal and external trapping of excitons (11a, b) with increasing styrene concentration the PE radical formation via (11a) will be suppressed and, therefore, grafting onto the matrix becomes inefficient.

From the ionic part of the styrene transients information on the reactivity of the ionic precursors in the PE matrix can be obtained. In distinction to the liquid state, styrene cations were formed only at relatively high concentrations that speaks for a fast fragmentation of the PE parent ions as already formulated for by reaction (1). As found in liquid state experiments for alkene cations *(5)* also the PE olefin type cations undergo only a relatively slow charge transfer to styrene (15).

$$>C^{\cdot}\text{-}C^{+}< + ST \longrightarrow ST^{\cdot +} + \text{-}\overset{|}{C} = \underset{|}{C}\text{-} \qquad (15)$$

In molten PE the trapped electrons have a lifetime much shorter than or electron pulse. But with high styrene concentrations they are scavenged, here as styrene anions.

$$e_{tr}^{-} + ST \longrightarrow ST^{\cdot -} \qquad (16)$$

In a similar way as for grafting the process of the oligobutadiene-sensitized PE crosslinking has been analyzed *(18)*. As intermediates radicals and cations were found.

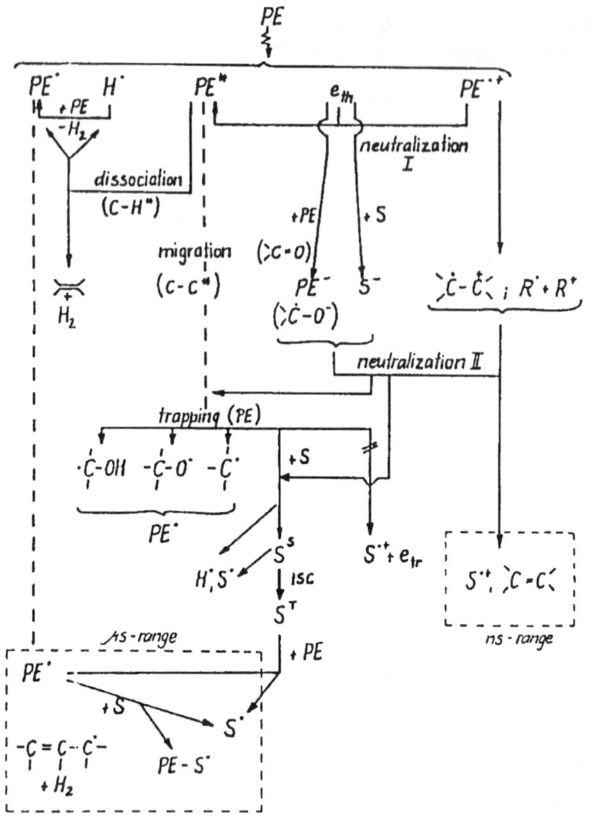

Figure 8. Radiation-induced primary processes in molten PE doped with a scavenger S. The processes proceed within the picosecond time range with the exception of the marked areas.

Conclusions

The pulse radiolysis experiments with molten PE samples enabled the observation of a lot of species and processes. Figure 8 gives a survey on the radiation-induced transient processes in the system PE / scavenger. Generally, PE excitons are precursors of the most part of matrix radicals as well as of the scavenger ones. As cations PE olefin ions were observed being products of the parent ion fragmentation. By electron scavenging and charge transfer product anions and cations are formed. Scavenger radicals are generated via exciton trapping (fast process) and reaction of massive PE radicals (ms time scale).

The observation of the mentioned species marked in Figure 8 by the dashed line areas may be of interest from the point of basic research (7) and also under more applied aspects as, e.g., the analysis of processes of polymer ageing, the description of modes of stabilizer action (14) and the better understanding of distinct processes of polymer modification.

Acknowledgments. The author gratefully acknowledges financial support of the LEUNA WERKE-AG (Germany) and scientific consultation by Drs. L. Stephan and T. Taplick.

Literature Cited

(1) Johnson, G. R. A.; Willson, A.; *Radiat. Phys. Chem.* **1977**, *vol 10*, p 89

(2) *Bertram, D.; Brede, O.; Helmstreit, W.; Mehnert, R.; Z. Chem. (Leipzig)* **1975**, *vol 15*, p 125

(3) Hargittai, P.; Stenger, V.; Földiak, G.; *ZfI-Mitteilungen (Leipzig)* **1984**, *vol 98*, p 514

(4) Brede, O.; Hermann, R.; Wojnarovits, L.; Stephan, L.; Taplick, T.; *Radiat. Phys. Chem.* **1989**, *vol 34*, p 403

(5) Mehnert, R; Brede, O.; Cserep, Gy.; *Radiat. Phys. Chem.* **1985**, *vol 26*, p 353

(6) Patridge, R. H.; *J. Chem. Phys.* **1970**, *vol 52*, pp 2485, 2491, 2501

(7) Brede, O.; Hermann, R.; Helmstreit, W.; Taplick, T.; Stephan, L.; *Makromol. Chem., Macromol. Symp.* **1988**, *vol 18*, p 113

(8) Brede, O.; Naumann, W.; *Radiat. Phys. Chem.* **1988**, *vol 32*, p 475

(9) Brede, O.; Wojnarovits, L.; Stephan, L.; Taplick, T.; *Poly. Deg. and Stab.* **1989**, *vol 24*, p 257

(10) Brede, O.; Radiation-Induced Processes in Polyethylene (a Collection of Papers), *ZfI-Mitt. (Leipzig)* **1987**, *vol 132*, pp 1-231

(11) Dole, M.; *Polymer Plast. Technol. Eng.* **1979**, *vol 13*, p 41

(12) Brede, O.; Wojnarovits, L.; *Radiat. Phys. Chem.* **1990**, *vol 36*, p 239

(13) Brede, O.; *Radiat. Phys. Chem.* **1987**, *vol 29*, p 369

(14) Brede, O.; *Proceedings of the 12th Intern. Conf. on Advances in the Stabilization and Controlled Degradation of Polymers,* Luzern / Switzerland, May 21-23, 1990, State University of New York, N.Y., USA, **1990**, pp 29-47; *ZfI-Preprint* No. 82, March 1990

(15) Friese, K.; Michler, G.; Steinbach, H.; Hamann, B.; Runge, J.; *Angew. Makromol. Chem.* **1986**, *vol 141*, p 185

(16) Brede, O.; Luther, B.; *Isotopenpraxis (Leipzig)* **1989**, *vol 25*, p 239

(17) Brede, O.; Bös, J.; Helmstreit, W.; Mehnert, R.; *Radiat. Phys. Chem.* **1982**, *vol 19*, p 1

(18) Brede, O.; Beckert, D.; Wunsch, K.; Hößelbarth, B.; Specht, W; Tannert, F.; *Radiat. Phys. Chem.* **1989**, *vol 34*, p 415

RECEIVED April 17, 1991

Chapter 6

Radiation Effects on the Structure and Properties of Poly(vinylidene fluoride) and Its Ferroelectric Copolymers

Andrew J. Lovinger

AT&T Bell Laboratories, 600 Mountain Avenue, Murray Hill, NJ 07974

Electron- and γ-irradiation of copolymers of vinylidene fluoride with tri- or tetra-fluoroethylene has been shown to induce solid-state ferroelectric-to-paraelectric transitions. Their Curie temperatures decrease with increasing radiation dose, while at the higher doses full amorphization is obtained as a result of cross-linking. Poly(vinylidene fluoride) homopolymer shows only amorphization without an intervening Curie transition. A structural model is presented to explain this behavior, and the thermal, dielectric, and dynamic mechanical properties of the irradiated polymers are discussed.

The radiation chemistry of polymeric materials is a rapidly expanding field because of the extensive modification of properties which it renders possible. This may be a result of chemical changes (*e.g.*, cross-linking or chain scission) as well as structural effects (*e.g.*, on crystallinity, molecular conformation, or interchain packing). When crystalline polymers are irradiated by X-rays or electrons, they typically undergo cross-linking through formation and subsequent recombination of free radicals, as typified by polyethylene. Rarely (*e.g.*, in polyoxymethylene) chain scission is the dominant effect, leading to progressive reduction in molecular weight and eventual degradation to nonpolymeric materials. These aspects of ionizing radiation on crystalline polymers have been reviewed thoroughly (see, *e.g.* ref. *1* and *2*).

Among such crystalline polymers, poly(vinylidene fluoride), $-(CH_2-CF_2)_n-$ [abbrev. PVF_2], and its copolymers continue to attract high interest because of their exceptional electroactive properties. These include piezo-, pyro-, and ferro-electricity, and are leading to important applications in areas such as electromechanical transducers, infrared-, motion-, and impact detectors, microphones, hydrophones, speakers, and medical devices. Copolymers with

0097–6156/91/0475–0084$06.00/0

trifluoroethylene, $-(CHF-CF_2)-$ [abbrev. F_3E], are particularly promising because they have higher crystallinities, piezoelectricities, and electromechanical coupling coefficients. Another reason is that they can be crystallized directly into their polar phases without requiring mechanically induced transformations as does PVF_2. The structures, properties, processing, and applications of PVF_2 and its copolymers have been summarized recently in a textbook (*3*) and in a number of reviews (*e.g.*, *4-7*).

The effects of irradiation on PVF_2 and its copolymers have been investigated only recently, and interest in them is growing because of the extraordinary findings obtained. Among these are major structural changes, including ferroelectric-to-paraelectric transformations at room temperature controllable by radiation dose. These effects on structure and properties are summarized in this report.

Irradiation of PVF_2

Poly(vinylidene fluoride) was found (*8*) to be quite sensitive to radiation-induced degradation. Typical changes in the electron-diffraction pattern of the electrically active polar β-form of PVF_2 are seen in Figure 1 as a function of irradiation dose. Here and in Figures 2 and 3 the actual electron doses were measured in C/m^2. The *absorbed* dose is a function of the stopping power of the particular material, which depends upon chemical composition, density, and film thickness. For common carbon-based polymers, 1 C/m^2 is equivalent to *ca.* 25-40 Mrad (*1*); for PVF_2 and its copolymers discussed here, Macchi *et al.* (*9*) found an equivalence of *ca.* 33-38 Mrad. As is seen in Figure 1, the crystallographic reflections become progressively weaker and more diffuse during irradiation, until their eventual replacement by amorphous haloes. This is indicative of a cross-linking process but occurs at doses much smaller than is typical for polymers (*e.g.*, *ca.* one-third of that for polyethylene). No substantial differences were found (*8*) in the rates or doses of this radiation-induced cross-linking among the α, β, and γ polymorphs of PVF_2.

Remarkable phenomena were observed for this polymer at much *lower* doses. Wang (*10*) showed that for γ-ray doses of up to 50 Mrad there was an improvement in thermal stability of polar $\beta-PVF_2$, as well as in its piezoelectric properties. Specifically, a slower piezoelectric decay that rendered the films usable to higher temperatures was associated with the slight formation of cross-links, which was not sufficient to cause amorphization but only to impede the molecular relaxation of the dipole-containing groups. For similar reasons, an enhanced dimensional stability was also observed at high temperatures (*10*). Pae and co-workers (*11*) have performed a similar study using 175-keV electrons from a Van de Graaf accelerator at doses up to 50 Mrad, this time on the anti-polar α-phase of PVF_2. They found that the limited introduction of cross-links leads to a lowering in melting point as expected, but also detected a significant increase in crystallinity. They ascribed the latter to chain scission, which could have increased the mobility of molecular chains, thus leading to further recrystallization (*11*).

Irradiation of PVF_2 Copolymers

The effects of γ- and electron irradiation have also been explored for copolymers of PVF_2 with trifluoroethylene (F_3E) and tetrafluoroethylene (F_4E). For the

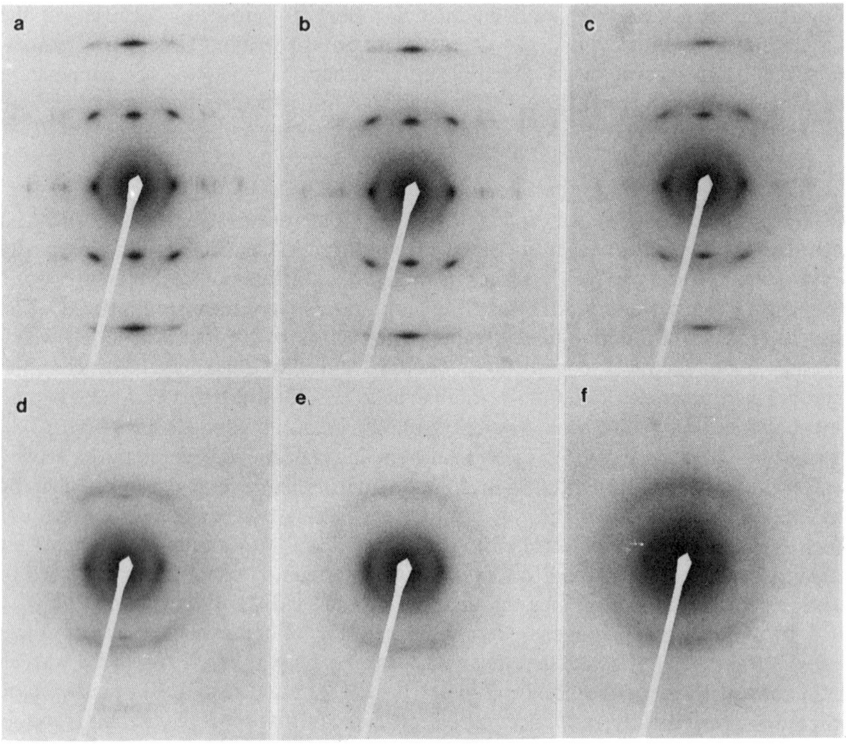

Figure 1. Changes in the electron-diffraction pattern of the piezoelectric β-phase of PVF_2 that had been uniaxially oriented (fiber axis vertical) and subjected to 100 keV electrons at 25°C. (a) initial exposure, (b) 9 C/m^2, (c) 20 C/m^2, (d) 25 C/m^2 , (e) 32 C/m^2, and (f) 38 C/m^2. Reproduced from ref. 8. Copyright 1985 American Chemical Society.

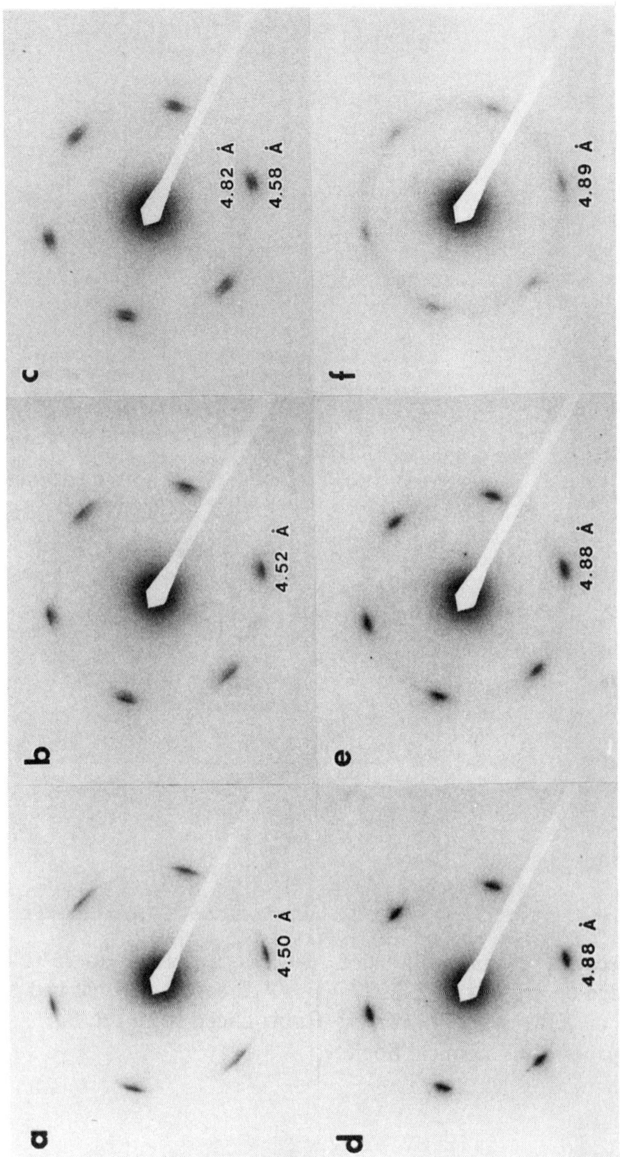

Figure 2. Effects of electron irradiation on the selected-area electron-diffraction pattern from a single crystal of a 73/27 mol % VF_2/F_3E copolymer, grown at 135°C and examined at 25°C at 100 keV. (a) initial exposure, (b) after 3 C/m^2, (c) 4 C/m^2, (d) 6 C/m^2, (e) 9 C/m^2, (f) 14 C/m^2. Reproduced from ref. 8. Copyright 1985 American Chemical Society.

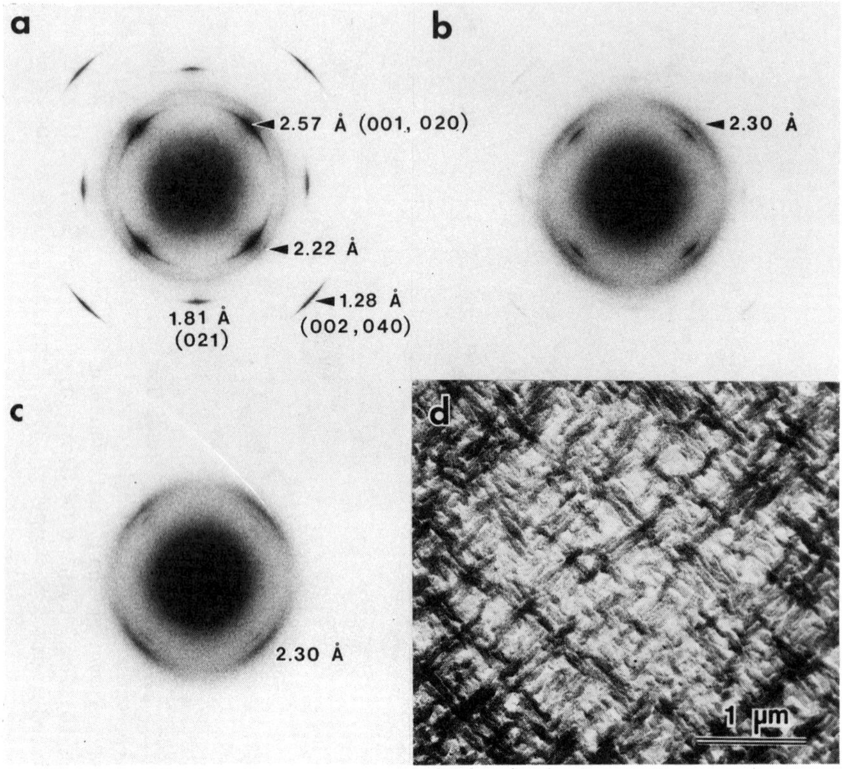

Figure 3. Consecutive electron-diffraction patterns (recorded at 25°C and 100 keV) from edge-on crystals of a 73/27 VF_2 / F_3E copolymer that had been grown epitaxially on KBr as seen in part d. Reproduced from ref. 8. Copyright 1985 American Chemical Society.

VF_2 / F_3E copolymers, compositions containing 50-80 mol % VF_2 have been studied extensively. These have the advantage that they generally crystallize directly into the ferroelectric β-phase of PVF_2, since the trifluoroethylene units promote adoption of the *trans* conformation (*12, 7*). Among tetrafluoroethylene copolymers, an 81/19 mol % VF_2 / F_4E composition has been studied in some detail, even though others also adopt the ferroelectric β-structure (*13, 14*). Contrary to PVF_2 homopolymer, both of these copolymers were found to undergo major and highly unusual crystallographic transformations upon irradiation. These are summarized below in terms of structure and properties.

Structural Transformations in PVF_2 Copolymers. The discovery of electron-induced structural transformations in VF_2 / F_3E copolymers was reported in ref. *8*. It was shown that single crystals of electron copolymers undergo solid-state transformations during electron irradiation at room temperature in the manner depicted in Figure 2. At extremely low doses (Fig. 2b and c) new reflections appear at a higher interplanar spacing but with the same azimuthal disposition as the original ferroelectric-phase reflections. At a slightly higher dose, the new reflections increase dramatically in both intensity and sharpness, as the original reflections disappear (Fig. 2d). With further irradiation, the new reflections begin to fade, leading to an amorphous, cross-linked structure (Fig. 2e and f). It was concluded *(8)* that the ferroelectric lattice is replaced by a paraelectric one since the spacings of the new reflections corresponded to those of the paraelectric lattice that had previously been obtained only at high temperatures *(12,15)*. The much higher intensity and reduced arcing of these paraelectric-phase reflections (Fig. 2d) compared to their ferroelectric counterparts (Fig. 2a) is *exactly opposite* to the usual degradative effects of electron irradiation. The greater perfection of this radiation-induced paraelectric lattice was also demonstrated by dark-field electron microscopy *(8)*.

However, the true paraelectric nature of the *individual* macromolecules could not be ascertained from this evidence, since all the reflections in Fig. 2 were of the hk0 type, *i.e. inter*molecular. For this reason, the *conformational* changes accompanying electron irradiation were probed after crystallizing the copolymer lamellae *on edge, i.e.* with the chains parallel to the substrate. These are seen in Figure 3d, together with the corresponding consecutive changes in the electron-diffraction pattern during irradiation (Fig. 3a-c). Comparison of Figures 3a and 3c shows that the sharp reflection from the intramolecular all-trans repeat (at 2.57 Å) is replaced by a broad and diffuse reflection at 2.30 Å, which had been shown earlier *(12,15)* to typify the disordered conformation of the paraelectric phase; this conformation consists of irregular sequences of TG, $T\bar{G}$, and TT groups. In this manner, this unique electron-induced ferroelectric-to-paraelectric transformation at room temperature was established both intra- and inter-molecularly *(8)*. It is also noteworthy that this transformation occurs at extremely low doses, which are *ca.* 20-25 times smaller than those needed to impart the usual electron-induced changes (*i.e.*, amorphization) to typical polymers (*e.g.*, polyethylene).

Similar changes were also observed for other vinylidene-fluoride copolymers using different irradiation sources and different detection techniques. Specifically, Odajima and co-workers *(16)* utilized a ^{60}Co source to expose thicker (20-40 μm) films of a 65/35 mol % VF_2 / F_3E copolymer to γ-rays. As seen in

Figure 4, irradiation at 120 Mrad caused the major intermolecular reflection of the ferroelectric phase (*i.e.*, the combined {200, 110} at *ca.* 19.50° 2θ) to disappear in favor of the equivalent paraelectric peak (at *ca.* 18.2°), in excellent agreement with the electron-microscopic results from the same copolymer *(8)*. For a lower dose of 60 Mrad (Fig. 4b), only an incipient transformation was found by irradiation at ambient temperature; however, exposure at 120°C (*i.e.*, with the polymer in the paraelectric state) was more effective in inducing this structural conversion. This conclusion was apparently also supported by IR spectroscopy, which was reported *(16)* to show an increasing absorption of a new band at 1710 cm^{-1} (attributed *(16)* to $-CF=CH-$ bonds) with temperature of irradiation.

For lower VF_2 contents, the behavior is more complex because the room-temperature interchain peak consists of two components that have been ascribed *(17,18)* to a coexistence of ferroelectric and paraelectric phases (an orthorhombic packing of the ferroelectric phase has also been proposed *(19)*. As was shown by Odajima, *et al.* *(20)* (see Figure 5a), the higher-angle one of these two peaks (consistent with the ferroelectric lattice *(17,18)*) continuously decreases in intensity as both peaks shift to a lower 2θ angle (characteristic of the paraelectric lattice) with increasing γ-ray dose. The partial transformation to the paraelectric phase at intermediate doses (*e.g.*, 80 Mrad) could be fully completed by subsequent heating (see Figure 5b).

Such transformations have also been documented for VF_2 / F_3E copolymers richer in vinylidene fluoride (*e.g.* 78 mol %) *(8,20)*, as well as for VF_2 copolymers with *tetra*fluoroethylene *(8)*. The central question following these findings is, of course, why all of these copolymers undergo this extraordinary radiation-induced polymorphic transformation, while the homopolymer itself does not. This may seem surprising particularly in terms of the very strong chemical and structural similarities between the two materials. As seen in Figure 6, the chains have the same (trans) conformation in their ferroelectric phases, as well as the same chain packing. The major difference lies in the expansion of the *a* and *b* axes of the unit cell by the presence of the randomly added tri- (or tetra-) fluoroethylene units. An explanation was proposed in ref. *8* based upon this lattice expansion (see Figure 7). In $\beta-PVF_2$ the chains are held at their closest possible contacts along the *b*-direction through their strong dipolar alignment. This renders rotation away from the trans conformation energetically unfavorable. On the contrary, the closest contacts between the copolymer chains will be at the randomly located $-CHF$ or $-CF_2$ groups, providing increased separation to the vinylidene fluoride units (see Figure 7). This should allow expansion of the effective cross-sectional area of the chains during the early stages of electron irradiation by internal rotations to the G and $\bar{G}$ conformations (as occurs commonly through introduction of defects or stresses), leading to adoption of the paraelectric packing. This expansion was confirmed by the X-ray studies of Odajima *et al.* *(16)*, and the above model *(8)* was supported by the analysis of Daudin and co-workers *(21)*.

Properties of Irradiated PVF_2 Copolymers. Thermal, mechanical, and electrical properties of irradiated PVF_2 copolymers have been the subject of very recent, but increasing, interest. Of these, the dielectric properties were first investigated by Odajima and co-workers *(16,20)* for samples irradiated with ^{60}Co γ-ray beams or 2 MeV electron beams. Typical results are shown in Figure 8. The unirradiated

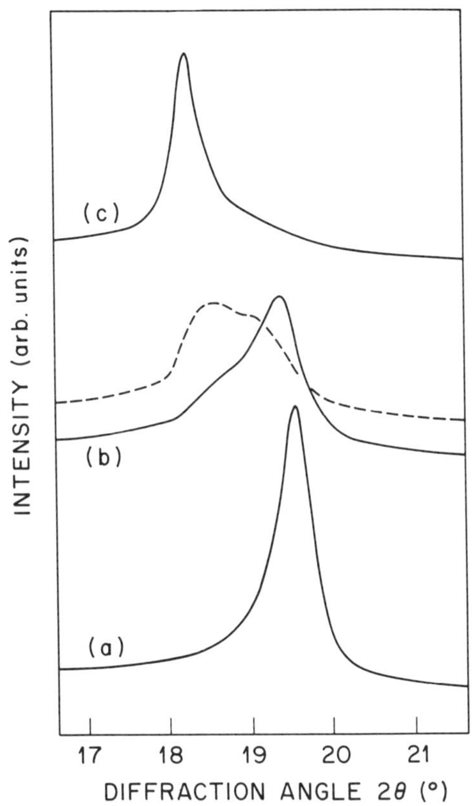

Figure 4. X-ray diffraction profiles of a 65/35 mol % VF_2 / F_3E copolymer (a) prior to irradiation, (b) after exposure to 60 Mrad at room temperature (solid line) or at 120°C (dotted line), and (c) after exposure to 120 Mrad. Reproduced with permission from ref. 16. Copyright 1985 Japanese Journal of Applied Physics.

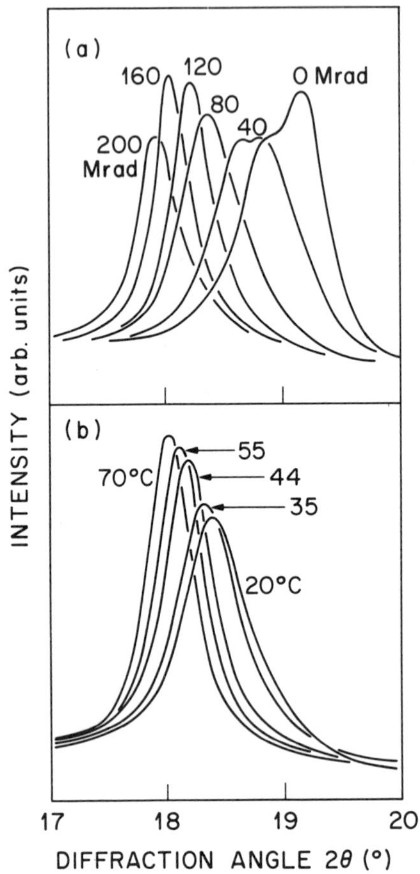

Figure 5. (a) X-ray diffractograms of the intermolecular peak of a 52/48 mol % VF_2/F_3E copolymer exposed to γ-irradiation at ambient temperature to the indicated doses. (b) Effect of heating of this copolymer, irradiated at 80 Mrad, to the indicated temperatures. Reproduced with permission from ref. 20. Copyright 1985 IEEE.

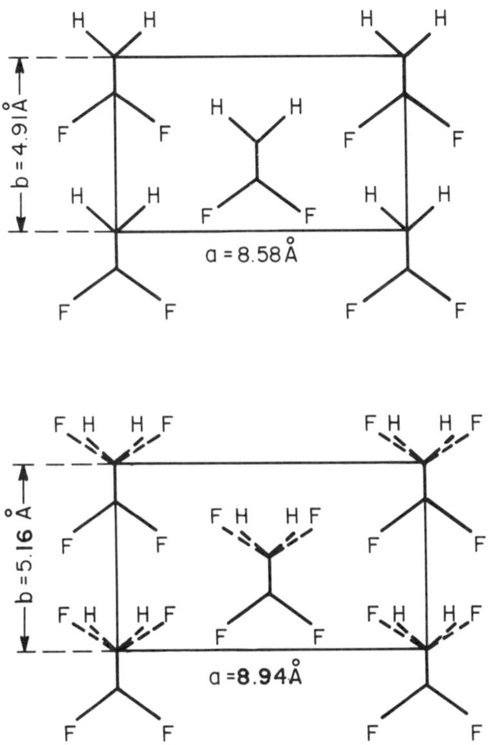

Figure 6. Projected cell bases of the ferroelectric phases of β−PVF_2 (top) and of a 73/27 mol % VF_2 / F_3E copolymer (bottom). Reproduced from ref. 8. Copyright 1985 American Chemical Society.

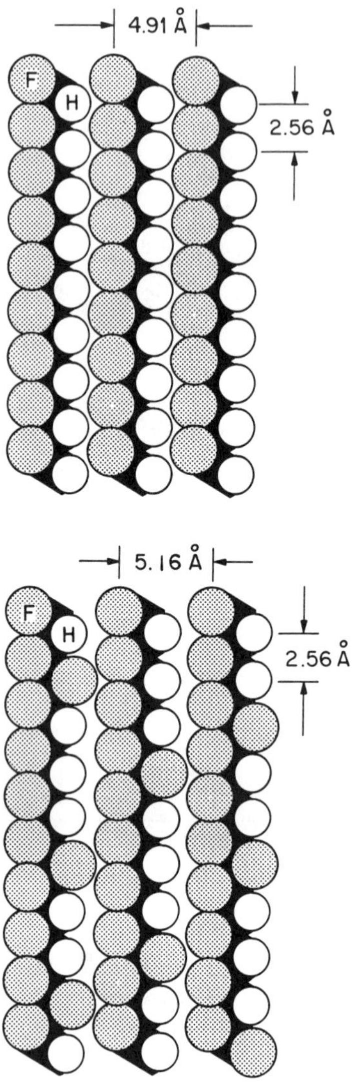

Figure 7. Schematic models of intermolecular packing on the bc planes of the ferroelectric phases of β−PVF_2 (top) and of a 73/27 mol % VF_2 / F_3E copolymer (bottom). Reproduced from ref. 8. Copyright 1985 American Chemical Society.

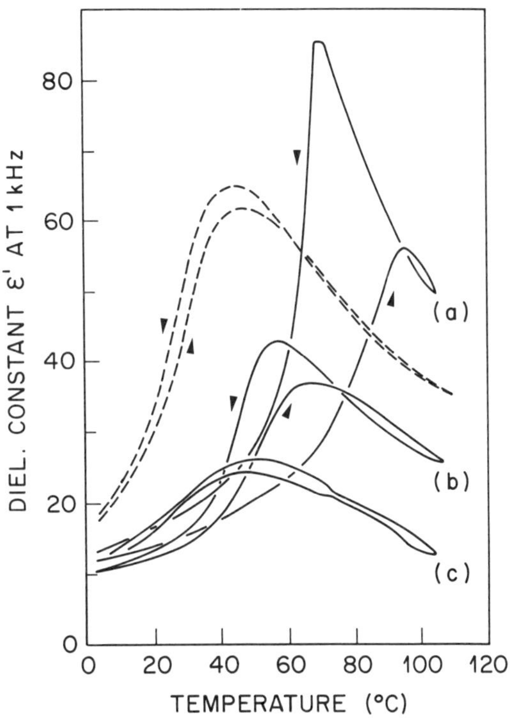

Figure 8. Variation of dielectric constant with temperature during heating and cooling of a 65/35 mol % VF_2 / F_3E copolymer irradiated at room temperature and (a) 0 Mrad, (b) 60 Mrad, and (c) 120 Mrad. The dotted curve is for irradiation at 120° C and 40 Mrad. Reproduced with permission from ref. 20. Copyright 1985 IEEE.

polymer (curve *a*) exhibits the well-known large increase in ε' at the ferroelectric-to-paraelectric transition, as well as the hysteresis between heating and cooling curves (characteristic of the first-order nature of this transition). After irradiation at 60 Mrad (curve *b*), the peak temperatures, heights, and hysteresis are all drastically reduced, and by 120 Mrad (curve *c*) the hysteresis has essentially disappeared and the peaks have merged and broadened. All these features are consistent with adoption of a paraelectric lattice during increasing irradiation and with disappearance of the dielectric anomaly associated with the Curie transition. Additionally, irradiation at higher temperature (120°C) caused disappearance of the hysteresis at much lower doses while retaining high values of the dielectric constant (see again Fig. 8). In general, subsequent dielectric studies by Daudin, *et al.* *(21-23)* were in agreement with these findings (excepting the reported *(20)* increase in ε' after irradiation at high temperature) and provided many useful extensions. In particular, they showed that the downward shift of the Curie temperature with radiation dose could not be extended below *ca.* 0°C, and analyzed it on the basis of a competition with the glass-transition which occurs at these temperatures *(21,23)*.

Thermal studies utilizing differential scanning calorimetry (DSC) were reported in ref. *20, 23* and *24*. Typical findings are illustrated in Figure 9. As seen by comparing the two curves, irradiation clearly lowers the temperatures of both the Curie and melting peaks, reduces the enthalpies of both transitions, and broadens the temperature ranges over which these processes occur. The Curie peak decreases much more rapidly with dose than does the melting endotherm, as seen in Figure 10. The reduction in Curie temperature with irradiation dose has been studied quantitatively by Macchi and Legrand *(9,24)*. They discuss three mechanisms contributing to this depression of T_c: (a) introduction of polarization defects, which is responsible for the major part of the reduction in T_c; (b) diminution of crystallite sizes; and (c) expansion of the crystalline lattice. Of these, the third was not detected for 3 MeV electrons *(9)*, the second had only a small effect, and the first was also diminished in magnitude.

Dynamic mechanical investigations were also carried out by Macchi *et al.* *(25)* using flexural and torsional modulus. A plot of the dissipation factor (tan ϕ) versus temperature from the latter study is seen in Figure 11. Noteworthy characteristics are the increase in the glass-transition (β) and sub-glass transition (γ) peaks following electron exposure. This is consistent with the increase in the amorphous content of the polymer. The α-peak is also enhanced; its origin is generally attributed to relaxation of defects at the interior or at the surfaces of crystalline lamellae *(26,27)*. At the same time, the Curie temperature (which shows the expected hysteresis in each cycle of Figure 11 is diminished both during heating and during cooling in the irradiated specimens.

Conclusions

A number of studies have shown that exposure of vinylidene fluoride copolymers with tri- or tetra-fluoroethylene to electrons or X-rays has unique effects on their ferroelectric nature and properties. It lowers the Curie transition to temperatures near or below ambient, the decrease becoming more pronounced with increasing radiation dose. While at the higher doses the specimens become eventually cross-

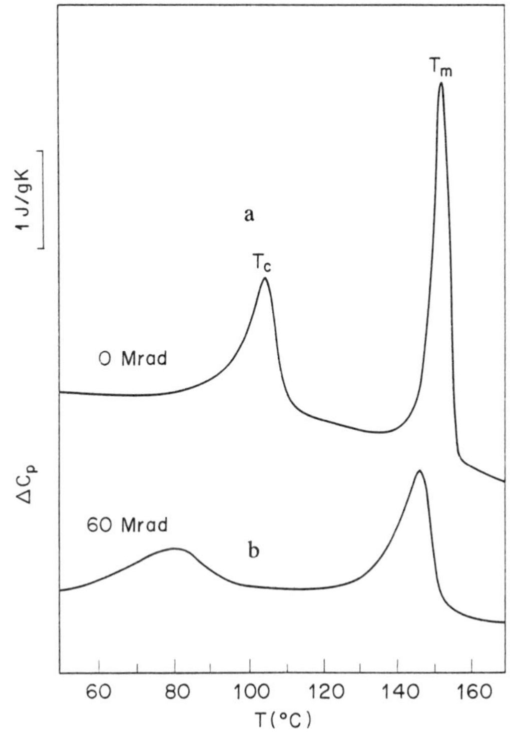

Figure 9. DSC curves of a 70/30 mol % VF_2 / F_3E copolymer (a) prior to irradiation, and (b) after irradiation by 3 MeV electrons to 60 Mrad. Adapted from ref. 9.

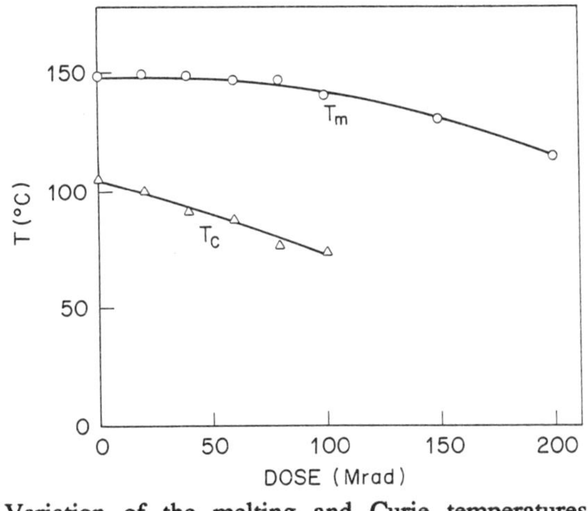

Figure 10. Variation of the melting and Curie temperatures (T_m and T_c, respectively) with radiation dose for 3 MeV electrons. Adapted from ref. 24. Copyright 1990 Elsevier Science Publishers.

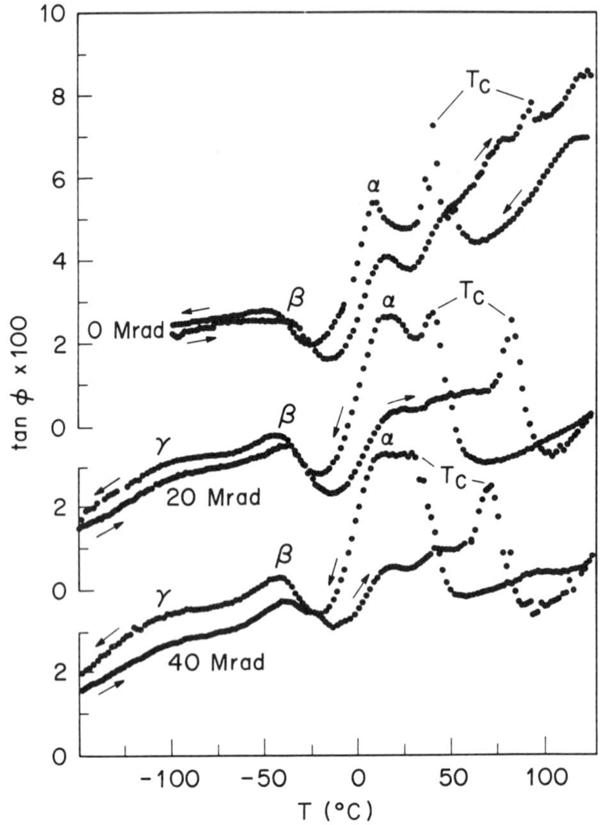

Figure 11. Variation of the mechanical dissipation factor (at 1 Hz) with temperature during heating and subsequent cooling of a 70/30 mol % VF_2 / F_3E copolymer irradiated by 3MeV electrons to the indicated doses. Adapted from ref. 25. Copyright 1990 Elsevier Science Publishers.

linked, the lowest doses yield a paraelectric phase of increased intermolecular perfection. On the other hand, PVF_2 homopolymer does not show any structural effect of irradiation other than the expected amorphization at the higher doses. The difference between homo- and co-polymers has been attributed to the increased interchain separation of the polar vinylidene fluoride segments by the bulkier tri- or tetra-fluoroethylene units. This should allow easier rotation of the VF_2 segments away from their ferroelectric *trans* conformation during irradiation, leading to subsequent adoption of the paraelectric packing. The thermal, mechanical, and dielectric properties of the irradiated copolymers are consistent with these structural changes.

Acknowledgments

I am very grateful to Drs. J. F. Legrand, F. Macchi, and A. Odajima, as well as to the Institute of Electrical and Electronics Engineers, Elsevier Publishers, and the Japanese Journal of Applied Physics, for permission to adapt and reproduce Figures 4, 5, and 8-11.

Literature Cited

1. Grubb, D. T. *J. Mater. Sci.* **1974**, *9*, 1715.
2. Keller, A. In *Developments in Crystalline Polymers-1*; Bassett, D. C., Ed.; Applied Science: London, 1982, chapter 2.
3. *The Applications of Ferroelectric Polymers*, Wang, T. T.; Herbert, J. M.; Glass, A. M., Eds.; Chapman & Hall: New York, 1988.
4. Lovinger, A. J. In *Developments in Crystalline Polymers-1*; Bassett, D. C., Ed.; Applied Science: London, 1982, chapter 5.
5. Lovinger, A. J. *Science* **1983**, *220*, 1115.
6. Furukawa, T. *Phase Transitions* **1989**, *18*, 143.
7. Tashiro, K.; Kobayashi, M. *Phase Transitions* **1989**, *18*, 213.
8. Lovinger, A. J. *Macromolecules* **1985**, *18*, 910.
9. Macchi, F. Doctoral thesis under the direction of Prof. J. F. Legrand, Université de Grenoble-1, BP 87, 38402 Saint Martin d' Hères Cédex, France 1990.
10. Wang, T. T. *Ferroelectrics* **1982**, *41*, 213.
11. Pae, K. D.; Bhateja, S. K.; Gilbert, J. R. *J. Polym. Sci., Polym. Phys. Ed.* **1987**, *25*, 717.
12. Lovinger, A. J.; Furukawa, T.; Davis, G. T.; Broadhurst, M. G. *Polymer* **1983**, *24*, 1225 and 1233.
13. Lovinger, A. J.; Johnson, G. E.; Bair, H. E.; Anderson, E. W. *J. Appl. Phys.* **1984**, *56*, 2412.
14. Lovinger, A. J.; Davis, D. D.; Cais, R. E.; Kometani, J. M. *Macromolecules* **1988**, *21*, 78.
15. Tashiro, K.; Takano,K.; Kobayashi, M.; Chatani, Y.; Tadokoro, H. *Polymer* **1983**, *24*, 199.
16. Odajima, A.; Takase, Y.; Ishibashi, T.; Yuasa, K. *Jpn. J. Appl. Phys.* **1985**, *24* (Suppl. 2), 881.

17. Lovinger, A. J.; Davis, G. T.; Furukawa, T.; Broadhurst, M. G. *Macromolecules* **1982**, *15*, 323.
18. Horiuchi, T.; Matsushige, K.; Takemura, T. *Jpn. J. Appl. Phys.* **1986**, *25*, L465.
19. Tashiro, K.; Takano, K.; Kobayashi, M.; Chatani, Y.; Tadokoro, H. *Polymer* **1981**, *22*, 1312.
20. Odajima, A.; Takase, Y.; Ishibashi, T.; Yuasa, K., In *Proc. 5th Int. Symp. Electrets*; Sessler, G. M., and Gerhard-Multhaupt, R., Eds., Heidelberg: 1985, p. 643 (© 1985 IEEE).
21. Daudin, B.; Dubus, M.; Legrand, J. F. *J. Appl. Phys.* **1987**, *62*, 994.
22. Daudin, B.; Dubus, M.; Macchi, F.; Legrand, J. F. *Ferroelectrics* **1988**, *81*, 357.
23. Daudin, B.; Dubus, M.; Macchi, F.; Legrand, J. F. *Nucl. Instr. and Meth. Phys. Res.* **1988**, *B32*, 177.
24. Macchi, F.; Daudin, B.; Legrand, J. F. *Nucl. Instr. and Meth. Phys. Res.* **1990**, *B46*, 324.
25. Macchi, F.; Daudin, B.; Hillairet, J.; Lauzier, J.; N'goma, J. B.; Cavaille, J. Y.; Legrand, J. F. *Nucl. Instr. and Meth. Phys. Res.* **1990**, *B46*, 334.
26. Nakagawa, K.; Ishida, Y. *J. Polym. Sci.-Polym. Phys. Ed.* **1973**, *11*, 1503.
27. Yano, S. *J. Polym. Sci. A-2* **1970**, *8*, 1057.

RECEIVED January 11, 1991

Chapter 7

Radiation and the Crystals of Polyethylene and Paraffins

G. Ungar[1] and A. Keller[2]

[1]School of Materials, University of Sheffield, Sheffield S10 2T7, United Kingdom
[2]H.H. Wills Physics Laboratory, University of Bristol, Bristol BS8 1TL, United Kingdom

The modes in which crystallinity in polyethylene and its n–alkane oligomers is disrupted by ionizing radiation are described. The "polyethylene" mode differs essentially from the "paraffin" mode: in the former cross–links are randomly distributed through the crystal lattice, while in the latter a separate liquid phase emerges and becomes the preferential location for new cross–link formation. Efficient radiation energy transfer over micron distances across the crystal must occur in n–alkanes. The change–over from the "paraffin" to the "polyethylene" mode is gradual and occurs in the chain length region of 80–90 C–atoms. The role and the nature of the radiation–induced columnar liquid crystalline phase in polyethylene is also discussed.

Ample evidence exists that irradiation–produced cross–links in solid polyethylene are primarily located in the non–crystalline phase (for a review see ref. [*1*]). Nevertheless, at high absorbed doses crystallinity is reduced and eventually lost at 2–3 Grad. Thus, e.g. high density polyethylene becomes transparent as refractive index fluctuations vanish. While such high doses seem almost out of reach for gamma irradiation, the loss of crystallinity due to electron beam damage at equivalent irradiation levels poses serious problems to electron microscopists since diffraction contrast rapidly fades during observation of polyethylene and other organic crystals.

In the following we summarize our past work and present new results on radiation–induced changes to the crystal lattice and the mechanism of lattice destruction in polyethylene and its model compounds, n–alkanes. The work which has lead to the recognition of a previously unsuspected long–range active site migration in alkane crystals is also covered.

0097–6156/91/0475–0101$06.00/0

Effect of Cross-Linking on the Crystal Lattice of Polyethylene

Loss of Crystallinity. Figure 1 shows the crystallinity of bulk linear polyethylene, slowly crystallized (s) and quenched (q), as a function of absorbed gamma radiation. While crystallinity is seen to decrease steadily, the dose dependence of melting temperature is more complex, as illustrated in Figure 2 [*2*].

Hexagonal Mesophase and the Pressure–Dose Superposition. Melting proceeds in two stages for highly irradiated polyethylene: in the first transition orthorhombic *(o)* crystals transform into the so-called "hexagonal phase" *(h)* which, as recently recognized, is a columnar liquid crystal (Ungar, G. *Polymer*, submitted) with the characteristic two-dimensional order [*3*]. Only in the second transition is the isotropic melt *(l)* attained. The dose dependence of the *o–h* transition temperature is marked by dashed lines in Figure 2, and the isotropization temperature (*h–l* and the direct *o–l* melting) by full lines. The crystal lamellae in the three samples in Figure 2 differ in their thickness, this being largest in the slowly crystallized bulk (ca. 30 nm) and smallest in the solution-grown single crystals (ca. 11 nm).

The hexagonal phase turns out to be of the same type as that occurring in unirradiated polyethylene subjected to hydrostatic pressures above 4 kbars [*4*]. In fact it was found that pressure and irradiation act in synergism to bring about the hexagonal mesophase. This is illustrated in Figure 3 which shows the dependence of transition temperatures (*o–l*, *o–h* and *h–l*) on irradiation dose for three different pressures for bulk polyethylene irradiated at 85°C [*5*]. As pressure increases, the triple point is seen to move to lower doses. Similarly, in a p–T diagram, the triple point moves to lower pressures as irradiation dose increases; consequently, at around 500 Mrad, the hexagonal phase becomes stable at 1 bar.

In qualitative terms, the reason behind the dose–pressure superposition effect described above is the following. Irradiation-induced cross-links significantly increase the free enthalpies of both the crystal, G_o, and the isotropic liquid, G_l. The free enthalpy of the mesophase, G_h, is also affected to some extent, but the behaviour of G_o and G_l is dominant, as shown below. The increase in G_o is due to the substantial lattice strain energy – a calculation by Guiu and Shadrake [*6*] puts it as high as 2.5×10^{-19} J per cross-link (compare with a heat of fusion of 2.2×10^{-19} for a molecule of n-alkane $C_{40}H_{82}$). On the other hand, the increase in G_l is due to the reduction in configurational entropy upon network formation.

A schematic free enthalpy – temperature diagram for irradiated polyethylene is shown in Figure 4. G_o^0, G_h^0 and G_l^0 (bold lines) refer to zero dose, while thinner lines correspond to irradiation doses 1 to 5. G_o^0 is taken as reference and thus coincides with the abscissa axis. The free enthalpy curves are drawn in such a way that their intersections outline the transition temperature – dose behaviour (dashed lines) qualitatively similar to that observed (cf. Figure 2 and Figure 3, bottom). The similarity becomes readily apparent if the free enthalpy diagram is rotated clockwise by 90 degrees.

In terms of the diagram in Figure 4, the effect of pressure is to increase the free enthalpies of all three phases, according to the relationship $dG = Vdp - SdT$, with V, S, p and T being, respectively, specific volume and entropy, pressure and temperature. Pressure exerts the largest effect on raising G_l, which can again be

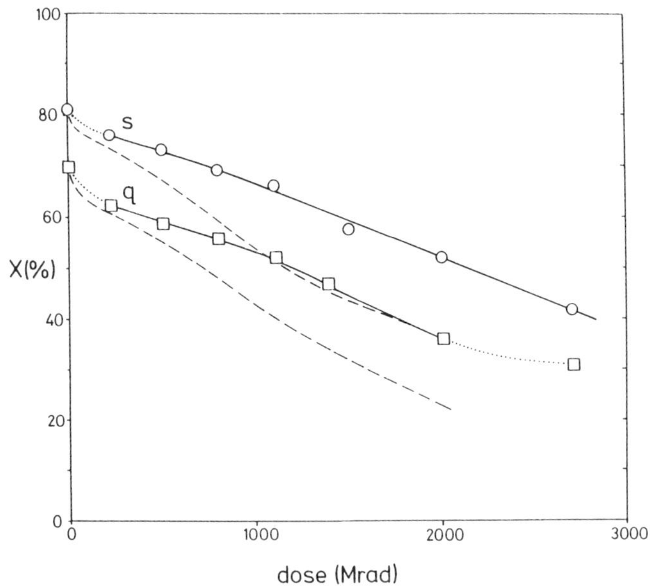

Fig. 1 Full lines: X–ray crystallinity (after Hermanns and Weidinger) *vs.* dose for as–irradiated bulk polyethylene (s = slowly crystallized, q = quenched). Dashed lines: crystallinity from heat of fusion.

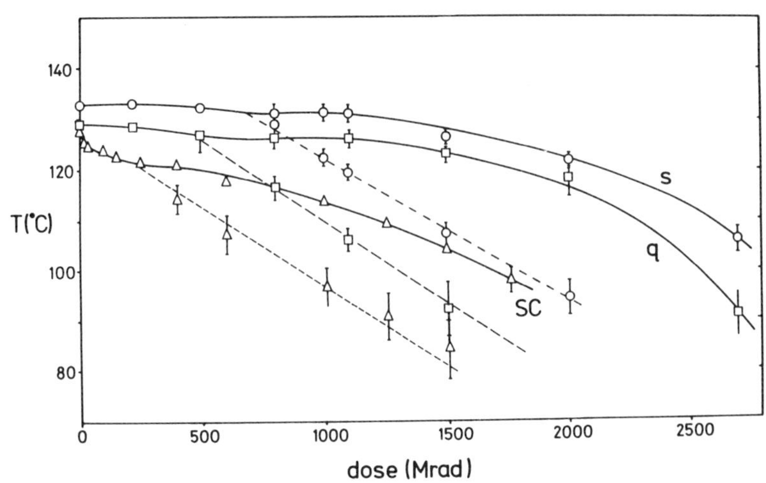

Fig. 2 Melting (full lines) and orthorhombic – hexagonal transition temperature (dashed lines) *vs.* dose for slowly crystallized (○) and quenched bulk (□) and for single crystals grown at 85 °C (△) (Reproduced with permission from ref. *2*. Copyright 1980 IPC Business Press).

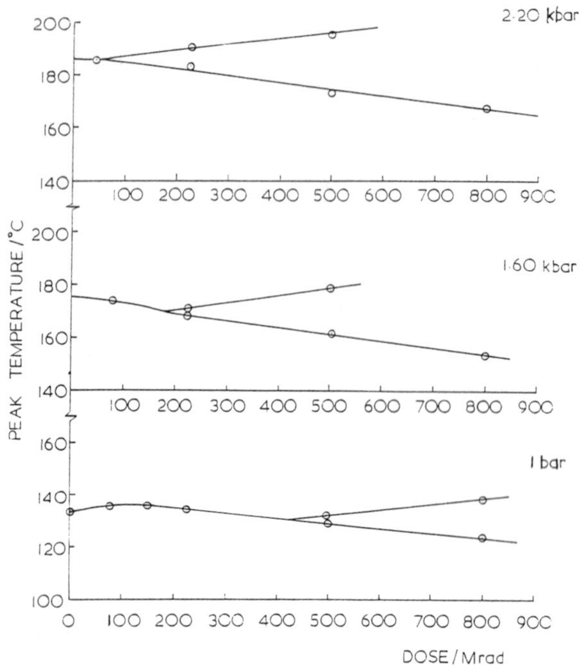

Fig. 3 *o–l*, *o–h* and *h–l* transition temperatures vs. dose, measured at hydrostatic pressures of 1 bar, 1.60 kbar and 2.20 kbar (slowly crystallized bulk polyethylene, gamma irradiation at 85°C) (Reproduced with permission from ref. 5. Copyright 1985 IPC Business Press).

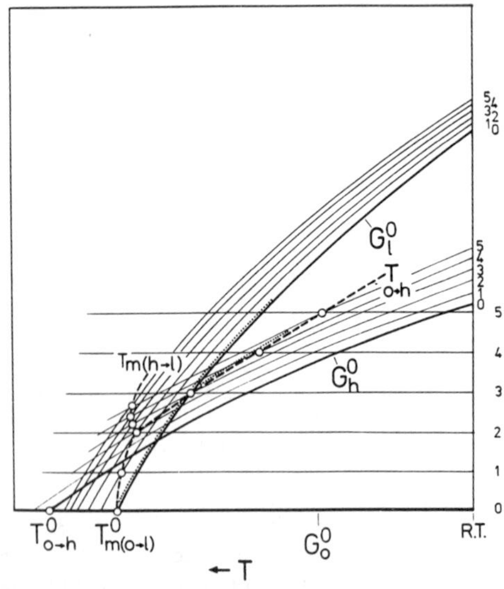

Fig. 4 Schematic free enthalpy *vs.* temperature diagram for irradiated polyethylene. With increasing dose (denoted 0 to 5) the transition temperatures follow the dashed lines (cf. Figures 2 and 3).

traced to the reduction in entropy, similar to the effect of network formation. Thus pressure helps shift the triple point to lower irradiation doses, as observed. It should be noted, however, that the synergistic effect of hydrostatic pressure is only present because the columnar mesophase is a high-entropy low-volume phase [*4,5,7*].

The Nature of the Mesophase. It had been postulated that there is considerable conformational disorder in the hexagonal phase in polyethylene [*8*]. NMR and IR spectroscopies are the appropriate techniques for the determination of the type and amount of conformational defects. Unfortunately neither technique is well suited for high pressure studies, so that the pressure-induced hexagonal phase has not been examined by these methods. Fortunately, the obstacle does not arise in the case of irradiated polyethylene. Thus a temperature dependent Fourier-transform IR study was carried out at atmospheric pressure on slowly crystallized (*s*) and quenched (*q*) bulk polyethylene samples irradiated with 800 Mrad at 85°C [*7*]. The DSC thermograms of these samples, displaying the double melting endotherms, are shown in Figure 5.

There are two common types of conformational defects in long hydrocarbons, the gauche-trans-gauche sequence (GTG and GTG*) and the gauche-gauche (GG) sequence, which have distinct localized IR-active vibration modes [*9*]. These give rise to absorptions at 1306 cm^{-1} and 1352 cm^{-1}, respectively. Figure 6 shows the temperature dependence of the 1306 cm^{-1} absorbance (GTG+GTG*). Sample s0 has not been irradiated and acts as reference. It melts directly into the isotropic liquid and the one-step increase in the 1306 cm^{-1} absorption illustrates this. On the other hand, the double endotherm melting of irradiated samples s800 and q800 (Figure 5) is mirrored also in a two-step increase in the 1306 cm^{-1} absorbance (Figure 6).

In contrast, the behaviour of the 1352 cm^{-1} absorbance (GG defects) is quite different. Here the defect concentration increases in only one step, simultaneously with the second endotherm, i.e. the *h-m* transition [*7*]. The implication is that no GG defects are present in the hexagonal phase, but there is a considerable amount (3 to 4 in every 100 C-C bonds) of GTG or GTG* defects (for details see ref. [*7*]). This difference in the temperature dependence of the two defect bands can be easily understood if it is assumed that all the defects responsible for the 1306 cm^{-1} absorption in the hexagonal phase are of the GTG* type. The latter defects (so-called kinks) do not alter the overall trajectory of the all-trans chain (see Figure 7), and could thus be incorporated within the hexagonal array of essentially extended chains (or columns) with little energetic penalty. On the other hand, GG defects would bend the chain by ca. 90 degrees and would not be viable in a lattice of any kind.

The formation of the hexagonal phase in polyethylene is promoted by the incorporation into the crystal lattice of any constitutional defect and not just cross-links. Thus a high concentration of methyl branches [*2*] or chlorine substituents [*10*] in ethylene-propylene or ethylene-vinylchloride copolymers can by itself induce the *h* phase. Furthermore, even in low propylene content orthorhombic EP copolymers, much lower irradiation doses are required for *h* phase formation than in the case of pure polyethylene [*11*].

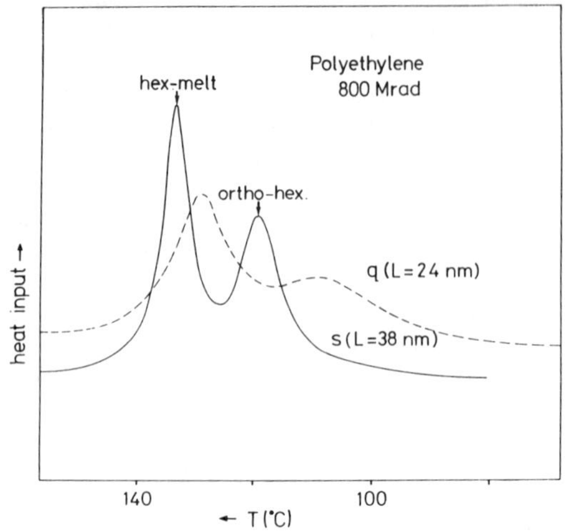

Fig. 5 DSC thermograms of the two bulk polyethylene samples *s* and *q*, gamma-irradiated with 800 Mrad at 85°C (Reproduced from ref. 7. Copyright 1986 American Chemical Society.)

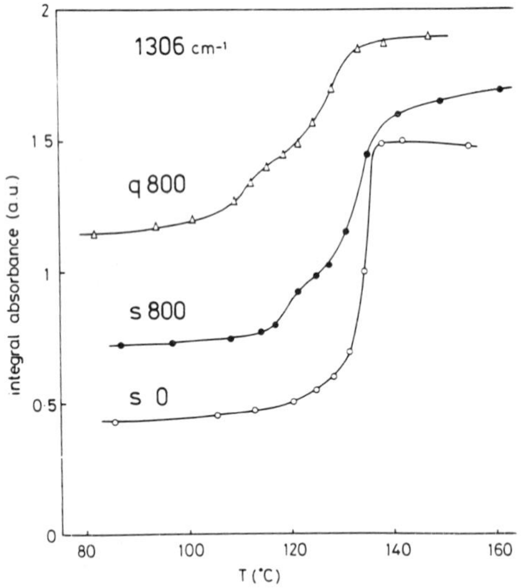

Fig. 6 Temperature dependence of the 1306 cm^{-1} GTG/GTG* band intensity for bulk polyethylene irradiated with 800 Mrad (samples *s* and *q*) and for unirradiated sample *s*. The curves in the diagrams are normalized to the same intensity in the melt and displaced vertically by 0.2 units for clarity (Reproduced from ref. 7. Copyright 1986 American Chemical Society.)

Effect of the Mesophase on Irradiation Response. In the context of irradiation effects, the importance of the *h* phase is in its high molecular mobility compared to the crystalline state. The cross-linking efficiency (G-value) of the mesophase is expected to be nearer that of the amorphous (or isotropic liquid) phase than that of the crystal. This expectation is clearly corroborated by experiments on n-alkanes. There it was found [*2*] that the cross-linking efficiency in the "rotator" phase, which is in many respects similar to the *h* phase in polyethylene, is at least as high as in the liquid.

The intervention of the *h* phase may account for the following two observations concerning the final loss of crystallinity at high irradiation doses: (i) that the lifetime of thick crystal is somewhat longer than that of thin ones, and (ii) that the final loss of crystallinity is preceded by a steep drop in melting (*h–m*) temperature (Figure 2). The proposed interpretation of both effects is that accelerated cross-linking commences as the *o–h* transition temperature approaches the irradiation temperature and molecular mobility increases. From Figure 2 it is clear that this approach occurs at a lower dose for thinner crystals, since the *o–h* transition temperature is lower. Such dependence of T_{o-h} on lamellar thickness is simply a thermodynamic consequence of the lower surface free energy of the hexagonal mesophase, σ_e^h, compared to that of the orthorhombic crystal, σ_e^o. The transition temperature for lamellae of thickness l is given by [*12*]:

$$T_t = T_t^0 \left[1 - \frac{2(v^o\sigma_e^o - v^h\sigma_e^h)}{l\,\Delta h_t} \right]$$

where T_t^0 is the equilibrium transition temperature of infinitely thick lamellae, v^o and v^h are specific volumes of the two phases, and Δh_t is the transition enthalpy per unit mass. We note that no change in l takes place upon the transition as the polymer is densely crosslinked at radiation doses in question.

Destruction of Crystallinity in n-Paraffins

Phase Separation Upon Gamma-Irradiation. Although n-alkanes are often considered and studied as models for polyethylene, they behave quite differently from their polymer counterpart regarding the mode of crystallinity destruction on irradiation. We have investigated this phenomenon in some detail, both by applying ^{60}Co gamma-irradiation [*13*] and electron microscope beam [*14–16*]. In addition to samples of polyethylene with different crystal morphologies, the materials studied included n-alkanes of different chain lengths, ranging from 20 to 94 C-atoms, as well as a long-chain α,ϖ-dicarboxy-n-alkane with ca. 100 C-atoms, denoted DPE. The latter was obtained by ozone degradation of polyethylene single crystals, and is not as strictly monodisperse as n-alkanes.

An illustration of the difference in irradiation behaviour between paraffin and polyethylene crystals is given in Figure 8, where changes in the unit cell parameters *a* and *b* are plotted as a function of gamma irradiation dose received at 40°C [*13*]. The crystal lattice in both polyethylene single crystals (the same is true for the bulk) and in the long chain diacid becomes greatly distorted. This is attributed mainly to cross-link formation in the lattice. The unit cell increases in the *a* direction, meaning that hexagonal symmetry (with $a/b=\sqrt{3}$), is approached.

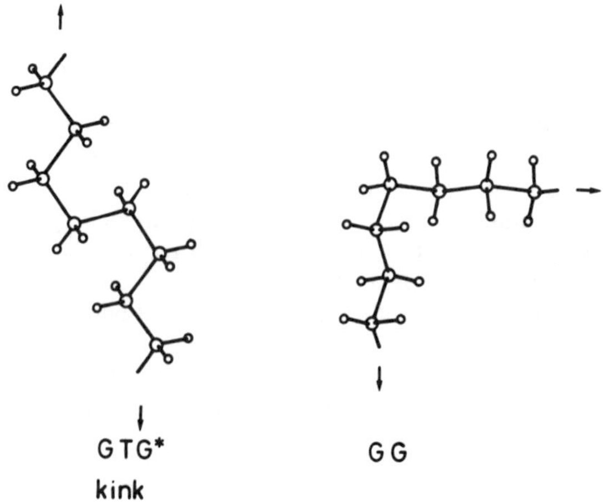

Fig. 7 Conformations of a GTG* and a GG defect, linked to an all-trans chain at either side. (Reproduced from ref. 7. Copyright 1986 Amer. Chem. Soc.).

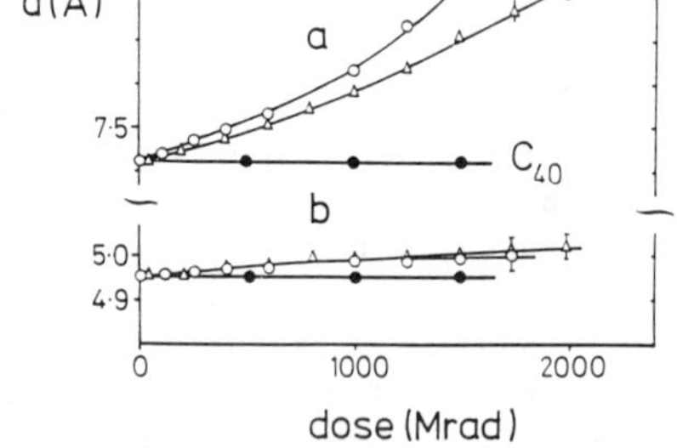

Fig. 8 Change in unit cell parameters *a* and *b* with dose of gamma irradiation. SC: polyethylene single crystals grown at 85°C; DPE: dicarboxylic acid; C_{40}: n-tetracontane (Reproduced with permission from ref. *13*. Copyright 1980 IPC Business Press).

In contrast, the unit cell in n–tetracontane ($C_{40}H_{82}$) does not change at all, until crystallinity is completely lost around 2 Grad. Still shorter paraffins behave in this respect like $C_{40}H_{82}$.

The diffraction linewidth parallels the change in lattice spacings: in n–alkanes $C_{40}H_{82}$ and shorter, where spacings do not change, the lines remain sharp. This indicates that the crystal lattice, although reduced in overall amount, remains largely free of defects. It also means that the phase separation is on a large scale: small crystalline domains would cause line broadening; on the other hand, point defects (isolated Angstrom–scale regions of disorder), while leaving sharp diffraction peaks, would not produce an amorphous halo with a discrete scattering maximum.

To help understand the difference in radiation response of crystal lattices in polyethylene and paraffins, it is instructive to compare the melting thermograms of the irradiated samples described above. Figure 9 shows thermograms of unirradiated and irradiated DPE (polyethylene behaves similarly) and paraffin $C_{40}H_{82}$. The difference in melting behaviour of these two compounds is as striking as the difference in dose dependence of their lattice parameters (Figure 8). The endotherm of the 800 Mrad irradiated DPE remains sharp but the melting point is substantially reduced (by >20°C). On the other hand, the "peak melting temperature" in equally irradiated paraffin is reduced by only 8°C, but the endotherm is substantially broadened. The shape of the latter is typical of two–phase melting of an impure compound where the impurities are soluble in the liquid but insoluble in the solid phase.

Further Evidence for Phase Separation. X–ray diffractograms of irradiated paraffins show the appearance of a diffuse peak, characteristic of the amorphous phase, superimposed on the crystalline diffraction pattern – see Figure 10. As either dose or temperature is raised the diffuse halo increases at the expense of the crystalline reflections. The change with temperature is fully reversible.

All this is evidence that phase separation occurs in irradiated $C_{40}H_{82}$, with the two phases being (i) pure perfect crystal and (ii) a liquid containing the bulk of irradiation products. For $C_{40}H_{82}$ it turns out that the uncrosslinked, or "monomer" fraction, as measured by GPC, coincides almost exactly with crystallinity at room temperature as determined from the heat of fusion (Ungar, G.; Stejny, J.; Keller, A., unpublished results). Thus the liquid phase at room temperature is almost entirely made up of cross–linked material.

The combination of GPC values for the cross–linked fraction and the shapes of the thermograms in Figure 9 (bottom) allows one to construct an equilibrium binary phase diagram uncrosslinked – crosslinked $C_{40}H_{82}$, which realistically describes the phase separation (liquidus curve) for different doses and temperatures – see Figure 11.

Recent high–resolution solid–state ^{13}C NMR spectra of gamma–irradiated $C_{20}H_{42}$ and $C_{21}H_{44}$ by Toriyama *et al.* [*17*] indeed confirm that a mobile phase is produced by irradiation. While there is general agreement between these authors and ourselves on the main issues, their interpretation of the mobile phase differs somewhat from ours: they regard it as being confined to isolated "glassy islands", containing only a few molecules, and dispersed within the crystal lattice. It may be argued, however, that such amorphous occlusions are likely to produce severe

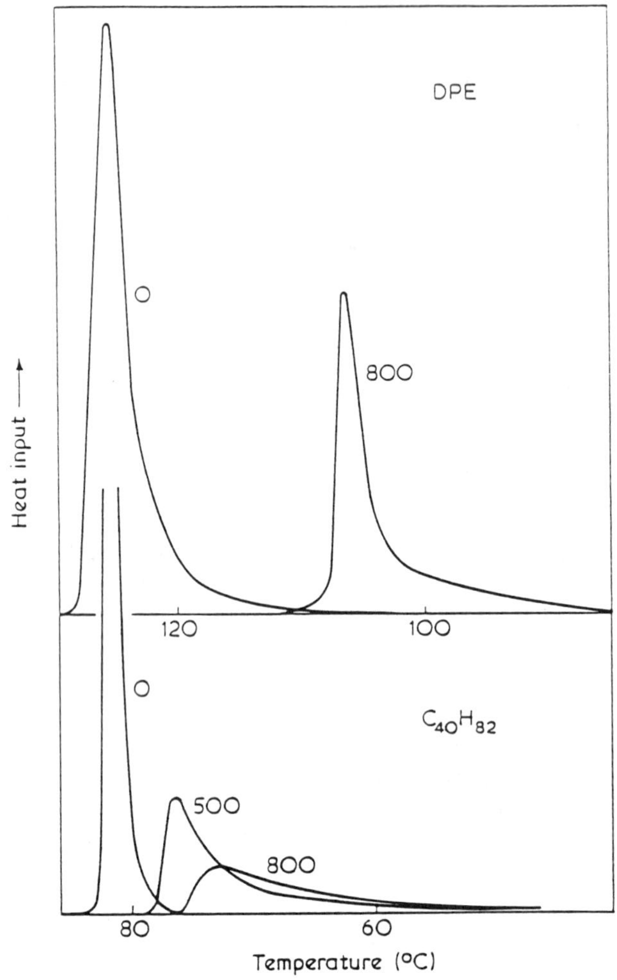

Fig. 9 Top: DSC melting traces of an unirradiated and an irradiated (800 Mrad at 85°C) sample of degraded polyethylene (dicarboxy–n–alkane). Bottom: DSC traces of an unirradiated and two irradiated (500 and 800 Mrad at 40–45°C) samples of n–$C_{40}H_{82}$. Thermograms are normalized to the same sample mass (Reproduced with permission from ref. *13*. Copyright 1980 IPC Business Press).

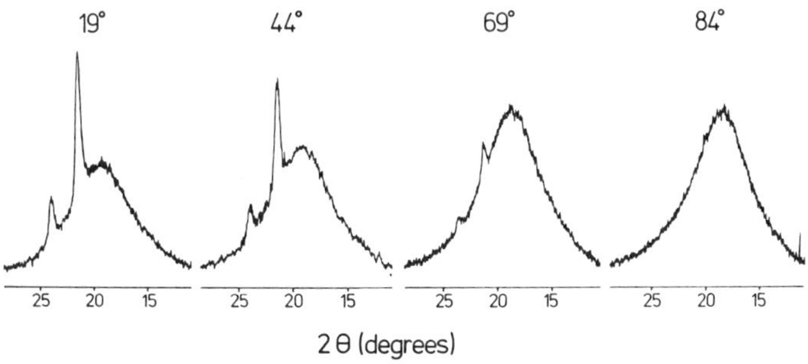

Fig. 10 Series of X-ray powder diffractograms of gamma-irradiated $C_{40}H_{82}$ (800 Mrad) recorded at increasing temperatures.

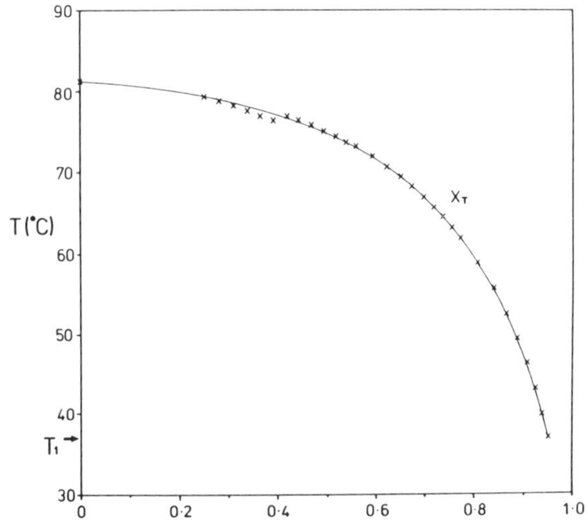

Fig. 11 Binary phase diagram for partially cross-linked $C_{40}H_{82}$ constructed after DSC curves and GPC data for the 219 and 500 Mrad samples. X is the weight fraction of cross-linked paraffin, X_T (the liquidus) is the composition of the liquid in equilibrium with the solid. The latter contains no cross-links as judged from the absence of any lattice distortion. Details of the construction will be described elsewhere.

lattice strain which could only be released by rejection of the underlying chemical defect (e.g. cross–link) into a completely separate phase. Indeed, our X–ray diffractograms already show a distinct amorphous halo for paraffins irradiated to comparable doses (90 Mrad). Even if full phase separation did not occur in as–irradiated crystals, it would have certainly taken place after melting and recrystallization (see above). However, the NMR spectra of recrystallized irradiated alkanes reportedly did not differ from the spectra of the crystals prior to melting [*17*].

We have found no evidence for phase separation either in polyethylene or in DPE. Here the cross–links are distributed through the crystal lattice resulting in a uniformly depressed melting point. Melting is of the "homophase" type (Figure 9, top) rather than "heterophase" as in shorter paraffins (Figure 9, bottom).

Primary or Secondary Phase Separation. The phase separation between crosslinked and uncrosslinked paraffin, described in the preceding section, has been established "post mortem", and it is not clear from the experimental evidence presented how this separation came about. It is certain that, mobility permitting, cross–links tend to be excluded from the crystal lattice of shorter paraffins. This tendency has been demonstrated on mixtures of n–alkane $C_{27}H_{56}$ and the model "H–shaped" molecules of 1,1,2,2–tetra(tridecyl)ethane, which contains two 27 C–atoms chains linked in the middle [*18*]. The "H" molecules did not co–crystallize with either the orthorhombic (*o*) or the rotator phase (Ungar, G.; Stejny, J; Keller, A., unpublished results).

According to the above, even if cross–links had initially formed at random throughout the crystal, phase separation could have occurred subsequently. It should be noted, however, that in the case of polyethylene or DPE crystals no amount of annealing could eliminate radiation–induced lattice distortions: once formed, the cross–links have always remained within the crystals.

Irradiation in the Electron Microscope. In order to elucidate the origin of phase separation in irradiated paraffins, we resorted to electron microscope (EM) experiments [*14–16*]. The advantage of EM is in its provision for simultaneous irradiation and monitoring, both of the appearance of the crystals in diffraction contrast, and of the changes in lattice spacings and distortions using electron diffraction. 100 keV electrons were used. The beam current density and specimen temperature were accurately controlled and measured.

The main question posed originally was answered early on in the course of EM experiments. The emergence of the second phase in the form of radiation–induced amorphous domains was visualised with remarkable clarity in images of paraffin crystals like those in Figure 12. The oval "holes" in the thin crystals are in fact non–diffracting amorphous droplets which appear at an early stage of irradiation. With increasing dose these droplets grow in size but not in numbers. Their growth is at the expense of the crystal which, otherwise, remains relatively free of defects.

The time–scale of EM irradiations was considerably shorter (minutes to hours) than that of gamma irradiation (weeks to months), and observable phase separation on a micron scale (Figure 12) often became established in a matter of one or a few seconds. Thus molecular segregation subsequent to cross–link formation

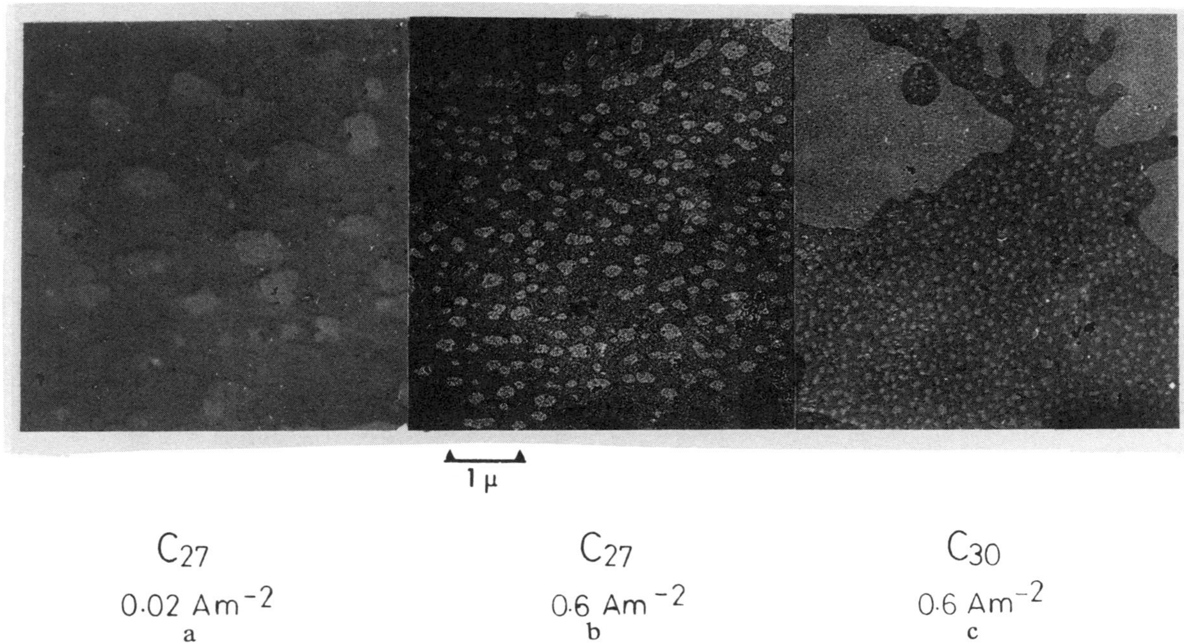

Fig. 12 Examples of droplet formation in n–alkane crystals irradiated in EM: (**a** and **b**) crystals of $C_{27}H_{56}$ exposed to an electron current density 0.022 A/m^2 (**a**) and 0.56 A/m^2 (**b**). (**c**): crystal of $C_{30}H_{62}$ exposed to electron current 0.56 A/m^2. An exposure of 1 Coulomb/m^2 (1 C/m^2) has been estimated to correspond to an absorbed dose of 25 Mrad for 10 nm thick crystals [*22*]. (Reproduced with permission from ref. *16*. Copyright 1991 Pergamon Press).

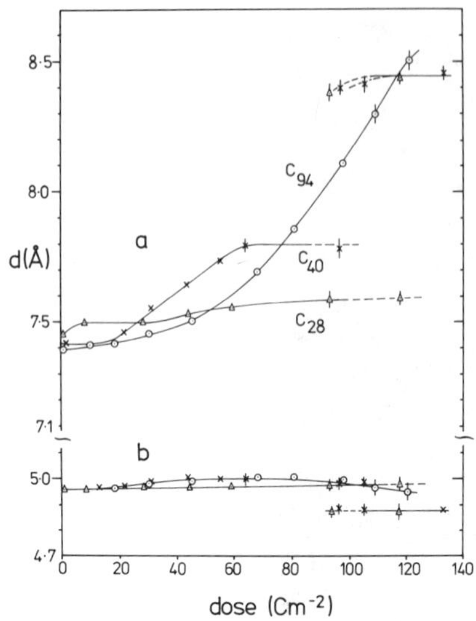

Fig. 13 Dose dependence of unit cell parameters *a* and *b* for paraffin $C_{28}H_{58}$ (Δ), $C_{40}H_{82}$ (x) and $C_{94}H_{190}$ (o) irradiated at room temperature. Electron current density 0.1 A/m^2. The new spacings corresponding to *a*=8.4A and *b*=4.9A appearing at around 100 C/m^2 are those of the hexagonal rotator phase (Reproduced with permission from ref. [*14*]. Copyright 1980 IPC Business Press).

(secondary segregation) cannot account for the observed phase separation. Cross–links and, possibly, other defects must actually be forming *in situ* within the newly created droplets. As shown in Figure 12, the lower the dose rate the larger the scale on which phase separation occurs. Extrapolation to the very low dose rates of gamma irradiation would thus suggest a very thorough spatial separation between the pure alkane and irradiation products, as indeed observed.

Since the initial energy deposition is random, a highly mobile active intermediary must exist in irradiated paraffin crystal, capable of intermolecularly traversing the lattice over micron distances and inducing cross–linking in the liquid. The rate of this migration is estimated [*14*] to be at least ten orders of magnitude higher than the diffusion rate of alkyl radicals within a polyethylene crystal [*19–21*]. Thus the active species in question cannot be alkyl radicals, which are commonly perceived as precursors of cross–links (also see below). The nature of the active migrating species still remains unknown, but we have accumulated by now a reasonable amount of information which might, in combination with other techniques, at least narrow the choice of likely candidates.

As has been shown, the distinction between the behaviours of paraffins on the one hand and polyethylene and DPE on the other is very clear in the case of gamma irradiation. However, this distinction is more blurred in electron irradiation, as illustrates in Figure 13. At one end, for paraffin $C_{27}H_{56}$, the unit cell parameters hardly change at all on irradiation. At the other end, in the long alkane $C_{94}H_{190}$, and also in DPE and in polyethylene, the lattice gradually distorts as in gamma irradiation of these same materials. However, $C_{40}H_{82}$ shows an intermediate behaviour: the lattice parameters change up to a point, and then remain constant. Here the effect of irradiation temperature is very important: if it is raised from 25°C to 60°C, $C_{40}H_{82}$ behaves like $C_{27}H_{56}$ at 25°C, i.e. negligible distortion occurs. Alternatively, at sub-zero temperatures $C_{40}H_{82}$ behaves like polyethylene.

Additionally, the following general rule was found to apply to electron-irradiated hydrocarbon compounds: the lack of sizeable change in lattice parameters is invariably accompanied by the observation of distinct non-diffracting droplets. Thus, e.g., at room temperature droplets are seen in $C_{27}H_{56}$ (Figure 12) but not in $C_{40}H_{82}$. On the other hand, large droplets occur in $C_{40}H_{82}$ at 60°C. It should be noted, however, that even at temperatures very close to the melting point no droplets of the above type are seen in $C_{94}H_{190}$.

The effect of the three explored variables, viz. temperature, dose rate and chain length, on the extent to which phase heterogeneity does ("paraffin-like" behaviour) or does not occur ("polyethylene-like" behaviour) is summarized in Table I.

Long-Range Active Site Migration. Most of the observed features of paraffin irradiation can be rationalized in terms of the following description: (1) The active sites in question can, under certain conditions, move rapidly and over large distances across the orthorhombic crystal lattice common to both paraffins and polyethylene. (2) Disorder, such as that found in the liquid or even in the "rotator" phase of n-alkanes, appears to inhibit the migration, as suggested by recent experiments [*16*]. (3) The active species terminate with formation of cross-links, but this termination occurs mainly upon the arrival at an existing liquid droplet. (4) Diffusion of the active species is thermally activated; its rate increases with temperature.

In Figure 14 we present a schematic illustration of the droplet formation in paraffin crystals. At the inception of irradiation the concentration of active sites increases uniformly throughout the lattice, producing occasional cross-links. Where cross-links are sufficiently abundant to raise the free energy above a critical value, the crystal will melt locally and the embryo droplet will form. This now serves as a sink for incoming active sites within a given capture radius, determined by the sites' diffusion rate. Thus the active site concentration in the vicinity of a droplet is kept low; no cross-links and hence no new droplets form in the crystal once the initial population is established. The droplet separation will be determined by their capture radius, i.e. by the sites' diffusion rate. This, in turn,

Table I. Effect of Chain Length, Irradiation Temperature and Dose Rate on Irradiation Behaviour of n-Alkanes

"Paraffin-type" behaviour		"Polyethylene-type" behaviour
	Increasing chain length →	
	← Increasing irradiation temperature	
	Increasing dose rate →	

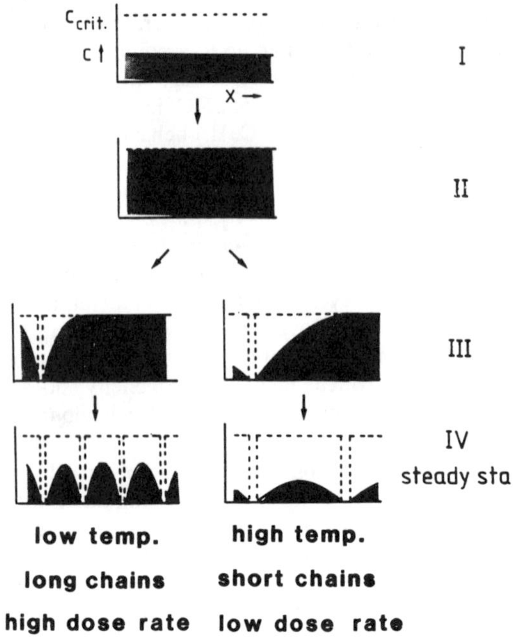

Fig. 14 Development of concentration profiles of active sites along a space coordinate x in the lamellar plane of an irradiated paraffin crystal. Broken line marks the critical concentration for droplet formation; dents in c_{crit} mark the droplets. (Reproduced with permission from ref. [*15*]. Copyright 1983 Pergamon Press).

will depend on temperature and chain length as already described. Furthermore, it is clear from the steady state dynamics of the process that a high dose rate (high input of sites) would result in a high density of droplets (cf. scheme on the left, Figure 14, and observation on $C_{27}H_{56}$, Figure 12).

At this point it is appropriate to mention the revealing experiments by Clough [*23*] on irradiated mixed crystals of perdeuterated and protiated n–tetracosane. Clear evidence for alkyl radical migration by hydrogen (or deuterium) hopping was obtained from the observation, by mass spectroscopy, of isotopic exchange occurring during irradiation. From the extent of the exchange, or "scrambling", it was deduced that, on average, there are 14.3 hops per radical pair before termination. Assuming random walk, this means an average migration distance of 6.3 Å for an alkyl radical. This is clearly not sufficient to account for the long–range active site migration, producing liquid domains microns apart, described above.

It should be remarked that formation of liquid droplets and certain other features of "paraffin–type" irradiation behaviour are not observed in alkanes with 25 or less C–atoms [*16*]. This has been attributed to the fact that even a relatively low dose of irradiation at ambient temperature induces a phase transition from the ordered crystal to the transient orientationally–disordered and mobile "rotator" crystal. The long–range active site migration is absent in the rotator phase, and it may be that alkyl radical migration by hydrogen hopping becomes the dominant process. It is thus possible that isotopic exchange experiments on an alkane longer than the $C_{24}H_{50}$ used by Clough would show comparatively less scrambling, if the non-radical long-range migration process replaces hydrogen hopping.

Active Site Migration in Long and Short Chain Hydrocarbons. Finally, we address the question of why polyethylene and paraffins behave differently. More precisely, why is the long-range active site migration not operative for chains with ca. 90 C-atoms or more? The experimental fact is that the temperature interval of droplet formation decreases with increasing chain length: it is at least 50°C for $C_{27}H_{56}$ [*16*], only 20°C for $C_{65}H_{112}$ (Ungar, G.; Hill, M.J.; Ogawa, Y., unpublished), and nil for $C_{94}H_{190}$ [*14*]. We propose that this convergence is the result of the site diffusion rate being linked to lattice mobility, whose activation energy increases approximately linearly with lamellar thickness (chain length). On the other hand, chain length dependence of melting point of paraffins levels off at higher molecular weights. Hence $C_{94}H_{190}$ and polyethylene melt before their lattice mobility becomes high enough for efficient radiation energy transfer.

Conclusion

There is solid experimental evidence for the existence of two different mechanisms of radiation-induced crystal destruction in hydrocarbon chain systems. The "polyethylene mode" differs essentially from the "paraffin mode": in the former cross-links are randomly distributed through the crystal lattice, while in the latter a separate amorphous phase (liquid or gel) forms and becomes the preferential location for new cross-link formation. Efficient radiation energy transfer over micron distances across the crystal must be invoked to account for the observed features of the "paraffin mode". The nature of the fast and long-range migrating species involved is not yet known, but it is unlikely that they are alkyl radicals.

The change-over from the "paraffin" to the "polyethylene" mode is gradual and occurs in the chain length region of 80-90 C-atoms, depending on irradiation temperature and dose rate.

At high doses the gradual accumulation of cross-links in the crystal lattice of polyethylene induces one-dimensional melting, manifested as a transition from the three-dimensional orthorhombic crystal to a two-dimensionally ordered columnar liquid crystal phase with pseudohexagonal symmetry. Final amorphization follows shortly after this transition, as cross-link formation is enhanced in the mobile columnar phase. The inverse relationship between radiation sensitivity and crystal thickness can be explained by a lower stability of the crystalline relative to the columnar phase in thinner lamellae; hence the dose required to induce the transition decreases with decreasing crystal thickness.

Literature Cited

1. Keller, A. in *Developments in Crystalline Polymers;* Bassett, D.C., Ed.; Applied Science: London, 1981; p. 37.
2. Ungar, G.; Keller, A. *Polymer* **1980,** *21,* 1273.
3. Yamamoto, T. *J. Macromol. Sci.-Phys.* **1979,** *B16,* 487.
4. Bassett, D.C. In *Development in Crystalline Polymers;* Bassett, D.C., Ed.; Applied Science: London, 1982; Ch. 2.
5. Vaughan, A.S.; Ungar, G.; Bassett, D.C.; Keller, A. *Polymer* **1985,** *26,* 726.
6. Guiu, F.; Shadrake, L.G. *Phil. Mag., Ser. A* **1980,** *42,* 687.
7. Ungar, G. *Macromolecules* **1986,** *19,* 1317.

8. Tanaka, H.; Takemura, T. *Polym. J. (Tokyo)* **1980,** *12,* 355.
9. Maroncelli, M.; Strauss, H.L.; Snyder, R.G. *J. Chem. Phys.* **1985,** *82,* 2811.
10. Gomez, M.A.; Tonelli, A.E.; Lovinger, A.J.; Schilling, F.C.; Cozine, M.H.; Davis, D.D. *Polym. Prepts., Amer. Chem. Soc.* **1989,** *30,* 317.
11. Marks, B.S.; Carr, S.H. *J. Polym. Sci. - Polym. Phys. Ed.* **1985,** *23,* 1563.
12. Marchetti, A.; Martuscelli, E. *J. Polym. Sci. - Polym. Phys. Ed.* **1976,** *14,* 323.
13. Ungar, G. *Polymer* **1980,** *21,* 1278.
14. Ungar, G.; Grubb, D.T.; Keller, A. *Polymer* **1980,** *21,* 1284.
15. Ungar, G.; Grubb, D.T.; Keller, A. *Radiat. Phys. Chem.* **1983,** *22,* 849.
16. Ungar, G.; Hill, M.J. *Radiat. Phys. Chem.* **1991,** *37,* 37.
17. Okazaki, M.; Toriyama, K. *Chem. Phys. Lett.* **1990,** in press.
18. Bennett, R.L.; Keller, A.; Stejny, J. *J. Polym. Sci. - Polym. Chem. Ed.* **1976,** *14,* 3021.
19. Seguchi, T.; Tamura, R. *Rep. Prog. Polym. Phys. Jpn.* **1971,** *14,* 565.
20. Shimada, S.; Maeda, M.; Kashiwabara, H. *Polymer* **1977,** *18,* 25.
21. Grimm, H.J.; Thomas, E.L. *Polymer* **1985,** *26,* 27.
22. Grubb, D.T. *J. Mater. Sci.* **1974,** *9,* 1715.
23. Clough, R.L. *J. Chem. Phys.* **1987** *87,* 1588.

RECEIVED April 22, 1991

Chapter 8

Gamma Radiolysis of Styrene-*co*-Methyl Acrylate Copolymers

An Electron Spin Resonance Study

Raymond Kellman[1], David T. J. Hill[2], D. S. Hunter[2], James H. O'Donnell[2], and Peter J. Pomery[2]

[1]Department of Chemistry, San Jose State University, San Jose, CA 95192–0101
[2]Polymer Materials and Radiation Group, Department of Chemistry, University of Queensland, Brisbane, Queensland, 4072 Australia

Gamma radiolysis of poly(styrene-*co*-methyl acrylate) was carried out at 77K and 300K. Radical yields were determined by ESR spectroscopy using double integration. At 77K G[R•] values decreased with increasing styrene content and a styryl protective effect was observed. At 300K results were complicated by T_g effects. Photobleaching at 77K and annealing experiments allowed structural elucidation of the radical anion and neutral radical species formed on radiolysis.

The rapidly expanding use of polymers as materials in hostile environments, in microlithography and in medical applications makes an understanding of the fundamental processes involved in radiation induced polymer degradation an incumbency. The Polymer and Radiation Group at the University of Queensland has for some time pursued a broad investigation of the fundamental relationship between molecular structure and radiation sensitivity of polymers and copolymers in the solid state (*1,2*) . As part of this program we were interested in the degradation of poly(styrene-*co*-methyl acrylate) under high-energy radiation (*3*). In this paper we wish to report results of an ESR study on the formation and fate of radical species produced on gamma irradiation of these copolymers.

Experimental

The styrene-methyl acrylate copolymers as well as the styrene and methyl acrylate homopolymers used in this

0097–6156/91/0475–0119$06.00/0

study were prepared by bulk, AIBN initiated polymerization which has been described in detail previously (*3*). Six copolymers which varied in composition from 12 to 80 mole % styrene were prepared and used (see Table I). Composition was determined by elemental microanalysis and H1 FT-NMR spectroscopy (Joel GX-400). Multiple integration of the C_6H_5- aromatic styryl signal and the $-OCH_3$ acrylate signal were averaged.

Table I. Composition and Tg's of Poly(styrene-*co*-methyl acrylate)s

no.	mole % STY	Tg, C
1	0.0	13
2	12	26
3	18	42
4	29	56
5	43	68
6	60	77
7	80	91
8	100	105

a- Determined by DSC; 0 to 125C @ 10C/min.

Polymer samples subjected to radiolysis were prepared by packing the solid polymers to a depth of ca. 3 cm in Spectrosil-A grade quartz ESR tubes (i.d. 2.87 mm). The samples were evacuated overnight(<1.0 $\times 10^{-5}$ Torr) and sealed. Gamma irradiations were carried out at 300K and 77K using a ^{60}Co source (GammaCell-220) with a dose rate of 0.58 Kgy/ hr. Samples were examined by ESR spectroscopy at several dose intervals between 0.0 - 10 Kgy. In some cases, samples were taken to a higher total dose of ca. 100 KGy. When removed from the Gammacell, samples were in all cases maintained at 77K.

Radical concentrations, [R•], and radical yields, G[R•], were determined by ESR spectroscopy utilizing a Strong Pitch standard (3.0×10^{15} spin/cm) using a Brucker ER-200D spectrometer. Spectra were determined at the temperature of irradiation, *viz.*, 300K or 77K using a Klystron frequency of 9.25 GHz, microwave power of 6.3 μW (45 dB) and magnetic field strength of 0.33 Tesla. Acquisition parameters were typically 0.3300 Tesla center field, 20-25 mT sweep width and 0.2 mT modulation amplitude.

Annealing experiments were carried out in 20°C increments from 77K to 355K. Samples were heated at a rate of 10K/min and allowed to thermally equilibrate for 5.0 min at each temperature before the ESR spectrum was determined. Spectra were recorded at 77K after each increment in temperature to eliminate Boltzman distribution effects. Annealing to 355K eliminated all radical species from the samples.

Photobleaching the irradiated samples at 77K was

carried out *in situ* in the ESR cavity using a 1000 W Hg/Xe lamp and the appropriate cut-off filters (>575 nm or >495nm). The samples were exposed for periods of *ca.* 10-30 sec., the ESR spectrum was recorded and doubly integrated. When no further change in integrated area could be detected, photolysis was stopped and the spectrum recorded. Spectral subtraction gave difference spectra from which the species eliminated could be identified.

Discussion and Results

Styrene-methyl acrylate copolymers are of particular interest because they contain both an aromatic, radiation-resistant, crosslinking styryl unit and an aliphatic, radiation-sensitive acrylate unit which is subject to chain

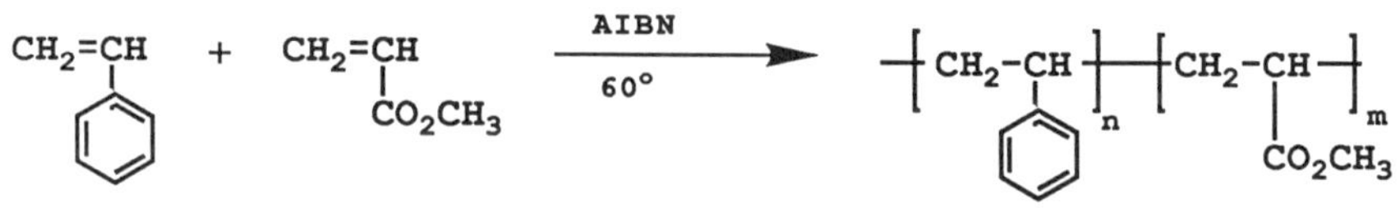

1 2 3

scission (4). Thus, one is able to examine the relative influence of these two species on the radiation behavior of these copolymers. Our objective was to determine radical yields, G[R•], for styrene-co-methyl acrylate polymers and to elucidate the structure and behavior of the radical species using ESR spectroscopic analysis.

This experimental approach afforded G[R•] values which showed a strong dependence on copolymer composition. At 77K, radical concentrations were observed which increased linearly with dose from 0.0 to 10 Kgy for all samples studied. [R•] and G[R•] values decreased in a regular way as the styrene content of the copolymer increased from 0 mole % through 100 mole % (See Figure 1 and Table II), which demonstrates the enhanced radiation resistance provided to the copolymers by the styryl units. However, a plot of G[R•] versus copolymer composition (Figure 3) clearly shows that radical yields were smaller than expected on the basis of copolymer composition. Figure 3 shows a distinctly concave relationship between copolymer composition and radical yield. That is, radiation resistance of each copolymer is greater than expected for a given styrene content. This indicates that the styryl units afford a definite protective effect to the copolymer. This observation is consistent with previous work reported from these and other laboratories (3). The aromatic ring is capable of absorbing the high energy associated with ionizing radiation without covalent bond rupture that would otherwise lead to scission and/or crosslinking.

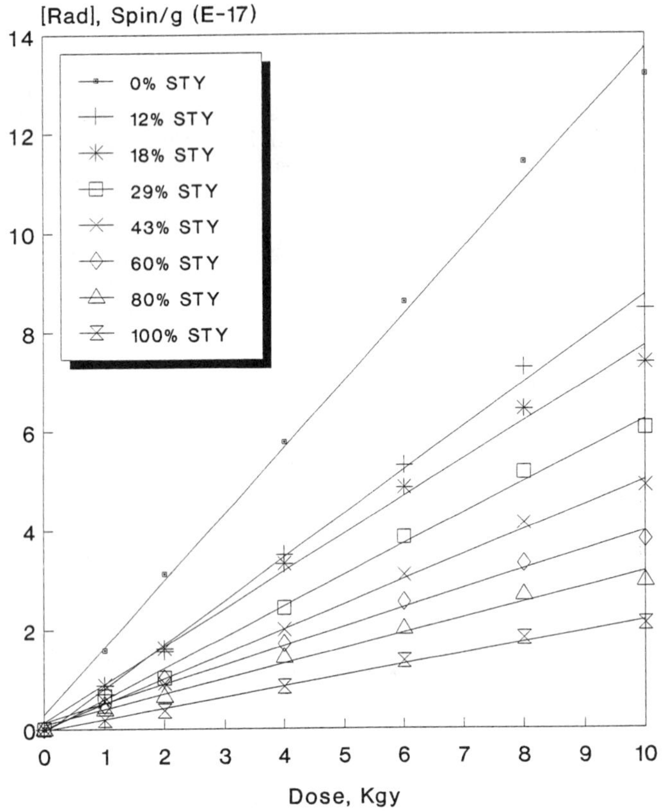

Figure 1. [R•] concentration versus dose (0-10.0 Kgy) for gamma radiolysis at 77K.

At 300K [R•] and G[R•] values were overall lower than those observed for radiolysis at 77K. Further, these values did not vary regularly with

Table II. G[R•] for Poly(styrene-*co*-methyl acrylate)s [a]

Mole % STY	*G[R•] 300K*	*G[R•] 77K*
0.0	0.877	2.76
12	1.33	1.69
18	1.39	1.58
29	1.10	1.15
43	0.889	0.942
60	0.641	0.819
80	0.396	0.676
100	0.140	0.404

a- Values for Total radical and radical anion concentrations

copolymer composition (see Figures 2 and 4 and Table II).Copolymer **2** (12 mole % STY) and homopolymer **1** (PMA) showed radical yields much lower than expected. One would have expected an increasing trend in the radical yields as the styrene content in the copolymer decreases to zero. In fact, PMA displayed a G[R•] value of 0.877, lower even than copolymer **5** (43 mole % STY; G[R•]= 0.889). These values are not consistent with styrene content and can not reflect the relative radiation resistance of these copolymers. However, the ambient temperature for irradiation (310K) is greater than the glass transition temperature (T_g) for polymers **1-3**. Consequently, these samples being above their T_g's, have enhanced segmental chain mobility which allows radical reactions to occur more readily than would be observed in the glassy state. This results in the observed radical concentrations and G (values) being lower than expected. For those polymers (*viz.*, **4-8**) irradiated at 300K and below their glass temperatures, behavior under radiolysis was analogous to results at 77K (see Figures 5 & 6). Radical yields for these polymers at 300K were observed to decrease with increasing styrene content. However, the distinct protective effect observed at 77K and attributed to the styryl units was not discernible for copolymers **4-8** within experimental error limits at 300K.

In all cases, the G[R•] values were determined from the slope of a plot of [R•] versus dose using a linear least squares treatment. Furthermore, the G[R•] values reported (see Table II) are those for <u>total</u> free radical content, which includes substantial contributions from radical anion species as well as from neutral free radicals.

The photolysis of the irradiated radical-containing samples at wavelengths of >575 nm and >495 nm has been reported to eliminate radical anions species selectively from complex mixtures of radical anions and neutral free

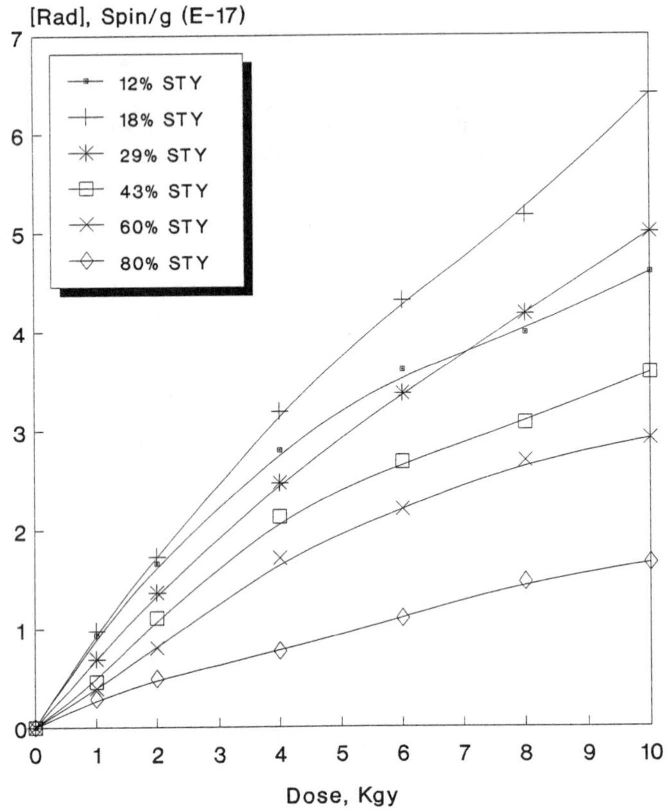

Figure 2. [R•] concentration versus dose (0-10.0 Kgy) for gamma radiolysis at 300K.

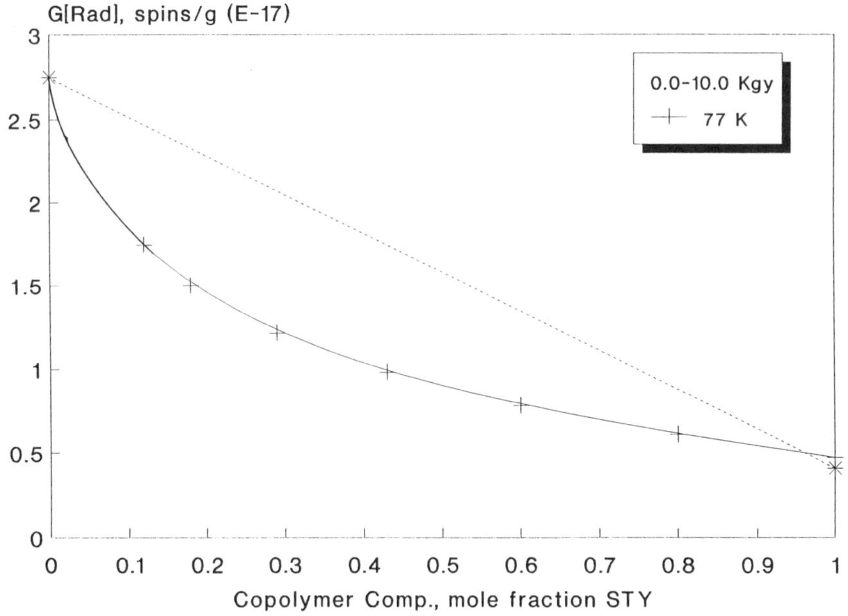

Figure 3. G[R•] versus copolymer composition for gamma radiolysis at 77K.

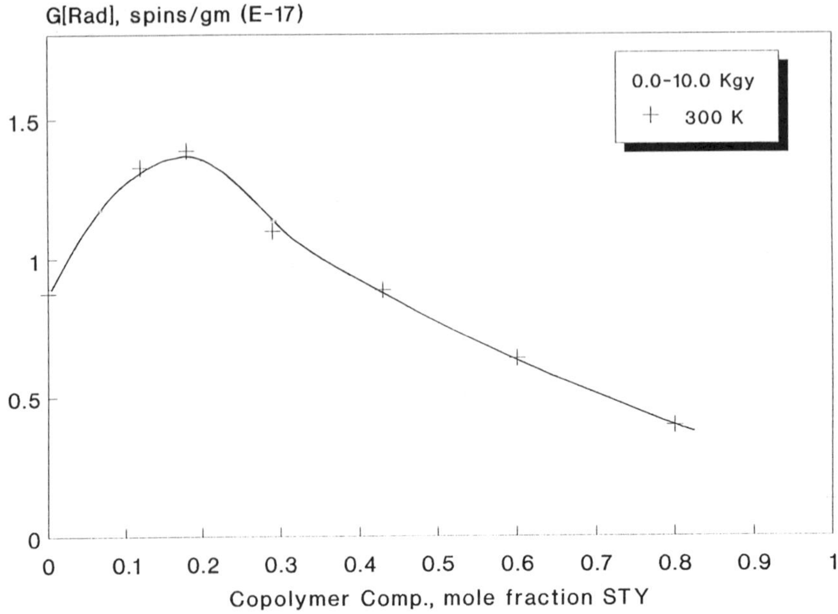

Figure 4. G[R•] versus copolymer composition for gamma radiolysis at 300K.

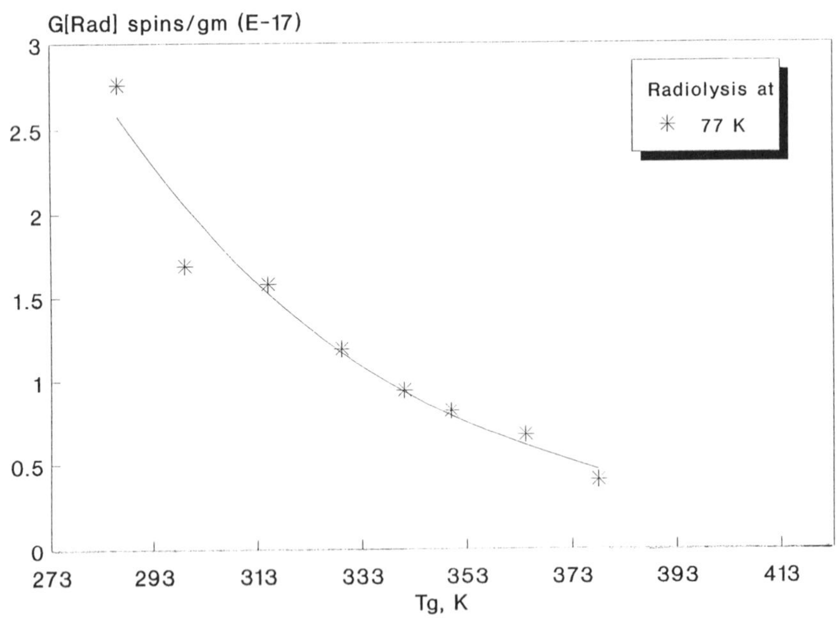

Figure 5. G[R•] versus T_g (K) for gamma radiolysis at 77K.

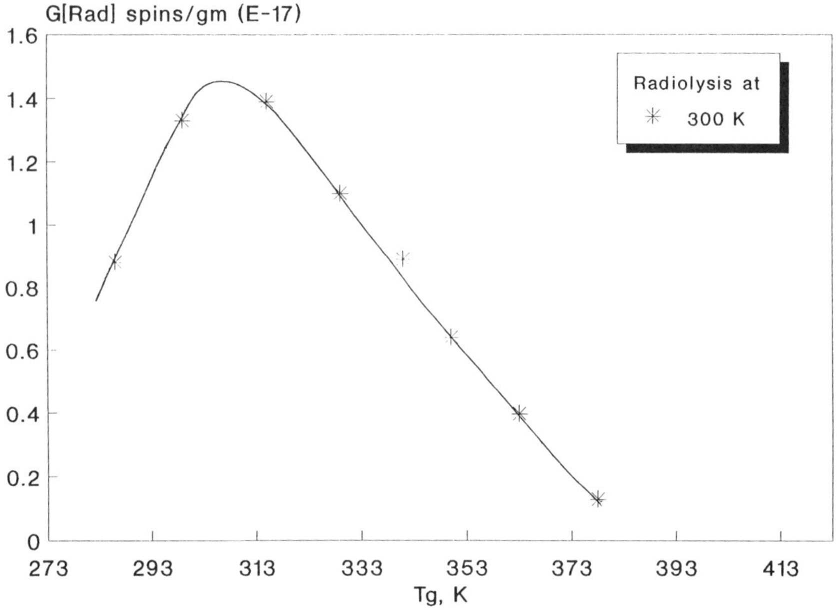

Figure 6. G[R•] versus T_g (K) for gamma radiolysis at 300K.

radicals formed on radiolysis (*2*). In this study, photolysis at 77K of several samples led to the loss of the styryl and carbomethoxy radical anions. These radical anions show characteristic broad singlets (line widths *ca.* 8 G) in the ESR. Loss of these species was evidenced by the disappearance of this broad singlet from the ESR spectrum. Copolymer samples which had been gamma irradiated and maintained at 77K were photolyzed initially with > 575nm until [R•] was constant (25 min) and then with > 495nm (25 min) to ensure complete photobleaching. The spectra were recorded and doubly integrated after photolysis at each wavelength. Figures 7 and 8 show the results of photobleaching polymers **1** and **5**, respectively. The initial spectra are dominated by a rather broad, featureless singlet which reveals little of the hyperfine splitting due to the complex radical mixtures present. However, photolytic elimination of the pendant carbomethoxy, $-CH_2CH(CO_2CH_3)^{\bullet}-$, and pendant styryl, $-CH_2CH(C_6H_5)^{\bullet}-$, radical anions removes the singlet (Fig 7b & 8b) from the spectra. The final spectra which show somewhat more discernible hyperfine structure are those of a mixture of the neutral free radical species formed on radiolysis. Verification of the species lost or gained on photolysis are shown in the difference spectra (Fig. 7b & 8b) obtained by subtracting the spectra before and after photobleaching.

Evidence that the styryl radical anion contributes a significant proportion of the observed intensity at 77K is shown in Table III. For example, 55% of the total radical signal in polystyrene is attributable to this species. In the copolymers studied the styryl radical

Table III. Percent loss of radical anions on photobleaching

%STY	% Loss of [R•] >595nm	>495nm	Total
100	50	5	55
43	30	12	42
28	25	13	38
18	17	10	27

anion is formed in relatively high concentration when compared to the mole % styryl units present along the polymer chain. Interestingly, the copolymers which display the highest protective effect (Figure 3) at 77K also give rise to relatively high concentrations of styryl radical anion. This may be in part responsible for the protective effect observed and attributed to the aromatic monomer unit. Photolysis at >575 nm eliminated only the styryl radical anion from the copolymer samples. However, at wavelengths >495 nm the pendant carbomethoxy radical anion, $-CH_2CH(CO_2CH_3)^{\bullet}-$, and for PMA formyl radical, •CHO (12.8 mT), are lost as well (see Figure 7b).

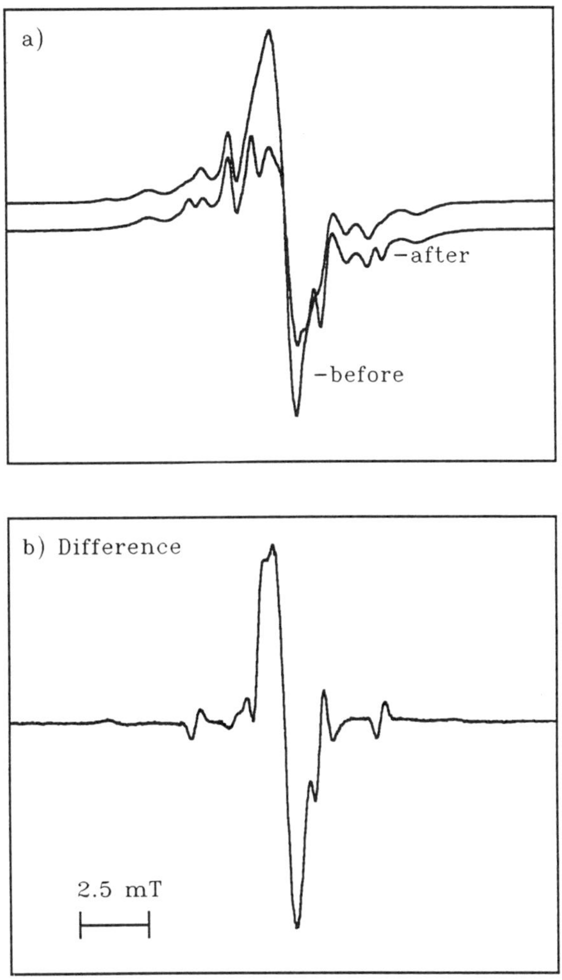

Figure 7. a) Photobleaching (>495 nm) of PMA(**1**) at 77K; b) Difference spectra of **1** before and after photobleaching.

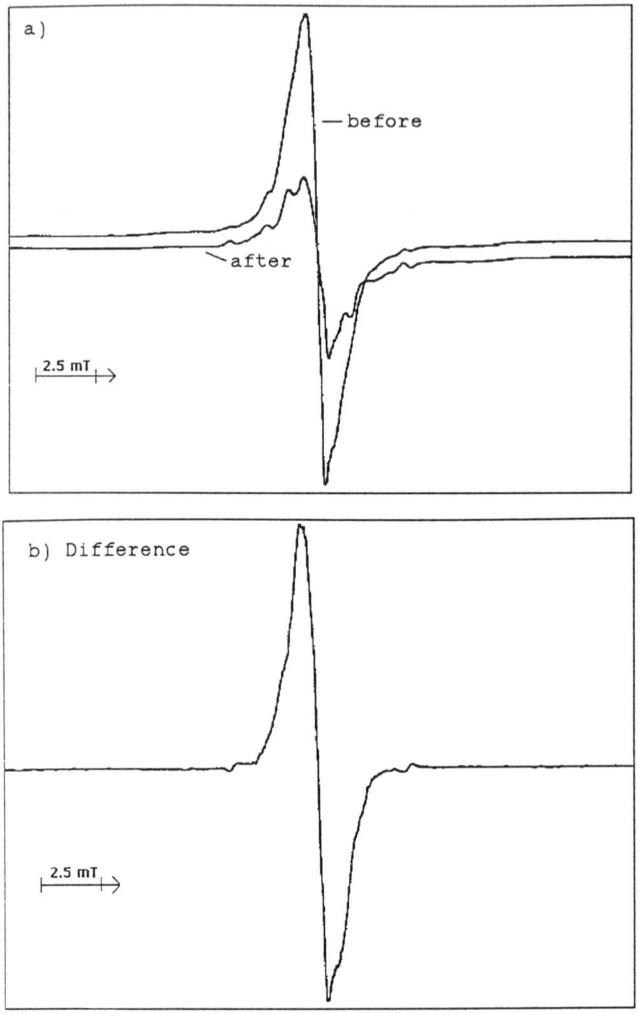

Figure 8. a) Photobleaching (>495 nm) of 43% STY copolymer (**5**) at 77K; b) Difference spectra of **5** before and after photobleaching.

Also, at the higher wavelengths spectral subtraction clearly indicated the slow formation of the methyl radical. The methyl radical gives rise to the inverted quartet (2.27 mT; line width of .33 mT) superimposed on the broad singlet due to carbomethoxy radical anion (Figure 7b). This is probably due to the photolytically induced homolytic cleavage of the carbomethoxy radical, and/or radical anion.

Annealing experiments were carried out on those samples irradiated at 77K. Each sample was warmed from 77K to <u>ca.</u> 355K in 20° increments and the spectrum recorded after each temperature increment. Spectral subtraction afforded difference spectra and allowed identification of the radical species lost on warming. For example, PMA gamma irradiated and subsequently photobleached at 77K showed loss of $\cdot CH_3$, $\cdot$CHO , and $\cdot CO_2CH_3$ radicals on warming to 115K (see Figure 9a & b). From 115K to 215K warming caused the loss of some minor and unidentified radical species along with a major triplet due to the radical localized on the methyl group of the carbomethoxy moiety ($-CO_2CH_2\cdot$). Above 200K presence of a doublet from the expected methyl acrylate propagating radical, a product of chain scission, became apparent from difference spectra. This radical persisted to 295K. In the copolymers investigated (see Table IV) besides the radicals derived from methyl acrylate degradation, styrene units undergo hydrogen atom addition to the phenyl ring to give the cyclohexadienyl radical as the dominant species (see Figure 10a&b). Also, hydrogen atom loss from the beta carbon of the styryl units gives rise to the secondary radical along the polymer backbone (5).

Conclusions

Gamma radiolysis of poly(styrene-*co*-methyl acrylate) copolymers at 77K affords radical yields, G[R•], which decrease in a regular way with increasing styrene content. Examination of the relationship between G[R•] and copolymer composition reveals an enhanced radiation resistance above that which one would predict based on copolymer composition. This is due to the ability of the aromatic styryl pendant group to dissipate the absorbed energy without subsequent energy loss through bond cleavage. At 300K radical termination reactions are faster and G[R•] values correspondingly lower. However, these results are complicated by the effects of T_g for copolymers **1-3**. That is, G[R•] exhibits a maximum, when plotted against composition due to the fact that radiolysis is carried out above T_g for these materials.

Photobleaching experiments demonstrated at the styryl radical anion is the dominant species formed on radiolysis at 77K, responsible for 55% of the radicals in polystyrene, at least 30% of [R•] in **5** (43 mole % STY)

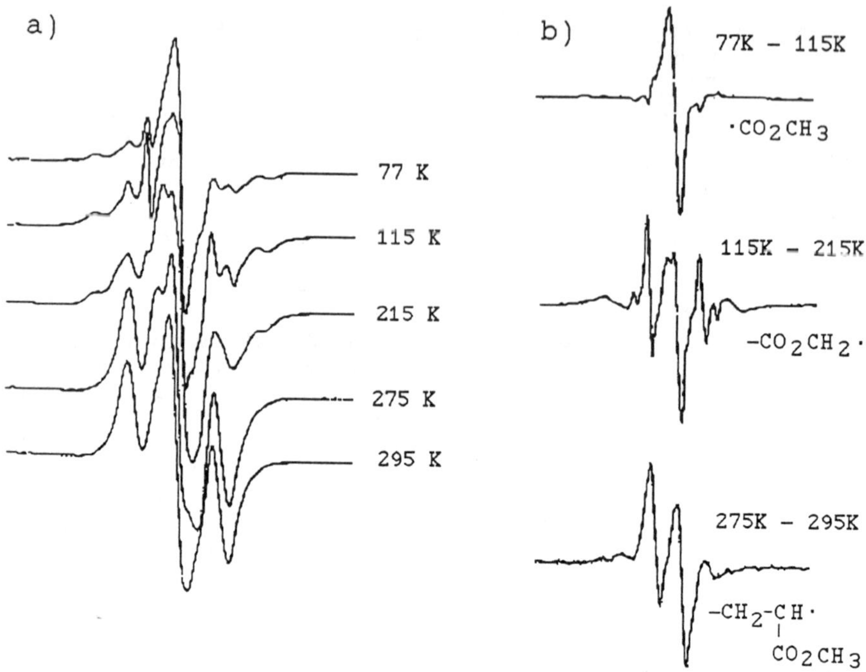

Figure 9. a) ESR spectra of PMA as a function of temperature; b) difference spectra.

Table IV. Disappearance Temperatures of Radicals from Gamma Radiolysis at 77K

115 K	$\cdot CH_3$ $\cdot CHO$ $\cdot CO_2CH_3$	$\sim CH_2-CH\sim$ (with $\cdot CO_2CH_3$ substituent)
200 K	$-CO_2CH_2\cdot$	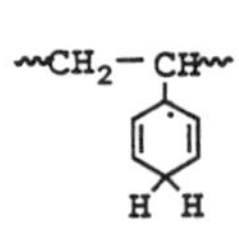
295 K	$\sim CH_2-\dot{C}(H)(CO_2CH_3)$	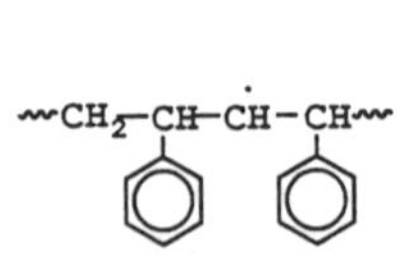
>295 K	$\sim CH_2-\dot{C}(CO_2CH_3)-CH_2\sim$	

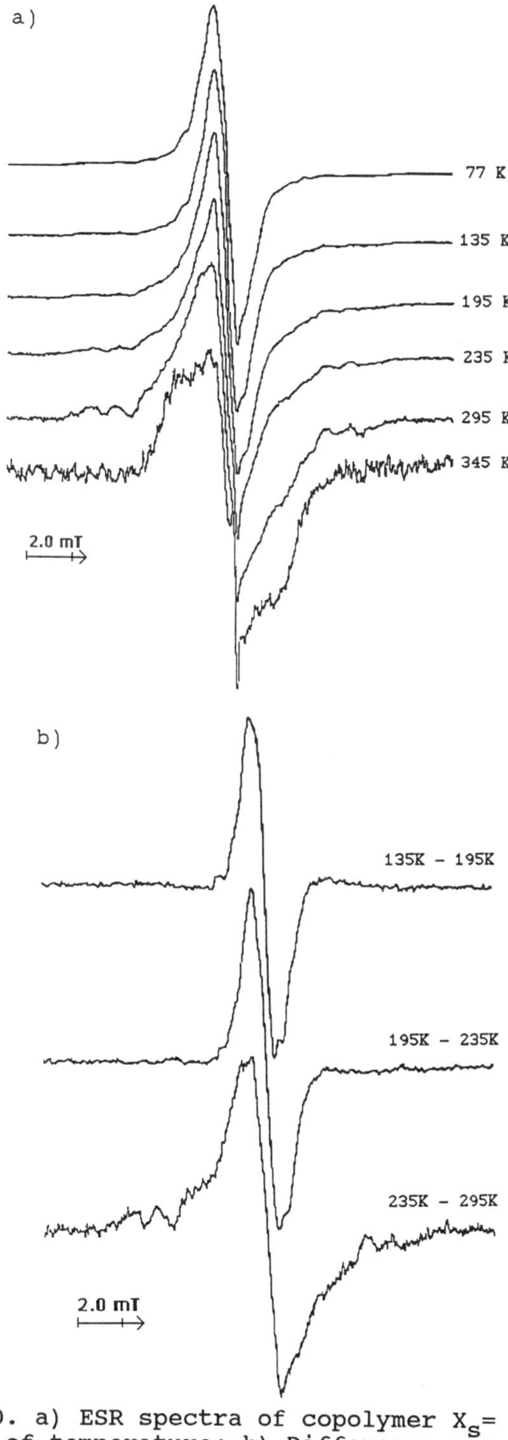

Figure 10. a) ESR spectra of copolymer X_S= 0.43 as a function of temperature; b) Difference spectra.

and 25% of [R•] in **4** (29 mole % STY). In those copolymers rich in methyl acrylate the carbomethoxy pendant group radical anion was present in significant amounts. The post photolysis ESR spectra at 77K revealed the presence a complex mixtures of free radicals which have been partially elucidated by warm-up experiments and spectral subtraction techniques (See Table IV). The styrene units along the polymer backbone gave rise to cyclohexadienyl and 2^0 backbone radicals but not to 1^0 or benzylic radicals which would result from chain scission. On the other hand, under radiolysis the methyl acrylate units did suffer chain scission to give the 2^0 propagating radical. Additionally, homolytic bond cleavage resulted in backbone radicals and pendant group radicals which are capable of crosslinking.

Acknowledgments

The authors (especially grantee RK) wish to thank the Fulbright Foundation, the Australian-American Educational Foundation for financial support of this work.

References

1- D.J.T. Hill, J.H. O'Donnell, and P.J. Pomery, "Fundamental Aspects of Polymer Degradation by High-Energy Radiation," in *Materials for Microlithography*, L.F. Thompson, C.G. Wilson and J.M.J. Frechet (Eds.), **ACS Symposium Series, 266**, 125-150, (1984).
2- D.J.T. Hill, J.H. O'Donnell, and P.J. Pomery, "Applications of ESR in Polymer Chemistry," in *Electron Spin Resonance 1985*, **9**, 223.
3- W.K. Busfield, J.H. O'Donnell, and C.A. Smith, *J. Macro. Sci.- Chem. 1982*, **A17(8)**, 1263.
4- J.H. O,Donnell in *The Effects of Radiation on High-Technology Polymers*, E. Reichmanis and J.H. O'Donnell (Eds.), **ACS Symposium Series, 381**, 1, 1989.
5- R.W. Garrett, Ph.D. Thesis, University of Queensland, 1985, Brisbane, Australia.

RECEIVED July 11, 1991

Chapter 9

Effects of Ultraviolet and Vacuum Ultraviolet Irradiation on Fluorinated Ethylene–Propylene Copolymers

An Electron Paramagnetic Resonance Study

Soon Sam Kim and Ranty H. Liang

Space Materials Science and Technology Section, Jet Propulsion Laboratory, California Institute of Technology, Pasadena, CA 91109

Fluorinated ethylene propylene copolymer (FEP) samples were irradiated by an excimer laser source, KrF (249 nm), ArF (193 nm) or F_2 laser (157 nm), and observed formations of two radical species; a terminal (I) or side chain radical (III), $-CF_2\bullet$, and a main chain radical $- CF_2 - \dot{C}F - CF_2 -$ (II). Room temperature EPR spectrum of (I) and (III) showed a strong triplet structure with a hyperfine splitting of 16.7 G and the radical (II) showed a doublet of quintet structure with hyperfine splitting of $a_\alpha = 42$ G and $a_\beta = 22$ G. Observations of the same radical species were reported in the literature by γ-ray or electron irradiation of polytetrafluoroethylene or FEP samples. It seems the same radicals are involved in the initial stages of degradation processes of fluorinated polymers by γ-ray, electron, UV or VUV irradiations.

As a part of materials evaluation program for space applications, we studied the effects of ultraviolet (UV) and vacuum ultraviolet (VUV) irradiation on several polymer materials. It is well known that UV and VUV induced degradation process of polymeric materials involve photogenerated radical species in the matrices as initial steps. One of the polymeric materials we studied is fluorinated ethylene propylene copolymer (FEP) (see Figure 1). FEP is one of the candidate materials for space applications, for

0097–6156/91/0475–0135$06.00/0

example, as thermal control blankets. Long term stability of these materials in space has practical importance for the integrity of space missions.

There have been numerous publications on the study of radicals generated after γ-ray or electron irradiation of fluorinated polymeric materials by electron paramagnetic resonance (EPR); polytetrafluoroethylene (PTFE) (1-6) and with FEP (7-9). However, study of UV and VUV effects on these materials was rare (5,10).

It is the aim of this study to obtain molecular level information for the initial stages of degradation processes, and we describe the results of our study on photogenerated radicals of FEP after excimer laser irradiation by EPR spectroscopy.

Experimental

Samples of FEP (Du Pont, FEP 6000L) (propylene mole ratio, ~14%, Tg = 260-280°C), of a size ~0.4 x 1.8 x 0.15 cm thick and weight of ~0.2 gr, were irradiated by an excimer laser source (Lambda Physik 101E), either as KrF laser (249 nm, 250 mJ/pulse), ArF(193 nm, 200 mJ/pulse) or as F_2 laser (157 nm, 10 mJ/pulse). A xenon lamp (ILC Technology, LX300UV, $\lambda \geq 250$ nm) was also used for UV irradiation.

In order to capture possible short lived radical species, the FEP samples were irradiated at liquid N_2 temperature as shown in Figure 2. For VUV irradiation, a special arrangement with 45° beam turning mirror (Acton Research Corp) was used. The sample was placed on a wire screen in contact with liquid N_2. During irradiation, nitrogen gas was blown (35 SCFH) through the optical path to minimize absorption of VUV by 0_2. After irradiation, the sample was then dropped to the bottom of the dewar for EPR studies. For room temperature UV or VUV irradiation, the FEP sample was placed in a cell with LiF window (see Figure 3) and irradiated under vacuum. After irradiation, the sample was dropped to the bottom of the cell and EPR study was made.

A commercial EPR spectrometer (Varian Century Series E115) operated at 9.12 GHz was used. Spectral simulation was obtained by using a software package supplied by Varian (E-900).

Results and Discussion

At liquid N_2 temperature, after KrF laser irradiation (20 min. at 15 Hz), we observed EPR spectra with a strong central peak with wings at 223 G apart (see Figure 4). As the sample was warmed by pouring out liquid N_2 and by blowing room temperature nitrogen gas at the top of the dewar, the wing structures coalesced into the central peak. At room temperature

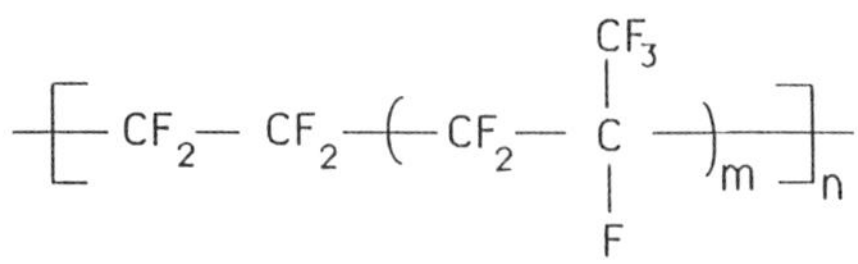

Figure 1. Chemical structure of fluorinated ethylene propylene copolymer (m= ~0.17).

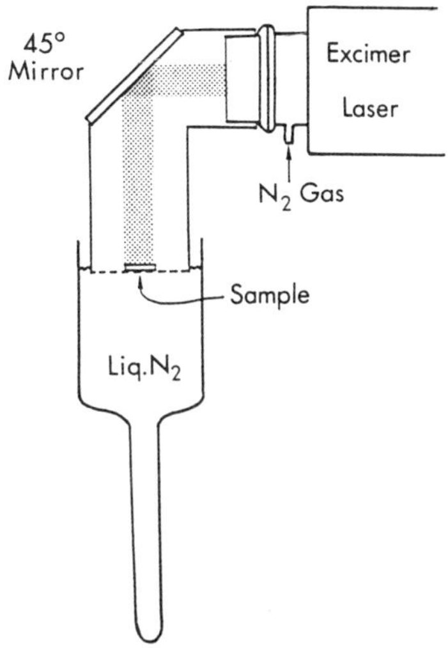

Figure 2. Experimental apparatus used for irradiation of samples with an excimer laser at liquid N_2 temperature.

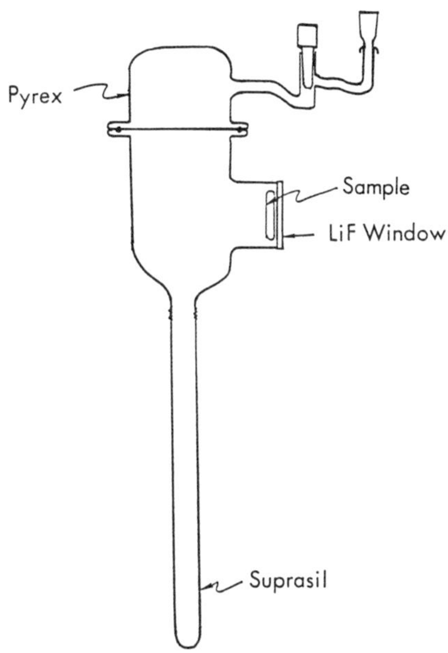

Figure 3. Experimental apparatus used for irradiation of samples with UV or VUV sources at room temperature.

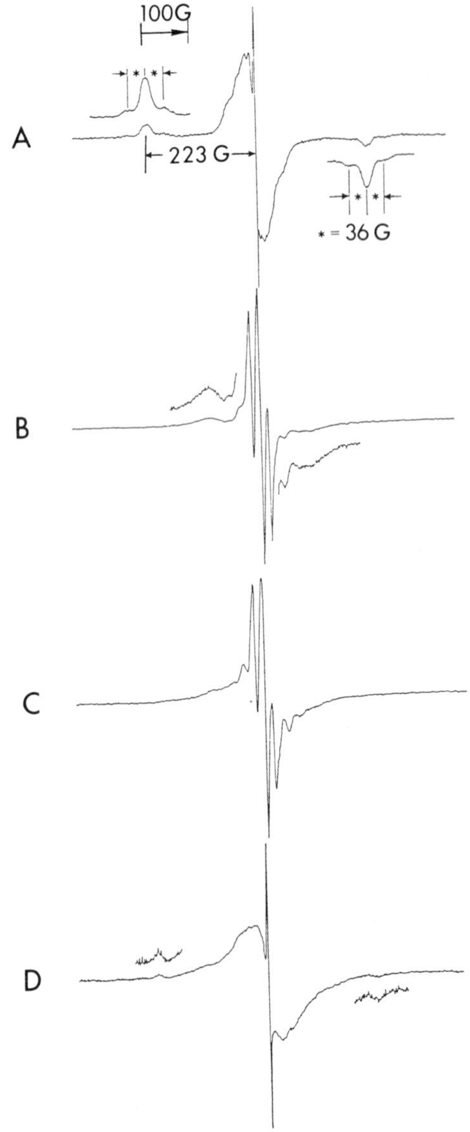

Figure 4. EPR spectra obtained after KrF laser irradiation (20 min., at 15 Hz) of a FEP sample at 77°K. EPR scans were made; A, at 77°K; B, warmed to ~0°C; C, at 23°C; D, cooled back to 77°K.

we observed only the central peak with hyperfine structures. When cooled back to liquid N_2, the wing structure returned.

Details of the central hyperfine structure at room temperature is shown in Figure 5. When the FEP sample was irradiated at room temperature in a cell under vacuum by the F_2 laser, we observed an EPR spectrum with 7-line pattern (Figure 5, B). Also, when FEP sample was irradiated in the same cell with a xenon lamp, we observed mainly of a triplet structure with 16.7 G splitting (Figure 5, C). By comparing the relative peak heights of the spectrum shown in A of Figure 5, one can deduce it is a composite of B and C, i.e., from two different radical species (5), one with triplet structure from two equivalent fluorine atoms (a = 16.7 G), the other is a doublet of quintet structure with one α-fluorine (a_α = 42 G) and 4 equivalent β-fluorine atoms (a_β = 22 G). Simulated spectra with the hyperfine coupling constants are shown in Figure 6. The observed one matches with the simulated composite spectrum.

EPR spectra with triplet structure and doublet of quintet structure were also observed in cases of γ or election irradiated PTFE (1-6) and also from FEP (7-9). The triplet structure was assigned to terminal - $\underset{\bullet}{C}F_2$ radicals (I), where as the doublet of quintet structure was assigned to a main chain radical (II), with larger hyperfine coupling constants, a_α = 86 ~ 90 G and a_β = 33 ~ 34 G (5,8).

- CF_2 - $CF_2\bullet$ (I) - CF_2 -$\underset{\bullet}{C}F$ - CF_2 - (II)

$$\begin{array}{c} \bullet CF_2 \\ | \\ -CF_2 - CF - CF_2 \end{array} \text{ (III)} \qquad \begin{array}{c} CF_3 \\ | \\ -CF_2 - \underset{\bullet}{C} - CF_2 - \end{array} \text{ (IV)}$$

We also assign the observed triplet and doublet of quintet structure to the same radicals, -$CF_2\bullet$ and -CF_2 -$\underset{\bullet}{C}F$ - CF_2- (II), respectively. In the case of -$CF_2\bullet$ radical, it could have originated from either sources, side chain -CF_3 unit, (III), or chain terminals, (I), as in the case of PTFE (11). For the FEP case, Iwasaki et. al. (10) chose the side chain radical configuration (III) over the chain terminals (I). They drew this conclusion based on the observation that the orientation dependence of α-fluorene hyperfine splitting for FEP is different from PTFE and also by the absence of β-fluorine hyperfine structure in the wing peaks at 77°K. From consideration of steric interaction between fluorine atoms in side CF_2 group and those in chain CF_2 groups of radical (III), they reasoned that the p-orbital of the unpaired electron on the carbon atom should be parallel to the chain axis. This configuration would result in a small β-fluorine coupling or lack of clear hyperfine structure in the wing peaks.

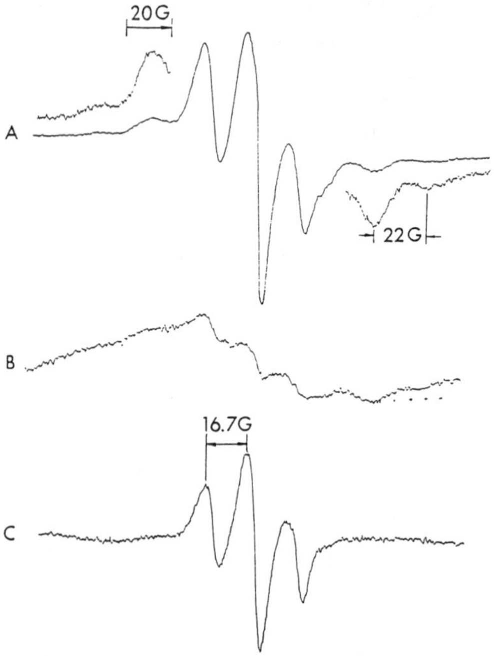

Figure 5. EPR spectra of FEP samples; A, the central hyperfine structure at 23°C after irradiation by KrF laser at 77°K; B, after F_2 laser irradiation at 23°C; C, after xenon lamp irradiation at 23°C.

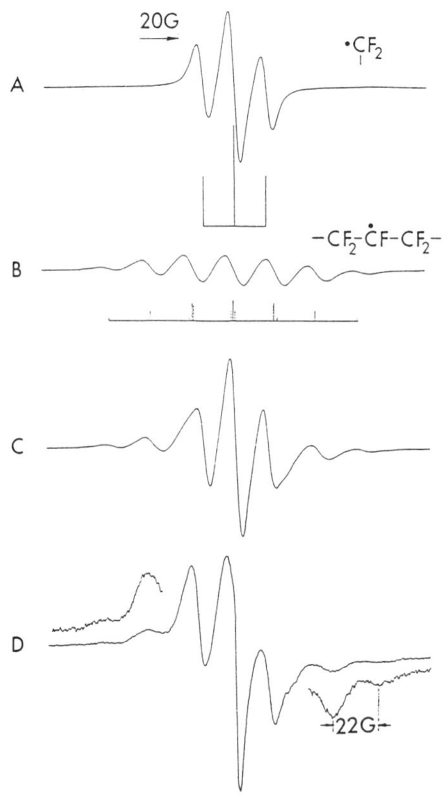

Figure 6. Simulated EPR spectra; A, -CF_2 radical (a = 16.7 G); B, radical (II) (a_α = 42 G, a_β = 22 G); C was obtained by addition of A and B at the relative radical concentrations of 1:4 ratio. D, the observed spectrum.

However, in our experiments we observed a partially resolved hyperfine structure in the wing peaks at 77°K (see Figure 4, A), and this could be assigned to two equivalent β-fluorene splitting, a_β(77°K) = 36 G (11). If we assume the radical (III) would not show any structure in the wing peaks as discussed by Iwasaki et al (10), a spectral overlap for radicals (I) and (III) will show a wing structure similar to what we observed in Figure 4 A; triplet with larger central peak. Most probably both radicals, (I) and (III), contribute to the observed triplet spectrum at room temperature (Figure 5 C) and the spectrum observed at liquid N_2 temperature with 223 G splitting (Figure 4 A).

The composite EPR spectrum shown in Figure 5A was also observed by irradiation of FEP samples by ArF laser under moderate vacuum (~10^{-4} mmHg) as shown in Figure 7. The radicals have very long life time as indicated in the figure caption. However, the central triplet spectrum from -CF_2 radical gradually changed to that of peroxy radical (see Figure 7, B and C) whereas the 7-line structure from the main chain radical did not show such change in 17 hours. It shows that the two radical species have different reactivities toward oxygen present in the cell, the primary radicals (I) and (III) reacting faster than the secondary radical (II). After 12 days, both radicals converted to the peroxy radicals. The rate of conversion will depend on the oxygen concentration, at higher oxygen pressure, the rate will be much faster.

Iwasaki et al (8,10) reported observation of the tertiary radical (IV), from their γ-irradiation experiments of FEP samples. They also reported a conversion of radical (IV) to side chain radical (III) after UV illumination at 77°K. Since the radicals in our experiments were generated by UV sources, it may be justified for not observing the tertiary radical (IV) from our samples. The hyperfine coupling constants of FEP samples we obtained from the spectra at 77°K and room temperature are listed in Table 1 and compared with those of Iwasaki et al (8,10).

It is reported (12,13) that hyperfine constant for the α-fluorine has large anisotropy. After electron irradiation of oriented PTFE, Ovenall (4) observed a variation of α-fluorine splitting from 91 G to a minimum of 25 G, and for β-fluorine from 33.5 G to a minimum of 28 G for the chain radical (II). The observed differences in the hyperfine coupling constants between irradiated PTFE and FEP may be explained by possible differences in the geometric arrangements of α and β fluorine atoms in the polymer chains (10,11), rotational motion of α C-F bonds around the chain axis at room temperature (4,8), the crystallinity of samples, and in the case of FEP, copolymer ratios of tetrafluoroethylene and hexafluoropropylene.

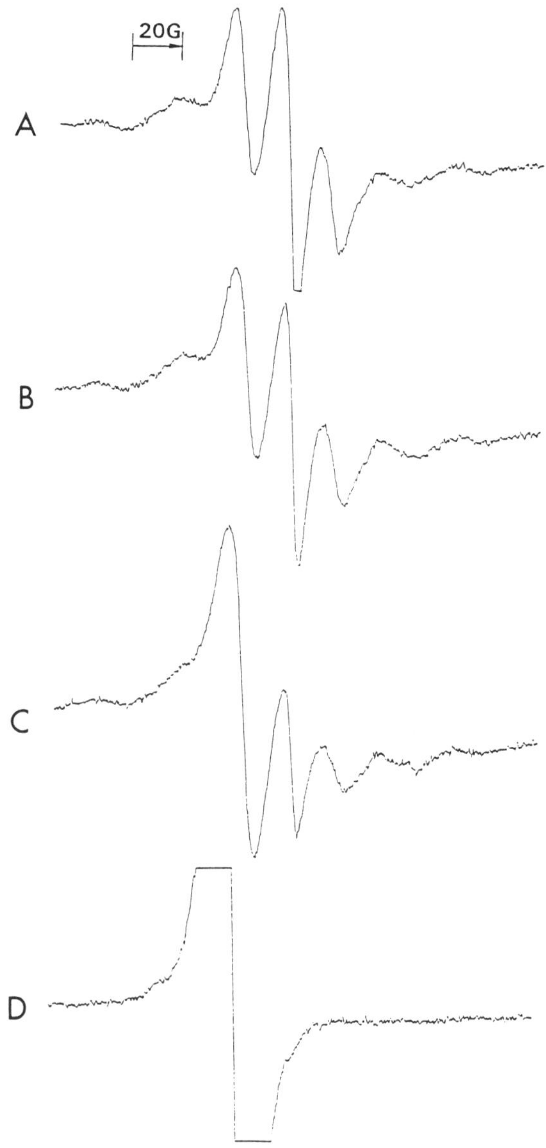

Figure 7. EPR spectra obtained after irradiation of FEP samples by ArF laser (10 min, 15 Hz) at room temperature under moderate vacuum (~10^{-4} mm Hg). A. right after ArF irradiation, B. after 2 hours, C. 17 hours and D. 12 days. Notice the gradual change of central triplet structure to that of peroxy radical. After 12 days, all the radicals converted to peroxy radicals.

Table 1. Hyperfine coupling constants[1] of photogenerated radicals from FEP

	Iwasaki et al (8,10)	this work
irradiation source	γ	UV and VUV
	(I) a = 16 G	(I) and (III) a_α(77°K) = 223 G (I) a_β(77°K) = 36 G
	(III) a_α(77°K) = 22.5 G a = 16 G	(I) and (III) a_α = 16.7 G
	(II) a_α = 86 G a_β = 33 G	(II) a_α = 42 G a_β = 22 G
	(IV) a_β = 28.5 G	not observed

[1] The hyperfine coupling constants are isotropic values measured at room temperature except indicated as 77°K.

The hyperfine coupling constants for the chain radical (II) obtained by Iwasaki et al (8,10) show larger values, a_α = 86 G and a_β = 33 G, compared with the values measured in this work, a_α = 42 G and a_β = 22 G. Their FEP samples (Teflon 100X) had 11 mole % of propylene contents with 40-60% crystallinity. The samples used in this work (FEP 6000L) have ~14 mole % of propylene contents with less crystallinity, 38.6% (14). It seems our smaller hyperfine coupling constants are a result of more isotropic averaging of the chain radical in the less crystalline environment.

Compared with the chain radical, the terminal (I) or side chain radicals (III), - $CF_2\bullet$, experience less restriction in the motional averaging at room temperature, and less sensitive to the crystallinity of the samples. Thus their hyperfine coupling constants matches well with those reported for PTFE (1-6) or FEP (8,10).

Acknowledgment

The work was supported by NASA.

References and Notes

1. Iu.D. Tsuetkov, N.N. Bubnov, M.A. Makul'skii, Iu.S. Lazurkin, Dokl. Akad. Nauk SSSR, 122, 1053 (1958).
2. H.N. Rexroad and W. Gordy, J. Chem. Phys. 30, 399 (1959).
3. N. Tamura, J. Phys. Soc. Japan, 15, 943 (1960).
4. D.W. Ovenall, J. Chem. Phys. 38, 2448 (1963).
5. S. Siegel and H. Hedgpeth, J. Chem. Phys. 46, 3904 (1967).
6. K. Toriyama and M. Iwasaki, J. Phys. Chem. 73, 2919 (1969).
7. R.E. Florin and L.A. Wall, J. Research, National Bureau of Standards A. Phys. Chem., 65A, 375 (1961)
8. M. Iwasaki, K. Toriyama, T. Sawaki and M. Inoue, J. Chem. Phys. 47, 554 (1967).
9. V.I. Muromtsev, R.A. Asaturyan and I.G. Akhvlediani, Usp. Khim. 40, 312 (1971).
10. M. Iwasaki and K. Toriyama, J. Chem. Phys. 47, 559 (1967).
11. K. Toriyama and M. Iwasaki, J. Phys. Chem. 73, 2919 (1969).
12. R. Lontz and W. Gordy, J. Chem. Phys. 37, 1357 (1962).
13. M. Iwasaki, K. Toriyama and B. Eda, J. Chem. Phys. 42, 63 (1965).
14. Personal communication with Mr. Ralph B. Williams from DuPont de Nemours & Co., Wilmington, Delaware, U.S.A. The crystallinity of FEP was estimated from the following equation which utilizes the difference in densities between amorphous phase (2.04 gr/cc) and crystalline phase (2.30 gr/cc) of FEP; % Crystallinity = (1/2.04 - 1/S) x 100/0.0554 in which S = specific gravity of FEP sample (2.133 gr/cc) used in the experiment.

RECEIVED August 5, 1991

Chapter 10

Temperature Dependence of the Radiation Chemistry of Polymers

R. Wayne Garrett[1], David J. T. Hill, Tri T. Le, Karen A. Milne, James H. O'Donnell[2], Senake M. C. Perera, and Peter J. Pomery

Polymer Materials and Radiation Group, Department of Chemistry, University of Queensland, Brisbane, Queensland, 4072 Australia

The temperature dependence of main-chain scission and mass loss by depropagation to monomer has been determined for γ-irradiation of poly(methyl methacrylate), poly(phthalaldehyde), poly(isobutene) and poly(α-methyl styrene). Radiation greatly enhanced the thermal degradation. The yield of main-chain scission also increased markedly with temperature, particularly when the calculation of G(scission) was corrected for the decreasing dose rate due to the mass loss. The behaviour of these polymers was compared with polystyrene and poly(arylene sulfone).

The changes in the properties of polymeric materials that occur when they are subjected to high-energy radiation, including γ-rays and electron beams, are a result of chemical reactions initiated by the radiation. The chemical reactions may be classified into (1) main-chain scission, (2) crosslinking, (3) formation of small molecular products, and (4) modification of the chemical structure of the polymer. An objective of fundamental studies of the radiation chemistry of polymers is to determine the rates of these reactions.

There is increasing interest in the utilization of polymer materials in radiation environments which are significantly above ambient temperatures, including aerospace applications (*1*). The deterioration in the properties of polymers may be markedly increased by relatively small rises in temperature.

The rates of chemical reactions normally show a positive temperature dependence, which is due to the activation energy barrier and can be described by an Arrhenius relationship. In polymers, the activation energy may show considerable variation and reflect both chemical and physical effects on the reaction. Also, the temperature dependence of reactions in polymers is likely to show discontinuities at first and second order transitions, especially at the glass transition, Tg, and crystalline melting, Tm, temperatures. The reaction rates may be quite different in crystalline and amorphous polymer, and between the glassy and rubbery amorphous states. There is increasing interest in the properties of the interfacial material between the amorphous and crystalline regions, which may have non-crystalline structure.

[1]Current address: Australian Nuclear Science and Technology Organization, Menai, New South Wales, 2234 Australia
[2]Corresponding author

0097–6156/91/0475–0146$06.00/0

The effects of temperature on the radiation chemistry of polymers has not been studied extensively. Charlesby and Moore (*2*) found that the yield of scission, G(S), in poly(methyl methacrylate) increased from 1.6 at 0 °C to 3.8 at 180 °C, and followed an Arrhenius relationship without any discontinuity at Tg. However, Wundrich (*3*), in a study of PMMA over a wider temperature range down to 77 K, found three distinct temperature regions with different activation energies. Wundrich also studied poly(isobutene) and obtained similar results (*4*).

Polymers which give high values of G(S) at ambient temperature frequently undergo depropagation to monomer during thermal degradation. Both types of behaviour have been attributed to the large steric strain in these polymers, which usually arises from a quaternary carbon atom in the repeat unit of the main chain and the related low enthalpy of polymerization. Thus, initiation by main-chain scission and degradation by depropagation are both favoured in these polymers.

Bowmer and O'Donnell (*5*) have shown that main-chain scission occurs on irradiation of poly(olefin sulfone)s with a high value for G(S) at temperatures from -196 to 100 °C, and that depropagation to the two monomers increases rapidly above the ceiling temperature for propagation/depropagation equilibrium.

In this paper we report measurements of the reduction in molecular weight and sample mass for γ-irradiation in vacuum of several polymers which have high G(S) values at ambient temperature, apparently do not undergo crosslinking, and degrade thermally by depropagation to produce mainly monomer. The behaviour of these polymers is compared with polystyrene and poly(arylene sulfone).

Experimental

Preparation. Poly(methyl methacrylate), PMMA, was prepared by free radical polymerization in bulk at 70 °C with AIBN initiator. The polymer was precipitated twice from chloroform into methanol and dried under vacuum with progressive heating from 30 to 100 °C. The polymer was ground to a fine powder under nitrogen at 77 K. Poly(styrene) was prepared similarly.

Poly(phthalaldehyde), PTA, was provided by Dr Hiroshi Ito of IBM Almaden Research Center and was prepared by anionic polymerization at -50 °C, and end-capped with acetaldehyde.

Poly(α-methyl styrene), PAMS, was made by cationic polymerization at -100 °C with BF3-ether in CH2Cl2 as initiator.

Bisphenol-A polysulfone, PSF, was provided by Union Carbide Corporation (Udel P-1700); it was purified by reprecipitation from chloroform into methanol and dried under vacuum.

Irradiation. The polymers were evacuated for 24 hours and sealed in glass ampoules for low temperature experiments. For irradiations at temperatures when mass loss became significant, samples of powdered polymer were contained in small, open tubes which were placed in larger tubes and evacuated continuously during irradiation. The tubes were placed in an aluminium block heater, which maintained a constant temperature within ± 1 °C.

The samples were irradiated with ^{60}Co γ-rays at about 5 kGy/h in an AECL Gammacell or in the Gatri radiation cell at the Australian Nuclear Science and Technology Research Laboratories for higher temperatures with continuous evacuation. Fricke dosimetery was used to determine the dose rates.

Characterization. The mass loss was determined by weighing the samples in the small, open sample tubes before and after irradiation. Molecular weights were obtained by gel permeation chromatography using calibration with polystyrene standards and appropriate corrections. The purity of the unirradiated polymers was checked by ^{1}H and ^{13}C NMR.

Results and Discussion

Poly(styrene). Values for Mn and Mw of the polydisperse polystyrene sample, before and after irradiation to a series of doses, were derived from the GPC chromatograms. The radiation chemical yields for scission, G(S), and crosslinking, G(X) per 16 aJ of energy absorbed, were calculated from the slopes of plots of 1/M versus dose according to equations 1 and 2. Equation 1 applies to any initial molecular weight distribution, MWD, but equation 2 is only valid for a most-probable MWD, when Mw/Mn = 2.

$$1/Mn(D) = 1/Mn(0) + A[G(S) - G(X)]D \qquad (1)$$

$$1/Mw(D) = 1/Mw(0) + B[G(S) - 4G(X)]D \qquad (2)$$

where the dose, D, is in gray and $A = 1.04 \times 10^{-10}$, and $B = 5.18 \times 10^{-11}$.

The variation in G(S)-G(X) with temperature from 30 ^{o}C to 160 ^{o}C is shown in Figure 1. Crosslinking is predominant at 30 ^{o}C, but scission increases with increasing temperature. However, there is no significant discontinuity at Tg (100 ^{o}C) apparent within the accuracy of the data.

Poly(styrene) is known to undergo thermal degradation above 250 ^{o}C to yield mainly monomer. Radiation was found to enhance the rate of thermal degradation, corresponding to an increase in temperature of about 100 ^{o}C. The rates of mass loss at 292 ^{o}C (Thermal) and 240 ^{o}C (5 kGy/h) are shown in Figure 2. Analysis of the volatile product and the residue has confirmed the formation of styrene monomer and of dimer, trimer and higher oligomers. It is uncertain whether the oligomers are formed during the depropagation, e.g. through an intra-molecular abstraction (back-biting) reaction, or by polymerization of monomer initiated by irradiation or by radical or ionic species trapped in the polymer.

Bisphenol-A Polysulfone. This commercial, high-performance, high-temperature polymer is an amorphous thermoplastic with a substantial aromatic content in the backbone chain. It differs from polystyrene which has substituent phenyl groups and a lower Tg of 100 ^{o}C, compared with 190 ^{o}C for PSF. Crosslinking predominates in PSF for irradiation in vacuum at 30 ^{o}C, but the G values are comparable to polystyrene, as might be expected. The yields of crosslinking and scission can be conveniently derived from Charlesby-Pinner plots of $s + s^{1/2}$ versus 1/D, according to equation 3, where s is the soluble fraction, determined by extraction of the irradiated polymer, provided crosslinking occurs by a H-linking mechanism, and $C = 4.82 \times 10^{9}$.

$$s + s^{1/2} = G(S)/2G(X) + C/[G(X)Mn(0)D] \qquad (3)$$

The variation of the soluble fraction with dose for irradiation of PSF under vacuum at 35, 125 and 220 ^{o}C is shown in Figure 3. The gel dose decreases substantially over this temperature range. The ratio of scission to crosslinking is indicated by the limiting soluble fraction. It is remarkable that crosslinking increases markedly above Tg and scission decreases to zero, which is in contrast to the behaviour of polystyrene. It is likely that the onset of depropagation above Tg in polystyrene stabilizes main chain scission, whereas depropagation does not occur in PSF and the chain radicals have greater mobility for

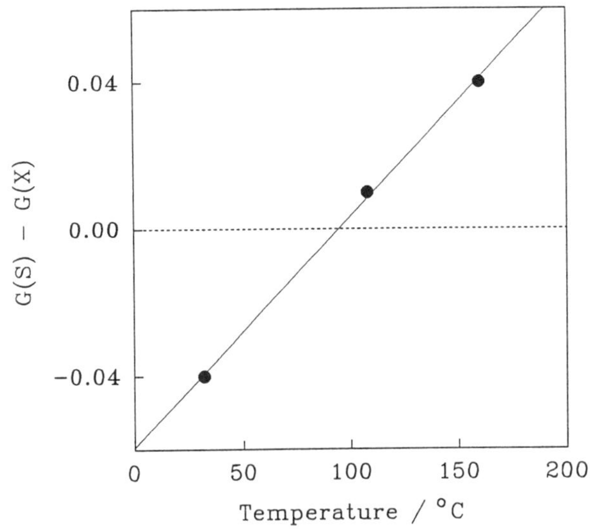

Figure 1. Temperature dependence of the net rate of scission, G(S) - G(X), in polystyrene for γ-irradiation in vacuum.

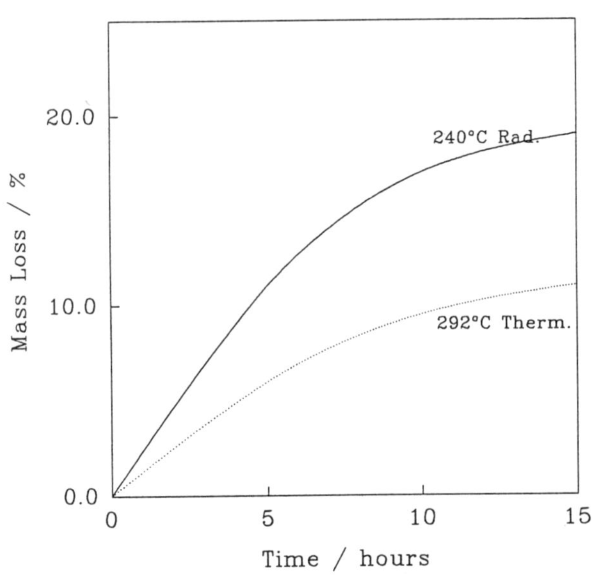

Figure 2. Radiation enhancement of the thermal degradation of polystyrene; mass loss with time at 240 ^{o}C (radiation), and 292 ^{o}C (thermal).

recombination reactions, including crosslinking. Crosslinking could also occur by a Y-linking mechanism, which would be enhanced above Tg.

Poly(methyl methacrylate). Irradiation of PMMA at ambient temperatures causes main-chain scission. Values of G(S) have been reported in the range 1-2, and apparently crosslinking does not occur. Charlesby and Moore (*2*) have found that G(S) increases with irradiation temperatures above ambient as shown in Figure 4. Wundrich (*3*) have reported increasing values of G(S) from 77 K with discontinuities at second order phase transitions.

We have measured the mass of small, powdered samples of PMMA during irradiation under vacuum and found that there was negligible mass loss up to 116 ^{o}C, but that it increased rapidly at higher temperatures. Above 200 ^{o}C the rate of depropagation was extremely rapid and may well have been determined by physical factors. Thermal degradation starts to become significant above about 250 ^{o}C, so that the enhancement by irradiation is very significant at 200 ^{o}C.

Measurements have been made by GPC of the change in molecular weight of these PMMA samples during irradiation. Corrections have to be made for the decreasing dose to the sample on account of the loss of sample through depropagation. G(S) has been found to increase very rapidly above 150 ^{o}C, and zip lengths are short compared with the degree of polymerization (DP) of the polymer.

Poly(phthalaldehyde). The ceiling temperature for the propagation/depropagation equilibrium of the reaction phthalaldehyde - poly(phthalaldehyde) is -40 ^{o}C in the liquid state. Therefore, the polymer molecules must be end-capped with acetaldehyde; otherwise the polymer undergoes slow depropagation on standing at ambient temperature.

The polymer has been investigated by Ito and Willson (*6*) as a resist for microelectronic applications in which photolysis or radiolysis of a sensitizer catalyses depropagation to monomer. Irradiation of poly(phthalaldehyde) in the absence of a sensitizer causes main-chain scission and the new chain ends undergo depropagation. Measurements of molecular weight (determined by GPC) and weight loss have been reported for γ-irradiation at ambient temperature (*6*).

In the present work, the effect of irradiation temperature on G(S), G(-monomer) and the zip length for depropagation have been investigated. The decrease in Mn with dose from irradiation at 25 ^{o}C of low mol. wt. PTA are shown in Figure 5 and compared with the results of Ito and Willson for PTA of higher molecular weight.

PTA is inherently unstable even at sub-ambient temperatures and therefore any radiation or thermal formation of propagating radicals will lead to depropagation. Consequently, the polymer is very sensitive to radiation enhancement of thermal degradation.

Polyisobutene. PIB is a polymer with a low Tg of -70 ^{o}C and a ceiling temperature of 175 ^{o}C. It is believed to undergo only main-chain scission on irradiation. Wundrich (*4*) has determined values of G(S), assuming that crosslinking does not occur, from solution viscosity measurements, for irradiation over a range of temperatures.

We have used GPC to derive Mn and Mw values of samples irradiated from 77 K to 350 K, and derived G(S) from equation 1.

Poly(α-methyl styrene): Compared with poly(methyl methacrylate) and poly(isobutene), poly(α-methyl styrene), PAMS, has a relatively low ceiling temperature of 60 ^{o}C for polymerization of the pure liquid monomer at 1

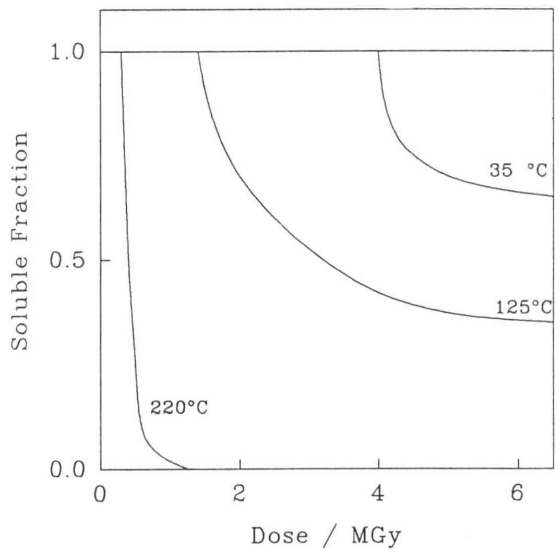

Figure 3. Dose dependence of the soluble fraction of bisphenol-A polysulfone for irradiation at 35, 125 and 220 ^{o}C.

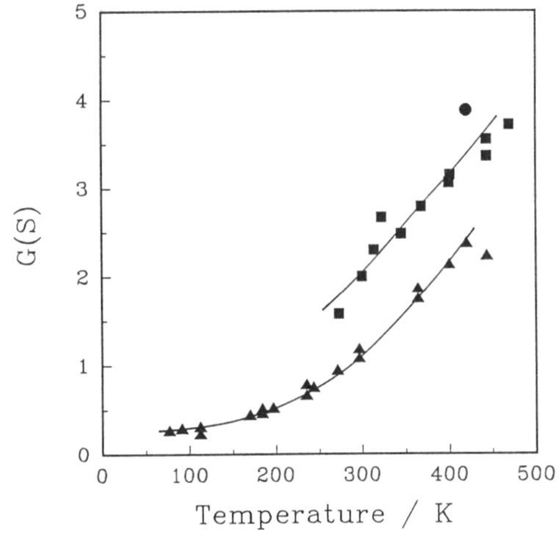

Figure 4. G(S) values for γ-irradiated poly(methyl methacrylate) in the temperature range -196 to 200 ^{o}C. (■) Charlesby and Moore - ref. 2; (▲) Wundrich - ref. 3; (●) present work.

atmosphere pressure (*7*). The Tg is 180 °C. The thermally-induced depropagation of PAMS has been studied extensively (*8-11*) and reported to occur at a significant rate at temperatures greater than 230 °C, with the only product of the depropagation being monomer. However, by comparison with the thermally-induced depropagation, the radiation-induced depropagation has received little attention.

Here we report a study of the radiation chemistry of PAMS at three temperatures (25 °C, 80 °C and 163 °C). The mass loss and molecular weight of polymer after various periods of irradiation have been measured and the yields of chain scission and loss of monomer have been determined.

The polymer (initially 50 mg, Mn=60,000 g/mol) was irradiated under vacuum (dose rate = 2.8 kGy/h) at each temperature. The mass of the polymer was monitored over a period of approx. 30 hours. At the two lowest temperatures, no significant loss in mass was observed during the period of irradiation, but at the highest temperature, the mass loss was approx. 25 %. GC/MS studies showed that the mass loss was caused by depropagation of the polymer to monomer. The variation in the mass of polymer was found to follow first-order kinetics, as demonstrated in Figure 6.

The yields for loss of monomer, G(-M), can be calculated from a plot of mass loss versus absorbed dose (Figure 7), to yield a value of 320 ± 45 at 163 °C, indicative of a chain reaction.

Because no significant depropagation occurs at 25 °C and 80 °C, the value of G(S) can be calculated from the linear relationship between 1/Mn and absorbed dose. These plots are shown in Figure 8 and yield values for G(S) of 0.29 ± 0.01 at 25 °C and 0.48 ± 0.06 at 80 °C.

At 163 °C, the value of G(S) cannot be determined directly from the 1/Mn versus dose plot, because the mass of polymer varies during irradiation. A second complication arises as a result of the depropagation process, because a chain scission within a zip length, z, of a chain end will not produce an increase in the number of polymer chains present in the sample. In order to deal with this problem, the variation in the molecular weight of the polymer (Mn and Mw) and the loss in mass have been analysed over a range of absorbed dose (0-50 kGy) using a Monte-Carlo simulation of chain scission and depropagation. The details of this simulation are discussed elsewhere (*12*). The Monte-Carlo simulation returns values for G(S) and z, which were found to be 1.8 and 400, respectively, at 163 °C.

In Figure 9 the values for G(S) and z are plotted against the irradiation temperature. The results clearly show that both G(S) and z rise substantially as the temperature increase above 80 °C. Poly(α-methyl styrene) can thus undergo radiation-induced depropagation at temperature approximately 100 deg. lower than that at which thermally-induced depropagation occurs at a significant rate.

Conclusions

Radiation enhances the rate of thermal degradation of polymers which undergo depropagation to monomer to an extent equivalent to a temperature increase greater than 100 °C. Complementary measurements of mass loss and reduction in molecular weight enable values to be derived for the rates of main-chain scission, G(S), and monomer loss, G(-monomer), provided that appropriate corrections are made for the decreasing dose rate due to the continuing mass loss. It is then possible to determine the average zip length for depropagation. The very high G(S) values obtained at high temperatures suggest that radiation energy supplements thermal energy in the initiation process.

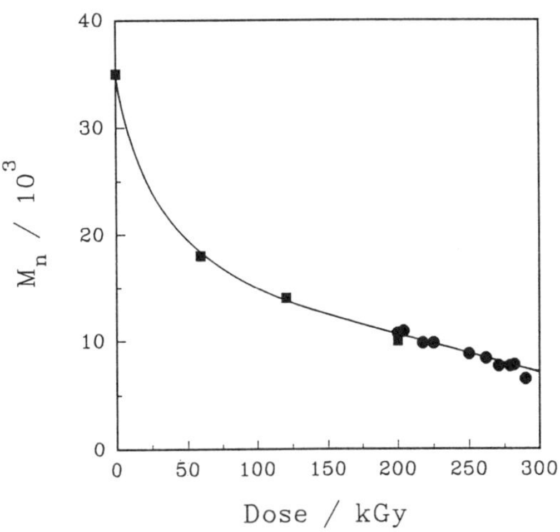

Figure 5. The decrease in Mn of poly(phthalaldehyde) with radiation dose for initial high and low molecular weight polymer. (■) high mol. wt. - ref. 6; (●) low mol. wt. - present work.

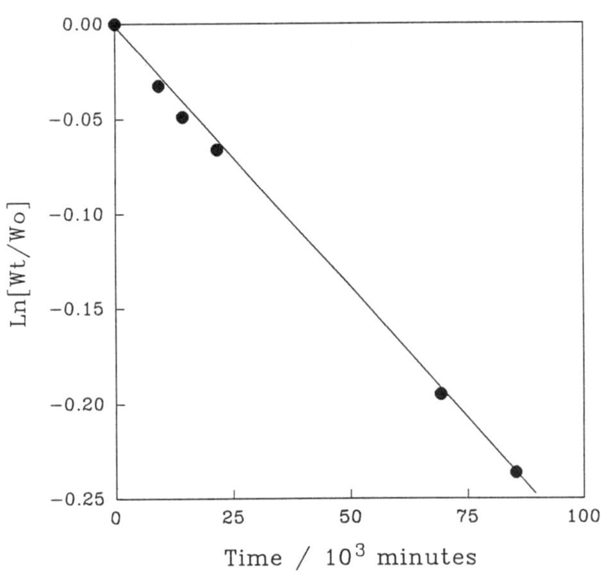

Figure 6. Mass loss of poly(α-methyl styrene) irradiated under vacuum at a temperature of 163 °C; examination of first-order kinetics.

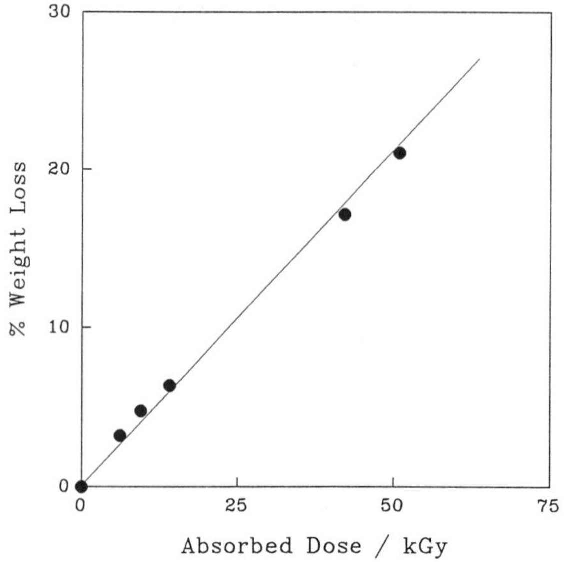

Figure 7. Percent mass loss of poly(α-methyl styrene) with absorbed dose of radiation under vacuum at a temperature of 163 ^{o}C.

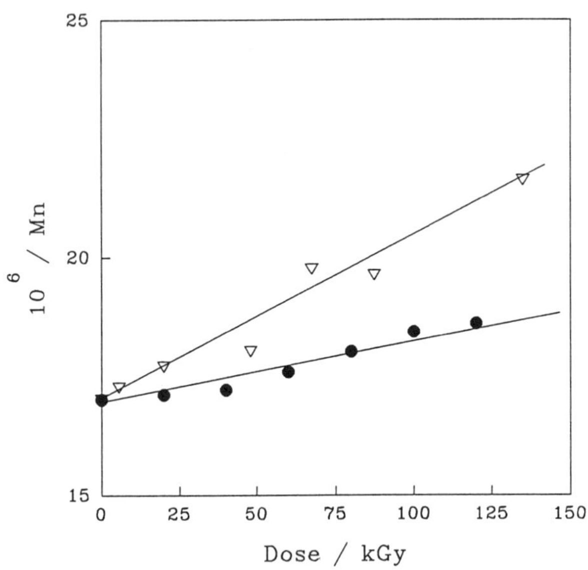

Figure 8. The variation in the value of 1/Mn for poly(α-methyl styrene) with absorbed dose of radiation under vacuum at temperatures of (●) 25 ^{o}C and (▽) 80 ^{o}C.

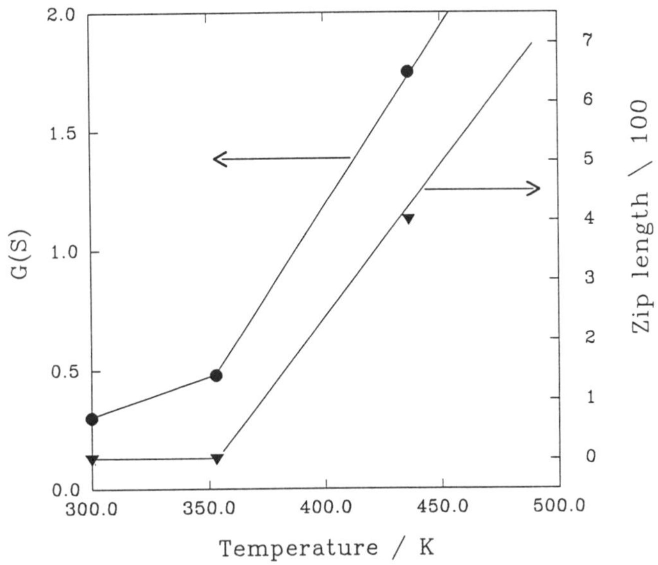

Figure 9. Temperature dependence of the values of (●) scission, G(S), and (▼) zip length, z, for γ-irradiation of poly(α-methyl styrene) in vacuum.

Acknowledgments

The authors thank the Australian Research Council, and the Australian Institute of Nuclear Science and Engineering for supporting their research and the Australian Nuclear Science and Technology Organization for providing irradiation facilities, Mr David F. Sangster for advice, Dr Hiroshi Ito for providing the sample of PTA and Dr G.C. Willson for encouragement.

Literature Cited

1. O'Donnell, J.H. In *The Effects of Radiation on High-Technology Polymers*; Reichmanis, E.; O'Donnell, J.H., Eds.; Symposium Series 381; Amer. Chem. Soc.; Washington, DC, 1989, p 1.
2. Charlesby, A.; Moore, N. Int. *J. Appl. Rad. Isotopes* **1964**, *15*, 703.
3. Wundrich, K. *J. Polym. Sci., Polym. Chem. Ed.* **1964**, *11*, 1293.
4. Wundrich, K. *Europ. Polym. J.* **1971**, *10*, 341.
5. Bowmer, T.N.; O'Donnell, J.H. *J. Macromol. Sci. Chem.* **1982**, *A17*, 243.
6. Ito, H.; Willson, C.G. *Polym. Eng. Sci.* **1983**, *23*, 1012.
7. Kilroe, J.G.; Weale, K.E. *J. Chem. Soc.* **1960**, 3849.
8. Brown, D.W.; Wall, L.A. *J. Phys. Chem.* **1958**, *62*, 848.
9. Bywater, S.; Blank, P.E. *J. Phys. Chem.* **1965**, *69*, 2987.
10. Roestamsjah, L.A.; Wall, L.A.; Florin, R.; Aldridge, M.H.; Fettes, L.J. *J. Res. Natl. Bureau Stand.* **1978**, *83*, 371.
11. Guaita, M.; Chiantore, O. *Polym. Degrad. Stab.* **1985**, *11*, 167.
12. Garrett, R.W.; Hill, D.J.T.; Le, T.; O'Donnell, J.H.; Pomery, P.J. In *Polymers for Microelectronics Science and Technology*; Tabata, Y; Mita, I.; Nonogaki, S., Eds.; VCH Publishers; Federal Republic of Germany, 1990, in press.

RECEIVED March 22, 1991

Chapter 11

Degradation of Poly(methyl methacrylate)

Deep UV, X-ray, Electron-Beam, and Proton-Beam Irradiation

J. A. Moore and Jin O. Choi

Polymer Science and Engineering Program, Department of Chemistry, Rensselaer Polytechnic Institute, Troy, NY 12180–3590

The chemical changes in PMMA films caused by irradiation with deep UV, x-ray, electron, and proton beams were studied by gel permeation chromatography, FT-IR, UV, and NMR spectroscopy. Quantitative analyses of spectral changes demonstrated a 1:1 correspondence between the disappearance of ester groups and the generation of double bonds in the polymer chain by all types of radiation. The ratio of main chain scission to changes in the number of ester groups and the generation of double bonds for deep UV was very close to the quantum yield of main chain scission of PMMA as reported in the literature. High energy radiation was about ten times more efficient than deep UV in causing main chain scission with removal of fewer ester groups. Four different types of unsaturated bonds were identified in NMR spectra of degraded PMMA films. In additon, conjugated unsaturated bonds were observed in UV-irradiated film at high doses. Polymer stereochemistry and state (solution or solid) exert influence on the course of the degradation process.

In semiconductor device fabrication, microlithography is the process of transferring a mask pattern to a radiation-sensitive polymer system (resist) and then into an underlying insulating thin film (*1,2*). Resist materials play an essential role in the fabrication of microdevices and integrated microcircuits. Photoactive and radiation-sensitive materials can be classified as positive or negative on the basis of their response to radiation. Depending upon the chemical nature of the resist material, the exposed areas become more soluble in some developing solvents than the unexposed areas, thereby producing a positive tone image of the mask. Conversely, in a negative-working material, the exposed areas become less soluble after exposure to radiation. There are two different ways of making a resist system responsive to radiation: by causing the functional groups in the resist to change or by changing the molecular weight of the polymer (main chain scission or crosslinking)(*3*).

Poly(methyl methacrylate) (PMMA) is a common plastic used in a variety of applications. Recently, it has become very important as a resist for advanced microlithography (*1-5*). In addition to deep UV exposures (*6-9*), high energy

0097–6156/91/0475–0156$10.25/0

radiations, such as electron beam (*10-14*), x-rays (*15-19*) and proton beam, (*20-22*) generate a positive pattern at very high resolution in thin films of PMMA. However, the sensitivity of PMMA is more than 10^{-5} C/cm^2 at an e-beam energy of 25 KeV, and PMMA is also not highly sensitive to deep UV, x-ray and proton beam radiation. Though PMMA has been very useful in the laboratory as a medium for the generation of test structures and devices, it is too "slow" to be practical in modern device manufacture. One of the important requirements for high resolution lithography is the ability to expose resist films on wafers at high speed to enhance the throughput of the lithographic process.

A wide range of PMMA analogs and copolymers have been synthesized and characterized in attempts to find a resist with the desirable processing properties of PMMA but with sufficient sensitivity ("speed") for use in the production of masks or actual devices (*1-5, 23*). Work to improve the speed of PMMA analogous systems has concentrated primarily on the synthesis of methacrylate polymers where the methyl group at the a-position of PMMA has been replaced with an electron withdrawing group (-CN, -CF_3, -Cl, etc.). Changing the ester group from methyl to a fluorinated alkyl also improves the lithographic speed (*3*). A "terpolymer" resist with methacrylic acid units that form anhydride units shows considerably improved performance (*4,5*).

When methacrylate films are exposed to a sufficiently intense dose of radiation, such as ultraviolet, x-ray, e-beam or proton beam, they are damaged by chain scissions and form volatile fragments. These changes profoundly affect the dissolution rate in methacrylate resist films. The increased solubility in the exposed area is caused, in part, by the degradation of PMMA by various types of radiation. The development of a good image in a thin film of resist is possible if the PMMA resist is irradiated with an appropriate dose. This appropriate dose is dependent on the intrinsic radiation sensitivity of the methacrylate resist.

From the view of degradation effects in PMMA film, damage to the polymer gives rise to the formation of free radicals (*1, 2*) other active intermediates (*24*), and microvoids (*5, 25*), which are generated even at doses lower than the dose required for good image generation by the conventional processes. These active intermediates can be used to initiate a polymerization with a new monomer to produce a copolymer with properties different from those of the initial polymer (*26*). The modification of irradiated PMMA with acrylic acid (AA) was first reported by Gazard et al. (*27*) who showed that AA monomer could be chemically bonded to damaged PMMA after exposure to an electron beam at a dose of 10^{-8} C/cm^2. The modified area becomes less soluble in non-polar solvents than the unirradiated area as hydrophilic monomers are incorporated. Therefore, this method induces a negative lithographic tone. Their work was repeated with various types of radiation sources to study the modification process and to determine if this process was general with respect to the type of radiation used (*28, 29*). Hydrophilic comonomers were introduced selectively into the irradiated areas of the film and chemically bonded to, and modified the solubility of, the irradiated areas. Thus, a high level of solubility alteration can be attained for a relatively small radiation dose (*29*). It was reported that appropriate solvents develop negative images generated at doses on the order of 1,000 times less than those needed to form positive images in PMMA when the irradiated PMMA is exposed to acrylic acid (AA) vapor prior to solvent development. During the course of that work it became apparent that the degradative pathways which control the sensitivity of PMMA as a resist were neither simple nor completely understood. As a result, we undertook, concurrently with the on-going attempts to enhance the sensitivity of PMMA, the task of trying to understand the details of the degradative process.

Radiation-Induced Degradation of PMMA

The degradation of PMMA films caused by irradiation with deep UV, x-ray, e-beam and proton beams encompasses main chain scission and the formation of gaseous products by cleavage of pendent methyl ester groups. These processes generate positive images because of increased solubility in the exposed area. Many gaseous products formed from PMMA films upon irradiation have been studied. These gases easily evaporate from films at room temperature even at atmospheric pressure. Because the formation of gaseous products and main chain scissions result in a decrease in the number of methyl ester groups and the generation of unsaturated bonds in damaged films, respectively, the degradation should involve changes of specific functional groups which can be detected spectroscopically. Fourier transform-infrared (FT-IR) and ultraviolet (UV) spectroscopy were used in this work to measure directly for the first time the changes in PMMA films after exposure to various radiations. Nuclear magnetic resonance (NMR) spectroscopy was used to characterize the influence of polymer stereochemistry on the nature of the double bonds produced in the degradation process.

The UV and FT-IR absorption spectra of different film thicknesses of PMMA were measured and the extinction coefficient(s) were determined for use in quantitative analysis of the degradation process. The absorption band of pendent methyl ester groups appears between 190 to 260 nm, and its maximum absorption ($lambda_{max}$) is at 216 nm, which is consistent with previous work (*9*). The absorbance was found to be proportional to film thickness with an extinction coefficient (E_{max}) at 216 nm of 150 l/mole-cm. In the FT-IR spectrum of PMMA film, absorption bands of the pendent methyl ester group were also found to be proportional to film thickness. The extinction coefficient of the carbonyl stretching band at 1730 cm^{-1} was found to be 365 l/mole-cm.

The molecular weights of PMMA film (about 1 μm thick) irradiated with four types of radiation were obtained by using gel permeation chromatography with polystyrene reference standards, and the number of main chain scissions was determined by using the equation, $Mn^{o}/Mn - 1$ = number of main chain scissions (*30*). Figure 1 is a plot of the number of main chain scissions vs. incident radiation dose. The linear plot, which is consistent with previous work (*1, 31*). indicates that the number of backbone scissions of PMMA is proportional to the incident dose from deep UV, x-ray, electron and proton beams.

As seen in the FT-IR spectra of PMMA irradiated at increasing doses (Figure 2), the change in specific bands on a transmittance scale is very small and difference techniques were required to follow the degradation quantitatively. Figure 3 shows a typical difference spectrum obtained before and after UV irradiation. The difference spectra obtained upon irradiation of PMMA with X-ray, electron and proton beams are entirely analogous. The decrease of the carbonyl stretching band at 1730 cm^{-1} is plotted as a function of incident dose in Figure 4. The major gaseous products formed from PMMA under UV and gamma-ray irradiations are methyl formate, methane and methanol (derived from $\cdot CH_3$), CO, and CO_2 along with many other products (*31 - 34*). Although the qualitative and quantitative analysis of gaseous products and their yields vary widely in the reported values, the formation of gaseous products evolved from pendent methyl ester groups should always be reflected in the changes in intensity of carbonyl (C=O) bands and C-O bands in the FT-IR spectra. It can be seen from the data reported here that the decrease in carbonyl band intensity is, within experimental error, linearly dependent on the incident dose of the four different radiations, indicating that the number of cleavages of pendent ester groups depends on the incident radiation dose. The removal of ester groups caused by a particular incident dose was calculated from the FT-IR difference spectra according to:

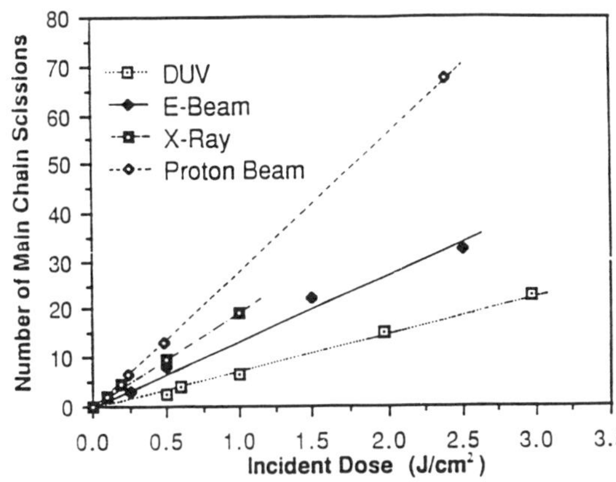

Figure 1. Number of Chain Scissions by GPC vs. Radiation Dose.

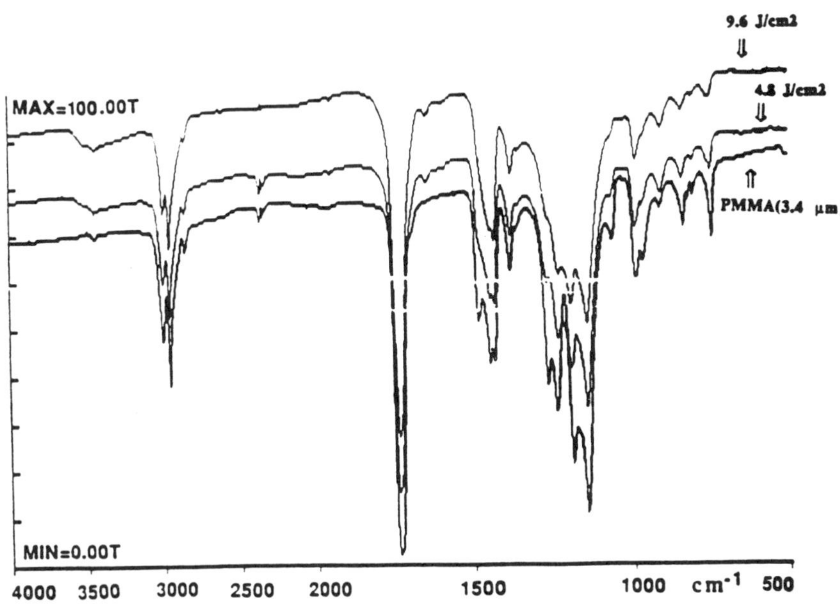

Figure 2. FT-IR Spectra of Native and UV-Irradiated PMMA.

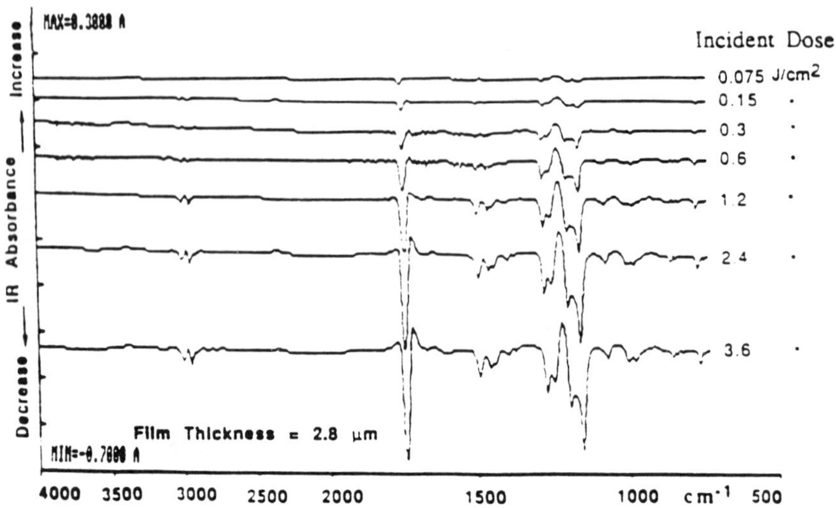

Figure 3. Difference FT-IR Spectra of Native and UV-Irradiated PMMA.

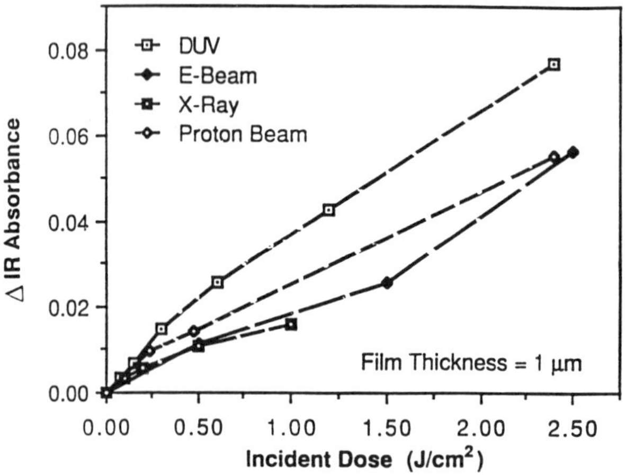

Figure 4. Decrease of IR Carbonyl Absorbance vs. Radiation Dose.

$$\Delta A = E \times T \times \Delta C$$

where ΔA = the change in absorbance at 1730 cm^{-1}
E = 365.3 l /mole-cm
T = PMMA film thickness
ΔC = change in concentration of ester groups in the PMMA film

and the results are presented in Table I.

Figure 5 shows a new UV absorption ($lambda_{max}$ = 195 nm) which appeared in the UV-irradiated PMMA film. This behavior is attributed to the generation of unsaturated bonds, the characterization of which by bromination, UV, IR and NMR spectroscopy will be discussed in detail below. The difference UV spectrum before and after irradiation with UV is shown in Figure 6. The difference spectra obtained upon irradiation with x-ray, electron and proton beams are entirely analogous. The absorbance at 195 nm which develops in the irradiated films is plotted in Figure 7. The increase of absorbance is a complex function of the incident dose. At the initial stages of radiation, the change is proportional to the incident dose. The concentration of unsaturated bonds in the irradiated films was calculated from the UV difference spectra as described above, where, in this case

ΔA = change in absorbance at 195 nm
E = extinction coefficient of C=C bonds at 195 nm
($10^{3.8}$ l /mole-cm (*35*)
T = PMMA film thickness
ΔC = change in concentration of generated C=C bonds

The number of unsaturated bonds in the irradiated PMMA film caused by the four types of radiation is given in Table I. The number of C=C bonds generated in the films

Table I. Main Chain Scission, Ester Group Removal and Unsaturated Bond Generation per 100 Monomer Units

	DUV 4~6 eV 0.6 J/cm^2	E-Beam 25 KeV 20μC/cm^2 (0.5 J/cm^2)	X-Ray 0.8 ~ 1.9 KeV 1 J/cm^2	Proton-Beam 300 KeV 1x10^{13}H$^+$/cm^2 (0.48 J/cm^2
Main Chain Scission (MCS)	0.22	0.46	1.08	0.75
Ester Removal (C=O)	8.53	2.59	3.68	3.22
Ratio (MCS / C=O)	0.026	0.178	0.293	0.233
C=C generation	7.06	2.24	3.72	2.83
Ratio (MCS / C=C)	0.031	0.205	0.290	0.265

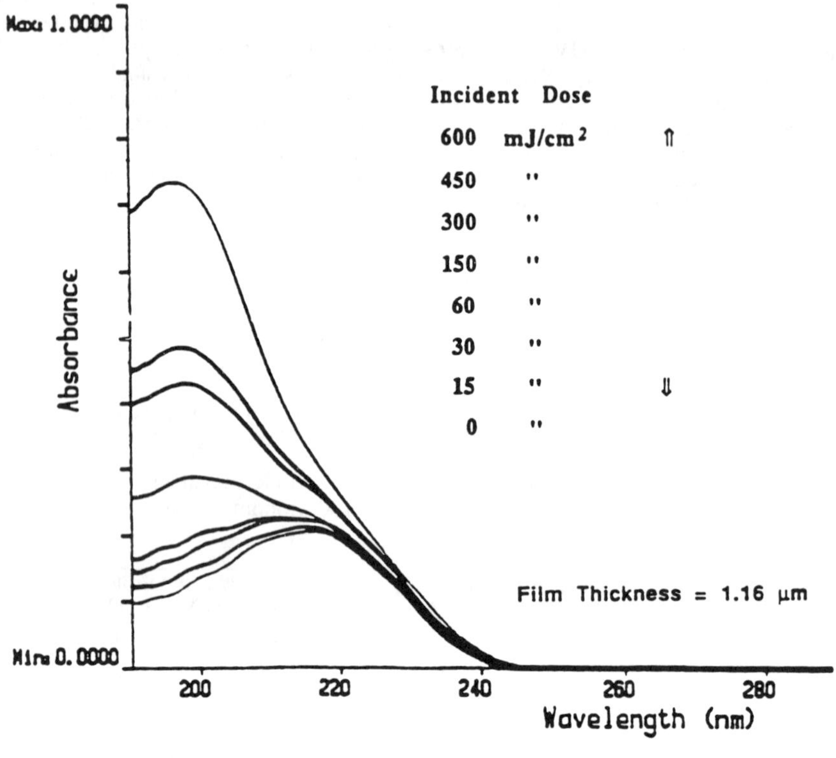

Figure 5. UV Spectra of UV-Irradiated PMMA Films.

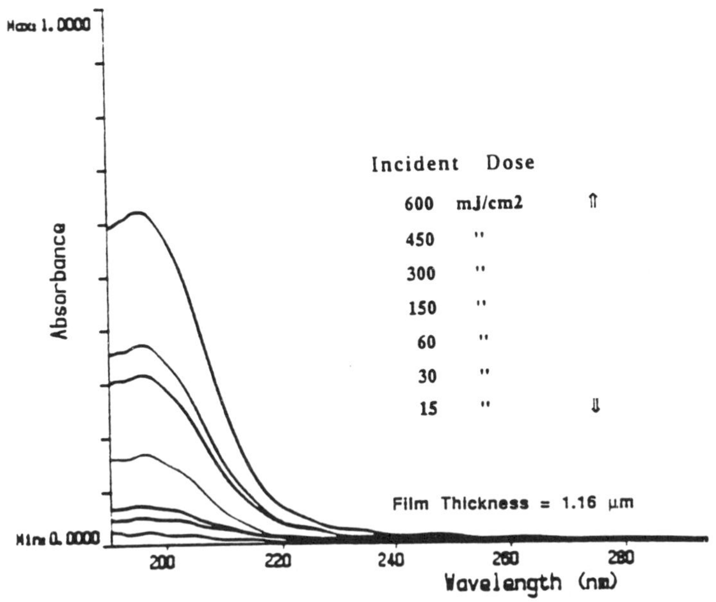

Figure 6. Difference UV Spectra of Native and UV-Irradiated PMMA.

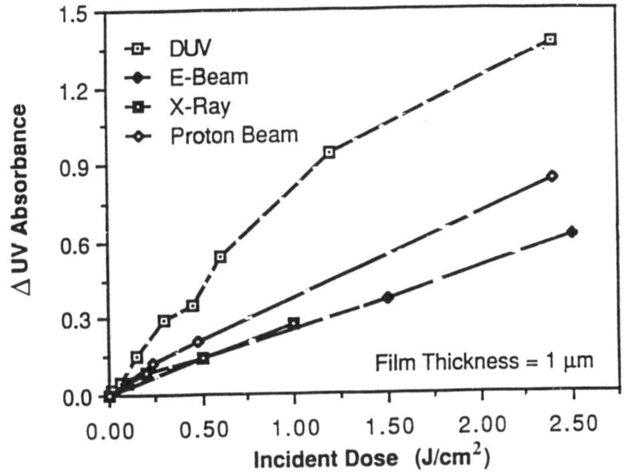

Figure 7. Increase of New Band at 195 nm vs. Radiation Dose.

is close to that of C=O groups removed by the same incident dose, and this result indicates that each removal of an ester group generates one C=C bond by main chain scission or by abstraction of hydrogen from methylene or a-methyl groups. The ratio of main chain scission to the change in the number of specific functional groups obtained with different types of radiation is compared in Table I. Quantum yields (main chain scissions/absorbed photons) for PMMA film, when irradiated with 254 nm light in air at room temperature, have been reported as 0.02 ~ 0.03 (*31, 36*). This ratio for our deep UV data, 0.026, is very close to the quantum yields of main chain scission obtained in air. Shultz's study (*37*) found the quantum yield for PMMA main chain scission caused by irradiation at 214 ~ 229 nm to be 0.03. This result indicates that the first n-p* transition possible for pendent esters at these shorter wavelengths does not enhance main chain scission relative to that obtained by exposure at 254 nm. Gupta observed that every photon of 254 nm light absorbed results in the degradation of an ester group (*31*). If it is assumed that all the absorbed photons from the UV source used in this experiment cause removal of ester side groups, the ratio for deep UV irradiation (0.026 in Table I) should equal the quantum yield for main chain scission.

It is apparent that high energy radiations, such as x-ray, e-beam and proton beam, are much more efficient than deep UV exposure in causing main chain scission with removal of fewer ester groups as can be seen from the results presented in Table I. The ionization process (Figure 8 path II) of PMMA by high energy radiation is very efficient for inducing backbone chain scission compared to the photo-excitation process (path I). Most of the changes in intensity of the carbonyl band caused by deep UV irradiation can be related to the production of methyl formate (quantum effieciency for HCO_2CH_3 = 0.92), which could be generated by hydrogen abstraction from alpha-methyl or methylene groups (path VI). In contrast, high energy radiation gives much larger amounts of gases such as CO, CO_2, CO_4 and methanol. This difference can be rationalized by considering differences in stability of the polymeric fragments generated by irradiation (Figure 8 path II, IV, VII and VIII). Ionization, followed by expulsion of a methyl formate radical would leave behind a relatively stable tertiary cation which is unlikely to undergo electrophilic abstraction processes. The methyl formate radical can therefore diffuse from the reaction cage and decompose to CO_2, CO, $\cdot CH_3$, $CH_3O\cdot$.

The ratio of main chain scission to the change of specific functional groups for e-beam is less than those for x-ray and proton beam in Table I. A possible rationale for this observation is that excess electrons trapped in the sample during e-beam irradiation can convert the cation-radicals generated by path II to excited state species (similar or identical to those formed in UV irradiation) path III, and decreasing the ultimate amount of main chain scission as evidenced in the smaller ratio of MCS/C=O as shown in Table I. X-ray radiation is seen to be the most efficient at causing main chain scission of PMMA among the four different types of radiation investigated in this work.

From the data in Table I, it can be seen that x-ray is about ten times more effective than deep UV in causing main chain scission by loss of one ester side chain. It has been reported that the ratio of main chain scission to pendent ester scission of PMMA by 1.24 MeV ^{60}Co gamma-radiation was about 1 (*33*). Thus, the value of MCS/C=O is largest for gamma-ray, followed by x-ray and UV in approximately the proportion of 50 : 10 : 1. It may be concluded that the observed differences in efficiency arise from the differences in energy of the radiation types (see Experimental), which give rise to different chemical processes as described above.

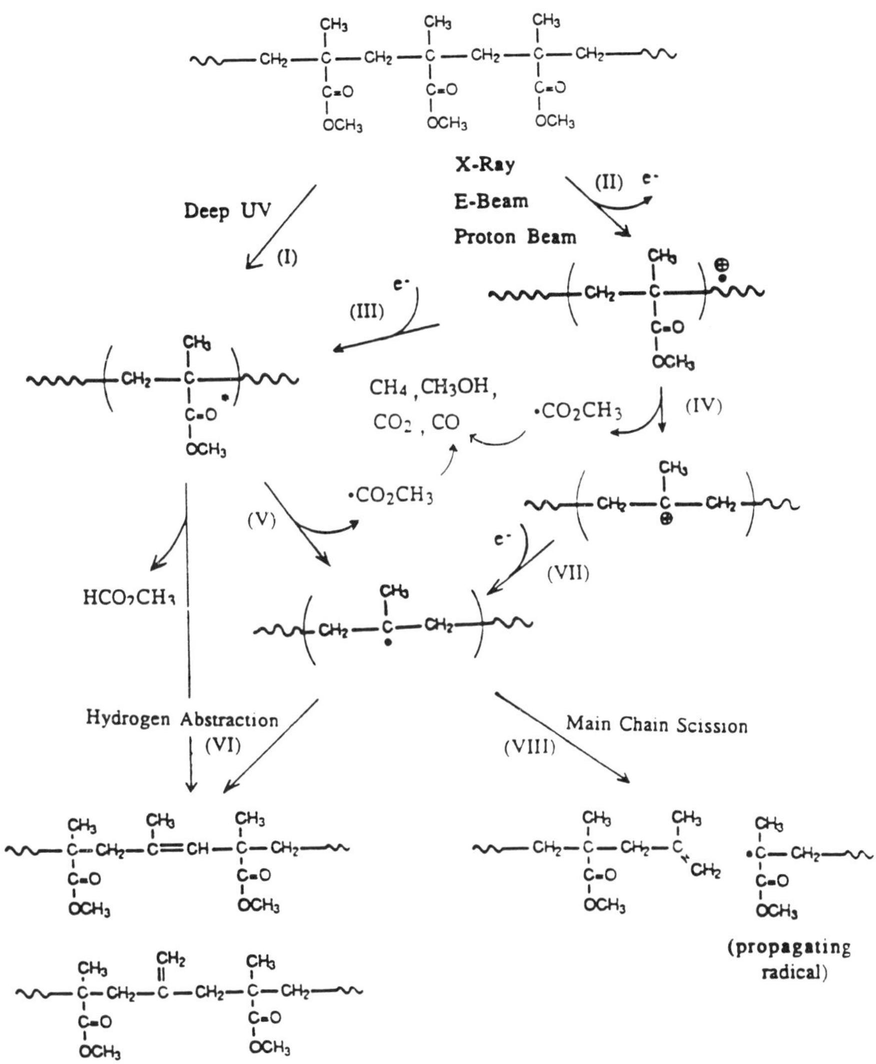

Figure 8. Radiation-Induced Degradation Processes in PMMA.

NMR Characterization of Radiation-Degraded PMMA Film

Although the degradation of PMMA by radiation must be associated with changes in specific functional groups in the films, most of the product analyses have concentrated on identifying the small gaseous molecular fragments produced and have paid relatively little attention to the detailed nature of the polymeric fragments. The quantitative analysis of spectroscopic changes in damaged film described above demonstrated that loss of each ester side chain corresponds to the generation of one unsaturated bond in the polymer chain by all types of radiation. The earlier results with UV and FT-IR spectroscopy are extended here to include characterization by NMR spectroscopy of the variety of unsaturated bonds formed in PMMA upon irradiation and an attempt is made to correlate these results with the degradation mechanism that arises from the different types of radiation.

Figure 2 shows the FT-IR spectra of virgin PMMA and of UV-irradiated films. The difference spectrum before and after UV exposure is shown in Figure 3 (a). The absorption bands at 2997, 2936, 1730, 1484, 1446, and 1265 ~1145 cm^{-1} were reduced in intensity upon loss of pendent methyl ester groups (*38, 39*). In addition, an increased absorption band appeared in the region 3500 ~3100 cm^{-1}, which was probably caused by photo-oxidative degradation of PMMA (*30*). The carbon-carbon double bonds associated with the cleavage of methyl ester groups and main chain scission in the PMMA films appear at 1650 cm^{-1}. Bands at 2852, 1215 and around 1730 cm^{-1}, increased in intensity because of shifts of methylene and carbonyl bands from the generation of unsaturated bonds. PMMA films exposed to x-ray, e-beam and proton beams showed the same changes in IR spectra as those exposed to UV light except for the photo-oxidation effect during the irradiation.

Exposure of PMMA films ($lambda_{max}$ = 216 nm) to four different radiation sources generates a new absorption band in the UV spectrum ($lambda_{max}$ = 195 nm), which was unchanged after baking at 120 ^{o}C for 30 minutes, as determined by UV spectroscopy. Methyl methacrylate monomer has a $lambda_{max}$ of 203 nm, Figure 9. Reaction of the carbon-carbon double bonds with bromine was used to test the presence of unsaturated bonds in the damaged PMMA (*40*). When the films were exposed to gaseous bromine for three hours at room temperature, the new absorption band generated by UV-irradiation disappeared, as shown in Figure 10. This result indicates that PMMA films exposed to UV, x-ray, e-beam and proton beam contain unsaturated bonds attached to polymer chains.

The changes in IR and UV spectra of irradiated films were proportional to the low incident doses (Figure 7). When PMMA films were exposed to UV light at much higher doses, the decrease in carbonyl group intensity remained approximately linear with dose, as shown in Figure 11. At the same time, absorption at longer ultraviolet wavelengths appeared while the intensity at 195 nm slowly increased and then was reduced, as shown in Figure 12. These results indicate that conjugated unsaturated bonds are formed at high doses, as shown in Scheme I.

To characterize the unsaturated bonds in the irradiated films, H^1 and C^{13} NMR (200 MHz) spectroscopy proved to be very useful. The changes in the NMR spectra recorded at normal intensities were not large enough to enable identification of the generated structures. However, as shown in Figure 13, the amplified NMR spectra showed several resonances of hydrogen atoms attached to sp^2 - hybridized carbon atoms, which disappeared after exposure to bromine. The intensity of these peaks was dependent on the incident dose of UV light. Similar peaks are apparent in the H^1 NMR spectra of irradiated samples caused by high energy radiation (Figure 14).

From the reported NMR spectra (*41 - 44*) of poly(isoprene) and model compounds containing similar unsaturated units, such as 2-methyl-1-butene and 2-

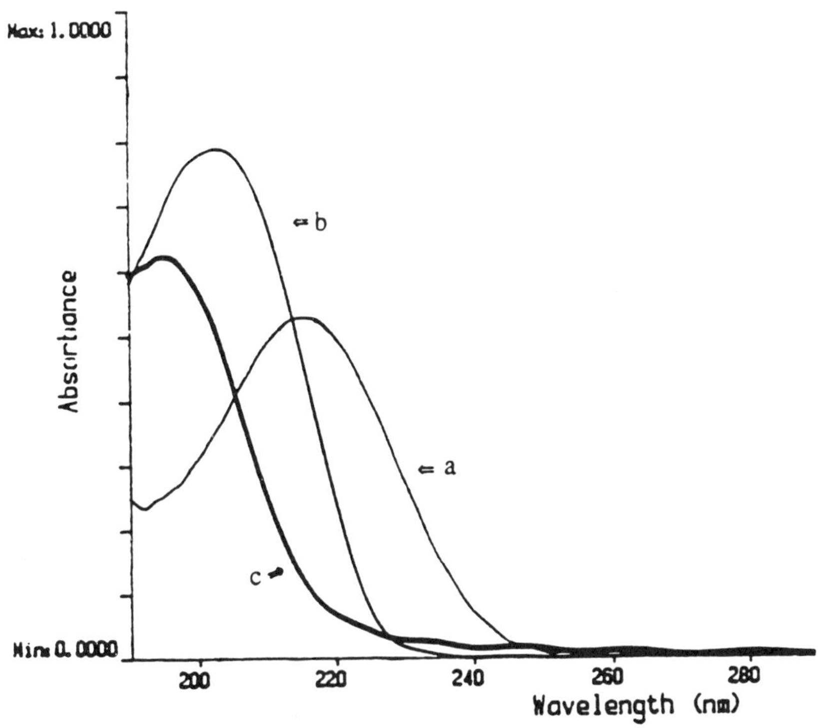

Figure 9. UV Absorption Spectra of: (a) PMMA Film (b) Methyl Methacrylate (c) Difference UV Spectra of Native and UV-Irradiated PMMA.

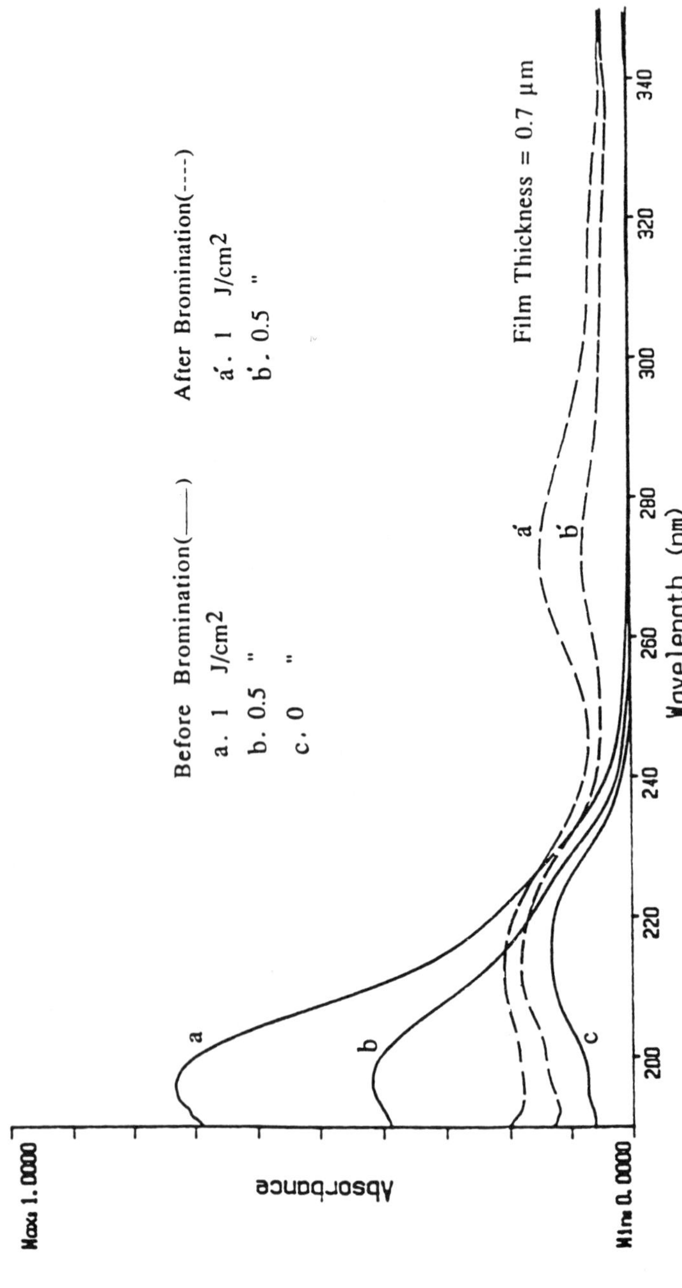

Figure 10. UV Spectra of UV-Irradiated PMMA Film before, and after Exposure to Bromine.

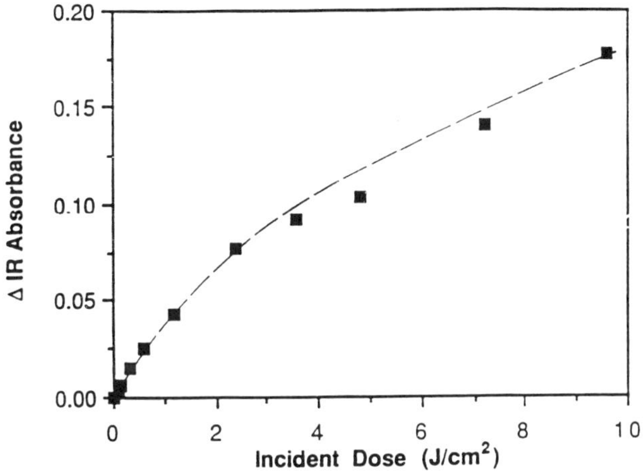

Figure 11. Decrease of IR Carbonyl Absorbance vs. UV Dose.

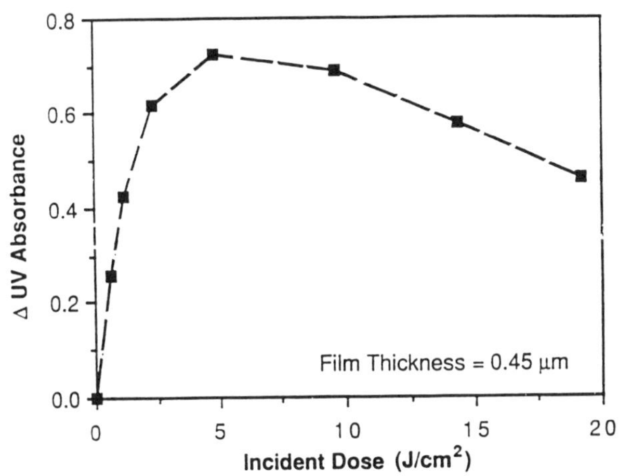

Figure 12. Increase of 195 nm Band vs. UV Dose.

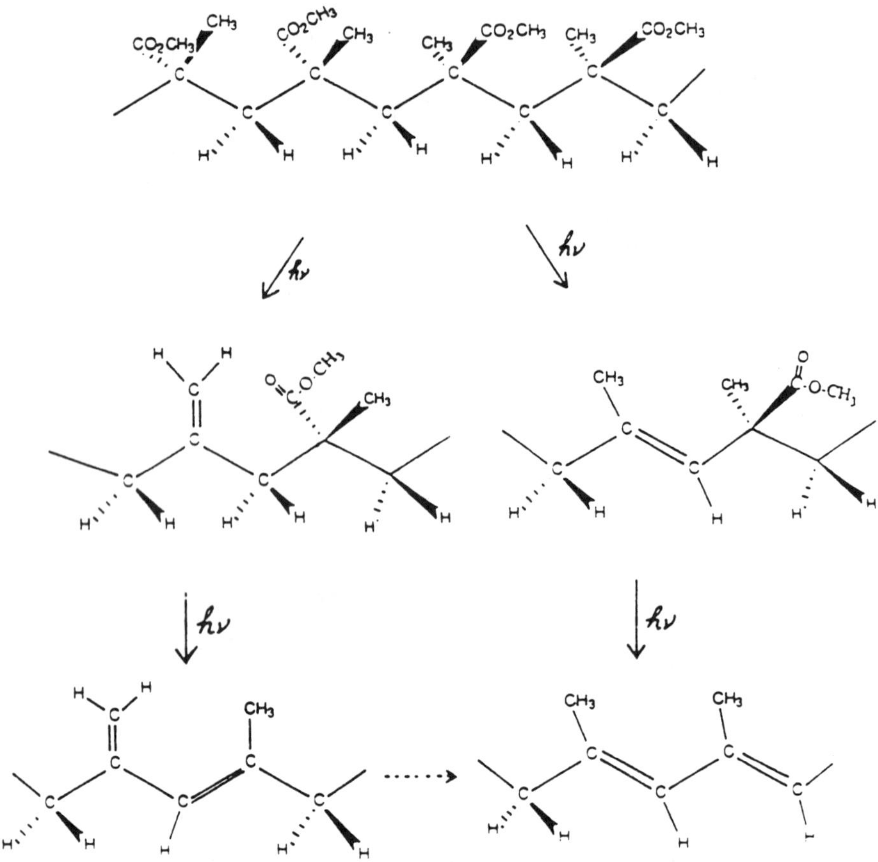

Scheme I. Generation of Conjugated Unsaturation in PMMA.

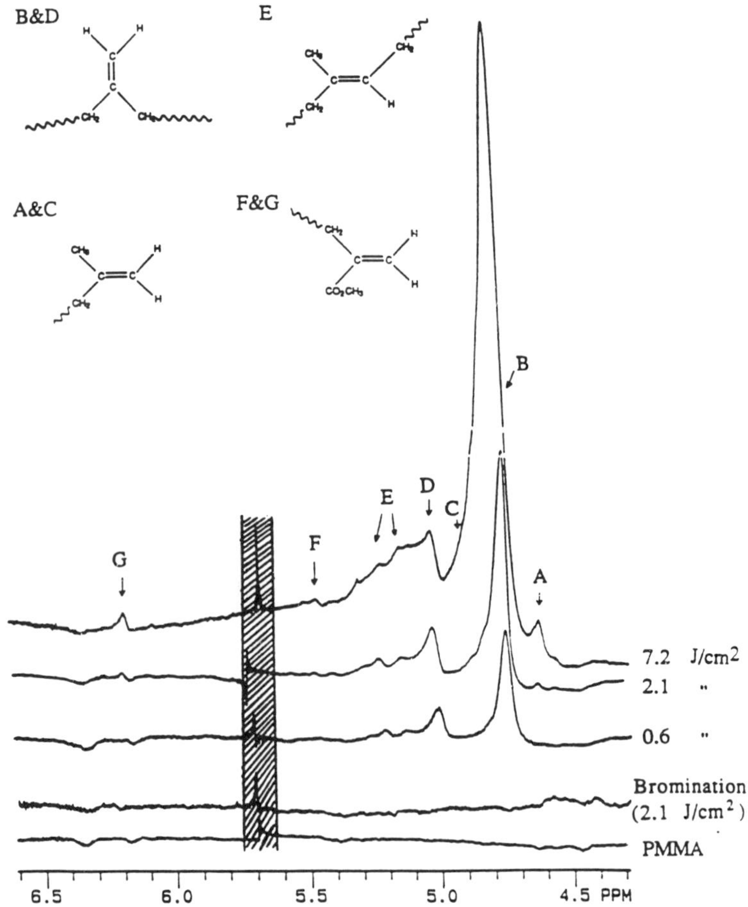

Figure 13. PMR Spectra (X17.5) of UV-Irradiated PMMA.

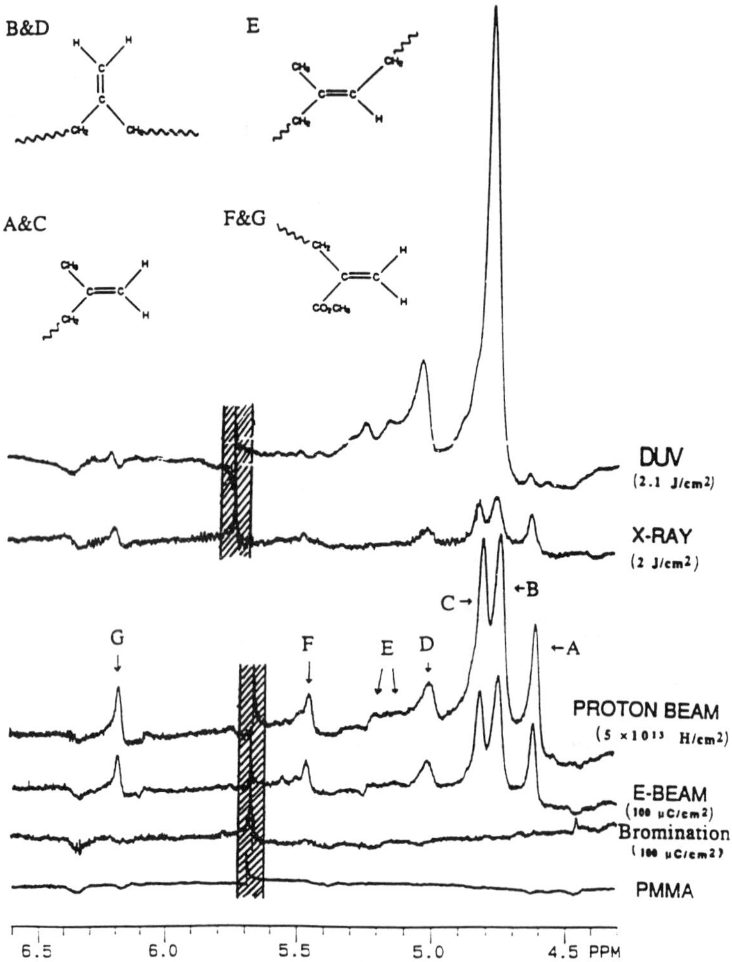

Figure 14. PMR Spectra (X35) of Radiation-Damaged PMMA.

methyl-1-hexene, the broad peaks around 5.14 and 5.22 ppm are assigned to vinylene hydrogen atoms and peaks at 4.6~5.1 ppm should be vinylidene protons generated by main chain scission and hydrogen abstraction from the alpha-methyl group.

The unsaturated bonds are probably generated by main chain scission (Reaction III) and hydrogen abstraction (Reaction I and II), as shown in Figure 15. The ratio of main chain scission to cleavage of pendent methyl ester groups was about 0.03 for UV radiation and about 0.25 for high energy radiation from GPC measurements, IR and UV difference spectra . In Figure 14, the peak intensity at 4.62 and 4.82 ppm is almost the same for high energy radiation and in Figure 13, the 4.82 ppm peak for UV radiation is probably so small that it is overlapped with the 4.75 ppm peak. From the NMR peak intensity of unsaturated bonds, the ratio of A+C to B+D+E was about 0.05 with 2.1 J $/cm^2$ UV exposure and about 0.5 for high energy radiation. This ratio is almost doubled compared with the previous results of main chain scission to methyl ester cleavage from GPC measurement and IR difference spectra. It is expected that this difference probably arises from the recombination of polymeric fragments, which is expected from the observed saturation of ESR peak intensity of propagating radicals (*31*) and peroxy radicals (*45, 46*) with increasing radiation dose. From this point, the peaks at 4.62 ppm and 4.82 ppm are assigned to protons of vinylidene groups generated by main chain scission.

In Table II, (*47, 48*) it can be seen that the dissociation energies of CH_3-H and CH_3O-H bonds are higher than those of secondary hydrogen atoms.

Table II. Bond Dissociation Energy

Bond	D_{298}, Kcal/mole
CH_3-H	104
$CH_3CH_2CH_2$-H	98
$(CH_3)_2CH$-H	94.5
CH_3O-H	102
CH_3OCO-H	92.7

These values indicate that the $CH_3\cdot$ and $CH_3O\cdot$ fragments can abstract hydrogen from a-methyl or methylene groups in a cage of PMMA matrix as shown in Scheme II. On the other hand, the formate radical ($\cdot CO_2CH_3$) is not sufficiently reactive to abstract hydrogen from a-methyl or methylene groups (Table II). However, the electronically excited carbonyl group could eject methyl formate (the quantum yield of methyl formate = 0.97) via a concerted process (Norrish type III reaction). Hydrogen abstraction (I and II) in Figure 19 can occur by a radical reaction or a process similar to the Norrish type II and III reactions (*49, 50*) via an excited state without main chain scission. Further work to clarify this pathway is in progress.

With respect to the bond dissociation energies of C-H bonds in a-methyl (primary) and methylene (secondary) groups, as well as with respect to the number of hydrogen atoms on each, the methylene group in the backbone chain seems more likely to undergo a hydrogen abstraction reaction. However, in the NMR spectrum of UV-irradiated PMMA, the ratio of vinylidene to vinylene groups is about 9 : 1, despite the

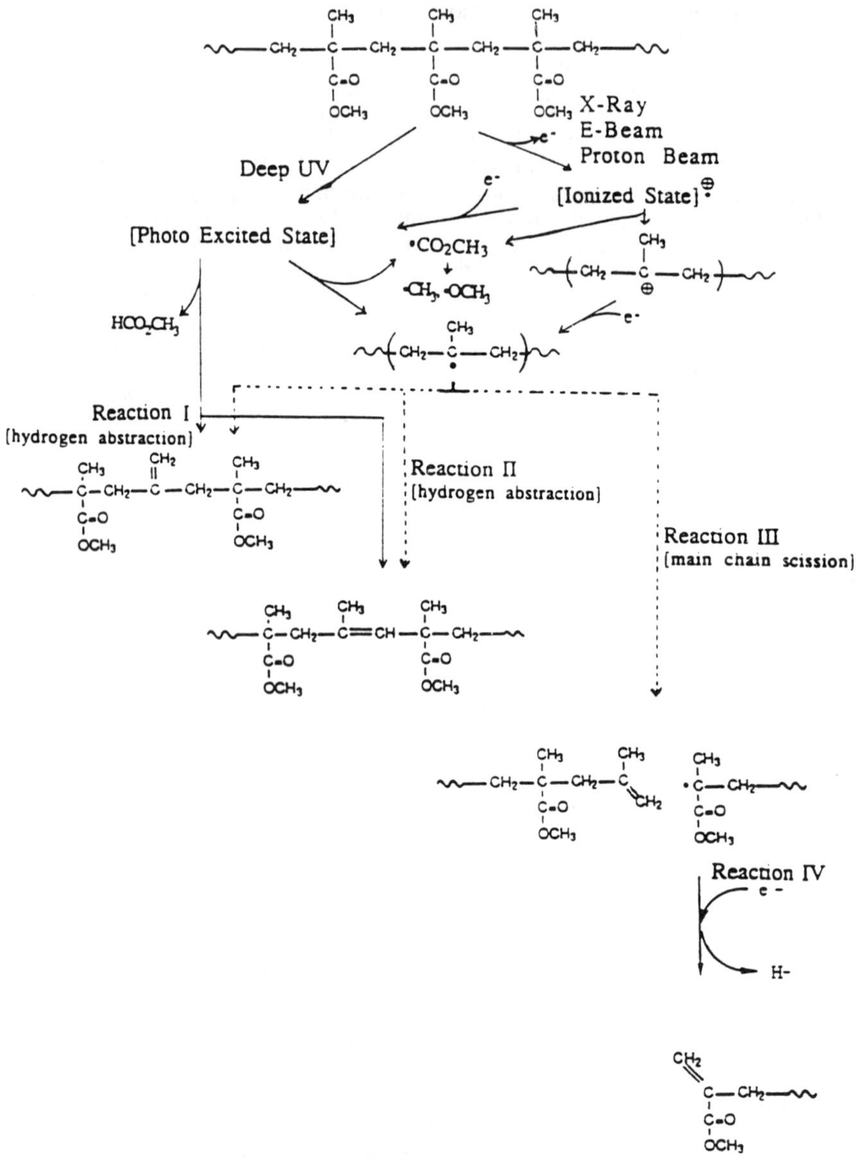

Figure 15. Generation of Unsaturation in Irradiated PMMA.

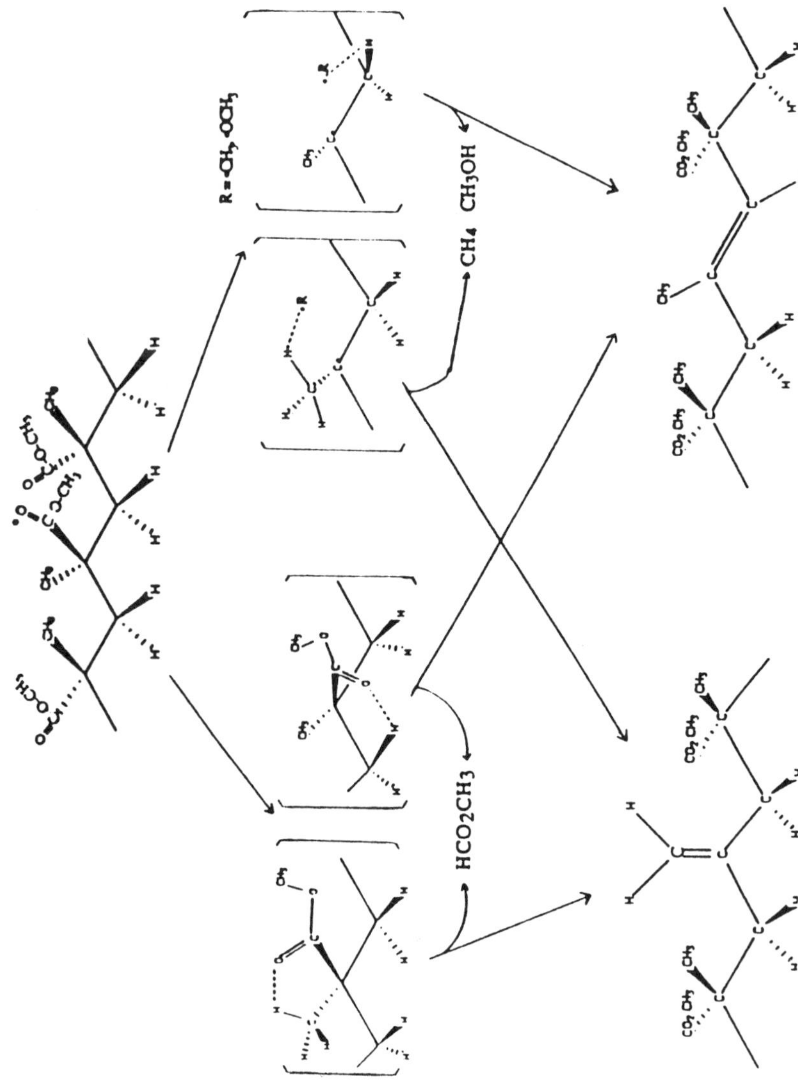

Scheme II. Generation of Backbone Unsaturation in PMMA.

fact that reaction at the methylene sites is favored on enthalpic and statistical grounds. A possible influence which may favor the formation of vinylidene double bonds arises from the effect of hybridization on chain mobility, because irradiation of PMMA is carried out in a solid film at room temperature. Vinylidene groups introduce only one sp^2-hybridized carbon atom into the chain while vinylene groups require two sp^2-hybridized carbon atoms. The resultant hindrance to free rotation stiffens the chain and decreases rotational entropy, thereby favoring vinylidene over vinylene structures. Further, vinylidene groups do not generate excessive steric interactions with the a-methyl and methyl ester groups while vinylene would experience steric repulsion. These reasons may be the partial basis of the preference for hydrogen abstraction from a-methyl groups caused by UV irradiation of PMMA film.

From the ^{1}H NMR peak intensities of the PMMA which was studied (*51, 52*) the content of rr (racemic-racemic), mr (meso-racemic) and mm (meso-meso) triad sequences in Du Pont Elvacite 2041 was found to be 63, 33, and 4 %, respectively. On the formation of vinylidene groups (v) in the triads of PMMA by hydrogen abstraction from a-methyl groups , rr, mm, and mr triad units give rise to two different types of segments (rv and mv), Scheme III.

The ratio of mm + $\frac{1}{2}$ mr to rr + $\frac{1}{2}$ mr triads is about 1 : 4 from the NMR spectrum before irradiation of PMMA, assuming that the mr triad has the same possibility to change to the mv or rv segments. In the degraded PMMA, the new peaks at 0.95 and 1.14 ppm appear in Figures 16 and 17. The relative height of these two peaks is about 1 : 4. These two peaks are assigned to alpha-methyl hydrogen atoms next to vinylidene groups: rv (from rr and $\frac{1}{2}$ mr triads) at 0.95 ppm and mv (from mm and $\frac{1}{2}$ mr triads) at 1.14 ppm. The hydrogen atoms of vinylidene groups in the triad sequences are also expected to be associated with the stereochemically structures, Scheme IV.The ratio of mm+rr to mr triads is about 1 : 2 from the NMR spectrum of unirradiated PMMA. The ratio of the peak intensity at 5.0 and 4.75 ppm is about 1 : 3~4 in the NMR spectra of the irradiated PMMA . From these data these peaks are assigned to the hydrogen atoms of vinylidene groups: 4.75 ppm (from mm and rr triads) and 5.0 ppm (from mr triads).

In the NMR spectra in Figure 14, the small peaks at 5.46 and 6.19 ppm are assigned to protons on unsaturated bonds (*53, 54*), which have been proposed to be generated by Reaction IV in Figure 15 (*34*). This process is almost unnoticeable in UV-irradiated samples (Figure 13) as contrasted with the results obtained with high energy radiation.

When a solution of PMMA in methylene chloride or acetonitrile was exposed to UV light, the hydrogen atoms of unsaturated bonds, the peak positions of which are different from those generated in solid film, appeared in the H^1 NMR spectrum and ^{13}C NMR spectrum, as shown in Figure 18 and 19. It was reported that in solution, the quantum yields of chain scission can be 0.2 ~ 2.0, but in the solid state, chain scission can be an order of magnitude lower because of recombination of radicals in immobilized chains (*55*). This large difference can be explained by the fact that in the solid film there is a cage effect such that, following scission, reactive radicals are held in proximity and may recombine rapidly. In solution, the probability of recombination is much diminished and the number of effective scissions is higher. The magnetic resonance spectra show that several types of unsaturated bonds are apparently generated in PMMA solution by UV-radiation. In the ^{13}C NMR spectrum of UV-degraded PMMA film in Figure 19, the peaks at 128.9 and 161.9 ppm are assigned to the sp^2 carbon atoms of vinylidene groups generated by hydrogen abstraction from

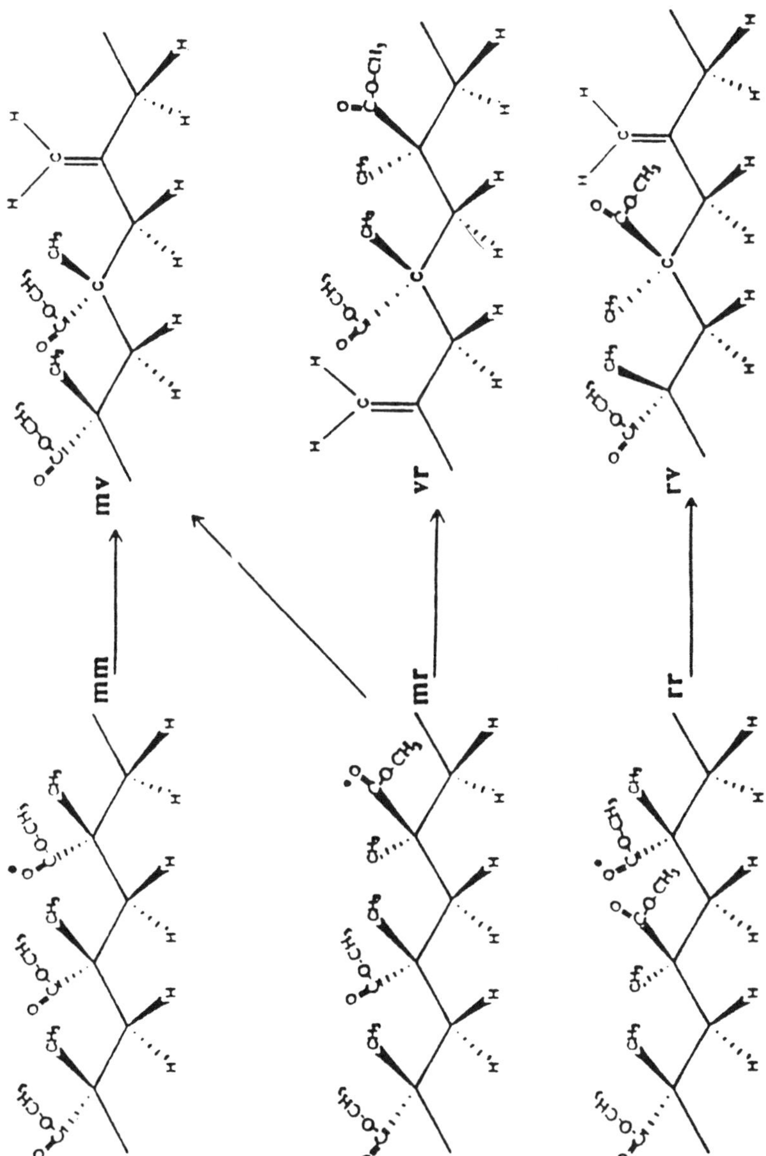

Scheme III. Local Environments of Vinylidene-Containing Triads.

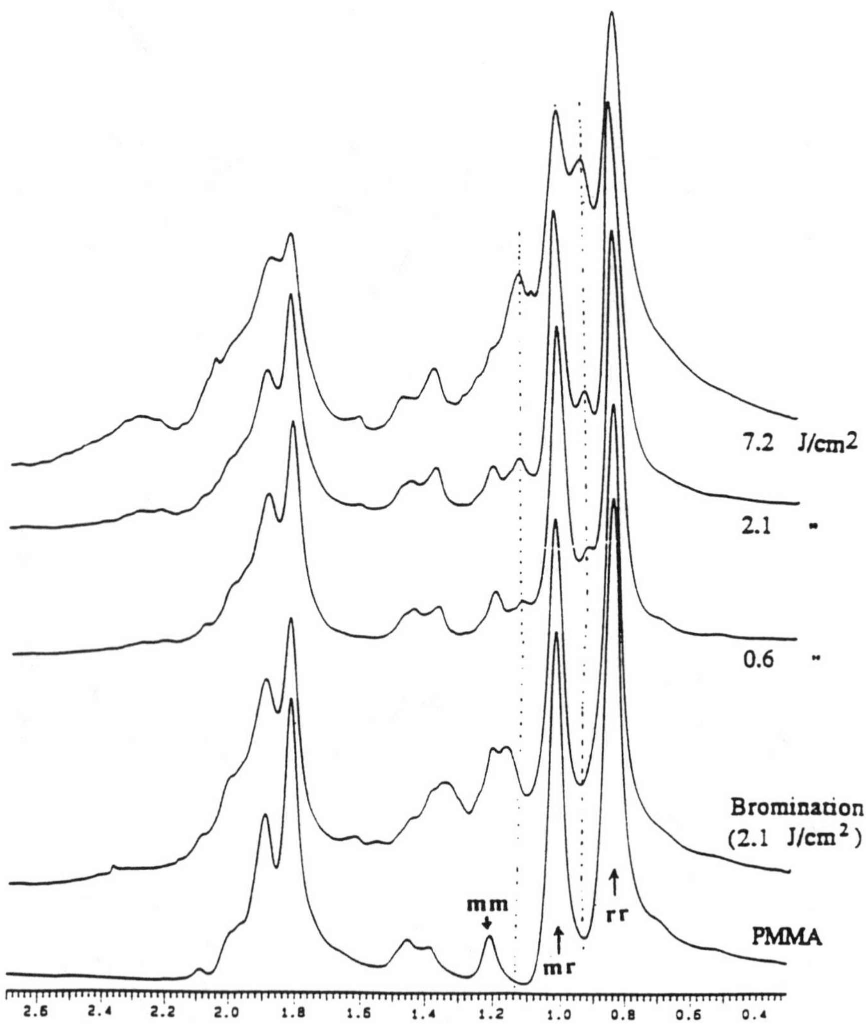

Figure 16. PMR Spectra (X3) of UV-Irradiated PMMA.

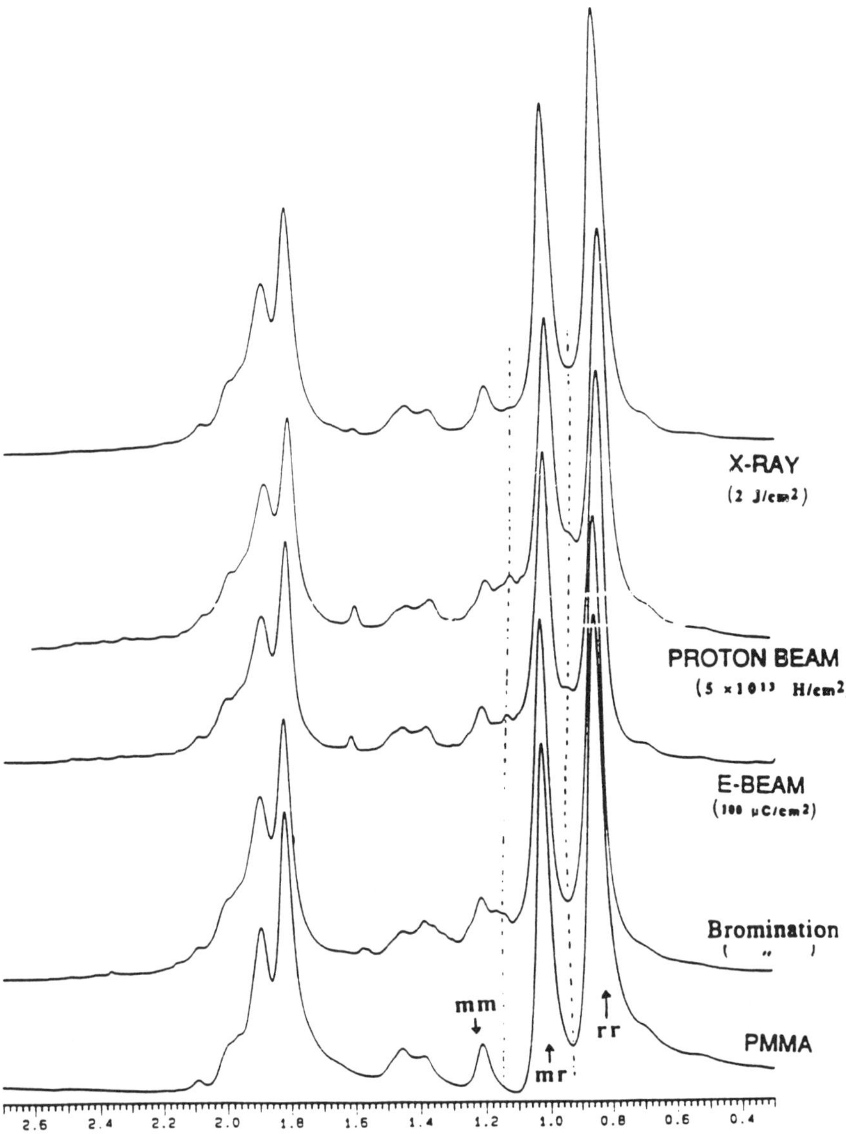

Figure 17. PMR SPectra (X3) of Radiation-Damaged PMMA.

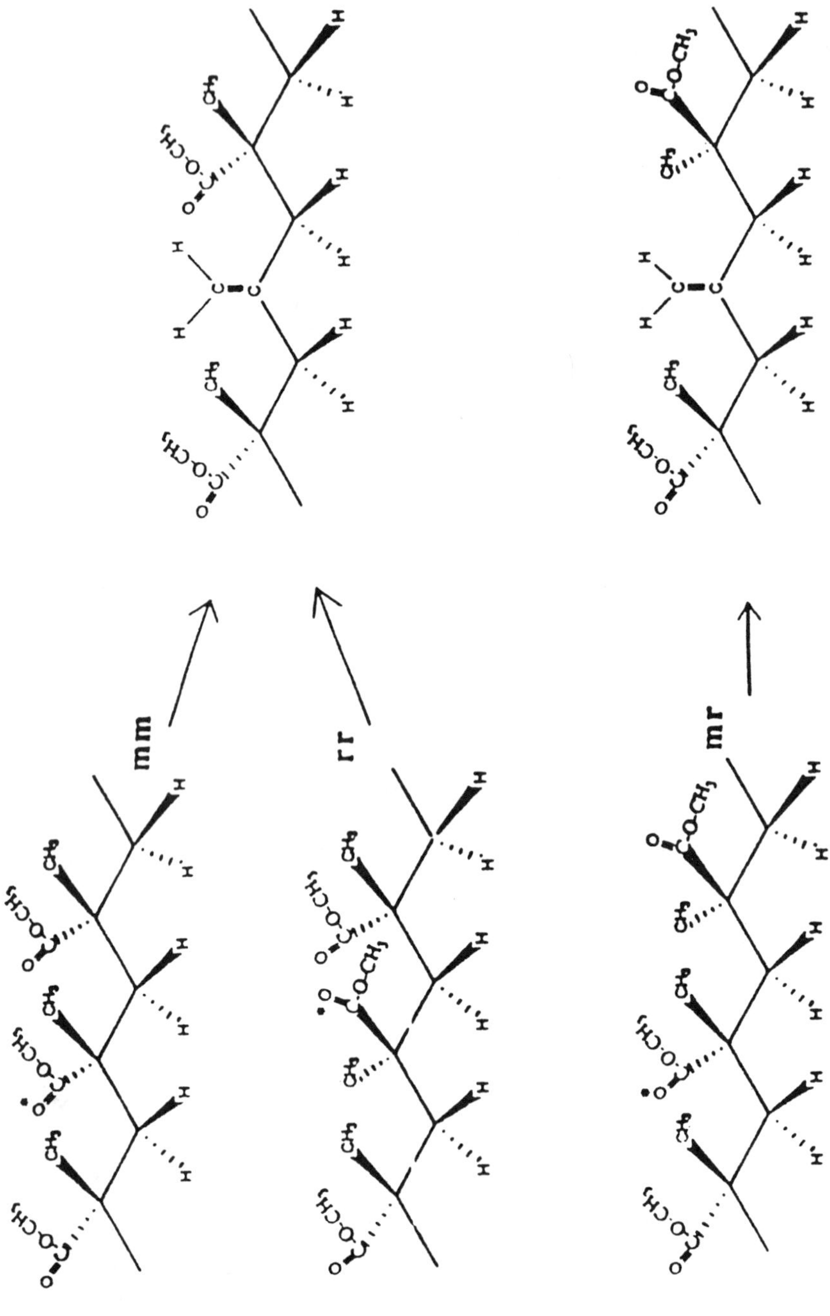

Scheme IV. Vinylidene Groups in Triads of Irradiated PMMA.

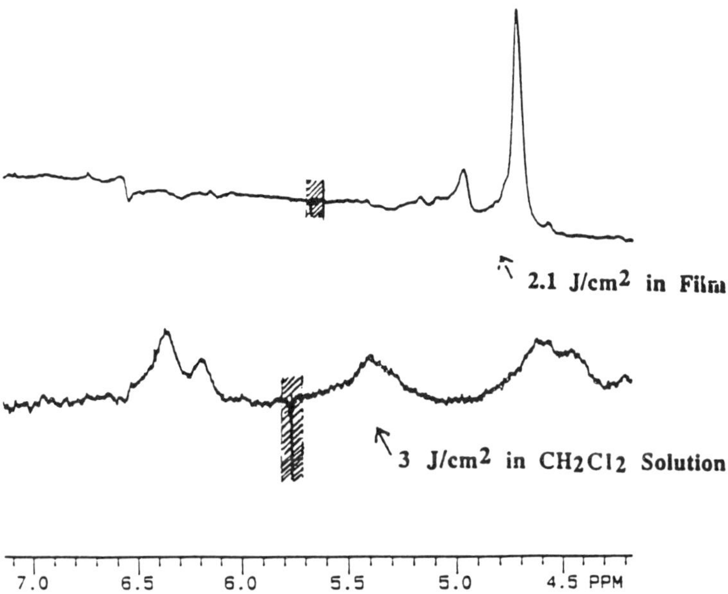

Figure 18. PMR Spectra (X17.5) of UV-Irradiated PMMA Film and Methylene Chloride Solution.

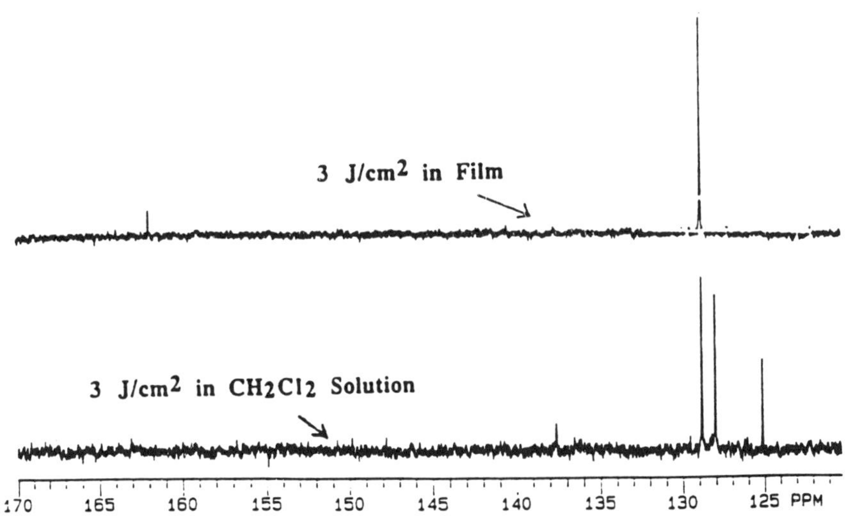

Figure 19. CMR Spectra (X3) of UV-Irradiated Film and Methylene Chloride Solution.

alpha-methyl groups (Reaction I in Figure 15). Other types of unsaturated bonds are presumably not apparent in this spectrum because they are present at too low concentrations to be detectable by ^{13}C NMR spectroscopy under these conditions.

To help explain the degradation mechanism of PMMA, the formation energy (ΔH_f) and the most stable conformational isomers of model compounds for the stereochemical sequences (rr, mr and mm) of PMMA and degraded segments were determined with the MINDO/3 (*56*) program and with the Macromodel program employing MM2 (*57*), respectively. The most stable conformational isomers of the model compounds determined by the Macromodel program are shown in Figure 20.

The data in Table III indicate that the relative portion of triad sequences of PMMA (Du Pont Elvacite 2041) prepared by radical polymerization corresponds to the order of the formation energy of the model compounds.

Table III. Formation Energy of Model Compounds for Degraded PMMA

Model Compounds	Formation Energy (Kcal/mole)	Fraction (%) of Triads
rr (syndio)	0.48	63%*
mr (hetero)	10.63	33%*
mm (iso)	19.40	4%*
vinylidene	3.40	90%**
vinylene	-36.85	10%**

*Elvacite 2041
**UV-Irradiated PMMA

As expected, the data in Table III show that the formation energy of vinylene segments, $R_1R_2C{=}CHR_3$, is less than that of vinylidene segments, $R_1R_2C{=}CH_2$. Thus, based on thermodynamic considerations, the hydrogen abstraction reaction in the photodegradation of PMMA should produce more vinylene groups. However, as described above, the ratio of vinylidene to vinylene groups in the H^1 NMR spectrum of UV-irradiated PMMA is about 9 : 1. Probably, the observed formation of vinylidene groups by hydrogen abstraction is preferred kinetically. It should be noted that these preliminary calculations neglect long range steric interactions expected to be important in polymers.

In the most stable conformational isomers of the model compounds, the distance of closest approach between the hydrogen atoms of the a-methyl or methylene groups and the oxygen atom of the middle carbonyl group was calculated with the Macromodel program. From the data in Table IV, the hydrogen atoms of the a-methyl group are placed much closer to the carbonyl oxygen atom than to those of the methylene group in the rr and mr triad sequences, which constitute the major fraction (96 %) in PMMA prepared with a radical initiator. Therefore, most of the hydrogen abstraction reaction should occur between the excited carbonyl and a-methyl groups. This result may be one reason that the vinylidene group is the major segment produced in the UV-irradiated atactic PMMA film. In the structure of mm triad models at the

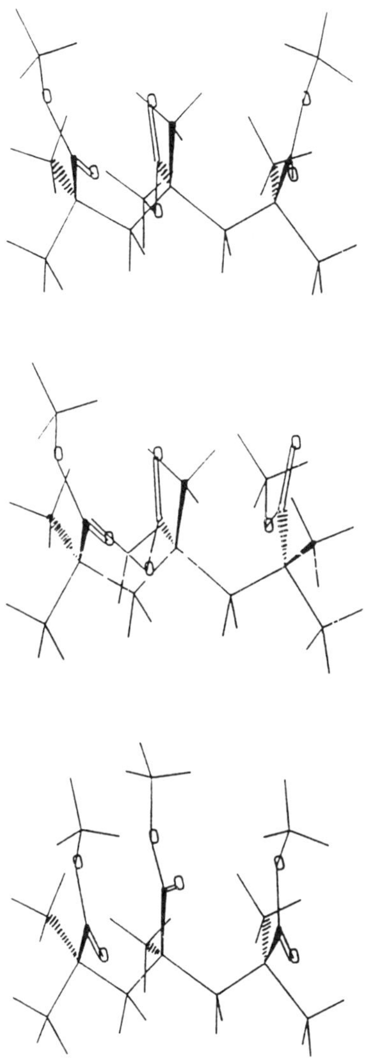

rr

mr

mm

Figure 20. The Most Stable Conformational Isomers of PMMA Models.

minimum energy state in the MM2 program (Figure 21 and Table IV), the carbonyl groups are located closer to the methylene hydrogen atoms than to the alpha-methyl hydrogen atoms.

Table IV. Distance (Å) Between Hydrogen and Oxygen Atoms of the Middle Carbonyl Group in the Model Compounds

Model Compounds	Distance (Å)		Fraction (%) of Triads
	α-methylene	methylene	
rr	2.730	3.819	63%
mr	2.695	3.725	33%
	2.765	3.855	
mm	4.023	3.042	4%

It was suspected that a vinylene fragment in the internal polymer chain might be the major unsaturated bond in isotactic PMMA film (mm : mr : rr = 70 : 21 : 9) irradiated with UV light. However, the amplified NMR spectra of isotactic samples (Figure 22) shows several resonances of hydrogen atoms of unsaturated bonds with peaks appearing at 4.5 ~ 5.3 ppm, which are very similar to those in atactic samples, and the other peaks at 5.5 ~ 6.5 ppm originate from the degradation of the isotactic structure.

In the UV degradation of isotactic PMMA film, a concerted process similar to the Norrish Type II and III reactions does not occur easily because of the location of the carbonyl groups. Probably the cleavage of methyl ester groups similar to the Norrish Type I reaction is also favorable to produce methyl and methoxy radicals which can abstract a hydrogen atom from a neighboring polymer chain, form the alkyl radicals I and II, and initiate b-bond scission, as shown in Figure 23. Hiraoka observed differences in the ESR spectra of syndio, iso, and atatic PMMA irradiated with UV light (*34*). It was also reported that, in addition to the propagating radical, the alkyl radicals I and II (Figure 23) would be formed in PMMA under the influence of different agents: chemical, radiation, photochemical and mechanical (*58, 59*). The hydrogen atoms of A and B (Figure 23), which could also be generated through a concerted process, are expected to appear at 4.5 ~ 5.3 ppm in the NMR spectra. The most probable unsaturated fragments with resonances at 5.5 ~ 6.5 ppm are expected to be C, D, E (Figure 23) in the degraded isotactic sample. The regenerated radicals would also be able to start the same sequence of reactions. Todd has suggested that radicals could eject methyl formate radical to form a stable chain end (*60*), which indicates that the regenerated radicals would also be able to start the same sequence of reactions:

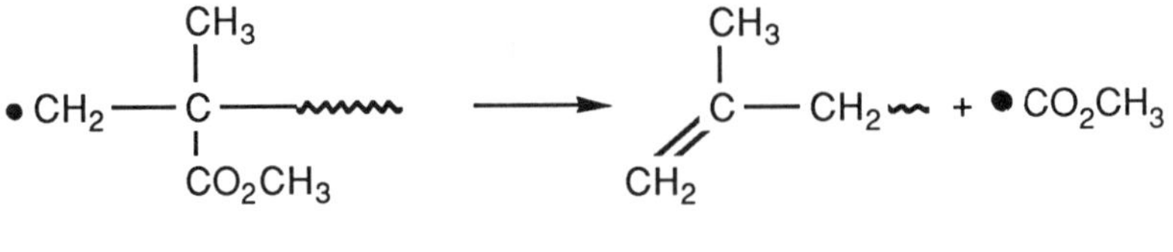

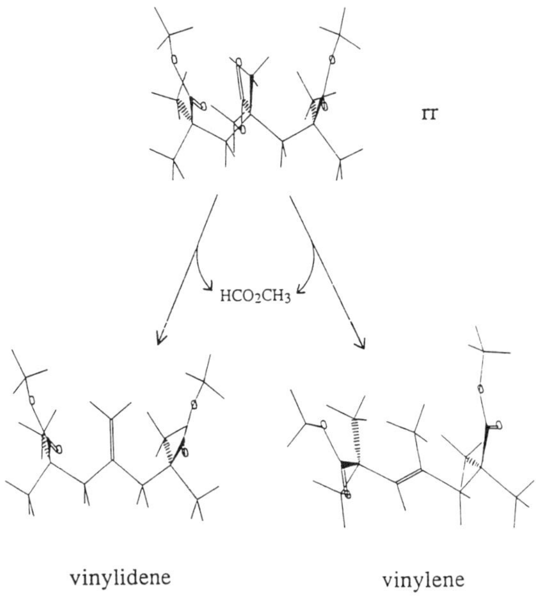

Figure 21. The Most Stable Conformational Isomers of Model Degradation Products.

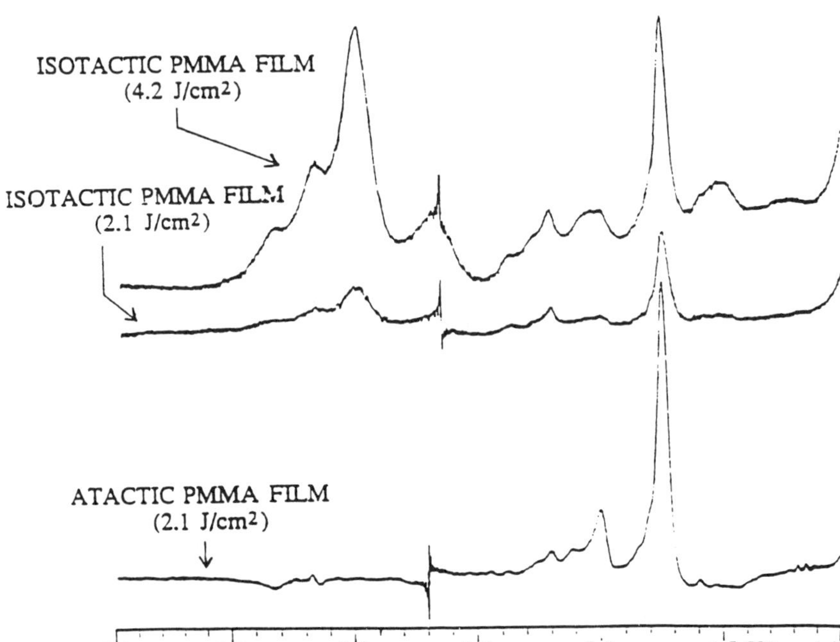

Figure 22. PMR (X25) of UV-Irradiated Isotactic PMMA Film

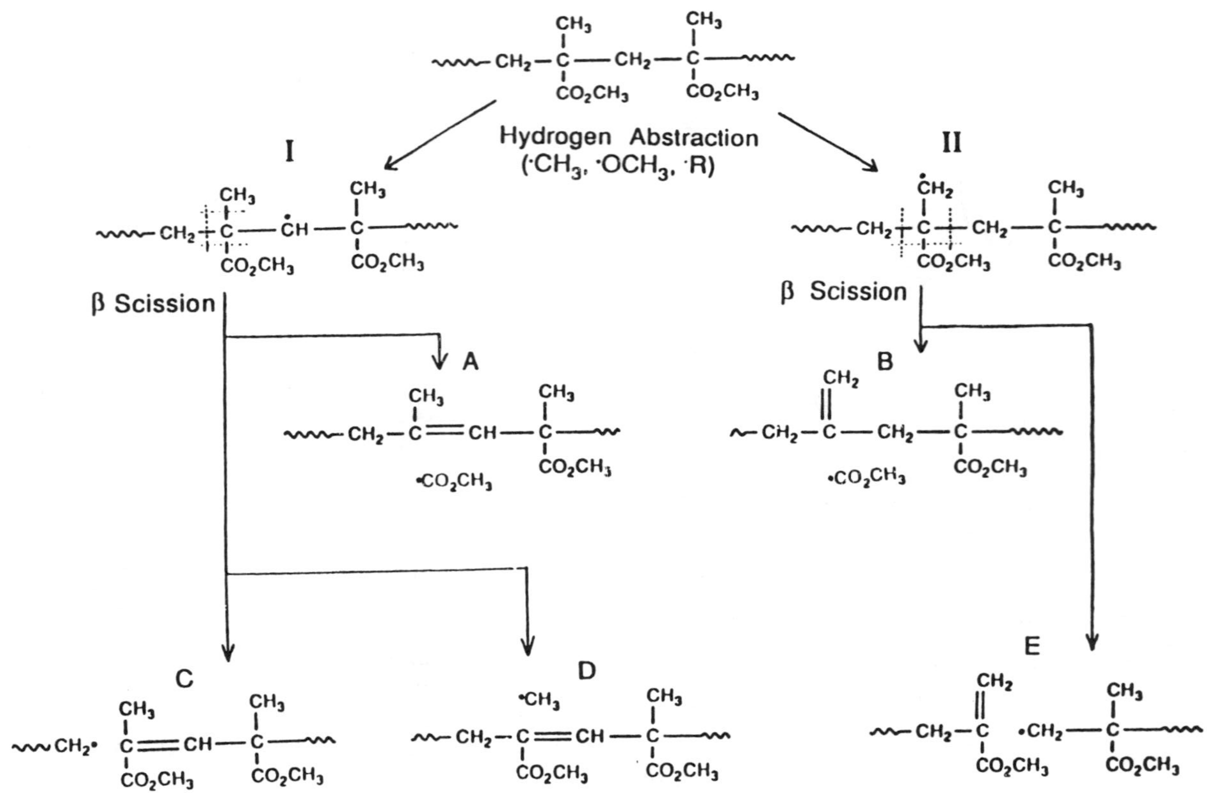

Figure 23. Degradation Pathways for Alkyl Radicals in UV-Irradiated Isotactic PMMA.

It was reported that the racemization of isotactic PMMA occurred upon ultraviolet and gamma-ray irradiation, as shown below (*61, 62*).

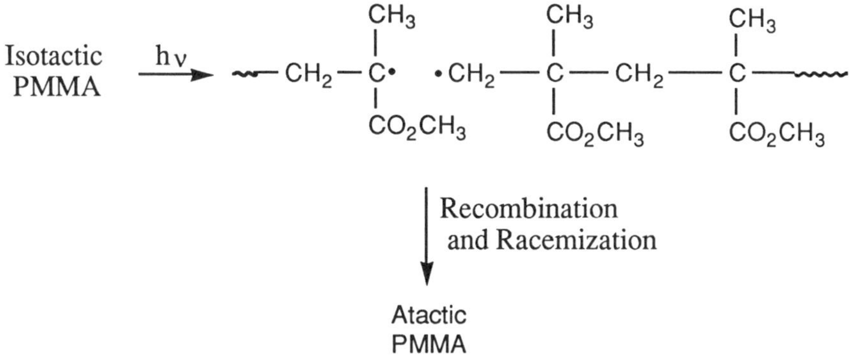

The racemization yield was calculated from the decrease of the content of meso pairs (dd or ll). However, a change of the meso content (mm + $\frac{1}{2}$ mr) was not observed in the ^{1}H NMR spectra of UV-degraded isotactic PMMA (Figure 24) in this work. On the other hand, the peak of the alpha-methyl group (mm triad) seems to be overlapped with the peak of the alpha-methyl group next to the vinylidene bond (from mm and $\frac{1}{2}$ mr triads) generated by hydrogen abstraction (Reaction I in Figure 15) as described before. Although the racemization of isotactic PMMA was observed at much higher UV dose than in this experiment, its occurrence under these conditions is questionable. In isotactic PMMA irradiated at much higher UV dose, conjugated unsaturated bonds are expected to be formed rather than racemization.

Conclusions

The chemical changes in poly(methyl methacrylate) caused by irradiation with deep UV, x-ray, electron and proton beams were studied by gel permeation chromatography, FT-IR, UV and NMR spectrosopies and are found to be much more complex than was previously apparent. Quantitative analyses of spectroscopic changes demonstrated a 1:1 correspondence between the diaappearance of methyl ester groups and the generation of double bonds in the polymer backbone by all types of radiation used in this study. The ratio of main chain scission to changes in the number of ester groups and unsaturated bonds was compared to determine the characteristics of degradation of PMMA by the different types of radiation. High energy radiation (x-ray, e-beam and proton beam) was about 10 times more efficient than deep UV in causing main chain scission with removal of fewer ester groups in processes which decrease the ultimate sensitivity of PMMA as a microlithographic resist. This ratio (approximately 0.03) for deep UV was very close to to the quantum yield for main chain scission of PMMA as reported in the literature.

Most of the reactions leading to loss of ester groups (about 97 %) in atactic PMMA film (which approximates a syndiotactic configuration) involved hydrogen abstraction from alpha-methyl and methylene groups leading to the generation of vinylidene groups (major) pendent from the chain and vinylene groups (minor) within the backbone, respectively. When films of isotactic PMMA, or solutions of atactic PMMA, were exposed to deep UV, unsaturated bonds which had peak positions and intensities in the NMR spectra *different* from those found for irradiated atactic PMMA

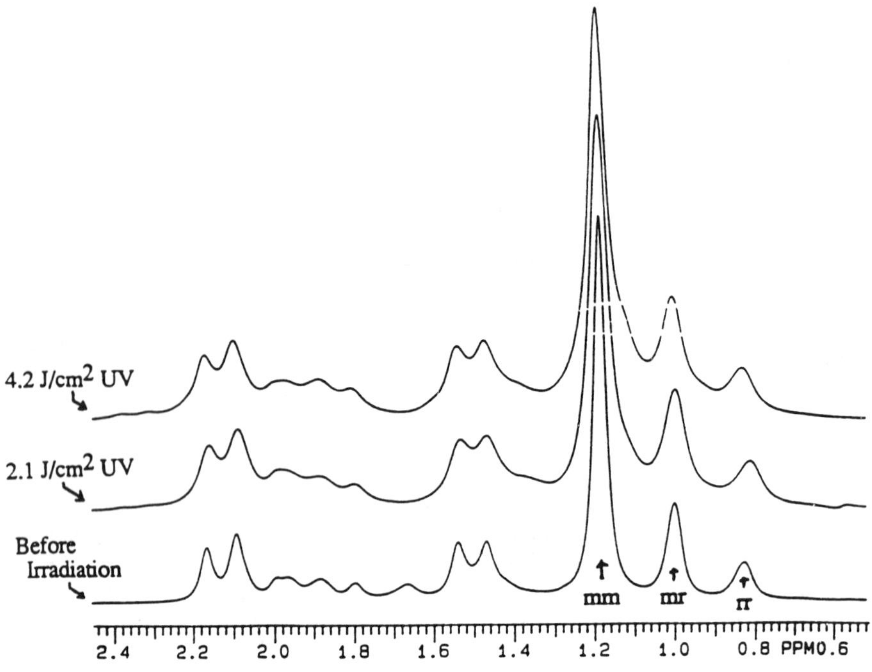

Figure 24. PMR Spectra of UV-Irradiated Isotactic PMMA.

films, appeared in the NMR spectra. Work to elucidate further the apparent influence of stereochemistry and constraint in a solid matrix on the outcome of the photoprocesses operative in this system is currently underway.

Within the limits of sensitivity of our instruments at this time, we find no indication of the radiation-induced racemization which had been reported previously, rather we see a tendency for the initially formed isolated double bonds to rearrange to form sequences of conjugated double bonds at higher doses.

Experimental

Materials. Atactic PMMA, Du Pont Elvacite 2041, was purified by reprecipitation from toluene solution into methanol and dried in a vacuum oven (oil pump at 50 ^{o}C) for one day. The dry PMMA (M_n = 1.76 x 10^5 g/mole from GPC measurement with polystyrene reference stantards) was dissolved in chlorobenzene to make 7.5 weight % solution. Isotactic PMMA (M_v = 2.5 x 10^4 g/mole, Polysciences, Inc.) was used without any purification.

Radiation Sources. Light from a high pressure 500 W mercury-xenon lamp, HTG Serial DUV-500 (Advanced Radiation Corp., CA) was used without a filter, in air, for deep UV irradiation. The deep UV portion of the radiation spectrum from this lamp was between 230 and 255 nm. The intensity of UV light was measured with a Black-Ray shortwave ultraviolet intensity meter J-225 (UVP, Inc., CA).

The x-ray irradiations were carried out on the IBM beam line using synchrotron radiation from the National Synchrotron Light Source (NSLS) at Brookhaven National Laboratory using the vacuum ultraviolet (VUV) storage ring (*15*). The VUV storage ring provides a high intensity source of collimated soft x-rays for which the effective energy spectrum lies between about 800 and 1800 eV (between 7 and 15 Å) (*16*).

Electron beam (25 KeV) exposures were made with an IBM EL-2 lithography tool at the Center for Integrated Electronics, RPI, Troy, NY. The system utilizes a dual channel deflection system and a variable shaped spot (*63*). The spot is positioned using raster scan deflection. The normal working beam current of the EL-2 electron column is 2.0 µA. The spot size is 0.25 µm x 0.25 µm. An accelerating voltage of 25 KeV was used for all electron beam exposures.

Proton beam irradiations were conducted with a 900 KeV H_3^+ beam from a Dynamitron Accelerator at the State University of New York, Albany, NY. It can provide a maximum proton yield of 600 µA at 4 MeV and the beam energy can be continuously varied from 0.2 MeV to 4 MeV.

Analysis of Irradiated PMMA Film

Atactic or isotactic PMMA was dissolved in chlorobenzene and spin-coated onto silicon wafers, quartz or sodium chloride disks and then baked at 150 ^{o}C for 30 minutes to remove solvents completely.

Irradiated PMMA films on silicon wafers were dissolved in tetrahydrofuran and analyzed by gel permeation chromatography (UV detector at 229 nm) using 500, 10^3, 10^4, 10^5, and 10^6 Å columns with polystyrene reference standards to measure changes in molecular weight. UV spectra of films on quartz disks before and after exposure to various radiation sources were obtained with a Perkin Elmer Lambda 4C UV spectrophotometer. FT-IR spectra of films on NaCl disks before and after irradiations were obtained with a Perkin Elmer 1800 Fourier Transform Infrared Spectrophotometer. Difference UV and FT-IR spectra were obtained by using a Perkin Elmer 7500 computer. Irradiated PMMA films on silicon wafers were dissolved in $CDCl_3$ and analyzed with Varian XL-200 NMR spectrometer (pulse width = 7 µsec, delay time = 2 sec, temperature = 40 °C, and acquisition number = 1000 ~ 7000). Irradiated PMMA films were exposed to bromine gas at room temperature for 3 hours and analyzed with UV and NMR spectrophotometers.

A PMMA solution (10 to 20 %) in methylene chloride or acetonitrile was exposed to UV light. The incident dose was 3 J /cm^2. The irradiated PMMA was dried in a vacuum oven (oil pump) at 50 °C for one day, dissolved in $CDCl_3$ and analyzed by NMR spectrometry.

The heat of formation (ΔH_f) of the model compounds was calculated with the MINDO/3 Program (*121*) on an IBM 3090 Model 200S, running the Michigan Terminal System (MTS). Molecular mechanics calculations were performed on a MicroVax II computer using the Macromodel program supplied by Columbia University. The force field used was MM2 which is based on the molecular mechanics developed by Allinger (*122*). The Steepest Descent minimization procedure was used until the first derivative of the potential energy fell below 1.0. Then, Block Diagonal Newton Raphson Minimization was chosen to obtain the lowest conformational energy of the model compounds.

Acknowledgment

It should be apparent that these results are the fruit of many collaborative interactions. The names of those involved are too numerous to repeat here and are given in the references to our previous work in this area. The unstinting efforts of the technical staff of the Major Instrumentation Center of the Department of Chemistry (Dr. H. Schwartz) and of the Center for Integrated Electronics (M. Bourgeois, D. C. King, N. King and D. Pulver) deserve special recognition. Financial support for the initial stages of this work was provided by the Semiconductor Research Cooperative and the IBM Corporation. Dr. Wayne Moreau served as the liason between IBM and RPI and was a wellspring of ideas and suggestions which helped to initiate this work.

Literature Cited

1. Willson, C. G. In *Introduction to Microlithography;* Thompson, L. F.; Willson C. G.; Bowden, M. J., Eds.; ACS Symposium Series #219; American Chemical Society, Washington, DC, 1983; pp. 87-159.
2. Moreau, W. M. *Semiconductor Lithography: Principles, Practices, and Materials;* Plenum Press: New York, NY, 1988.
3. Daly, R. C.; Williams, J. W. R. *Polymer News* **1986,** *11*, 164.
4. Moreau, W. M. *Proc. SPIE* **1982,** *33*, 2 .
5. Moreau,W. M. *Optical Engr.* **1983,** *22(2)*, 181.
6. Moreau, W. M.; Schmidt, *P. Electrochem. Soc. Ext. Abstr., 138th Meeting* **1970,** 459.
7. Lin, B. J. *J. Vac. Sci. Technol.* **1975,** *12(6)*, 1317.
8. Lin, B. J. *IBM J. Res. Develop.* **1976,** *20*, 213.
9. Mimura,Y.; Ohkubo, T.; Takeuchi,T. Sekikawa, K. *Jap. J. Appl. Phys.* **1978,** *17(3)*, 541.
10. Greeneich, J. S. *J. Electrochem. Soc.* **1974,** *121(2)*, 1669.
11. Greeneich, J. S. *J. Electrochem. Soc.* **1975,** *122(7)*, 970.
12. Deckman , H. W.; Dunsmur, J. H. *J. Vac. Sci. Technol.* **1983,** *B1(4)*, 1166.
13. Harada, K.; Sugawara, S. *J. Appl. Polym.Sci.* **1987,** *27*, 1441.
14. Hatzakis,M. *IBM J. Res. Develop.* **1988,** 32*(4)*, 441.
15. Silverman, J. P.; Haelbich, R. P.; Grobman, W. D.; Warlaumont, J. M. *Proc. SPIE* **1983,** *448,* 50.
16. Silverman, J. P.; Haelbich, R. P.; Grobman, W. D.; Warlaumont, J. M.*Proc. SPIE* **1983,** *393*, 99.

17. Sotobayshi, H.; Asmussen, F.; Thimm, K.; Schnabel,W.; Betz,H.; Einfeld, D. *Polym. Bull.* **1982,** *7*, 95.
18. Spiller, E.; Feder, R. In *X-Ray Optics: Application to Solids* ; Oueisser, H.-J., Ed.; Springer-Verlag, New York, NY, 1977; pp. 35-91.
19. Yamada, H.; Hori, M.; Morida, S.; Hattori, S. *J. Electrochem. Soc.* **1988,** 135(4), 966.
20. Brault, R. C.; Miller, L. J. *Polym. Engr. Sci.* **1980,** *20(6)*, 1064.
21. Hall, T. M.; Wagner , A.; Thompson, L. F. *J. Vac. Sci. Technol.* **1979,** 16(6), 1889.
22. Tung, N. C. *J. Electrochem. Soc.* **1984,** 131(9), 2152.
23. Willson, C. G.; Ito, H.; Miller, D. C.; Tessier, T. G. *Polym. Engr. Sci.*, **1983,** *23(18)*,1000.
24. Tarro, R. M.; Warden, J. T.; Corelli, J. C.; Moore, J. A.; Steckl, A. J.;Kumar, S. In *Microcircuit Engineering 84*, Heuberger, A.; Beneking, H. Eds.; Academic Press, London, 1985; p. 537.
25. Ouano, A. C. *Polym. Engr. Sci.* **1978,** *18(4)*, 306.
26. Chapiro, A. *Radiation Chemistry of Polymeric Systems*, Interscience Publishers, New York, NY; 1962.
27. Gazard, M.; Duchesne, C.; Dubois, J. C.; Chapiro, A. *Polym. Engr. Sci.* **1980,** *20(16)*, 1069.
28. Liu, W.-T.; Corelli, J. C.; Steckl, A. J.; Moore, J. A.; Silverman, J. *Appl. Phys. Lett.* **1984,** *44(10)*, 973.
29. Moore, J. A.; Choi, J. O.; Kim, S. Y.; Corelli, J. C. *Proc. Electrochem. Soc., 172nd Meeting* **1987,** .
30. Dickens, B.; Martin, J. W.; Waksman, D.; *Polymer* **1984,** *25*, 706.
31. Gupta, A.; Liang, R.; Tsay, F. D.; Moacanin, J. *Macromolecules* **1980,** 13, 1696.
32. Abraham, R. J.; Whiffin, D. H. *Trans. Farad. Soc.* **1958,** *54*, 1298.
33. David, C.; Fuld, D.; Geuskens, G. *Makromol. Chem.* **1970,** *139*, 269.
34. Hiraoka, H. *IBM J. Res. Dev.* **1977,** *21*, 121.
35. *Organic Electronic Spectral Data* ; Ungnade, H. E., Ed.; Interscience Publishers Inc., New York, NY, 1960, Vol. II, p. 57, 76.
36. Fox, R. B.; Issacs, L. G.; Stokes, S.; *J. Polym. Sci.* **1963,** *A1*, 1079.
37. Shultz, A. R.; Frank, P.; Griffing, B. F.; Young, A. L.; *J. Polym. Sci., Poly. Phys.*,**1985,** *23*, 1749.
38. Bohn, P. W.; Taylor, J. W.; Guckel, H.; *Anal. Chem.* **1981,** *53*, 1082.
39. Nagai, H. *J. Appl. Polym. Sci.* **1963,** *7*, 1697.
40. Simako, S. J.; Bryan, S. R.; Griffis, D. P.; Murray, R. W.; Linton, R. W. *Anal. Chem.* **1985,** *57*, 1198.
41. *Spectres RMN des Polymeres: ^{1}H and ^{13}C NMR Spectra of Polymers,* Pham, Q.-T.; Petiaud, R., Eds.; Editions SCM: Paris, France,1980, Vol. 1.
42. *Sadtler Standard Spectra, Carbon-13 NMR Collection* : Sadtler Research Laboratories; Philadelphia, 1976.
43. *Sadtler Standard Spectra, Proton NMR Collection;* Sadtler Research Laboratories; Philadelpia, PA; 1967.
44. *The Aldrich Library of NMR Spectra Edition II* ; Pouchert, C. J., Ed.;Aldrich Chemical Company, Inc.: Milwaukee, Wisconsin, 1983.
45. Tarro, R.; M.S. Thesis; Rensselaer Polytechnic Institute, 1985.
46. Pivovarov, S. P.; Muller, S.; Vasilevskaja, L.; Messerle, P. E.; Lorkowski, H.-J. *Acta Polymerica* **1987,** *38*, 114.
47. Benson, S. W. *J. Chem. Ed.* **1965,** *42*, 502.
48. Kerr, J. A. *Chem. Rev.* **1966,** *66*, 465.
49. Calvert,J. G.; Pitts, Jr., J. N. *Photochemistry,* John Wiley & Sons, Inc.: New York, NY 1966; pp 384, 434.

50. Roffey, C. G. *Photopolymerization of Surface Coatings*; John Wiley & Sons: New York, NY,1982; pp 73, 74.
51. Bovey, F. A. *Accounts Chem. Res.* **1968,** *1*, 175.
52. Ferguson, R. C. *Macromolecules* **1969,** *2(3)*, 237 .
53. Hatada, K.; Kitayama, T.; Masuda,E. *Polym. J.* **1986,** *18(5)*, 395.
54. Kashiwagi, T.; Inaba, A.; Brown, J. E.; Hatada,K.; Kitayama,T.; Masuda, E. *Macromolecules* **1986,** *19*, 2160.
55. MacCallum, J. R.; Schoff, C. K. *Trans. Faraday Soc.* **1971,** *67*, 2383, 2372.
56. Bingham, R. C.; Dewar, M. J. S.; Lo, D. M. *J. Am. Chem. Soc.* **1975,** *97*, 1285. These calculations were carried out in conjunction with Prof. K. J. Miller.
57. Burkert, U.; Allinger, N. L. *Molecular Mechanics*; ACS Monograph; American Chemical Society: Washington, D C, 1982.
58. Butyagin, N. L. *Dokl. Chem. Proc. Acad. Sci. USSR (Engl. Transl.)* **1965,** *165*, 1061.
59. Tabata, M.; Yamakawa, H.; Takahashi, K.; Sohma,J. *Polym. Degrad. Stab.* **1979,** *1*, 57.
60. . Todd, A. *J. Polym. Sci.***1960,** *42*, 223.
61. David, C.; Verhasselt, A.; Geuskens,G. *Polymer* **1968,** *9*, 289.
62. Geuskens,G.; Fuld, D.; David, C. *Makromol. Chem.* **1972,***160*, 347.
63. Giuffre, G. J.; Marquis, J. F.; Pfeiffer, H. C.; Stickel, W. *J. Vac. Sci.Technol.* **1979,** *16(6)*, 1644.

RECEIVED February 20, 1991

Chapter 12

Characterization of Macromolecular Morphology by Pulsed NMR Spectroscopy

A. Charlesby

Silver Spring, Eagle Lane, Watchfield, Swindon, Wiltshire, SN6 8TF England

The pattern of the spin–spin relaxation T_2 derived from NMR measurements can provide detailed information on morphology, mobility, and other physical properties of polymeric and other macromolecular systems. For a low-molecular-weight, flexible polymer, the spin–spin relaxation time T_2 is related directly to molecular weight, temperature, concentration in solvents, and free volume and, in a simple quantitative manner, to viscosity. At radiation doses sufficient to form a partial network, the T_2 relaxation curve then consists of two components T_{2l} and T_{2s} that provide characteristic data on the network and nonnetwork fractions in crosslinking polymers. High-molecular-weight, flexible polymers show a T_2 pattern characteristic of a dynamic network even in the absence of permanent crosslinks, and this pattern is ascribed to the presence of entanglements, whose number and temperature dependence can therefore be derived by NMR. The importance of these dynamic entanglements to such properties as viscosity, modulus, and melting can be outlined quantitatively. T_2 data can also be used to follow curing of polyesters, crosslinking of polymers in aqueous solutions, radiation protection offered by additives, crystallization rates under various conditions, oxidation reactions, and other reactions that may affect molecular morphology and mobility. The NMR technique may also be extended to radiation changes in molecular arrangement in biopolymers. Many physical properties of polymers can be related to the concentration of these entanglements plus crosslinks as deduced from T_2; this information is quite different from that obtained from spin lattice relaxation T_1.

Nuclear magnetic resonance (NMR) has long been established as a very powerful method of chemical analysis also applicable to macromolecular systems. However, its potential use for the determination of the morphology of such systems is far less appreciated. By using pulse techniques, the spin–spin relaxation time T_2 can be measured under a variety of conditions, and whose determination provides a most valuable means

0097–6156/91/0475–0193$07.25/0

of nondestructive analysis of many of the more physical features of such systems, such as molecular weight and mobility, crystallinity or glassy structure, and melting characteristics of a partially crosslinked or entangled network. It is often possible to measure quantitatively the effect of radiation on these and other properties of the polymer and, where these effects are known, to interpret the meaning of the observed NMR changes. Once the meaning of the NMR changes has been elucidated, it becomes possible to use these NMR patterns as a sensitive method of tracing and measuring radiation effects in other systems or, alternatively, to apply the information gathered by the NMR of a series of irradiated specimens to the quantitative assessment of similar properties in a range of unirradiated or chemically modified materials.

The pulsed NMR method involves the examination of the spin–spin lattice relaxation of small specimens, many previously irradiated, and placed in the static magnetic field of the spectrometer under suitable conditions such as temperature or solvent. An electromagnetic pulse at right angles perturbs the nuclear spin system, and its subsequent relaxation can be followed from the induced signal it produces in the surrounding coil. The spin–spin relaxation curve is measured as a function of time, often by using the Carr–Purcell sequence or free induction decay (FID). A typical relaxation decay is in the form $A(t) = A(0) \exp(-t/T_2)$ (Lorentzian) or $A(t) = A(0) \exp(-t^2/T_2^2)$ (Gaussian) or some intermediate. For many polymeric systems, the relaxation curve might consist of the superposition of two or more of these decay sequences, yielding useful information on the relative amounts of different components in the polymeric specimens examined. To distinguish between values of T_2 for different systems, subscripts l (long), s (short), and c (very short; *see* below) have been used. Decay times T_2 are typically of the order of a fraction of a second, a millisecond, and for some a microsecond.

Low-Molecular-Weight Polymers

The simplest system to analyze is a linear, long-chain polymer with a molecular weight of typically 10,000 or less and measured in a homogeneous, flexible state. In this case the T_{2l} relaxation is a typical Lorentzian exponential, with T_{2l} of the order of about 100 ms to several seconds. The parameter T_{2l} is found to depend on molecular weight, M, approximately as follows (Figure 1):

$$T_{2l}^{-2} = B \cdot M_n$$

This simple relation has been verified for a series of polyisobutylene specimens subjected to various radiation doses and for which the molecular weight can therefore be readily calculated from 5200 to 88 (Table 1). For other polymers, polydimethylsilicone and linear molten polyethylene (paraffin), a similar relation has been derived with a slightly different coefficient. It is not known if the difference is significant.

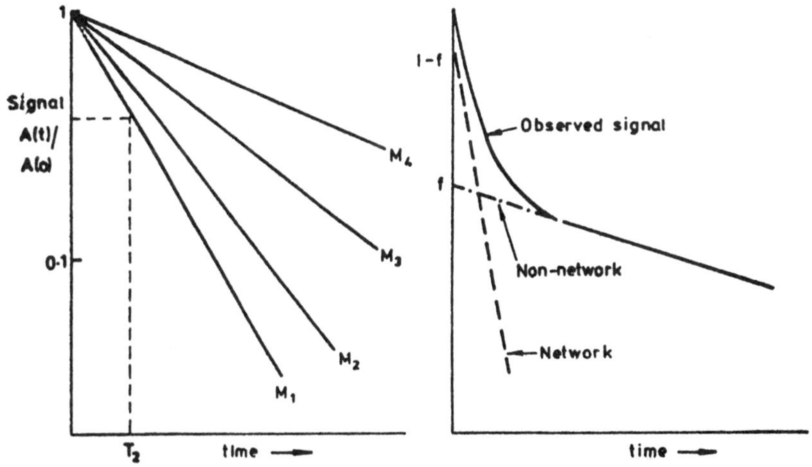

Figure 1. (Left) Exponential decay T_{21} for low-molecular-weight polymers. T_{21} varies with MW. $M_1 > M_2 > M_3 > M_4$; f = 1 (only for non-network. (Right) Superposition of two exponential decays T_{21} and T_{2s}. Total signal is due to their sum. Symbols: — · —, slow component T_{2L}; — — —, fast component T_{2S}; ——, total signal. f = non-network; 1 − f = network. (Reproduced with permission from *Rad. Phys. Chem.* **1985,** *26,* 467. Copyright Pergamon)

Table 1. Values of $T_{21}M^{0.5}$ for Low-Molecular-Weight Irradiated Polyisobutylene (one T_{21} only; 150 °C)

R (kGy)	M_n	$T_{21} \times 10^{-3}$ *(s)*	$T_{21}M^{0.5}$
0	5200	52	3.75
21.6	4603	62	4.21
75	3586	72	4.31
328	1752	96	4.02
780	915	125	3.8
1400	553	158	3.72
1500	505	168	3.77
2900	283	210	3.53
4270	195	270	3.77
9670	88	400	3.75

The dependence of T_{21}^{-2} on molecular weight M is similar to that linking bulk viscosity and M (for a low-molecular-weight polymer) so that T_{21}^{-2} and viscosity appear to measure related quantities, some form of molecular mobility. The variation of the parameter B with temperature shows an approximately similar activation energy as for self-diffusion. If this similarity is verified for a number of other polymers, it would confirm that T_{21} is determined primarily by the mobility of individual molecules within the bulk.

The preceding discussion refers to a pure polymer below a critical molecular weight, as described below. We may now consider the change in T_{21} in the presence of a solvent (Figure 2). This has been studied in the case of low-molecular-weight, monodispersed *cis*-polyisoprene in various concentrations of tetrachloroethylene as solvent. It is concluded that T_{21} depends primarily on the free volume available to it for molecular motion, as well as molecular weight (as described above). This free volume v_f is determined by that of the polymer and also by that provided by the solvent (and therefore dependent on its concentration), as well as by the temperature, which can also increase this free volume. Thus one can summarize the situation:

$$\log T_{21} = \log A' + B''/v_f \text{ or}$$

$$= \log A' + B''/(T - T_\infty)$$

where the temperature T_∞ agrees with that used in describing rheological properties. The parameter A′ is directly related to molecular weight (T_{21} is proportional to $M_n^{-0.5}$).

It may therefore be concluded that this interpretation of measurements of T_{21}, based on a series of quantitative measurements of irradiated low-molecular-weight polymer, can be extended to the understanding of other types of rheological measurements and, because it is a fairly simple

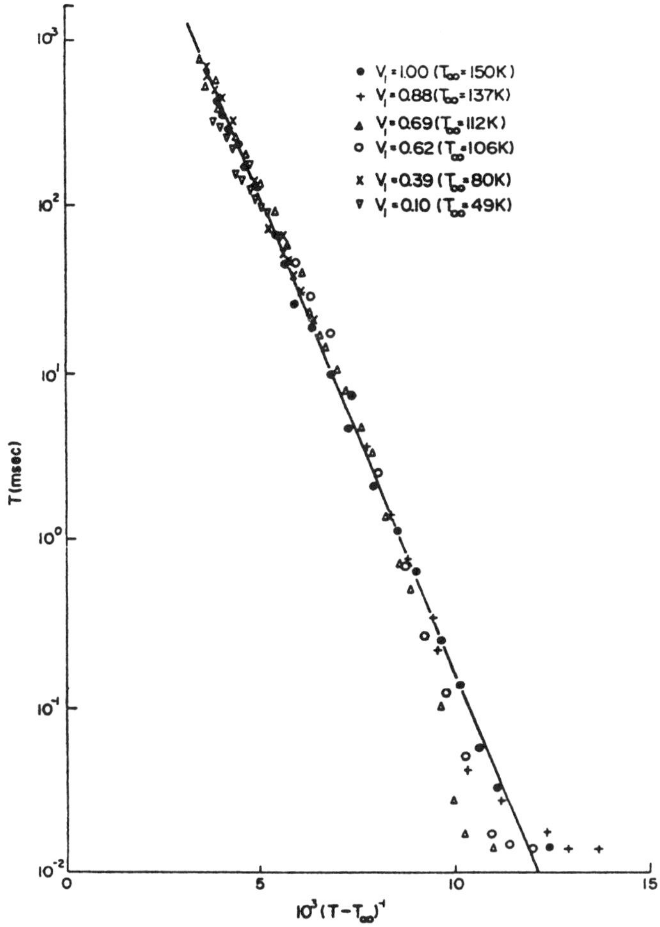

Figure 2. Monodispersed *cis*-polyisoprene. T_{21} depends on free volume due to temperature and solvent (tetrachloroethylene) - polymer fraction v_1 - as well as molecular weight 7200. (Reproduced with permission from *Eur. Polymer J.* **1981,** *17,* 681. Copyright Pergamon)

technique, might indeed be used to evaluate properties in unirradiated polymers. Furthermore it is available for the investigation of many radiation effects in macromolecules. There remain a number of points to elucidate for which more data is required: How far are these promising but still tentative conclusions valid for other polymers and solvents? What is the effect of a more direct interaction between polymer and solvent? How far can the values of B′, B′, v_f, T_∞, etc., be related to the chemical nature of the polymer units? What occurs in more complex systems such as emulsions, etc.?

Low-Molecular-Weight Polymers Crosslinked by Radiation

For many polymers, the effect of crosslinking on their T_{21} values differs greatly from the situation for polymers that scission, and the system goes through several distinct phases as the crosslink density increases. At this stage we consider a flexible polymer molecule.

1. A slight increase in average molecular weight and viscosity, with a corresponding decrease in T_{21}, but the effect is small; random crosslinking of a linear polymer results only in the formation of a limited fraction of star-shaped molecules, whose viscosity is increased marginally.

2. A drastic change at a critical concentration of crosslinks when there is an average of one crosslinked unit per weight average molecule (δ = 1) and an incipient network or gel begins to form.

3. Beyond this *gel* point ($\delta > 1$), the irradiated polymer splits into two very distinct fractions; an insoluble network structure or gel and a residue of free molecules plus a small fraction of star-shaped molecules. This dramatic difference produces a drastic change in polymer properties and is usually traced by lengthy Soxhlet extraction procedures, but it can be followed by NMR measurements of T_2.

Once the network is present, the total spin–spin T_2 relaxation curve consists of the sum of two exponential decays with characteristic times T_{21} and T_{2s}. The complete decay may be written in the form

$$A(t)/A(0) = f \cdot \exp(-t/T_{21}) + (1 - f) \cdot \exp(-t/T_{2s})$$

In the equation, f corresponds to the soluble fraction s, and (1 − f) corresponds to the insoluble, gel fraction, as otherwise determined by solubility measurements. Because the two values of T_2 are usually very different, $T_{2s} << T_{21}$, a simple plot of the decay curve on single log paper allows the sol fraction to be determined immediately by extrapolation of the longer part of the curve to t = 0 (*see* Figure 1, right side).

The new T_{2s} is directly related to the value of M_c, the average molecular weight between adjacent crosslinks, and hence inversely proportional to the crosslink density in the network (but *see* below). The significance of the old T_{21} value in this more complex structure has not been fully investigated, but one would expect it to relate to the mobility of the free molecules (plus stars) caught up within the network, or to those outside it (when the network is fully swollen and cannot accommodate more free molecules). In such systems one might even anticipate several types of T_{21}: from separate molecules trapped within the network, from those free outside this restraining influence, and possibly even from those linked in some way to the surface of the network. In some cases an extra component has been reported. This aspect has not yet been followed up, but it might provide a promising method of evaluating surface adsorption in certain systems.

High-Molecular-Weight Polymers and Networks by Entanglements

The situation with high-molecular-weight, flexible polymers (typically $M_w >> 10^4$) is very different and for a good reason. Even in the absence of any radiation or other sources of crosslinking and with a completely soluble material, such polymers still produce a bifunctional relaxation T_2 curve, just as if the specimen were in fact crosslinked (Figure 3). The explanation is that the molecules are indeed linked together by some form of temporary association, termed *entanglement,* whose lifetime is sufficiently long for it to behave as the equivalent of a crosslink on the time scale of the NMR measurements, though not sufficient for the far longer term extraction procedure. Measurement of such a high-molecular-weight polymer can therefore provide a direct determination of the number of such entanglements and even an estimate of their lifetime. Their temporary presence may even account in part for such physical behavior as creep under stress as they slowly disappear to reform elsewhere.

The appearance of a two-phase relaxation curve in an unirradiated and noncrosslinked polymeric system first occurs at a molecular weight corresponding to an average of one effective entangled unit per weight average molecule, the same condition needed for the incipient formation of a permanent network by crosslinks. Experimental NMR measurements confirm this interpretation. The NMR values of $1 - f$ representing an apparent network fraction for an irradiated polymer of high molecular weight follow those for a similar crosslinked low-molecular-weight polymer but displaced by a virtual dose equivalent to the density of entangled units. Thus T_{2s} is related to M_t where $1/M_t = 1/M_c + 1/M_e$. Here M_t represents the effective total due to true, permanent crosslinks c plus those due to entanglements e. The transition from "low" to "high" molecular weight would occur at the weight average molecular weight corresponding to an average of one entangled unit per molecule.

Some interesting results emerge from these investigations. One is that the density of entanglements Ne per gram of polymer is independent of molecular weight (to within the accuracy of these measurements). For a low-molecular-weight-polymer with A/M molecules per gram (A = Avogadro) and with δ_e = Ne ÷ A/M entangled units per weight average molecule

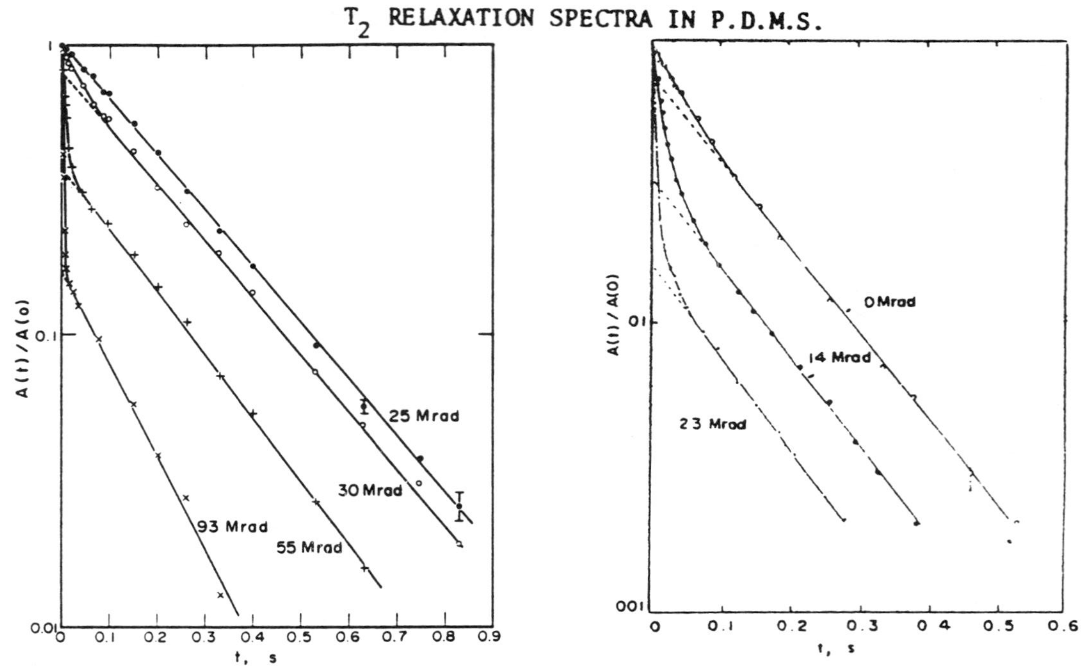

Figure 3. Experimental decay curves. (Left) Low-molecular-weight PDMS (6000); no network and no T_{2s} up to 25 Mrad; then increasing network fraction with dose (about 85% at 93 Mrad). (Right) High-molecular-weight PDMS (26,000); network already present at 0 dose (r_e) due to entanglements; further increased by permanent crosslinks due to dose. (Reproduced with permission from *Rad. Phys. Chem.* **1976,** *8,* 557; **1977,** *10,* 62. Copyright Pergamon)

less than unity, the required minimum of one per molecule for network formation is not met. For polymers of high molecular weights, δ_e = Ne ÷ A/M $\geq$ 1; thus a temporary dynamic network can be formed. The critical parameters are the number of entanglements of adequate lifetime and the average molecular weight. The transition might also be related to the change in viscosity from M^1 to $M^{3.4}$ (Figure 4).

Unlike crosslinks produced chemically or by irradiation, these entanglements are temperature sensitive, and their density decreases with rising temperature. This is readily demonstrated by comparing the NMR of a low-molecular-weight polymer irradiated to give the same NMR pattern as a high-molecular-weight polymer of identical chemical structure but only held together by its entanglements (Figure 5). As the temperature is raised, there is little change in the NMR of the low-molecular-weight specimen, whereas for that from the high-molecular-weight, purely entangled polymer, the NMR pattern changes with an increase in the f value as the nonnetwork fraction increases until at a certain temperature f = 1 (Figure 6). At this point the number of entangled units equals the number of molecules. It is therefore possible to determine the number of effective entanglements per gram and their variation with temperature.

Many mechanical properties of flexible polymers depend directly on the density of these entanglements and the formation of even a temporary network; an example might be cold flow and the variation of strength with molecular weight, temperature, and time. An example is uncrosslinked polystyrene of various molecular weights (Figure 7). Below the glass temperature the modulus is determined by stiff chains (T_{2c}) but as the temperature is raised, molecules acquire adequate mobility, and for low-molecular-weight polystyrene, the polymer goes directly from glass to liquid (no flexible network is formed because there is less than one entangled unit per molecule at T_g). For a polymer of much higher molecular weight, a plateau is reached almost independently of molecular weight, corresponding approximately to M_e = A /Ne. As the temperature is further raised and Ne decreases, there is a fall in T_{2s} until a point is reached when M_e becomes comparable to M_w and the temporary network falls rapidly as does the modulus until there is no network left and hence no elastic modulus. For a polymer of higher molecular weight, a higher temperature is required to reach this low density per molecule, as is indeed observed. This pattern of behavior, determined directly from the NMR measurements of T_2, agrees quantitatively with that deduced from mechanical measurements of modulus of a range of polystyrenes of different molecular weights, measured over a range of temperatures. The major difference between the two sets of measurements, mechanical and NMR T_{21}, are the temperatures of failure due of course to the different time scales involved, 20 min and less than 1 s, leaving far longer for disentanglement in the mechanical tests.

High-Molecular-Weight Polymers That Scission on Irradiation

The transition from a low-molecular-weight polymer with only a T_{21}

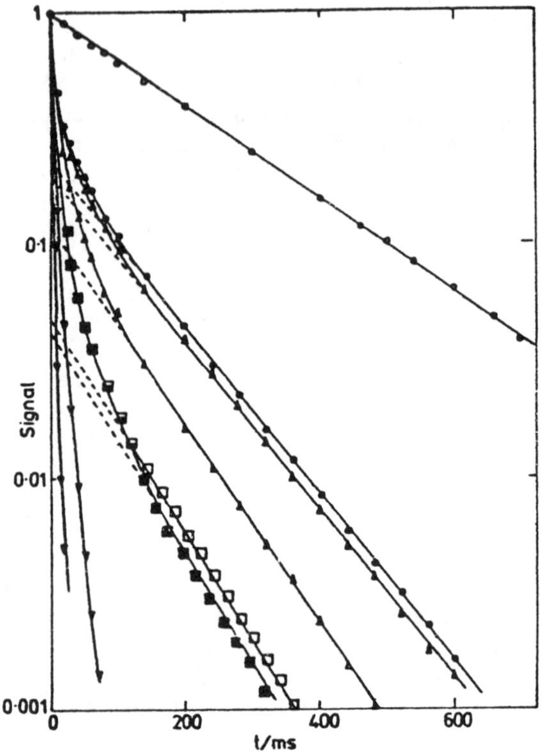

Figure 4. Polyethylene at 153 °C. ○, Low-molecular-weight M_w = 7000; only a non-entangled component T_{2l}; ∇, polyethylene fraction M_w = 53,000, high-molecular weight T_{2s}; ▼, crosslinked network only T_{2s}. Other curves for commercial products with entangled T_{2s} and non-entanlged T_{2l}: ●, Rigidex 50; Δ, Marlex 6009; ▲, Alkathene 403; □, Alkathene 334; and ■, DFDS 6600. (Reproduced with permission from *J. Polymer Sci. (Polymer Lett.)* **1978,** *16,* 340. Copyright Wiley)

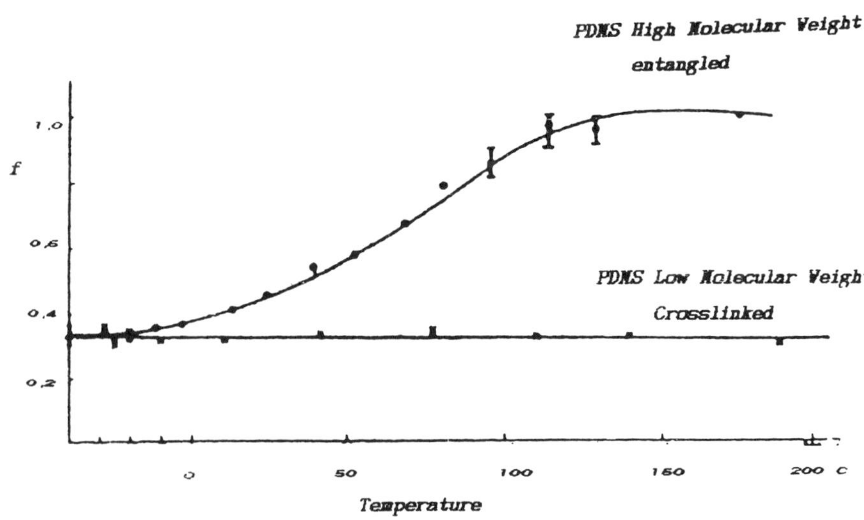

Figure 5. Temperature effect on entanglements but not on crosslinks (35% at room temperature). Low-molecular-weight, crosslinked PDMS stays 35% as temperature rises. High-molecular-weight sample loses effective entanglements and becomes a fluid at about 140 °C. (Reproduced with permission from *Proceedings of the Sixth Radiation Research Congress,* **1979,** 345. Copyright Japanese Association for Radiation Research)

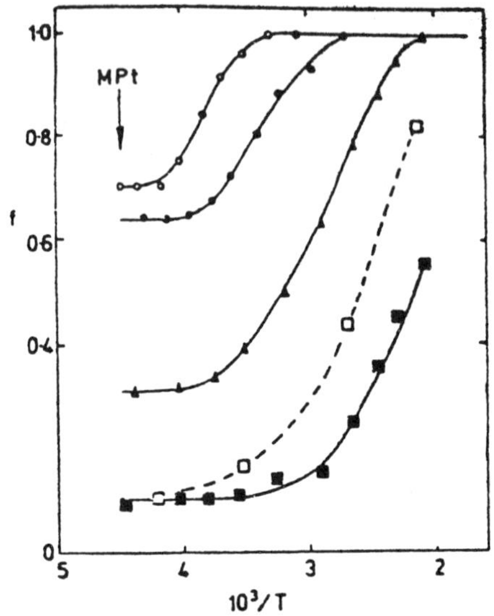

Figure 6. f represents the sol nonnetwork fraction and corresponds to molecules that are not entangled into the network. As the temperature is raised, the density of effective entanglements decreases. This temperature effect applies to entanglements but not to crosslinks. For low-molecular-weight PDMS, the entanglement density sinks to unity, one per weight average molecule (f = 100%) at 300K, whereas for a much higher molecular weight PDMS, a temperature above 500K may be needed. This type of plot can be used to determine the temperature dependence or lifetime of these entanglements and hence the rheological behavior. Note the effect of filler particles. Symbols: ■, filled PDMS; □, unfilled PDMS; ▲, ●, and ○, unfilled PDMS of decreasing molecular weight. (Reproduced with permission from *J. Polymer Sci. (Phys. Lett.)* **1978,** *16,* 1050. Copyright Wiley)

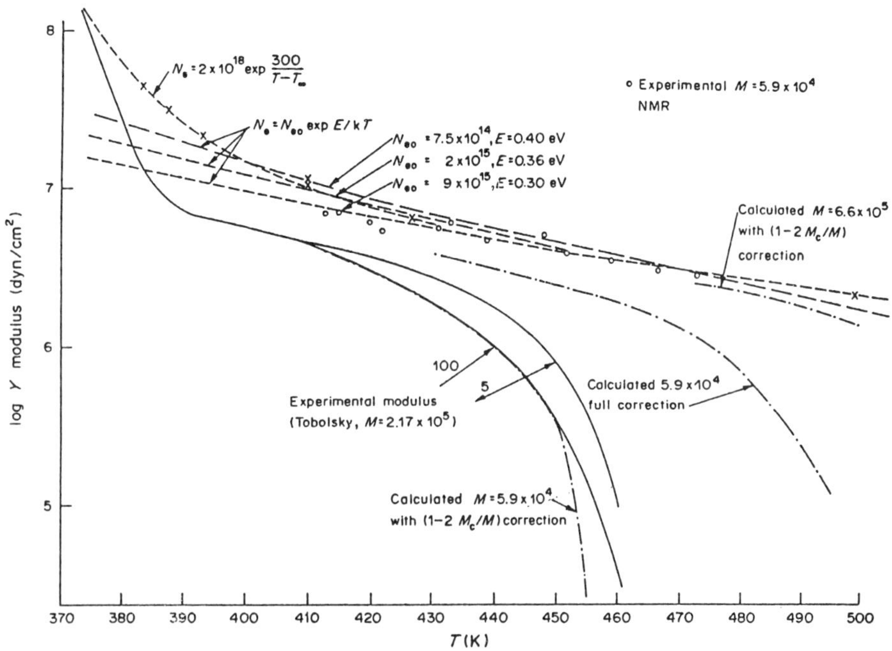

Figure 7. Comparison of elastic modulus of polystyrene from rheology (Tobolsky, A. V. *Properties and Structures of Polymers,* Wiley, 1960 pp 72–75) and from pulsed NMR (Charlesby and Jaroskiewicz, *Europ. Polym. J.* **1985,** *21*(1), 55). Both analyses lead to similar values of E above T_g (log E = 6.6) until the entanglement density per molecule does not greatly exceed unity when end corrections become very important and the polymer eventually becomes a viscous liquid. The temperature at which this occurs depends on the molecular weight of polystyrene and the time available for an entanglement to disengage, which is much longer in rheological measurements. Symbols: ○, individual values deduced from f vs $10^3/T$ curve (NMR, experimental); -----, calculated from N_c values, not corrected for end-effects; ——, rheology, experimental; — · —, calculated from N_c values with correction for molecular weight. (Reproduced with permission from *Eur. Polymer J.* **1985,** *21,* 62. Copyright Pergamon)

component (no network due to crosslinks or entanglements) to a higher molecular weight polymer of identical chemical structure but with a T_{2s} network can be demonstrated very clearly by analyzing the opposite transition, by using a noncrosslinked high-molecular-weight polymer (e.g., polyisobutylene) with a T_{2s} spectrum corresponding to a network formed by entanglements. On irradiation this polymer scissions until its molecular weight sinks to the level at which $\delta < 1$, when only a T_{21} NMR relaxation curve remains. The intermediate stages in which both T_{21} and T_{2s} are present can then be measured and compared with those calculated from the doses. The agreement is excellent and confirms the accuracy of the explanations presented here.

There remains one apparent discrepancy that requires an explanation. This has been found and leads to a somewhat different approach to a widely accepted equation for the elastic modulus of a partly crosslinked elastic network.

The modulus E of an elastic network is usually expressed as $E = 3\rho RT/M_c$ if end effects are ignored, with M_c, the average molecular weight of a segment between adjacent crosslinks, being inversely proportional to the crosslink density (or to radiation dose with appropriate corrections). This is also observed with pulsed NMR of crosslinking polymers, because T_{2s} is proportional to M_s and varies inversely as dose. However, the NMR results for polyisobutylene do not follow this pattern (Figure 8). With higher doses and increasing scission between crosslinks, the average value of M_c of the remaining segments must diminish, simply because the longer segments are more likely to suffer scission and the average of the remainder M_c must diminish, as should T_{21}. In fact the opposite is observed. The network fraction diminishes, but the value of T_{21} increases, as would correspond to an *increase* in M_c. Similar results are obtained for the elastic modulus; this decreases with increased random scissions, although M_c has decreased. The explanation must be that modulus and also $1/T_{2c}$ depend not on the inverse length or mass of the segments between crosslinks, as given in the formula, but on the number of such segments, large or small, per unit volume. Scission reduces their number, and E and $1/T_{2s}$ both decrease.

This explanation in terms of number of segments rather than their size may also help to explain reinforcement with certain fillers; their presence within the network could increase the number of effective segments.

Partially Crystalline or Glassy Polymers

So far we have summarized some of the T_2 NMR data for a flexible polymer, whether free, entangled or crosslinked. One can now outline a few results for a partially crystalline polymer, in this case polyethylene.

At temperatures below the crystalline melting point, the NMR spin–spin relaxation shows two components. One of these would correspond to T_{2s}, corresponding to the length of mobile chains between crosslinks but, in this case, most probably the length of chains caught between adjacent crystalline regions.

The other component, of far shorter time (~10 μs) and denoted by T_{2c}, follows a Gaussian pattern exp $\{-t^2/T_{2c}{}^2\}$. This may best be ascribed to motion of extremely short lengths within the very restricted crystalline region. The separation of the complete decay spectrum into these two components can therefore measure not only the degree of crystallinity but also the lengths of the more flexible chains in the amorphous regions serving to link the crystals together. Indeed, this NMR technique has been used to follow the rate of crystallization of polyethylene at various temperatures (Figure 9). The results obtained agree with those obtained by another method (volume change). Because of the inherent advantages of the NMR technique, this rate can be followed over a much wider time scale, and proposed mechanisms can be compared. The effect of radiation on this behavior can also be investigated. How far does crosslinking interfere with crystallization reactions?

As the temperature is raised, both T_{2c} and T_{2s} increase because of thermal expansion and more free volume. The degree of crystallinity can be followed from the ratio of these components up to close to the melting point, when T_{2c} rapidly disappears. At this point, T_{2s} also increases very rapidly, from about 0.1 to 10 ms, and when all crystals have melted, the increase proceeds far more slowly. T_{2s} represents at first the segment length of the amorphous chain, linked at each end into crystallites. As these melt, the length of such segments increases until all the crystals have melted, and the segment length is only limited at its ends by the entangled (or crosslinked) units. Further temperature rises only reduce the effective number of these links slowly; corresponding to this, there is a slower further increase in T_{2s}.

In addition to the T_{2s} component of about 10 ms well above the melting point, two new components now appear: a long component of about 50 ms, probably T_{2l} due to free low-molecular-weight molecules, and a new component T_{2x} of about 2 ms whose origin is as yet unexplained. Is it some form of structure in the liquid state?

The importance of morphology during irradiation has also been examined. Different T_2 patterns are obtained from quenched or slowly cooled polymer. Perhaps more surprising is that different patterns are obtained for specimens of identical molten polyethylene depending on their previous degree of crystallinity. Some form of memory is retained in the liquid, dependent on its previous morphology prior to melting.

This NMR technique of analysis can therefore provide quantitative determination of important features of the melting process. It might also be expected to tell us about the occurrence of radiation-induced crosslinking or scission in crystalline and amorphous regions. An example is PTFE (polytetrafluoroethylene), which scissions on irradiation with a small increase in the degree of crystallinity; this is greatly enhanced on remelting, because shorter molecules are less likely to become linked into several crystals. The great reduction in these link molecules in the amorphous fraction completely changes the cohesion of such specimens.

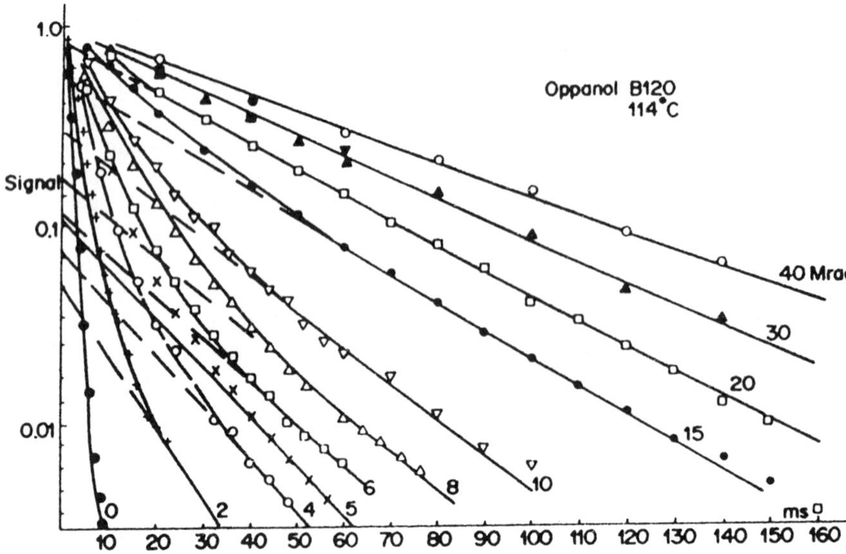

Figure 8. Scission of high-molecular-weight polyisobutylene leads to its transformation from an entangled network, to a partial network and free molecules, and then to free molecules only (T_{2s} to T_{21}). (Reproduced with permission from *Rad. Phys. Chem.* **1982,** *19,* 157; **1985,** *26,* 469. Copyright Pergamon)

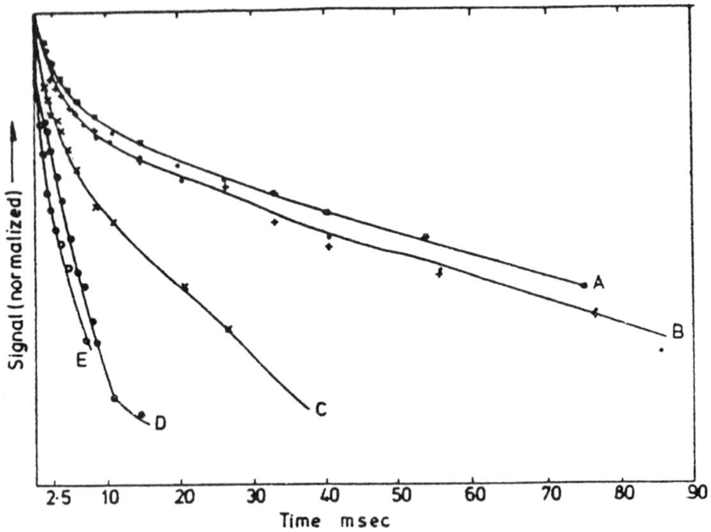

Figure 9. Polyethylene: effect of rate of cooling and of radiation at 47 Mrad. A, Marlex as received (0 Mrad); B, Marlex slow cooled, reheated, or quenched (0 Mrad); C, Marlex slow cooled (45 Mrad); D, Marlex ice quenched (45 Mrad); E, Marlex as received (47 Mrad). (Reproduced with permission from *Rad. Phys. Chem.* **1979,** *14,* 929. Copyright Pergamon)

Determination of T_2 for the polymer irradiated in air reveals a long component due to an increased low-molecular-weight component owing to oxidation (Figure 10). It may be possible to evaluate the length of these shortened chains by means of their T_{21} relaxation curves.

Mechanical Behavior Due to Network Formation by Entanglements

Many mechanical properties of flexible polymers depend directly on the density of these dynamic entanglements and the formation of even a temporary network. Their lifetime and density would influence such properties as cold flow and the variation of strength with molecular weight, temperature, and time. An example is uncrosslinked polystyrene of various molecular weights. Below the glass temperature, the modulus is determined by stiff chains (T_{2c}), but as the temperature is raised, molecules acquire adequate mobility, and the possible formation of a dynamic network due to entanglements can play a very significant role. For low-molecular-weight polystyrene with less than one entangled unit per weight average molecule at T_g ($\delta < 1$), no flexible network can be formed, and the polymer goes directly from glass to liquid. At higher molecular weights, δ can exceed 1, and the polymer can form a dynamic network with a plateau in its elastic modulus E, which is almost independent of molecular weight; $E = 3\rho RT/M_e$ and $M_e = A/Ne$ approximately.

As the temperature is further raised, Ne decreases more rapidly than rising T; there is a net fall in T_{2s} until a point is reached when M_e becomes comparable to M_w (δ approaches 1), and the temporary network fraction falls rapidly as does the modulus until there is no network left ($\delta < 1$)and hence no elastic modulus. For a polymer of higher molecular weight, a higher temperature is required to reach this low density per molecule, as is indeed observed. This pattern of behavior, determined directly from the NMR measurements of T_2, agrees quantitatively with that deduced from mechanical measurements of modulus for a range of polystyrenes of different molecular weights, measured over a range of temperatures. The major difference between the two sets of measurements, mechanical and NMR T_2, are the temperatures at which the networks disappear. This is of course due to the different time scales involved, 90 min and less than 1 s, leaving far longer time available for disentanglement in the mechanical tests.

One can visualize the same explanation for polymer behavior under stress: In some regions, many entanglements of finite life may disappear; part of the individual polymer molecule under stress can then flow until equivalent entanglements reform elsewhere to restrict flow for a further limited period. When the external stress is removed, recovery to a random configuration can be retarded by these new entanglements.

Reinforcement by Fillers

In many polymer applications, mechanical properties can be greatly

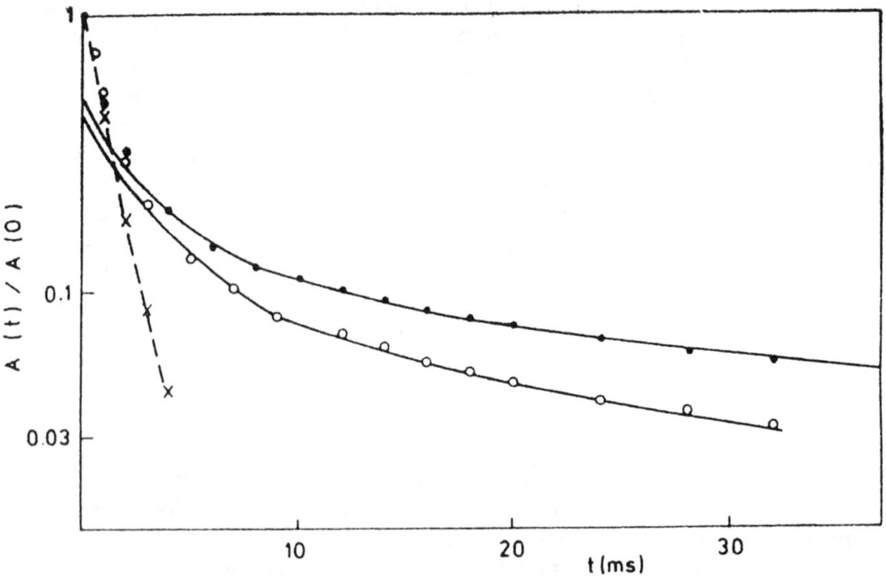

Figure 10. Polyethylene: irradiation in air at 150 °C. Radiation doses of 1 Mrad (○) and 10 Mrad (●) give T_{21} component due to shorter length products. × = 0 Mrad. (Reproduced with permission from *Rad. Phys. Chem.* **1978,** *11,* 87. Copyright Pergamon)

enhanced by the incorporation of filler particles such as fine silica or carbon black. The question then arises as to the nature of the interaction. Is there a chemical link, or does the particle divert the direction of the stress? Preliminary measurements on a range of such samples seem to indicate that no network structure is set up, and the T_2 relaxation time is merely reduced, as would be expected from an increased mass or molecular weight. This technique requires more detailed attention.

The NMR behavior of trifunctional (T) and tetrafunctional (H) crosslinking systems has been found to be very different, even when the molecular weights between successive links are similar. The role of plasticizer and enhanced crosslinking additives may also be investigated, because they may reduce temporarily or permanently the values of M_c. The number of such bridges and their permanence have been ascertained and related to such factors as concentration of components. Other factors to be considered are dose, dose rate, temperature, and solvent or diluent.

Radiation of Polymers in Solution

Very interesting results have been obtained from the irradiation of aqueous solutions of certain polymers, where a swollen crosslinked structure may be obtained by the direct absorption of radiation by the polymer, and indirectly by the radiation products of the solvent. The crosslinking reaction and the formation of a swollen gel network is greatly enhanced at lower concentration. By using heavy water, the sequences involved in these reactions can be followed by NMR, and the importance of concentration, dose rate, and reactions within particles can be determined. The transition from a macrogel to a microgel structure as the initial concentration is reduced can be examined. The effect of additives in such systems can also be ascertained. For example, polyethylene oxide in D_2O has been studied at concentrations of 0.2–10% (Figure 11). In the presence of 0.05% thiourea, there is a 10-fold protection, that is, the dose required to produce a given NMR change is increased by this factor. The ability to measure such changes accurately and at relatively low doses for hydrophilic polymers with extremely low concentration of additives or at different pH, etc., opens up the possibility of applying the technique in active biological systems, where very high molecular weight systems are irradiated in an aqueous environment and in the presence of additives and where configuration may be of great importance

Polymerization and Grafting

Comparatively little work has been published on the changes in T_2 during polymerization, when one might expect to see evidence of the presence of at least two components. Such changes in the T_2 relaxation have been used to follow the course of polymerization of methyl methacrylate and methacrylic acid. Here three different T_2 values are involved: a relatively long Lorentzian decay (L1) with T_2 of about 1 ms, a very short, Gaussian

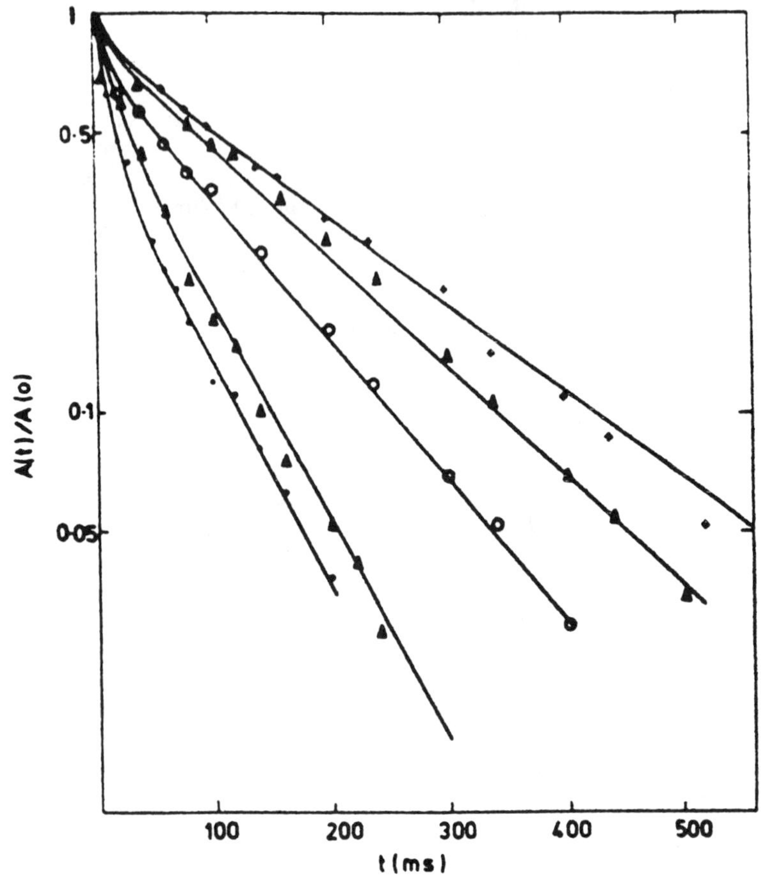

Figure 11. Irradiation of a 4.5% water solution of polyethylene oxide (MW = 340,000; M_n = 51,000) in the presence of 0.05% thiourea acting as a radiation protector. The increased slope at 0, 25, 38, and 130 kGy shows increased crosslinking and eventually partial formation of a crosslinked network. In the absence of thiourea, approximately the same effect results from a one-tenth smaller dose (12 kGy). Thus on a weight basis thiourea is one thousand times as effective. Symbols: +, 0 Mrad with thiourea; Δ, 2.5 Mrad with thiourea; ○, 3.8 Mrad with thiourea; ●, 13 Mrad with thiourea; ▲, 1.2 Mrad without thiourea. (Reproduced with permission from *Rad. Phys. Chem.* **1980,** *15,* 402. Copyright Pergamon)

decay (G1) with T_2 about 20 μs, and an intermediate Lorentzian (L2) of about 100 μs. When only the monomer is present, the long component (L1) is seen. As polymerization proceeds, the shortest component (G1) appears, and at a later period, the intermediate component (L2) also appears. At the same, time the longest T_2 shows a drastic drop by a factor of two or three.

The fraction of the total signal due to the shortest component (G1) correlates closely with the polymer yield as determined by a direct method and could be ascribed to a glassy polymer. It can reach 80–90%. The fraction due to the monomer and later possibly growing low-molecular-weight polymer (L1) decreases to about 5%. The meaning of the intermediate component (L2) is unknown, but after an initial rise, it remains steady at about 20%. The interpretation of these components (L1 and G1) agrees with that previously derived for low- and high-molecular-weight polymers, but L2 seems to be another example of the extra component already seen in a viscous liquid, as in molten polyethylene. One wonders whether it represents an additional and unsuspected component of polymer under these conditions or an artifact of the technique of analysis by NMR. The fact that it is only observed under certain polymer conditions inclines one to the former view.

The technique may also be useful in following the curing of unsaturated polyester resins or other chain reactions. In a series of polyester–styrene mixtures subjected to small radiation doses, two T_2 relaxations were observed. The longer T_{2l} could be correlated with the soluble fraction. Although a T_{2s} component was observed in a system with no gel fraction after extraction, this could well be due to some form of shorter term interaction. Up to 3 kGy no gel point was observed, whereas at 6 kGy there were two components over a limited temperature interval and a single one at a higher temperature. At 9 kGy two T_2 components were found. All this behavior is similar to that due to temperature-dependent entanglements near the gel point.

One might also like to see this method used to follow radiation-induced grafting reactions. How far can one separate the T_2 relaxation of the main chain and of the chemically different side chain?

Conclusions

At this stage it would appear that the analysis of the NMR T_2 pattern can provide a most valuable method of studying aspects of macromolecules more closely related to conformation than to detailed chemical structure. This fact is of great importance because many of the most interesting properties of such systems derive from their overall molecular morphology and mobility rather than the detailed chemical structure of the units. Even the hydrophobic properties of an umbrella depend primarily on its shape.

Only a small part of this field has been explored in any detail, and this relates primarily to the formation and destruction of network systems, either by chemical links or by what are here termed entanglements. The precise nature of these entanglements is far from clear, and any appropriate form of association between adjacent molecules may be envisaged, which leads to the formation of a network under the same quantitative conditions as are required for permanent crosslinks. However many of their basic properties are readily derived by this NMR method. They behave as temporary links between adjacent chains; their number is independent of molecular weight over the range of polymers already studied; their number appears to vary in a well-defined manner with temperature and free volume, either because their number truly changes or, more likely, their lifetime decreases at higher temperatures so that even with the same geometrical arrangement fewer are observable in the time limit set by the equipment. These dynamic entanglements may perhaps act as the precursors to solidification into a crystal or glass.

As presented here, these entanglements are related to the value of M_e or by analogy to M_c. It may be that a wider relation with the whole network will be required. In a solution with more free volume and greater chain mobility, entanglements would become very short lived and their average number would be insufficient for network formation. Only T_{21} relaxation would then be expected, except at low temperatures. Indeed one can determine the temperature dependence of the lifetime of these entanglements. Their numbers in swollen polymeric matrices would also be of interest.

The apparent existence of some form of structure in the liquid state as indicated by some unexpected new patterns of relaxation could be of great interest. It would agree with the persistence of some form of memory in liquid polymer. Many of the most valuable applications of radiation are concerned with macromolecules such as polymers and biopolymers, and modification of their arrangement can have the most profound effect even if the number of purely chemical reactions is minute. Even in a simple polymer with a molecular weight of 10^6, a one-bond change per molecule at an unspecified site can completely alter its behavior; biological systems can be far more sensitive. The technique outlined in this summary appears to provide an excellent technique, as yet almost unexplored, of investigating such morphological rather than chemical changes in polymeric and biological macromolecules.

The influence of minute amounts of certain additives can have a great influence in irradiated systems, and the resultant changes in morphology can be readily revealed by pulsed NMR T_2. One would be glad to see this method examined for possible use in molecular and morphological studies in biophysics, where many important changes may arise from such modifications despite the minuteness of the chemical changes involved. Close analogies exist between the T_2 values and rheological data. Because the interpretation of much of this NMR data is clear, its introduction into physical, and even mechanical, studies may be beneficial.

Acknowledgment

Much of the earlier work described in this chapter was carried out in my old laboratories at the Royal Military College of Science, Shrivenham, UK. Because of my retirement on grounds of calendar age, my work there has ceased, leaving unsolved a number of intriguing questions. I have outlined a few of these questions in this chapter in the hope that scientists in other laboratories will wish to meet this challenge. Needless to say, I would be glad to collaborate in such investigations.

References

General

1. Charlesby, A. Use of pulsed NMR in measurement of radiation effects in polymer. *Rad. Phys. Chem.* **1979,** *14,* 919.
2. Charlesby, A.; Folland, R. Use of pulsed NMR to follow radiation effects in long-chain polymers. *Rad. Phys. Chem.* **1980,** *15,* 393.
3. Charlesby, A.; Folland, R. Pulsed NMR technique for studying radiation effect in macromolecules. *Sixth Int. Congress Rad. Res.* (Tokyo), **1979,** 336.
4. Charlesby, A. Characterisation of polymers using pulsed NMR. *Tihany 5th Symposium.* Akad. Kiado Budapest, 1989, p 843.
5. Charlesby, A. Use of pulsed NMR to study radiation effects. *Z. Fl. Mitt.* **1984,** *97,* 24.
6. Charlesby, A. Radiation effects in macromolecules; their determination with pulsed NMR. *Rad. Phys. Chem.* **1985,** *26*(5), 463.
7. Charlesby, A. Use of pulsed NMR to determine morphology of macromolecules. *J. Radioanal. Nucl. Chem.* **1986,** *101*(2), 401.

Polyethylene

1. McCall, D. W.; Douglas, D. C.; Anderson, E. W. *J. Polym. Sci.* **1962,** *59,* 301.
2. Charlesby, A.; Kafer, P.; Folland, R. Very high molecular weight polyethylene. *Rad. Phys. Chem.* **1978,** *11,* 83.
3. Bridges, B. J.; Charlesby, A.; Folland, R. Crystallization kinetics from melt and solution. *Proc. Roy. Soc. Lond.* **1979,** *A367,* 343.
4. Folland, R.; Charlesby, A. Effect of previous history on NMR relaxation of polyethylene. *Eur. Polym. J.* **1979,** *15,* 953.
5. Folland, R.; Charlesby, A. Entanglement effects on NMR spin–spin relaxation. *J. Polym. Sci.* **1978,** *15,* 339.

6. Kamel, I.; Charlesby, A. NMR relaxation in solid and molten polyethylene structures. *J. Polym. Sci.* **1981,** *19,* 803.

Polystyrene

1. Charlesby, A.; Jaroskiewicz, E. M. Entanglement and network formation in polystyrene: viscoelastic behavior from pulsed NMR. *Eur. Polym. J.* **1985,** *21,* 55.

Polydimethyl siloxane (PDMS)

1. Folland, R.; Charlesby, A. Radiation-induced crosslinking and gel formation in linear PDMS. *Rad. Phys. Chem.* **1976,** *8,* 555.
2. Folland, R.; Charlesby, A. Crosslinking and entanglements in high-molecular-weight linear PDMS. *Rad. Phys. Chem.* **1977,** *10,* 61.
3. Folland, R.; Steven, J. S.; Charlesby, A. *J. Polym. Sci.* (Pub. Phys. Edn.), **1978,** *16,* 1041.
4. Charlesby, A.; Folland, R. Pulsed NMR studies of irradiated PDMS. *Tihany 4th Symposium on Radiation Chemistry.* Akad. Kiado Budapest, 1977, p 335.
5. Charlesby, A.; Folland, R.; Steven, J. S. Analysis of crosslinked and entangled polymer networks. *Proc. Roy. Soc. Lond.* **1977,** *A335,* 189.

Polyisoprene

1. Folland, R.; Charlesby, A. Pulsed NMR of *cis*-polyisoprene 1. *Polymer* **1979,** *20,* 207.
2. Folland, R.; Charlesby, A. Pulsed NMR of *cis*-polyisoprene 2. *Polymer* **1979,** *20,* 211.
3. Charlesby, A.; Bridges, B. J. Pulsed NMR of *cis*-polyisoprene solution T_3 and T_2 relaxation, free volume, viscosity relationships. *Eur. Polym. J.* **1981,** *17,* 645.

Polyisobutylene (PiB)

1. Charlesby, A., Steven, J. H. Molecular weight determination of irradiated PiB by NMR. *Rad. Phys. Chem.* **1976,** *8,* 719.
2. Charlesby, A.; Bridges, B. J. Entanglement density in irridiated PiB. *Rad. Phys. Chem.* **1976,** *19,* 497.

Polymerization

1. Kurotu, T. Pulsed NMR investigation on polymerization I: methyl methacrylate. *Polym. J.* (Japan) **1986,** *18*(11), 859.

2. Kurotu, T. Pulsed NMR investigation on polymerization II: bulk polymerization of methacrylic acid. *Polym. J.* (Japan), **1987,** *19*(2), 285.

3. Kurotu, T. et al. Pulsed NMR study of bulk polymerization of methacrylic acid. *Memoirs of National Defense Academy* **1988,** *28*(1), 53.

4. Kurotu, T. et al. *Polym. J.* (Japan), **1982,** *14,* 149.

Polyester Curing

1. Andreis, M.; Veksli, Z.; Ranogajec, F.; Hedvig, P. Radiation-induced crosslinking in polyester–styrene systems. *Polymer* **1989,** *30,* 1498.

RECEIVED May 3, 1991

Technological Applications of Radiation to Polymers

ADVANCES IN THE EFFECTIVE USE OF AN ELECTRON BEAM (E-beam) to graft substances onto polymeric surfaces and for bulk cross-linking are exemplified in reports on polyethylene and polystyrene, respectively. A promising application of E-beams to directly form polymeric microspheres is also described. The use of E-beam and X-ray technologies in conjunction with contemporary chemical knowledge is addressed in chapters on X-ray and E-beam high-sensitivity applications. Coupling chemical concepts with radiation effects in polymeric substrates is illustrated in two chapters on controlled reactions on tailored chains and on thermal marking of amorphous substrates, respectively. Among the non-traditional applications of electromagnetic radiation is the growing use of microwave heating to effect chemical reactions, including direct formation of high-performance fiber-reinforced composites; two chapters address this topic. Finally, in the important area of health products, Chapters 17 and 18 describe the applications of radiation in the production of hydrogels and in the development of radiostable absorbable polyesters for use as the first radiation-sterilizable absorbable systems, respectively.

Chapter 13

Electron-Beam Irradiation of Polymerizable Polyaphrons

A Method for Making Polymeric Microspheres

Dale C. McHerron and Garth L. Wilkes

Polymer Materials and Interfaces Laboratory, Department of Chemical Engineering, Virginia Polytechnic Institute and State University, Blacksburg, VA 24061

This study illustrates a new approach to the formation of cross-linked polymeric microspheres by incorporating trimethylol-propane propoxylate triacrylate (a radiation sensitve monomer) into a polyaphron system and exposing it to electron beam radiation. Polyaphrons are essentially a collection of droplets encapsulated in a very thin aqueous film dispersed in an aqueous matrix. This film gives extreme stability to the system which allows for extremely high dispersed phase concentrations. The formation of polyaphrons requires the presence of a surfactant in both the dispersed (monomer) and continuous (aqueous) phases. The effects of monomer phase concentration, monomer and aqueous phase surfactant concentration, aqueous phase surfactant type, and preparation temperature were studied to gain an understanding of the necessary conditions for the formation of polymeric microspheres with a narrow size distribution. Polymeric microspheres ranging from 2 to 40 microns in diameter were formed with monomer concentrations as high as 67 volume percent. without any aphron bridging occuring or need for agitation to keep the monomer phase dispersed in the aqueous matrix.

Polymeric microspheres have been produced by a variety of different techniques. Emulsion polymerization of vinyl polymers is probably the most well known process (*1*). More recently, several investigators have described the formation of polymeric microspheres using radiation processes. Yoshida et al. (*2*) have reported on the formation of polymeric microspheres by gamma irradiation of a homogenous solution of diethylene glycol dimethacrylate in methyl orthoformate. Rembaum et al. (*3*) also demonstrated the formation of microspheres by a gamma radiation emulsion polymerization technique.

The investigation reported in this paper illustrates the formation of cross-linked polymeric microspheres utilizing a polyaphron system and electron

0097–6156/91/0475–0220$06.00/0

beam radiation. This method posesses several advantages over past or current techniques that have been reported. The polyaphron system facilitates formation of microspheres in an aqueous medium with monomer concentrations as high as 67 volume percent. In addition, electron beam radiation allows for reaction times on the order of seconds, as opposed to gamma radiation, where reaction times are on the order of hours. Hence, electron beam radiation is well suited for continuous web line processing, which would be easily adapted to this application for polymeric microsphere production.

At the onset of this work, the formation of polymeric particulates was attempted using trimethylolpropane propoxylate triacrylate (a radiation curable monomer) in an emulsion system coupled with electron beam radiation. This scheme posed numerous processing difficulties and the resulting product was quite disperse in size. In order to circumvent these problems, an attempt was made to incorporate the monomer into a polyaphron system, which, as will be shown in this paper, is a successful approach.

Polyaphrons were discovered by Sebba, who also developed all of the current theory for these systems and coined the term "aphron" from the Greek word for foam (*4*). In addition, he invented a continuous polyaphron generator, which may make it possible to design a continuous process for the formation of solid polymeric microspheres, although this scheme has yet to be reduced to practice. Sebba's work deals with strictly liquid systems, although he does mention the possibility of making polymerizable aphrons, which is the focus of this work.

According to Sebba, an aphron is a phase that is encapsulated in a thin soapy film (*5*). Polyaphrons (a collection of a large number of aphrons) are similar to foams in that they consist of a continuous aqueous phase with a second phase (of greater volume) dispersed throughout the aqueous matrix. Instead of a gas being dispersed, as with a foam, an immiscible liquid constitutes the dispersed phase. Each droplet of this immiscible liquid is encapsulated in a thin soapy film which imparts extremely high stability to the system. This stability allows for the existence of high concentrations of the dispersed phase with no coalescence or need for agitation. Obviously, aphrons are structurally very different from emulsion droplets. Emulsion systems consist of two phases with an electrical double layer at the phase interface, which stabilizes the droplets. On the other hand, aphron systems contain three phases (continuous, dispersed, and encapsulating phases), where the third encapsulating film phase imparts some unique and desirable characteristics to the system.

The mechanism of aphron formation requires the ability of each phase to spread on the other. This is achieved through the presence of a surfactant in both phases. In order for an aphron to form, the liquid to be dispersed must spread on an aqueous surface. A large aqueous surface area can be obtained by foaming the water prior to contact with this second phase. This large surface area allows the liquid to spread into a very thin film on the aqueous interface. As the film becomes very thin, and the contact angle becomes low enough, the aqueous phase spreads over this liquid film. As the film spreads, local disturbances eventually cause the film to break up into tiny globules and these globules become encapsulated by the aqueous phase. Hence, each thin film forms a single layer of aphrons (*6*).

As is evident from the proposed aphron formation mechanism, the creation and resulting size distribution of a system of polyaphrons will be strongly dependent on the spreading phenomena that occur when two immiscible phases come in contact with each other. Aphron formation requires the ability of each phase to spread on the other, as previously mentioned. Thus, the surfactant concentration in both phases should be sufficient such that the resulting spreading coefficients (or surface pressures) are approximately equal. If the coefficents are substainially different, one of the phases will have difficultly spreading on the other, resulting in inefficient (or total lack of) aphron formation. If the nonpolar phase has a significantly higher spreading coefficient than the water phase, it will be able to spread easily on the water surface. However, the water will not be able to spread over the top of this phase as easily, which is necessary for aphron formation. On the other hand, if the spreading coefficient of the water phase is significantly higher than the nonpolar phase, there will be a large aqueous surface area, but the nonpolar phase will have difficulty spreading on it, due to the high surface pressure (compared to its own surface pressure.)

The above discussion suggests many factors that will affect the formation of aphrons and the underlying spreading phenomena that are occuring, some of which include (*7*): (a) nature and concentration of the water soluble surfactant, (b) nature and concentration of the nonpolar phase soluble surfactant, (c) equilibrium surface pressure on the water surface (related to (a) above), (d) preparation temperature, and (e) viscosity of the phases. While this list is by no means a complete listing of all the factors affecting spreading, it does provide a basis for this study.

This paper addresses the effects of the aqueous surfactant type, monomer and aqueous phase surfactant concentration, and the formation temperature on a polyaphron system containing a radiation sensitive monomer as the dispersed liquid phase which is subject to electron beam radiation to form crosslinked polymeric microspheres. Obviously, due to the nature of the aphron formation process, a distribution of aphron sizes will be produced. The objective of studying these variables is to determine what conditions may be utilized to obtain a narrow distribution of polymeric microspheres.

Experimental

Materials. (1) Distilled water, (2) Trimethylolpropane propoxylate triacrylate (TMPPTA) (Henkel Corp.), (3) Sodium dodecyl benzene sulphonate (NaDBS) water soluble anionic surfactant (Tennessee Chemical Co.), (4) Sodium didodecyl disulfonated diphenyloxide (Dowfax 2Al) water soluble anionic surfactant (Dow Chemical Corp.), (5) Alkyloxypolyethyleneoxyethanol (Tergitol 15-S-3) oil soluble surfactant (Union Carbide Corp.).

Aphron Formation. A specified amount of aqueous surfactant solution is placed in a closed vessel and shaken vigorously to create a foam. Several drops of the monomer (with the specified amount of surfactant) is added to the foam and shaken vigorously for approximately 20 seconds. This process is repeated until the desired concentration of monomer is obtained. However, after several milliters of monomer has been added, it is possible to add increasing amounts of monomer in a single step. Aphron formation requires the presence of a large aqueous surface area. However, the surface of an aphron works just as well as an aqueous film in this respect. Therefore, once a significant number of aphrons have been produced, increasing amounts of monomer may be added at one time. For reasons to be discussed, the final concentration of monomer dispersed in the aqueous solution was 67% for most samples prepared in this study. In order to gain an understanding of how various factors affect the particulate formation, several variables were systematically manipulated to determine their influence on the final product. The following variables were considered: (a) Monomer phase concentration. Four aphron samples were prepared with a 1.0% NaDBS aqueous solution and 0.4% Tergitol 15-S-3 in TMPPTA in a monomer to water ratio of 1:1, 2:1, 3:1, and 9:1 to determine the effect of monomer concentration on the resulting product. (b) Monomer phase surfactant concentration. Surfactant concentrations of 0.04, 0.07, 0.4, 0.8 and 1.0 volume % of Tergitol 15-S-3 in TMPPTA were used with various aqueous solutions to determine the effect of the monomer phase surfactant concentration on the final product. (c) Aqueous phase surfactant type and concentration. Aqueous solutions containing 0.5% and 1.0% of NaDBS or Dowfax 2A1 were prepared for use as the continuous phase of the aphron systems to determine what effect this has on the final product. (d) Temperature. Various aphron samples containing different combinations of the specified aqueous and monomer solutions were prepared at 25°C and 75°C to determine the effect of formation temperature on the final product since the viscosity and surface tension of the phases are dependent on this parameter.

Radiation Curing. The aphron samples were cured to form crosslinked polymeric microspheres under the following conditions. A small amount of solution was placed on an inert, semicrystalline polyester substrate, PET, and spread to a thickness of 5 mil (maximum thickness allowed to ensure uniform radiation dosage) using a wet film applicator. Samples that had been made more than a day or two prior to curing tended to settle out, forming a clear aqueous layer on the surface of the polyaphron phase. This aging posed no problem; the liquid was simply shaken to restore homogeniety and ensure a representative sampling. All wet film samples were cured at ambient temperatures under a nitrogen atmosphere with electron beam radiation. The electron beam source utilized was an Energy Sciences Inc. CB150 Electrocurtain operating at 175KV, with a beam current of 2.4 milliamps and a conveyor speed of 40 ft/min. This resulted in a total radiation dose of 4 megarads. This dosage is slightly above the manufacturer's recommended dose for film curing (3 megarads). After curing, the particles were washed in acetone, filtered, washed in water, filtered, then washed in acetone and filtered again, then placed under vacuum for at least 24 hours prior to analysis.

Analysis. Particle analysis was performed using a Cambridge Instruments Stereoscan 2000 scanning electron microscope and a Shimadzu SA-CP3 centrifugal particle size analyzer to obtain information on the size, shape, and size distribution of the crosslinked microspheres. The particle size analyser is based on the sedimentation method for particle size determination and measures the particle concentration photometrically. Accelerated sedimentation of small particles is facilitated by a rotating sample holder which exerts a centrifugal force on the colloid solution. All distributions shown represent an average of three runs made on the analyzer, which were performed using a 30 volume % solution of gylcerol in distilled water with Dowfax 2A1 as a dispersant. To obtain an indication of the error associated with these distributions, three solutions of one polyaphron sample were prepared and analyzed. In addition, three polyaphron samples of the same composition were prepared and analyzed. This allowed the error associated with the aphron preparation and the particle analysis to be qualitatively separated.

Surface tension values were measured using the Du Nouy ring method of surface tension determination. In order to obtain a quantitative measurement of the relative foam stability of the aqueous surfactant solutions, 10 milliters of each solution was placed in a capped 100 ml graduated cylinder and shaken vigorously to create a foam. The foam was allowed to sit for twenty seconds and the volume of drained water was measured as a qualitative indication of this property.

Results and Discussion

As previously mentioned, the formation of polymeric particulates was first attempted by introducing TMPPTA into an emulsion with glycerol, water, and Dowfax 2A1 constituting the continuous phase. The emulsion was subjected to electron beam radiation to form solid polymeric microspheres. This system required extensive stirring prior to processing to sufficiently disperse the monomer phase and the monomer concentrations had to be kept fairly low (about 5 volume %). The resulting product was quite disperse in size (see Figure 1) and required extensive purification, which resulted in the production of a large amount of by-product (i.e. the reaction medium). As mentioned earlier, these difficulties were eliminated by developing a polyaphron system with TMPPTA incorporated into the dispersed phase.

Electron beam irradiation of the polyaphron samples did result in the creation of polymeric microspheres, but their formation was strongly dependent on the monomer concentration. Figures 2a-d illustrate the effect of monomer concentration on the formation of microspheres. At monomer concentrations of 50 and 67 volume %, polymeric microspheres are produced with very little (if any) production of other geometric structures (see Figures 2a&b) In addition, there does not appear to be any difference in the size distribution of particles between these two concentrations. However, as the monomer concentration is increased to 75%, the production of irregular fragments (in addition to microspheres) becomes significant (Figure 2c). In fact, at a monomer concentration of 90%, irradiation of the sample results in the formation of a continuous film (Figure 2d). Apparently, as the monomer concentration approaches 75% or higher, the aphrons become so close to one another that the encapsulating soap film around the individual aphrons does not prevent monomer in one aphron from coalescing with monomer in neighboring aphrons, under the reaction conditions specified in this study. It should be noted that at a concentration of approximately 75% of the dispersed phase, Sebba (*6*) has reported that liquid aphrons begin to lose their spherical character and take on a polyhedral structure due to the close packing of aphrons. In light of this, the production of irregular fragments at a monomer concentration of 75% should become evident. Due to the lower yield of microspheres and an increase of undesirable structures created at this monomer concen-

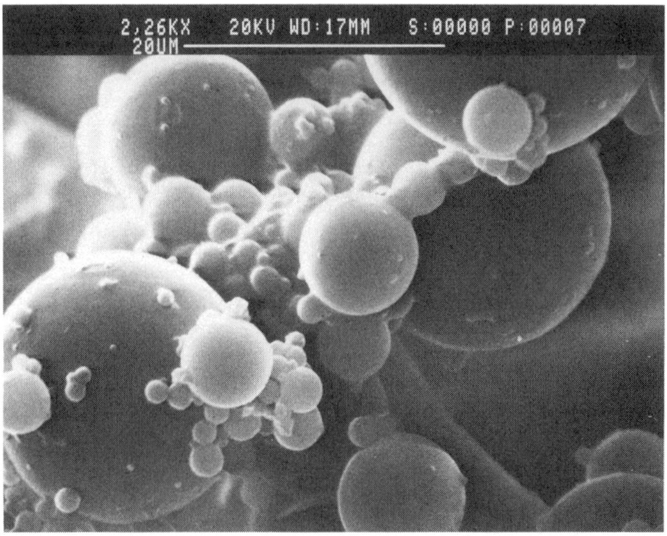

Figure 1. Polymeric microspheres formed via irradiation of an emulsion system containing TMPPTA.

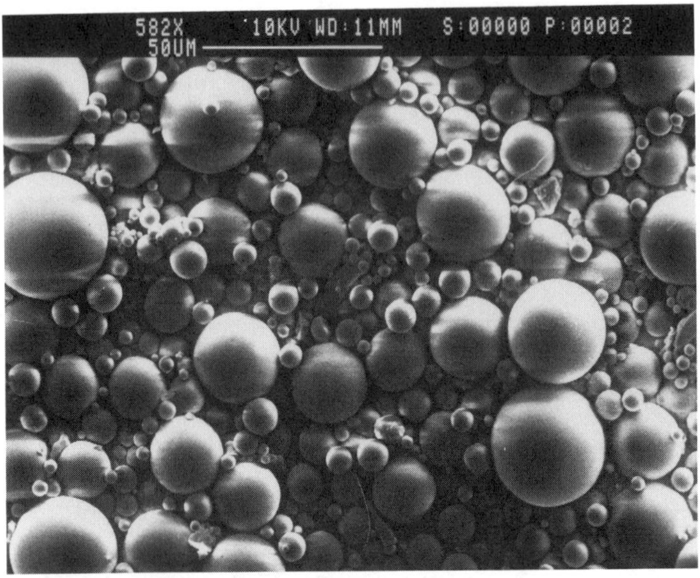

a

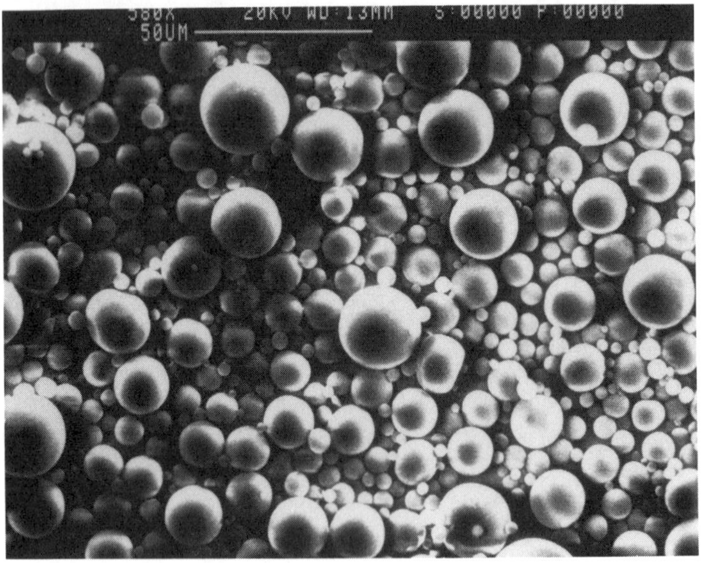

b

Figure 2a-d. Polymeric microspheres obtained by irradiation of a polyaphron system containing (a) 50%, (b) 67%, (c) 75% and (d) 90% TMPPTA dispersed in an aqueous matrix.

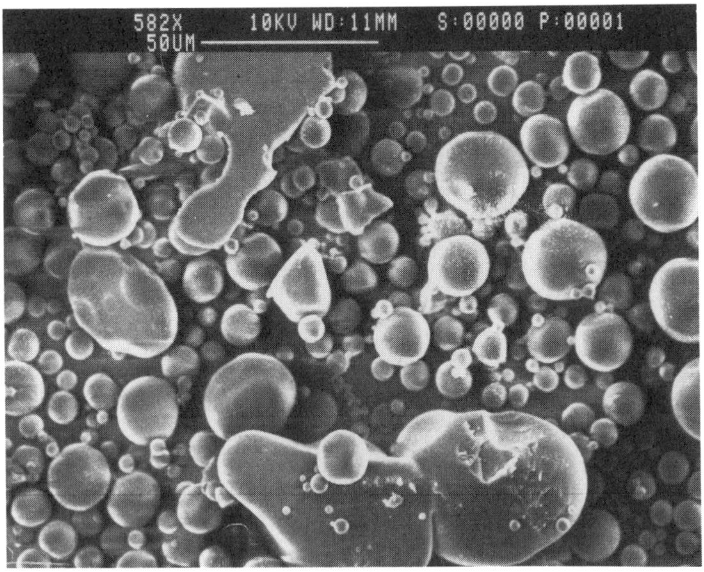

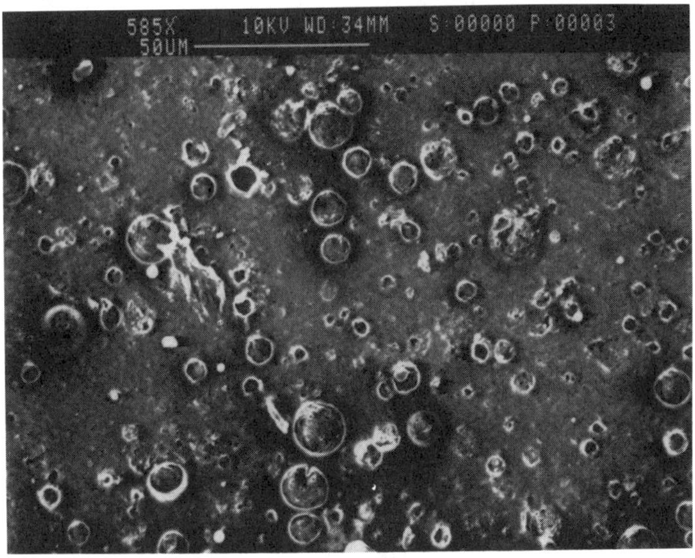

d

Figure 2. Continued.

tration, all subsequent aphron samples to be discussed were produced using a monomer concentration of 67 volume %.

By varying the concentration of the monomer phase surfactant, it is possible to change the size distribution of the resulting microspheres. Although the end points of the size distribution (i.e. the largest and smallest spheres produced) are not dramatically affected by the monomer phase surfactant concentration, the relative distribution of particle sizes between the end points is significantly dependent on this variable (see Figure 3). It should be noted that an error analysis was performed on the sample containing 1.0% Tergitol 15-S-3 in TMPPTA. The results show an overall average error of ±14% for all data points greater than 10 cumulative %. Below this value the percent error was somewhat greater, due to the small average value used in the calculations. This represents the error associated with both the aphron preparation and the size analysis. The average error associated with the size analysis alone was determined to be ±12%. Although this error analysis was performed on only one composition, it is indicative of the error associated with all the distributions shown.

As illustrated in Figure 3, varying the monomer phase surfactant concentration from 0.04 to 1.0% results in a minimum average particle size at 0.4%. Increasing the surfactant concentration in the monomer phase from 0.04 to 0.4 volume % results in a significant decrease in the average particle size, which may be attributed to the Gibbs effect. The Gibbs effect applies to thin films or small volumes which contain a surfactant, and can be described as follows. If a fixed volume of a liquid contains a surfactant and the surface area is increased (as in spreading), then surfactant molecules will move to the surface in accordance with the Gibbs adsorption isotherm and reduce the surface tension of this new surface. However, if the surfactant concentration becomes so low that the liquid surface already contains a significant portion of the total amount of surfactant, then there may not be enough surfactant available to reduce the surface tension of this new surface. The net effect is to produce a restoring force which resists further thinning of the liquid surface. Hence, as the surfactant concentration is increased, the monomer phase can form thinner films prior to breaking up and being encapsulated in the soap film, thus forming smaller aphrons. However, as the surfactant concentration is increased further (to 0.8 and 1.0 volume %), the average particle size increases.

The increase in average particle size at the higher monomer phase surfactant concentrations may have several origins. One possibility is that at high concentrations, the surfactant may be forming complex micellar structures in the monomer phase which may be very stable and not break down readily as the surface area is increased. Thus, the surfactant molecules may not be able to diffuse to the surface, allowing the Gibbs effect to occur. A second possible explanation is related to the Marangoni effect. This effect is similar to the Gibbs effect except that it is not restricted to thin films. If a liquid contains a surfactant and its surface area is increased, this new surface area will have a greater surface tension than the surface which was already present. This is a non-equilibrium condition because the Gibbs adsorption isotherm requires that surfactant molecules must diffuse to the surface from the bulk to reduce the surface tension to its equilibrium value. Although this diffusion time is very small, the momentary increase in surface tension

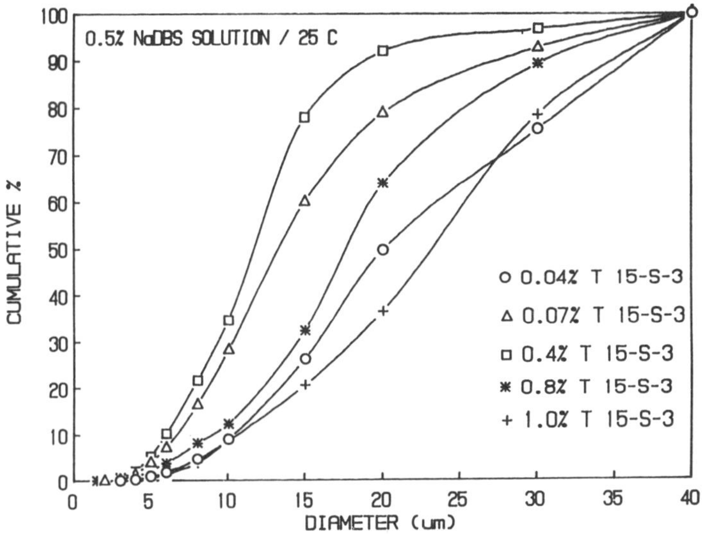

Figure 3. Effect of the Tergitol 15-S-3 concentration in TMPPTA on the particle size distribution of microspheres formed at 25°C with a 0.5% NaDBS solution as the continuous phase.

produces a restoring force preventing the extension of the surface, thus causing a surface elasticity. If the concentration of the surfactant is very high, this diffusion time will be very small. Hence, the surface will have minimal elasticity, and the film may become unstable. This instability could cause the monomer film to break up and become encapsulated by the soap film before it is able to thin sufficiently, thus producing larger aphrons. A third possibility is that the high surfactant concentration produces a spreading coefficient that is substantially "out of balance" (as discussed earlier) with the spreading coefficient of the aqueous phase thus causing inefficient formation of aphrons due to the hindered ability of the aqueous phase to spread over the monomer phase.

Figure 4a illustrates the variation of the particle size distribution with monomer phase surfactant concentration for a series of polyaphrons formed using the Dowfax 2A1 aqueous surfactant. Note that there is no apparent trend here with respect to particle size distribution and the monomer phase surfactant concentration (as was discussed prevously for the polyaphron systems utilizing the NaDBS aqueous surfactant). This indicates that other factors may be present here which essentially override the ability of the monomer phase surfactant concentration to control the resulting particle size distribution. The reduced foamability of the Dowfax solution (compared to the NaDBS solution - see Table I) may account for this apparent lack of control. The Dowfax solution produces less aqueous surface area for the monomer to spread on which may affect the aphron formation process. However, if the preparation temperature is increased to 75°C, the trends discussed with the systems containing the NaDBS solution is established (see Figure 4b). Table I shows that the main effect of the temperature increase is to reduce the viscosity and surface tension of the monomer phase. Both of these physical property changes will improve the ability of the monomer to spread, so the surface area produced with the Dowfax solution may now be sufficient for the Gibbs effect (or other effects previously discussed) to govern the aphron formation process.

Table I: Some physical properties and their temperature dependence of the surfactant solutions used in this study

Material	Apparent Surface Tension(dyne/cm)		Viscosity (cps)		Foam Stability (vol%)[1]
	25°C	75°C	25°C	75°C	
TMPPTA[2]	37	32	140	30	--
NaDBS[3]	33	32	--	--	25
Dowfax[3]	33	32	--	--	50

1. volume percent of water drained in 20 seconds after foaming solution.
2. all Tergitol concentrations.
3. 0.5 volume percent aqueous solution.

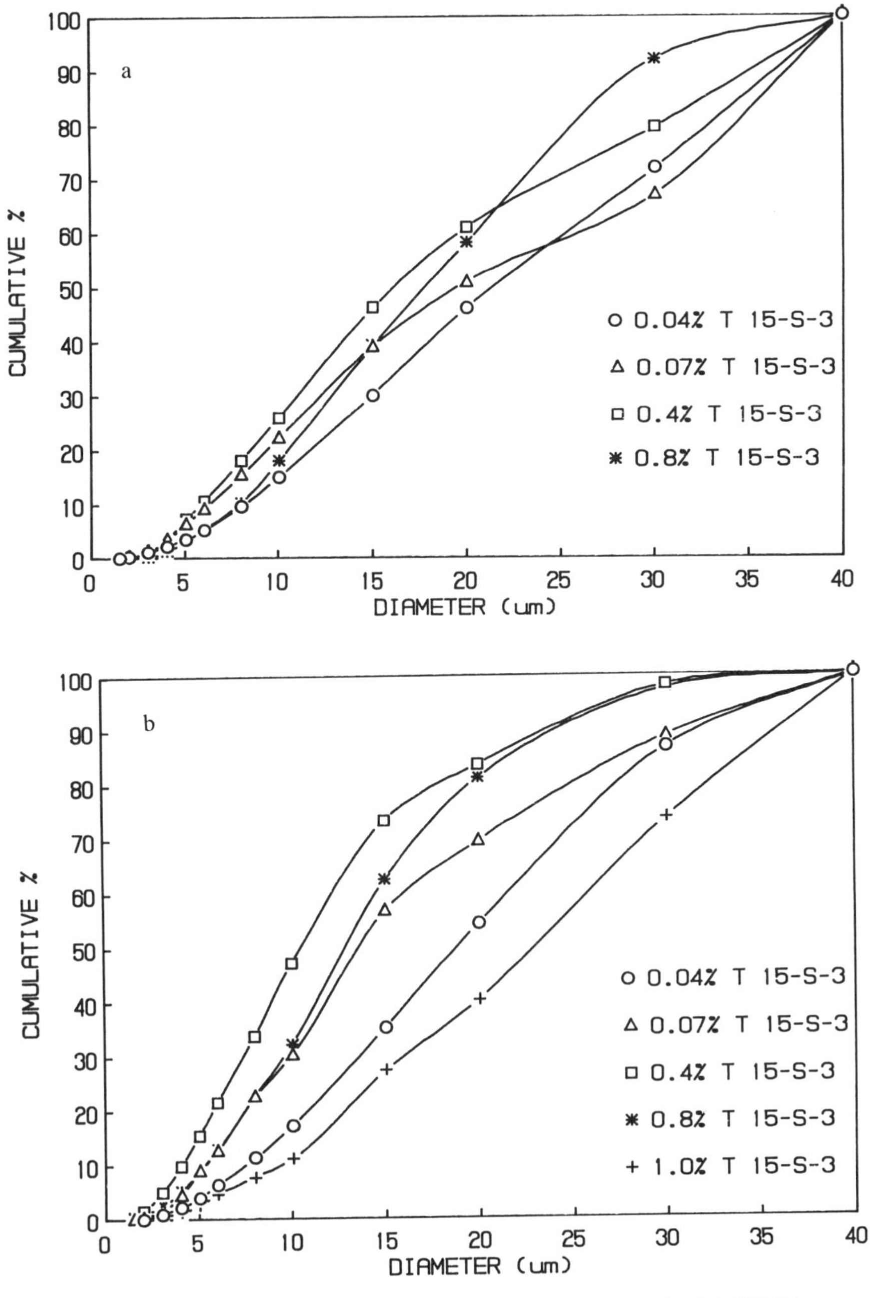

Figure 4a-b. Effect of the Tergitol 15-S-3 concentration in TMPPTA on the particle size distribution of microspheres formed at (a) 25°C and (b) 75°C with a 0.5% Dowfax 2A1 solution as the continuous phase.

These points lead one to consider the effects of the aqueous phase surfactant type and concentration on particulate formation and properties. Figure 5 illustrates these effects on the particle size distribution for a series of aphron systems formed at 75°C with a Tergitol 15-S-3 concentration of 0.07% in the monomer phase. An increase in the aqueous surfactant concentration increased the average particle size for both NaDBS and Dowfax 2A1 surfactants. Increasing the surfactant concentration produces a higher surface pressure on the aqueous film, which increases the resistance to spreading of the monomer film. This will produce a thicker monomer film, which will lead to the formation of larger aphrons when the film breaks up and becomes encapsulated in a thin soapy film. However, if the Tergitol 15-S-3 concentration is increased to 1.0%, the average particle size decreases with increasing aqueous phase surfactant concentration. This trend reversal at high monomer phase surfactant concentrations strongly supports the fact that a "balance" between the spreading coefficients (or spreading pressures) is necessary for the efficient production of small aphrons.

SEM analysis of the aphrons has shown some differences in the shape of the resulting polymeric microspheres that appears to be related to the aqueous phase surfactant type. It should be noted that SEM sample preparation was facilitated by using double sided adhesive tape to secure the specimen to the SEM stub. Some microspheres may appear to be semicircular, where actually they are imbedded in the adhesive layer. Figures 6a&b shows SEM micrographs of polymeric microspheres formed with 0.5% Dowfax 2A1 and 0.5% NaDBS solutions respectively, with equal concentrations of Tergitol 15-S-3 in the monomer phase. The micrographs show the presence of some indentations on the surface of the larger microspheres formed with the NaDBS solution. These distortions were not found on all the samples prepared from the NaDBS solution, but they were present to a much larger extent than the microspheres produced with the Dowfax 2A1 solution. Sebba (*6*) has reported that increasing the ionic strength of the encapsulating film (i.e. higher surfactant concentrations or strengths) increases the attractive forces between aphrons and the smaller aphrons will tend to adhere to the surface of larger aphrons and distort their sphericity. This is a likely cause of the surface distortions seen on the microspheres formed with the NaDBS solution. However, it is expected that some shrinkage of the monomer phase occurs during curing as a result of the polymerization through double bond moieties. This shrinkage can induce stresses which may affect the curvature of the larger microspheres to a greater extent, due to the larger volume, but this does not account for the apparent surfactant type dependence. Hence, we discount this latter explanation as the cause for the lack of particle sphericity.

As previously discussed, increasing the aphron preparation temperature produces a significant decrease in the monomer viscosity and surface tension, as well as a slight decrease in the surface tension of the aqueous solution. As illustrated in Figure 7, this temperature increase causes a decrease in the average particle size over the entire range of monomer phase surfactant concentrations studied. The reasons for this should be obvious. A reduction in the monomer surface tension will promote more spreading since the surface energy is lower. The same holds true for the aqueous phase, but to a much lesser extent. A reduction in the monomer viscosity obviously promotes better

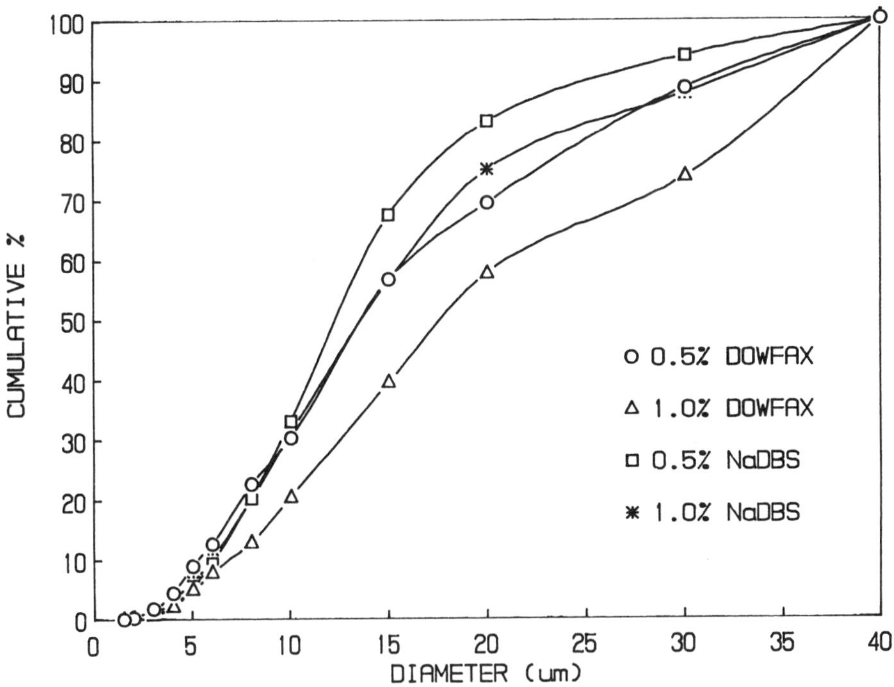

Figure 5. Effect of the aqueous phase surfactant type and concentration on the size distribution of microspheres formed at 25°C with 0.07% Tergitol 15-S-3 in TMPPTA.

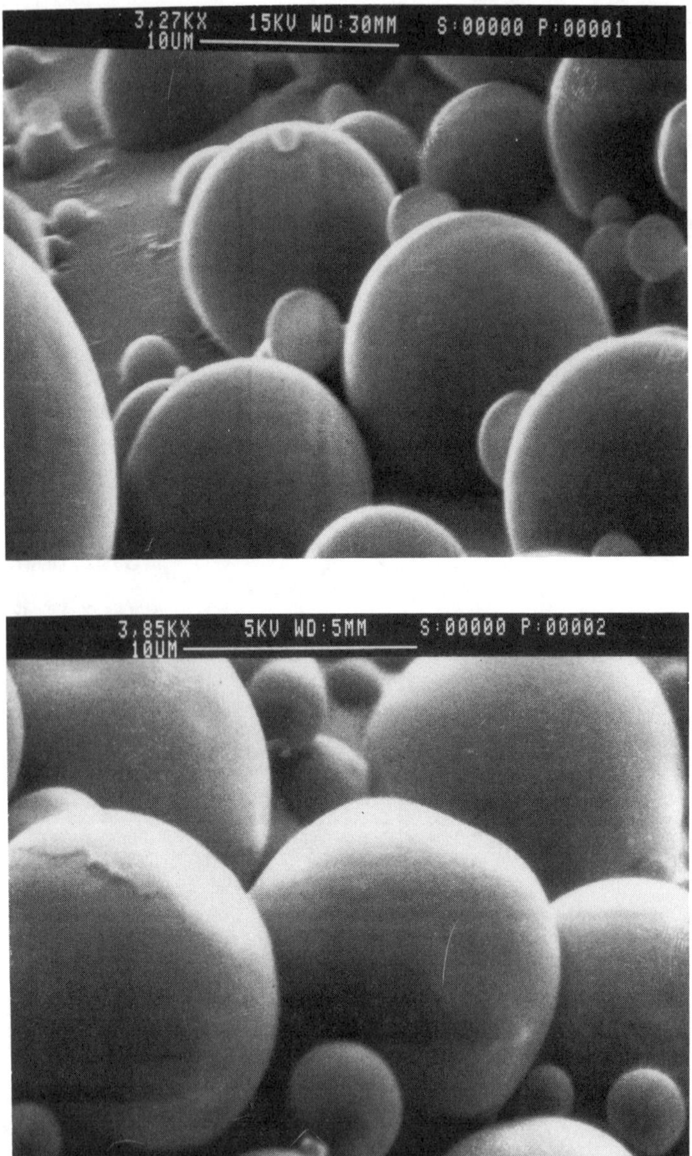

a

b

Figure 6. SEM micrographs depicting the effect of the aqueous phase surfactant type on the shape of the resulting microspheres.

spreading, but this will also reduce the diffusion time of the surfactant molecules to the surface as spreading occurs. This may cause the monomer film to become stable for a longer period of time (allowing for more film thinning) as the surfactant molecules become depleted from the bulk and the diffusion paths (or diffusion times) become longer. It should be noted from Figure 7 that the reduction in particle size with increasing temperature appears to be dependent on the monomer phase surfactant concentration. The reasons for this are not entirely clear, but it may be related to the relative balance of spreading coefficients that exists in each system and the extent to which temperature may affect this condition.

Since the objective of this investigation has been to gain an understanding of the conditions necessary for the production of small polymeric microspheres with a narrow size distribution, it is worthy to highlight the best and worst cases found in this respect, as illustrated in Figure 8. The "best case" has an average particle size of approximately 7 microns with 90% of the particles between 3 and 18 microns. These microspheres were produced using 0.4% Tergitol 15-S-3 in TMPPTA and a 0.5% NaDBS aqueous solution prepared at 75°C. The "worst case" distribution has an average particle size of 24 microns with 90% of the particles between 5 and 38 microns. These particles were formed at 25°C with 0.04% Tergitol 15-S-3 in TMPPTA and the 0.5% Dowfax 2A1 aqueous solution. In contrast, a similar "worst case" (distribution not shown) was obtained by forming microspheres at 25°C with 1.0% Tergitol 15-S-3 in TMPPTA and a 0.5% NaDBS aqueous solution. The average size of these microspheres was 23 microns with 90% of the particles between 7 and 40 microns. These best and worst cases illustrate some of the conditions that will promote (and hinder) the production of polymeric microspheres with a narrow size distribution. The best case illustrates that a higher preparation temperature facilitates a narrower distribution. The two worst cases further support the need for a delicate balance between the surfactant concentrations in both phases; either too much or too little surfactant in the monomer phase can have a detremental effect on the production of small microspheres with a narrow size distribution.

Summary and Conclusions

This work has illustrated the successful incorporation of TMPPTA into a polyaphron system for the production of crosslinked polymeric microspheres using electron beam radiation. Spherical particulate formation is facilitated at a solids content as high as 67 volume %, with complete conversion and no agitation necessary. There are several parameters that affect the particle size distribution which include: preparation temperature, the relative concentrations of the monomer and aqueous phase surfactants, and the aqueous phase surfactant type. By careful manipulation of these variables, the average particle size can be controlled between approximately 6 and 25 microns. While this paper illustrates the use of TMPPTA in a polyaphron system, it should be realized that a variety of different materials may be used in this scheme to produce microspheres with various physical properties.

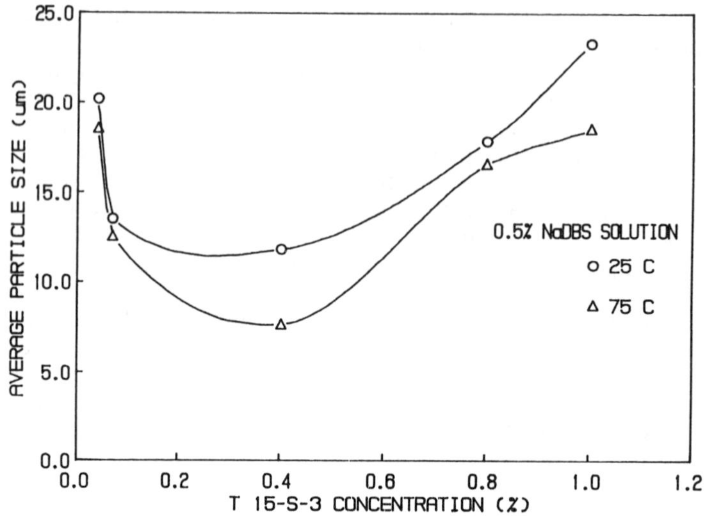

Figure 7. Effect of aphron formation temperature on the average particle size of microspheres formed with varying amounts of Tergitol 15-S-3 in TMPPTA.

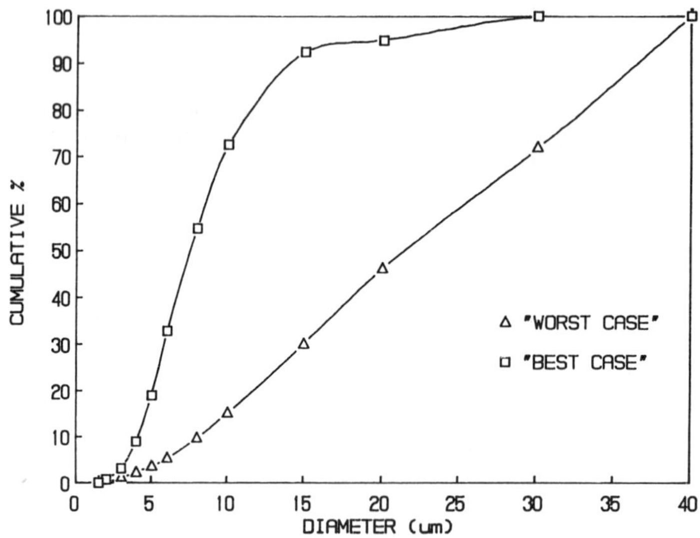

Figure 8. Particle size distributions illustrating the "best case" and "worst case" microspheres obtained in this study.

Acknowledgments

The financial support of 3M Corporation for this work is gratefully acknowledged. The authors would also like to thank Dr. James Wightman, for his helpful discussion of the results.

Literature Cited

1. Piirma,I. *Emulsion Polymerization*; Academic Press: New York,NY,1982.
2. Yoshida,M.; Asano,M.; Kaetsu,J.; Morita,Y. *Rad. Phys. Chem.* **1987**,*30*,39.
3. Rembaum,A.; Yen,R.C.K.; Kempner,D.H.; Ugelstad,J. *J. Immun. Meth.* **1982**,*52*,341.
4. Sebba,F. *Foams and Biliquid Foams-Aphrons*; John Wiley and Sons: New York,NY,1987;pp 49.
5. Sebba,F. *Foams and Biliquid Foams-Aphrons*; John Wiley and Sons: New York,NY,1987;pp 50.
6. Sebba,F. *Chem. Ind.***1984**,*10*,367.
7. Sebba,F. *Foams and Biliquid Foams-Aphrons*; John Wiley and Sons: New York,NY,1987;pp 91.

RECEIVED January 14, 1991

Chapter 14

Acrylic-Acid-Grafted Polyethylene by Electron-Beam Preirradiation Method

Some Effects of Reaction Variables

Junji Harada[1], Rey T. Chern[2], and Vivian T. Stannett[2]

[1]Tsukuba Research Laboratories, Mitsubishi Paper Mills, Ltd., 46 Wadai, Tsukuba, Ibaraki, Japan
[2]Department of Chemical Engineering, North Carolina State University, Raleigh, NC 27695–7905

High density polyethylene (HDPE) and low density polyethylene (LDPE) samples preirradiated with an accelerated electron beam (accelerated for 1-64 Mrad) were grafted with acrylic acid under various conditions. The effects of irradiation dose, reaction time, monomer concentration and inhibitor concentration on the grafting were studied. According to the DSC analysis on dry grafted samples, the graft reaction occurs both in the amorphous region and at the boundary of crystallites. The asymptotic grafting increased exponentially with the preirradiation dose. The grafting rate and asymptotic grafting value of HDPE are much higher than those of LDPE. The grafting of the 64 Mrad irradiated HDPE reached 26,000% within 2 hours of reaction.

It is well known that graft polymerization can be initiated by many methods, for example, high energy (gamma ray, electron beam) irradiation, plasma treatment, ultraviolet light radiation, decomposition of chemical initiators and oxidation of the polymers. Among these methods, electron beam irradiation may be the most convenient and most effective for industrial usage because of its high irradiation dose rate, ease in generating active sites in many kinds of polymers, effective penetration into the polymer and moderate initiation reaction conditions (room pressure and temperature). Electron beam irradiation induced graft copolymerization can be used to modify either the surface or the bulk of the polymer. Many factors affect the modification reaction, they include the radiation chemistry of the substrate polymer (i.e. crosslinking or degradation type), extent of swelling of the grafted polymer by the monomer solution, the electron beam acceleration voltage, the grafting reaction time and temperature, and many others.

By graft modification of the polymer surface, one can change the polymer's wettability (1,2), adhesion (3,4), printability, metalization, anti-fog properties (5), anti-statics properties (6-8), and biocompatability (9-13). Bulk graft modification of the polymer has been used to improve water absorbency (14-16), fire retardancy (17), and to produce battery separators (18-26), ion exchange

0097–6156/91/0475–0238$06.00/0

and ion trap materials (8,27-30), separation membranes (29,31,32), deodorant materials, immobilized enzymes (33,34) and so on.

Two standard methods of radiation grafting were developed early in the field and are still the main techniques available. These are the mutual (direct, simultaneous) and preirradiation (indirect, consecutive)methods. In this study, we used only the preirradiation method. This method can achieve effective grafting reactions, especially for a highly reactive monomer, such as acrylic acid, because the grafting reactions are carried out with little homopolymerization which is one of the main disadvantages of the mutual grafting method.

This paper will consider the acrylic acid grafting reaction to polyethylene film by the electron beam preirradiation method. Acrylic acid is one of the most reactive monomers and grafted poly (acrylic acid) is expected to impart ion exchange and ion trap capabilities, wettability, water absorbent properties, anti-static properties, permselectivities and so on. Polyethylene is a typical crosslinkable polymer by irradiation and is not swollen by water or acrylic acid. Several papers have been published about radiation induced grafting of acrylic acid onto polyethylene (18-25). Ishigaki et al., studied the preirradiation grafting using Mohr's salt as an inhibitor, and produced comparatively low grafting materials which are suitable as battery separators (23-25).

The present paper deals with comparatively high temperature and redox reagent free grafting reactions, especially concerning the effects of irradiation dose, reaction time, monomer concentration and other reaction conditions.

Experimental

Materials High density polyethylene (HDPE) film (d=0.963g/cm^3, 95μm thick, crystallinity=70%) and low density polyethylene (LDPE) film (d=0.926g/cm^3, 95μm thick, crystallinity=46%) were used. Polytetrafluoroethylene (PTFE) film (100μm thick) was used. Acrylic acid (Aldrich, including 200ppm hydroquinone monomethyl ether as a stabilizer) was used without any purification. Vacuum distilled acrylic acid was used only for the experiment on the effects of the Mohr's salt. Mohr's salt (ammonium iron (II) sulfate hexahydrate) was ACS reagent grade (Aldrich). All acrylic acid aqueous solution concentrations were expressed in volume percentage.

Irradiation An electron curtain type EB accelerator (Energy Sciences Inc., CB 175/15/180) was used. The irradiation was carried out at 175 KV accelerated voltage, and under a nitrogen atmosphere (oxygen concentration level was 50-100 ppm). The irradiation dose rate range was 3.4 to 13.5 Mrad/sec. Actual irradiation dosage to the film was measured by absorbance change of the cellulose triacetate film (Fuji film, FTR-125) with the CTA film dose reader (Narumi Co. FDR-01). According to the dose-depth profile at 175KV electron energy level of this system (35), the irradiation dosage was expected to be almost constant (>95%) throughout the sample thickness (<100μm).

Grafting Reaction All monomer solutions were aqueous and were bubbled with nitrogen in the Erlenmeyer flasks for at least 30 minutes to degas the solution before the grafting reaction. The irradiated polyethylene samples (2cm x 2cm) were put into the reaction flask which was kept in an isothermal water bath, immediately after irradiation or after at most 14 days storage in liquid nitrogen. Nitrogen bubbling was continued during the grafting reaction. After reaction, the grafted samples were taken out and washed by flowing hot water for about 24 hours, followed by vacuum drying for 48 hours at 55°C. By this washing method, the grafted polymer weight reached an almost constant value (at most 2 % weight excess relative to the complete extraction result 2 days water reflux and 2 days methylalcohol reflux). The grafting (%) was calculated by the next equation,

$$\text{Grafting (\%)} = \frac{100 \times (\text{After grafting weight} - \text{Original weight})}{(\text{Original weight})}$$

Thermal and Microscopy Analyses A differential scanning calorimeter (Parkin-Elmer DSC 7) was used to measure the melting endotherm of polyethylene under nitrogen purge and at 10°C/min heating rate.

A scanning electron microscope (JEOL, JXA-840) was used to observe the surfaces and cross sections of the grafted films. The grafted samples were neutralized by a 1% potassium hydroxide solution to enhance the contrast between the grafted layer and the ungrafted layer.

Results and Discussion

Effects of Mohr's Salt Figure 1 shows the effects of the Mohr's salt on the acrylic acid grafting. Vacuum distilled acrylic acid was used for this experiment. The grafting ratio dramatically decreases with the increase of Mohr's salt concentration and tended to reach a constant value. Above 500ppm Mohr's salt concentration, the grafting seems to be independent of the irradiation dose because little apparent grafting difference exists between low and high irradiation dose samples.

According to the X-ray microanalysis, the grafting reaction occurred almost equally throughout the film depth after two hours of reaction. It appears that the Mohr's salt cannot easily penetrate into the ungrafted part of the sample. As the grafting front advances into the polyethylene film, the grafted region behind the front becomes accessible to the Mohr's salt, leading to annihilation of the reactive free radicals and therefore the termination of the grafting reaction in the region. As soon as the grafting fronts from both sides of the film sample meet in the center of the film specimen, essentially all the reactive free radicals disappear almost immediately, independent of the radiation dose. This is probably why the grafting ratio changes little with irradiation dose at high concentrations of the Mohr's salt, as shown in Figure 1.

Moreover, assuming that the GR• value (the yields of the free radicals for 100eV of energy absorbed) of polyethylene is 3 (36), one would expect the maximum amount of reactive free radicals to be 108 μmol in a 500 mg sample exposed to 16 Mrad irradiation. This amount is of the same order of magnitude as the amount of the ferrous ion present (190 μmol, 500ppm Mohr's salt in 150ml of solution).

It is known that copper metal chips inhibit the homopolymerization of acrylic acid during the distillation purification of the monomer. As shown by the dotted line in Figure 1, copper metal chips (5g/150ml) which were dropped into the reaction flask five minutes before the grafting reaction was started, worked as an inhibitor similar to the Mohr's salt and the effect was almost the same as that of 10000 ppm Mohr's salt .

Monomer Concentration Figure 2 shows the typical effect of monomer concentration. The reaction were carried out for 120 minutes at 70°C immediately after electron beam irradiation of the HDPE. Over the concentration range up to 90%, the grafting ratio increases with monomer concentration, but between 90% and 100% monomer concentration it decreases dramatically. The reason for the low grafting ratio at 100% may be due to the low solubility of acrylic acid in the poly (acrylic acid) which is grafted on the surface of the polyethylene films. This low solubility retards further penetration of acrylic acid monomers to the reactive sites inside the bulk of the polymer samples. It is clear that the reactant solution must be able to swell the grafted layer or the bulk polymer if high grafting is to be achieved.

Over the whole monomer concentration range, higher irradiation dose yielded higher grafting except for the 16Mrad sample at acrylic acid concentrations below 30%. This exception is believed to be related to the crosslinking effect to be described later.

Pop-corn reaction (rapid, uncontrolled homopolymerization) happened for samples irradiated at high doses when the monomer concentration was 80% and 90%. To avoid pop-corn homopolymerization and to achieve rapidly high grafting one should use a monomer concentration between 50 and 60%.

Reaction Time and Reaction Rate As shown in Figure 3, the grafting of the comparatively low dose (1-16 Mrad) samples increase with reaction time and asymptote to constant values after 40-120 minutes reaction time. The polyethylene sample was observed to start swelling 10 to 20 minutes after the reaction was started. At this point the grafting was above about 130%. The grafting reaction occurs very rapidly in the case of HDPE. The grafting reached 50% of the asymptotic value within 15 minutes for all irradiation doses (1,2,4,8,16 Mrad), and 90% within 60 minutes.

Figure 4 shows the sample after different reaction times (the transparent sample is the original HDPE at 0 min., and the white sample is a piece of paper of the same size, included here as a reference). After 5 minutes of reaction, the surface layer of the HDPE sample was grafted by the acrylic acid, and its color turned white, the grafting was 69% at this point. The sample started to swell after about 10 minutes of reaction and its surface roughened. The sample turned transparent again after roughly 25 minutes of reaction. The sample remained square throughout the whole course of the reaction.

The short time grafting data for the 8 Mrad irradiated HDPE are plotted in Figure 5. Both the grafting and the changing physical dimension of the samples are shown. Below 8 Mrad, the crosslinking effects are presumably negligible, as described later. The grafting reaction rate (the slope) obviously changed after 6 minutes and the sample area also started to increase. At this point, the grafting reached almost 130% and the grafting front met at the center of polyethylene film as shown Figure 6.

According to the logarithmic plots of the grafting rate against the irradiation dose (1-8 Mrad), the grafting rate in initial reaction stage (before grafting fronts meet at the center of the film) is proportional to $[\text{Dose}]^{0.92}$ as described in the previous paper (37). This nearly first-order dependence on the irradiation dose appears to suggest that the amount of reactive free radicals at the beginning of the grafting reaction is proportional to the preirradiation dose. On the other hand, the reaction order in the second reaction stage (presumably after grafting fronts meet at the center of the film) the grafting rate is almost proportion to the square root of the irradiation dose. It suggests that the graft polymerization inside of polyethylene can be treated as a steady state reaction, typical of radical polymerization.

The Effects of Crosslinking Figure 7 shows the crosslinking effects on the grafting of HDPE. Similar effects were observed for LDPE. Polyethylene is a typical irradiation-crosslinkable polymer. Experiments with polytetrafluoroethylene (PTFE), a chain-scission type upon irradiation, are included in Figure 7 for comparison. While the acrylic acid grafting of PTFE increased with the increase in the irradiation dose, the grafting of HDPE showed a maximum and decreased at higher doses. This decrease of grafting may be due to the lower monomer diffusion rate into the crosslinked polymer matrix and lower swelling. Obviously, highly irradiated HDPE was crosslinked because samples irradiated above 16 Mrad had non-extractable gel fractions after 30 hours toluene extraction (38) as shown Table 1.

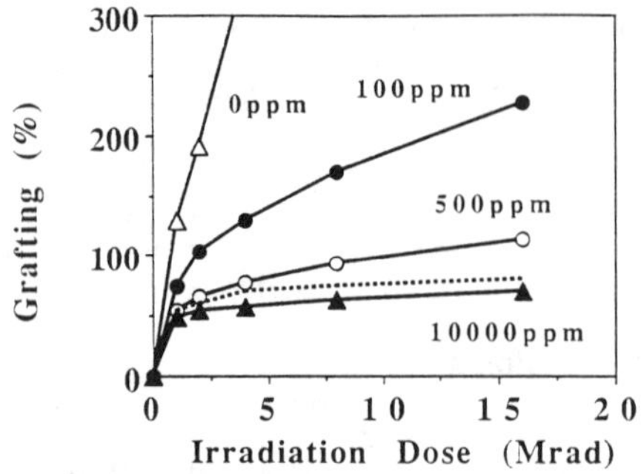

Figure 1 Effect of Mohr's salt
HDPE: 95μm, grafting temperature: 70°C, reaction time: 120 minutes, 50% acrylic acid solution, Dotted line: Copper metal chips (5g/150ml).

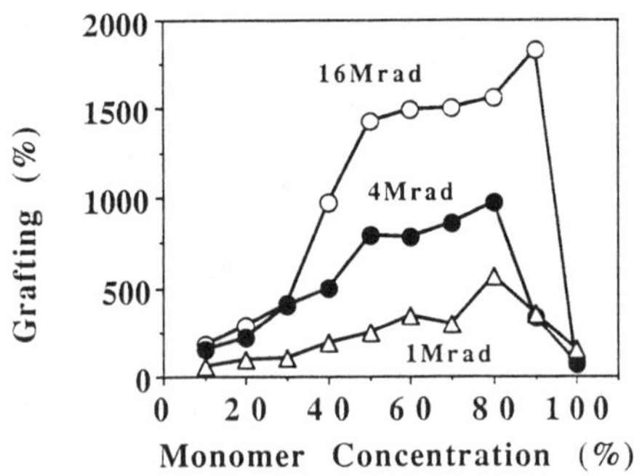

Figure 2 Effect of monomer concentration
HDPE: 95μm, grafting temperature: 70°C, reaction time: 120 minutes.

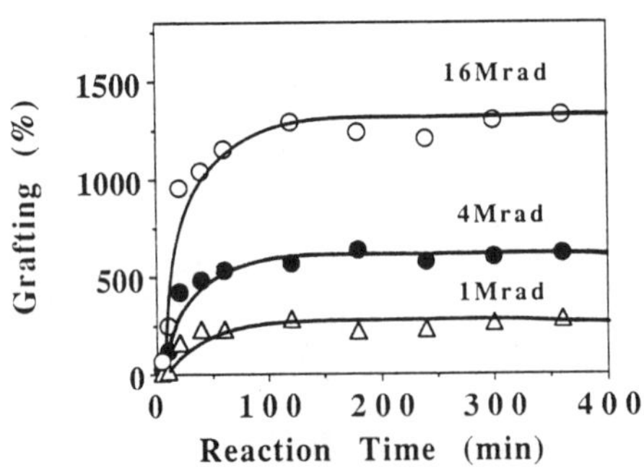

Figure 3 Effect of reaction time
HDPE: 95μm, grafting temperature: 70°C, 50% acrylic acid solution.

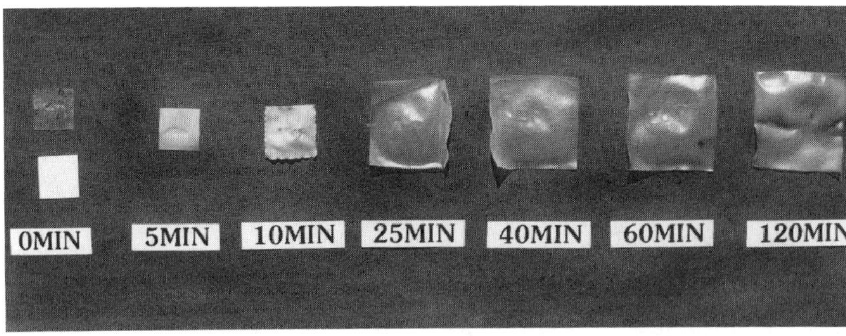

Figure 4 Acrylic acid grafted HDPE
HDPE: 95μm, grafting temperature: 70°C, irradiation dose: 16Mrad, 50% acrylic acid solution.

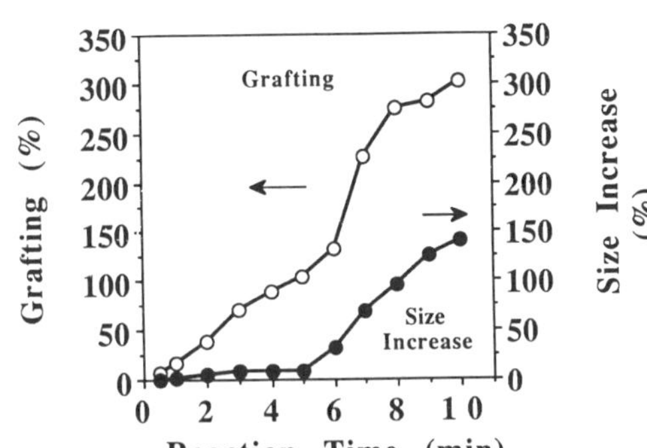

Figure 5 Grafting and size increase
HDPE: 95μm, grafting temperature: 70°C, irradiation dose: 8 Mrad, 50% acrylic acid solution.

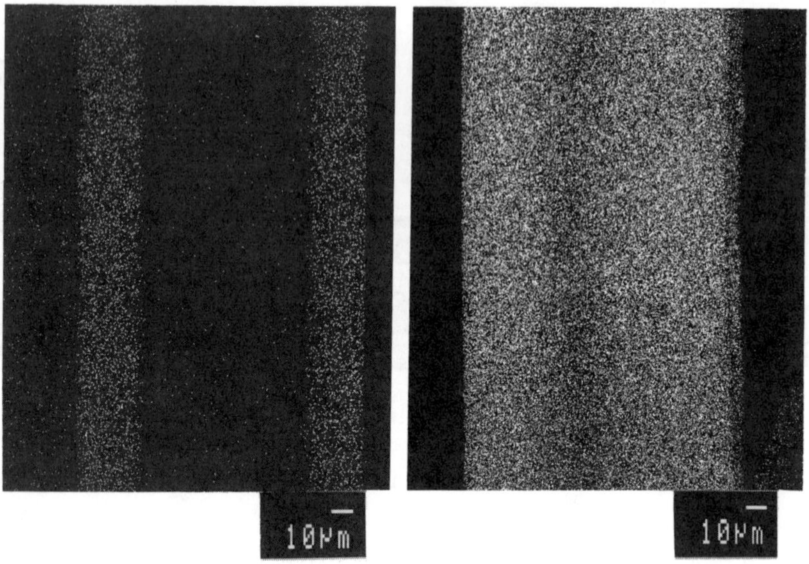

Figure 6 Cross section X-ray microanalysis photography
HDPE, grafting temperature: 70°C,
(left) irradiation dose: 4 Mrad, grafting 35%,
(right) irradiation dose: 2 Mrad, grafting 131%.

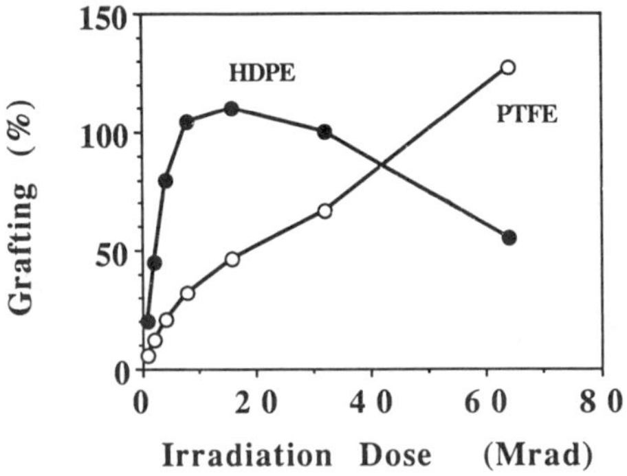

Figure 7 Short time grafting reaction
● HDPE: 95µm, grafting reaction :70°C, 5 minutes,
○ PTFE:100µm, grafting reaction :70°C, 10 minutes.

Table 1 Gel fraction of irradiated HDPE

Dose (Mrad)	8	16	32	48	64	96	128
Gel (%)	soluble	43	66	71	78	84	89

When the grafting is above the swelling line (ca. 130%) , the relationship between the grafting and the irradiation dose is close to linear as shown Figure 8.

Asymptotic Grafting Ratio As presented earlier, when the grafting was plotted versus reaction time, an asymptotic value was reached at long reaction times for each dose. These asymptotic values are plotted versus radiation dose in Figure 9. Clearly, the data can be fitted to the following equation:

$$\text{Asymptotic grafting (\%)} = K \times [D]^{b} \qquad \text{Eq.1}$$

where K and b are constant, and [D] is the irradiation dose (Mrad) .
The b value is about 0.50 for HDPE and 0.37 for LDPE. K is about $10^{2.7}$ for HDPE and $10^{2.4}$ for LDPE. Eq.1 is valid for the HDPE data below 32 Mrad and for the LDPE data above 2 Mrad. This equation makes it possible to estimate the asymptotic grafting at a given irradiation dose.

The asymptotic grafting above 32 Mrad is much higher for HDPE than the value calculated from this equation and the reaction is more complicated.

High Irradiation Dose HDPE samples irradiated at high doses (48 Mrad, 64 Mrad) exhibited another behavior distinctive from samples irradiated at lower doses. As mentioned in a previous paragraph, the "short-time" grafting of the high dose irradiated HDPE samples are smaller than that of the 8 Mrad irradiated sample. However, at longer reaction times the grafting of the high-dose samples surpassed those of the low-dose samples (Figures 10 and 11). The grafting rate of the high dose polyethylene increased continuously after 20 minutes whereas the grafting rate of the lower dose polyethylene decreases. The grafting of the 64 Mrad irradiated HDPE reached 26000% after 120 minutes reaction and had not yet reached an asymptotic value as observed in the lower dose samples.

The physical size of the 64 Mrad sample changed from 2cm x 2cm to 20cm x 20cm, and the thickness changed from 95μm to 650μm after 120 minutes of grafting reaction. This film is essentially pure polyacrylic acid (HDPE is less than 0.4% by weight) and is uniform, transparent, elastic (wet situation), high in tensile strength and insoluble in hot water and hot methanol.

Comparison Between HDPE and LDPE The grafting kinetics for HDPE and LDPE are compared in Figure 12. The grafting rate and the asymptotic grafting of HDPE are much higher than those of LDPE, even at the initial stage of reaction. Similar results were observed at other irradiation doses. Only 15 minutes was needed for HDPE grafting to reach half of the asymptotic grafting over the whole range of irradiation dose (1-16 Mrad) whereas 50 minutes was needed for LDPE.

Since the diffusion of acrylic acid in HDPE is expected to be smaller than that in LDPE, the results in Figure 12 seem to suggest that the diffusion rate of monomer may not be a controlling factor in determine the rate of the grafting for the systems studied here. HDPE can trap more radicals than LDPE because of its higher crystallinity (70% compared with 46%). The larger asymptotic grafting of HDPE (ca. 1300% for HDPE vs. 900% for LDPE at 16 Mrad) can be attributed to the higher crystallinity of HDPE leading also to a higher trapped radical concentration.

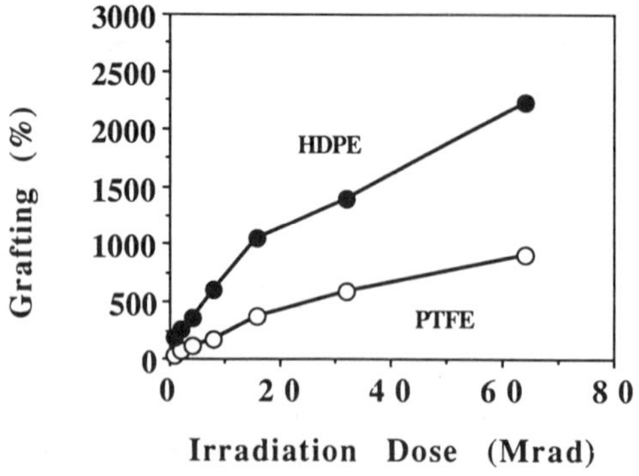

Figure 8 Long time grafting reaction
● HDPE: 95μm, grafting reaction :70°C, 40 minutes,
○ PTFE:100μm, grafting reaction :50°C, 120minutes.

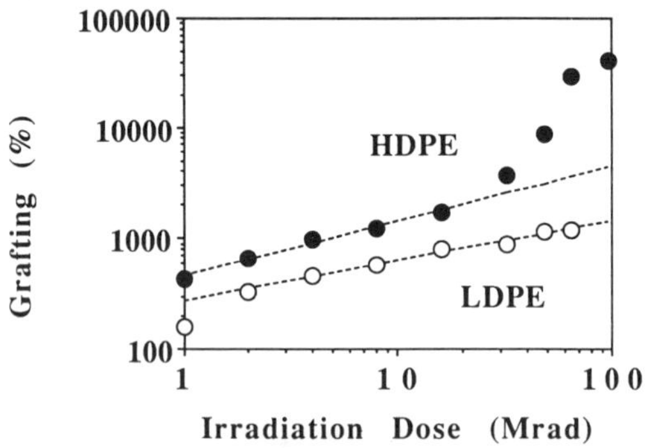

Figure 9 Asymptotic grafting
● HDPE: 95μm, ○ LDPE: 95μm, 50% acrylic acid solution, grafting temperature: 70°C, reaction time: 120 minutes.

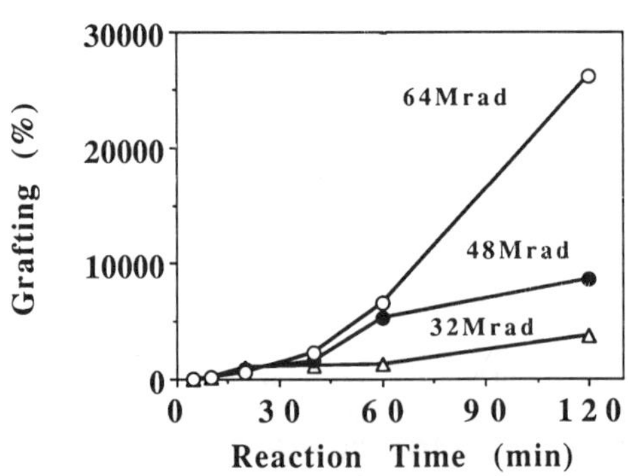

Figure 10 Grafting reaction of high dose irradiation
HDPE: 95μm, grafting temperature: 70°C,
50% acrylic acid solution.

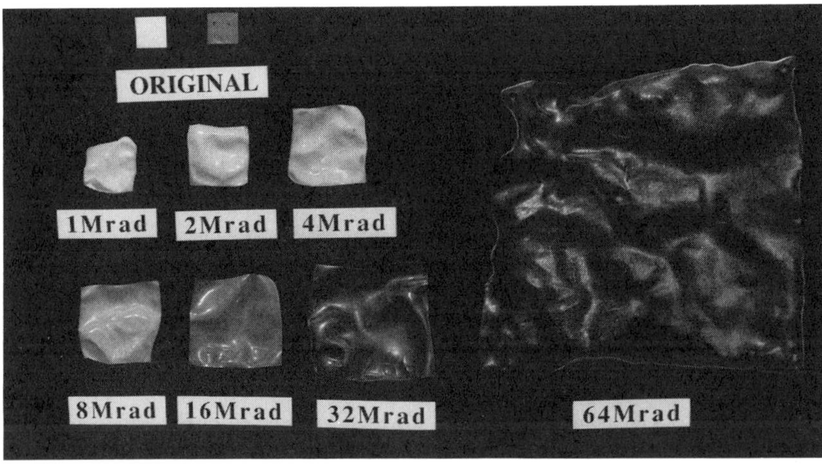

Figure 11 Grafted samples of HDPE
HDPE: 95μm, grafting temperature: 70°C,
reaction time: 120 minutes, 50% acrylic acid solution.

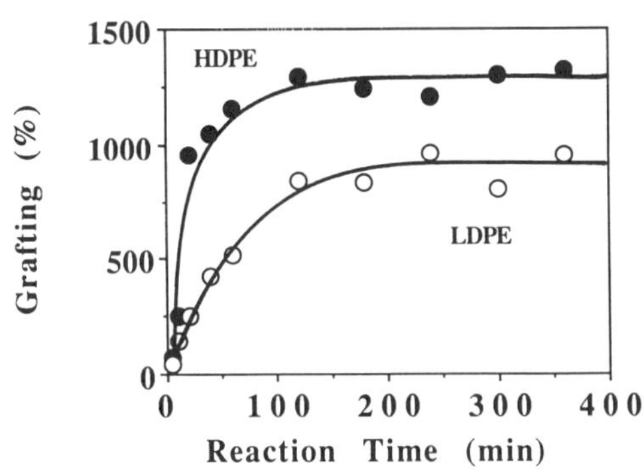

Figure 12 Comparison between HDPE and LDPE.
● HDPE: 95μm, ○ LDPE: 95μm, irradiation dose: 16Mrad,
grafting temperature: 70°C, 50% acrylic acid solution.

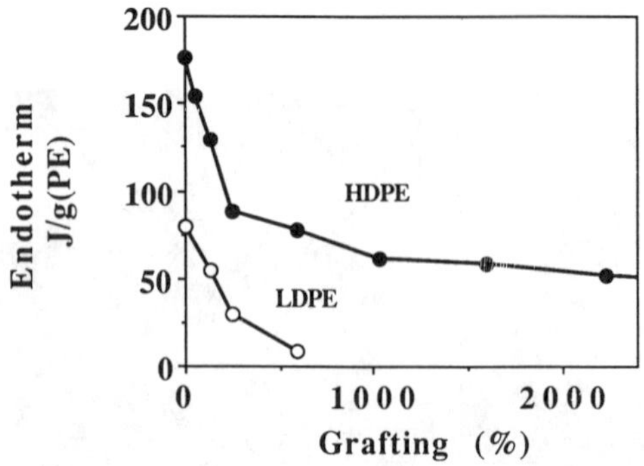

Figure 13 DSC analysis.
Melting endotherm per unit weight of polyethylene
● HDPE, ○ LDPE, heating rate: 10°C/minute.

DSC Analysis Figure 13 shows the plot of grafting and the melting endotherm (per unit weight of PE) by DSC. It indicates that the amount of crystallinity decreases with increasing grafting ratio. One can tentatively conclude that the grafting reaction occurs both in the amorphous region and the boundary of the crystallites initially, then crystallites are destroyed gradually by subsequent grafting reactions.

For HDPE, the crystallinity drops to half of the original value at 250% grafting, and one third at 2300% grafting. The rate of decrease in crystallinity slows down at about 250% grafting. These data suggest that the later stage of the grafting reaction occurs in grafted region rather than the remaining crystalline region in the sample.

Conclusions

The acrylic acid grafting reaction on preirradiated polyethylene films proceeds gradually from the film surface towards the center of the film. The asymptotic grafting achieved in the absence of redox reagent depends on the preirradiation dose while the grafting is independent of the irradiation dose in the presence of Mohr's salt.

When the effect of crosslinking is negligible, the asymptotic grafting increased exponentially with increasing irradiation dose. Both the grafting rate and the asymptotic grafting of HDPE are much higher than those of LDPE. The grafting reaction seems to occurred in both the amorphous region and the boundary of the crystallites.

Highly irradiated HDPE could be changed to almost pure poly (acrylic acid) by grafting reaction. We intend to report the characterization, the physical and mechanical properties, the ion trap capabilities, the wettabilities, the water absorbency, and the permselectivity in a future paper.

Safety Consideration

Electron processors such as the one we used is an electron energy source of very high intensity but the distance which the electrons can penetrate is very small, e.g.

about 45 cm in air at 200 KeV energy. However, secondary radiations such as X-rays or "Bremsstrahlung" are also generated. Users are protected from secondary radiations by shielding lead walls (ca. 6-12 mm thick). The EB accelerator which we used was designed to reduce the exposure level of the operator to at most 0.1 mrem/hour, and is classified as self-shielded. Although the users are protected by such self-shielded system, they should nevertheless wear a film badge when they operate the device, and they should check the radiation level of the machine routinely.

Acknowledgement

We would like to thank Energy Sciences Inc. for access to their electron beam accelerator. The microscopy analyses provided by the Analytical group of the Mitsubishi Paper Mills, Ltd. Central Research Laboratories are also highly appreciated.

References

1)Shiraishi, N; Williams, J. L.; Stannett, V. T. *Radiat. Phys. Chem.*, **1982**, *19*, 73.
2)Suzuki, M.; Tamada, Y; Iwata, H.; Ikada, Y. *Physicochemical Aspects on Polymer Surface;* Mittal, K. L. Ed., Plenum, NY, 1983, Vol 2; pp923-941.
3)Simpson, J. T. *Radiat. Phys. Chem.*, **1985,** *25*, 483.
4)Wang, U. *Radiat. Phys. Chem.*, **1985,** *25*, 491.
5)Japan Kokai Tokkyo Koho; 54-148547.
6)Kaji, K.; Hatada, M.; Yoshizawa, I.; Kohara, C.; Komai, K. *J. Appl. Polym. Sci.*, **1989**, *37*, 2153.
7)*Annual Report of the Osaka Laboratory for Radiation Chemistry Japan Atomic Energy Research Institute,vol 20* JAERI-M 88-272, **1988**, 29.
8)Omichi, H.; Chundury, D.; Stannett, V. T. *J . Appl. Polym. Sci.*, **1986**, *32*, 4827.
9)Andrade, J. D. *ACS Polymer Preprint* , **1972,** *13*, 290.
10)Rather, B. D.; Hoffman, A. S. *J. Appl. Polym. Sci.*, **1974**, *18*, 3183.
11)Lyman, D. J. *Trans. Am. Soc. Art. Int. Org.*, **1972**, *18*, 19.
12)Chapiro, A.; Millequant, M.F.; Jendrychowska-Bonamour, A. M.; Lerke, Y.; Sadurni, P.; Domurado, D. *Radiat. Phys. Chem.*, **1980,** *15* , 423.
13)Sasaki, Y.; Ranter, B. D.; Hoffman, A. S. *ACS Polymer Preprint*, **1975**, *16,* 435.
14)Japan Patent 53-46199.
15)Vitta, S. B.; Stahel, E. P.; Stannett, V. T. *J. Macromol. Sci. A*, **1985**, *22*, 579.
16)Vitta, S. B.; Stahel, E. P.; Stannett, V. T. *J. Appl. Polym. Sci.*, **1986,** *32,* 5799
17)Shiraishi, N; Williams, J. L.; Stannett, V. T. *Radiat. Phys. Chem.*, **1982,** *19*, 79.
18)Charlesby, A.; Fydelor, P. J. *Radiat. Phys. Chem.*, **1972,** *4,* 107.
19)Guimson, C. *Radiat. Phys. Chem.*, **1979,** *14,* 841.
20)Lawler, J. P.; Charlesby, A. *Radiat. Phys. Chem.*, **1980,** *5,* 595.
21)Sidorova, L. P.; Aliev, A. D.; Zlobin, V. B.; Aliev, R. E.; Chalkh, A.E.; Kabanov, V. Y. *Radiat. Phys. Chem.*, **1986,** *28,* 407.
22)Grushevskaya, L. N.; Aliev, R. E.; Kabanov, V. Y. *Radiat. Phys. Chem.*, **1990,** *36,* 475.
23)Ishigaki, I.; Sugo, T.; Senoo, K.; Takayama, T.; Machi, S.; Okamoto, J.; Okada, T. *Radiat. Phys. Chem.*, **1981**, *18*, 899.
24)Ishigaki, I.; Sugo, T.; Takayama, T; Okada, T.; Okamoto, J.; Senoo, K. *J. Appl. Polym. Sci.*, **1982,** *27,* 1033.

25)Ishigaki, I.; Sugo, T.; Takayama, T; Okada, T.; Okamoto, J.; Machi, S. *J. Appl. Polym. Sci.*, **1982,** *27,* 1043.
26)Tanso, S.; Yoshida, S.; Senoo, K.; *Yuasa Jiho,* **1985,** *59*, 35.
27)Omichi, H.; Okamoto, J. *J. Appl. Polym. Sci.*, **1985,** *30,* 1277.
28)Okamoto, J.; Sugo, T.; Katakai, A.; Omichi, H. *J. Appl. Polym. Sci.*, **1985,** *30,* 2967.
29)Furusaki, S.; Okamoto, J.; Sugo, T.; Saito, K. *Chemical Engineering,* **1987,** 521.
30)Saito, K.; Yamada, S.; Furusaki, S.; Sugo, T.; Okamoto, J. *J. Membrane Sci.*, **1987,** *32,* 307.
31)Okamoto, J. *Membrane,* **1989,** *14*, 277.
32)Pimg, Z. H.; Nguyen, Q. T.; Clement, R.; Neel, J. *Membrane Sci.*, **1990,** *48,* 297
33)Hoffman, A. S. *Intern. Meet. Radiat. Process.,* **1989**, Tokyo, 297.
34)Kaetsu, I. *Radiation Technology for Immobilization of Bioactive Materials,* **1987,** Beijing, 153.
35)Nablo, S. V.; Frutiger, W. A. *Radiat. Phys. Chem.*, **1981**, *18*, 1023.
36)Ungar, G. *J. Mater. Sci.*, **1981**, *16*, 2635
37)Harada, J.; Chern, R. T.; Stannett, V. T. ; *RadTech'90 in North America,* **1990,** *1*, 493.
38)Kampouris, E. M.; Andreopoulos, A. G. *Biomaterials,* **1989,** *10,* 206.

RECEIVED May 1, 1991

Chapter 15

Electron-Beam Curing of Aramid-Fiber-Reinforced Composites

C. B. Saunders[1], A. Singh[1], V. J. Lopata[1], S. Seier[1], G. D. Boyer[1], W. Kremers[1], and V. A. Mason[2]

[1]Whiteshell Laboratories, AECL Research, Pinawa, Manitoba, Canada
[2]Chalk River Laboratories, AECL Research, Chalk River, Ontario, Canada

Electron-beam (EB) processing of aramid fiber-reinforced composites uses ionizing radiation, specifically high-energy electrons or X-rays produced by an electron accelerator, to initiate polymerization and cross-linking reactions in suitable matrices, curing the polymer and enhancing its specific physical and chemical properties. Many benefits have been identified for using EB processing for composites rather than thermal curing, including ambient curing temperatures, reduced curing times, and improved material handling and resin stability. EB-curable matrices have been developed with properties comparable to thermally cured epoxies and polyesters. This paper examines the effect of irradiation dose and dose rate on the curing of acrylated epoxies and discusses selected properties of aramid fabric-acrylated epoxy laminates.

Aramid or aromatic polyamide fibers are used to reinforce several polymer matrices, producing composite materials with high tensile strengths and moduli, fatigue resistance, damage tolerance and thermal stability *(1)*. Applications for aramid fiber-reinforced composites are found in several industries, including aircraft and aerospace, military, marine and sporting goods *(2)*.

Electron beam (EB) processing of composites involves using electrons as ionizing radiation to initiate polymerization or cross-linking reactions in suitable substrates, thereby enhancing specific physical and chemical properties. Industrial EB processing of plastics and composites has been an established industry for 45 years *(3)*. Currently, there are about 600 EB accelerators being used worldwide to process plastics *(4)*.

0097–6156/91/0475–0251$06.00/0

The benefits of EB curing fiber-reinforced products, include *(5)*:

1. Curing at ambient temperature. Tooling/mandrel materials with appropriate thermal expansion coefficients must be used in composite processing methods to produce a component to strict dimensions while minimizing internal stresses. The tool expands and contracts during the thermal curing cycle, often at a different rate than the molded composite. These movements can alter the dimensions and produce excessive internal stresses in the cured product, which can decrease its strain-to-failure and fracture toughness *(6)*. EB curing at ambient temperatures eliminates thermal expansion in both the tool and the product, thus reducing dimensional changes and internal stresses in the final product.

2. Reduced curing times for individual components. A typical EB-curable composite can be cured with a maximum dose of about 50 kGy (50 kJ/kg). The product throughput for an accelerator is a function of the machine's power. A 50-kW electron accelerator can provide this dose to about 1.8 Mg/h of material, assuming that at least 50% of the energy provided by the accelerator is actually absorbed by the product being irradiated. This production speed is several times higher than for thermal curing with a typical autoclave or oven, even though the products are cured one at a time rather than in large batches.

3. Improved material handling. EB processing is a continuous operation and components can be EB-treated immediately after they are produced. This continuous operation makes production scheduling and inventory control easier and reduces the number of identical molds needed to manufacture products economically, as compared to using a batch process such as thermal curing with autoclaves or ovens. EB processing is also better suited for short production runs because parts are cured one at a time.

4. Improved resin stability at ambient temperature. The shelf-life of EB-curable resins can be much longer than the shelf-life of formulations for thermal curing because EB-curable formulations do not normally auto-cure at room temperature, making low-temperature storage unnecessary.

5. Reducing the amount of volatiles produced. Thermal curing of composites often produces volatile degradation products that can be hazardous and require proper control procedures *(6)*. EB processing eliminates the production of thermal degradation products, though very small amounts of gases such as hydrogen and carbon dioxide may be produced *(7)*. The disposal of these gases would not cause problems

and would require much less effort than that required to deal with the volatiles from thermal curing.

6. Better control of the energy-absorption profile for a component, allowing greater design flexibility. In conventional processing, materials with vastly different thermal curing cycles cannot be combined in a single product. Since electron beams can be easily manipulated, different EB-curable materials can be used together in the same product and each component can be given a different dose, if required.

Many traditional fabrication methods, including filament/tape winding, resin transfer molding, pultrusion and hand lay-up, can be combined with EB processing. Filament/tape winding is particularly well-suited because it requires limited or no external pressure (high pressure requirements complicate the EB process) and high winding rates can make efficient use of the available energy.

The constraint of EB processing for manufacturing composite structures is that EB-curable (free-radical-initiated) matrix systems are required, but none have as yet been widely qualified for high-performance aerospace or aircraft applications. Although the polyester systems commonly used for winding consumer products can be cured via free-radical mechanisms, the epoxy formulations currently being used in advanced composites for aircraft, aerospace, and naval applications are generally unsuitable for EB curing *(8)*. For many products, epoxies provide an almost unbeatable combination of handling characteristics, processing flexibility, composite mechanical properties, and acceptable cost *(1)*. Most epoxies polymerize by a cationic mechanism under the influence of high-energy radiation. This process is inhibited by trace amounts of water *(8)*. However, equivalent EB-curable polymer systems are being designed using appropriate resin formulations; for example, by acrylating the terminal epoxy groups of the epoxy resin *(8)*.

Research in EB-curable composites over the past 15 years, together with the recent availability of industrial, high-energy, high-power accelerators, has made possible the industrial production of filament-wound, fiber-reinforced motor cases for aerospace applications *(9)*. Research has been conducted on a number of topics, including matrix development, sizings and interface properties, EB effects on fiber properties, process development, process control, and radiation safety and dosimetry. This paper describes our research program to study EB-curable aramid fiber-reinforced composites. The program objective is to design and manufacture EB-curable composites that meet mechanical and physical property specifications for selected applications.

Materials and Sample Preparation

Table I gives the common forms of acrylated epoxies, polyesters and urethanes for composite manufacturing *(9)*. The resin system selected for our initial experiments was an epoxy diacrylate. Size exclusion chromatography (SEC) showed that the resin was 60% dimer with a molecular weight of 908 g/mol, with 36% trimer and higher oligomers, 3% monomer and 1% acrylic acid and hydroquinone, an ultraviolet stabilizer. Table II gives the range of properties presently available for both epoxy and EB-curable acrylated epoxy polymers *(10)*. The glass transition temperature (T_g) of the difunctional EB-curable acrylated epoxy is up to 150°C, while the tetrafunctional forms have a T_g up to 200°C.

Aramid fibers, specifically Kevlar fibers, are poly (p-phenyleneterephthalamide) (PPD-T). These fibers were selected for composite preparation because of their radiation stability, a consequence of the highly aromatic structure *(11)*, and because Kevlar is the fiber of choice for many of the products that are suitable for EB processing *(2)*. The suitablity of a product depends on the product size and shape, and the required curing conditions, such as pressure and temperature. A plain-weave Kevlar 68 fabric was selected for our initial experiments. Table III lists some of the fabric/fiber specifications *(1)*.

A solvent process, using acetone, was employed to impregnate the aramid fabric to a resin loading of 40% (by mass). The fibers were not chemically treated prior to impregnation. Chemical treatment increases the expected adhesion between the selected polymer and the fibers *(12)*. Laminates containing 10-plys, all in the same fiber orientation (±1°), were prepared using standard hand lay-up methods, including vacuum-bagging to remove volatiles produced during EB treatment and to eliminate oxygen, a free radical scavenger. Table IV gives the experimental curing conditions. A laboratory-scale pressure chamber can be used to EB cure the laminates. The pressure limit for our chamber is 5 MPa at 100°C and the maximum sample thickness is 15 mm. Samples of the neat resin were also EB treated under vacuum, at various dose rates, for gel fraction analyses.

Process Considerations

The electron beam penetration limit must be considered when determining the applicability of EB processing for composites. Electron penetration is a function of beam energy and product density. A 10-MeV beam can penetrate about 40 mm in unit-density material, or 40 kg/m^2 with one-sided irradiation and 90 mm (90 kg/m^2) with two-sided irradiation, with the top and bottom surface doses being equal *(13)*. The penetration limit is inversely proportional to density. The EB must also penetrate any vacuum bag and/or pressure vessel if these are required for manufacturing the component. This penetration limit makes it necessary to design the product and the pressure vessel so that the

Table I. Chemical Forms of EB-Curable Resins [9]

Resin Type	*Chemical Formula*
Difunctional Acrylate Epoxy	$CH_2=CH-C(=O)-O-[CH_2CH(OH)-CH_2\ -(C_6H_4)-\ C(CH_3)_2\ -(C_6H_4)-\ CH_2-CH(OH)-CH_2-O-C(=O)-CH=CH_2$
Tetrafunctional Acrylated Epoxy	$R_2N-Z-NR_2$ where R may be $-CH=C(CH_3)-COO-CH_2-C(OH)=CH_2$ and Z may be aromatic or cycloaliphatic
Acrylated Polyesters	$CH_2=CH-C(=O)-O-(CH_2)_6-[O-C(=O)-(CH_2)_4-C(=O)-O-(CH_2)_6]-O-C(=O)-CH=CH_2$
Acrylated Urethanes	$CH_2=CH-C(=O)-O-CH_2-CH_2-O-C(=O)-NH-(C_6H_3-CH_3)-NH-C(=O)-O-CH_2-CH_2-O-C(=O)-CH=CH_2$

Table II. Typical Properties of Epoxies and EB-Curable Acrylated Epoxies [10]

Property	*Epoxies*	*Acrylated Epoxies*
Ultimate Strength		
(MPa): Tensile	28-90	50-82
Compressive	103-172	NA[a]
Modulus (GPa):		
Tensile	2.4-4.1	2.3-4.2
Compressive	2.8-4.1	NA
Elongation at Break (%)	3 - 6	1.5-8.2
Coef. of Linear Thermal Expansion (μm/(m·°C))	45 - 65	NA
Glass Transition Temperature (°C)	100-175	53-200

[a]Not available from the supplier.

Table III. Kevlar 68 Fabric Specifications/Fiber Properties [1]

Specification/Property	*Value*
Weave Type	Plain
Fabric Thickness	
10-ply	1.6 mm
Fiber Tensile	
Strength	3600 MPa
Modulus	110 GPa
Fiber Density	1440 kg/m^3

product can be fully penetrated and cured by an EB. Products with specific densities greater than 40 kg/m^2 can also be irradiated by modifying the 10-MeV electron accelerator to produce X-rays. Figure 1 gives the penetration of both 10-MeV electrons and X-rays in a typical aramid fiber composite (density = 1500 kg/m^3). The absorbed 10-MeV electron dose initially increases because of an electron scattering effect, reaching a maximum at a penetration depth of about 25 kg/m^2.

Table IV. EB Processing Conditions

Dose Range	0 to 100 kGy
Dose Rate	1 to 1500 kGy/h
Beam Energy	10 MeV
Beam Power	1 kW
Beam Scan Width	0.6 m
Curing Pressure	600 kPa
Curing Temperature	25°C

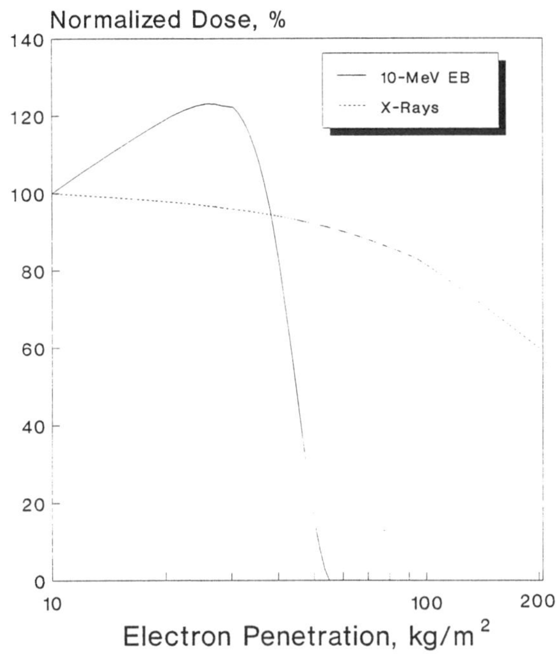

Figure 1. Penetration Limit of 10-MeV Electrons and X-Rays In Aramid Fiber Composites (Density = 1500 kg/m³)

Figure 2 gives the penetration limit for a 10-MeV EB in aramid composites (density = 1500 kg/m³) as a function of the pressure required during processing. This limit was calculated from the energy lost by the beam penetrating a titanium window thick enough to withstand the pressure and 100 mm of air at the specified pressure before hitting the product. The beam energy is effectively reduced from 10 to 8.4 MeV at the required pressure of 1 MPa. The penetration limit for the beam is therefore reduced from 27 to 24 mm for one-sided EB treatment, and from 60 to 50 mm for two-sided treatment.

Thermal energy is added to material during EB processing by radiation-induced exothermic reactions and by the net energy of the

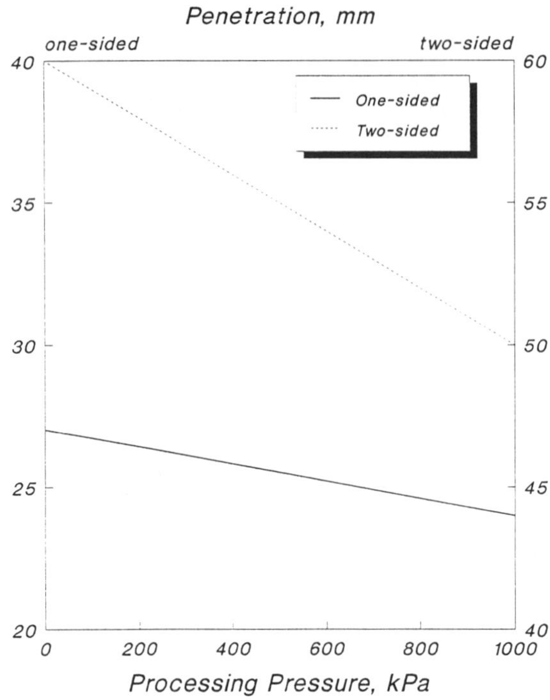

Figure 2. Penetration Limit of 10-MeV Electrons in Aramid Fiber Composites as a Function of Gauge Curing Pressure (Density = 1500 kg/m³)

absorbed radiation *(14)*. The final temperature of an irradiated object depends on its shape, specific heat, thermal conductivity and the electron dose and dose rate. In composites thicker than 2 cm, local heating can occur, leading to composite damage because the energy absorption properties of the reinforcements are different than those of the matrices *(14)*. Excessive heating in composites can be avoided by proper selection of the resin formulation to minimize the thermal energy released by exothermic reactions, reducing the required curing dose by using cross-linking promoters and fractional irradiation techniques *(15)*. Converting to low-dose-rate X-rays also reduces or eliminates heating problems. Multiple passes are normally required to deliver the needed energy to any large component, which allows the product to be maintained at an acceptably low temperature (< 50°C) *(9)*.

An electron accelerator is a source of penetrating radiation, and must therefore be operated in a shielded room. Shielding is usually supplied by concrete with an earth berm often supplementing the concrete. The penetrating radiation of concern is not the electrons, but the Bremsstrahlung (X-rays) and neutrons produced when the electrons impinge in materials. The efficiency for Bremsstrahlung production increases with the atomic number (Z) of the irradiated material.

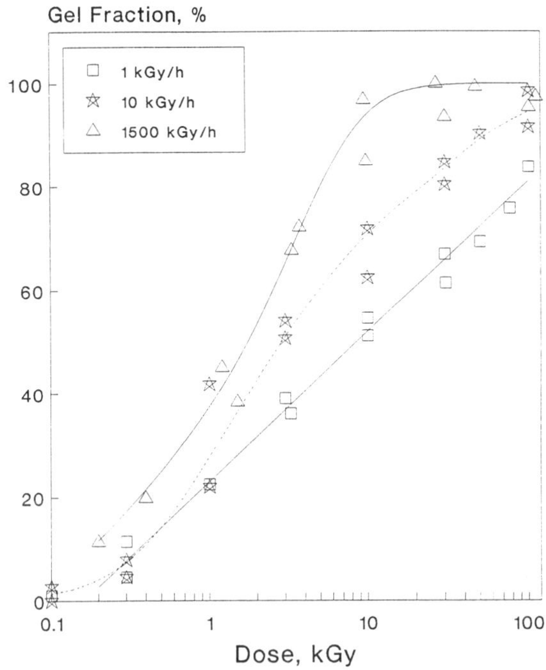

Figure 3. Effect of Dose Rate on the EB Curing of Epoxy Diacrylate Polymer

Facilities designed to shield against Bremsstrahlung using concrete are generally more than adequate to shield against neutrons.

Results and Discussion

Figure 3 plots the gel fraction of our EB treated polymer matrix as a function of dose at three dose rates, 1 kGy/h (X-rays), 10 kGy/h (X-rays), and 1500 kGy/h (EB). Gelation begins at a dose of about 0.1 kGy over the entire dose rate range. At any specific dose up to 100 kGy, the gel fraction in the cured polymer increases with increasing dose rate. The final properties of a cured composite are a function of the degree of cross-linking in the matrix polymer, and are therefore affected by any large changes in dose rate, such as when going from EB to X-ray treatment to obtain the necessary beam penetration. The magnitude of these gel fraction changes must be known so the required dose can be delivered at each selected dose rate to ensure uniform properties throughout an entire product.

Table V gives some of the mechanical properties of our 10-ply laminates, EB-cured at 600 kPa to a dose of 50 kGy in the absence of oxygen, and comparable thermally cured aramid fabric-epoxy laminates *(1)*.

As expected, the tensile properties of the EB-cured laminate compare favourably with the thermally cured material. The tensile properties are primarily fiber-dependent and EB treatment (50 kGy) does

Table V. Typical Tensile and Compression Properties of Kevlar 68 Fabric-Epoxy Laminates[a] [1]

Property	*Cure Type*	
	EB-Cure Epoxy	*Thermal Epoxy*
Tensile Strength (MPa)	590	525
Tensile Modulus (GPa)	34	27
Compressive Strength (MPa)	89	180
Compressive Modulus (GPa)	33	27

[a]A 10-ply laminate; 600-kPa curing pressure; 60% fiber mass.

not affect the properties of the fibers. The compression properties of the EB-cured laminate are lower than the thermally cured material because they depend largely on the polymer properties and the adhesion between the fibers and the matrix. The fibers used to prepare the EB-cured material had no coupling agents applied to them to increase the fiber-matrix bonding *(6)*. Research is continuing to study the properties of EB-curable resin formulations and the interface chemistry to select sizings/coupling agents to improve the compression properties of EB cured laminates.

X-ray diffraction analysis of the EB-cured polymers confirmed that the polymers were amorphous *(16)*. The volatile content in each EB-cured polymer was below 2%. Gas chromatography and mass spectroscopy showed that the volatiles released during EB curing were primarily low-molecular-weight hydrocarbons and some carbon dioxide *(16)*. Carbon dioxide was released from samples irradiated under either oxygen or nitrogen, suggesting that atmospheric oxygen takes little or no part in the formation of CO_2. The CO_2 is a radiolytic degradation product.

Summary

Advanced composites, including aramid fiber-reinforced composites, are used extensively for a variety of demanding structural applications. The benefits of using EB-curing rather than thermal curing for aramid fiber-reinforced composites include curing at ambient temperature, reduced curing times for individual components and fewer volatiles. The gel point of the EB-cured epoxy diacrylate polymer was about 0.1 kGy and was unaffected by dose rate. The gel fraction in the cured polymer increases with increasing dose rate at any dose selected up to 100 kGy. The final properties of a cured composite are therefore affected by any large changes in dose rate, such as when going from EB to X-ray treatment to obtain the necessary beam penetration, and the delivered dose must be increased to compensate for this change. The tensile properties of a 10-ply, EB-cured, aramid fiber laminate were comparable to the properties of thermally cured aramid fiber laminates.

Literature Cited

1. *Composites;* Dostal, C.A., Ed.; ASM International: Metals Park, Ohio; 1987.
2. *Advanced Thermoset Composites;* Margolis, J.M., Ed.; Van Nostrand Reinhold Co.: New York, N.Y., 1986.
3. Menezes, T.J.; Nablo, S.V. *Radiat. Curing.* **1985,** *12,* 2, pp. 2-9.
4. Saunders, C.B. *Radiation Processing in the Plastics Industry: Current Commercial Applications.* Atomic Energy of Canada Limited Report, AECL-9569, 1988.
5. Saunders, C.B.; Singh, A. *The Advantages of Electron Beam Curing of Fibre-Reinforced Composites.* Atomic Energy of Canada Limited Research Company Report, RC-264, 1989. Unpublished report available from SDDO, AECL Research, Chalk River Laboratories, Chalk River, Ontario K0J 0J0.
6. Weeton, J.W.; Peters, D.M.; Thomas, K.L. *Engineer's Guide to Composite Materials;* ASM International: Metals Park, Ohio, 1987.
7. Bradley, R. *Radiation Technology Handbook;* Marcel Dekker: New York, N.Y., 1984.
8. Dickson, L.W.; Singh, A. *Radiat. Phys. Chem.* **1988**, 31, 4-6, pp. 587-593.
9. Beziers, D. *Electron Beam Curing of Composites*; Proc. of 35TH International SAMPE Symposium, Anaheim, CA, 1990.
10. *Modern Plastics Encyclopedia;* Juran, R., Ed.; McGraw Hill: New York, N.Y., 1990, pp. 516-517.
11. Bly, J.H. In *Modern Plastics Encyclopedia;* Juran, R., Ed.; McGraw-Hill: New York, N.Y., 1990.
12. Chen, M.L.; Ueta, S.; Takayanagi, M. *Polym. J.* **1988,** *20,* 8, pp. 673-680.
13. Dickson, L.W.; McKeown, J. *Radiation Interactions with Linac Beams;* in Proc. of Working Meeting on Radiation Interactions, Leipzig, 1987.
14. Zagorski, Z.P. *Radiat. Phys. Chem.* **1985,** *25,* pp. 291-293.
15. Saunders, C.B.; Carmichael, A.A.; Lopata, V.J.; Singh, A. *Physical and Mechanical Characterization of Radiation-Curable Carbon Fibre Composites;* Proc. of 9th Annual CNS Conference, pp. 480-486. Canadian Nuclear Society: Toronto, Ontario, 1988.
16. Pigliacampi, J.J.; Riewald, P.G. In *Modern Plastics Encyclopedia;* Juran, R., Ed.; McGraw-Hill: New York, N.Y., 1990.

RECEIVED February 4, 1991

Chapter 16

Gamma-Ray Irradiation of Polystyrene in the Presence of Cross-Linking Agents

Delia López[1] and Guillermina Burillo[2]

[1]Escuela de Ciencias Químicas, Universidad Autónoma de Puebla, Pue., México, C. Postal 1613
[2]Instituto de Ciencias Nucleares, UNAM, Ciudad Universitaria, México, D. F. 04510

The radiation induced crosslinking of polystyrene is difficult because of its aromatic units, and the polymer degrades when irra diated in air. Therefore, the effects of gamma-ray irradiation on polystyrene in the presence of polyfunctional monomers such as trimethyl propanetrimethacrylate (TMPTM), dimethyleneglycol bis-allylcarbonate (DEGBAC) and propyleneglycol bis-allylcarbonate(PGBAC). These monomers readily polymerize by the ra diation with small dose.
In the case of DGBAC and DPGBAC (allyl monomers) the dose required for crosslinking of polystyrene was higher than that with TMPTM (vinyl monomer).
The relationship between doses and % of gel formed as well as radiochemical yields for gelation and degradation for those poly functional monomers will be presented.

A great number of works on the solid state irradiation of polymers have been reported, many of them aimed at obtaining crosslinked polymers with improved physical properties; however not all the polymers irradiated readily undergo crosslinking mainly and predominate main-chain degradation. Polystyrene (PS) for example does not undergo crosslinking when irradiated in air but it crosslinks when irradiated in vacuum, although crosslinking yield (Gc) is very low; this high resistance to radiation induced changes, is due to the protective effect of the benzene rings in the molecule. Charlesby[(1)] reported that the incipient gel formation (rg) occurred with 270 KGy and the crosslinking yield is only 0.03; the PS is considered to be one of the polymer difficult to be crosslinked by irradiation therefore it is important to improve this effect in the presence of polyfunctional monomers.

0097–6156/91/0475–0262$06.00/0

There are many factors on the effects of PS irradiation: the rg can be changed depending on the initial molecular weight from 50,000 to 3x10^6 [2]. It was found that rg depend on the age of the PS samples, the old ones need higher doses than freshly prepared material. Burillo and Ogawa[3], found that pressure on the samples before irradiation reduces lagerly the rg and Schults[4], shown that high dose rate decrease rg until 10 Mrad, (100) KGy), even in the presence of air.

It must be emphasized that polymers are very seldom used in this pure polymeric form and usually contain plasticizers, antioxidants, stabilizers, etc. to facilitate the processing of these materials; no additive can be considered to be "inert" with regard to radiation, it will be appreciated that the additives can influence the stability of a polymer toward radiation. Certain additives (sensitizers) can accelerate the radiation crosslinking and thereby reduce the irradiation dose requiered to achieve a specific degree of crosslinking.

In this paper we studied the effect of radiation on the crosslinking, first, in the presence of an external pressure and then with the addition of crosslinking agents.

EXPERIMENTAL SECTION

SOLVENTS AND POLYFUNCTIONAL MONOMERS

The Solvents of "reagents grade" were used as received, but those of uncertain grade were purified according to literature method, the high purity monomers (purity >98%) as trimetehyl propanetrimethacrylate (TMPT) and diethyleneglycol bis-allyl carbonate (DEGBAC) were supplied by Ancomer Limited and Akso-Chemie, they were used as received. Propyleneglycol bis-allylcarbonate (PGBAC), was synthesized according to the literature and purified by molecular distillation or low temperature crystallization.

The technical grade styrene were purified twice by fractional distillation just before polymerization. AIBN was crystalized from toluene solution.

POLYMERIZATION.

Different $\bar{M}n$ polystyrene (PS) samples were obtained in our laboratory by bulk or emulsion polymerization using AIBN or $K_2S_2O_8$ as initiator. The intrinsic viscosities in benzene at 30°C were determined and the number-average molecular weights of PS calculated from the Mark-Houwink equation.[6] Their molecular weight distribution were obtained from SEC (GPC).

IRRADIATION OF PS SAMPLES.

Compressed disks were obtained by external pressure applied on the solid Purified polymers with hydraulic press.

Samples of PS($\bar{M}n = 41300$) and polyfunctional monomers were prepared with monomers in solution with a non solvent of PS (MeOH). When PS were well wetted the MeOH was exhaustively evaporated. One Gammabean 651 PT from atomic Energy of Canada with 50000 Ci(initial) was used as the source of ^{60}Co γ-radiation. Dose rates from 2.6 to 10.5 KGy/h were utilized. Samples for irradiation: powder, pressure disks or mixtures of polystyrene-polyfunctional monomers were irradiated in a vacuum sealed pyrex tubes at different dose and dose rates. The percentage of gel formed by irradiation of PS, were determined by extracting the soluble fraction with benzene or Toluene. The irradiated samples of PS-polyfunctional monomers mixtures were extracted with THF.

Sol and gel fractions were characterized by viscometry, IR, or SEC. Perkin Elmer 1600 Ft, infrared spectrometer and a model 501 Waters Associates liquid chromatograph with two columns of ultrastyragel were used. Dose for incipient gel (rg) was estimated by extrapolation in a plot of % gel versus radiation dose at 0% gel, and by a plot of log (s $+\sqrt{s}$) against log dose extrapolated until s $+ \sqrt{s} = 1$, because it showed more linear behaviour. The radiochemical yields of degradation Gd and crosslinking Gc were calculated using the Charlesby Pinner equation[1].

RESULTS AND DISCUSSION.

It was studied by Burillo and Ogawa[3,7] that pressure applied to vinyl polymers before irradiation and by (millinchuck et al[8-9]), that one external pressure applied to the polymers in the moment of the irradiation, increases the crosslinking and decrease the chain degradation. We found that it is also true in case of PS, but even the increase is considerable, the % of gel formed were not comercially important at the range of molecular weigh (413000 - 579000), doses (0 - 500 KGy) and dose rate (0.46 - 22 KGy/h) studied. We observed that the % of gel formed increase with the increase in pressure of the samples until one optimun pressure and then at higher pressure the % of gel decrease. The optimun pressure in case of polystyrene was 7.5 ton/cm^2 in different conditions of $\bar{M}n$, dose rate(I), and irradiation dose as shown in Figure 1.

In table I we show radiochemical yield of crosslinking Gc and degradation Gd, as well as dose for incipient gel in different conditions for the compressed disks of PS with different $\bar{M}$n. When the irradiation was performed at 7.5 ton/cm^2 pressure; we observe that dose for incipient gel decrease but Gc and Gd are independent of molecular weight and depend on dose rate irradiation.

Gc values found in literature for different polystyrene samples and radiation sources ($\bar{e}, \gamma$, nuclear reactor) are between 0.19 to 0.049. In case of irradiation of samples under previous pressure, the G_c values increases and we found values from 6.7 to 15.2, and doses for incipient gel formation decreases from 270 KGy for powder samples until 62-5 KGy in cases of compressed samples.

The effect of external pressure applied to polymer is to decrease the interchain distance to facilitate the combination of macroradicals. The optimum pressure is caused by the balance of two factors; the interchain distance and the polymer chain mobility, which decrease with increase in pressure. Further increase in pressure beyond the optimum pressure inhibits the chain motion. In any case the percent of gel obtained are not commercial useful, so we tried to improve crosslinking by addition of crosslinking agents.

Table I. Gc, Gd and rg values in polystyrene irradiation

$\bar{M}n$	$I = KGy/h$	$rg(KGy)$	Gc	Gd	Gd/Gc
41300	0.46	114.9	6.7	24.7	3.7
41300	8.80	100.0	15.2	30.4	2.0
57900	0.46	62.5	6.8	25.9	3.8

IRRADIATION OF CROSSLINKING AGENTS.

Various types of crosslinking agents have been employed to lower the dose required to achieve a sufficient degree of crosslinking. They are different kinds of polyfunctional monomers; one of them is a vinyl monomer whose chemical reactivity is more enhanced than that of the allyl monomers. There are:

One vinyl polyfunctional monomer, the trimethylpropanetrimethacrylate (TMPTM) and two allylcarbonates: diethyleneglycol bis-allylcarbonate (DEGBAC) and propyleneglycol bis-allylcarbonate (PGBAC), was studied. Figure 2.

They were sealed in vacuum and irradiated at a dose rate of 2.6 to 10.5 KGy/h, the amount of gel formed were determined by extracting the soluble fraction with THF. Figure 3 shows the % of gel formed at different doses and dose rate for TMPTM in the presence of air and vacuum irradiation for two different dose rates. At a dose rate of 10.5 KGy/h, we found rg=1 KGy for air irradiation, but in vacuum rg is only about 0.1 KGy independent of dose rate.

The crosslinking of both allyl monomers by γ irradiation is shown in the fig. 4. We can see that at the same dose rate of 2.6 KGy/h the rg of DGBAC is about 56 KGy while the rg of PGBAC is 65KGy. However the rg values of the allyl monomers are closest to each other. For both of them the amount of gel formed at 250 KGy is over 70 %.

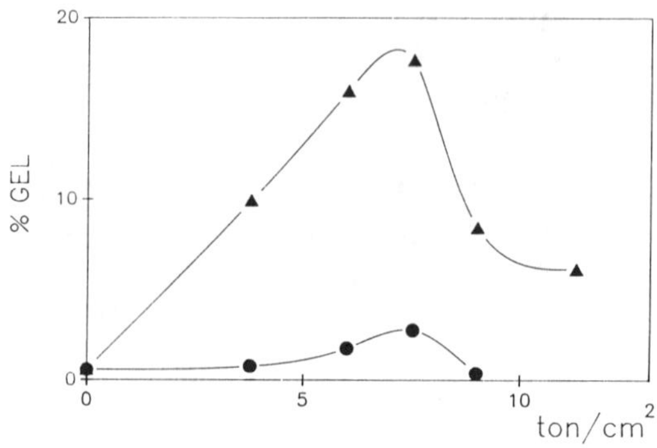

Figure 1. % gel obtained at different pressures; • • • 200 KGy, I = 0.46 KGy/h, $\bar{M}n$ = 41300; ▲▲▲ 400 KGy, I = 12 KGy/h, $\bar{M}n$ = 350 000; adapted from ref. 3.

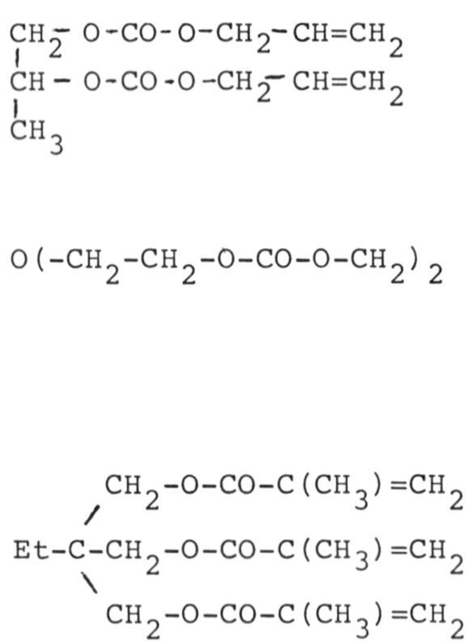

PGBAC

PM = 244.2

$C_{11}H_{16}O_6$

DEGBAC

PM=274.26

$C_{12}H_{18}O_7$

TMPTM

PM=338.4

$C_{18}H_{26}O_6$

Figure 2. Crosslinking Agents.

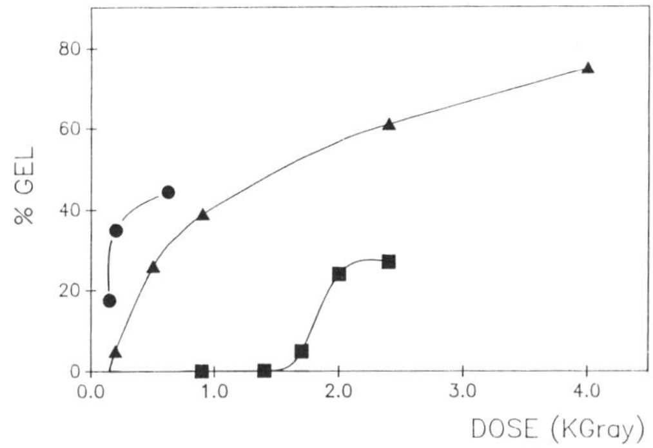

Figure 3. % gel obtained at different dose rates for TMPTM; ● ● ● I = 2.45 KGy/h (vacuum), ▲▲ 10.5 KGy/h (vacuum); ■ ■ ■ 10.5 KGy/h (air).

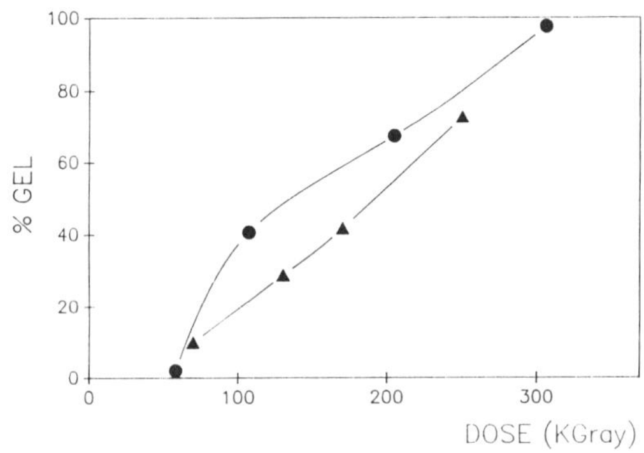

Figure 4. % gel formed for : ● ● ● PGBAC; ▲▲ DGBAC.

The results shown that the more reactive of them the TMPTM polymerizes at too lower doses them the least reactive allyl carbonates monomers. But the rg values of the allyl monomers are not too different with regard to the rg value of the PS here studied.

IRRADIATION OF POLYSTYRENE CROSSLINKING AGENTS.

Polystyrene powder mixed with the polyfuntional monomers studied before was crosslinked by gamma irradiation at different doses and dose rates.

Figure 5, shows the relationships between gel percent and irradiation dose for PS ($\bar{M}n = 41300$) with different amounts of TMPTM. We can see that, the higher the percentage of added monomer the higher the percentage of gel formed. In all mixtures most of the gel is formed at about 30 KGy, although the % of gel formed is determined by the amount of TMPTM in the mixture. It is also interesting to point out that all of the irradiated mixtures present the same rg, which is about 0.2 KGy like that of the pure TMPTM thereby, in this case the rg is independent of the composition of the mixture and it is also independent from the dose rate used.

In Order to get higher Gc, lower rg and higher ratio PS/additive, we tried another polyfunctional monomers with higher rg than TMPTM. Figure 6 shows the percent of gel formed by the irradiation of PS added with DGBAC and PGBAC in amounts of 20% at a dose rate of 0.26 KGy/h.

In spite of the similar behavior of the two allylcarbonate monomers, their % gel obtained in the mixture PS/monomer was two different; when the relationship between percent of gel and the irradiation dose for PS/20% DGBAC in vacuum was observed, the % gel found was low in all cases, even for 300 KGy and the rg observed is about 8.3 KGy. The % gel obtained for the mixture PS/20% PGBAC irradiated in vacuum was higher than in case of DGBAC and TMPTM, about 70% at a dose of 250 KGy. In table II, the rg for the three monomers and mixture of PS-monomere was given.

Table II. rg in KGy on the compound studied

	Monomer	*PS − monomer*
TMPTM	0.10	0.20
DGBAC	56.00	96.00
PGBAC	65.00	70.00
PS	−−	270.00

Decrease on Dose for incipient gel (rg) is important in relation with the PS without crosslinking agents, the lowest value was founded in case of TMPTM,

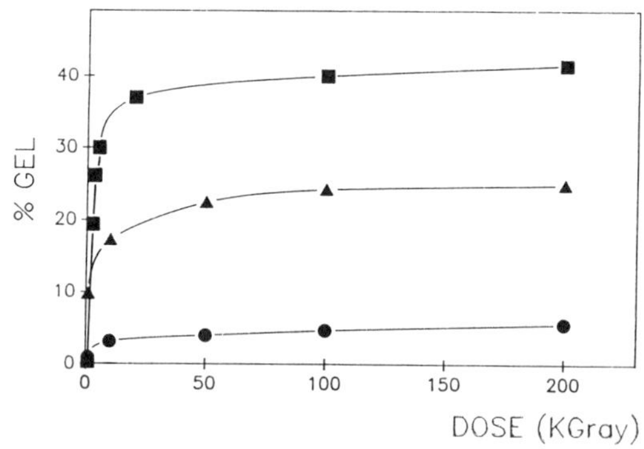

Figure 5. Relationship between % gel and the irradiation doses for PS-TMPTM with different amounts of TMPTM; ● ● ● 5%, ▲▲▲ 20%, ■ ■ ■ 33%.

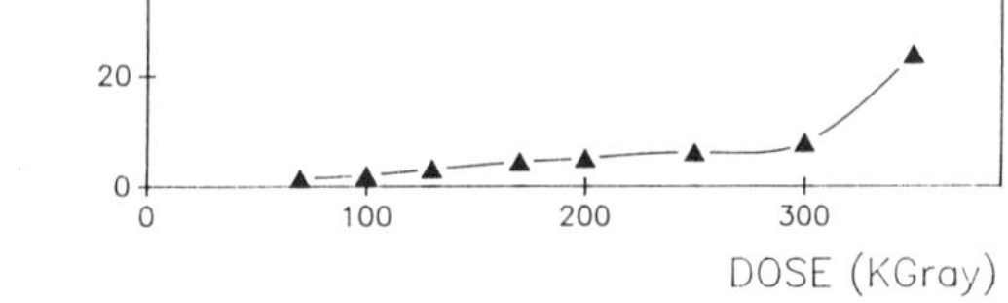

Figure 6. Relationship between % gel and doses; ▲▲▲ PS-DGBAC; ● ● ● PS-PGBAC.

but the maximum % of gel was obtained for the mixture PS-PGBAC. Other PS-PGBAC system and other sources of irradiation will be studied in order to low the doses, dose rate and crosslinking rate for this system. Beside, It will be studied the physical changes of the irradiated PS-PGBAC because of the individual allylcarbonate monomers casting in "Engeenering Polymers". It has been reported (10) that in the case of some trifunctional monomers as ATMPT, it exhibits a high gel fraction at low doses, and as the absorbed dose further increases, the gel fraction, again start increasing; it could be explained because at doses below 200 KGy only two of the tree functional groups reacts and the third participates at higher doses. Therefore highER doses most be studied for the PS-TMPTM system.

LITERATURE CITED.

1.- A. Charlesby and S. H. Pinner, Proc. R. Soc. A, 244, 367. (1959)
2.- L. A. Wall and D. W. Brown; the journal of physical chemistry, 61, No. 2, 129 (1957).
3.- G. Burillo and T. Ogawa; Radiat, Phys. Chem., 25, No. 1-3, 338-388 (1985).
4.- Shultz, A. R., Ruth, P. J. and Rathmann, G. B., J. Polymer Sci., 22, 495 (1956).
5.- Irving E. Muskat and Franklin Strain; U. S., Pat. 2370 565; Feb. 27, (1945).
6.- Brandrup, J., and E. H. Immergut: "Polymer Handbook", IV. 18, Interscience, New York, (1966).
7.- G. Burillo, G. Canizal and T. Ogawa; Radiat. Phys. Chem., 33, No. 4, 351 (1989). Int. J. Radiat. Appl. Inst. Part C.
8.- Millinchuk V. K., Klinshpont E. R. and Kiryukhin V. P., Radiat. Phys. Chem., 21 (6), 519 (1983).
9.- Millinchuk V. K., Klinshpont E. R. and Kiryukhin V. P., Radiat. Phys. Chem., 28 (3), 331 (1986).
10.- T. Sawasaki and A. Nojiti, Radiat. Phys. Chem. 31, No. 4-6, 877 (1988).

GENERAL REFERENCES.

- Joseph E. Wilson, "Radiation Chemistry of Monomers, Polymers, and Plastics"; Marcel Dekker, Inc. N. Y., 1974.
- A. Charlesby. Atomic Radiation and Polymers. Pergamon Press. Great Britain (1966).

RECEIVED July 2, 1991

Chapter 17

Hydrogel Dressings

Janusz M. Rosiak

Institute of Applied Radiation Chemistry, Technical University of Lodz, Wroblewskiego 15, 93–590 Lodz, Poland

Hydrogel dressings are the most promising class of temporary coverings of burn wounds. They contain over 90% of water and accelerate a cure. These dressings may be also applied for the healing of all other kinds of skin damages in which humid medium is favourable. A new method of production of hydrogel dressings as well as their physical and medical properties have been described. The mechanism of radiation formation of hydrogels and the role of intermediate products of water radiolysis in the process of polymerization and crosslinking of N-vinylpyrrolidone and acrylamide has been discussed.

Hydrogels are most often defined as two-component systems where one of the components is a hydrophilic polymer, insoluble in water because of existing three-dimensional network joining its chains, and the second one is water. These systems may swell in water up to a equilibrium state and retain their original shape. The interactions responsible for water sorption by hydrogels include the processes of hydration (it is connected with the presence of such chemical groups as *–OH*, *–COOH*, $-CONH_2$, *–CONH–*, $-SO_3H$), and those related to the existence of capillary areas and differences in osmotic pressure. The forces which make hydrogel dissolution impossible are connected with the existence of covalent bonds between individual polymer chains although they may also have a character of electrostatic or hydrophobic interactions. Because the polymer as well as water interpenetrate it is not possible to distinguish a dispersed phase in swelled hydrogels as it is in the case of suspensions or emulsions.

Crosslinking reactions lead to the formation of hydrogels in which individual polymer chains are connected by covalent bonds being only decomposed in the case of a total macromolecule destruction. Physical hydrogels (reversible gels, pseudogels) form also a three-dimensional network but individual polymer chains are connected by weak hydrogen bonds or electrostatic interactions. In this case it is usually possible convert them into homogeneous solutions. The classical example of such hydrogels are gelatin, agar etc.

0097–6156/91/0475–0271$08.25/0

The ability to absorb water and ions, without loss of shape and mechanical properties, is a very essential feature of hydrogels. Such a properties exhibit many body organs like e.g. muscles, tendons, cartilage and intestines. Hydrogels show usually good biocompatibility in the contact with blood, body fluids and live tissues. Hence, they are often used as contact lenses, burn wound dressings, artificial cartilagesor membranes for hemodialysis. They are used to coat the materials being applied in the contact with the live body e.g. coating of the surface of catheters, electrodes, and vascular prostheses etc. A comprehensive revue of hydrogels application in medicine and pharmacy can be found in the monograph published in parts during 1986-1987 (1).

Rising interest in using hydrogels to biomedical purposes begins from the paper of Wichterle and Lim published in 1960, and devoted to radiation crosslinking of hydroxyethyl polymethacrylate and its properties (2). Use of ionizing radiation for the formation of biomaterials has a number of advantages. They list:

- initiation, proceeding and termination of reactions leading to the hydrogel formation take place in the absence of chemical initiators (residual, highly toxic superoxide products have to be removed from the final hydrogel obtained by chemical processes),
- whole process ie. gel formation and simultaneous sterilization can be realized in ready to "serve" packages,
- process of crosslinking can be carried in a broad temperature range without the necessity of heat input to the reaction system,
- polymer crosslinking proceeds also in prefrozen solutions. After completing irradiation procedure final products brought to the room temperature retain original desired shape,
- initiation and termination of chemical reactions is achieved simply by introducing or removing the material from the field of irradiation,
- requested physical and/or chemical properties of the final product can be achieved by adjusting intensity and type of radiation, irradiation time (dose), modification of composition of initial batch, etc.

Directions of the research devoted to the radiation formation of biomaterials are presented in a number of revues (3-6). These investigations are also included in the research programme co-ordinated by the International Atomic Energy Agency in Vienna (7,7a).

Formation of hydrogels by means of ionizing radiation can be simply (but fairly accurately) explained as a result of mutual recombination of macroradicals. In consequence of radiation interaction with matter, intermediate, reactive products such as ions, excited states and free radicals are formed. These species undergo various mutual processes but might react also with host molecules or in general with surrounding medium. If radicals located on different polymer chain are favorably positioned they might undergo recombination. As a result of this process new, covalent bond between these chains is formed. If the amount of these new bonds is sufficiently high (from the statistical point of view it is equal to the number of macromolecules) then insoluble fraction i.e. the gel appears in the system. Further irradiation of the system under consideration increases the amount of gel, although a part of macromolecules (or their fragments) may be still left unbound (sol).

Because of the presence of water as a solvent, the process of radiation induced hydrogel formation proceeds in more complicated way. Indirect effect due to reactions of products of water (solvent) radiolysis with macromolecules and solvent enhanced mobility of polymer chains influence the overall process.

Hydrogel Dressing HDR

Healing of wide and serious burn wounds can be often executed only by the transplantation. Auto transplantation (the skin taken from other parts of the body) or implantation of the skin cultured *in vitro* from fibroblasts of the patient are practically the only possible solutions. There is usually significant, up to few weeks, lapse of time between injury and performing transplantation. For this period the wounds should be properly protected against the infection. The dressings used for protection and healing the wounds can be divided into three groups:

1. conventional - gauze, bandages impregnated with ointments or silicon oils etc.
2. biological - the skin taken from cadavers, amnions, animal skin, sponges, collagen mats, dextran mats etc.
3. synthetic - i.e. formed from synthetic polymers and used as foams, ointments, sponges, membranes, fabrics etc.

A common feature of all dressings is that they are used periodically. The frequency of changing the dressings is connected with their properties such as the time for "wearing out" or the defence reactions of the organism (all "foreign" biological systems are antigenic factors).

In the literature there are some proposals concerning properties of "ideal" dressing used for healing burn wounds (8). Although there is no general consensus such a dressings should obviously have following properties: absorb effectively the body fluids, act as a efficient barrier against bacteria, adhere well to the wound and healthy skin, exhibit high elasticity but also mechanical strength, show good transparency, enable the oxygen to penetrate through the volume of dressing to the wound surface, enable the control of drug dosage, offer good handling ie. easy placement and replacement. In addition it should be sterile, easy to store, stable, generally available and relatively cheap (8-10). Some authors believe that not far from this ideal standard is a hydrogel dressing called "Geliperm" (Geistlich and Sons Co., Switzerland) produced according to the technology patented in many countries, including Poland, by Max Planck Society (9,11). This dressing is produced by chemical polymerization and crosslinking of acrylamide and methylene-bis-acrylamide in aqueous solutions containing some additives like polysaccharides and proteins.

The process of forming polymer network proceeds in two stages. In the first stage an aqueous solution of monomers is heated at 56 *°C* for the period of 30 minutes, and in the second one the batch is conditioned at 4 *°C* for 24 hours. From the point of view of the applied technology the most critical stage is the process of multiple extraction of the reaction mass with the purpose to remove completely unreacted monomer - acrylamide. This compound is a toxic one, believed to act mainly on the central nervous system and is listed as a potential carcinogen (12). During extraction procedure, hydrogel swells strongly, therefore the hydrogel foil has to be further dried, before cutting it into the pieces of requested size. Sterile condition has to be maintained during swelling, cutting and packaging operations. In spite of the fact that a ready dressing contains about 95% water and the cost of raw-materials is rather small, the complicated technology of production brings its price to significant level.

The original way of production of hydrogel wound dressings (trade name HDR) developed in Poland allowed to eliminate the disadvantages of "Geliperm" technology without loosing any of essential properties of the final product (13).

The first stage of this technology consists in the preparation of aqueous solution of dressing components. A combination of synthetic polymers as well as natural ones such as: polyvinylpyrrolidone (PVP), polyacrylamide (PAA) or their monomers, polyethylene glycol and agar is a usual case. After mixing components at the temperature of 60 °C clear, homogeneous brownish solution is formed.

In the second stage of this technology the moulds, which are also used as final packages for the ready-made dressings, are filled with solution to the requested depth. After solidification of the solution (several minutes, at room temperature) the moulds are tightly sealed in the foil non permeable for air and bacteria. Thin layer of pseudo-gel (unstable gel, which after heating will again form a solution) in the mould is thus formed.

In the final stage of this process, afore-mentioned products, assembled in commercial packages are treated with an ionizing radiation (a beam of high energy electrons from the accelerator or gamma rays from e.g. ^{60}Co). The dose of radiation, 25 kGy, is sufficient to sterilize the material and insures the formation of a stable three-dimensional network consisting of polymer chains joined together by covalent bonds. After sterilization, the pseudo-gel (encased in commercial packages) turns into sterile hydrogel dressing.

The technology of producing HDR dressings presents some advantages in comparison with the known chemical processings. They are as follows:

1. the technology is simple, easy and clean (there are no side-products such as wastes, sewage, fumes etc.) and all chemical substances present in HDR are safe for humans,
2. there is no need of maintaining special, sterile rooms, but the final product is completely sterile,
3. the process can be carried out as a continuous one (production line made up of package extruding press, automatic feeder and an accelerator) or as a batch process based on the separation a sites for preparing solutions, pseudo-gel packing and sterilization.

Some variations of this technology have been worked out, and the flexibility of manufacturing methods used is very promising, enabling them to be adapted also for producing other biomedical products.

Mechanism of Formation of Polyacrylamide and Polyvinylpyrrolidone Hydrogels

The process of hydrogel dressing formation is essentially due to radiation crosslinking of acrylamide or N-vinylpyrrolidone. The presence of other components i.e. agar and polyethylene glycol do not change the mechanism of these processes (under the action of radiation, agar is degraded and polyethylene glycol - crosslinked). Because the reactions proceeding in aqueous solutions are mainly initiated by intermediate products of the radiolysis of water, attention has been devoted to their contribution in the polymerization process.

The Role of Intermediate Products of Water Radiolysis in the Process of Polymerization and Crosslinking. Irradiation of aqueous monomer solution leads to the formation of crosslinked structures. The degree of acrylamide and N-vinylpyrrolidone conversion as a function of the absorbed dose is presented in Figure 1a,b. Data shown in this figure refer to the oxygen free solution, saturated with purified argon prior to irradiation. As the dose absorbed by the system increases, there is a rise in the degree of conversion of both monomers, although yield of their polymerization is different.

In order to confirm that the process of acrylamide and N-vinylpyrrolidone polymerization proceeds according to the radical mechanism, comparison of experimentally estimated rate of polymerization with the values resulting from respective theoretical model has been made.

The process of radiation polymerization proceeding according to the radical mechanism involves following reaction steps: initiation, propagation and termination of growing chains. It follows from the general theory of radical polymerization that the rate of initiation is directly proportional to the dose rate (I), and the rate constant of this reaction depends on the radiation yield of radical formation in the monomer itself (direct effect) and in the water (indirect effect):

$$k_1 = G_M[M] + G_S[S] \tag{1}$$

where $[M]$ and $[S]$ denote monomer and solvent concentration, while G_M , G_S , radiation yield of radical formation in monomer and solvent (radiation yield, G, is the amount of a definite product or the value of a parameter changing as a result of absorbing 100 eV energy by the system). Taking into account several stages of the process of radical polymerization i.e. initiation, chain propagation and assuming that chain termination takes place only as a result of mutual recombination of macromolecules one can get a general equation defining the overall polymerization rate in solution (22):

$$V_P = k_2\, k_3^{-0.5} I^{0.5} G_M^{\;1.5}\left(1+\frac{G_S[S]}{G_M[M]}\right)^{0.5} \tag{2}$$

k_2 and k_3 represent rate constants for propagation and termination processes.

In Figure 2 solid line represents the general solution of the equation 2, assuming the ratio of the yield of radical formation as presented in this figure (14,15). According to the theory of radical polymerization, the rate of this process should increase with content of monomer and reach maximum value at the concentration of monomer in the range of 60-70%. Further increase in the monomer content in the system leads to the decrease in the polymerization rate. It is to be noted that the shape of the theoretical curve (in higher range of concentrations) is determined essentially by the ratio of radiation yield of radical formation in the monomer and in the solvent (the value for N-vinylpyrrolidone has estimative character only). As it follows from the comparison of experimental data (points) and theoretical dependence (solid curves, Figure 2) the equation 2 describes with a good accuracy the course of radiation polymerization of both monomers. It seems to confirm a priori made assumption that polymerization proceeds according to radical mechanism. Similar conclusions can be inferred from the

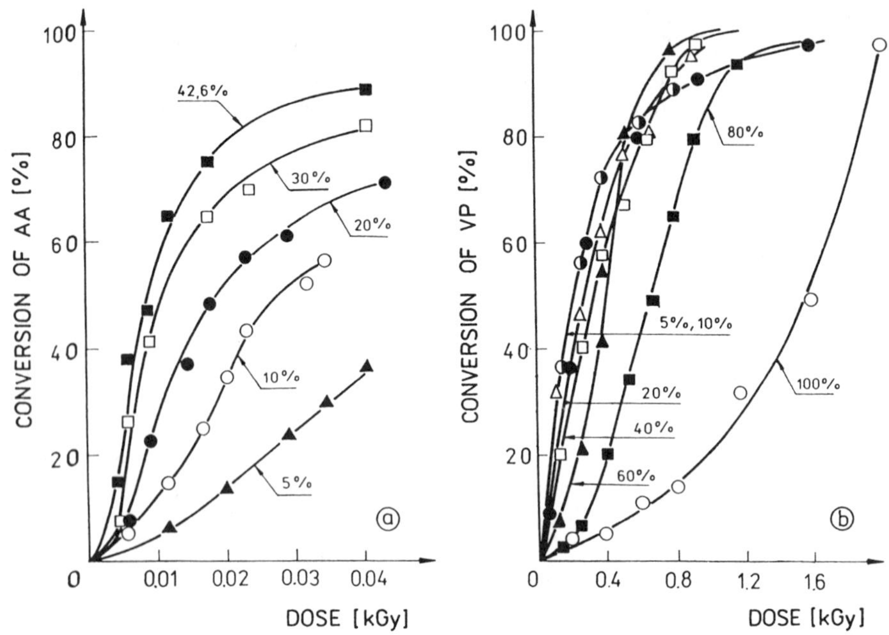

Figure 1. Radiation polymerization of acrylamide (a) and N-vinylpyrrolidone (b) in aqueous solutions saturated with argon (concentrations are marked in the figure). Dose rate (a) - 0.1 Gy/s; (b) - 0.23 Gy/s.

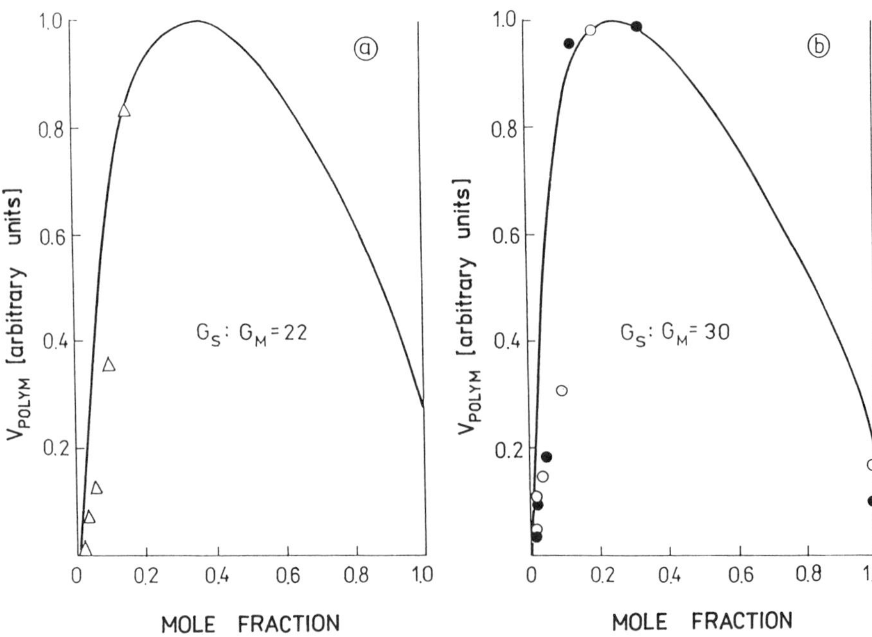

Figure 2. Relative rate of polymerization of acrylamide (a) and N-vinylpyrrolidone (b) in aqueous solutions saturated with argon. Solid curves - theoretical dependence according to the equation (2), points - experimental values (black circles - on the basis of the paper [16]).

results of other authors (16) (in Figure 2b they are presented as darken points - original experimental data have been recalculated according to the equation 2).

Polymerization rate increases along with growing water content in the irradiated system. This effect is due to the fact that polymerization may be initiated not only by radicals being formed as a result of direct action of radiation on the monomer (direct effect), but also by intermediate products of the radiolysis of water (indirect effect):

$$M \xrightarrow{\gamma} R^{\cdot} \quad (3)$$

$$S\,(H_2O) \xrightarrow{\gamma} S^{\cdot}\,(OH^{\cdot}, H^{\cdot}, e_{aq}) \quad (4)$$

$$S^{\cdot} + M \rightarrow R^{\cdot} \quad (5)$$

It is to be noted that radiation yield of the reactions (4) and (5) is much higher than that of (3). For individual intermediate products of the radiolysis of water, yields amount respectively: e_{aq} - solvated electron (G = 2.8), $OH^{\cdot}$ - hydroxyl radical (G = 2.7), $H^{\cdot}$ - hydrogen atoms (G = 0.5) (15).

The role of individual products of the radiolysis of water on the initiation ofpolymerization was evaluated in series of the experiments with scavengers which introduced to the solution react selectively with particular transient species.The results are presented in Figure 3 and Table I.

Nitrous oxide, in neutral solution (pH ≈ 7), reacts effectively only with hydrated electron. Equivalent concentration of hydroxyl radicals is then formed according to the reaction (6)($k_6 = 9\times10^9\ M\,s^{-1}$) (15):

$$e_{aq} + N_2O \xrightarrow{k_6} N_2 + OH^{\cdot} + OH^{-} \quad (6)$$

Substitution of hydrated electrons by hydroxyl radicals does not affects the polymerization rate (curves 2 and 3, Figure 3). It implies that both eaq and $OH^{\cdot}$ radicals initiate the process of monomers polymerization.

Methanol is known as a good scavenger of $OH^{\cdot}$ radicals as well as of $H^{\cdot}$atoms,

$$CH_3OH + OH^{\cdot}(H^{\cdot}) \rightarrow {}^{\cdot}CH_2OH + H_2O\,(H_2) \quad (7)$$

One can thus expect that in the presence of methanol the rate of polymerization would be markedly hampered - this effect is however nearly negligible (curves 4 and 2 , Figure 3) . Comparison of two competing reactions (8) and (9) enables to estimate "inhibitory" effect of methanol due to scavenging of $OH^{\cdot}$ radicals:

$$OH^{\cdot} + CH_3OH \xrightarrow{k_8} H_2O + {}^{\cdot}CH_2OH \quad (8)$$

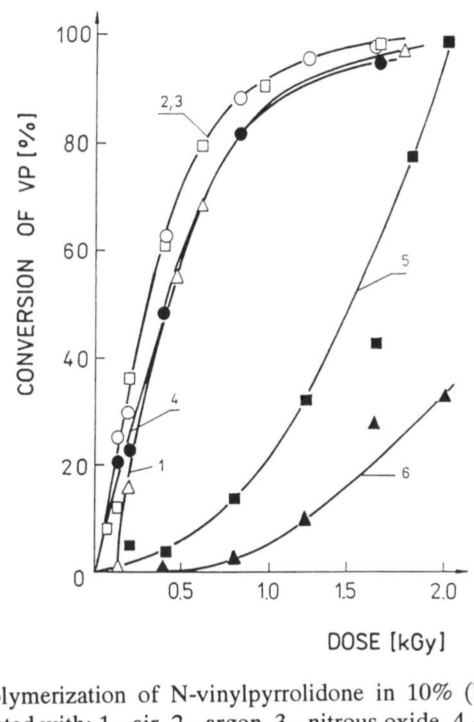

Figure 3. Polymerization of N-vinylpyrrolidone in 10% (by weight) aqueous solution saturated with: 1 - air, 2 - argon, 3 - nitrous oxide, 4 - argon (with methyl alcohol at the concentration 2 M). Liquid monomer saturated with: 5 - argon, 6 - air.

$$OH^{\cdot} + VP \xrightarrow{k_9} OH\text{–}VP^{\cdot} \tag{9}$$

Rate constants of the reactions (8) i (9) amount to: $8.4\times10^{8}\ M\,s^{-1}$ and $7.0\times10^{9}\ M\,s^{-1}$ respectively (15,17). Assuming that the reactions (8) and (9) are of the second order, at the applied concentration of methyl alcohol equal to 2 M, the initiation reaction (9) will proceed still about 4 times faster than the competing one (8). Such a result and the fact that hydrated electrons, which also can initiate polymerization are not scavenged by methyl alcohol may explain the observed but rather small effect of CH_3OH on the rate of this process. Further data confirming participations of e_{aq} , $OH^{\cdot}$ and $H^{\cdot}$ in the initiation of polymerization of N-vinylpyrrolidone and acrylamide are presented in Table I.

Table I. The rate of radiation polymerization of acrylamide (5% by weight, I = 0.09 Gy/s) and N-vinylpyrrolidone (10% by weight, I = 0.16 Gy/s) in the presence of scavengers

$Cd(NO_3)_2$	[M]	0	0.01	0.049	0.1	-
V [%/min]	AA	4.53	4.51	4.41	4.22	-
V [%/min]	VP	1.11	1.00	0.90	0.74	-
Metanol	[M]	0	2	4	8	12
V [%/min]	VP	1.11	1.06	0.82	0.70	0.12
Isopropanol	[M]	0	0.11	0.65	1.31	3.27
V [%/min]	AA	4.53	4.16	3.93	3.67	3.51

Cd^{+2} ion is a highly effective scavenger of hydrated electrons - the respective rate constant is equal to $k = 5.6\times10^{10}\ M\,s^{-1}$ (18). Lower polymerization rates of both monomers in the presence of Cd^{+2} are in agreement with accepted contribution of e_{aq} in the initiation process.

The increase in the concentration of methyl alcohol or introducing isopropyl alcohol (even better scavenger of $OH^{\cdot}$and $H^{\cdot}$ (15)) radicals causes a distinct decrease in the rate of monomer conversion .

Oxygen present in the system exhibits inhibitory effect on the process of polymerization of pure, liquid monomer (curves 5 and 6, Figure 3) but in aqueous solutions this effect is relatively small. Oxygen molecules interact with radicals being formed in the system (as a result of the radiolysis of solvent and monomer). Therefore total amount of "active species" being able to initiate the process of polymerization is diminished. In diluted aqueous solutions, the amount of radicals being formed is so high, that inhibitory action of oxygen has only a small effect on the course of polymerization (small inductive effect - curve 1, Figure 3). However when oxygen dissolved in the system ($\approx 10^{-3}\ M$) has been "used up", the process of polymerization proceeds with approximately the same rate as in aqueous solutions, previously saturated with argon.

After reaching 80-90% monomer conversion, the rate of polymerization rapidly decreases. Molecular weights of polymers separated from solutions are quite similar apart from their initial saturation with argon, nitrous oxide or addition of methyl alcohol. Further irradiation of these solutions must lead to the increase in the molecular weight of existing polymers assuming they undergo crosslinking. This predicted increase can only appear as a result of joining different macroradicals. Such a case is illustrated in Figure 4. Polyvinylpyrrolidone, isolated from 10% solutions containing different scavengers (dose about 1 kGy, the degree of conversion about 90%) has in all the cases similar molecular weight ($M_v \approx 10^6$). Further irradiation causes increase in the molecular weight of PVP in all solutions, but the rate of this process depends essentially on the additives present in the system.

Joining of polymer chains is caused by macroradicals recombination. Macroradicals can be formed either as a result of direct action of radiation on a polymer itself, or by the reaction of the intermediate products of the radiolysis of water with macromolecules. The curve 2 (Figure 4) describes the dose dependence of polyvinylpyrrolidone molecular weight increase in the solutions saturated with argon no the dose.The double increase in the concentration of hydroxyl radicals, in the solutions saturated by nitrous oxide (curve 3, Figure 4), is the obvious reason for a distinct increase in the growth of molecular weight. Methyl alcohol present in the solution (curve 4, Figure 4), which reacts with $OH^{\cdot}$ and $H^{\cdot}$ radicals causes however very strong decrease of this process. Under these conditions the process of crosslinking is limited entirely to the reaction of macroradicals being formed in the polymer by direct action of ionizing radiation (see Table II). These results allow to state that in the irradiated aqueous solutions, $OH^{\cdot}$ and $H^{\cdot}$ radicals are responsible for the reactions leading to the formation of macroradicals, whereas the hydrated electron does not take part in these reactions. Similar results pointing out the lack of e_{aq} contribution in the reactions of macromolecules linking have been obtained for the solutions of acrylamide (19).

Oxygen presence in the irradiated solutions has only small decreasing effect on the growth the molecular weight (curve 1, Figure 4). Reaction of oxygen with macroradicals leads finally to the rupture of macromolecules into smaller fragments and thereby to the decrease in the molecular weight (25).

Summing up this part of discussion one can state that polymerization and crosslinking of N-vinylpyrrolidone and acrylamide macromolecules proceeds according to the radical mechanism. All intermediate products of water radiolysis take part in the initiation of the process of monomers polymerization whereas in the crosslinking only hydroxyl and hydrogen radicals are active.

Network Formation. The course of the process of crosslinking and the properties of the gel formed depend to a high degree on the method of forming crosslinks. In the systems under study gel formation is preceded by the process of polymerization. Development of a network may be therefore based on the formation of four- or three functional sites of macro-molecules joining. The first ones are formed as a result of joining the radicals located on the chains (except end groups). This process is defined as classical crosslinking or simply crosslinking. The second sites are formed when the radical located on the terminal group of the chain recombine with any macroradical (it is called endlinking). Depending on the way of network formation, the amount of the gel being formed for comparable doses is different. By analysis of the amount of gel (g) and sol (s) formed in the system one can draw the conclusions on the mechanism of network formation.

Mathematical Analysis of the Results. The most known and popular mathematical description of the course of crosslinking process is the equation of Charlesby and Pinner:

$$s + \sqrt{s} = 0.5\,\lambda + 0.5\,(4-\lambda)\,(D_g\,/D) \tag{10}$$

where s is the content of sol in the system (s = 1 - g), Dg and D - the dose absorbed by the system, λ is the ratio of the radiation yield of degradation and crosslinkingprocesses (20). The plot of the equation 10 is a straight line with intercept equal to 0.5 λ. If no degradation occurs, then the straight line crosses the origin of coordinates (λ = 0). Most often however the process of crosslinking is accompanied by splitting the main chain. For this case λ ≠ 0. Although equation 10 is a very convenient way of presenting the results it has a rather limited range of application. The relation can be used in the case of polymers, which before irradiation show a random distribution of molecular weights, with the distribution width equal to $R_0 = M_w\,/M_n = 2$. The other equation defining the amount of sol contained in the system has been worked out by Inokuti and may be applied for polymers characterized by other types of molecular weights distribution, without any limitations to its width (21). This equation looks as follows:

$$(1-g)\,x\,(2g+\lambda)^3 = x\lambda^2(2g+\lambda) + 4\lambda g\,/u_1 - 4\lambda g\int_{(2g+\lambda)x}^{\infty} F(z)dz + 4\lambda^2 xg^2(2g+\lambda)F[(2g+\lambda)x)] \tag{11}$$

where g denotes the content of the gel in the system ($g = 1-s$), x - density of crosslinking, that is the number of crosslinks for a base unit in the polymer chain, u_1 - is the number average polymerization degree of the initial polymer, and $F(z)$ - a Laplace transform of the distribution function of molecular weights $m\,(P,0)$ of the initial polymer.

The distribution of molecular weights is defined in such a way that $m\,(P,0)\,dP$ represents the number of polymer fractions of a polymerization degree contained in the range P to $P + dP$.

$F\,(z)$ is defined by the equation:

$$F(z) = \int_0^{\infty} m(P,0)e^{-zP}dP \quad (z \geq 0) \tag{12}$$

Function $m\,(P,0)$ has been chosen in such a way that following equation is fulfilled:

$$\int_0^{\infty} Pm(P,0)dP = 1 \tag{13}$$

It can be concluded from Inokuti's equation that the amount of gel being formed in the system under the same experimental conditions may be different depending on the molecular weights distribution of the initial polymer. The plot of the equation 11 is represented in the Charlesby-Pinner coordinate system, $s + \sqrt{s} = f(D_g/D)$, by lines with a shape depending on the type of molecular weights distribution and its width. The plot for the random distribution (where $R_0 = 2$ for the initial polymer) is a straight line, and the Inokuti equation is reduced to expression 10. The solutions of the equation 11 have been worked out for different distributions of molecular weights, and are presented in a tabular form (22). In the case of the Wesslau's distribution valid is the dependence $M_n M_z / (M_w)^2 = 1$, where M_n, M_w and M_z denote respectively a number average, weight average and average molecular weight of the polymer. The distribution width is determined by a parameter β defined by the dependence $M_w / M_n = exp(0.5\beta^2)$.

Maximum amount of gel, which can be formed in the system does not depend on the type of molecular weights distribution or its width and is defined by the equation:

$$g_{max} = 0.5\,[1-\lambda+(1+2\lambda)^{0.5}] \tag{14}$$

The value g_{max} may be estimated experimentally on the base of gel-sol analysis. It allows to estimate the ratio of the radiation degradation and crosslinking.

The way of working out the results of sol-gel analysis is based on adjusting theoretical relation obtained as a solution of the equation 11 to fit the experimental results.

Sol-gel Analysis. Inokuti's equation, describing the course of the crosslinking process has been deduced taking into account following assumptions:

a) radiation yields of crosslinking and degradation are constant and independent of the dose
b) intramolecular crosslinking proceeds to a negligible degree
c) degradation and crosslinking processes proceed entirely independently
d) formation of a network proceeds in a classical way i.e. by the formation of four-functional joining points only (no endlinking).

For acrylamide and N-vinylpyrrolidone solutions under investigation as well as for many other systems subjected to irradiation, the assumptions a, b and c are assumed to be generally true. The assumption d can be used as a test of the mechanism of network formation. When experimental results may be described with good accuracy by the Inokuti's equation then network formation follows the classical mechanism.

In Figure 5 the dependence of sol fraction on the absorbed dose is presented. On the ground of Inokuti's equation one can determine the maximum angle, between the line joining experimental points, and a straight line $s + \sqrt{s} = 2$ (corresponding to the conditions just before the gelation point). Depending on the type of the distribution of polymer molecular weights, this angle may have a maximum value: $\Theta = 63°$ for the most probable distribution and the distribution according to Schulz-Zimm or $\Theta = 56°$ for Wesslau distribution.

In Figure 5 afore-mentioned theoretical angles are shown together with solid line, representing initial course of crosslinking. The experimentally obtained values were also put on, indicating the sol content in the system for : a - acrylamide solutions,

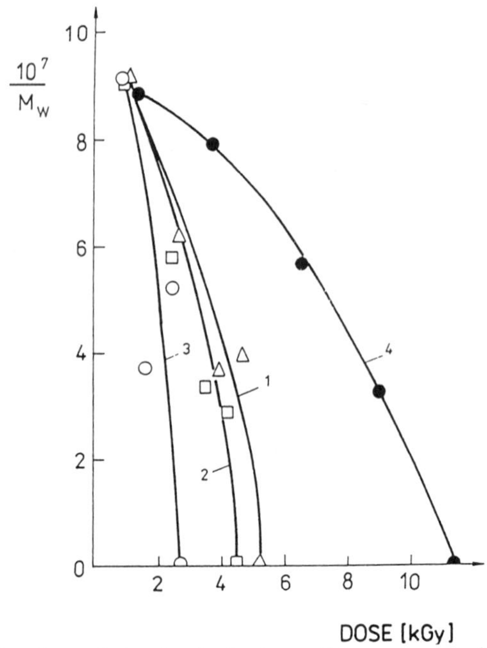

Figure 4. Molecular weight of polyvinylpyrrolidone separated from 10% aqueous solution saturated with: 1 - air, 2 - argon, 3 - nitrous oxide, 4 - argon, with the content of methyl alcohol at the concentration 2 M.

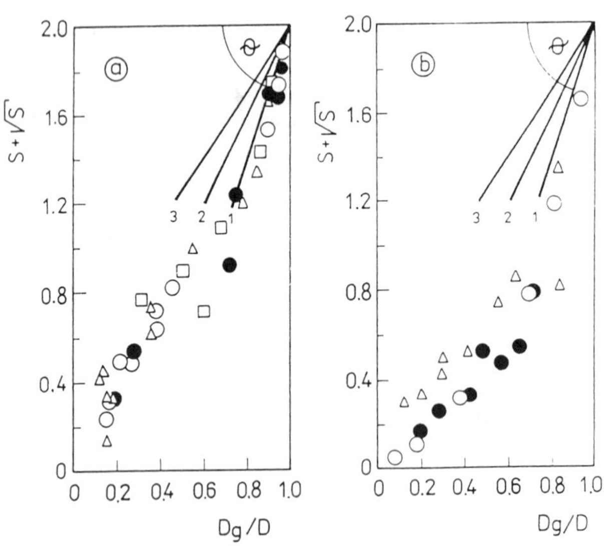

Figure 5. The content of sol fraction in the system: a - aqueous acrylamide solutions saturated with argon at the concentrations as follows: (a) **O** - 42.6%, Δ - 30%, □ - 20%, ● - 5% (by weight). (b) - aqueous 10% solutions of N-vinylpyrrolidone saturated with: **O** - argon, ● - nitrous oxide, Δ - with access of air . Solid lines - theoretical content of sol fraction for the distribution of molecular weights : 1 - uniform, 2 - according to Schulz-Zimm, 3 - according to Wesslau.

b - N-vinylpyrrolidone solutions. Solid line 1 represents the initial course of the function for the case when the initial distribution of molecular weights was monodisperse i.e. $M_w / M_n = 1$.

As it follows from the data presented in Figure 5, the initial slopes of crosslinking curves determined experimentally are greater than predicted theoretically. This relation implies that crosslinking does not proceed according to the classical mechanism but probably as an endlinking. However, one should take into account the fact that irradiated solutions contain initially monomer and not a polymer, and part of the energy absorbed by the system is used in the processes of polymerization. This part of the energy (by convention it was accepted that it is the dose corresponding to 80% conversion of a monomer into polymer) should be subtracted from the gelation dose and dose absorbed by the system. If this fact is not taken into consideration one may arrived with erroneous interpretation (23).

In Figure 6 are shown the results of the gel-sol analysis for the system containing different amounts of acrylamide, corrected for polymerization dose. The values marked in the plots by points were determined experimentally, and solid lines were plotted on the ground of Inokuti,s equation. These curves have different shape being dependenton the concentration of the monomer in the initial solution and the presence of oxygen. The best agreement between experimental data and theoretical dependence has been found for the distribution of molecular weights of Wesslau's type. The width of the molecular weights distribution of polyacrylamide which undergoes crosslinking is different and depends on the concentration for the initial solution - for 5% solution it amounts to $M_w / M_n = 1.1$, while for 42.6% solution it is equal to 17.8.

The presence of oxygen in the irradiated solution brings about that alongside of crosslinking, scission of the main chain of polyacrylamide proceeds. The ratio of the degradation yield to the yield of crosslinking amounts to 0.7 (for 5% solution) and 0.5 for the system containing 20% acrylamide. It indicates that in the solutions containing oxygen a sol fraction is always present independent of the absorbed dose. Maximum content of gel fraction determined on the basis of the equation 14 for 5% and 20% amounts to: $g_{max} = 0.93$ and $g_{max} = 0.96$, respectively.

For the solutions with a constant acrylamide content, the ratio of the degradation yield to that of crosslinking increases with the increase of the dose rate (Figure 7). It may be caused by the enhancement in the yield of crosslinking or by the depression in the yield of main chain breaking. Higher dose rate leads to higher temporary concentration of macroradicals, and therefore increases the rate of crosslinking to higher extend than that for the reaction of the polymer radicals with oxygen (leading finally to degradation). These competitive processes can be presented schematically as:

$$P^{\cdot} + P^{\cdot} \rightarrow crosslinking \qquad V_{CROSS} = k_s[P^{\cdot}]^2 \tag{15}$$

$$P^{\cdot} + O_2 \rightarrow degradation \qquad V_{DEG} = k_d[P^{\cdot}][O_2] \tag{16}$$

The process of gel formation in 10% aqueous solution of N-vinylpyrrolidone is shown in Figure 8. The symbols a, b, c and d refer to the solutions which have been saturated with argon, air, nitrous oxide or with argon (methyl alcohol added) prior to

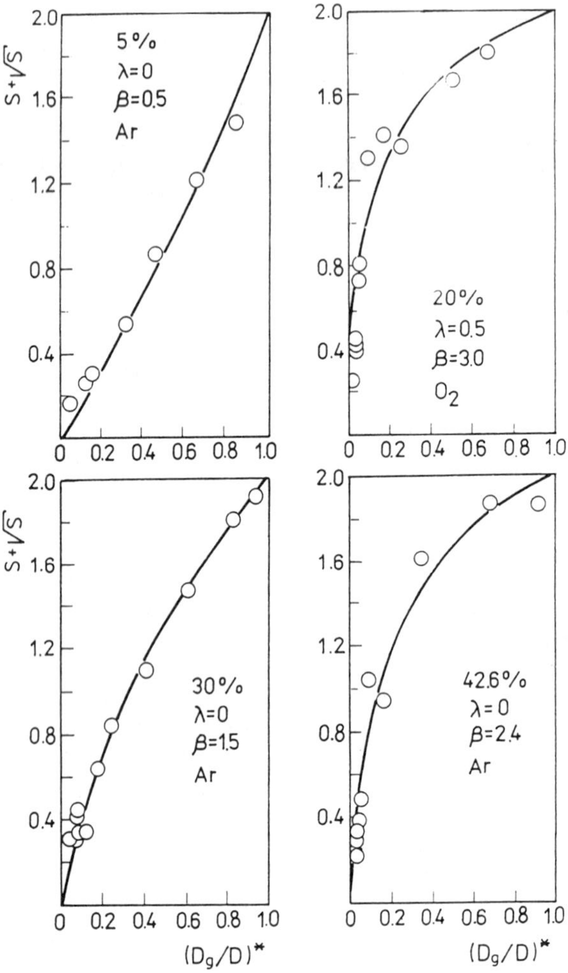

Figure 6. The content of sol fraction in irradiated acrylamide solutions (concentrations are shown in the figure). Parameters α, β and solid lines - on the basis of Inokuti's equation, for the distribution of molecular weights of Wesslau's type.

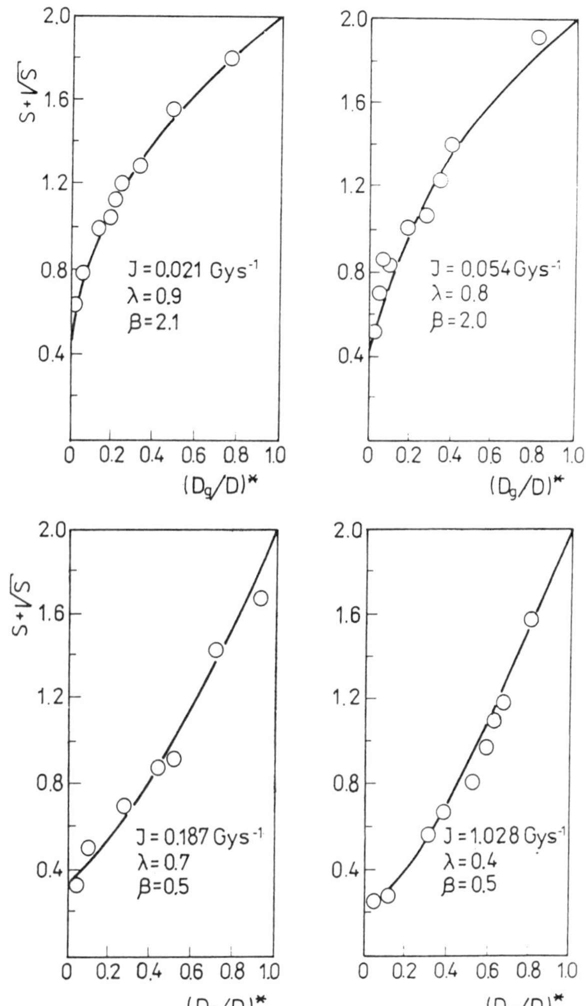

Figure 7. The content of sol fraction of 5% aqueous acrylamide solutions irradiated at different dose rates (in the presence of air). Parameters α, β and solid lines as in Figure 6.

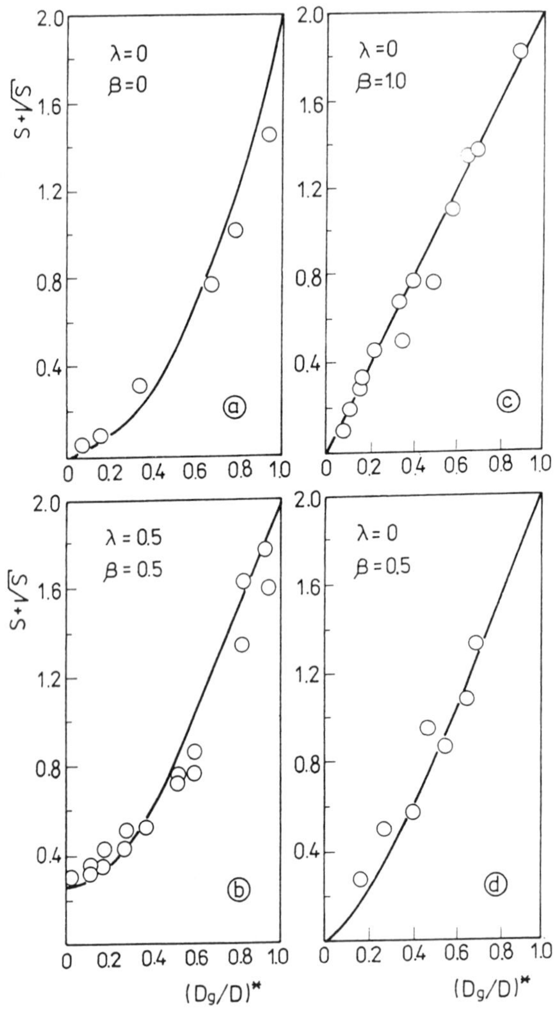

Figure 8. The content of sol fraction in 10% aqueous solutions of N-vinyl-pyrrolidone saturated with: a - argon, b - air, c - nitrous oxide, d - argon (solution contained methyl alcohol at concentration 2M). Parameters α, β and solid lines as in Figure 6.

irradiation. The experimental values may be described with a good accuracy by the relation reported by Inokuti (11). In all the cases the best agreement was obtained for Wesslau's type of distribution of molecular weights. In the absence of oxygen in the system, the parameter λ is equal to zero what means that crosslinking is not accompanied by main chain scission. Therefore, for respectively high doses absorbed by the system no soluble fraction can be found.The entire polymer forms the network (gel). In the presence of oxygen crosslinking is accompanied by rupture of the main chain, similarly as it has been found for solutions containing acrylamide ($g_{max} = 0.96$). For 10% N-vinylpyrrolidone solutions the parameter β is contained in the range 0 - 1. It indicates that the distribution of molecular weights for the polymer undergoing crosslinking is relatively narrow - the ratio M_w / M_n varies respectively from 1 to 1.7.

In all the cases of crosslinking being analyzed, the experimental values may be described with sufficiently good accuracy by the Inokuti's equation (after correction for the dose used up for polymerization). Such an agreement indicates that crosslinking of both polymers proceeds according to the classical mechanism, i.e. by joining of two radicals located in the polymer main chain, and not through end groups of this chain.

Radiation Yield of Hydrogel Crosslinking. Radiation yield of hydrogel crosslinking in aqueous solutions of N-vinylpyrrolidone and acrylamide has been determined by the ROCH method (24). On the basis of gravimetric determinations of gels separated from irradiated systems, the concentrations of effective bonds, V_e, were calculated. The examples of such dependence have been presented in Figure 9 a,b. As seen in all analyzed cases, there is linear plot of experimental values. In such a case it is possible to calculate straightforward the yield of crosslinking. Corresponding values referring to hydrogels being formed in N-vinylpyrrolidone solutions are presented in Table II. On the basis of the results obtained one can state that the yield of crosslinking in aqueous systems is many times higher compared with pure monomer. Reactions of intermediate products of the radiolysis of water with polymer chains leading to the formation of macroradicals are responsible for observed high yield. Results shown in Table II enable to estimate the contribution of this indirect effect to the formation of crosslinks.

Table II. Radiation yield of hydrogel crosslinking obtained in aqueous N-vinylpyrrolidone solutions. Dose rate I = 0.23 Gy / s

System	G_x [1 / 100 *eV*]
1. 10 % aqueous solution	
a) Ar saturated	0.34
b) nitrous oxide saturated	0.44
c) in the presence of 2 M CH_3OH	0.04
d) with the access of air	0.26
2. 20% aqueous solution - Ar saturated	0.42
3. 40% aqueous solution - Ar saturated	1.32
4. Pure monomer Ar saturated	0.46

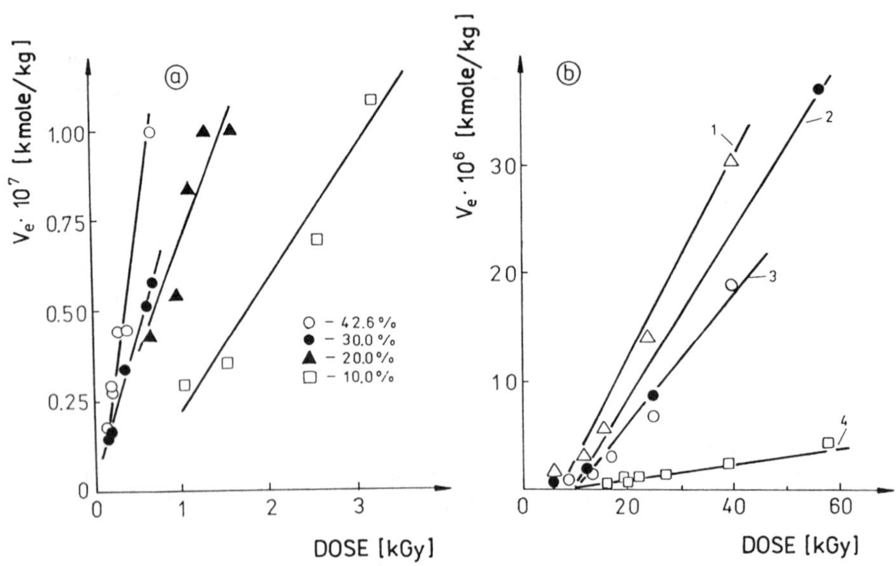

Figure 9. The concentration of effective chains, Ve, of hydrogels formed in aqueous solution: a - acrylamide saturated with argon; b - N-vinylpyrrolidone (10% w/w) saturated with: 1 - nitrous oxide, 2 - argon, 3 - air, 4 - argon (solution contained methyl alcohol at concentration 2M).

Rather low G-values suggest also that only a part of the radicals which are formed in the irradiated water react with the polymer. Most of them undergoes recombination leading to the formation of stable products of the radiolysis of water (H_2 , H_2O_2) or torecreation of H_2O. The lower is the concentration of the polymer in the system, the higher is the probability of mutual recombination of intermediate products of the radiolysis of water. Because in aqueous solutions indirect effect prevails, the yield of crosslinking decreases with the decreasing concentration of the polymer in the system.

Conversion of hydrated electron into hydroxyl radical in solutions saturated with N_2O is immediately manifested by the increase in the yield of crosslinking. This fact confirms that macroradicals involved in crosslinking are formed mainly in the reaction of macromolecules with hydroxyl radicals.

The presence of 2M of methyl alcohol diminishes the yield of radiation crosslinking to the limiting value corresponding to the direct effect only (see position 1c and 4, Table II). In other words, the number of crosslinks being formed is such as it would be if the radicals generated during the radiolysis of water could not form the macroradicals. This result can be accepted as an additional proof for previously formulated conclusion that hydrated electron does not take part in the reaction leading to the formation of crosslinking bonds (see previous text).

In the presence of oxygen in the system the yield of crosslinking decreases. As mentioned before (equation 16) the reaction of oxygen with macroradicals leads to the main chain scission, and in consequence brings about the net decrease of crosslinking yield. According to the mechanism presented a simple equation should be fulfilled:

$$G_x(O_2) = G_x(Ar) - 0.5\ G_s(O_2) \qquad (17)$$

Relation (17) links the yield of crosslinking G_x in the presence of oxygen and argon with the yield of degradation G_s . Equation 17 holds for a case if non-formation of one crosslink corresponds to such a conversions of two macroradicals, which results in the main chain rupture. From previous considerations it is known that the ratio of degradation and crosslinking yields is equal to 0.5 λ. Substituting corresponding values from the Table II to the equation 17 we arrive at $G_s\ (O_2) = 0.16$, what in turn allows one to calculate the parameter λ (λ = 0.47). This value agrees well with λ = 0.5 determined experimentally - on the ground of the gel-sol analysis. One should remember that the values of the radiation yield of crosslinking and degradation vary with the change in the concentration of polymer. For a pure monomer containing oxygen the yields of crosslinking and degradation determined on the basis of the equation 17 amount to $G_x\ (O_2) = 0.3$ and $G_s\ (O_2) = 0.29$, in good agreement with the value $G_x\ (O_2) = 0.29$ determined experimentally (24).

Using the equation 17 one can thus calculate the yield of degradation, which accompanies the radiation crosslinking when oxygen is present in the system.

Similar calculations of the yield of crosslinking by the ROCH method have been made for the solutions of acrylamide (Table III). In the solutions saturated with argon, the values of the yield of radiation crosslinking increase with the increasing concentration of the initial solution. The values are however much lower than in the case of corresponding N-vinylpyrrolidone solutions. It points to a lower yield of macroradical formation, contributing to the formation of polymer network. In the absence of oxygen in the system, no main chain scission occurs. Hence, the yield of radiation degradation is equal to 0.

Table III. Radiation yield of crosslinking of hydrogels formed in aqueous acrylamide solutions. Dose rate I = 0.1 Gy/s

System	G_x [1/100 *eV*]
5% aqueous solution	
a) argon saturated	0.01
b) containing oxygen	0.07* (0.0065**)
10% aqueous solution, argon saturated	0.03
20% aqueous solution	
a) argon saturated	0.08
b) air saturated	2.5* (0.06**)
30% aqueous solution, argon saturated	0.135

* calculated according to the ROCH method (dose rate 0.2 Gy/s)

** calculated on the basis of equation (17)

For the systems containing oxygen, the values of the radiation yield of crosslinking determined by the method ROCH are surprisingly high. In such systems, the process of crosslinking is accompanied by degradation ($\lambda \neq 0$, see previous text). It seems therefore reasonable to assumethat the yield of crosslinking should be smaller in comparison with corresponding values for the systems saturated by argon. In addition the values, G_x, calculated using the ROCH method are even higher than theoretically predicted upper limit (e.g. 20% solution, Table III). Yields of crosslinking determined by the ROCH method are therefore incorrect. The reason for this discrepancy is a high degree of chain branching of polyacrylamide being formed in the solutions containing oxygen. For 20% solution irradiated in the presence of oxygen, the parameter β determined on the basis of gel-sol analysis is equal to 3. It means that the ratio M_w / M_n for such polymers amounts to about 90. ROCH method can not be applied for systems where $M_w / M_n >> 2$. Correct values of crosslinking and degradation yields in such system can be obtained using the equation 17 and the values of parameter λ determined on the basis of the gel-sol analysis. The values for the yield of hydrogel crosslinking calculated in this way for 5% and 20% solutions are given in Table III (in brackets).

Some Properties of Hydrogel Dressings

Irradiation of aqueous solutions of vinylpyrrolidone and acrylamide with sterilization dose equal to 25 kGy leads to the formation of a crosslinked system containing more than 90% gel. These gels show high absorptivity of water. The foils which have been obtained are flexible, fully transparent, and do not dissolve in hot water. They are also easy and pleasant in handling. The essential drawbacks of such foils (obtained for 10% aqueous solutions of monomers) from the view points of their application as dressings are as follow:

1. the necessity of carrying out irradiation in leveled vessels (if this condition is not fulfilled, the foils of variable thickness are produced)
2. high rate of loosing water contained in the foils. After 8 - 10 h the foils were dry, and in consequence rigid and fragile,
3. increased (along with the diminished content of water in the foil) ability of adhering to the human skin and even the ability, especially in the case of acrylamide foils, to suck out body fluids (characteristic blood extravasation).

The above mentioned features of hydrogel systems are undesirable and even harmful from the point of view of demands concerning dressings. It does not mean however that such a hydrogels can not find other medical applications.

In order to avoid some troubles connected with a variable thickness of the product, it is necessary to add into the initial aqueous solution an appropriate component fixing desired, stable form (cf.agar, gelatin, alginic acid etc.).

In order to avoid such a situation that the hydrogel dressing might form a rigid, fragile plate after drying, it is necessary to add to initial solution a substance preventing from a complete loss of water (cf.polyethylene or polypropylene glycols).

Already a small water content is an excellent plasticizing agent which allows to preserve elasticity.

Manufacturing of hydrogel dressing can be divided into three basic operations:
1) preparation of aqueous solution of dressing components,
2) filling the mould with this solution,
3) irradiation of sealed mould with the dose of 25 kGy.

Production of HDR dressing can be realized as continuous or batch process.

The production line for continuous process in standard version would consist of:
1. extruding press to form moulds,
2. mixers for preparing the initial solution,
3. feeder to fill individual moulds (packages),
4. machine for closing the moulds,
5. electron accelerator.

The production lines of this type (excluding accelerator) are commonly found in the food industry.

The batch process can be carried out in different ways. The stage of irradiation can follows immediately filling the moulds or can be realized even few days apart. Both cobalt-60 gamma rays and electron beam from the matched accelerator can be used.

It is evident the gamma sources are the only ones which should be used at a suitably low dose rate if the solution is made up from monomers as initial raw-materials. In the case of preparing initial solutions from polymers both irradiation sources can considered. It is believed that electron accelerators are favoured due to lower costs of the processing carried out on commercial scale.

Mechanical properties of hydrogel foils obtained from the solutions of acrylamide and N-vinylpyrrolidone are similar indeed. Aqueous extracts obtained from theses foils show however essential differences in the interaction with microorganisms. Biological tests concerning the interaction of aqueous extracts from hydrogel foils on the microorganisms (e.g. Dapnia magna) proved that the extracts from polyacrylamide foils show some toxic action (due to residual unreacted monomer left). Extracts from foils obtained from N-vinylpyrrolidone (or from its polymer) are entirely non toxic. Hydrogel dressings HDR offered for medical applications are made therefore only from that last polymer. The dressing of a trade mark HDR-1 is from the N-vinylpyrrolidone is produced, whereas HDR-2 from already prepared polymer.

HDR dressings in the final form of sheet about 3.5 mm thick contain about 90% water. In spite of such a high content of water they retain capacity for its further absorption.The swelling rate and the degree of absorptivity depend on the content of individual components. In Figure 10 is shown the kinetics of swelling of hydrogel foils immersed in water at the temperature 295 K. Hydrogel was obtained by irradiation of N-vinylpyrrolidone solutions containing different amounts of polyethylene glycol (PEG) and agar. The change in the content of these two components has an essential effect on the course of swelling. Generally, the increase of the content of glycol leads to the higher sorption ability of foils, whereas the increase in the concentration of agar has opposite effect. A similar effects were observed for the sheets made by irradiation of polyvinylpyrrolidone solutions with a beam of electrons (Figure 11).

Mechanical properties of hydrogel dressings such as tensile strength and elongation at break varies not only with the content of the components but also depends on the absorbed dose. As it follows from the data shown in Figure 12 at higher doses the tensile strength increases but the elongation at break decreases. These changes are due to the higher degree of hydrogel crosslinking.

Hydrogel dressings because of their intrinsic content of water allow for permeation of oxygen (gases) to the wounds. Also solid substances placed on the upper surface can diffuse through the bulk of the dressings. Figure 13 shows the time profile of, a methyl red, dye penetration. Dye was placed on the dressing surface in the form of a powder.

Hydrogel dressings after passing rigorous biological tests have been subjected to intensive clinical examinations in some hospitals in Poland, namely in the Military Medical Academy (Lodz), Institute and Clinic of Traumatic Surgery (Wroclaw), and Central Hospital of Steel Industry (Siemianowice). On the ground of these examinations, the official permission has been issued to use HDR for healing the burn wounds as well as for other types of skin damages. They may be applied for healing burn wounds, trophic ulcerations, bedsores, post-operative wounds (gynecology, laryngology, plastic surgery). They may be used for the treatment of all kinds of skin damages in which humid medium is favourable.

On the basis of the laboratory and clinical investigations, and the observations of the course of healing one may summarize the advantages of HDR hydrogel dressings use.

They:

1. form an efficient barrier against bacteria from outside and prevent excessive loss of body fluids,
2. give access of oxygen to the wound,
3. show high flexibility and pleasant softness yet retain good mechanical strength. They can be used to cover even those parts of the body, which when immobilized could cause significant discomfort for the patient, e.g. face, joints, palms etc.,
4. show good adhesion to the wound but enable a painless change of dressing without disturbing the medical treatment,
5. provide the way for introducing the medicines into the wound without removing the dressings, by simply wetting the outside surface of the dressings with the drug

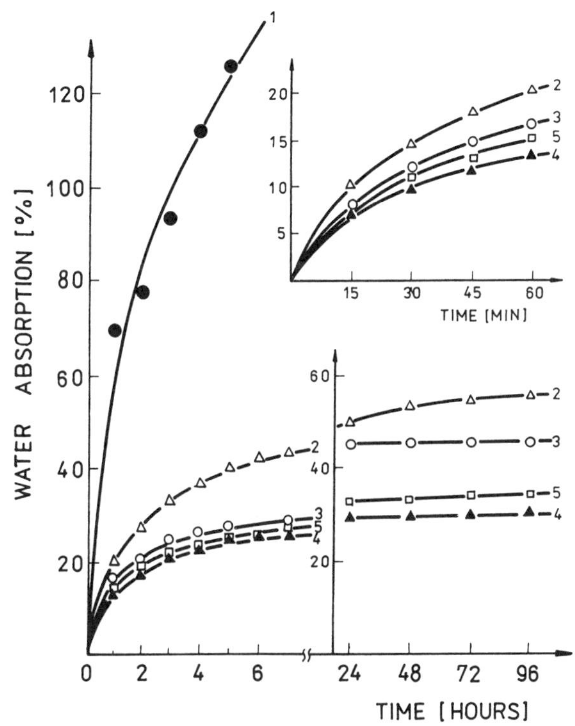

Figure 10. The sorption of water by hydrogel foils (310 K). The initial composition of dressings contained 10% of N-vinylpyrrolidone and: 1 - PEG 1%; 2 - PEG 1% plus agar 1%; 3 - PEG 1% plus agar 1.5%; 4 - agar 1.5%; 5 - PEG 2% plus agar 1.5%. The total dose 27 kGy (60Co).

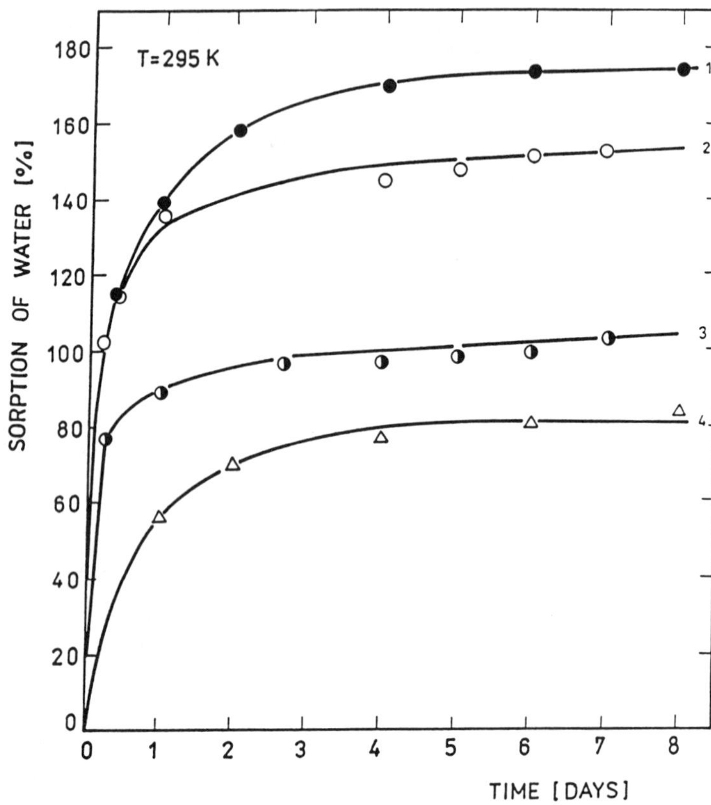

Figure 11. The sorption of water by hydrogel foils (6% PVP plus 1.5% PEG) containing different amounts of agar: 1 - 0.5%, 2 - 1%, 3 - 1.5%, 4 - 2%. The total dose - 26 kGy (electron accelerator).

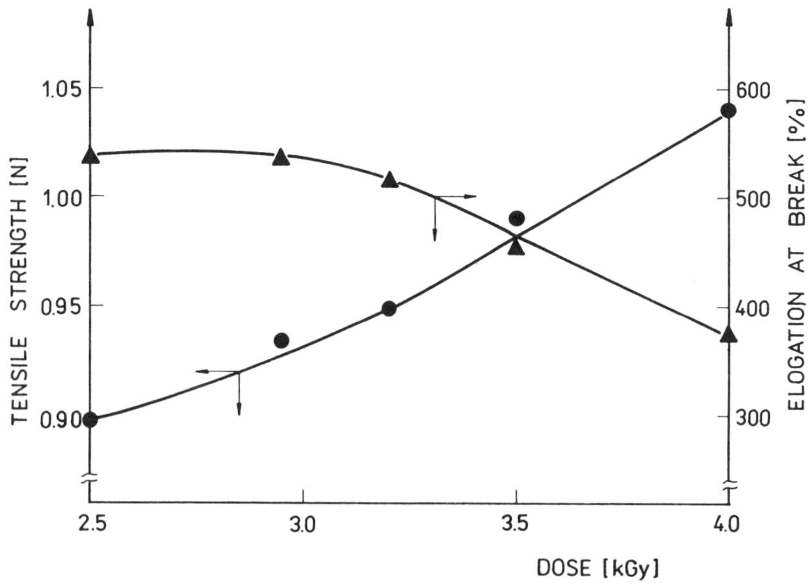

Figure 12. Mechanical properties of HDR hydrogel dressings.

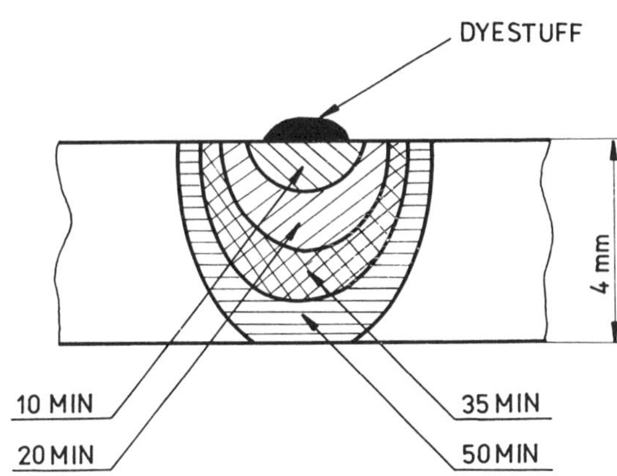

Figure 13. Time evolution of the penetration of dye, methyl red, through the hydrogel dressing.

solution. Drugs can also be introduced under the surface of the dressing by means of a syringe injection or by swelling its in the drug solution before placing on the skin,
6. absorb the exudates and bacterial toxins,
7. are non-antigenic and non-allergic,
8. soothe pain and ensure psychical "comfort" of a patient,
9. reduce the possibility of scars formation,
10. remain transparent. This allows the control of the course of healing without removing the dressings,
11. are sterile, easy in use and allow for storage at room temperature,
12. are definitely not expensive (the cost of raw materials for the production of the standard HDR dressing of a size 12 x 12cm is equal to about 0,15 US dollars).

Moreover, the HDR dressings may be used as:

a) oral agent detoxicating the organism,
b) therapeutic system for drug release at a controlled rate,
c) dressings for all the kinds of wounds and also in the form of tampons, suppositories etc.,
d) first aid dressings (in first-aid kit),
e) material for cosmetic purposes (eg. some beauty masks).

The above mentioned properties of HDR dressings and the simplicity of production technology allow to expect a broad range of their application in medical practice.

Acknowledgments

Author wish to express his gratitude to Professor W.Pekala, Dr.A.Rucinska-Rybus, Dr.J.Olejniczak, Dr.K.Burczak, Dr.S.Witkowski and Mrs.J.Ignaczak. Without their collaboration and assistance this work could not be done. Author would also like to thank for the support received from the International Atomic Energy Agency (RB/5509 and POL/1/010) and from the national sources (CPBP 01.19 and CPBR 05.08). Author would like to express his great appreciation to Polymer Division of American Chemical Society for giving him opportunity to present this paper at 200th National Meeting ACS, Washington D.C.,1990.

Literature Cited

1. *Hydrogels in Medicine and Pharmacy*; Peppas, N.A. Ed.; CRC Press, Inc.: Boca Raton, Florida; vol.I 1986, vol.II and III 1987.
2. Wichterle, O.; Lim, D.; *Nature* **1960,** *185*, 117.
3. Hoffman, A.S.; *Radiat.Phys.Chem.* **1981,** *18*, 323.
4. Kaetsu, I.; *Radiat.Phys.Chem.* **1981,** *18*, 343.
5. Rosiak, J.; Olejniczak, J.; Pękala, W.; *Polymers in Medicine* **1987,** *17*, 85.
6. Ikada, Y.; Mita, T.; Horii, F.; Sakurada, I.; Hatada, M.; *Radiat.Phys.Chem.* **1977,** *9*, 693.
7. *Radiation Technology for Immobilization of Bioactive Materials*; IAEA - TECDOC - 486; IAEA: Vienna, 1988.

7a.Rosiak, J.; Olejniczak, J.; Rucińska-Rybus, A.; Galant, S.; Burczak, K.; Pękala, W.; *ibid.*, p. 95.
8. *Burn Wound Coverings*; Wise, D.L. Ed.;CRC Press, Inc.: Boca Raton, Florida, 1984.
9. Quinn, K.J.; Courtney, J.M.; Evans, J.M.; Gaylor, J.D.S.; Reid, W.H.; *Biomaterials* **1985**, *6*, 369.
10.Rudowski, W.; Nasiłowski, W.; Ziętkiewicz, W.; Zienkiewicz, K.; *The Burn Wounds as a Research and Healing Problem.*; PZWL: Warsaw, 1984.
11.Polish Patent No 118 053 (Proprietor Max-Planck Gesellschaft, FRG), dated 1983.03.31.
12.*The Merck Index. Encyclopaedia of Chemicals and Drugs*; Merck and Company, Inc.: Rahway, New Jersey, 1976.
13.Rosiak, J.; Rucińska-Rybus, A.; Pękala, W.; Polish Patent No 151 581 and also: US Patent No 4,871,490; FRG Patent No 3744289; GDR Patent No 273200.
14.Chachaty, C.; Forchioni, A.; *J.Polym.Sci.* **1972**, *A-1, 10*, 1905.
15. *The Study of the Fast Processes and Transient Species by Electron Pulse Radiolysis*; Baxendale, J.M.; Busi, F. Eds.; Proc. NATO Adv. Stud. Inst., Capri, 7-18 September, 1981.
16.Davis, J.E.; Senogles, E.; *Aust.J.Chem.* **1981**, *34*, 1413.
17.Davis, J.E.; Sangster, D.F.; Senogles, E.; *Aust.J.Chem.* **1981**, *34*, 1423.
18.Swallow, A.J.; *Radiation Chemistry. An Introduction*; Longman Group Ltd.: London, 1973.
19.Burczak, K.; Ph.D.Thesis, Technical University of Łódź, 1988.
20.Charlesby, A.; *Atomic Radiation and Polymers*; Pergamon Press: New York, 1960.
21.Inokuti, M.; *J.Chem.Phys.* **1963**, *18*, 2999.
22.Dole, M.; *The Radiation Chemistry of Macromolecules*; Academic Press: New York - London, 1972.
23.Rosiak, J.; Burczak, K.; Czołczyńska, T.; Pękala, W.; *Radiat.Phys.Chem.* **1983**, *22*, 917.
24.Rosiak, J.; Olejniczak, J.; Charlesby, A.; *Radiat.Phys.Chem.* **1988**, *32*, 691.
25.Rosiak, J.; Olejniczak, J.; Pękala, W.; *Radiat.Phys.Chem.* **1990**, *36*, 747.

RECEIVED May 29, 1991

Chapter 18

A Polymeric Radiostabilizer for Absorbable Polyesters

Dennis D. Jamiolkowski and Shalaby W. Shalaby[1]

Ethicon, Inc., Somerville, NJ 08876

A new strategy was developed to prepare absorbable copolymers of acceptable absorption profile and radiostability. This entailed the preparation of a crystallizable segmented copolymer of poly(ethylene 1,4-phenylene-bis-oxyacetate) (PEPBO) and polyglycolide (PG) units with PG as the dominant component of the system. Incorporation of about 10 mole percent of the polymeric radiostabilizer (PEPBO) into the PG chain leads to high strength fibers and substantially improves the *in vivo* strength retention; the copolymer was shown to be absorbable.

Interest in synthetic absorbable polymers, which undergo chain scission and loss of mass in the biological environment, has grown considerably over the past two decades because of their increased applications in the biomedical and pharmaceutical areas (1-4). Medical products made of these polymers are usually sterilized by ethylene oxide (1-4) as opposed to gamma irradiation, a convenient and highly reliable method of sterilization. This is because of their susceptibility to degrade by high energy radiation; although losses in baseline physical properties can be modest, the rate at which physical properties are lost *in vivo* is generally unacceptible for irradiated, synthetic absorbable materials (exposure to a typical gamma radiation sterilization cycle, using a Co-60 source) (2,3,5-7) . The desire to develop biomedical and pharmaceutical products which are suitable for gamma sterilization provided an incentive to pursue studies in this area by Shalaby and coworkers since the early 1980's, who paid special attention to radiation sterilizable absorbable polyesters (6-13). This report deals with a member of a new class of polymeric radiostabilizers, which can be incorporated in the chains of absorbable polymers to produce radiation sterilizable, absorbable devices including surgical sutures.

Chemistry of the Polymeric Radiostabilizer and Copolymerization with Glycolide

The specific radiostabilizer used in these studies is a partially aromatic polyester with selected structural features to provide

[1]Current address: Department of Bioengineering, Clemson University, Clemson, SC 29634

0097–6156/91/0475–0300$06.00/0

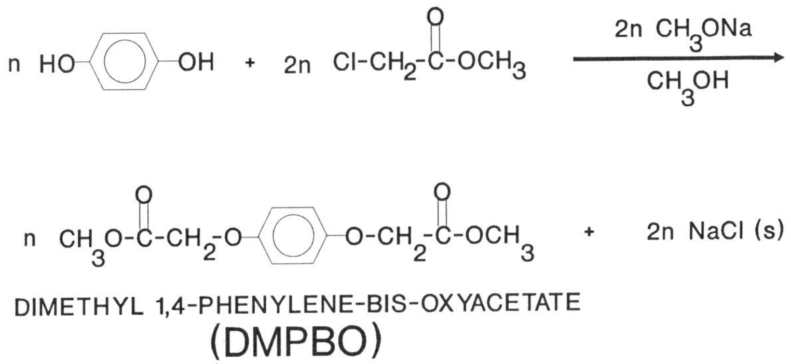

Figure 1. Synthesis of dimethyl 1,4-phenylene-*bis*-oxyacetate.

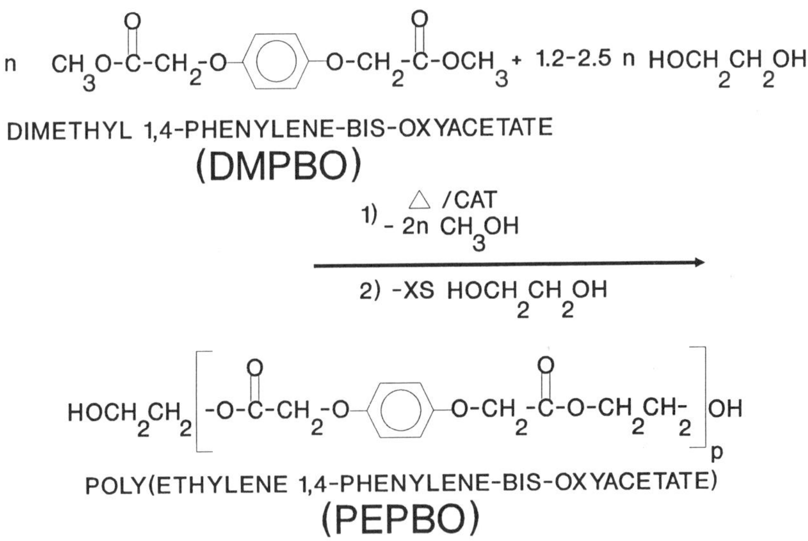

Figure 2. Preparation of PEPBO.

absorbability as well as radiostability. The requirements for radiostability has been satisfied by incorporating phenylene groups in the main chain (14). To improve absorbability, aroxyacetate linkages which have a chemical structure similar to glycolate esters are used to build the chain (15). To prepare the aromatic polyester, poly(ethylene 1,4-phenylene-bis-oxyacetate) (PEPBO), dimethyl 1,4-phenylene-bis-oxyacetate (DMPBO - prepared by O-alkylation of hydroquinone with methyl chloroacetate) was reacted with an excess of ethylene glycol in the presence of a catalytic amount of dibutyltin oxide. This α,ω-dihydroxy condensation polymer was then reacted with glycolide to produce a segmented copolymer of polyglycolide (PG) and PEPBO, which is referred to as modified PG or MPG. Reaction schemes for the syntheses of DMPBO, PEPBO and MPG are given in Figures 1-3. In addition, PEPBO was prepared in the presence of small amounts of glycolide to produce two PEPBO chains containing a small fraction of glycolate sequences for increased absorption. The unmodified PEPBO and two modified polymers are referred to as polymers I, II and III, respectively.

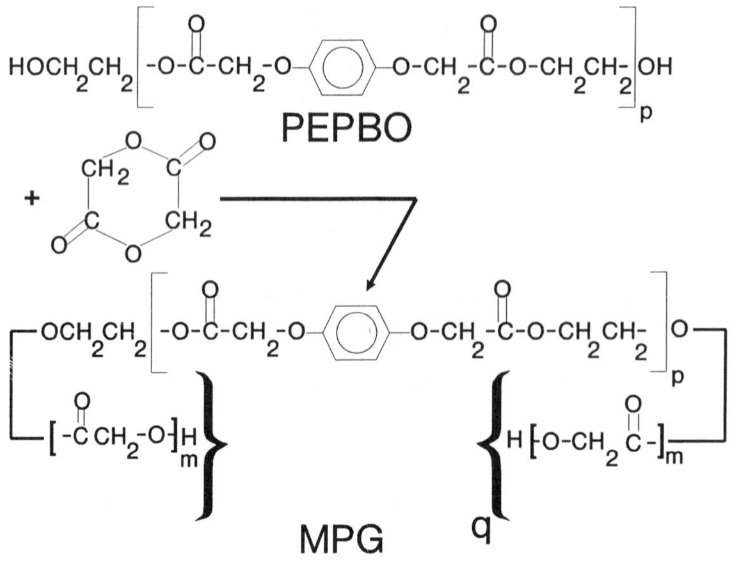

Figure 3. Preparation of MPG.

Experimental

Monomer Preparation. The preparation of dimethyl 1,4-phenylene-bisoxyacetate was carried out in a dry 5 liter, 3-neck round bottom flask equipped with an addition funnel, a nitrogen inlet, a mechanical stirrer, a reflux condenser with drying tube, a thermometer and a heating mantle. To this flask were charged 330.3 grams (3 moles) of hydroquinone, 651.1 grams (6 moles) of methyl chloroacetate, and 1722 mL of methanol. The contents of the flask were brought to reflux (approximately 68 °C) after an initial purge with nitrogen. A solution of sodium methoxide in methanol (1182 grams, 27.4 weight percent sodium methoxide, 6 moles) was charged into the addition funnel and, over the course of approximately one hour, allowed to slowly enter the refluxing reaction solution. After the addition was completed, the reaction mixture was refluxed an additional 17 hours during which time the reflux temperature dropped to 65 °C. The solution was filtered while hot (above 60 °C) to remove the sodium chloride. The filtrate was slowly cooled and material crystallized out of solution. The crystals were isolated by suction filtration and vacuum dried; 498.9 grams of crude product were obtained. The crystals were twice recrystallized from methanol using 4 mL of methanol per gram of dry crystals. The resulting dimethyl phenylene-bis-oxyacetate had a melting point of 99-101 °C (>55% yield). The structure of the monomer was verified by IR and ^1H NMR spectroscopic techniques. ^1H NMR (300 MHz, hexafluoroacetone sesquadeuterate(HFAD)/deuterobenzene (C_6D_6), δ in ppm): 3.53 [s, $-CH^3$], 4.20 [s, -CH2-], 6.74 [s, ϕH].

Syntheses of Unmodified and Modified PEPBO. Unmodified PEPBO (Polymer I) was prepared according to the following procedure. To a flame dried, mechanically stirred, 1 liter glass reactor, suitable for polycondensation reactions, were charged 127.1 grams of dimethyl 1,4-phenylene-bis-oxyacetate (0.5 moles), 62.1 grams of ethylene glycol (1.0 mole) and 9.0 milligrams of dibutyltin oxide (0.0071 weight percent based on the expected polymer weight). After purging the reactor and venting with nitrogen, the reactor was immersed in a silicone oil bath and connected to a gas supply to maintain nitrogen at 1 atmosphere of pressure. The stirred mixture was heated to and maintained at 160, 190, and 210°C for 2, 1, and 2 hours, respectively, during which time methanol along with some ethylene glycol was collected. The reactor was allowed to cool to room temperature. Some time later, the reactor was evacuated and heated; temperatures of 190, 210, and 220°C were maintained for 1, 1, and 2 hours, respectively. The collection of distillate was continued during the low pressure (less than about 100 microns) stage of the polymerization. The temperature was increased from 220 to 240°C over the course of 30 minutes and 240°C was maintained for 3 hours. The resulting polymer was isolated, ground and dried. The polymer had an inherent viscosity of 1.31 dL/g as determined in hexafluoroisopropyl alcohol at 25°C and at concentration of 0.1 g/dL. The structure of the polymer was verified by IR and ^1H NMR spectroscopic techniques. ^1H NMR (300 MHz, HFAD/C_6D_6, δ in ppm): 4.20 [s, -CH^2-], 4.39 [s, -CH^2CH^2-], 6.82 [s, ϕH].

(Note: The polymer may also be produced in a three stage polymerization wherein the diester, diol and a catalyst are heated under nitrogen at atmospheric pressure, in a molten state, followed by reaction at reduced pressure in a molten state to produce relatively low molecular weight species of the polymer. The low molecular weight material, upon crystallization and grinding into a relatively fine powder, is further polymerized at reduced pressure at temperatures below its melting point. The solid state polymerization increases the molecular weight significantly.)

Modified PEPBO (polymers II & III) were prepared by including glycolide as a reactant at the start of the above transesterification step of the polycondensation. The melting behavior of the copolymers was influenced not only by the composition but also by the reaction conditions, suggesting that the sequence distributions tended towards, but were not entirely random. Two examples of copolymers prepared by including minor portions of glycolide in the polymerization feed are polymers II and III. Polymer II was prepared by reacting 25.0 grams of dimethyl 1,4-phenylene-bis-oxyacetate (0.0983 moles), 13.2 grams of ethylene glycol (0.213 moles), 2.17 grams glycolide (0.187 moles) and 9.7 milligrams dibutyltin oxide (0.036 weight percent based on expected polymer weight). The stirred mixture was heated to and maintained at 180°C for 7 hours while under nitrogen at one atmosphere, followed by reaction at reduced pressure at temperatures of 180, 190, and 200°C for 2, 0.5, and 8 hours, respectively. To crystallize the copolymer, the temperature was reduced to 80°C for 3 hours followed by an annealing treatment of 130°C for 4 hours. The polymer was isolated, ground, and dried. It displayed an inherent viscosity of 0.60 dL/g. The molecular weight of the polymer was increased by conducting a solid state post-polymerization. The finely ground polymer was charged to a round bottom flask and the pressure reduced below about 100 microns; the flask was then immersed in a silicone oil bath at 80 and 135°C for 2 and 41 hours, respectively. The inherent viscosity of the copolymer increased to 1.22 dL/g. The structure of the polymer was verified by ^1H NMR spectroscopy. ^1H NMR (300 MHz, HFAD/C_6D_6, δ in ppm): 4.20 [s, -CH^2-], 4.38 [s, -CH^2CH^2-], 4.65 [s, glycolyl -CH2-], 6.82 [s, ϕH]. It was also shown by ^1H NMR spectroscopic analysis to have 16 mole percent glycolide-based moieties. The copolymer exhibited a major endothermic transition at 159°C (DSC; 20 °C/min) and 30% crystallinity as measured by x-ray techniques.

Polymer III, a copolymer, was prepared in a fashion similar used for Polymer II. The reactor was charged with 25.0 grams dimethyl 1,4-phenylene-bis-oxyacetate (0.0983 moles), 12.2 grams ethylene glycol (0.197 moles), 1.56 grams glycolide (0.0134 moles) and 4.9 milligrams of dibutyltin oxide (0.019 weight percent of the expected polymer weight). The stirred mixture was heated to and maintained at 180°C for 7 hours during which time the methanol was collected. The reactor was heated to 230°C while still under nitrogen to continue the polymerization and remove some excess ethylene glycol. The temperature was lowered to 200°C and the pressure reduced. Temperatures of 200, 220, and 240°C were maintained for 0.5, 1.5, and 2 hours, respectively, while continuing to remove distillates under reduced pressure. The polymer was isolated, ground, and dried under vacuum at room temperature. The polymer was found by ^{1}H NMR analysis to contain approximately 12 mole percent (5.9 weight percent) glycolide moieties. The inherent viscosity of the polymer was determined to be
0.95 dL/g. Hot stage optical microscopy of the bulk polymer revealed a melting transition below 140°C.

Synthesis of Modified Polyglycolide (MPG). The modified polyglycolide is prepared by reaction of glycolide with an unmodified sample of PEPBO to produce a segmented copolymer. Thus, a flame dried 1,000 mL round bottom flask outfitted with a vacuum tight, stainless steel mechanical stirrer and a hose connection was charged under dry, oxygen-free, conditions with 313.4 grams (2.7 moles) of glycolide and 75.7 grams of finely divided (passing a 10 mesh screen), amorphous, dry, poly(ethylene 1,4-phenylene-bis-oxyacetate) resin, prepared in the presence of 0.0197 weight percent of dibutyltin oxide as a catalyst, to an inherent viscosity of 0.63 dL/g. The reactor was purged and vented with nitrogen, immersed in a silicone oil bath, and connected to a gaseous supply to maintain a nitrogen atmosphere at a pressure of one atmosphere. The mixture was heated for about one half hour using a bath temperature of 120°C to melt the glycolide and start the dissolution of the polyester. The temperature was increased at the rate of 1.8 °C per minute to 150°C which was maintained for 8 minutes to continue the dissolution process. The heating bath was brought up in temperature to 195°C at an average rate of 1.5 °C per minute. Stirring was discontinued prior to reaching 195°C because of the viscous nature of the reaction mass. The resulting polymer crystallized and was maintained at 195°C for 8 hours. The polymer was isolated, unreacted glycolide was removed by heating the ground polymer to 110°C at 0.1 mm mercury for 16 hours. A weight loss of only 0.2% was observed indicating a high degree of conversion. The ground polymer was seived to remove particles less than 1 mm in diameter. Two hundred and twenty five grams of the polymer, free from fine particules, were produced.

The copolymer was analyzed by ^{13}C NMR for sequence distribution. The ^{13}C NMR carbonyl resonances of the glycolyl and PEPBO moieties at the block junctions are resolvable from the carbonyls within the blocks. Relative intensities of "junction" and "interior" carbonyls were used to determine sequence lengths for both the PEPBO and PG blocks. ^{13}C NMR (75.4 MHz, HFAD/C_6D_6, δ in ppm): 169.4 [PG "interior" carbonyl], 169.7 [PG "junction" carbonyl], 171.7 [PEPBO "junction" carbonyl], 172.2 [PEPBO "interior" carbonyl]. The ^{13}C NMR data indicated the chemical structure of the modified PG polymer, MPG, was that of a segmented copolyester as shown in Figure 3. ^{1}H NMR data indicated that the copolymer comprised 89.5 mole percent (79.7 weight percent) of polymerized glycolide moieties and 10.5 mole percent (20.3 weight percent) of the poly(ethylene 1,4-phenylene-bis-oxyacetate) moieties. The inherent viscosity of the polymer, after prolonged heating at 50°C to dissolve, was found to be 1.69 dL/g. The segmented copolyester could also be prepared entirely in the melt.

Polymer Extrusion and Fiber Properties. Polymers I and III were extruded at 160°C using a Model 3211 Instron Capillary Rheometer employing a 40 mil die, a shear rate of 213 sec^{-1}, an ice water quench and a take up rate of 27 feet per minute. The melt viscosities at this temperature were determined to be 3600 and 2200 poise, respectively. The extrudates were oriented in a glycerine draw bath in two stages as summarized in Table I. The fibers were subsequently annealed under tension. The physical properties of the fibers can be found in Table II.

Table I. Drawing Conditions of Fibers Made of Polymers I and III

	Fiber of Polymer I	Fiber of Polymer III
First Stage:		
Draw Ratio	6X	8X
Temperature, °C	52	53
Second Stage:		
Draw Ratio	1.5X	1.25X
Temperature, °C	90	65
Overall Draw Ratio	9X	10X

Table II. Physical Properties of Fibers Made of Polymers I and III

	Fiber of Polymer I		Fiber of Polymer III	
	Not Annealed	Annealed 65°C/16 hr	Not Annealed	Annealed 63°C/2 hr
Diameter, mil	5.8	5.7	5.4	5.5
Straight Tensile Strength, Kpsi	89.7	92.4	42.9	44.3
Knot Tensile Strength, Kpsi	---	75.0	40.3	40.5
Elongation to Break, Percent	24	---	55	40
Young's Modulus, 10^6 psi	---	2.40	1.06	1.59

The annealed monofilament fiber of Polymer III was exposed to 2.5 Mrads of gamma radiation. The diameter, straight and knot tensile strengths, elongation to break, and Young's modulus were determined to be 5.5 mil, 44.7 and 38.2 Kpsi, 40%, and 1.51 x 10^6 psi, respectively. Modified polyglycolide (MPG) was melt spun to produce monofilaments using the Instron Rheometer, as discussed above for Polymers I and III, at 245°C at a shear rate of 213 sec^{-1} employing a 40 mil die with an L/D ratio of 24. The extrudate was taken up through ice water and subsequently drawn in a glycerine draw bath in two stages: 5X at 49°C followed by a draw of 1.5X at 92°C. The inherent viscosity of the extrudate was found to be 1.52 dL/g, as measured at 25°C at a concentration of 0.1 g/dL in hexafluoroisopropyl alcohol. The drawn fibers were determined to be 33% crystalline, and upon annealing under tension at 112°C for 9 hours, were found to be 36% crystalline. The final fiber had a diameter of 7.0 mils. The properties of the fibers are provided in Table III.

Table III. Physical Properties of MPG Monofilaments

	Drawing 5X	Drawing 5X Followed By 1.5X	Drawn Fiber After Annealing
Tensile Strength, Kpsi	120	172	174
Knot Tensile Strength, Kpsi	73.4	104.0	117.0
Elongation at Break, %	53	12	11
Young's Modulus 10^6 psi	0.77	1.71	2.42

To produce multifilament yarn from braid construction of MPG, the polymer was melt spun in an extruder using a 16 hole die with the application of a spin lubricant to produce 7000 yards of 56.8 denier yarn (3.55 denier per filament). The extrusion conditions were as follows:

Melt temperature in block: 260°C
Melt temperature in die: 266°C
Chimney air temperature: 271°C
Throughput: 485 grams/hour

Orientation at a draw ratio of 5X with a feed roll temperature of 68°C and an annealing roll temperature of 91°C was used. The yarn was assembled into a braid made up of three core threads, 16 filaments each and 16 carrier threads, 16 filaments each. The final braid was hot stretched and annealed under tension. The braid denier was 1072 grams and the diameter was 13.5 mils. The braid had an elongation at break of 23%, a straight tensile strength of 106.0 Kpsi and a knot tensile strength of 67.1 Kpsi. A sample of polyglycolic acid was similarly extruded, braided and post-treated for use as a control. Portions of the braid were cut to appropriate lengths, placed in individual paper folders and subsequently into vented heat sealable metal foil envelopes. The materials were dried by *in vacuo* treatment at 50°C followed by sealing. The braids were sterilized by gamma irradiation by using standard industrial conditions; some sets of samples were subjected to double doses. The physical properties of both braids were determined and the results are given in Table IV.

Table IV. Physical Properties of Polyglycolide and MPG Sterilized Braids

	Fibers of MPG		Fibers of PG	
	Before Irradiation	After 2.89 Mrads	Before Irradiation	After 2.76 Mrads
Diameter, mils	13.6	13.4	14.1	14.2
Straight Tensile, Kpsi	105.9	100.0	130.7	113.2
Knot Tensile, Kpsi	65.0	61.2	74.0	59.7
Elongation at Break, %	24	22	19	16

Biological Properties of MPG Braided Sutures. The key properties of absorbable sutures, namely *in vivo* absorption and tissue reaction, and breaking strength retention were evaluated using braided multifilament yarns of MPG and PG (as a control) subjected to gamma sterilization. The sutures, following established protocol (16), were implanted intramuscularly and subcutaneously in rats to monitor their absorption and tissue reaction, and breaking strength retention (BSR) profiles, respectively. The absorption data are shown in Table V. Scores of the elicited tissue responses to the MPG braids were in the slight range with some zero scores after 91 and 119 days of implantation. Synthetic absorbable braided sutures would need to retain some breaking strength after 21 days *in vivo* to be considered efficacious for general surgery. The average breaking strength values in rats was determined on both MPG and PG braids (as a control) and respective data are summarized in Table VI.

Table V. Braided Sutures *In Vivo* Absorption Data: Percent Cross-Sectional Area Remaining

	Days Post-Implantation				
	5	70	91	119	182
MPG Braid Irradiated at 2.89 Mrads	100	92	49	1	0
PG Braid Irradiated at 2.76 Mrads	100	59.5	19.5	3	0

Table VI. *In-Vivo* Breaking Strength Retention Data of MPG and PG Braids

	Days Post-Implantation			
	0	7	14	21
MPG Braid, Non-Irradiated				
Pounds	9.51	11.63	9.96	9.23
Percent	100	122	105	97
MPG Braid, Irradiated 2.89 Mrads				
Pounds	9.43	10.63	9.82	6.76
Percent	100	113	88	72
MPG Braid, Irradiated 5 Mrads				
Pounds	8.58	9.44	7.80	4.71
Percent	100	110	92	55
PG Braid, Non-Irradiated				
Pounds	13.41	15.05	9.90	6.85
Percent	100	112	74	51
PG Braid, Irradiated 2.75 Mrads				
Pounds	12.76	11.06	4.91	0.00
Percent	100	87	38	0
PG Braid, Irradiated 5.33 Mrads				
Pounds	12.91	9.70	2.04	0.00
Percent	100	75	16	0

Discussion of Experimental Results and Conclusions

Using DMPBO, high molecular weight samples of the partially aromatic polyester PEPBO were obtained as a crystallizable fiber forming polymer. This is contrary to an earlier report (17) suggesting that PEPBO is an amorphous solid. It was in fact demonstrated that this polymer can be converted into drawn monofilaments with excellent tensile properties. Although these monofilaments underwent hydrolytic cleavage and absorption, the rate of *in vivo* absorption was much slower than that of commercially available absorbable fibers such as those of PG. Limited *in vivo* absorption studies indicated that these fibers retain most of their cross sectional area at one year. Due to its aromatic constituent, however, the expected resistance of these monofilaments to gamma degradation was documented.

Efforts to render PEPBO more absorbable by incorporating a small fraction of glycolate moieties through direct polymerization of DMPBO and ethylene glycol in the presence of glycolide led to a copolymer of poor fiber forming quality. This can be attributed to the random incorporation of glycolate units into the PEPBO chain which compromised its ability to crystallize. Therefore, a new strategy was developed to prepare absorbable copolymers of acceptable absorption profile and radiostability. This entailed the preparation of a crystallizable segmented copolymer of PEPBO and PG units with PG as the dominant component of the system. Thus, under controlled conditions, a partially aromatic polyester (PEPBO) was copolymerized with glycolide to produce a crystalline copolyester, MPG, which contains segments of polyglycolide and PEPBO, as shown by spectroscopic and thermal techniques. Incorporation of about 10 mole percent of the polymeric radiostabilizer (PEPBO) appears to impart radiostability to the polyglycolide matrix. This is quite evident when one compares the biological properties of fibers prepared from this copolymer and PG. It is worth note that segmentation of PG did not compromise the strength of resulting fiber. In fact, in addition to improving the *in vivo* strength retention substantially, the tensile strength of certain segmented fibers is extraordinarily high. The unexpected behavior of these fibers are being examined.

Acknowledgments

The authors wish to thank C. L. Ace for her NMR analysis and J. A. Persivale for his efforts in the absorption studies. Special thanks to S. C. Arnold for his thorough review of this manuscript.

Literature Cited

1. Vainiopaa, S.; Rokkanen, P.; Tormala, P. Prog. Polym. Sci. 1989, 44, 679.
2. Shalaby, S.W. Encyclopedia of Pharmaceutical Technology; Swarbrick, J. and Boylan, J.C., Eds.; Marcel Dekker, New York, 1988; Vol. 1, p 46.
3. Gupta, M.C.; Desmuth, V.G. Polymer 1983, 24, 827.
4. Tormala, P.; Vasenius, J.; Vainiopaa, S.; Laiho, Pohijonen, T.; Rokkanen, P.J. Biomed. Mater. Res. 1991, 25, 1.
5. Palmer, G.R.; Baier, R.E. Ann. Conf. Eng. Med. 1987, 29, 139.
6. Jamiolkowski, D.D.; Shalaby, S.W. Polym. Prepr. 1990, 32, (2), 329.
7. Bezwada, R.S.; Jamiolkowski, D.D.; Shalaby, S.W. Trans. Ann. Meeting Soc. Biomater. 1991, 14, 186.
8. Shalaby, S.W.; Jamiolkowski, D.D. U.S. Pat (to Ethicon, Inc.) 1984, 4,435,590.
9. Jamiolkowski, D.D.; Shalaby, S.W. Polym. Prepr. 1990, 32, (2), 327.
10. Bezwada, R.S.; Shalaby, S.W.; Jamiolkowski, D.D. U.S. Pat. (to Ethicon, Inc.) 1985, 4,510,295.
11. Bezwada, R.S.; Shalaby, S.W.; Jamiolkowski, D.D. U.S. Pat. (to Ethicon, Inc.) 1985, 4,532,928.
12. Koelmel, D.F.; Jamiolkowski, D.D.; Shalaby, S.W. U.S. Pat. (to Ethicon, Inc.) 1985, 4,559,945.
13. Kafrawy, A.; Jamiolkowski, D.D.; Shalaby, S.W. J. Bioact. Biocomp. Polym. 1987, 2, 305.
14. Shalaby, S.W. Macromol. Revs. 1979, 14, 406.
15. Shalaby, S.W. Unpublished work on the susceptibility of different linkages to undergo hydrolytic cleavage.
16. Shalaby, S.W. High Technology Fibers - Part A; Lewin, M.; Preston, J., Eds.; Marcel Dekker, New York, 1985, Chap.3.
17. Spanagel, E.E.; Carothers, W.H. J. Amer. Chem. Soc. 1935, 57, 926, 935.

RECEIVED July 31, 1991

Chapter 19

Chemistry and Manufacturing Requirements of X-ray Resists

J. W. Taylor[1–5], C. Babcock[2,4], M. Sullivan[2,5], D. Suh[2,5], and D. Plumb[5]

[1]Department of Chemistry, [2]Materials Science Program, [3]The Synchrotron Radiation Center, [4]Engineering Research Center in Plasma-Aided Manufacturing, and [5]Center for X-ray Lithography, University of Wisconsin—Madison, 3731 Schneider Drive, Stoughton, WI 53589–7192

The requirements placed on the polymer formulation chemistry for resists used by semiconductor manufacturing are severe when submicron features are to be printed. This forces some new strategies in resist molecular design, lithographic mechanism, formulation, measurement, and evaluation. Testing for compliance to these requirements utilizing e-beams and X-rays requires protocols which are still in the process of being developed. Commercial and experimental chemically-amplified resists, both positive-tone and negative-tone, appear to be able to meet the challenge of printing 0.25 μm minimum linewidth features, but the future thrust is to feature sizes of 0.15 μm and smaller. At these dimensions, control of processing parameters becomes critical because they provide the limits to manufacturing. The key ones for resists are described.

New polymer chemistry insight is required to design photoresists used for microcircuit manufacture, and several references provide an overview of the expectations (*1-4*). This problem is especially acute when the polymer system must be capable of printing submicron features down to 0.25 micrometer(μm or micron) or smaller in minimum linewidth, and the manufacturing requirements for throughput force the resist sensitivity to lower and lower values. Sources of exposure which will operate at the reduced linewidth dimensions include electron-beams, X-rays, deep UV photons, and ion-beams (*5*). Writing with e-beams is a relatively slow "serial-type" process, whereas X-rays, ion-beams, and deep UV photons can be used in a parallel mode, once a suitable mask is available to control the wafer exposure areas. X-rays rely on a mask

0097–6156/91/0475–0310$06.00/0

with different areas of contrast to define the required exposure pattern, whereas ion-beams may employ stencil-like masks. Recently in the US, Germany, and Japan, there have been substantial efforts to explore the reactions of resist systems to X-rays generated from an electron storage ring or from a laser-plasma source (*6-12*). From these efforts there have come a new generation of resists which circumvent the low sensitivities normally observed with conventional resists. This class of resists is termed "chemically-amplified". In equation 1 is shown the basic reactions for a conventional positive-tone resist.

$$\text{Inhibitor} + \text{Polymer} + \text{X-rays} \longrightarrow \text{Altered Inhibitor} + \text{Polymer} \qquad (1)$$

The chemically-amplified resists differ in that the reaction sequence is initiated by the initial absorption event, which generates an acid. As shown in equations 2 and 3, the acid can react catalytically with the original polymer or an additive

$$\text{Radiation-Sensitive Photoacid Generator} + \text{X-rays} \longrightarrow \text{Acid} \qquad (2)$$

$$\text{Acid} + \text{Polymer} (+ \text{Additive}) + \text{Heat} \longrightarrow \text{Altered Polymer} + \text{Acid} \qquad (3)$$

to alter the polymer and produce an image. This is also shown schematically in Figure 1 using the chemistry for a negative-tone chemically-amplified resist where the alteration consists of a cross-linking step. In this schematic the hexamethoxymethylmelamine is the cross-linking agent, which reacts with the polymer matrix after being activated by the photoacid generator. The cross-linked product is of greater molecular weight and is less soluble in the developer than the unexposed resist. With a different formulation, the acid might be used to depolymerize a cross-linked polymer or to reduce the effect of a dissolution inhibitor. Both of these latter steps lead to the positive-tone image by producing an altered polymer which is more soluble in the developer than the unexposed resist. Both processes are strongly dependent on radiation dose and time. Some negative-tone resists are also dependent on the post-exposure baking temperature and time because these two variables are used to control the kinetics of the acid reaction. There are a number of resist systems which follow these steps (*3,7-10,12-15*), and subsequent chapters in this book will detail more of their chemistry.

There also have been a number of comparisons of the differences in feature resolutions from different excitation sources - e-beam, X-ray, or ion-beam (*5,10,12*). One paper modeled the catalytic process for a negative-tone chemically-amplified resist (*13*). More importantly, however, there is a general protocol for evaluating and reporting resist performance (*16*).

The purpose of this chapter is to provide a background for the polymer requirements needed in microcircuit manufacture by concentrating on the results from experiments using X-rays from an electron storage ring at the Center for X-ray Lithography at the University of Wisconsin-Madison. This Center is a SEMATECH Center of Excellence and has benefitted greatly from advice and encouragement from industrial colleagues. In particular, at a Resist Workshop held at Wisconsin (and attended by representatives from AT&T, Intel, Olin Hunt, Rohm & Haas, IBM, Perkin-Elmer, Texas Instruments, and Hampshire Instruments), we attempted to develop a common system for reporting resist exposure data obtained from X-ray sources. This protocol is patterned, in part, along the lines proposed by Gary N. Taylor of AT&T and follows most of the suggestions for resist systems exposed by other sources (*16*).

At present, there are two exposure stations on the Aladdin Storage Ring at the Center for X-ray Lithography designed for these efforts (*17*), and shortly a fully-operational X-ray stepper beamline will be implemented. These stations permit evaluations of resist sensitivity as a function of exposure. The resolution characteristics are determined by exposure through an appropriate X-ray mask, development in an solvent, and evaluation of the resulting images by scanning electron microscopy(SEM). Prior to exposure, the resist is spin-coated onto the wafer with a uniform specified thickness, which usually ranges from 0.5 to 1.5 μm for various process steps.

Resist reactions are initiated by energy absorption events, but resist performance is measured by their ability to increase (or decrease) their dissolution rate in a developer. This means that any measurement of resist performance is coupled to the type, the developer concentration, development time, developer temperature, and method of developer application to the wafer surface. The chemically-amplified resists develop in the same way as conventional resists, except they have an additional step, as shown in equations 2 and 3 for which time must be allowed for the acid-catalyzed event to occur. Precisely timed baking on a vacuum hot plate may be required to achieve the necessary control of the reaction kinetics for some of the negative-tone resists. The development process is critical for some resists, and the developer normally is an aqueous base solution. In all the studies reported here, the wafers are immersed in the developer for a specified period of time and specified normality of the base.

Schematically, the process of exposure using an electron storage ring is shown in Figure 2. The X-rays, in the case of synchrotron radiation, come from the broad distribution of wavelengths produced by the relativistic electrons which lose energy as radiation after passing through a bending magnet. The radiation is processed in a beamline (*17*), which may utilize mirrors or filters to bring the radiation to the X-ray mask. In the case of the Center for X-ray Lithography exposure station ES-1, which has a 25 μm beryllium filter, the

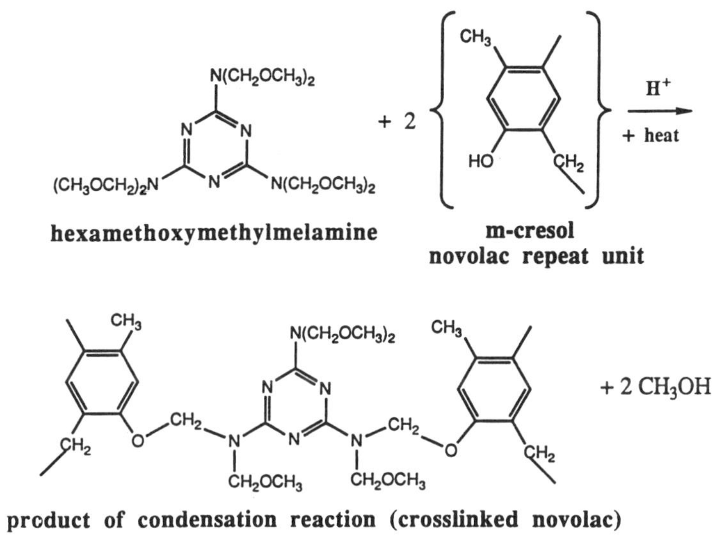

Figure 1. Schematic of the acid catalyzed reactions for chemically amplified negative-tone resists.

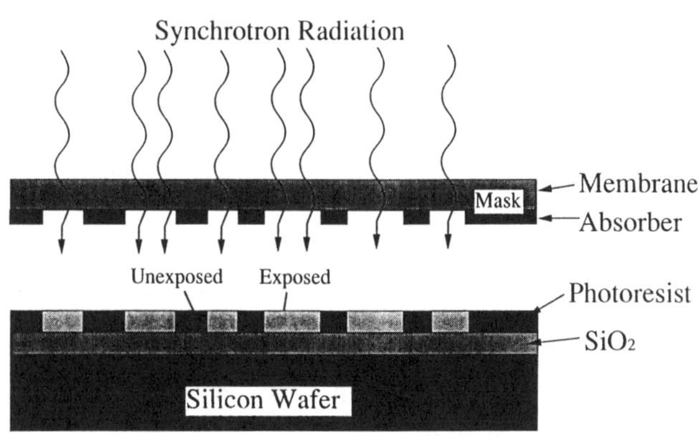

Figure 2. Schematic of the resist exposure process at the Center for X-ray Lithography.

radiation at an electron energy of 800 MeV is a broad but skewed band centered at 0.82 nm. At half-height, one side of the skewed band appears at 0.58 and the other is at 1.12 nm. Operation at 1,000 MeV (or 1.0 GeV) produces a higher energy band centered at 0.69 nm. The radiation passes through the X-ray mask before striking the resist-coated wafer surface. The actual spectrum observed by the resist is modified by the mask transmission and can be calculated if the composition, density, and thickness of the mask materials are known (*17,18*).

The mask consists of a semi-transparent membrane upon which a pattern has been constructed with a metal absorber. The membranes are approximately 1-2 μm in thickness and can be made of boron nitride, polysilicon, silicon nitride, silicon carbide, or diamond. The absorber is usually gold or tungsten. The membranes are not perfectly transparent but absorb 30-65 percent of the X-ray intensity (*19*). Likewise, the absorber is not a perfect absorber. For each combination of absorber and membrane, however, a mask contrast can be calculated as the ratio of the transmitted x-ray power through the unpatterned membrane to that through the membrane plus the absorber.

Semiconductor Manufacturing Requirements For Resist Systems

For polymer chemists not aware of semiconductor manufacturing requirements, it might be helpful to examine Table I, which outlines the desired resist performance. The following paragraphs will also illustrate how those requirements are measured to evaluate the prospects for new resist systems.

Lithographic Sensitivity. The lithographic sensitivity describes the sensitivity that the resist must have to yield the desired manufacturing throughput. It is determined by the combination of the mask and resist-coated wafer. A lithographic sensitivity requirement of 100 mJ/cm^2 arises from the manufacturing need to expose the complete wafer surface in one minute or less, to yield a throughput of 60 wafers/hr. Only a portion of the wafer is exposed at one time, and a device called a stepper is employed to move and align the exposure fields. This means that part of the one minute time per wafer exposure is assigned to stepper motion and alignment. Because the mask is located in front of the resist, the transmission of the mask membrane, 30-65 percent, must also be taken into account. For example, with a manufacturing lithographic sensitivity of 100 mJ/cm^2 and a 50 % transparent mask membrane, the actual resist sensitivity would have to be 50 mJ/cm^2 or smaller to meet this requirement.

The resist sensitivity is defined in terms of exposure using the units of mJ/cm^2 determined from a plot of normalized resist thickness(NT) plotted against the log of dose in mJ/cm^2. For the plot, the NT is the film thickness after development divided by the thickness of the unexposed film prior to development. (For chemically-amplified negative-tone resists this is modified slightly (*20*) by using the thickness after development divided by the thickness after post-exposure baking. If one were to be technically correct and use the protocols of Taylor (*16*), there should be two NT plots. One plot would use resist thickness data before exposure compared to after development; the second would be as described. Because the differences are small - indicating little thickness loss on baking - the post-bake thickness is the one normally used for reference with these resists.) For the log of dose, it is sometimes helpful to compare different X-ray sources. In order to do this, the actual wavelength, or exposure band, must be specified because shorter wavelength X-rays have a larger energy for each photon. At the shorter wavelengths, however, the absorption cross section decreases sufficiently rapidly that a more sensitive resist (lower dose) is required. Resist sensitivity can be expressed in units of J/cm^3 to avoid this problem, but this is not in general use by the resist community.

The protocol for resist evaluation calls for the complete NT versus the log of dose exposure curve to be shown. Such a plot for a conventional positive resist (*21-22*) is shown as Figure 3, and that for a chemically-amplified negative-tone resist is shown in Figure 4. It can be seen that the sensitivity for a positive resist, $D_p^{0.0}$, is normally given as the intercept in the NT versus log dose plot (Figure 3). For negative resists, there have been two definitions of sensitivity. The point at which 0.5 of the NT is reached provides one definition (see point showing 28.5 mJ/cm^2 on Figure 4); and the point at which the curve leaves the straight line provides another (point D_n^m, where m= 0.75 in Figure 4, gives a value of 30.7 mJ/cm^2). The second is a more stringent definition, but none of these answer the question of what exposure should be used to print a particular feature on the mask. These conditions have to be determined experimentally followed by examination with a SEM.

Lithographically Useful Sensitivity. When the resist system is actually used in the manufacturing environment, the dose required to print the mask dimension of a particular feature is usually the parameter desired. This may be called the dose to size (or energy to size). This dose is usually greater than one determined from the previous sensitivity definitions because of loss of exposure energy at the feature edge. The resist sensitivity required to print a particular feature, therefore, is the lithographically useful sensitivity. With conventional resists, the dose required to print a small feature may be 1.2 to 3 times larger than the sensitivity shown on Figure 3. With the chemically-amplified negative resists we have tested so far, however, we have found that the sensitivity measured at a NT value of 0.90 is a more reliable measure of the resist lithographically useful sensitivity. From Figure 4, one obtains 34.5 mJ/cm^2 at the NT=0.90 point. Thus, the 0.90 NT value is the most stringent definition and, in this case, is about 1.2 times the dose required for NT = 0.50. It, then, becomes the starting point for determining the dose required to print the features on the mask. Once that is determined for equal line and space 0.25

Table I. Desired X-ray Resist Properties for Semiconductor Device Manufacturing [a]

1.	Type	Positive- or Negative-Tone, Single Level
2.	Lithographic Sensitivity	Preferably < 100 mJ/cm^2
3.	Lithographically Useful Resist Sensitivity	< 40 mJ/cm^2
4.	Resist Contrast(γ)	> 3.0
5.	% Absorption/μm of Resist	8 to 50
6.	Resolution Between Lines/Spaces	0.08 to 0.25 μm
7.	Total Control of Linewidth	Better than ± 5 %
8.	Dose Latitude	Greater than ± 10 %
9.	Wall Profile	Greater than 85°
10.	Thermal Properties	T_g > 120 ° C T_d > 180 ° C

[a] Adapted from material supplied by G. N. Taylor(AT&T) at the Wisconsin Resist Workshop, 1989.

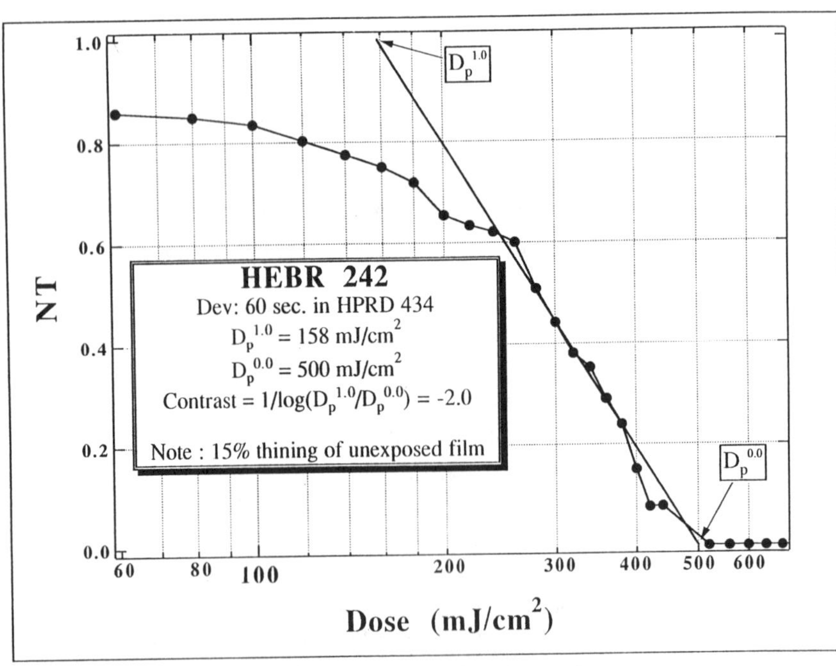

Figure 3. Sensitivity and contrast plot for Olin Hunt HEBR 242 positive-tone X-ray resist in HPRD-434 developer.

μm features by examining SEM pictures, that becomes the lithographically useful dose for the line and space features.

Resist Contrast. The resist contrast provides a measure of distinguishing between the exposed and the unexposed regions. The value of the contrast, gamma or γ, is defined as the slope of the straight line region on the plot of NT versus log dose in a manner similar to photographic film definitions of contrast. From Figure 3, one can calculate the contrast from 1/ log (dose at NT = 1.0, $D_p^{1.0}$, divided by the dose at intercept of line at NT = 0.00, $D_p^{0.0}$). The measured contrast in this case is -2.0. From Figure 4, the calculation of γ (using 1/log $D_n^{1.0}/D_n^{0.0}$) gives a resist contrast of 7.6. More recent resolution studies (*23*) with chemically-amplified negative-tone e-beam resists have suggested that there may better resolution correlation with γ when γ is measured using only two points - values at NT = 0.00 and NT = 0.90. For the data of Figure 4 and values of 24.5 mJ/cm^2 and 34.5 mJ/cm^2,respectively, the γ value would be 6.7 by this newer definition. The correlation with resolution suggests that the steepness of the slope as it approaches values of NT=0.90 may be an important parameter for resolution. It may also suggest that the dose increment needed to reach a NT=0.90, once the resist has received a sufficient dose to begin significant cross-linking, may be a better definition of the resist performance. None of these latter definitions are part of the protocol suggested by Taylor (*16*). Obviously, more data needs to be obtained to clarify the use of these definitions. In all cases, the best way to convey the resist performance is to publish the complete NT versus log dose curve. In that way subsequent definitions and correlations can be applied and evaluated because the data are available.

Absorption as Percent/Micron of Thickness. Because the initial chemistry of the resist is governed by its absorption of energy, the dose received by the top layer of resist should not differ greatly from that received at the bottom for the usual type of resists. The consequence of a substantial absorption would be to have no exposure at the bottom and a fully exposed resist at the top. In general, as one utilizes shorter and shorter wavelengths, the difference in absorption decreases. At 0.82 nm (with resists consisting primarily of carbon, oxygen, and nitrogen) the absorption can be calculated and shown to be on the order of 8-10 percent. The X-ray absorption step is amenable to modeling by application of the TRANSMIT program, which was developed by Professor Franco Cerrina and his students at Wisconsin (*18*). These programs provide a tool for the evaluation of the photoresist absorption as a function of atomic composition, photoresist density, and wavelength of the impinging X-rays. The computer program also considers contributions from the beam line optics and the absorption from filters, the mask carrier, the mask absorber, and any cooling gas which might be employed. This means that the total exposure system can be modeled, and variations in various mask compositions can be predicted. It is also possible to model other X-ray sources so that work at other facilities can be compared. Because both resist density and the resist atomic composition are required as input, it is possible to calculate what changes will occur in the absorption step with variations in resist substituents, provided the density, the source, and the mask conditions are fixed. In this fashion, resist atomic structure composition can then be designed to match the source X-ray wavelength and maximize the absorption (*24*).

With this modeling, however, the structural arrangement of the atomic constituents, which may control the subsequent chemistry, is not considered. This means that the structurally-driven resist reactions such as those kinetic reactions occurring with acid generated as a consequence of the absorption step, are not part of the model. These factors are much more difficult to determine and must be examined experimentally. In situ spectroscopic techniques can detail the functional group changes and give some insight into mechanisms. The application of Fourier Transform Infrared(FTIR) and mass spectrometry have been used for these purposes (*22*).

Resolution. The critical measure of the usefulness of a resist system is whether it will faithfully reproduce the features on the mask. Although there are excellent modeling approaches to predict what will be observed from a given set of conditions, the best test is to observe the resist image after it is developed on the wafer surface. There are a number of problems in insuring that the measurement step is accurate at these small dimensions, but the key step is to measure the wafer feature with a scanning electron microscope designed for that purpose.

Total Control of Linewidth. An earlier general "rule of thumb" for mask feature and wafer printed feature was to control the dimension to ± 10 % of the nominal value. More recent and tighter specifications reduce this to ± 5%. Thus a 0.25 μm feature would have a limiting total error of 25 nm (3 sigma). This total error budget would have a number of sources are associated with this deviation when a manufacturing stepper is used for the wafer exposure, including distortions of the mask while the exposure is in progress. The resist may be assigned a value on the order of 80 % of the total budget or < ± 4%. Any or all of the processing parameters may influence this requirement, and many measurements of linewidth have to be made, preferably with statistical experimental designs to reduce the number of experiments, to determine that the resist performance meets this stringent requirement.

Dose Latitude. This term refers to the variation of the measured resist linewidth with a change of the dose, while observing the same mask feature. What is desired here is a large variation in dose before any linewidth variation occurs. The total dose incident on the resist should be controlled by the beamline to better than 1-2 %. This means, therefore, that the rest of the resist processing parameters must be controlled, if this latitude value is met. A large dose latitude would be preferred, but the manufacturing requirement is to have a latitude of at least ± 10 % for the smallest feature size before the resist error budget for the linewidth is exceeded.

Wall Profile. One might expect that the "ideal" image on the wafer surface is a profile which produces a perpendicular wall from top to bottom in the resist, or a wall profile of 90°. In subsequent manufacturing steps, however, as long as the base width does not change, that expectation can be relaxed somewhat. Some processes, such as deposition and etching, may be improved at angles slightly less than 90° provided the base width remains at the desired dimension. It should be noted that the variation of five degrees on a 0.10 μm line is far more serious change in the appearance of the line profile than on a line of 0.50 μm linewidth because for a negative-tone resist, the flatness width at the top of the resist is reduced.

Thermal Properties. Because the actual construction of the microcircuits may involve a number of different steps (such as metal coating, etching, etc.) and will include a number of different masks and feature geometries, the resist system may be exposed to high temperatures during these processing steps. For that reason, the resist decomposition temperature is required to be high. This fixes the decomposition temperature, T_d, to be at least 180° C. The effect of the glass transition temperature, T_g, has been discussed by Thompson and Bowden (*25*). Basically, the glass transition temperature is a measure of the point at which the polymer may begin to flow. During the pre-bake step (after spin-coating and prior to exposure), solvent is being removed, and polymer movement is not a critical event. In fact, some strains introduced during the spin-coating step can be annealed with this heating step. After exposure, however, polymer motion may result in loss of feature resolution. One simple way to measure this is to determine any change in a feature after a timed exposure on a hot plate. For example, flow at 130° and no flow at 125 ^{0}C will suggest a thermal stability of 125^0 C. Fortunately, the cross-linking of the negative chemically-amplified resists result in thermal flow values ≥ 220^0 C. Requiring a glass transition temperatures at least on the order of 120° C represents a reasonable compromise (*20*) between the need for annealing and the need for retaining feature dimensions during further processing steps.

Other Properties. Not listed in Table I are the manufacturing requirements for etching and adhesion. Adhesion is critical for all resist polymers and strongly dependent on feature size. Etching characteristics may be important, depending on the particular processing step. Because the etching characteristics depend on the gases used to differentiate between the exposed and unexposed resist areas, it is not an easy requirement to state in advance.

Commercial Resist Systems Examples and the Resolution Observed with Them.

Poly(methyl methacrylate)(PMMA) is the highest resolution X-ray resist which has been described in the literature. The sensitivity of this resist, however, is too poor for it to be practical in semiconductor manufacturing, although it has applications for sensor construction and instrument calibration. The first US commercial conventional X-ray resist was manufactured by the Olin Hunt Company as HEBR 242. This is a dissolution-inhibition type resist containing two essential components: a photoactive diazonapthoquinone in a novolac resin matrix. In this case, the diazide is converted upon photoabsorption to an intermediate which eventually becomes an organic acid. Solvent development with a metal-ion free developer such as tetramethylammonium hydroxide dissolves away the exposed resist leaving the positive-tone image. We have measured sensitivities on the order of 500 mJ/cm^2 with this resist (Figure 3) and line resolution on the order of 0.25 μm. It performs well as a conventional X-ray resist, but the sensitivity is too low (higher dose required) to meet the throughput requirement.

The Rohm and Haas Company, in conjunction with Shipley, has developed a number of negative chemically-amplified resist systems. Their sensitivity and the resolution are, in part, determined by the control of a number of the processing variables. Among the major variables are: the post-exposure bake temperature, the post-exposure bake time on a vacuum hot plate, the developer normality, and the developer time. Figure 5 shows how the sensitivity varies with some of these conditions for XP-8933B, a resist system which utilizes a photoacid generator to provide crossing-linking to the novolac matrix. Lower developer normality of the developer and shorter development times are seen to yield better sensitivity. All the data shown are for a post-exposure bake temperature of 125^0 C and a bake time of 60 sec, whereas better resolution is seen at temperatures on the order of 110^0 C.

We have examined a number of these negative-tone formulations and find that resolution can be on the order of 0.25 μm or smaller with sensitivities on the order of 30-50 mJ/cm^2, see Figures 4 and 6, respectively. The resolution testing to date is limited by the mask features which are available on the test masks. Figure 6 shows one such resolution test provided by Shipley XP-90104C. This is a negative-tone chemically-amplified resist which shows 0.25 μm line resolution at a dose of 22 mJ/cm^2. Higher resolution masks are being written for these test purposes.

From Germany the Hoechst Company has produced two commercial chemically-amplified resists, Ray PF and Ray PN (*9*). The Ray PF is a positive-tone resist which uses the acid catalyst to provide depolymerization of an acetal-type polymer. We have less experience with this system to date but have observed resolution on the order of 0.25 μm and sensitivities on the order of

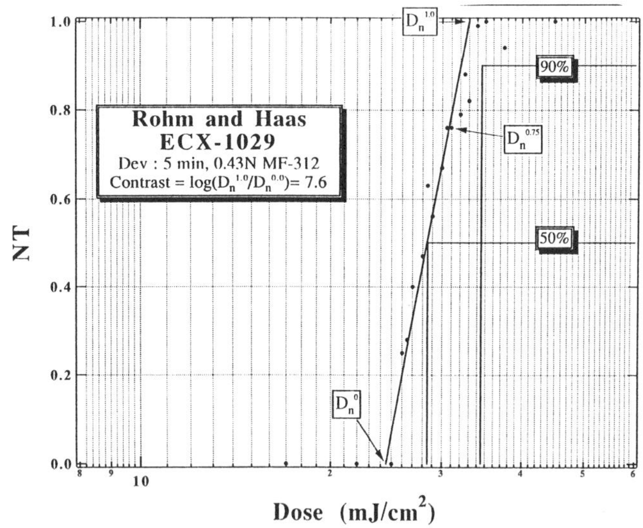

Figure 4. Sensitivity and contrast plot for Rohm & Haas ECX-1029 negative-tone chemically amplified resist in MF-312 developer.

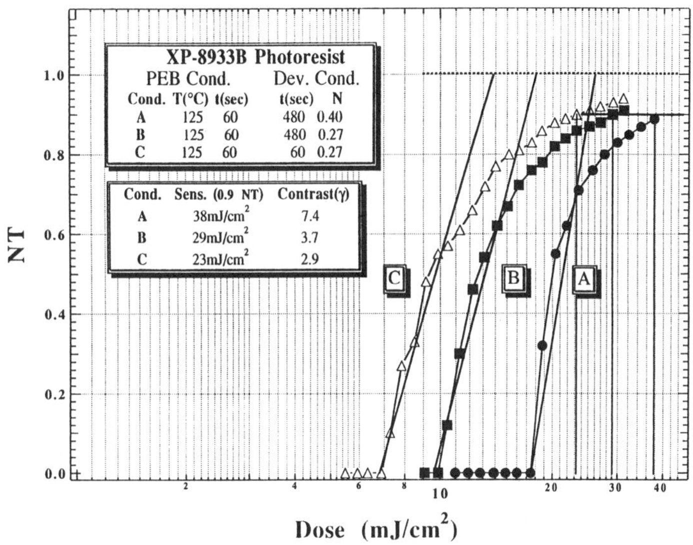

Figure 5. The influence of post-exposure baking conditions and development parameters on sensitivity and contrast for Rohm and Haas/Shipley XP-8933 B negative chemically amplified resist. (Figure modified from ref. 20 and used by permission from *J. Vac. Sci. Technol.*)

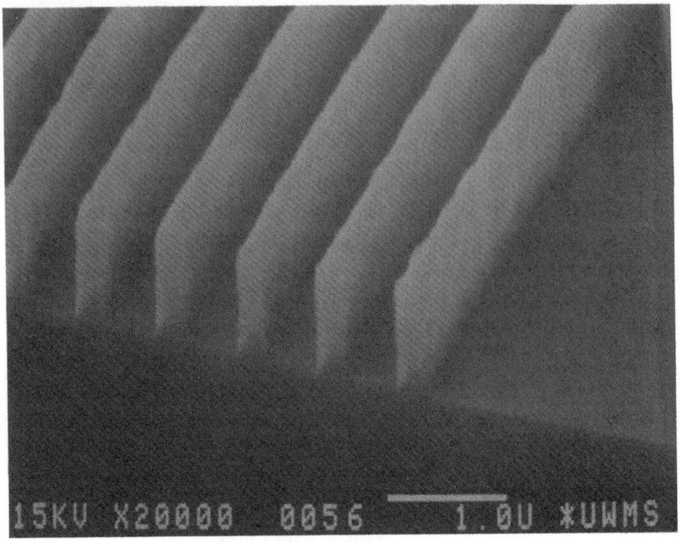

Figure 6. Resolution test pattern for Shipley XP-90104C negative-tone chemically amplified resist showing 0.25 μm lines at a dose of 22 mJ/cm^2.

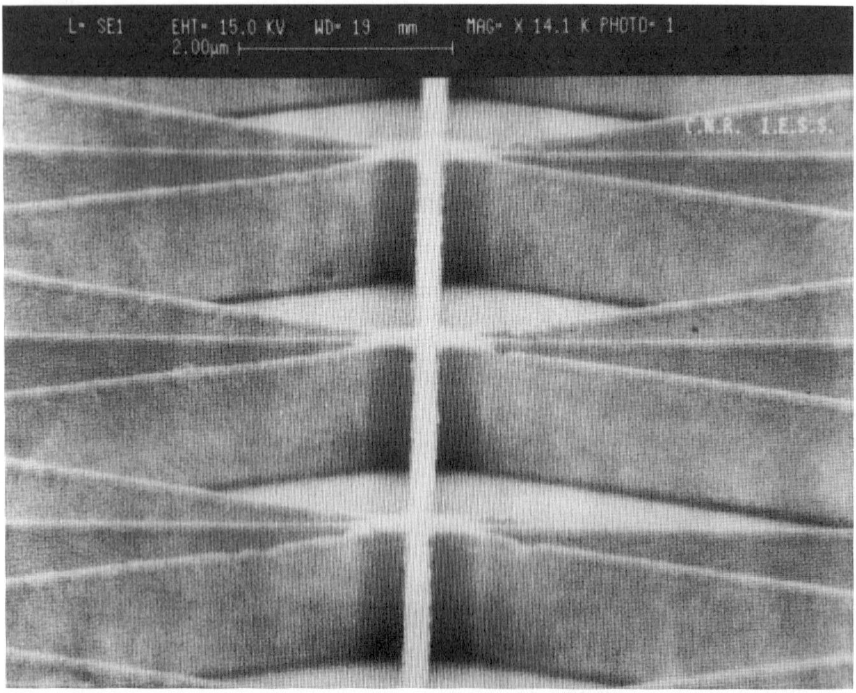

Figure 7. Resolution test on Hoechst Ray PF positive-tone resist showing 0.25 μm lines at a dose of 50 mJ/cm^2.

35-40 mJ/cm^2. Figure 7 shows the image produced through a test mask supplied by our colleagues at the IESS in Rome and shows a resolution of 0.25 μm with a dose of 50 mJ/cm^2.

Conclusions

There appear to be a small number of commercial resists which are available for X-ray exposure and which show resolution characteristics down to the 0.25 μm level. Some appear capable of printing smaller dimensions down to 0.15 μm, based on e-beam data. The next chapters will also illustrate a number of experimental resists which also show promise for high resolution X-ray and e-beam lithography. Evaluation of the resist performance for a manufacturing application, however, requires that a number of other factors be measured and evaluated. Foremost among these is the ability of the resist system to be developed within a broad range of processing conditions and still show critical dimensional control of the image linewidth. Other factors such as adhesion, stability, sensitivity to ambient conditions, and resistance to plasma etching are also important in the design of resist systems for manufacturing, but these requirements are not described here.

Acknowledgments

The Center for X-ray Lithography is supported by the SEMATECH Center of Excellence Program of the Semiconductor Research Corporation under Contract TN 856 CA, and the Department of Defense (DARPA) supplied funding for facilities support through Navy grant number N00014-89-J-2017. Partial personnel support was available from the National Science Foundation funding of the Engineering Research Center under grant number ECD-8721545. The Synchrotron Radiation Center is operated with support of the National Science Foundation under grant number DMR 8821625. The authors acknowledge the substantial help from the Olin Hunt Corporation, the Rohm & Haas Company, the Shipley Company, and the Hoechst Company in supplying resist samples; the contributions of Q. Leonard and M. Sisco of the Center for X-ray Lithography for their aid in the exposure and development evaluations; D. Mancini for the Olin Hunt HEBR 242 data; F. Cerrina, B. Lai, and F. Baszler for their aid with the TRANSMIT programs; and M. Gentili of Instituto Di Electtronica Dello Stato Solido (IESS) in Rome for the use of the high resolution masks. Finally, the authors acknowledge the substantial aid given by Gary N. Taylor of AT&T in the initial phases of the resist evaluation and in his sharing of experiences in resist testing and resist performance through the X-ray Resist Workshop.

Literature Cited

1. *Introduction to Microlithography*, Thompson, L. F.; Willson, C. G.; Bowden, M. J.; Eds., ACS Symposium Series 219, American Chemical Society: Washington, D. C., **1983**.
2. Reiser A.; *Photoreactive Polymers, The Science and Technology of Resists*, John Wiley and Sons, New York, **1989**, 184-185.
3. *Advances in Resist Technology and Processing VII*, Watts, M. P. C., Ed., SPIE Proceedings: The International Society for Optical Engineering: Bellingham, WA, **1990**, 1262.
4. Vollenbroek, F. A.; Spiertz, E. J.; in *Advances in Polymer Science 84, Electronic Applications*, Springer-Verlag, Heidelberg, **1988**, 85-111.
5. *Electron-Beam, X-ray, and Ion-Beam Technology: Submicrometer Lithographies IX*, Resnick D. J.; Ed., *SPIE Proceedings: The International Society for Optical Engineering*: Bellingham, WA, **1990**, 1263.
6. Heuberger, A.; *J. Vac. Sci. Technol.,* **1988**, *B6*, 107-121.
7. Lingnau, J.; Dammel, R.; Theis, J.; *Solid State Technol.,* **1989**, *32*, 105-111.
8. Conley, W. E.; Moreau, W.; Perrault S.; Spinillo, G.; Wood, R. L.; Gelorme, J. D.; Martino, R. M.; in *Advances in Resist Technology and Processing VII*, Watts, M. P. C.; Ed., *SPIE Proceedings: The International Society for Optical Engineering:* Bellingham, WA, **1990**, 1262, 49-60.
9. Dammel, R.; Lindley, C. R.; Pawlowski, G.; Scheunemann, U.; Theis, J.; *ibid.*, 378-391.
10. Suzuki, M.; Kaneko, T.; Saitoh, Y.; *J. Vac. Soc. Technol.,* **1989**, *B7(1)*, 47-54.
11. Peters, D.; Frankel, R.; *Solid State Technol.*, **1989**, *32*, 77-81.
12. Bruns, A.; Luethje, H.; Vollenbroek, F.; Spiertz, E.; *Microelectronic Eng.*,**1987**, *6*, 467-471.
13. Seligson, D.; Das, S.; Gaw, H.; *J. Vac. Sci. Technol.*, **1988**, *B6*, 2303-2310.
14. Blum, L.; Perkins, M. E.; Liu, H.; *ibid.*, 2280-2285.
15. deGrandpre, M.; Graziano, K.; Thompson, S. D.; Liu, H.; Blum, L.; in *Electron-Beam, X-Ray, and Ion-Beam Technology: Semiconductor Lithographies VII*, Yanof, A.; Ed., *SPIE Proceedings: The International Society for Optical Engineering*: Bellingham, WA, **1988**, 923, 158-168.
16. Taylor, G. N.; *Solid State Technol.*, **1984**, *27(6)*, 105-110.
17. Wells, G. M.; Lai, B.; So, D.; Redaelli, R.; Cerrina, F.; *Nucl. Instrum. Methods,* **1988**, *A266*, 278-287.
18. Baszler, F.; Hansen, M.; Cerrina, F.; *J. Vac. Sci. Technol.,* **1990**, *B8(6)*, 1529-1534 and references therein.
19. Wells, G. M.; Palmer, S.; Cerrina, F.; Purdes, A.; Gnade, B.; *ibid.*, **1990**, *B8(6)*, 1575-1578, and references therein.
20. Taylor, J. W.; Babcock, C. P.; Mancini, D. C.; Plumb, D.; Olsen, R. J., Fedynshyn, T.; *ibid.*, **1990**, *B8(6)*, 1547-1550.
21. Mancini, D. C.; Taylor, J. W.; Jayaraman, T. V.; West, R. J.; in *Advances in Resist Technology and Processing V,* MacDonald, S. A.; Ed., *SPIE*

Proceedings: The International Society for Optical Engineering: Bellingham, WA, **1988**, 920, 372-379.

22. Mancini, D. C.; Sullivan, M.; Taylor, J. W.; Bormett, R.; Jayaraman, T. V.; West, R.; in *Advances in Resist Technology and Processing VI*, E. Reichmanis; Ed., *SPIE Proceedings: The International Society for Optical Engineering:* Bellingham, WA, **1989**, 1086, 388-395.
23. Takahaski, H.; Wakui, K.; Herr, D. J. C.; Petersen, J. S.; Fedynyshyn, T. H.; Cronin, M. F.; in *Advances in Resist Technology and Processing VII*, Watts, M. P. C.; Ed., *SPIE Proceedings: The International Society for Optical Engineering:* Bellingham, WA, **1990** 1262, 330-342.
24. Taylor, J. W.; Babcock, C. P.; Mancini, D. M.; Suh, D.; Sullivan, M.; Plumb, D.; Abstracts of the 200th Meeting, American Chemical Soc., American Chemical Society: Washington, D. C., **1990**, Abstract Poly-0351.
25. Thompson, L.; Bowden, M. J.; *Introduction to Microlithography*, Thompson, L. F.; Willson, C. G.; Bowden, M. J.; Eds., ACS Symposium Series 219; American Chemical Society: Washington, D. C., **1983**, 161-214.

RECEIVED March 22, 1991

Chapter 20

Chemical Amplification for the Design of Sensitive Electron-Beam and X-ray Resist Systems

Hiroshi Ito

IBM Research Division, Almaden Research Center, International Business Machines Corporation, 650 Harry Road, San Jose, CA 95120–6099

Electron beam and x-ray lithographic technologies are capable of resolving sub-half-micrometer features and thus could support future ULSI fabrication. However, the economical operation of these high resolution technologies in the production environments demands extremely high resist sensitivity. In order to meet the requirement, we have proposed "chemical amplification." In chemically amplified resist systems, an initial radiochemical event induces a cascade of subsequent chemical transformations. We have chosen "acid" as catalytic species in conjunction with acid-labile polymer matrices. Triarylsulfonium salts are convenient radiochemical acid generators. In this paper are described acid-catalyzed deprotection to change the polarity and therefore solubility of matrix polymers and acid-catalyzed depolymerization. The negative systems include poly(p-t-butoxycarbonyloxystyrene), poly(t-butyl p-vinylbenzoate), and an alternating copolymer of α,α-dimethylbenzyl methacrylate with α-methylstyrene, which provide swelling-free negative images upon development with anisole. Selective silylation of exposed areas of these resist systems allows negative-tone development with O_2 reactive ion etching (RIE). The positive systems based on polyphthalaldehyde include a thermally developable O_2 RIE barrier resist utilizing poly(4-trimethylsilylphthalaldehyde) for the bilayer lithography and use of polyphthalaldehyde as a polymeric dissolution inhibitor for a novolac resin.

Microelectronics technology has been advanced at a remarkable rate for the past decade. The improvements of integrated circuit devices have resulted from the increase in the number of components per chip, which has been made possible by decreasing the minimum feature size on the chip. The feature size continues to shrink. Four-megabit dynamic random access memory (DRAM) devices are currently in production with minimum features in the 0.8 μm range. High numerical aperture step-and-repeat reduction projection tools operating at the conventional 436-365 nm (g-i) lines of the Hg arc lamp are used to produce today's state-of-the-art devices and it is generally believed that such tools in conjunction with conventional diazoquinone/novolac positive resists will be capable of producing chips with features as small as 0.5 μm (16 Mbit). Further reduction in feature size will require the introduction of a new lithographic technology.

0097–6156/91/0475–0326$06.00/0

Electron beams can be focused to a few nanometers in diameter and their wavelength is on the order of a few tenths of an angstrom. Therefore, the resolution achieved with electron beams is higher than that with photolithography. The e-beam resolution is, however, limited by forward scattering of the electrons in the resist film and back-scattering from the substrate. Soft x-rays with wavelengths ranging from 0.4 to 8 nm represent another alternative radiation source for high resolution lithographic imaging. The deposited energy is very uniform throughout the resist film and the diffraction effect can be ignored in x-ray exposures, especially at the shorter wavelengths, providing fine images with high aspect ratios even on topographic substrates.

E-beam and x-ray lithographic technologies are thus capable of resolving sub-half-micrometer features and could support future ULSI fabrication. However, the economical operation of these high resolution technologies in the production environments demands extremely high resist sensitivity. In order to meet the sensitivity requirement, we have sought a chemical method and proposed "chemical amplification" (*1-3*). In chemically amplified resist systems, an initial radiochemical event induces a cascade of subsequent chemical transformations. We have chosen "acid" as "catalytic" species in conjunction with acid-labile polymer matrices (*4-6*). Triarylsulfonium salts originally developed for photochemical curing of epoxy resins (*7*) are convenient radiochemical acid generators.

In this paper are described acid-catalyzed deprotection reactions to change the polarity and therefore the solubility of matrix polymers as well as acid-catalyzed depolymerization for the design of positive resist systems. The negative-working systems based on the polarity change include poly(p-t-butoxycarbonyloxystyrene) (PBOCST) (*1,2,8,9*), poly(t-butyl p-vinylbenzoate) (PTBVB) (*9,10*), and an alternating copolymer of α,α-dimethylbenzyl methacrylate (DMBZMA) with α-methylstyrene (MST) (*11*), which provide swelling-free negative images upon development with anisole. Selective silylation of exposed areas of these resist systems allows negative-tone development with oxygen reactive ion etching (RIE) (*12*). The positive-working systems based on acid-catalyzed depolymerization of polyphthalaldehyde (PPA) derivatives include self-(*3-6*) and thermal developments (*13,14*), a thermally developable oxygen RIE barrier resist utilizing poly(4-trimethylsilylphthalaldehyde) (PSPA) for the bilayer lithography (*14-16*), and the use of PPA as a dissolution inhibitor for a novolac resin (*16-18*).

Experimental

Materials. PBOCST was prepared by radical polymerization of p-t-butoxycarbonyloxystyrene (BOCST) with α,α-azobis(isobutyronitrile) (AIBN) or benzoyl peroxide (BPO) in toluene (*19*) or by reacting commercial poly(p-hydroxystyrene) (PHOST) (Maruzen Oil, Lyncur PHM) with di-t-butyl dicarbonate (Scheme I) (*9*). t-Butyl p-vinylbenzoate was prepared by reacting p-vinylbenzoyl chloride with lithium t-butoxide or by treating p-vinylphenylmagnesium chloride with di-t-butyl dicarbonate and then subjected to radical polymerization (Scheme I) (*10,20*). DMBZMA was synthesized by reaction between methacryloyl chloride and lithium α,α-dimethylbenzyl alkoxide and copolymerized with MST with AIBN in toluene at 60°C (*11*). PPA was obtained by anionic polymerization of phthalaldehyde with n-butyllithium (nBuLi) in tetrahydrofuran (THF) at −78°C (*3,4*). Poly(4-chlorophthalaldehyde) (PCPA) was prepared by cationic polymerization with BF_3OEt_2 in dichloromethane at −78°C of 4-chlorophthalaldehyde made from 4-chloro-o-xylene (*13,14*). 4-Trimethylsilylphthalaldehyde was synthesized from 4-chloro-o-xylene and converted to PSPA by anionic polymerization at −78°C with nBuLi or s-butyllithium (sBuLi) (*14-16*). The radiochemical acid generators employed in these studies were triphenylsulfonium hexafluoroantimonate (TPS), triphenylsulfonium trifluoromethanesulfonate (triflate, $Ph_3S^{+-}OTf$), and

4-thiophenoxyphenyl(diphenyl)sulfonium hexafluoroantimonate (TPPDPS), which had been synthesized according to the literature (*21,22*). The novolac resin used as a binder was a cresol-formaldehyde addition-condensation product. The resists were formulated using cyclohexanone or propylene glycol monomethyl ether acetate as a casting solvent.

Lithographic Imaging. Spin-cast resist films were baked on a hot plate at 100°C for 5 min. The e-beam exposures were carried out on an IBM VS-1 (20 keV) or VS-4 (25 keV) or on a Hontas-4 (25 keV) system. The x-ray exposures were carried out at Stanford Synchrotron Radiation Laboratory and also at Brookhaven National Laboratory. The exposed wafers were postbaked on a hot plate at 100, 130, or 160°C for 1-2 min. The negative imaging was achieved with use of anisole as the developer.

Selective silylation was carried out at 100°C with hexamethyldisilazane (HMDS, 100 torr) (*12*).

A Plasma Therm parallel-plate etcher was used for O_2 RIE pattern transfer with a stainless steel cathode maintained at 25 °C.

Measurements. IR spectra were recorded on an IBM IR/32 FT spectrometer using undoped Si wafers as a substrate. Molecular weight determination was made by gel permeation chromatography (GPC) using a Waters Model 150 chromatograph equipped with 6 μStyragel columns at 30°C in THF or by membrane osmometry using a Wescan 230 recording membrane osmometer in toluene at 30°C. Thermal analyses were performed on a Du Pont 1090 thermal analyzer at a heating rate of 5°C/min for thermogravimetric analysis (TGA) and 10°C/min for differential scanning calorimetry under inert atmosphere. Film thickness was measured on a Tencor alpha-step 200.

Acid-Catalyzed Deprotection

Poly(p-t-butoxycarbonyloxystyrene) (PBOCST). PBOCST can be prepared from monomer or from PHOST (Scheme I). This polymer is converted upon postbake at ca. 100°C to PHOST by reaction with radiochemically generated strong Bronsted acid, quantitatively releasing CO_2, isobutene, and a proton (Scheme II) (*1,2,8,9*). One proton cleaves 1000 tBOC groups under standard processing conditions, which corresponds to the acid diffusion distance of ca. 50Å (*23*). The deprotection reaction converts the lipophilic polymer to the hydrophilic and acidic polymer, providing either positive or negative imaging, depending on the polarity of the developer solvent (Scheme II). The resist is capable of resolving sub-half-micrometer features by e-beam or x-ray irradiation (*8,9*). Positive and negative images printed in the tBOC resist by the Brookhaven synchrotron x-ray radiation are presented in Figure 1. The incident dose required for 100% final thickness in the case of the negative development with anisole of the tBOC resist (1.1 μm thick film containing 15 wt% of TPS) was 13 mJ/cm^2 at 1.0-3.5 keV with a contrast (γ) of 15, demonstrating that high sensitivity and high contrast are not mutually exclusive (*8*). Sub-half-micrometer negative images (0.3/0.3 and 0.25/0.5 μm line/space) delineated in a tBOC resist containing 15 wt% of TPPDPS by e-beam exposure of 3 μC/cm^2 at 25 keV are presented in Figure 2.

The acid generation has been shown using poly(p-methoxystyrene) and 5 wt% of TPS to be almost linearly related to e-beam dose at 20 keV for doses less than 10 μC/cm^2 (*24*). It has been also demonstrated that the observed acid production ratio at 10, 15, and 20 keV is close to the Monte Carlo calculated ratios for energy deposition at these acceleration voltages (*24*). As the scanning electron micrographs in Figures 1 and 2 demonstrate, the resist does not suffer from swelling even in the negative development with anisole, whereas conventional negative resist systems based on crosslinking tend to exhibit limited resolution due to swelling

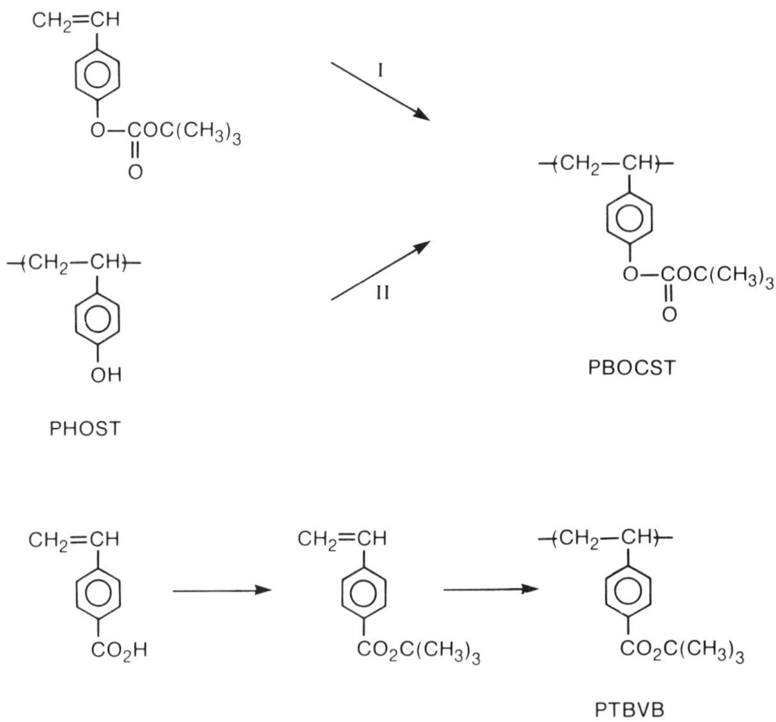

Scheme I. Preparation of PBOCST and PTBVB.

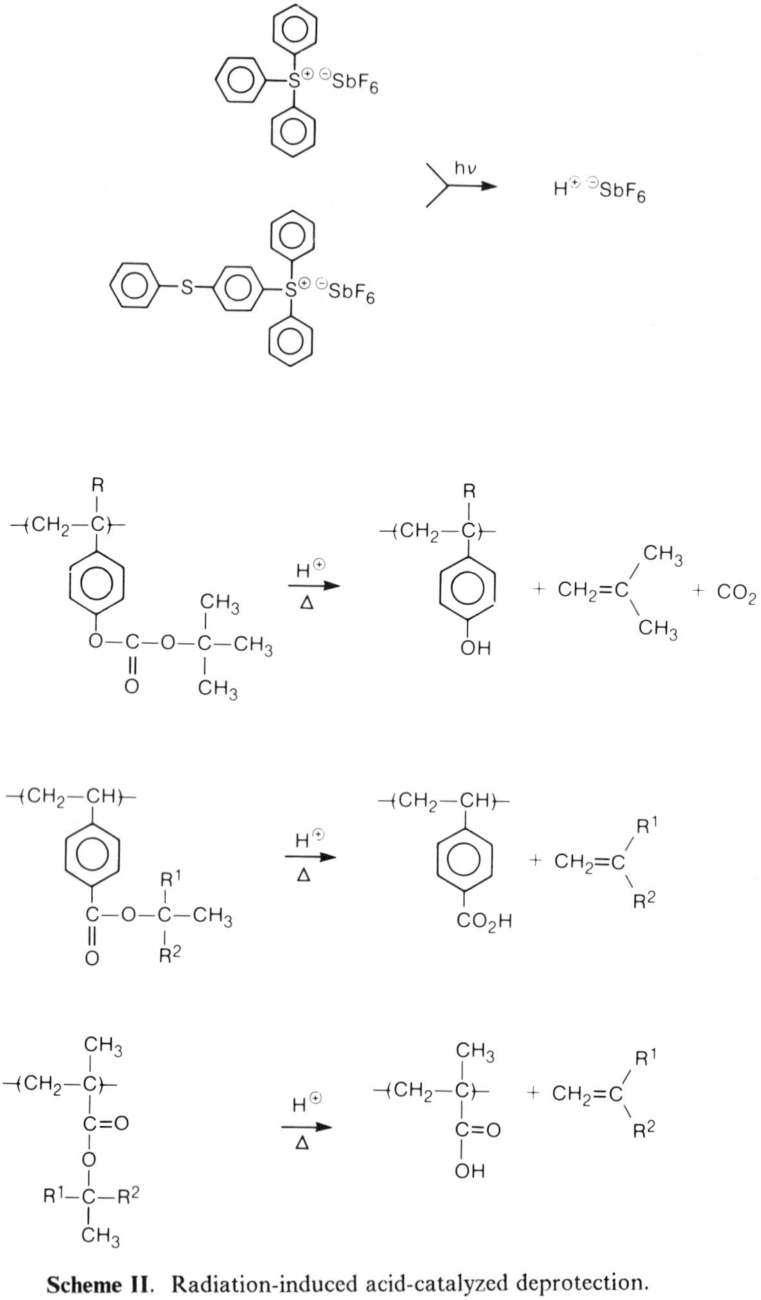

Scheme II. Radiation-induced acid-catalyzed deprotection.

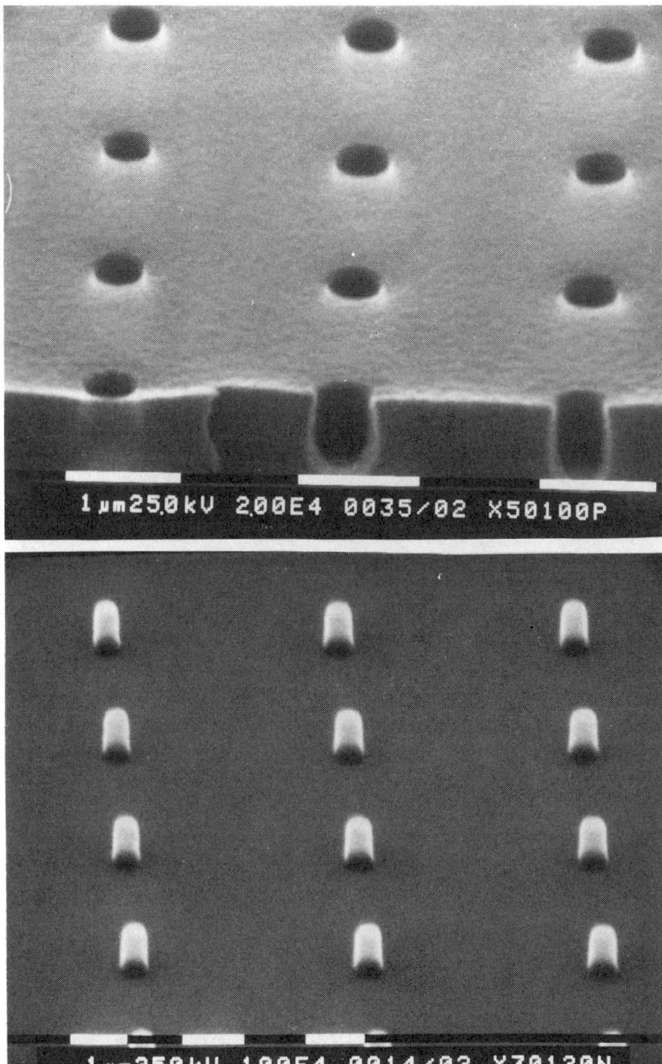

Figure 1. Positive and negative images of tBOC resist printed by x-ray radiation. (Reproduced with permission from Ref. 8. Copyright 1988 American Vacuum Society.)

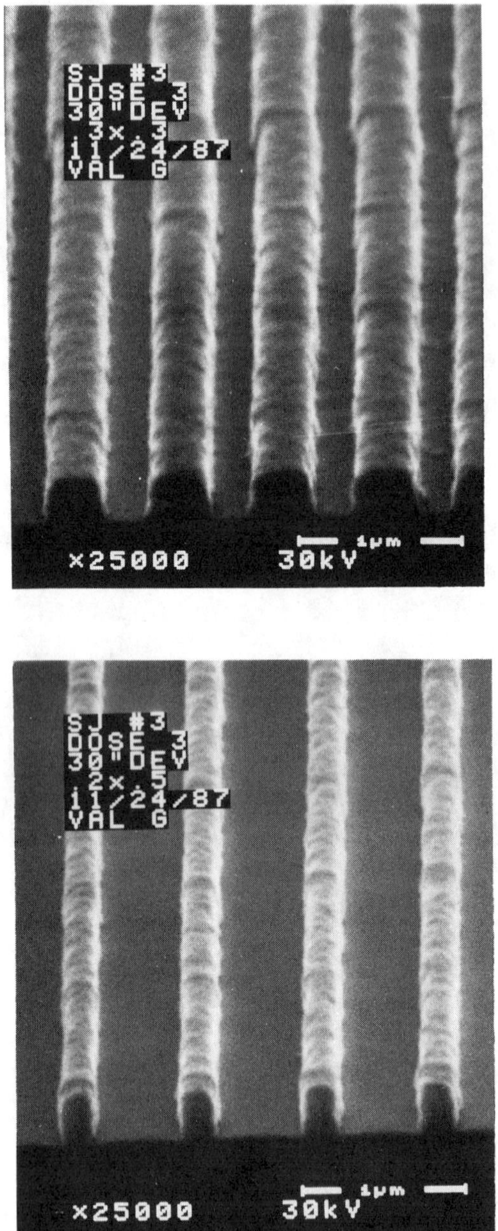

Figure 2. Negative images of tBOC resist printed by e-beam exposure (3 $\mu C/cm^2$ at 25 keV). (Reproduced with permission from Ref. 9. Copyright 1989 Society of Photo-Optical Instrumentation Engineers.)

during development with organic solvents. E-beam nanolithography with the tBOC resist has provided 19 nm resolution, indicating that chemical amplification and high resolution are not mutually exclusive (*25*). As the positive imaging mechanism does not rely on main chain degradation, the aromatic resist exhibits a good CF_4 plasma resistance in spite of its high radiation sensitivity.

The polarity change is equivalent to alteration of reactivity (Scheme III). Gaseous silylating reagents such as HMDS and N,N-dimethylaminotrimethylsilane react only with PHOST generated in the exposed regions to form poly(p-trimethylsilyloxystyrene). Thus, the large difference in O_2 plasma etch rates of the Si-containing exposed regions and the organic polymer in the unexposed areas can be utilized in the design of negative resist systems that can be developed with O_2 plasma (*12*). The negative images delineated in the tBOC resist by O_2 RIE after exposure to x-ray radiation (Brookhaven) are presented in Figure 3.

Poly(t-butyl p-Vinylbenzoate) (PTBVB). PTBVB can be readily obtained by radical polymerization of TBVB (Scheme I). This lipophilic polymer is converted upon postbake to poly(p-vinylbenzoic acid) (PVBA) by reaction with a radiochemically generated acid, quantitatively releasing isobutene and a proton (Scheme II). Development with anisole results in negative imaging and use of aqueous base as a developer provides positive imaging (*10*). In contrast to the tBOC resist, the PTBVB resist is not useful as a single layer deep UV resist due to a very high absorption of PTBVB and PVBA below 300 nm. However, the system containing 8.2 wt% of TPPDPS is a high sensitivity e-beam resist, providing 0.5-0.65 $\mu C/cm^2$ (20 keV) sensitivity for 50% film retention and γ of 3.8 in the case of the film thickness of ca. 0.7 μm (*9*). The typical imaging dose is ca. 0.7 $\mu C/cm^2$ (*9*). The e-beam sensitivity curve of the PTBVB resist is presented in Figure 4. Due to liberation of isobutene, the resist film shrinks upon postbake (130°C in this case) and no further thickness loss is observed during development with anisole as soon as the maximum shrinkage is attained during postbake.

In order to directly compare the intrinsic sensitivity of the t-butyl carbonate and the t-butyl ester, a copolymer of BOCST (42%) with TBVB (58%) was mixed with 3.0 mol% of TPPDPS, exposed to 0.5 and 2.5 $\mu C/cm^2$, and subjected to IR analysis as shown in Figure 5. The unexposed film exhibits two carbonyl absorptions at 1755 (carbonate) and 1715 cm^{-1} (benzoate) (Figure 5a). The 0.67-μm thick film exposed to 0.5 $\mu C/cm^2$ followed by postbake at 130°C showed ca. 90% tBOC conversion and ca. 50% t-butyl ester deprotection (Figure 5b), indicating that the t-butyl carbonate is more susceptible to acidolysis than the t-butyl benzoate. The tBOC deprotection is completed and the t-butyl ester conversion is very high at 2.5 $\mu C/cm^2$ as Figure 5c indicates. The lithographic sensitivity of the PTBVB resist is higher than that of the tBOC resist whereas the radiochemically-induced acidolysis is more facile on the tBOC group than on the t-butyl ester group, suggesting that conversion of PTBVB to PVBA provides a more pronounced solubility differentiation than the conversion of PBOCST to PHOST.

One attractive feature of the PTBVB resist is its high thermal stability (*9,10*). The negative images do not exhibit any thermal flow at 230°C owing to the high glass transition temperature ($T_g = 250°C$) of PVBA.

Alternating Copolymers of α,α-Dimethylbenzyl Methacrylate with α-Methylstyrene. Poly(methyl methacrylate) (PMMA) was the very first polymeric material to be used as an e-beam resist and provides a very high resolution. Its sensitivity, however, is very low, reflected by its low ^{60}Co γ-radiolysis G_s of 1.3. The sensitivity of the methacrylate-based resists can be readily improved by incorporating chemical amplification, especially by utilizing the α, α-dimethylbenzyl ester which is highly susceptible to A_{AL}-1 acidolysis (*26*). In order to make the methacrylate resists more attractive in device fabrication, however, two more problems have to be solved; low thermal stability and low dry etch resistance. Alternating copolymers of DMBZMA with MST prepared by radical

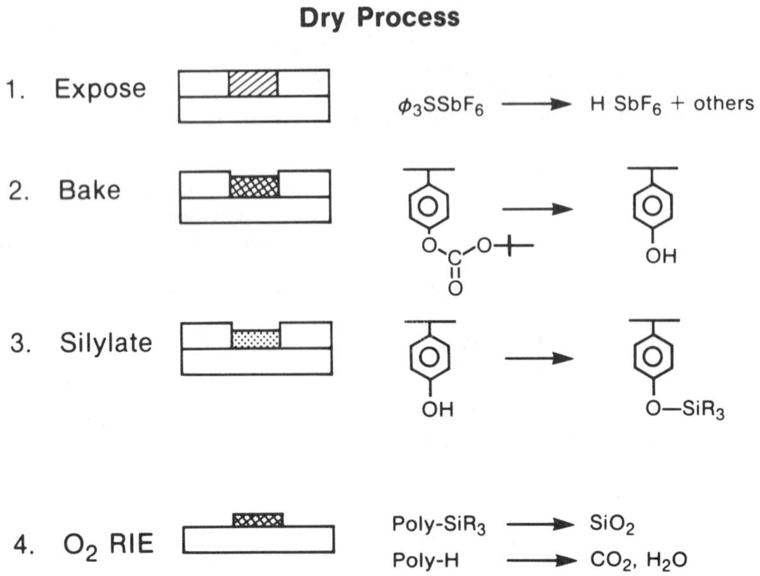

Scheme III. Selective silylation based on deprotection.

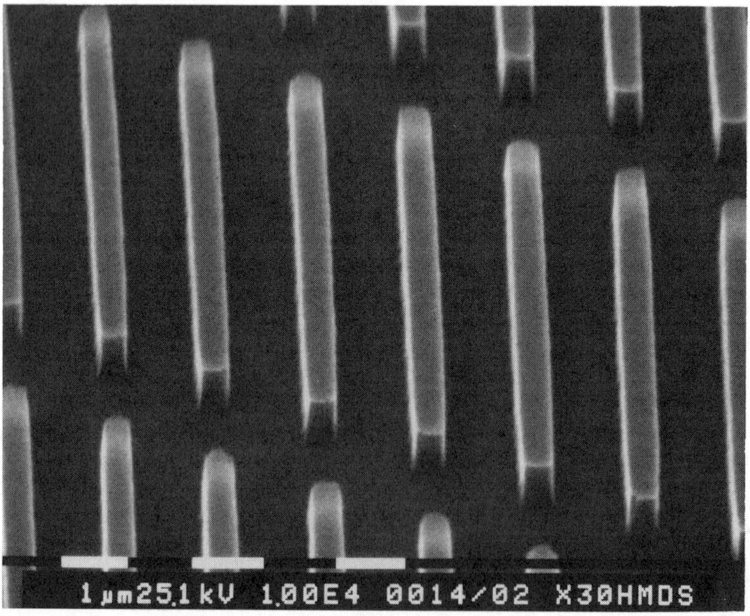

Figure 3. Negative images delineated in tBOC resist by x-ray irradiation, silylation, and O_2 RIE.

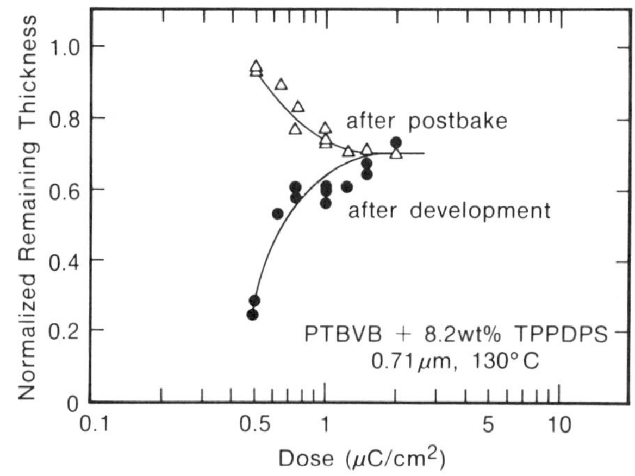

Figure 4. E-beam sensitivity curve of PTBVB resist at 20 keV. (Reproduced with permission from Ref. 9. Copyright 1989 Society of Photo-Optical Instrumentation Engineers.)

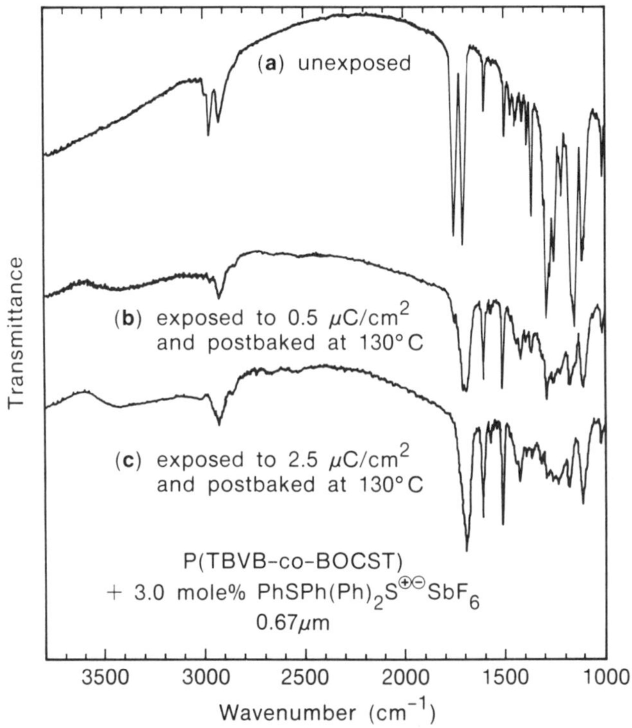

Figure 5. IR spectra of poly(BOCST-co-TBVB) resist exposed to e-beam radiation. (Reproduced with permission from Ref. 9. Copyright 1989 Society of Photo-Optical Instrumentation Engineers.)

copolymerization are highly stable in CF_4 plasma owing to incorporation of ca. 50% of the styrenic unit (*11*). The copolymer resist (0.4 μm thick) containing 9.3 wt% of TPPDPS exhibits 50 and 100% thickness retention at 0.3 and 0.7-0.8 μC/cm², respectively, of 20 keV e-beam radiation when developed with anisole after postbake at 100°C (Figure 6). In this system, the maximum concentration of poly(methacrylic acid) generated in the film is only 50% but the dual tone imaging is still possible. The negative images thus obtained are devoid of thermal flow up to 220°C. The high thermal stability is achieved by minimizing the intramolecular dehydration by isolating the methacrylate units in the copolymer through alternating copolymerization and thus by taking advantage of the high T_g of poly(methacrylic acid) (T_g = 228°C) and of poly(α-methylstyrene) (T_g = 168°C) (*11*). Hatada and co-workers have shown that poly(DMBZMA) is deesterified to a certain extent upon e-beam exposure and functions as a sensitive positive resist when developed with an alcoholic solution of alkoxide (*27*).

Acid-Catalyzed Depolymerization of Polyphthalaldehydes

Self- and Thermal Development. Aldehydes undergo anionic and cationic polymerizations, which are typically equilibrium reactions with low ceiling temperatures (T_c) and acetal bonds are labile toward acids. We have chosen PPA (*28*) for its high solubility. Aliphatic polyaldehydes are crystalline and rather intractable. PPA end-capped by acylation is stable thermally to ca. 200°C and then depolymerizes cleanly to the starting monomer. Addition of a small amount of an onium salt acid generator results in spontaneous positive-tone relief image formation upon irradiation, which we have termed "self-development" (Scheme IV) (*1,2*). The self-development was chosen to demonstrate the dramatic effect of chemical amplification on the resist sensitivity and lithographic processes but is not practical because the volatile products could contaminate expensive exposure tools. Reduction of the onium salt loading and exposure dose suppresses depolymerization during exposure but the positive development can be achieved simply by heating after exposure (thermal development). PCPA is particularly useful in the thermal development because the polymer does not exhibit any thermal flow at 160°C, providing high resolution positive images at 1 μC/cm² (20 keV) of e-beam irradiation (*13,14*). The residual non-volatile sulfonium salt can be removed by rinsing with isopropanol.

Copolymers of aliphatic aldehydes have been reported to undergo depolymerization upon e-beam exposure and to function as self-developing e-beam resists (*29*).

Polymeric Dissolution Inhibitor. Polymers that undergo efficient depolymerization do not possess high plasma resistance in general. In order to combine the high radiation sensitivity and high dry etch resistance, we have used the lipophilic PPA as a dissolution inhibitor for a novolac resin (*16-18*). The unsubstituted PPA is highly miscible (40/60, 60/40) with a cresol-formaldehyde novolac resin and inhibits the dissolution of the phenolic matrix polymer in aqueous base very efficiently at a low loading (<10 wt%) (*16-18*). The polymeric dissolution inhibitor is removed upon postbake from the exposed regions by acid-catalyzed depolymerization, which results in rapid dissolution of the exposed areas in aqueous base, providing positive images (Scheme V), that are resistant to CF_4 plasma due to the aromatic matrix resin. The e-beam sensitivity curve presented as Figure 7 indicates that the three-component resist based on a novolac resin containing 5.7 wt% of PPA and 8.5 wt% of TPPDPS has a sensitivity of ca. 1 μC/cm² (20 keV) and γ of 6.3. The thiophenoxy-substituted sulfonium salt is a much less efficient dissolution inhibitor than triphenylsulfonium salts (*30*) and therefore the high loading of 8.5 wt% does not affect the dissolution of the exposed regions but increases the sensitivity. Even the low PPA concentration of 5.7 wt% permits the full thickness retention of the

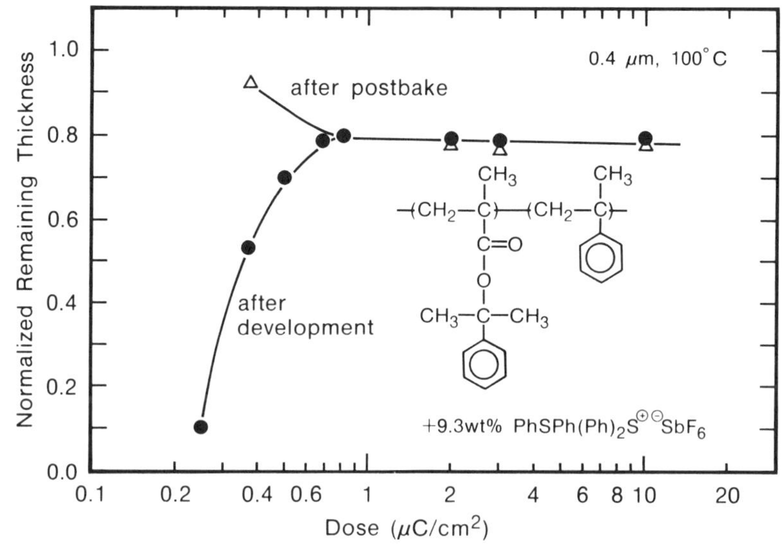

Figure 6. E-beam sensitivity curve of poly(DMBZMA-co-MST) resist at 20 keV.

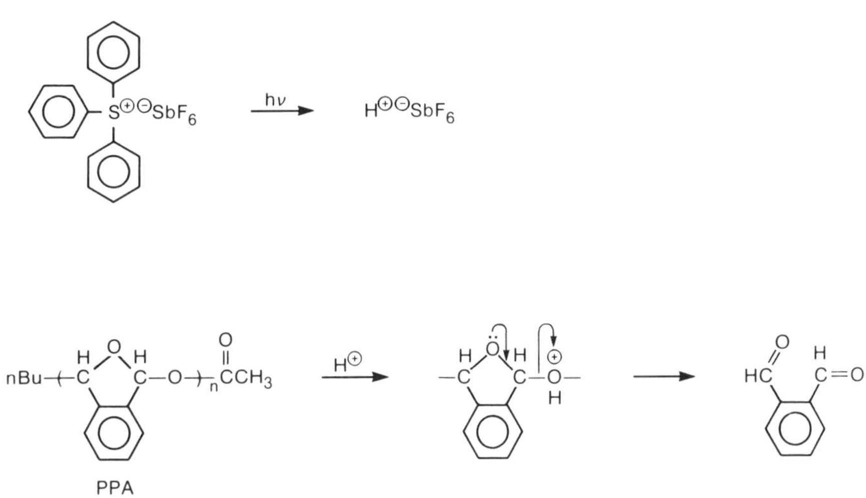

Scheme IV. Acid-catalyzed depolymerization of PPA.

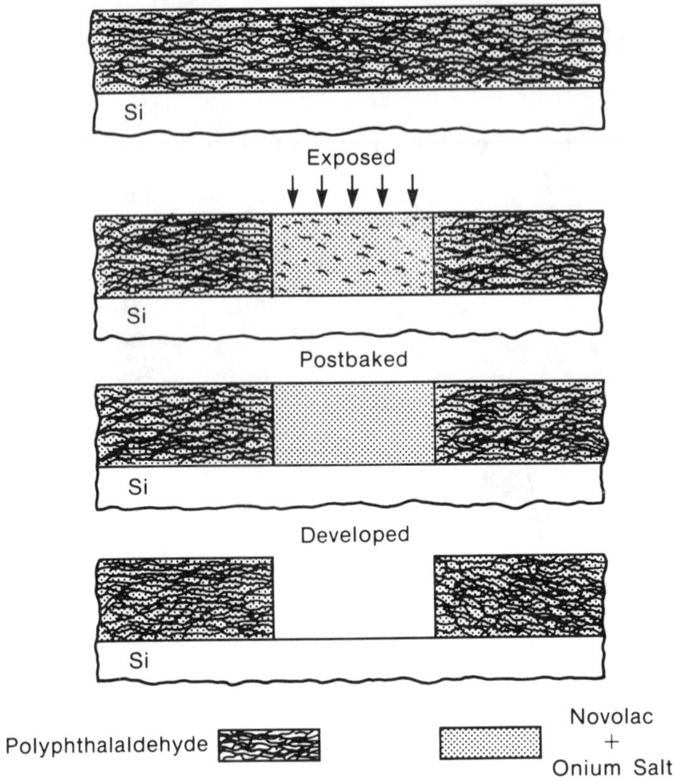

Scheme V. Three-component positive resist based on polymeric dissolution inhibitor.

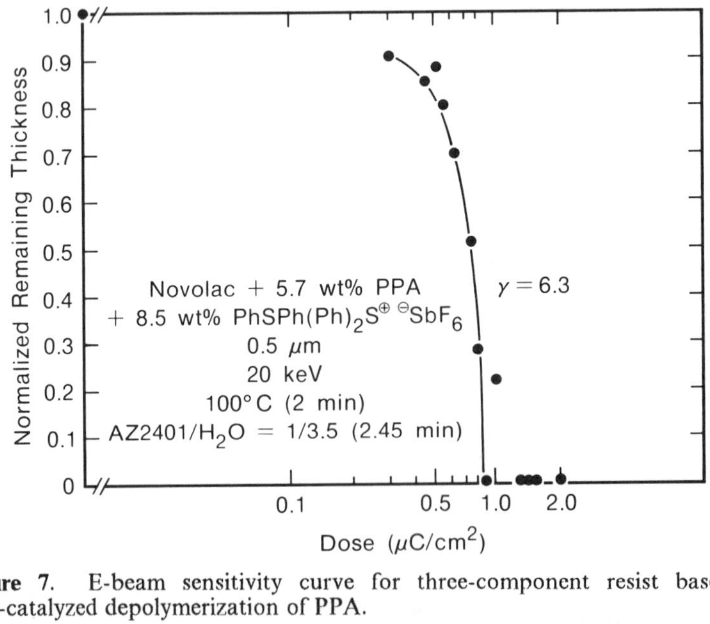

Figure 7. E-beam sensitivity curve for three-component resist based on acid-catalyzed depolymerization of PPA.

unexposed areas in AZ2401/H_2O = 1/3.5 (aqueous KOH solution) for at least 2.45 min. A scanning electron micrograph of positive 0.5-μm line/space patterns printed in the three-component e-beam resist is presented in Figure 8.

Small acetal and ketal compounds have been also used with a novolac resin in the acid-catalyzed chemical amplification scheme for e-beam and x-ray lithography (*31*). It has been reported that a copolymer of 3-trimethylsilylpropanal with 3-phenylpropanal functions as a polymeric dissolution inhibitor for phenolic resins in e-beam and x-ray imaging (*32*). Certain poly(olefin sulfones) have been also blended with novolac resins for the design of positive-tone, dry etch resistant, e-beam and x-ray resists (*33-35*).

Thermally Developable Oxygen RIE Barrier Resist. Another approach to rendering the highly sensitive PPA system practical is to incorporate Si into PPA (*14-16*). We have synthesized PSPA by anionic initiation in THF at −78°C. Thermal development provides positive images, which function as an O_2 plasma mask, allowing etching of an underlying organic polymer layer only in the exposed areas where the Si-containing PPA has been removed (Scheme VI). The PSPA resist (0.6 μm thick) containing 1.2 or 2.6 wt% of $Ph_3S^{+-}OTf$ can be completely developed by baking at 130-100°C the films exposed to 2.0 or 1.5 $\mu C/cm^2$ (20 keV) (*14*). The positive images thus obtained by heating can now be transferred to a 2-μm thick hard-baked novolac resin by O_2 RIE (20 sccm, 10 mtorr, 0.25 W/cm^2, −290 v), as Figure 9 demonstrates, without an appreciable thickness loss in the unexposed regions during etching. Thus, the thermal development is compatible with high resolution lithography and the acid-catalyzed depolymerization mechanism and good plasma resistance are not mutually exclusive.

Figure 8. Positive images delineated in three-component resist by e-beam exposure and aqueous base development.

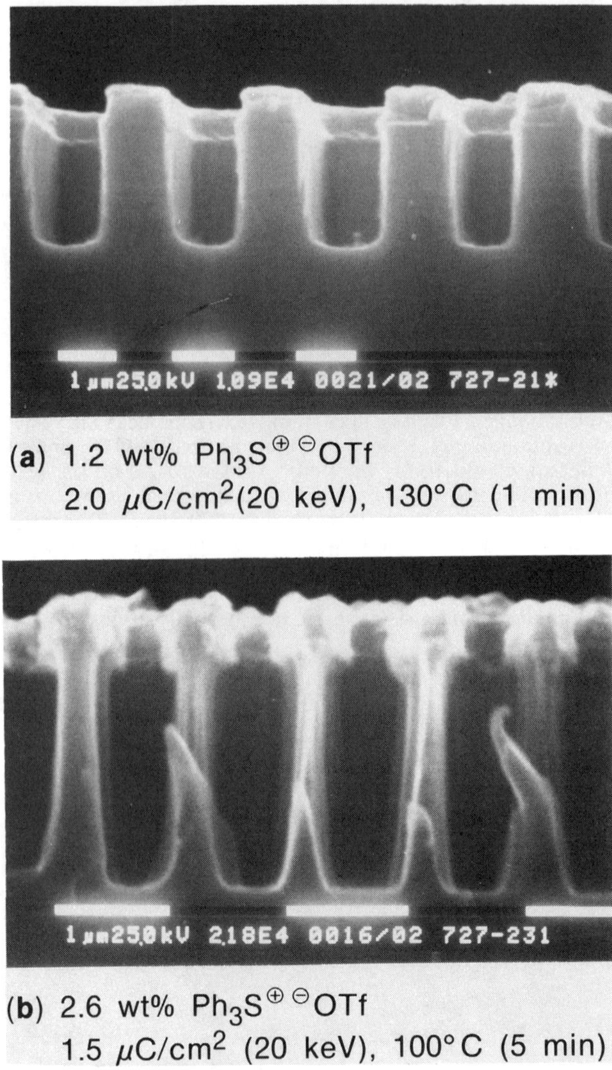

Figure 9. Positive PSPA/novolac bilayer images obtained by thermal development and O_2 RIE. (Reproduced with permission. Copyright 1989 Blackwell Scientific Publications.)

PPA containing two trimethylsilyl groups has been also utilized in a similar fashion for optical lithography (*36*).

Summary

Chemical amplification based on acid-catalyzed polymer reactions not only achieves high resist sensitivity but also provides a basis for designing an entire family of advanced resist materials possessing unique lithographic properties. Such acid-catalyzed resist systems are likely to play an important role in future ULSI fabrication that employs short wavelength lithographic technologies.

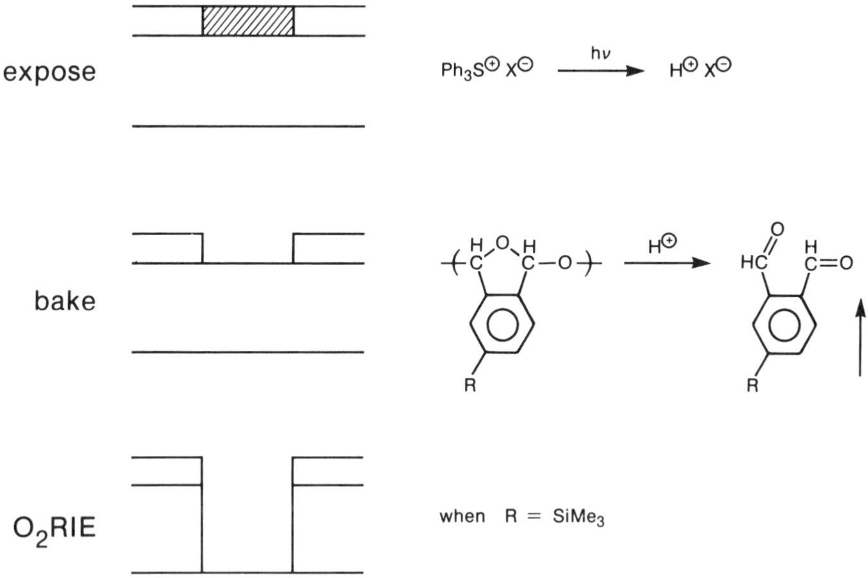

Scheme VI. Thermally developable oxygen RIE barrier resist for bilayer lithography.

Acknowledgment

The author thanks his co-workers whose names appear in the cited references for their contributions.

Literature Cited

1. Ito, H.; Willson, C. G.; Fréchet, J. M. J. *Digest of Technical Papers of 1982 Symposium on VLSI Technology* **1982**, 86.
2. Fréchet, J. M. J.; Ito, H.; Willson, C. G. *Proc. Microcircuit Eng.* **1982**, 260.
3. Ito, H.; Willson, C. G. *Technical Papers of SPE Regional Technical Conference on Photopolymers* **1982**, 331.
4. Ito, H.; Willson, C. G. *Polym. Eng. Sci.* **1983**, *23*, 1012.
5. Ito, H.; Willson, C. G. In *Polymers in Electronics* Davidson, T., Ed.; Symposium Series 242; American Chemical Society: Washington, D. C., 1984; pp. 11-23.
6. Ito, H. *Proc. KTI Microelectronics Seminar* **1988**, 81.
7. Crivello, J. V. In *UV Curing: Science and Technology*, Pappas, S. P., Ed.; Technology Marketing Corporation: Stamford, Connecticut, 1978; p. 23.
8. Seligson, D.; Ito, H.; Willson, C. G. *J. Vac. Sci. Technol.* **1988**, *B6(6)*, 2268.
9. Ito, H.; Pederson, L. A.; Chiong, K. N.; Sonchik, S.; Tsai, C. *Proc. SPIE* **1989**, *1086*, 11.
10. Ito, H.; Willson, C. G.; Fréchet, J. M. J. *Proc. SPIE* **1987**, *771*, 24.
11. Ito, H.; Ueda, M.; Ebina, M. In *Polymers in Microlithography*; Reichmanis, E.; MacDonal, S. A.; Iwayanagi, T., Eds.; Symposium Series 412; American Chemical Society: Washington, D. C., 1989; pp. 57-63.

12. MacDonald, S. A.; Ito, H.; Hiraoka, H.; Willson, C. G. *Technical Papers of SPE Regional Technical Conference on Photopolymers* **1985**, 177.
13. Ito, H.; Schwalm, R. *J. Electrochem. Soc.* **1989**, *136*, 241.
14. Ito, H.; Ueda, M.; Schwalm, R. *J. Vac. Sci. Technol.* **1988**, *B6(6)*, 2259.
15. Ito, H.; Ueda, M.; Renaldo, A. F. *J. Electrochem. Soc.* **1989**, *136*, 245.
16. Ito, H.; England, W. P.; Ueda, M. *J. Photopolym. Sci. Technol.* **1990**, *3*, 219.
17. Ito, H.; Flores, E.; Renaldo, A. F. *J. Electrochem. Soc.* **1988**, *135*, 2328.
18. Ito, H. *Proc. SPIE* **1988**, *920*, 33.
19. Fréchet, J. M.; Eichler, E.; Ito, H.; Willson, C. G. *Polymer* **1983**, *24*, 995.
20. Ito, H.; Ueda, M.; England, W. P. *Macromolecules* **1990**, *23*, 2589.
21. Crivello, J. V.; Lam, J. H. W. *J. Org. Chem.* **1978**, *43*, 3055.
22. Miller, R. D.; Renaldo, A. F.; Ito, H. *J. Org. Chem.* **1988**, *53*, 5571.
23. McKean, D. R.; Schaedeli, U.; MacDonald, S. A. *J. Polym. Sci., Part A, Polym. Chem.* **1989**, *27*, 3927.
24. MacDonald, S. A.; McKean, D. R. *J. Photopolym. Sci. Technol.* **1990**, *3*, 375.
25. Umbach, C. P.; Broers, A. N.; Willson, C. G.; Koch, R.; Laibowitz, R. B. *J. Vac. Sci. Technol.* **1988**, *B6*, 319.
26. Ito, H.; Ueda, M. *Macromolecules* **1988**, *21*, 1475.
27. Hatada, K.; Kitayama, T.; Danjo, S.; Tsubokura, Y.; Yuki, H.; Morikawa, K.; Aritome, H.; Namba, S. *Polym. Bull.* **1983**, *10*, 45.
28. Aso, C.; Tagami, S.; Kunitake, T. *J. Polym. Sci. Part A-1* **1869**, *7*, 497.
29. Hatada, K.; Kitayama, T.; Danjo, S.; Yuki, H.; Aritome, H.; Namba, S.; Nate, K.; Yokono, H. *Polym. Bull.* **1982**, *8*, 469.
30. Ito, H.; Flores, E. *J. Electrochem. Soc.* **1988**, *135*, 2322.
31. Dammel, R.; Lindley, C. R.; Pawlowski, G.; Scheunemann, U.; Theis, J. *Proc. SPIE* **1990**, *1262*, 378.
32. Nate, K.; Inoue, T.; Yokono, H.; Hatada, K. *J. Appl. Polym. Sci.* **1988**, *35*, 913.
33. Bowden, M. J.; Thompson, L. F.; Fahrenholtz, S. R.; Doerries, E. M. *J. Electrochem. Soc.* **1981**, *128*, 1304.
34. Shiraishi, H.; Isobe, A.; Murai, F.; Nonogaki, S. In *Polymers in Electronics*; Davidson, T., Ed.; Symposium Series 242; American Chemical Society: Washington, D. C., 1984; pp. 167-176
35. Ito, H.; Pederson, L. A.; MacDonald, S. A.; Cheng, Y. Y.; Lyerla, J. R.; Willson, C. G. *J. Electrochem. Soc.* **1988**, *135*, 1504.
36. Steinmann, A. *Proc. SPIE* **1988**, *920*, 13.

RECEIVED January 14, 1991

Chapter 21

Thermal Marking of Amorphous Poly(ethylene terephthalate)

C. M. Roland, J. P. Armistead, and M. F. Sonnenschein[1]

Chemistry Division, Code 6120, Naval Research Laboratory, Washington, DC 20375–5000

It is shown that exposure of amorphous films of poly(ethylene terephthalate) (PET) to infrared light from a CO_2 laser induces either crystallization or ablation, depending on the intensity of the radiation. The resolution of these thermal images is not only exempt from diffraction limitations, it is also found to be minimally affected by thermal diffusion. Crystallization and ablation can both serve as a basis for microlithography, providing a single step process producing high resolution images of good contrast and edge acuity. The extension of the technique to selective metallization of polymer films is demonstrated.

The use of polymeric materials as media for optical data storage has received much attention and a large body of relevant literature exists [1-7]. The desired attributes of an imaging process include the production of stable images which are high both in contrast and in resolution. High resolution enables higher storage density on the polymer surface. When absorption of a single photon causes an incremental change in the image intensity, the spatial resolution is diffraction limited. Hence "photon mode" processes rely on short wavelength radiation (ultraviolet, x-ray, and electron beam) to achieve high resolution. These approaches are employed in much photoresist technology [1-6].

A familiar commercial application of optical data storage is the compact disk (CD) [7]. Information takes the form of a series of pits in the surface of a polymer film, each depression having the same depth but being of different lengths and separations. The variation in the

[1]Current address: Dow Chemical Research Center, Walnut Creek, CA 94598

reflection of a light beam as it traverses alternately the smooth CD surface and the pits provides the data transcription. The pits in a CD are created by a stamping process; a master engraves onto the polymer substrate a negative relief. Polycarbonate is the current polymer of choice for CD's, due to its high glass transition temperature, toughness, relatively low cost, and the low birefringence obtained in injection moldings.

An alternative to writing mechanisms based on photon or mechanically induced alterations in the recording medium is the use of heat to effect a physical or chemical change. Thermal processes selectively raise the temperature to some critical point at which image formation commences. Because a single photon provides insufficient energy to induce this process, thermal methods are not diffraction limited. There is hence no *a priori* reason to employ short wavelength radiation; however, it is anticipated that thermal diffusion away from the directly heated regions will smear the image and thus limit resolution.

Crystalline - amorphous phase transformations are a potential basis for thermal lithographic techniques. The conversion of an amorphous polymer to the crystalline state alters the optical properties, introducing opacity and birefringence, and thus producing an image. A number of studies have employed the crystallization of small molecule species residing on or in a polymeric matrix [8-12]. The polymeric substrate itself can serve as the active medium. It must be initially amorphous, but yet highly crystallizable, in order to produce a high contrast image. Few polymers crystallize extensively above room temperature, while still crystallizing sufficiently slowly to be obtainable in the amorphous state. Poly(phenylene sulfide), poly(ether ether) ketone (PEEK), isotactic polystyrene, and poly(ethylene terephthalate) are among the polymers which meet this criterion.

An early demonstration of using the amorphous - crystalline phase change to impart an image involved poly(phenylene sulfide) [13]. Crystalline lines as narrow as 100 μm were produced on a film of the filled polymer ("Ryton") by the application of heat from a CO_2 laser. It is unknown whether elimination of the filler would slow down crystallization and thereby allow finer imaging.

A CO_2 laser was also employed to selectively crystallize PEEK films [14]. In these experiments image sizes were as large as a few mm, and the authors suggested the use of dyes to achieve high resolution. By placing the film on a substrate to slow the rate of cooling, the perfection of the crystallites was improved; the relationship of this to image quality was not addressed. It was reported that heating at a high rate for a short period rendered the PEEK amorphous, since ambient air quenching was sufficiently rapid to minimize recrystallization. An obvious advantage of using a physical change as the writing mechanism is that such a process is reversible, potentially allowing for multiple read/write cycles.

This reversibility was demonstrated in PET heated with an infrared laser, wherein crystallization and melting were successively executed by adjustment in the intensity or duration of the irradiation [15]. Directional crystallization upon cooling of polyethylene oxide from the melt was carried out by translating films through an infrared laser beam [16].

Although lacking such reversibility, a chemical change induced by heat can also serve as a lithographic method. Ablation at the surface of a polymer film, as well as other photochemical reactions, have been widely explored [17-28]. The use of an infrared laser to ablate oriented sheets of semi-crystalline PET ("Mylar") has been reported [29]. The texture arising from the presence of spherulites resulted in poor image quality. Photoetching and the photochemical degradation of poly(ethylene terephthalate) exposed to U.V. radiation has been extensively studied [30-34]. It was demonstrated that, even in the UV region, a thermal mechanism is operative at high radiation intensities [29,33]. High resolution imaging via ablation using infrared radiation has recently been reported [35].

Reviewed herein are ongoing efforts [15,35,36] to accomplish thermal microlithography in poly(ethylene terephthalate). Both physical and chemical changes can be induced by the application of heat. The utility of these processes for optical data storage is discussed.

Experimental

Experiments were carried out using extruded sheets (130 μm thick) of amorphous PET (intrinsic viscosity = 0.75 dl/g) obtained from Eastman Chemicals, Kingsport, TN. Qualitatively similar results were obtained when a higher molecular weight PET (from the Goodyear Tire and Rubber Co. with I.V.= 1.04) was utilized. Heating was accomplished by exposing the film to 10.6 μm radiation from a Coherent Model 42 CO_2 laser. The intensity profile of the laser beam was quite nonuniform; thus, average power measurements were of dubious value.

To create an image, the laser beam flood illuminated a mask comprised of gold patterned on a GaAs wafer. The mask lay on top of the polymer, whereby radiation was selectively prevented from impinging on the polymer surface by reflection. In an alternative arrangement, an image was produced by reflecting the laser light from an aluminum surface onto the PET film.

Micrographs were obtained with an AMR Model 1000 scanning electronic microscope (SEM) and with a hot stage equipped Zeiss optical microscope. Differential scanning calorimetry was conducted on samples cut from the films using a Perkin-Elmer DSC-7 at 20 deg per minute.

Results

Laser Induced Crystallization. In Figure 1 is displayed

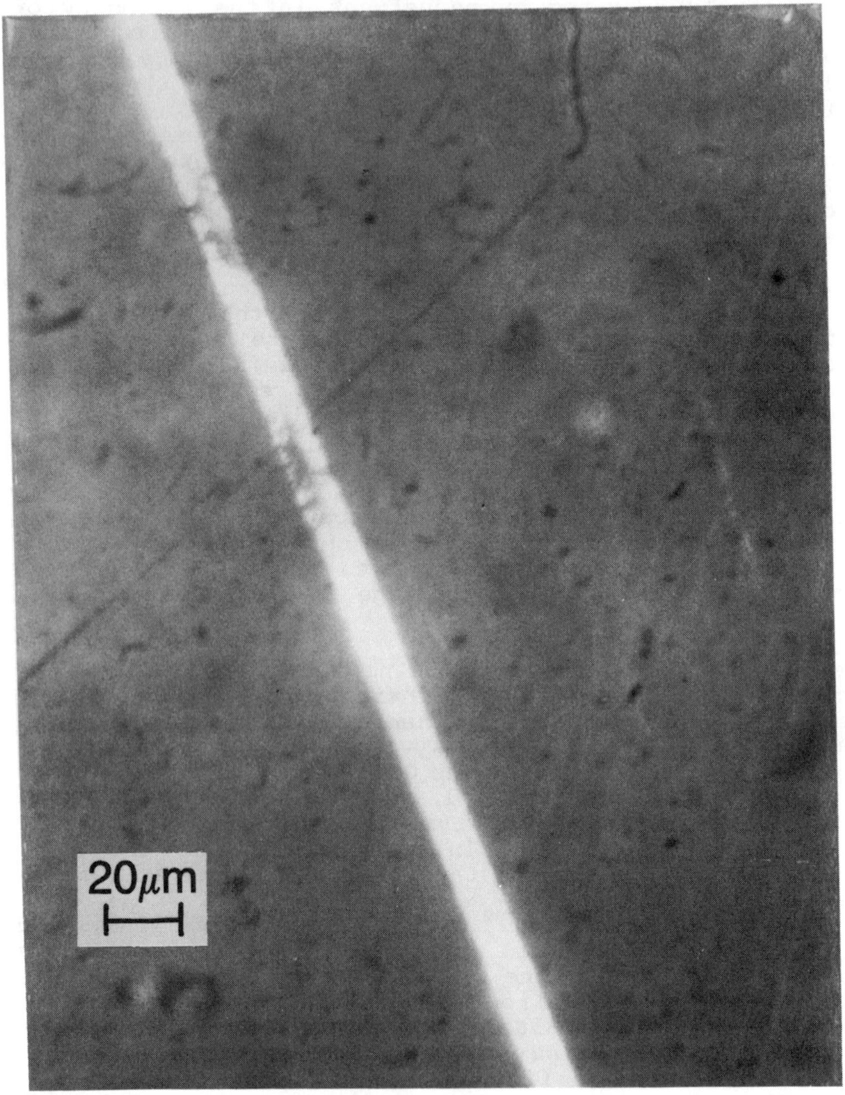

Figure 1. Crystallization line produced in amorphous PET by reflection of CO_2 laser radiation from an aluminum surface positioned near a film of PET. The irradiation time was 30 seconds.

amorphous PET film containing a crystalline image, roughly 10 μm wide, as seen through cross polarizers. This image was produced by heating the film with laser radiation (0.2 watts) reflected from a piece of aluminum. The localized crystallization is a consequence of heating the PET to above its crystallization temperature. The high birefringence associated with the crystal phase produces a distinct image through crossed polarizing filters (as illustrated in the figure). The crystallites are sufficiently large for significant light scattering, and consequently the images are also directly visible. Although there is a 10% reduction in mass density upon crystallization, image formation was not accompanied by any observable alteration of the surface topography of the films. From differential scanning calorimetry measurements it was estimated that irradiation induced a level of crystallinity equal to about 34% (using 140 J/g as the perfect enthalpy of fusion [37]). The crystalline phase melted at 245°C. This is equivalent to the melting point obtained when the PET was crystallized at various temperatures over a range of from 140°C to 230°C [15]. From experiments on a hot stage and microscope, it was determined that crystallization commences at an observable rate at temperatures as low as 125°C. The laser induced crystallization presumably transpires within this temperature range.

When a mask was used to define the image, the gold coating comprising the mask image was in contact with the polymer film. This probably reduces the temperature of the regions shielded from the radiation, since the gold functions as a heat sink. Representative images obtained by this method are seen in Figures 2 and 3.

With the mask in contact with the film, higher power levels (*circa* 1 watt) were necessary to induce crystallization, presumably to compensate for reflection losses at the interfaces (measured to be 30%) and heat conduction to the gold. The laser intensity profile was non-uniform, and as a result the extent of crystalline image formation was found to vary over the irradiated zones of the film, as can be seen in Figure 2. Note that individual spherulites are visible in the irradiated films. A cross-section of a laser crystallized film is displayed in Figure 4. It is observed that the crystalline phase extends completely through the film thickness. More opaque and more strongly birefringent images were found to have a greater concentration of spherulites (Figure 4). In no case was there evidence of any gradient or accumulation of spherulites toward the irradiated side of the films.

The absorptivity of the PET at 10.6 μm was equal to about 90 cm^{-1}. Assuming a heat capacity of 12 J/g-deg [37], and an amorphous density of 1.335 g/cm^{-3}, a 0.25 watt laser could provide sufficient energy to heat a one square mm section of the film to beyond 125°C in one second. This is faster than the actual time scale for the laser writing. The efficiency of the marking process could not be

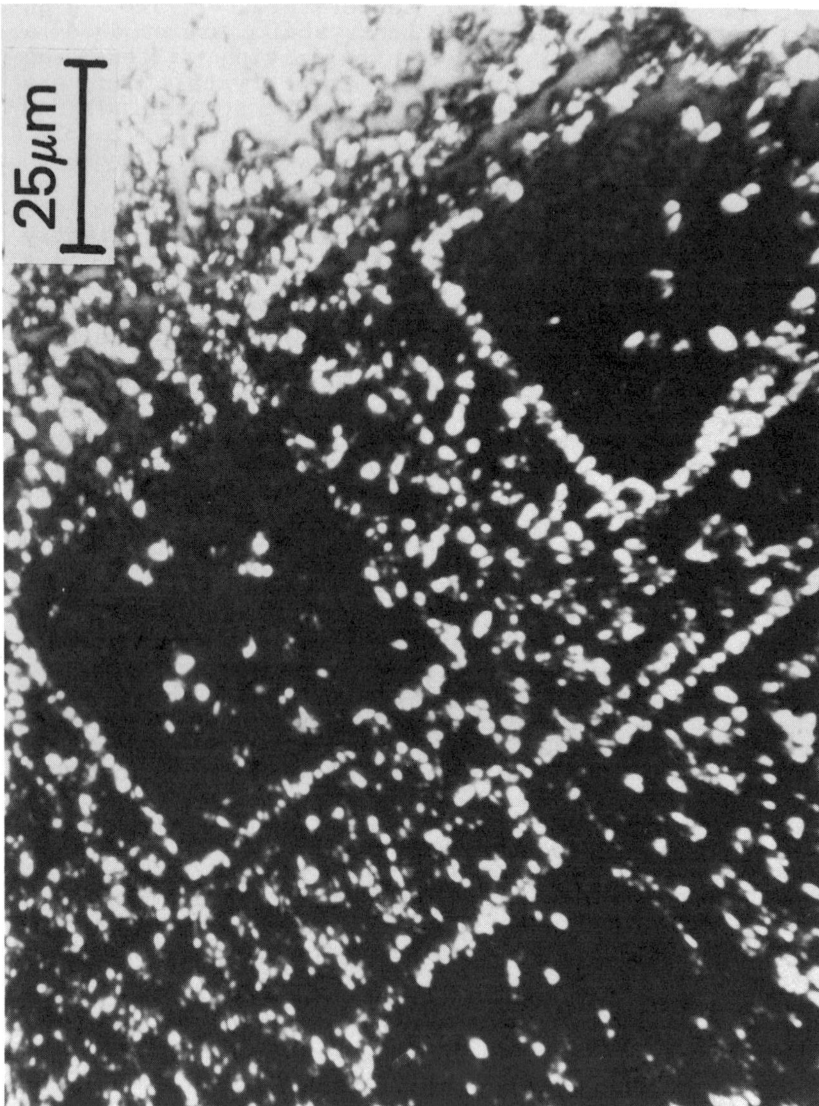

Figure 2. Amorphous PET film irradiated through a light field gold on GaAs mask for 20 s at low laser power. The metallized side of the mask was in physical contact with the film. Images obtained on the PET film were identical in size to those on the mask.

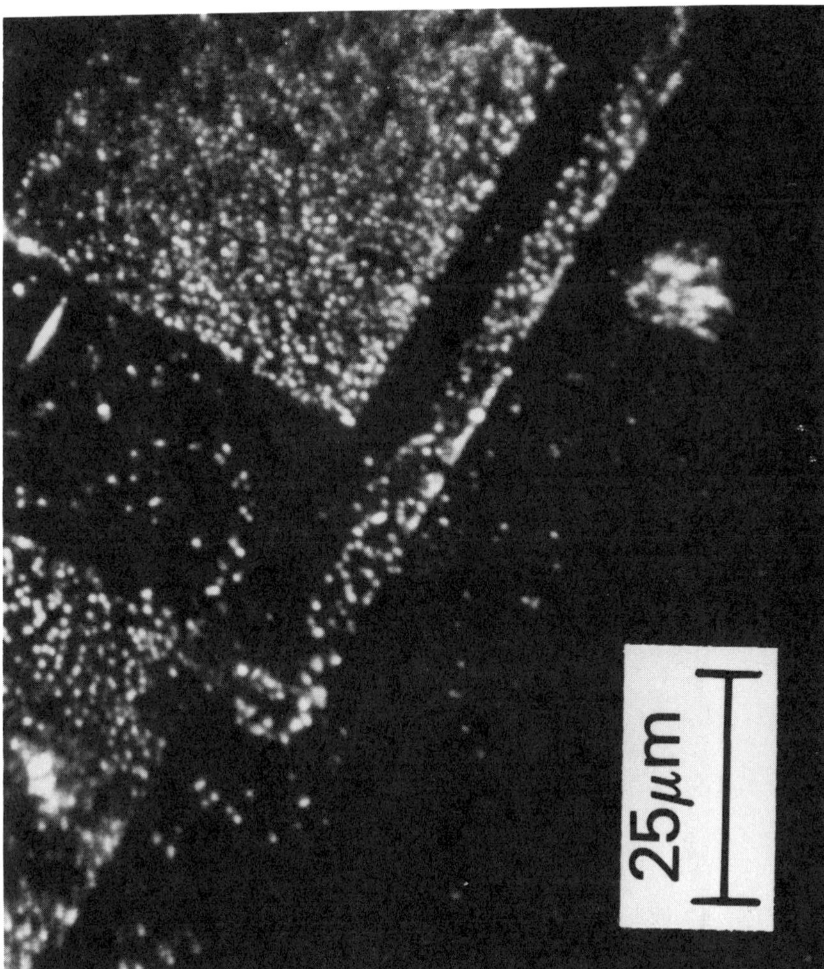

Figure 3. Results from irradiation of amorphous PET film through a dark field mask for 30 s.

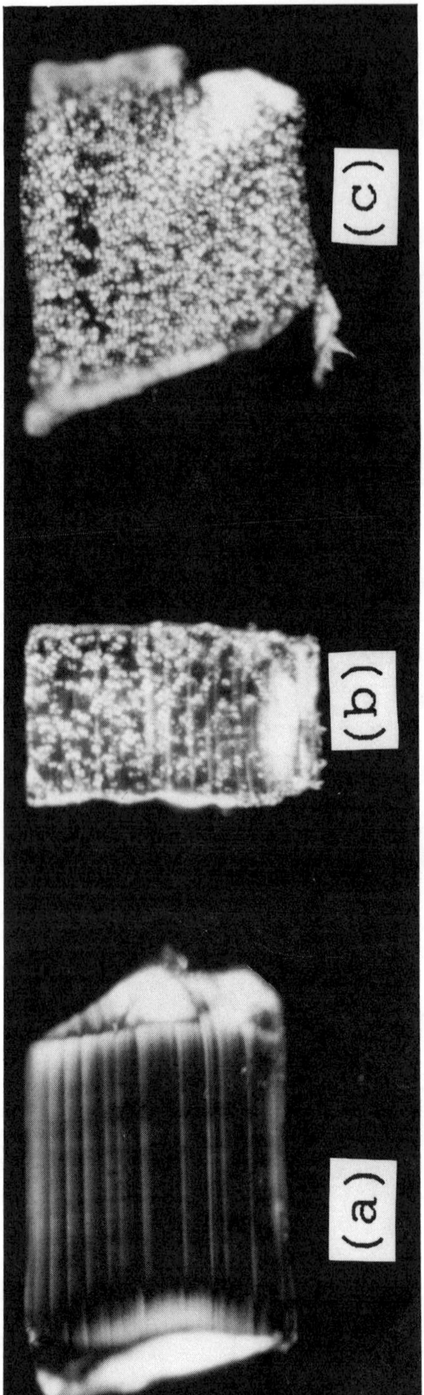

Figure 4. Cross-section of PET film (a) prior to exposure to the laser, wherein microtome marks can be seen; (b) after irradiation sufficient to give a weak crystalline image; (c) after prolonged irradiation that yielded a very opaque image.

quantified with the present experimental apparatus and no effort was expended toward optimizing this efficiency. A representative dose-response curve is shown in Figure 5. The non-linearity of the response is an attractive feature of thermal imaging processes.

Crystallization as a mechanism for microlithography is attractive only to the extent that high resolution marking can be achieved. As noted, the resolution of thermal imaging is not subject to wavelength constraints due to diffraction, and indeed features in Figures 2 and 3 are significantly smaller than 10 μm. As indicated by the data in Figure 5, there is a threshold level of radiation absorption below which the consequent temperature rise is insufficient to effect marking of the PET film. This threshold is beneficial for obtaining both good contrast and high resolution.

The resolution of a thermal process is expected to be governed by the diffusion of heat away from the directly irradiated regions. When optical imaging by means of an amorphous to crystalline phase transition was carried out with either polyphenylene sulfide [13] or poly(ether ether ketone) [14], the reported resolution was poor. The crystallization extended beyond the directly irradiated regions. Modeling of the heat flow has suggested that thermal diffusion away from the irradiated areas will smear the image unless extremely fast heating rates are employed [15]. Some representative results from these calculations are give in Table I. Listed is the temperature one micron away from the edge of a 10 micron wide irradiated area at the time a temperature of 120°C is attained by the material one micron within the edge. Different times correspond to different heating rates. Note that rapid crystallization commences in PET at roughly 120°C.

Table I. Calculated Temperatures in Region Adjoining Directly Irradiated Areas* [15]

TIME (msec)	1	10	100	1000
TEMPERATURE (C)	102	115	118	120

*At heating rates for which directly heated region reached 120°C.

A slow marking process is predicted to be unable to yield images as fine as 10 μm. The reports of low resolution (> 10^2 μm with polyphenylenesulfide [13], and > 10^3 μm with PEEK [14]) are consistent with such expectations. Given these considerations, the image dimensions apparent in

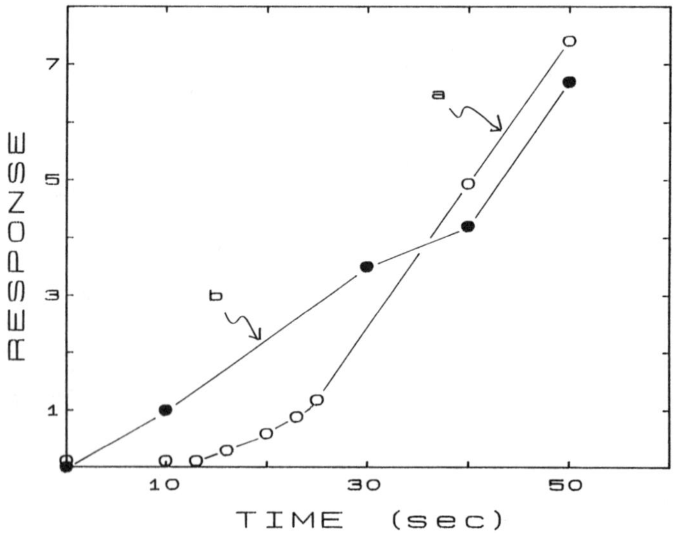

Figure 5. (a) The optical density (in arbitrary units) of a crystalline mark formed by irradiation at 0.2 watt. (b) The depth (in μm) of the ablation from exposure to 4 watts infrared laser radiation.

Figures 1-3 are surprising. Although the crystalline image quality is not superb, it should be noted that the extraneous crystallites are randomly situated. There is no suggestion that this "noise" is associated with the borders defining the images. It is certainly considered that more uniform films and a more uniform light source would improve the image quality.

Figure 6a shows a thermally crystallized film of PET on which the crystallization in a small region has been erased by exposing the film to the CO_2 laser radiation. The amorphous spot is transparent and non-birefringent, appearing through crossed polarizers as dark against the birefringent crystalline background. This erasure is reversible, as demonstrated in Figure 6b, which shows the "erased" spot following partial overwriting by exposure to less intense laser light. With further irradiation the film could be made completely opaque and birefringent, equivalent to its initial condition.

A disordering process such as melting can be accomplished significantly faster than the reverse operation of ordering the polymer segments into a crystalline phase; therefore, in principle, induced melting of crystalline film is potentially a faster marking process than crystallization of amorphous polymer. However, whereas crystallization commences at temperatures somewhat above the glass transition temperature, the reverse process requires heating to the higher temperature of melting, thus presumably requiring higher power input.

Ablation. As described above, at laser powers on the order of one watt or lower, crystalline images can be produced in amorphous PET, while very brief (< 1 sec) irradiation at slightly higher intensities effects melting. When the PET is exposed to significantly higher levels of the infrared radiation, however, a different process is observed. As seen in Figures 7-9, images produced by irradiation at several watts are three dimensional. Also, unlike the images produced by the amorphous-crystalline phase transformation, these images are not erasable. The more intense irradiation promotes significant chemical bond rupture in the PET. The byproducts of this decomposition are then expelled as fragments or vaporized. This ablation at the PET surface creates an image.

The dimension of the image in Figure 9 is less than 1 μm, and as seen in both Figures 8 and 9, edge acuity is quite good. Extraneous pits and debris are present on the original film. Although marks less than 1 μm in width can be produced, the masks have defects at this scale. The ultimate resolution achievable with this technique can not be determined presently.

The depth of the laser etching is proportional to the irradiation time (see Figure 5). A correspondence was also found between this depth and the weight loss measured for the films [35]. The steepness of the walls reflects a minimal extent of thermal diffusion, which coupled with the

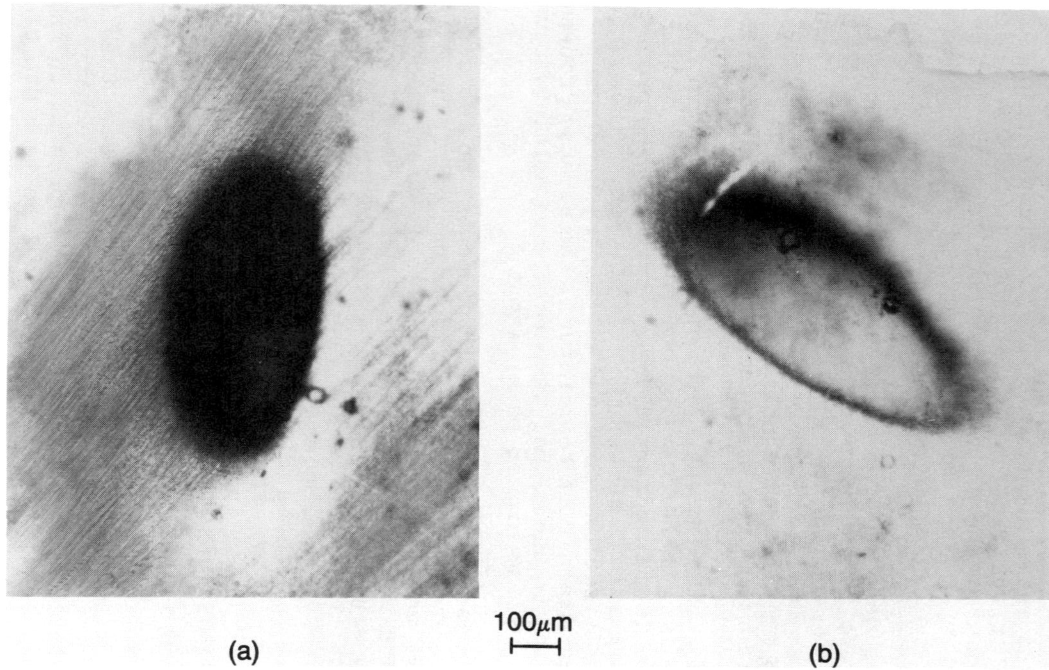

Figure 6. (a) Amorphous spot produced by laser melting in an initially crystalline PET film (as viewed through crossed polarizing filters). (b) The same spot after being partially overwritten through recrystallization via laser heating. After complete overwriting, no evidence of the original spot remained.

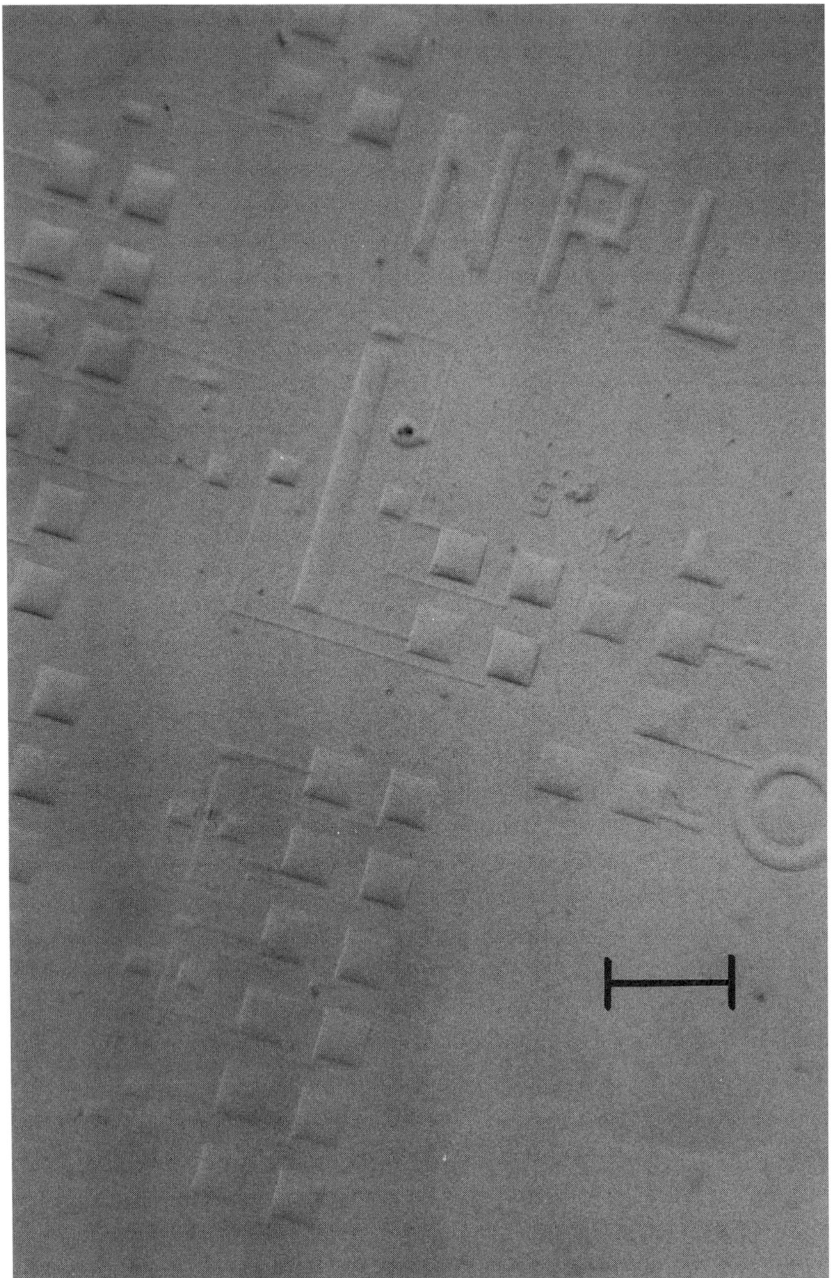

Figure 7. Optical micrograph of ablated images formed in amorphous PET by exposure to infrared radiation through a mask. The scale mark is 300 μm.

Figure 8. Scanning electron micrograph of a portion of Figure 7. The imperfections are due to the condition of the original PET film and to mask errors.

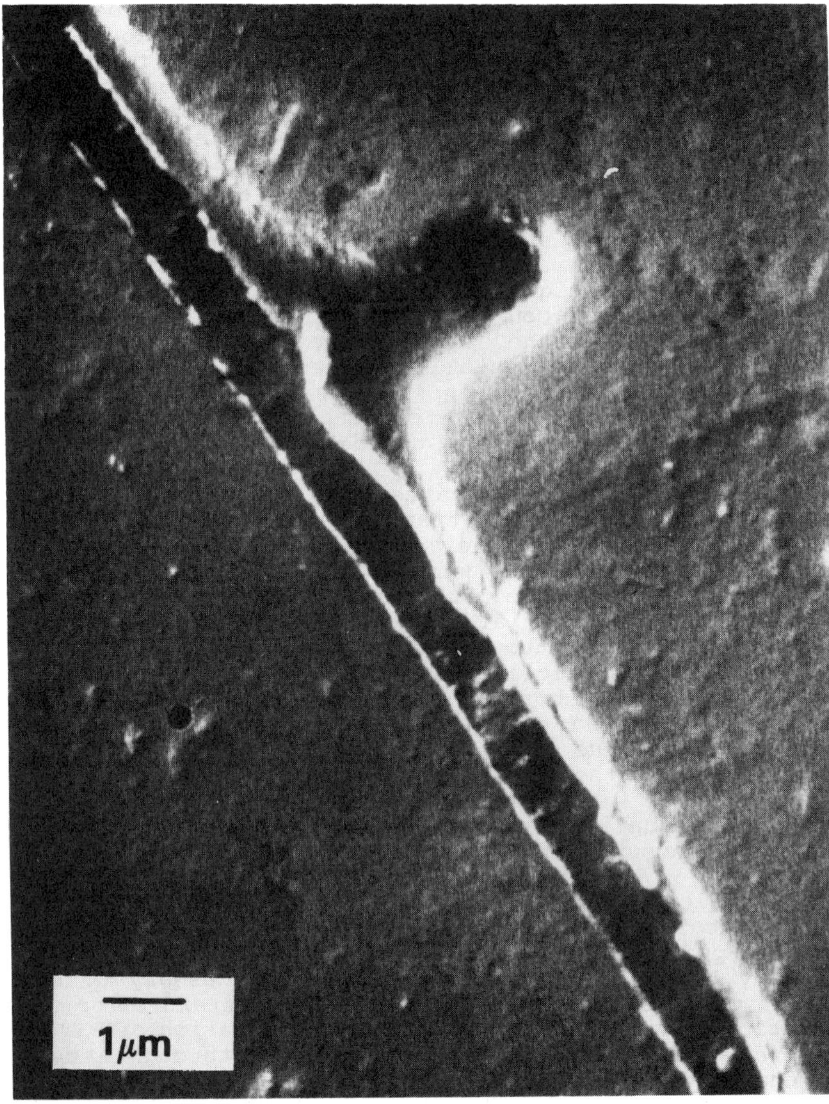

Figure 9. SEM of a portion of Figure 7. Notwithstanding the defect arising from an error in the mask, the edge acuity of the line is striking.

expected existence of a threshold for thermal ablation facilitates attainment of high resolution imaging.

When the lower power level associated with induced crystallization (< 1 watt typically) was continued well beyond the time required for crystallization, only melting, but no ablation, resulted. Ablation of the PET could only be achieved using radiation levels of at least a few watts. The PET under the ablated areas remains amorphous, as inferred from the transparency and absence of birefringence of the images (Figure 10a and 10b). When the ablated film is subsequently crystallized on a hot stage, both the unablated regions of the film and the imaged areas become opaque and birefringent (Figure 10c and 10d). Since the laser light is only attenuated roughly 20% upon passage through the film, it is expected that sufficient intensity is available for significant heating of the underlying material during the course of the ablation. Evidently a temperature above the crystalline melting point is attained, whereby no crystallinity results. For PET films as thin as used herein (130 μm), air cooling is sufficiently fast to prevent crystallization during cooling from the melting point down through the glass transition temperature.

Ablative imaging was attempted on semi-crystalline PET, which was nonetheless transparent and optically clear due to the small size of the crystallites. Consistent with previous work [29], only very coarse, ill-defined images could be produced. This might seem surprising since prior to ablation, the material presumably is heated above the crystalline melting temperature. Apparently a level of thermal energy sufficient for ablation is attained in a time period too short for thermal equilibration and consequent melting. High quality imaging requires initially amorphous PET.

The use of thermal ablation as an imaging method has been attempted with other polymers [35]. The efficiency and imaging quality varied widely, with the best overall results achieved with PET.

Metallization. It is known that PET will adhere to metals, and particularly to metal oxides, when raised above its melting point while in contact with the metal [38]. Actually, for amorphous PET it is only necessary to heat the polymer above the glass transition temperature for bond formation to transpire. Localized metallization of PET can therefore be effected by its irradiation through a mask. The underside of the PET is maintained in contact with the metal, for example in the form of a thin film, during exposure to the infrared light. After irradiation the metal in the unexposed regions is brushed or stripped away, leaving behind a metallized pattern. Representative results of this process are seen in Figure 11. The bonding of PET is superior to copper and aluminum than to gold. Attempts to remove the former result in cohesive failure of the polymer.

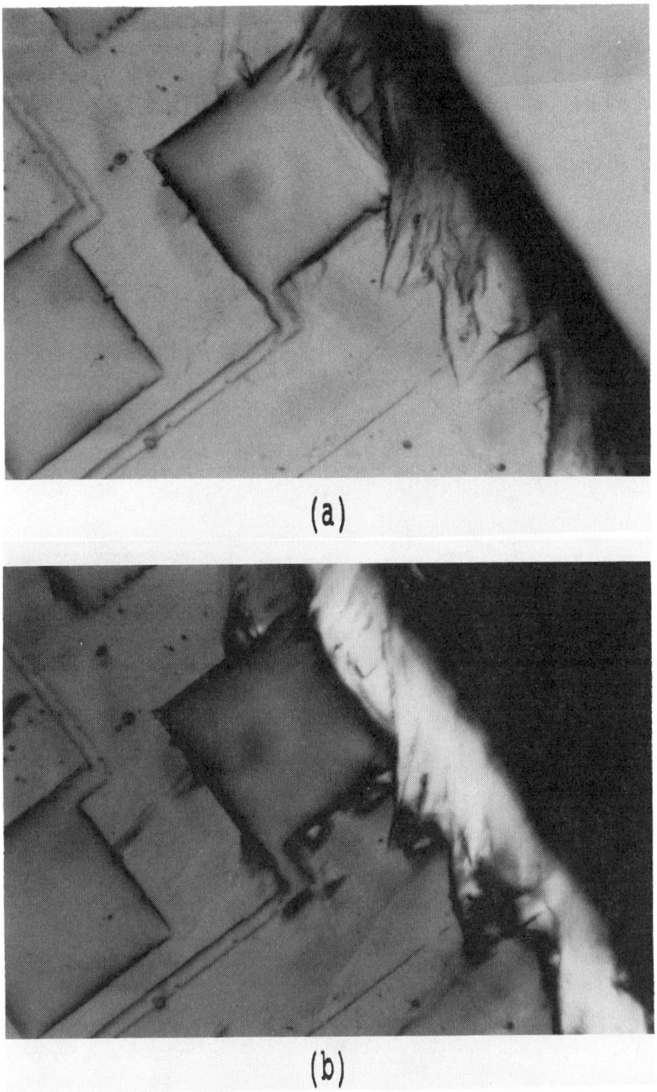

Figure 10. Optical transmission micrograph of ablated images (a) observed with unpolarized light, where edge scattering enables the images to be seen and (b) seen through crossed polarizers. The material lying under the ablated area is not birefringent. After heating at 2 degrees per minute to 127°C, the entire film becomes translucent and birefringent as respectively seen in (c) unpolarized and (d) polarized light.

Continued on next page

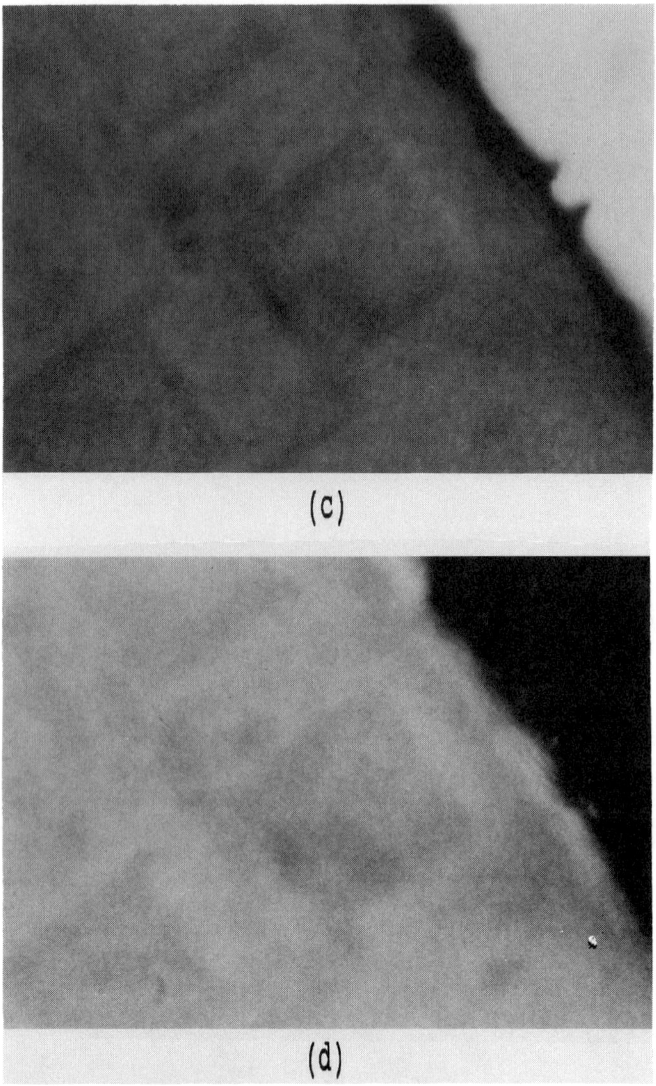

Continued from page 359.

Figure 10. Optical transmission micrograph of ablated images (a) observed with unpolarized light, where edge scattering enables the images to be seen and (b) seen through crossed polarizers. The material lying under the ablated area is not birefringent. After heating at 2 degrees per minute to 127°C, the entire film becomes translucent and birefringent as respectively seen in (c) unpolarized and (d) polarized light.

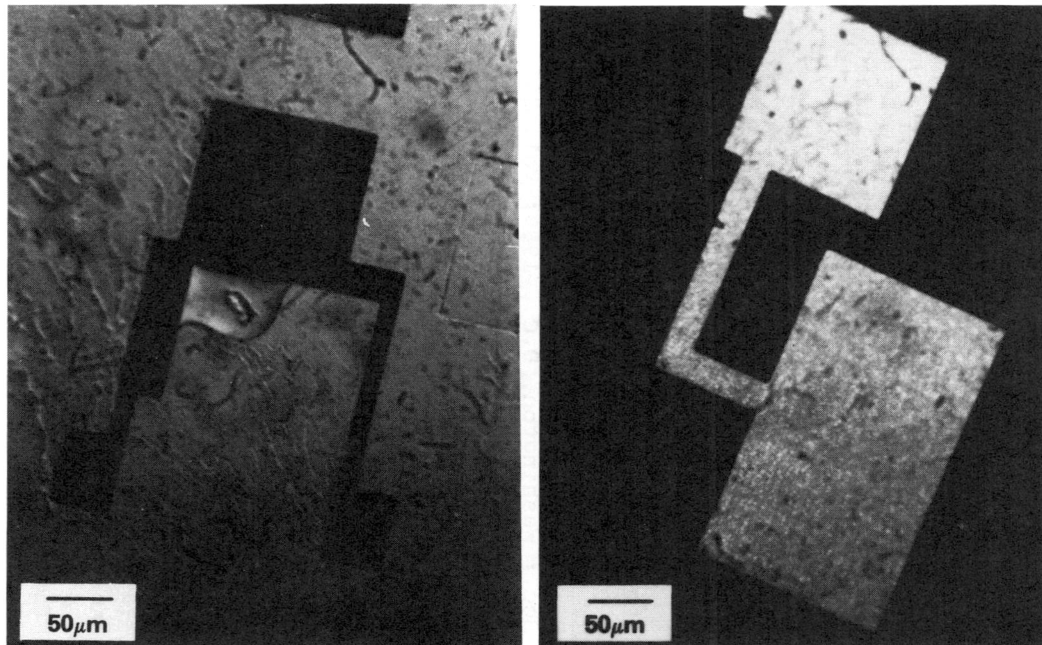

Figure 11. A gold image adhered to PET by laser irradiation through a light field mask and a PET film on which aluminum was bonded over all but selected regions. The latter were shielded from the irradiation by a dark field mask.

Summary

In anticipation of smearing by thermal diffusion, it is generally believed that photon processes are necessary for high resolution lithography. The present results with infra-red laser marking of PET, however, indicate that thermal methods have potential for high resolution optical data storage applications. Both radiation induced crystallization and ablation of PET film have been demonstrated. The two processes prevail at distinctly different levels of radiation intensity. While only the former process is reversible, ablation may offer advantages with regard to contrast and resolution. These techniques can be readily extended to achieve select metallization of the polymer film.

Acknowledgments

This work was sponsored by the Office of Naval Research. MFS acknowledges the National Research Council for an NRC-NRL postdoctoral associateship.

Literature Cited

1. The Effects of Radiation on High-Technology Polymers, Reichmanis, E.; O'Donnell, J., Eds.; ACS Symposium Series; Amer. Chem. Soc.; Washington, D.C., 1989; Vol. 381.
2. Symposium on Polymers in Microlithography; Poly. Mater. Sci. Eng., 1989, 60, 40.
3. Reichmanis, E.; Thompson, L.F. Chem. Rev. 1989, 89, 1273.
4. Electronic and Photonic Applications of Polymers; Bowden, M.J.; Turner, S.R., Eds.; Advances in Chemistry; Amer. Chem. Soc.: Washington, D.C., 1988; Vol. 218.
5. Symposium on Polymer in Information Storage Technology; Polymer Preprints, 1988, 29, 195.
6. Polymers for High Technology - Electronics and Photonics, Bowden, M.J.; Turner, S.R., Eds.; ACS Symposium Series; Amer. Chem. Soc.; Washington, D.C., 1987; Vol. 346.
7. Croucher, M. D.; Hopper, M. A. Chemtech 1987, 17, 426.
8. Werner, A.; Hibst, H.; Petermann, J. U.S. Patent #4,766,021 1988.
9. Takenaga, M.; Yamada, N.; Nishiuchi, K.; Akihara, N.; Ohta, T.; Nakamura, S.; Yamashita, Y. J. Appl. Phys. 1983, 54, 5376.
10. Ota, K. Japanese Patent #58,199,345A 1984.
11. Novotny, V.; Alexandru, L. J. Appl. Poly. Sci. 1979, 24, 1321.
12. Ovshinsky, S.R.; Fritzsche, H. IEEE Trans. Elec. Dev. 1973, ED-20, 91.
13. Rubner, M.F.; Cukor, P. United States Patent #4,486,463 1984.
14. Fang, S-J.; Salovey, R.; Allen, S.D. Poly. Eng. Sci. 1989, 29, 1241.

15. Sonnenschein, M.F.; Kotliar, A.M.; Roland, C.M. Poly. Eng. Sci. 1990, 30, 1165.
16. Yang, M.; Salovey, R.; Allen, S.D. J. Poly. Sci. Poly. Phys. Ed. 1990, 28, 245.
17. Chiello, E.O.; Garbassi, F.; Malatesta, V. Die Ange. Makro. Chem. 1989, 169, 143.
18. Muchova, M.; Pelzbauer, Z. J. Appl. Poly. Sci. 1989, 37, 817.
19. Limm, W.; Winnik, M.A.; Smith, B.A. Poly. Eng. Sci. 1989, 29, 911.
20. Kuder, J. E.; Taskier, H. T.; Hamer, G. J. Appl. Poly. Sci. 1988, 35, 1257.
21. Lub, J.; van Vroonhoven, F.C.; van Leyan, D.; Benninghoven, A. Polymer 1988, 29, 998.
22. Factor, A.; Ligon, W.; May, R.J. Macromolecules 1987, 20, 2461.
23. Cozzens, R.F.; Walter, P.; Snow, A.W. J. Appl. Poly. Sci. 1987, 34, 601.
24. Bell, A.E. J. Appl. Phys. 1982, 53, 3438.
25. Cornet, J. U.S. Patent #4360895 1982.
26. Maffit, K.N.; Robbins, W.B.; Wilson, R.F. European Patent #58496 1982.
27. Kawamura, Y.; Toyada, K.; Namba, S. Appl. Phys. Lett. 1982, 40, 374.
28. Howe, D.G.; Wrobel, J.J. J. Vac. Sci. Tech. 1981, 18, 92.
29. Andrew, J. E.; Dyer, P. E.; Forster, D.; Key, P. H. Appl. Phys. Lett. 1983, 43, 717.
30. Dunn, D.S.; Ouderkirk, A. J. Macromolecules 1990, 23, 770.
31. Day, M.; Wiles, D. M. J. Appl. Poly. Sci. 1972, 16, 175, 191, 203.
32. Kokai, F.; Saito, J.; Fujioka, T. Macromolecules 1990, 23, 674.
33. Srinivasan, R.; Braren, B. J. Poly. Sci. Chem. Ed. 1984, 22, 2601.
34. Srinivasan, R.; Lazare, S. Polymer 1985, 26, 1297.
35. Sonnenschein, M.F.; Roland, C.M. Appl. Phys. Lett. 1990, 57, 425.
36. Sonnenschein, M.F.; Roland, C.M. U.S. Patent #4,975,358 1990.
37. Mehta, A.; Gaur, U.; Wunderlich, B. J. Poly. Sci. Poly. Phys. Ed. 1978, 16, 289.
38. Brown, D.J.; Lamb, P.; Windle, A.H. J. Appl. Poly. Sci. 1988, 35, 279.

RECEIVED March 22, 1991

Chapter 22

Electromagnetic (Microwave) Radiation Effects on Terminally Reactive and Nonreactive Engineering Polymer Systems

S. C. Liptak, S. P. Wilkinson, J. C. Hedrick[1], T. C. Ward, and J. E. McGrath[2]

Department of Chemistry and NSF Science and Technology Center: High Performance Polymeric Adhesives and Composites, Virginia Polytechnic Institute and State University, Blacksburg, VA 24061–0212

Amorphous poly(arylene ether sulfones) of controlled $\overline{M}_n$ were synthesized by step polymerization of bisphenol-A phenolate with activated aromatic halides in the presence of m-aminophenol. Utilization of the amino terminated polysulfones (NH_2PSF) to co-cure with the diglycidyl ether of bisphenol A (DGEBA) and/or tetraglycidyl 4,4'-diaminodiphenylmethane (TGDDM) via thermal or microwave (EM) processing was conducted at temperatures of ~190-240°C. 4,4'-Diaminodiphenylsulfone (DDS) was employed as the curing agent. The bismaleimide systems (Matrimid 5292 A/B) were thermally or electromagnetically crosslinked through free radical reactions of the maleimide groups at temperatures of ~250°C. The influence of oligomer molecular weight on the thermal, mechanical, and morphological properties of the thermoplastic toughened epoxy networks were investigated by DSC, fracture toughness (K_{IC}) and SEM. Microwave processing can dramatically enhance the rate of network formation in a variety of systems. Morphological structure in thermoplastic toughened thermosets can be controlled via the microwave power utilized. Non-reactive thermoplastics which contain even modestly polar groups (e.g., PEEK, polyimides, etc.) can also be rapidily melt processed.

Radical ions which subsequently decay to free radicals are usually considered to be responsible for the crosslinking and scission reactions observed in polymers exposed to ionizing radiation. Nonionizing radiation, in contrast, is generated from that part of the electromagnetic spectrum for which the quantum energy is too low to ionize an atom (1). Microwave radiation is not energetic enough to cause ionization or chain scission. Thus, any acceleration of reaction rates observed for systems exposed to microwave radiation

[1]Current address: IBM Corporation, T. J. Watson Research Center, P.O. Box 218, Yorktown Heights, NY 10598
[2]Corresponding author

0097–6156/91/0475–0364$06.00/0

compared with thermal curing is probably due to more efficient energy transfer.

Polarization phenomena in homogeneous dielectric materials may be classified into three categories based on the nature of the atomic or molecular origins(2):

(1) Electronic polarization that results from a shift of electron orbits with respect to the center of the atom.
(2) Atomic polarization that results from a shift in relative positions, e.g., between the carbon and hydrogen atoms by stretching in response to an electric field.
(3) Dipole polarization that results from orientation of permanent dipoles under the influence of the electron field.

The mechanism of energy transfer in microwave heating is thought to occur by electric dipolar coupling of the radiation to permanent dipole moments in the polymer (Category 3), rather than by thermal conductivity as in conventional processing. Indeed, most molecules by their very nature are polarizable in an electric field. The degree of polarization and the energy required to achieve it control the loss factor or dissipation factor of a material. A material that is readily polarized by a small electric field has a high loss factor and is observed to be easy to heat.

Reactive systems inherently contain a large concentration of very mobile dipolar groups, and thus, are easy to heat via electromagnetic radiation. As the reaction proceeds though, the loss characteristic of the material decreases with the consumption of polar functional groups and the decreasing mobility of polarizable moieties, thereby necessitating a greater input of energy to maintain the "global temperature" of the reacting system. By the same reasoning, nonreactive systems with their associated negligible concentration of highly polar reactive endgroups and hindered dipolar rotation require higher input energies to attain efficient heat generation by dipolar relaxation mechanisms.

In our laboratories(3,6), and elsewhere, microwave radiation has been utilized to process both reactive thermosetting polymers and nonreactive thermoplastic(7-11) polymers. More specifically, Gourdenne et.al.(12-15) and colleagues at Michigan State University(16-17) have examined continuous and pulsed microwave processing of epoxy/amine systems. In these studies both groups have independently verified that energy transferred by a pulsed electromagnetic source is more efficient than by an energy equivalent generated continuous wave. During these investigations Hawley et al. made an important contribution to the field of microwave processing with their implementation of computer controlled pulsed microwave processing of epoxies(18). Supplementing the work conducted in the area of the epoxy/amine systems, Gourdenne(19) and Jullien(20) have investigated other step growth thermosetting type systems, such as the polyurethanes, processed via electromagnetic radiation at a frequency of 2.45 GHz. Additionally, Teffal and Gourdenne(21), in some of his earlier work, investigated microwave radiation activated radical polymerization of hydroxyethyl methacrylate.

Employment of the microwave processing technique possesses a number of potential advantages when compared to more traditional thermal methods. There is the potential for fast volumetric heating of thermoplastics, significantly reduced cure times in thermosetting systems, the ability to produce materials that may display lower residual stresses (22). Other reported advantages of EM processing include maintenance of mechanical properties (5,17,23-24), an increase in thermal properties and an improvement in adhesion at the interface of carbon fiber/epoxy composites(25) when compared to thermally processed samples. An additional reported benefit is that a reduction in chemical pollution occurs with some microwave processing applications (1).

In this investigation, the feasibility of processing high performance polymeric matrix resins for composite structures utilizing microwave radiation has been analyzed. The systems which were investigated include thermoplastic toughened epoxy resin networks, thermosetting bismaleimides, and nonreactive thermoplastics such as PEEK and Ultem. The relationships between processing conditions on the thermal, mechanical and morphological properties in these material systems were investigated.

EXPERIMENTAL

The thermoplastic modified epoxy resins were prepared by reacting DGEBA (diglycidyl ether of bisphenol A) epoxy resin (DER 332) with an amino functional poly(arylene ether sulfone) oligomer with a number average molecular weight of 16,000 gm/mole. The crosslinked network was prepared by reacting DGEBA in a 2:1 molar ratio with respect to the NH_2PSF oligomer and the curing agent, DDS (4,4'-diaminodiphenylsulfone) (Aldrich Chemical). The synthesis of the NH_2PSF thermoplastic modifier and its reaction into the epoxy resin network have been previously described (26,27).

Additionally, thermoplastic modified epoxy resins were prepared by reacting TGDDM (tetraglycidyl 4,4'-diaminodiphenylmethane) epoxy resin (Ciba-Geigy MY-721) with an amino functional PSF oligomer with a number average molecular weight of 14,700 gm/mole. In contrast with the other epoxy systems investigated, this crosslinked network was prepared by reacting TGDDM in a 1:1 molar ratio with respect to the NH_2PSF oligomer and the curing agent, DDS.

Bismaleimide resins based on 4,4'-bismaleimidodiphenylmethane (57% w/w) (Ciba-Geigy Matrimid 5292 A/B) were cured utilizing either microwave or thermal processes. The Matrimid 5292 resin was degassed with continuous stirring under vacuum at 130°C for 30 minutes prior to processing.

Various testing specimens could be prepared by pouring the homogeneous solution of either epoxy/modifier/curing agent or maleimide/hardener into preheated RTV silicone rubber molds. These samples were then cured via microwave or conventional thermal processing techniques into specimens of suitable dimensions for fracture toughness, dynamic mechanical thermal analysis, and stress-strain measurements.

The instrumentation utilized for the electromagnetic processing experiments is shown schematically in Figure 1. It consists of a Raytheon magnetron that generates continuous microwave energy at a frequency of 2.45 GHz. The power is transmitted through coaxial transmission lines to a circulator which protects the magnetron from reflected power by terminating it at a dummy load. A directional coupler is utilized to divert power so the incident and reflected radiation can be monitored via power meters. An 18 cm diameter,

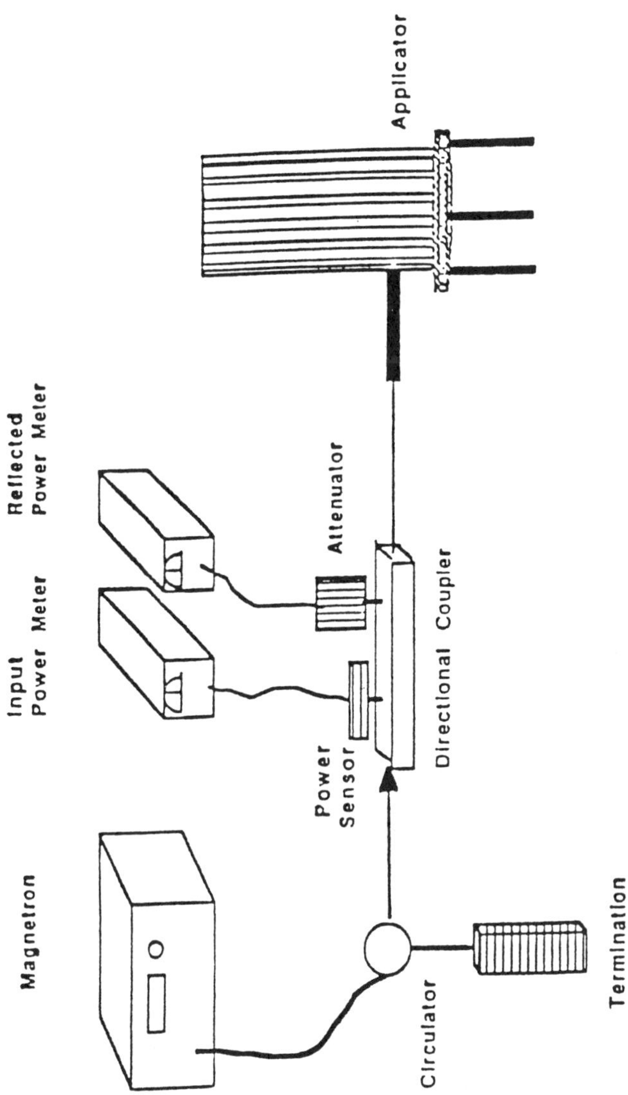

Figure 1. Schematic Representation of the Instrumentation Utilized for Electromagnetic Processing Experiments.

cylindrical brass cavity resonating in the TE_{111} mode was used for the electromagnetic processing. This single mode applicator has the distinct advantage over multimode cavities in that the impedance mismatch between the external circuit and the cavity due to changes of dielectric properties of the loaded material during processing can be avoided by tuning the cavity (16,7). Greater details of this instrumentation have been provided elsewhere (3,5).

Thermoplastic toughened epoxy systems consisting only of DGEBA were cured by a two stage process in the microwave cavity. The power introduced to the applicator was initially set at 20 watts to quickly attain the cure temperature. Gradually the power was reduced to 14 watts so as to maintain a constant cure temperature of ~240°C. Once the cure temperature had been attained, a processing time of 10 minutes was utilized to yield networks with ≥ 99% conversion (via DSC). The sample temperature was continually monitored during the course of the crosslinking reaction via a Luxtron (Model 750) fiber optic temperature probe.

Systems containing the more highly reactive TGDDM resin were reacted for longer times at lower microwave powers. For instance, epoxy networks incorporating 100 mole% TGDDM were first heated at 6 watts for ~10 minutes, then 8 watts for ~25 minutes, and finally 10 watts for ~20 minutes to effect a cure temperature of 195 - 200°C. As the TGDDM was "diluted" by the incorporation of DGEBA, higher processing powers could be realized. A 50/50 mole % DGEBA/TGDDM thermoplastic modified epoxy resin was cured by increasing the applied power to the cavity in a stepwise fashion. Overall, the power had been increased from 6 watts to 14 watts, by 1 watt intervals, every 4-8 minutes. Utilizing this cure schedule a higher cure temperature (~225 - 230°C) was achieved. The lowering of the dielectric loss of the materials, as the cure progressed, necessitated the increase in microwave power to maintain the cure temperature at optimal levels.

The Matrimid 5292 A/B resin was EM processed in a manner similar to that of the epoxy resin systems. To raise the temperature of the BMI/Diallyl resin to 200°C, 25 watts of power was employed. This temperature was maintained by continually tuning and detuning the microwave cavity. The samples were maintained at 200°C for 5 minutes and then raised to 250°C where again the temperature was held isothermally by tuning and detuning the cavity. For fracture toughness specimens the samples were held at 250°C for 40 minutes. Samples utilized in the extraction experiments were held at 200°C and 250°C for various lengths of time, again by tuning and detuning the cavity.

Extraction experiments were performed using chloroform solvent for 4 days. The samples were weighed before and after extraction when constant weights had been attained. The ratio of the weights after and before extraction gave the percent gel fraction remaining for the given length of time at a certain temperature. Samples for extraction that followed a thermal cure were performed in a forced air convention oven held isothermally at either 177°C or 200°C for particular time intervals.

RESULTS AND DISCUSSION

Thermoplastic Modified Epoxy Resin Networks of DGEBA: The polysulfone toughened epoxy resin networks of DGEBA cured with microwave radiation (MR) demonstrated essentially complete cure (~99% conversion by DSC) after the 10 minute processing period. Fracture toughness results and glass transition temperatures for the microwave and thermally processed epoxy resin networks as a function of increasing PSF incorporation are shown in Table I.

Table I. Fracture Toughness Properties and Glass Transition Temperatures of Polysulfone Modified Epoxy Resin Networks of DGEBA

	wt.% NH_2PSF*	$\overline{K}_{IC}x10^6(N/M^{3/2})$	Tg(°C)
Thermal cures:	0	0.5	238
	10	0.5	185, 230
	20	0.9	186, 230
	25	1.2	185, 229
	30	1.6	186, 227
	40	1.8	186, 229
Microwave cures:	0	0.6	221
	10	0.7	**, 221
	20	0.8	**, 214
	25	1.2	179, 210
	30	1.4	182, 217

* <Mn> = 16,000 gm/mole

** Not Detected Via DSC

These results demonstrate that fracture toughness increases continually with increasing weight percent of the PSF modifier for both thermal and MR processed materials. Furthermore, the K_{IC} values for the MR and thermally processed systems were essentially identical, possibly indicating that fracture toughness was independent of processing conditions.

DSC results demonstrated that two distinct Tg's occur in this phase separated system. The glass transition temperatures in the thermally cured networks appeared at ~185°C and ~230°C, for the PSF and epoxy resin phases, respectively, thus indicating complete phase separation. In contrast, the MR processed system exhibited depressed Tg's at ~180°C and ~220°C suggesting some degree of phase mixing.

Thermoplastic Modified Epoxy Resin Networks of DGEBA/TGDDM: The PSF toughened epoxy resin networks of 50/50 mole % DGEBA/TGDDM, as in the case of the 100 mole % DGEBA series, cured via MR also demonstrated essentially complete cure (~99% conversion by DSC) after the 50-55 minute processing cycle. In contrast, however, networks incorporating 100 mole % TGDDM cured by MR were found to contain unreacted species, as evidenced by a minor exotherm appearing in the DSC trace. Fracture toughness values and glass transition temperatures for the microwave and thermally processed epoxy resin networks as a function of 0,15 and 30 weight percent incorporation of the PSF modifier are shown in Table II. As expected, these results demonstrate that fracture toughness increases for microwave and thermal processed samples as the thermoplastic modifier was increased from 0 to 30 weight percent. Unexpectedly though, the fracture toughness results were comparable to the DGEBA systems of lower crosslink densities. It is also of interest to note that the presence of unreacted endgroups in the 100 mole % TGDDM systems did not compromise the fracture toughness data obtained. As in the previous series, essentially identical K_{IC} values for the MR and thermally processed system seems to indicate that fracture toughness was independent of the processing technique utilized.

Table II. Fracture Toughness Properties and Glass Transition Temperatures of Polysulfone Modified Epoxy Resin Networks of DGEBA/TGDDM

	wt.% NH_2PSF*	DGEBA/TGDDM(mole%)	$\overline{K}_{IC} x10^6 (N/M^{3/2})$	Tg(°C)
Thermal Cures:	0	0/100	0.3	**
	15	100/0	0.7	184, 233
	15	80/20	0.6	**, 223
	15	60/40	0.5	**, 225
	15	40/60	0.6	**, 234
	15	20/80	0.7	186, 252
	15	0/100	0.6	185, **
	30	100/0	1.6	186, 227
	30	80/20	1.3	186, 221
	30	60/40	1.5	187, 223
	30	40/60	1.6	187, 233
	30	20/80	1.7	190, 250
	30	0/100	1.6	188, **
Microwave Cures:	0	0/100	0.3	179
	15	50/50	0.5	185, 235
	15	0/100	1.0	188, **
	30	50/50	1.7	187, 232
	30	0/100	1.5	187, **

* <Mn> = 14,700 gm/mole; ** Not Detected Via DSC

DSC results again demonstrate that two distinct Tg's occur in this phase separated system. The glass transition temperatures in the thermally cured networks appeared at ~187°C and, depending on the molar ratio of DGEBA to TGDDM, from ~221°C to 252°C for the PSF and epoxy resin phases, respectively. In the MR processed specimens, the glass transition temperatures of the cured networks appeared at ~187°C and ~233°C, respectively, for the PSF and epoxy resin phases. The MR cured materials of this series show undepressed Tg's, thereby indicating in both microwave and thermal processed materials, that a well defined phase separated morphology had been developed.

Fracture surface electron micrographs (SEM) of the thermoplastic toughened epoxy resin networks are shown in Figures 2-4 for the thermally processed systems, respectively. The thermally cured networks exhibited a significant change in morphology as the weight percent incorporation of PSF was varied from 10 to 40 in the DGEBA systems and from 15 to 30 in the DGEBA/TGDDM systems. A continuous epoxy matrix with phase separated spheres of PSF was observed at low weight percent incorporations of the thermoplastic modifier. As the incorporation of PSF increased, however, phase inversion resulted, producing a continuous matrix of PSF surrounding discrete spheres of epoxy resin.

Morphological differences were observed for the DGEBA and DGEBA/TGDDM incorporated systems (thermally cured) at low and high compositions of the thermoplastic modifier. Whereas the phase separated PSF spheres, at low weight percent incorporation of PSF, were of uniform size (~0.6 μm) and distribution in the systems composed of DGEBA (Figure 2), those materials that employed TGDDM as the epoxy resin consisted of PSF spheres which varied in size (~0.5-1.9μm) and distribution (Figure 3). Additionally, the phase inverted morphology, exhibited at the higher loadings of thermoplastic modifiers, evolved into an occluded-type phase inversion (Figure 4) as the levels of TGDDM were increased. Also it was observed that small pockets of phase inverted material developed at 15 weight percent incorporation of the thermoplastic modifier when 100 mole percent TGDDM was utilized.

The morphology of the microwave processed systems of DGEBA was less defined. No distinct point of phase inversion was observed from the SEM micrographs as shown in Figure 5. Furthermore, the phase separated spheres of PSF at low incorporation were much smaller (~0.1 μm) than those observed for the thermally cured systems (0.6 μm) at the same level of incorporation. These observations suggested that the microwave cured systems of DGEBA were incompletely phase separated and possibly possessed some degree of phase mixing. At the higher weight percent loadings of thermoplastic modifier, microwave cures of systems incorporating TGDDM, rather than DGEBA, had better defined morphologies as shown in Figure 6. This possibly indicates the increase in cure time allowed phase separation to occur more fully. As in the case of the thermally cured material, pockets of phase inverted material were present in the 100 mole percent TGDDM system with 15 weight percent of the polysulfone modifier. These pockets appeared to be larger and more numerous in the MR cured specimens and possibly lend an explanation to the high K_{IC} values obtained for this set of samples.

Thermosetting Bismaleimides (Matramid 5292 A/B): A comparison of the dynamic mechanical spectra of both the thermal and microwave cured BMI resins (Figure 7) illustrates an increase in storage modulus after the initial drop for the microwave cured specimen. The absence of this peak in the spectra for the thermally cured resin suggests that the microwave cured material had not reached the same level of crosslink density as the thermally cured material. Any differences in crosslink density would be difficult to measure using swelling experiments on such highly crosslinked material, however the

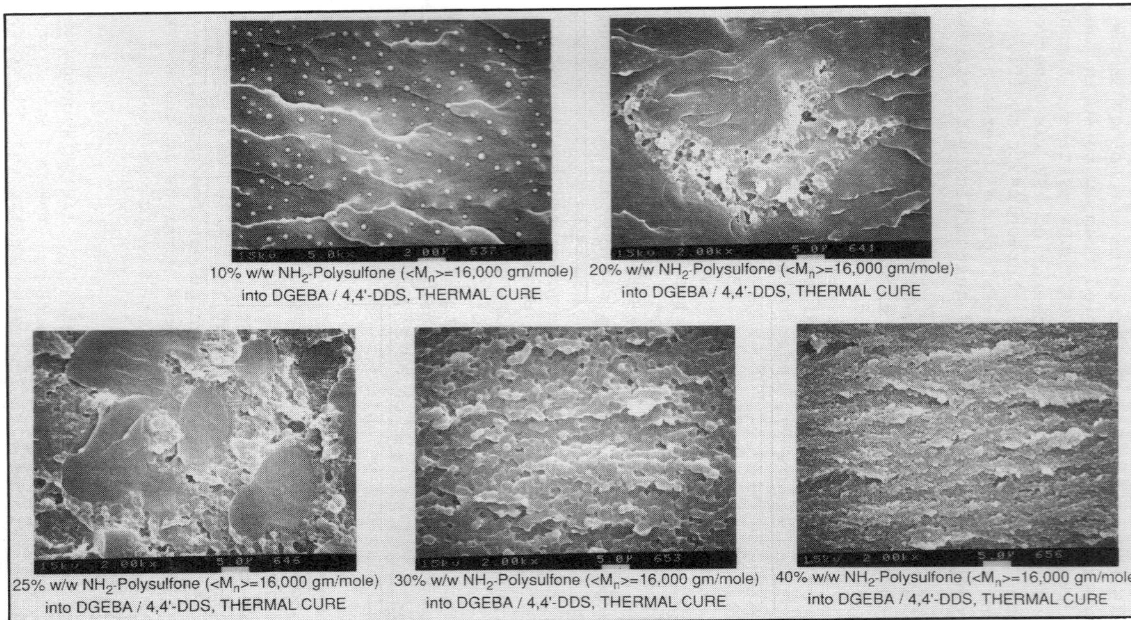

Figure 2. Fracture Surfaces of Thermally Cured Epoxy Resin (DGEBA)Networks.

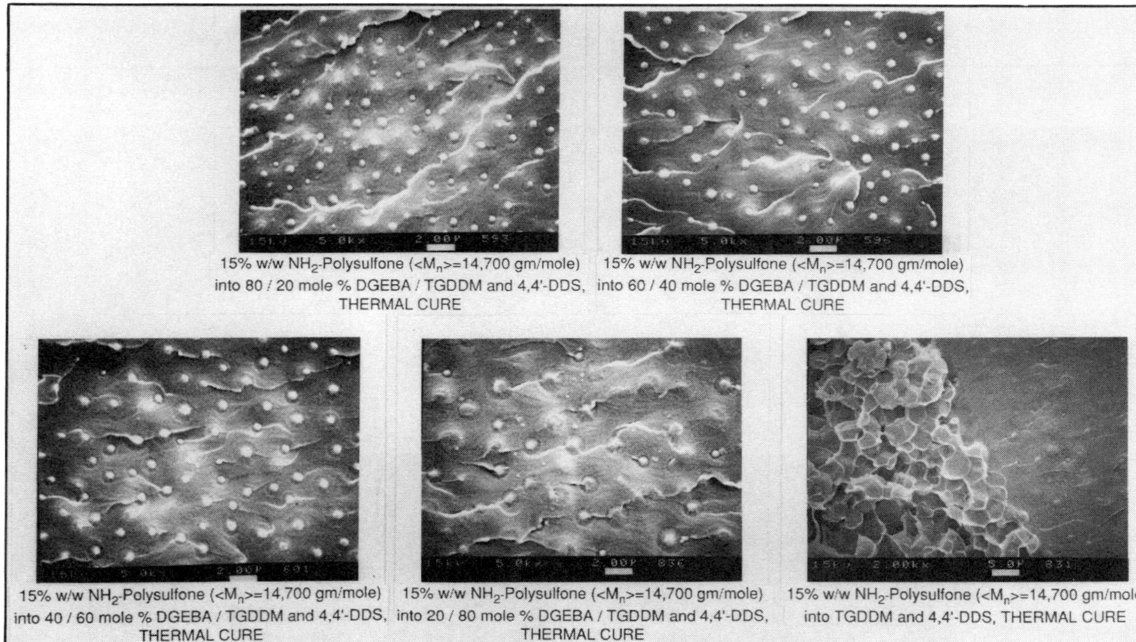

Figure 3. Fracture Surfaces of Thermally Cured Epoxy Resin (DGEBA/TGDDM) Networks with 15 Weight Percent Incorporation of Thermoplastic Modifier.

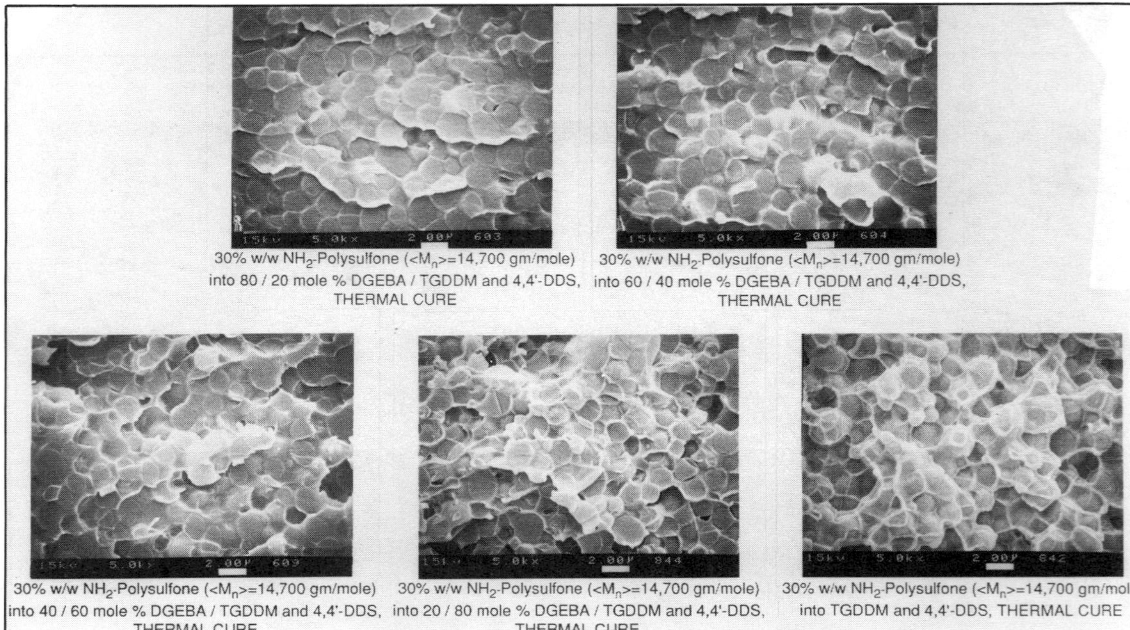

Figure 4. Fracture Surfaces of Thermally Cured Epoxy Resin (DGEBA/TGDDM) Networks with 30 Weight Percent Incorporation of Thermoplastic Modifier.

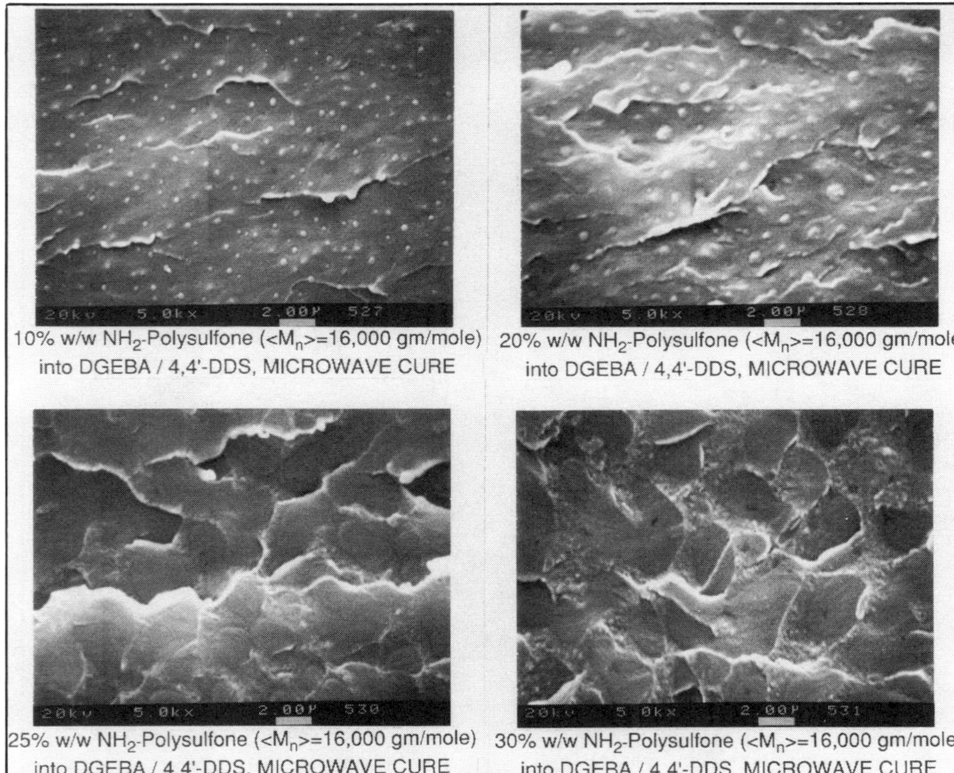

Figure 5. Fracture Surfaces of Microwave Cured Epoxy Resin (DGEBA) Networks.

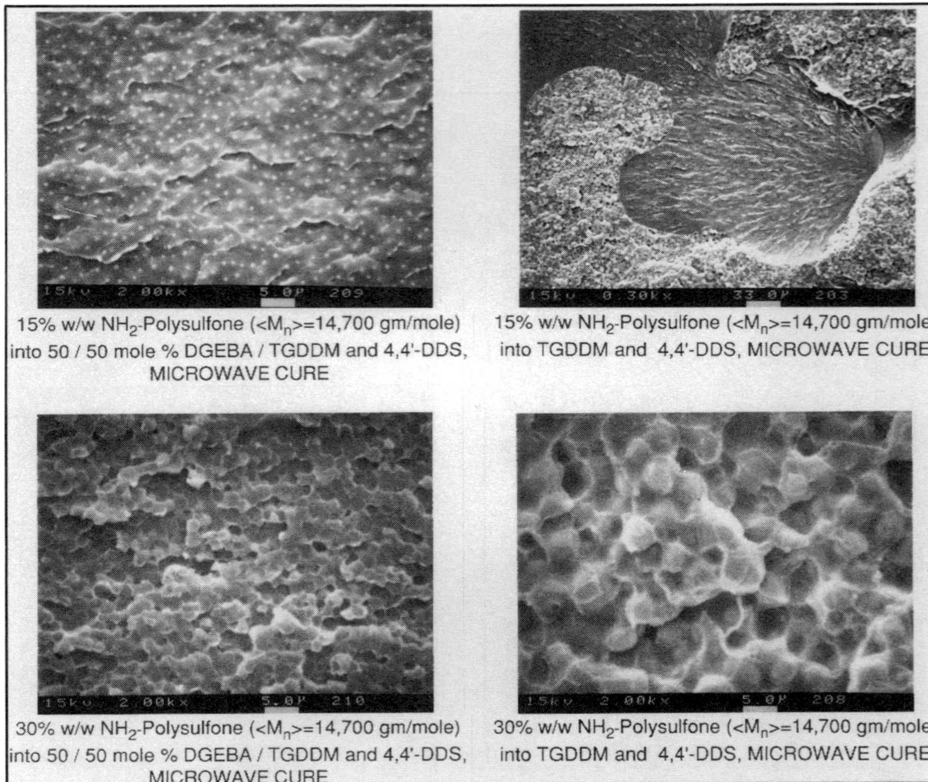

Figure 6. Fracture Surfaces of Microwave Cured Epoxy Resin (DGEBA/TGDDM) Networks with 15 and 30 Weight Percent Incorporation of Thermoplastic Modifier.

effect of the apparent difference in crosslink density on fracture toughness was studied. Table III indicates that within experimental error the fracture toughness was not affected.

Table III. Fracture Toughness Properties of Thermal and Microwave Processed Matramid 5292 A/B Systems

	$\overline{K}_{IC} x10^6 (N/M^{3/2})$
Thermal Cure	0.5
Microwave Cure	0.5

The interaction of microwaves with dipoles in crosslinking polymeric materials changes as their mobility is hindered by the growing network. Once the network has reached the point of gelation, chemical conversion proceeds through diffusion controlled reactions. The effect on the percent gelation due to isothermal cures by both microwave and thermal processes was studied using Soxhlet extraction techniques. Figure 8 illustrates the differences in the time taken to reach a particular level of gelation for the thermal and microwave processes. Such rapid improvements in reaching higher levels of gelation in shorter cure times was first reported by Hedrick et al.(28) where epoxy resins were used to form the thermosetting network.

Nonreactive Thermoplastics: Nonreactive poly(arylene ether ketone) or polyimide thermoplastics such as PEEK and Ultem were processed very effectively using microwave radiation as shown in Figures 9 and 10, respectively. Rapid temperature equilibrium was reached in short processing times due to the coupling of the electromagnetic radiation with the permanent dipole moments within the backbone of these systems. Rapid volumetric heating of this type can be very advantageous in future investigations on MR processing of thermoplastic composites.

CONCLUSIONS

Microwave radiation can be utilized to process both reactive (thermosetting) and nonreactive (thermoplastic) polymeric materials very effectively. Accelerated MR processing of thermoplastic modified epoxy resin networks was achieved with the retention of good mechanical properties. A greater reduction in processing time could be realized for the less reactive DGEBA containing systems than for the highly MR absorbing samples incorporating TGDDM.

In contrast to those samples cured via conventional thermal processing, networks that underwent electromagnetic processing were shown to contain novel morphologies. Systems which consisted solely of DGEBA as the epoxy resin were found to have a poorly defined phase inverted morphology, perhaps due to incomplete phase separation. In

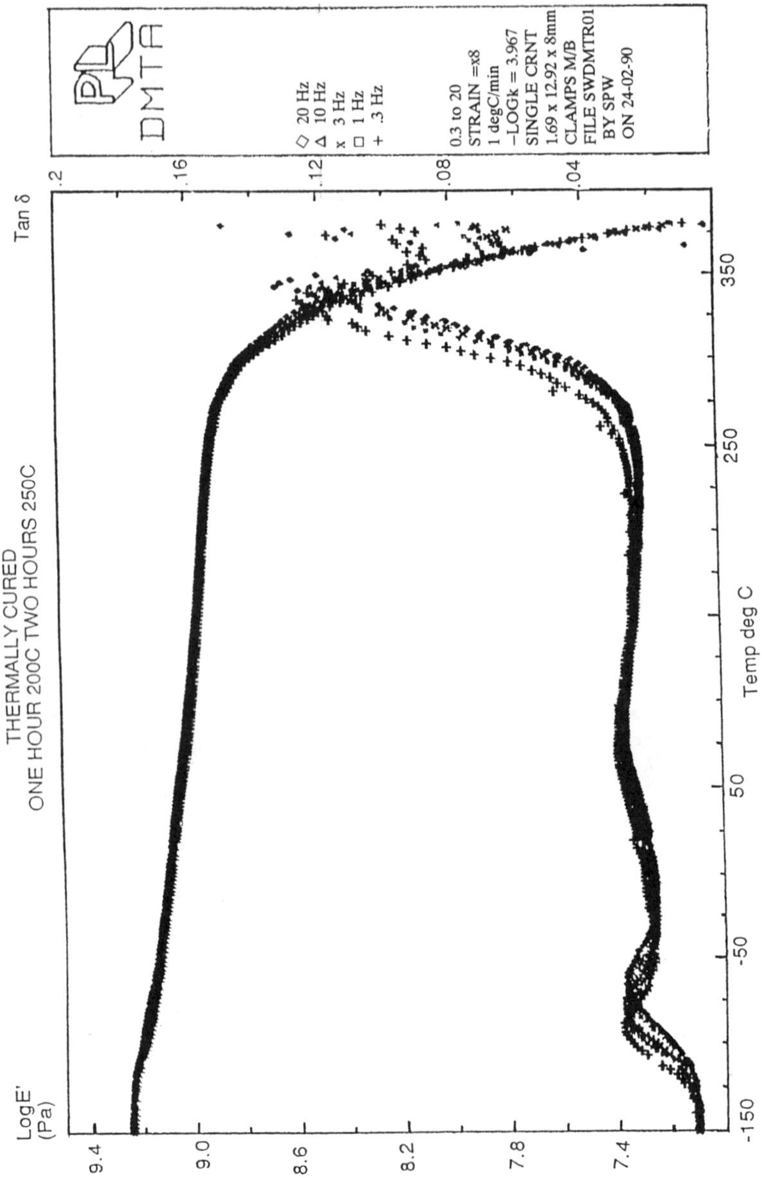
THERMALLY CURED
ONE HOUR 200C TWO HOURS 250C
LogE'
(Pa)
9.4
9.0
8.6
8.2
7.8
7.4
Tan δ
.2
.16
.12
.08
.04
-150
-50
50
250
350
Temp deg C
DMTA
◇ 20 Hz
Δ 10 Hz
x 3 Hz
□ 1 Hz
+ .3 Hz
0.3 to 20
STRAIN =x8
1 degC/min
-LOGk = 3.967
SINGLE CRNT
1.69 x 12.92 x 8mm
CLAMPS M/B
FILE SWDMTR01
BY SPW
ON 24-02-90

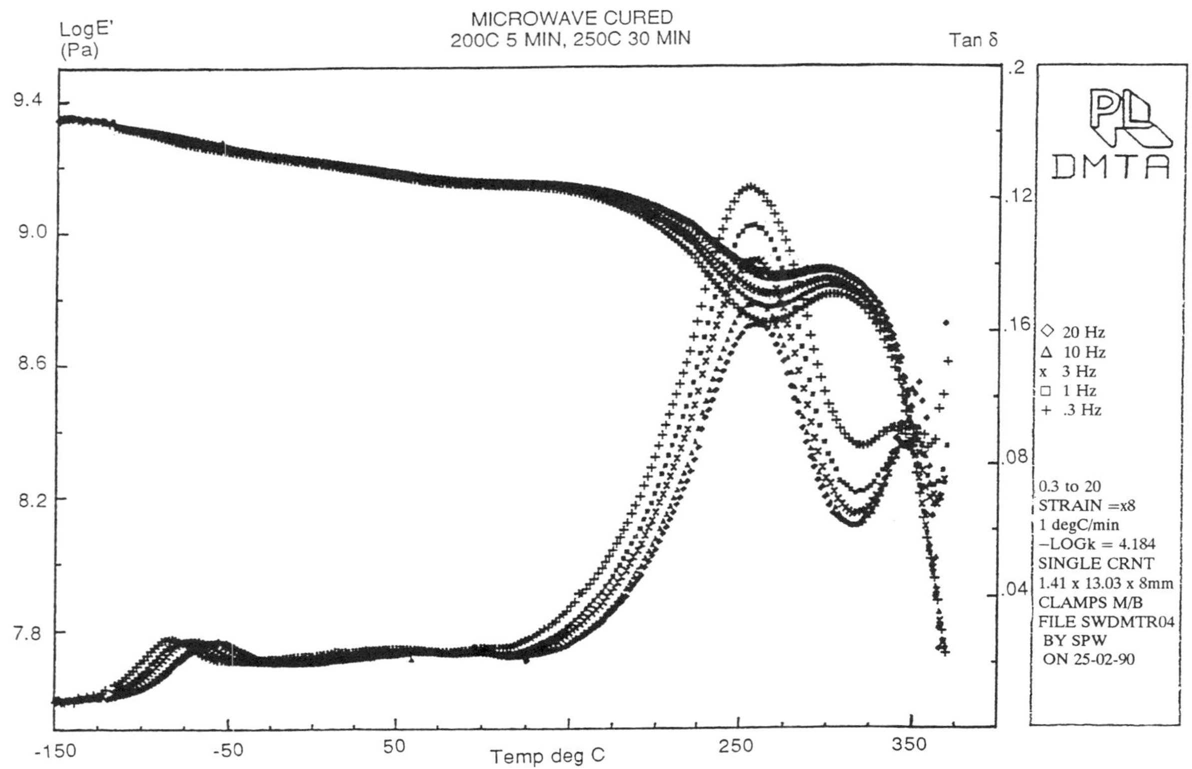

Figure 7. Dynamic Mechanical Thermal Analysis Spectra of Thermal and Microwave Cured Matramid 5292 A/B Thermosetting System.

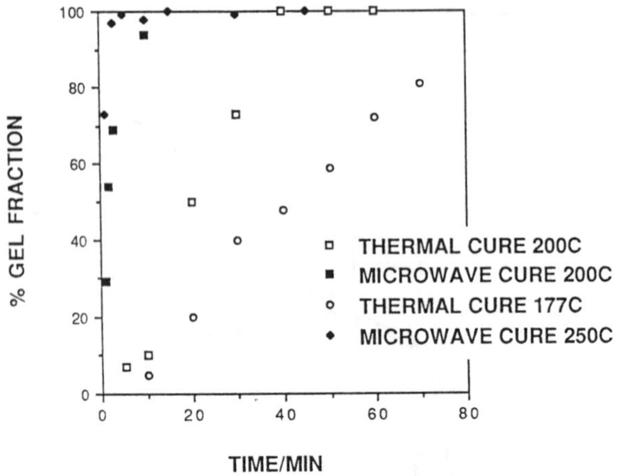

Figure 8. Time Required to Reach the Gel Point for Thermal and Microwave Cured Matramid 5292 A/B Systems.

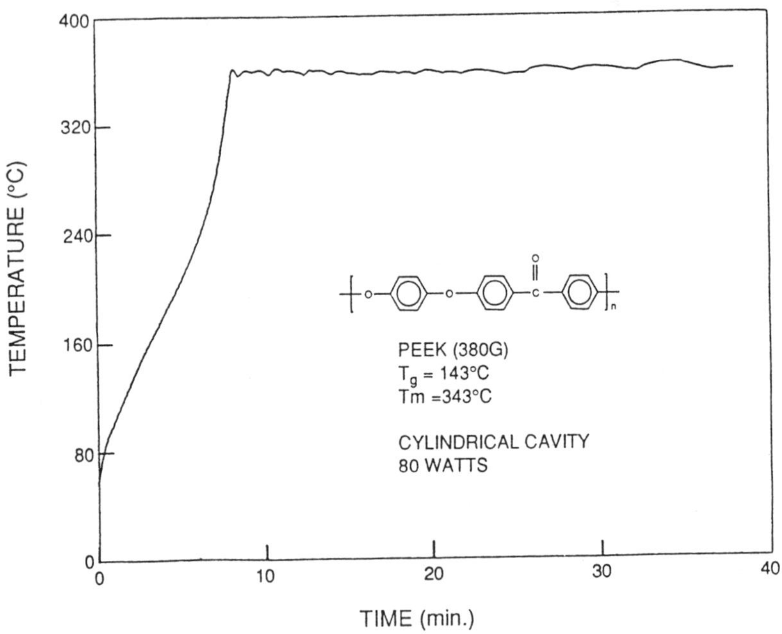

Figure 9. Microwave Processing of PEEK.

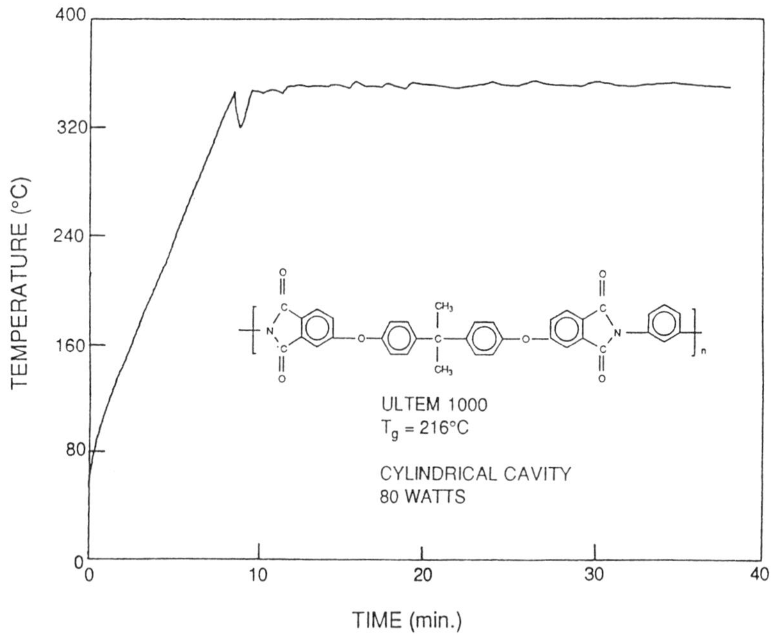

Figure 10. Microwave Processing of Ultem.

contrast, the systems incorporating TGDDM were shown to have a well developed phase inverted morphology, in this case, possibly due to the higher functionality of the epoxy resin, which would result in gel points at lower extents of reaction. At 15 weight percent loading of the thermoplastic modifier an unexpectedly high value of fracture toughness was obtained, again perhaps relating to the higher functionality of the TGDDM resin and the resultant larger pockets of phase inverted material being present in these specimens. The exact nature of this observation is the subject of a continuing study in our laboratories.

Initial results on the MR curing of thermosetting bismaleimides suggest further optimization of the processing parameters is needed to achieve full cure. The Matramid 5292 A/B system is also a prime candidate for thermoplastic toughening via amino or maleimide terminated poly(arylene ether sulfones) or poly(arylene ether ketones) and their subsequent cure by thermal or electromagnetic means.

Acknowledgments

The authors wish to acknowledge the support of this research by DARPA and administered through the AFML under contract #F33615-85-C-5153. The research samples of DER 332 from Dow Chemical Company and Matramid 5292 A/B resin obtained from Ciba-Geigy are also gratefully acknowledged. Utilization of the NSF S & TC facilities are also appreciated.

LITERATURE CITED

1. Osepchuk, J. M. in Encyclopedia of Chemical Technology, 3rd Edition; Grayson, M., Ed.; John Wiley and Sons: New York, NY, 1981, Vol. 15; pp. 494-522.
2. Matsuoka, S. in Encyclopedia of Polymer Science and Engineering; Kroschwitz, J. I., Ed.; John Wiley and Sons: New York, NY, 1987, Vol. 5; pp. 23-36.
3. Lewis, D. A.; Hedrick, J. C.; McGrath, J. E.; Ward, T. C. Polymer Preprints **1987**, 28(2), 330.
4. Hedrick, J. C.; Lewis, D. A.; Lyle, G. D.; Wu, S. D.; Ward, T. C.; McGrath, J. E. Polym. Mater.: Sci. and Eng. Preprints **1987**, 60, 438.
5. Lewis, D. A.; Hedrick, J. C.; Lyle, G. D.; Ward, T. C.; McGrath, J. E. in Microwave Processing of Materials; Sutton, W. H.; Brooks, M. H.; Chabinsky, I. I., Eds.; Materials Research Society Symposium Proceedings; Publishers Choice Book Mfg. Co.: Mars, PA, 1988, Vol. 124; pp. 181-188.
6. Rodrigues, D. E.; Wilkes, G. L. Polymer Preprints **1989**, 30(2), 227.
7. Asmussen, J.; Lin, H. H.; Manring, B.; Fritz, R. Rev. Sci. Instrum. **1987**, 58(8), 1477.
8. Chen, M.; McGrath, J. E.; Ward, T. C. Polym. Mater.: Sci. and Eng. Preprints **1989**, 60, 443.
9. Jow, J. Ph.D. Thesis, Michigan State University, East Lansing, MI, 1988.

10. Manring, E. B. M.Sc. Thesis, Michigan State University, East Lansing, MI, 1988.
11. Chen, M. Ph.D. Thesis, Viginia Polytechnic Institute and State University, Blacksburg, VA, 1989.
12. LeVan, Q.; Gourdenne, A. Eur. Polym. J. **1987**, 23(10), 777.
13. Beldjoudi, N.; Bouazizi, A.; Douibi, D.; Gourdenne, A. Eur. Polym. J. **1988**, 24(1), 49.
14. Beldjoudi, N.; Gourdenne, A. Eur. Polym. J. **1988**, 24(1), 53.
15. Beldjoudi, N.; Gourdenne, A. Eur. Polym. J. **1988**, 24(3), 265.
16. Jow, J.; Hawley, M. C.; Finzel, M.; Asmussen, J.; Lin, H. H.; Manring, B. IEEE Trans. Microwave Theory Tech. **1987**, MTT-35(12), 1435.
17. Singer, S. M.; Jow, J.; DeLong, J. D.; Hawley, M. C. SAMPE Quarterly **1989**, 20(2), 14.
18. Jow, J.; DeLong, J. D.; Hawley, M. C. SAMPE Quarterly **1989**, 20(2), 46.
19. Silinski, B.; Kuzmycz, C.; Gourdenne, A. Eur. Polym. J. **1987**, 23(4), 273.
20. Jullien, H.; Valot, H. Polymer **1985**, 26, 506.
21. Teffal, M.; Gourdenne, A. Eur. Polym. J. **1983**, 19(6), 543.
22. Metaxas, A. C.; Meridith, R. J. Industrial Microwave Heating; Peter Peregrinus Ltd.: London, 1983.
23. Hedrick, J. C.; Lewis, D. A.; Ward, T. C.; McGrath, J. E. Polymer Preprints **1988**, 29(1), 363.
24. Liptak, S. C.; Wilkinson, S. P.; Hedrick, J. C.; Joseph, W. D.; Ward, T. C.; McGrath, J. E. Polymer Preprints **1990**, 31(2), 320.
25. Jow, J.; Hawley, M. C.; DeLong, J. D. Proceedings of the American Society for Composites; Technomic: Lancaster, PA, 1988; 4th Technical Conference, pp. 305-312.
26. Hedrick, J. L.; Yilgor, I.; Wilkes, G. L.; McGrath, J. E. Polymer Bulletin **1985**, 13, 201.
27. Jurek, M. J.; McGrath, J. E. Polymer **1989**, 30, 1552.
28. Hedrick, J. C.; Lewis, D. A., Lyle, G. D.; Wu, S. D.; Ward, T. C.; McGrath, J. E. Polym. Mater.: Sci. and Eng. Preprints **1989**, 60, 438.

RECEIVED June 7, 1991

Chapter 23

Dielectric Properties at Microwave Frequencies of Epoxy Undergoing Cure

Eva Marand, Kenneth R. Baker, and Jack D. Graybeal

Department of Chemistry, Virginia Polytechnic Institute and State University, Blacksburg, VA 26061

It is well known that the kinetics and the reaction mechanism of epoxy curing determine the final network morphology and ultimately the physical properties of the epoxy resin. In recent years, the introduction of microwave processing of thermosetting materials has prompted the need for comprehensive microwave dielectric data, necessary to properly interpret the cure mechanism induced by electromagnetic radiation. In situ measurements at 2.45 GHz of the complex dielectric constant of an epoxy formulation undergoing thermal cure have been compared with theoretical predictions, based on changes in the dipole concentration and the changes in the relaxation time of the epoxy matrix with extent of cure. It has been shown that the rate of decrease of the dielectric constant primarily reflects the rate of crosslinking within the resin.

Measurements of dielectric properties have been used to monitor chemical reactions for more than sixty years (1). Typically, most of these studies were and still are carried out in the conveniently accessible audio frequency range below 100 MHz, where most polymers and polymer composites exhibit relaxation dispersions at room temperature. Very few investigations have actually focused on the microwave dielectric properties of polymeric materials (2, 3). In this frequency region, the primary molecular response to the applied electric field is that of orientational polarization. Even though some polymers exhibit relaxation dispersions at high frequencies, for the most part, the losses observed often have their origins outside the microwave region.

In the past several years, however, there has been a growing demand for good microwave dielectric data, sparked by new technological developments. Fortunately, advances in instrumentation have made such measurements more accurate and more convenient. There are also certain advantages in using higher frequency dielectric measurements to study chemical reactions. At lower frequencies, below 100 MHz, the interpretation of dielectric data in terms of the true physical properties of the reacting system is hampered by interfacial and electrode polarization. At microwave frequencies, where one employs waveguides, such effects are eliminated and hence the dielectric data truly reflects the chemical and morphological state of the system.

Our study focuses primarily on the measurement of the dielectric properties at microwave frequencies of epoxies under going thermal cure. The mechanism of epoxy curing, which is known to significantly affect the properties of the forming glassy network, has long been the subject of detailed investigations (1, 4, 5). In recent years,

0097–6156/91/0475–0384$06.00/0

research efforts have been initiated in the area of microwave processing of epoxy materials (6-10). Although microwave radiation has been shown to significantly accelerate the curing of epoxy resins, there remains no systematic and comprehensive examination of the chemistry and mechanism of microwave curing. It is not known, for example, whether the preferential orientation of the dipoles in the microwave field, as opposed to random orientation in the thermal field, could alter the rates of reactions in the curing process, leading to a different network morphology and properties. The radiative power absorbed by the material, causing a temperature rise, depends on the dielectric properties of the material. Hence, comprehensive microwave dielectric data is necessary in order to properly interpret the cure mechanism induced by the electromagnetic radiation.

In addition to establishing a data base for future microwave curing studies, our goal is to understand the changes, invoked in the dielectric properties by curing, from a fundamental point of view. We need to answer questions such as; To what extent do the dielectric properties reflect the changing chemical compositions of the sample, i.e. changes in polar group concentrations? and also; How much influence does the transforming polymer morphology have on the final dielectric properties?

Methods

The dielectric measurements have been performed at 2.45 GHz using an HP8510 Network Analyzer. A specialized waveguide cavity, resonant in the TE $_{103}$ mode, has been constructed in order to perform these measurements and to permit temperature control of the sample up to 250°C. Figure 1 shows a cross-sectional view of the rectangular waveguide cavity and the placement of the sample. The sample is heated by hot N_2 gas, whose temperature is controlled by the N_2 flow rate and the settings on a fluidized hot bath through which the gas is circulated. The nitrogen gas is

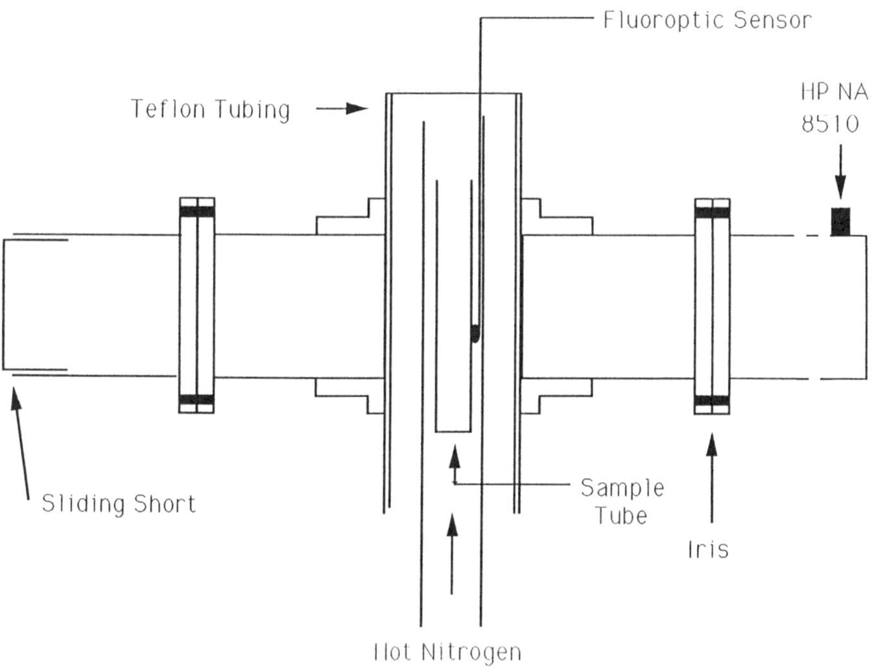

Figure 1. Cross-sectional View of the Rectangular Waveguide and Placement of the Sample

passed around the sample twice; first by the inner wall, then by the outer wall of the Teflon @ tubing, in order to minimize temperature gradients along the sample. An example of the sample temperature fluctuations as a function of time is shown in figure 2. The measurements are made using a cavity perturbation technique (11, 12). The cavity perturbation technique assume that the introduction of the sample into a resonant cavity causes only small perturbation in the fields inside. As a result, the solution for the fields inside the cavity when it contains the sample is approximated from the solution of the electromagnetic fields in the empty cavity. It is the nature of the particular electromagnetic fields that are able to exist in the cavity that determines the resonant frequency and the quality factor, Q. Initially, the resonant frequency and the Q of the empty cavity are measured. Empty, in this case means without the sample; all other components, such as the sample holder, temperature probe and teflon tubing which conducts the hot N_2 are in place and are assumed to contribute constant value to the measurements as the temperature is varied. After the sample is introduced, the changes in the resonant frequency and the Q yield the dielectric constant and the dielectric loss, respectively. The following equations are used to relate ε' and ε" to the cavity parameters ω_0, ω_s, Q_0 and Q_s.

$$\varepsilon'_s = 1 + \frac{1}{2}\frac{(\omega_o - \omega_s)}{\omega_o}\frac{V_{cav}}{V_{sam}} \qquad (1)$$

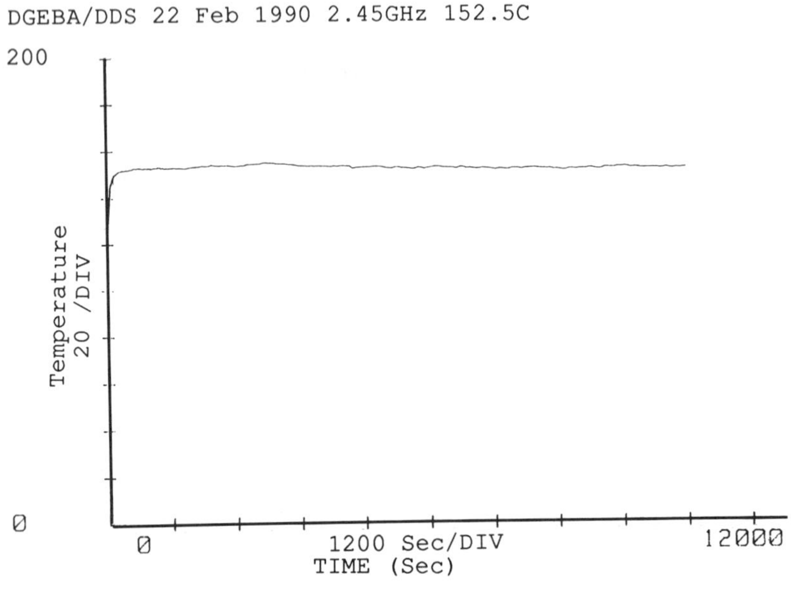

Figure 2. An example of sample temperature fluctuations as a function of time.

$$\varepsilon''_s = \frac{1}{4}\left(\frac{1}{Q_o} - \frac{1}{Q_s}\right)\frac{V_{cav}}{V_{sam}} \qquad (2)$$

where:

ε'_s = real part of the relative dielectric constant of the sample
ε''_s = imaginary part of the relative dielectric constant of the sample
ω_0 = resonant frequency of the empty cavity
ω_s = resonant frequency of the perturbed cavity
Q_0 = Q of the empty cavity
Q_s = Q of the perturbed cavity
V_{cav} = volume of the cavity
V_{sam} = volume of the sample

Both ε' and ε" are reported in units of permittivity of free space, ε_o, which has the value 8.85×10^{-14} Farads/cm. All measurements and calculations are computer controlled. The accuracy of the measurements was verified by comparison of the measured dielectric properties of Teflon @ at room temperature with literature values. The reported values of the dielectric constant and dielectric loss factor for Teflon@ at 1×10^8 Hz are 2.1 (13) and 4×10^{-4} (14), respectively. Our measurements at 2.45 GHz were $1.96 \pm .05$ and $2 \times 10^{-4} \pm 1 \times 10^{-4}$, respectively.

The epoxy system consisting of diglycidylether of bisphenol -A (DGEBA) and 4, 4-diphenyldiamine sulfone (DDS) was investigated. The DGEBA (DOW Chemical) was recrystallized and dried for two days under vacuum at room temperature to remove water and remaining volatile solvents. The 4, 4-DDS (Aldrich) was recrystallized in $MeOH/H_2O$ and dried under vacuum for two days at room temperature. The 4, 4-DDS was added to degassed DGEBA at 120°C in stoichiometric ratio (1:2) and allowed to dissolve. The epoxy system was then transfered by a preheated pipette into a preheated NMR tube.

Infrared curing studies were conducted on a Nicolet 5 SXB FTIR spectrometer using a SpectraTech High Temperature Cell. Changes in the characteristic epoxide absorption peak at 915 cm^{-1} were measured, corrected for thickness fluctuations using a reference band at 1507 cm^{-1}, and normalized. Although the infrared measurements were not conducted simultaneously with the dielectric measurements, they were, nevertheless, carried out under the same conditions.

Results and Discussion

There are a number of different reactions which may occur during a curing process illustrated in figure 3. These are drawn schematically below.

1) 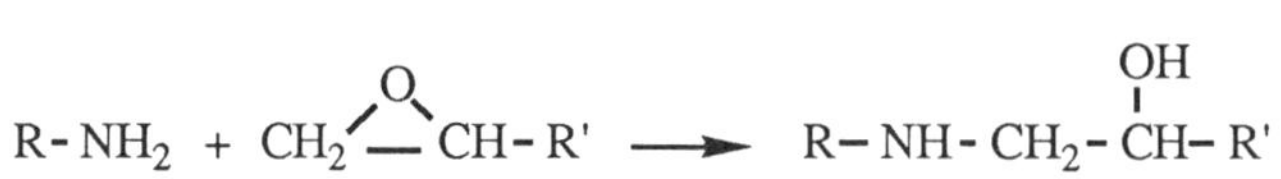

2) 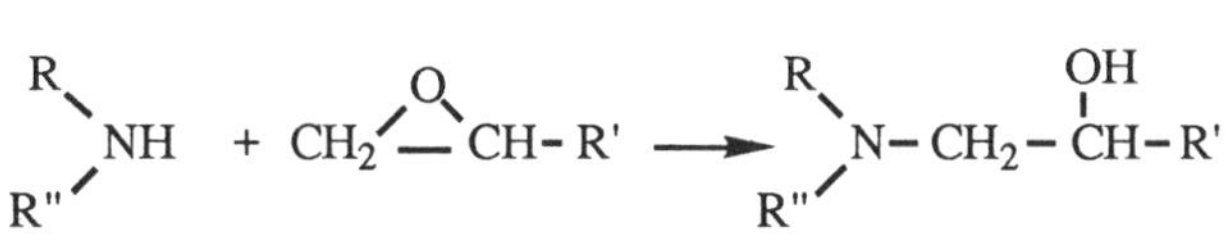

3) 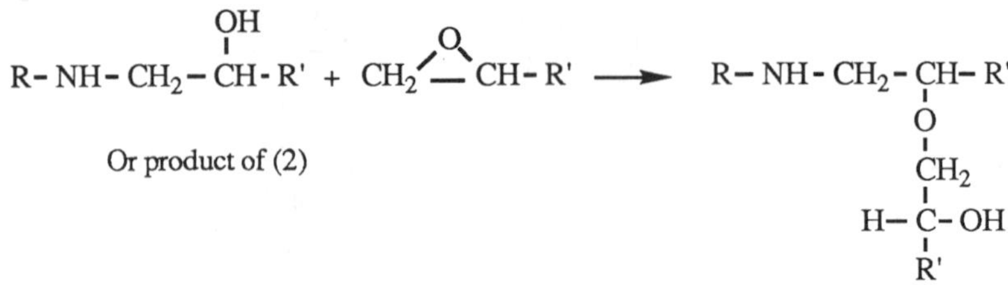

The epoxy moieties of the DGEBA can react either with the primary or the secondary amine to form an OH adduct, which can later react with another epoxide ring in an etherification reaction to cross-link the resin. Here, we have assumed that the homopolymerization of the epoxy is negligible (1). The results of FTIR experiments are shown in figure 4. Examination of the epoxide fraction profile for various temperatures shows that there are three regions present. The first region is marked by the adduct reactions which occur prior to the etherification reaction and is characterized by a slow rate of epoxide depletion and exhaustion of the N-H groups. An example of the change in the N-H concentration as measured by the area of peak 3234 cm^{-1} is shown in figure 5. The second region corresponds to the crosslinking reaction which is marked by a rapid decrease in the epoxide concentration. The third region occurs when the sample vitrifies and the reaction of the epoxide groups becomes diffusion limited. As the temperature is increased, the rates of these reactions increase, giving rise to higher slopes. However even at higher temperatures, one can still identify the three regions. Similar observations have been made in DGEBA/Imidazole curing studies (15).

The dielectric constant and the dielectric loss factor as a function of curing temperature and time are shown in figures 6 and 7. These two properties reflect the ability of the molecular dipole moments in the material to follow the oscillations of an applied electric field. While the dielectric constant represents the component of immediate alignment of the dipole moments in the direction of the electric field, the dielectric loss factor reflects the retardation incured in the rearrangement. Hence, the changes in the dielectric properties should reflect not only the alterations in molecular dipole moments as a result of chemical reactions taking place, but also the rapid decrease in molecular mobility that occurs during the formation of a cross-linked network.

Initially, the dielectric properties of the epoxy system increase as the sample temperature reaches that of the pre-heated sample holder, then decrease with the onset of curing. The magnitudes of the dielectric properties, particularly at the beginning of the curing process, are larger at higher curing temperature. This effect is diminished as the reaction proceeds, although the influence of temperature is still observed even in fully cured samples. This effect is illustrated in figure 8 for a sample cured at T=180°C for three hours.

Figures 9 and 10 show the dielectric properties of the pure components, DGEBA and DDS, as a function of temperature. Comparison of the dielectric properties of the epoxy undergoing cure with the dielectric properties of the pure components at corresponding temperatures suggests that the diamine contributes substantially to the magnitude of ε' and ε'' of the epoxy system. This is in agreement with previous kinetic studies which have calculated the dipole moments of the reacting polar groups as 7.6×10^{-30} C-m for the epoxide, 14.8×10^{-30} C-m for the primary amine and 12.6×10^{-30} C-m for the reacted amine (1). The dipole moment of the hydroxyl group may be slightly higher than that of the epoxy ring (16).

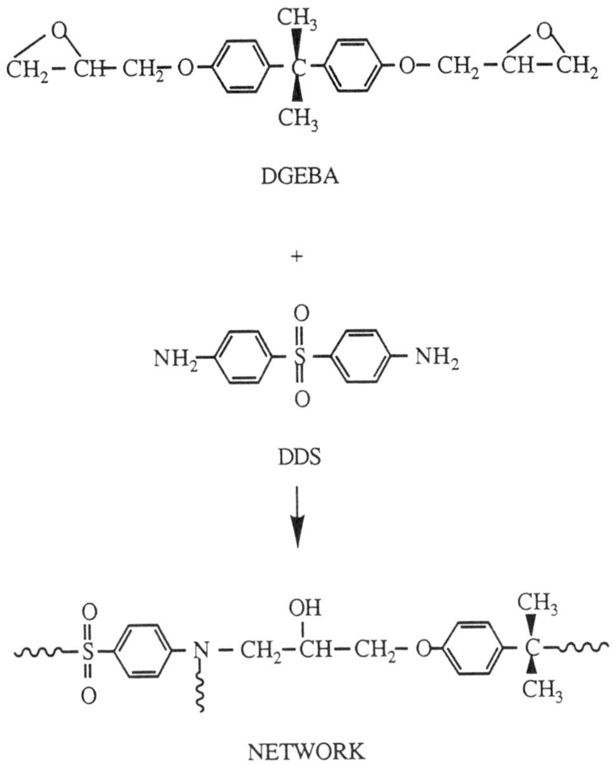

Figure 3. DGEBA/DDS Curing Process

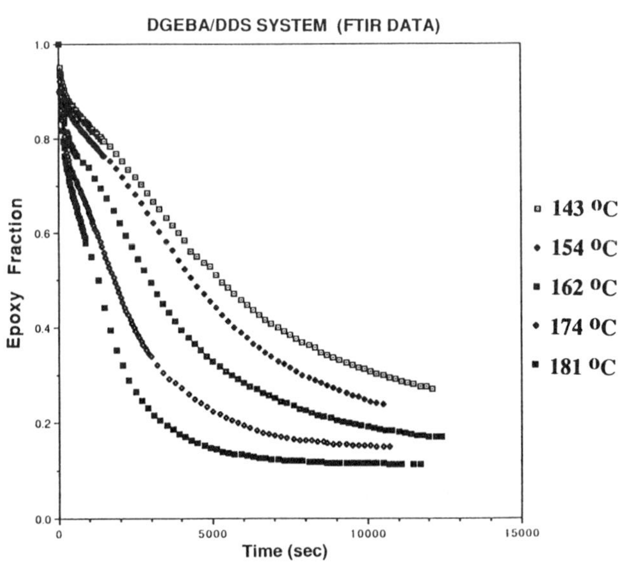

Figure 4. Fraction of Epoxy Groups as a Function of Time and Temperature (FTIR Data)

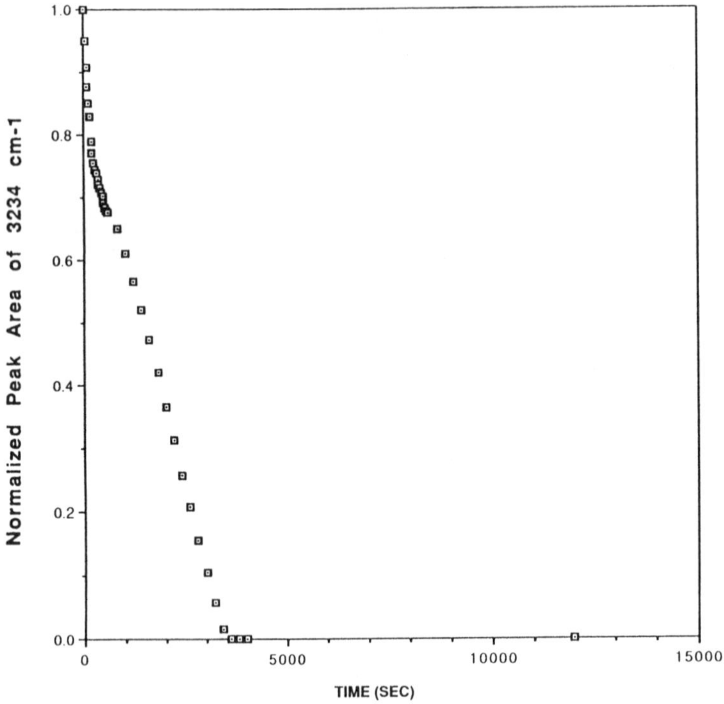

Figure 5. Decrease in the N-H Group Concentration as a Function of Time at 162°C (FTIR Data)

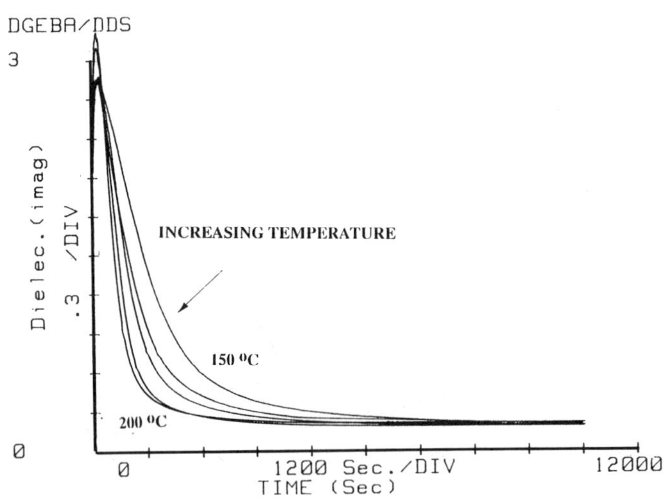

Figure 6. The Dielectric Constant as a Function of Time and Curing Temperature

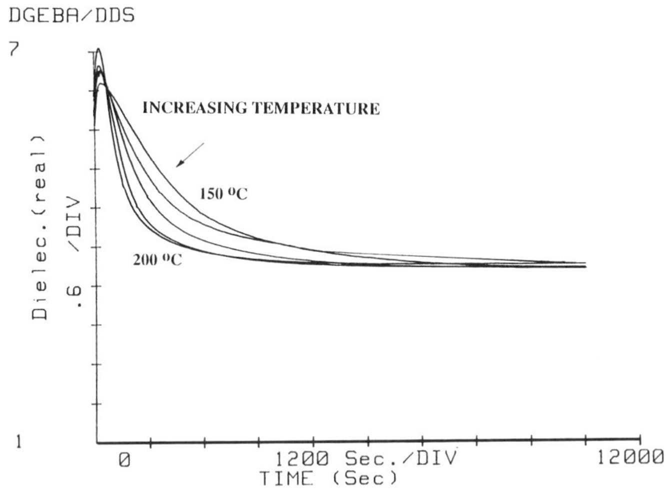

Figure 7. The Dielectric Loss Factor as a Function of time and Curing Temperature

Figures 11 and 12 show the relationship between the epoxy fraction and the dielectric constant and the dielectric loss factor, respectively. These relationships were obtained by cross-correlating the FTIR transmission data and the dielectric data using time as a common denominator. Both curves are characterized by an initial rapid decrease of the dielectric properties with a slow change in dielectric properties above 59% extent of reaction (extent of reaction = 1 - epoxy fraction). We wish to explain this behavior based on theoretical considerations. Unfortunately, the quantitative theories that relate the relaxed dielectric constant to the concentrations and dipole moments of the various polar end groups are insufficient to deal with polymers, where intermolecular and intramolecular forces, as well as entanglements play a role (17, 18). However, we deem it reasonable to assume that the dielectric properties will scale with $\Sigma N_i (\mu_i)^2$, where N_i and μ_i are respectively, the number of polar groups of particular species i and their respective dipole moments. Based on our previous consideration of the reaction chemistry, the quantity $\Sigma N_i (\mu_i)^2$ must vary with the extent of reaction, α, as

$$\Sigma N_i (\mu_i)^2 = A' (1 - \alpha) + B' \alpha + C \tag{3}$$

Where the A' constant represents primarily the dipole moment contribution of the epoxy groups and primary amine groups, as well as some contribution from secondary amine groups and hydroxyl groups. The B' constant represents primarily the contributions of the hydroxyl groups and the secondary and tertiary amine groups. Based on our previous discussion concerning the relative magnitude of the various dipole moments, we can safely assume that A' > B'. In this case, equation 3 can be reduced further to

$$\Sigma N_i (\mu_i)^2 = N = N_o - k\alpha \tag{4}$$

Where N_o and k are constants having units of $(C\text{-}m)^2$. Hence, the dielectric properties should decrease linearly with the extent of reaction if ε' and ε" depend primarily on the concentration of the polar end groups. The fact that this is not the case suggests that

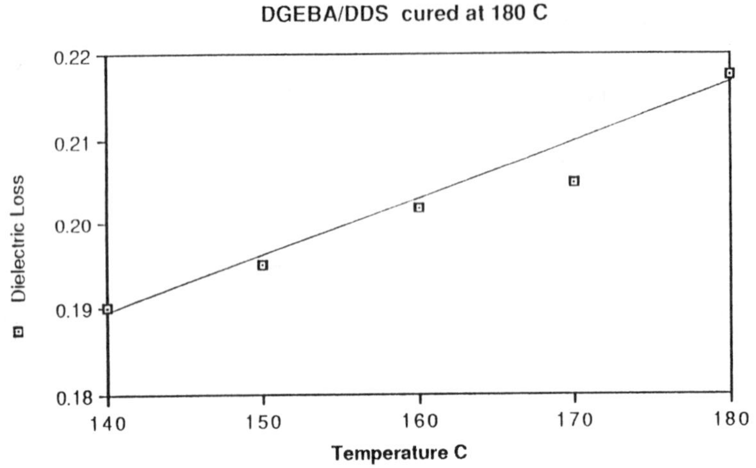

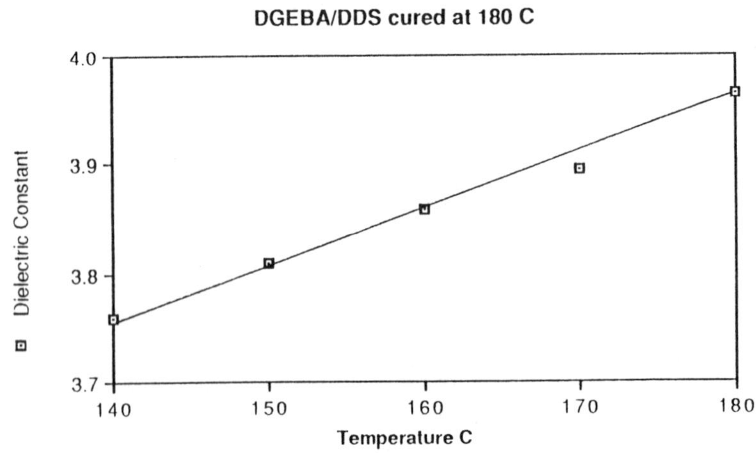

Figure 8. The Temperature Dependance of the Dielectric Properties of a fully Cured Sample (T=180°C for three hours)

the dielectric properties must also reflect the changes in the relaxation time, τ, which will increase with the extent of reaction as the cross-linking restricts the polymer matrix. A typical equation used to establish the relationship between the dielectric properties and the relaxation time is the Debye equation (eg. 5 and 6) (17, 18).

$$\varepsilon'(\omega) = \varepsilon_\infty + \frac{\varepsilon_s - \varepsilon_\infty}{1+\omega^2\tau^2} \tag{5}$$

$$\varepsilon''(\omega) = \frac{\varepsilon_s - \varepsilon_\infty}{1+\omega^2\tau^2}\,\omega\tau \tag{6}$$

where ε_s = static dielectric constant

ε_∞ = instantaneous dielectric constant

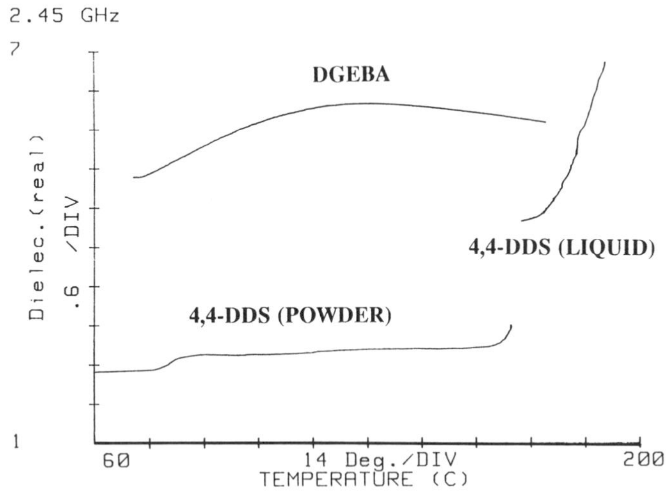

Figure 9. The Dielectric Constant of the Pure Components as a Function of Temperature

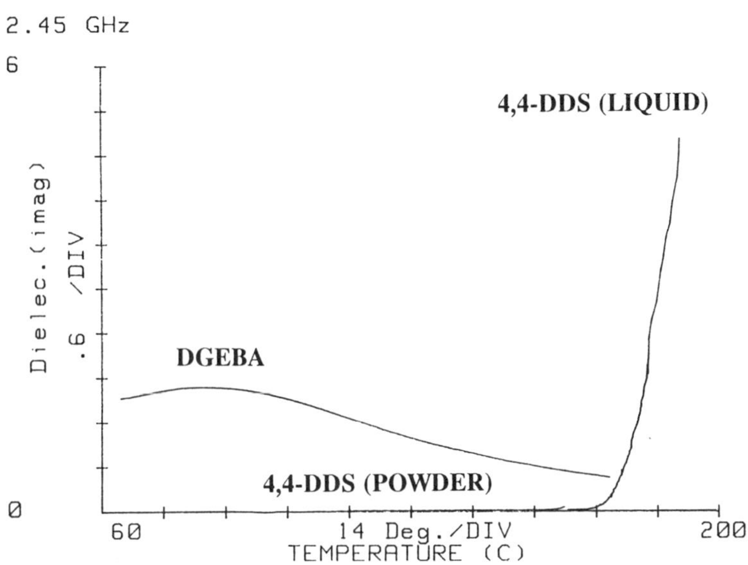

Figure 10. The Dielectric Loss Factor of the Pure Components as a Function of Temperature

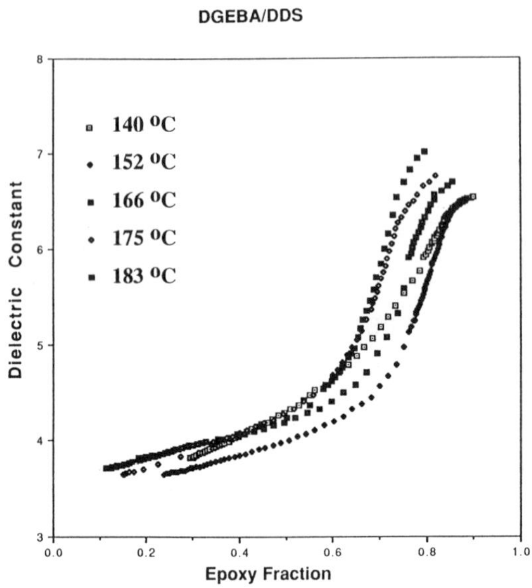

Figure 11. Relationship between the Dielectric Constant and the Epoxy Fraction

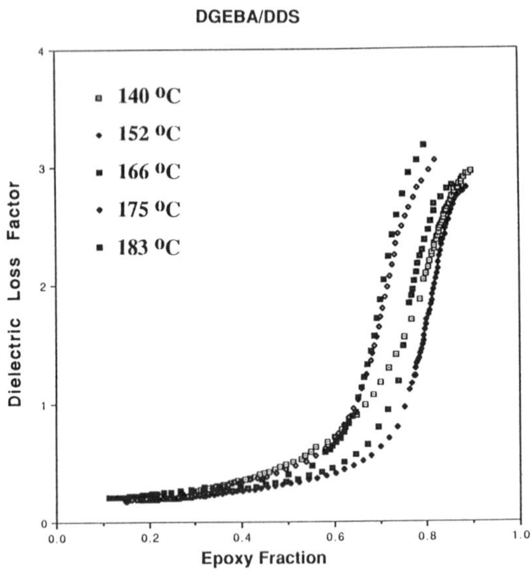

Figure 12. Relationship between the Dielectric Loss and the Epoxy Fraction

Because, in a curing reaction, the relaxation time will vary from about 1 x 10^{-8} sec to 1 hour over the course of an isothermal curing experiment (5), the contribution of the $(1/1 + \omega^2 \tau^2)$ term will diminish with time. Therefore, at long times, ε' will approach ε_∞ and ε" will approach zero. This is what we observe experimentally.

Because omega is large, $\omega = 15.4 \times 10^9$ sec $^{-1}$, we assume that $\omega^2\tau^2 > 1$ and we let

$$\varepsilon' = N\left(a_o + \frac{a_1}{\tau_2}\right) \tag{7}$$

$$\varepsilon'' = a_2\left(\frac{N}{t}\right) \tag{8}$$

where a_o, a_1 and a_2 are constants.

Because the relaxation time will change with the extent of reaction, α, a model for the curing reaction of epoxy resins developed by S. Matsuoka et. al (5) has been used in our analysis;

$$\tau = D\exp\left[\frac{C(T)}{1-(A + B\,\alpha)}\right] \tag{9}$$

This model is derived from the Adam and Gibbs (19) formula for relaxation phenomena in the equilibrium state. The relationship considers cooperative motions of molecular segments in domains of varying sizes, where the total configurational entropy, S, is the factor that determines the critical size for such domains,

$$\ln \tau = C + \frac{\Delta \mu\, s^*}{k\,T\,S} \tag{10}$$

Here, Δμ is the potential energy hindering the cooperative rearrangement of a smallest possible molecular unit, having itself a configurational entropy, s*. The total configurational entropy, S, is assumed to be equivalent to the quantity $s^*(1 - T_2/T_f)$, where T_2 is the temperature below which the equilibrium state cannot be reached, T_{2o} and $T_{2\infty}$ are the values of T_2 at t=0 and t =∞, respectively, and T_f is a fictive temperature defined as the temperature at which the liquid state has been frozen to form a given glassy state (5). Furthermore, T_2 varies with the degree of cure as,

$$T_2 = T_{2o} + \alpha(T_{2\infty} - T_{2o}) \text{ (5)} \tag{11}$$

Hence, the constants in equation 9 can be expressed as
$C(T) = [\Delta\mu/kT]$, $A = T_{2o}/T_f$ and $B = (T_{2\infty} - T_{2o})/T_f$

Substituting equations 4 and 9 in equation 7 and 8, obtain

$$\varepsilon'(\alpha) = (N_o - k\,\alpha)\left\{ a_o + \frac{a_1}{\left[\exp\left(\frac{C(T)}{1 - A + B\,\alpha}\right)\right]^2} \right\} \tag{12}$$

$$\varepsilon''(\alpha) = (N_o - k\,\alpha)\left\{ \frac{a_2}{\exp\left(\frac{C(T)}{1 - A + B\,\alpha}\right)} \right\} \tag{13}$$

We have used equations 12 and 13 depicting the relationship between the dielectric properties and the extent of reaction to fit our data. In case of polymers, the hindrance energies of internal rotations, $\Delta\mu$, are expected to be few kilocalories per mole (19). Hence, we assume that C(T) = 1 for all purposes. We also know that the sum of A + B must be less than 1, because $(T_{2\infty})/T_f < 1$ (5). Finally, a_o (1 - k) is the value of the dielectric constant at long curing times, when letting N_o be normalized to 1, (since each sample begins with the same stoichiometry).
Using a least squares fitting routine (20) with a five parameter fit, we calculate that

Dielectric Constant	140 C	150 C	160 C	170 C	180 C
k	0	.067	0	.03	.137
α_o	2.24	3.86	3.48	3.72	4.24
α_1	1.65×10^5	1.98×10^5	1.75×10^5	2.18×10^5	2.36×10^5
A	.805	.81	.805	.806	.798
B	.054	.088	.097	.074	.110

The fitted and original data are shown in figures 13–17. Similarly, using a four parameter fitting routine, we can fit the dielectric loss factor data. An example of this at T = 140°C is shown in figure 18.

Conclusion

The above numbers suggest that the decrease in the microwave dielectric properties of the DGEBA/DDS epoxy system undergoing cure is predominantly dictated by the nature of the polymer superstructure, and not as much by the changes in the polar end group concentrations. This is indicated by the very low value of the k constant, which reflects the relative contribution by the polar end groups to the dielectric properties.

Based on our results, it seems reasonable to suggest that the use of microwave energy for driving reactions may be useful in systems where the network morphology is not rigid and the dipoles are free to move, giving rise to a relatively high dielectric loss. In the DGEBA/DDS system, the dielectric properties remained nearly constant above 50% extent of reaction, the dielectric loss being close to zero. Thus at higher conversion rates, the microwave field may not be as efficient as thermal heating in curing the resin.

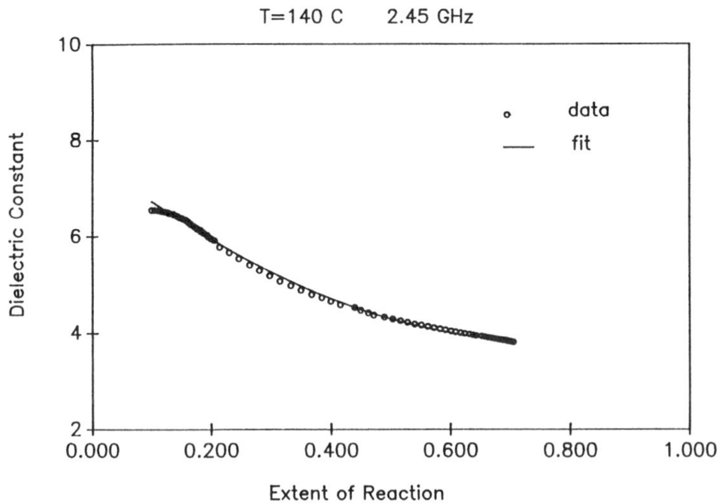

Figure 13. The dielectric constant data and their fit.

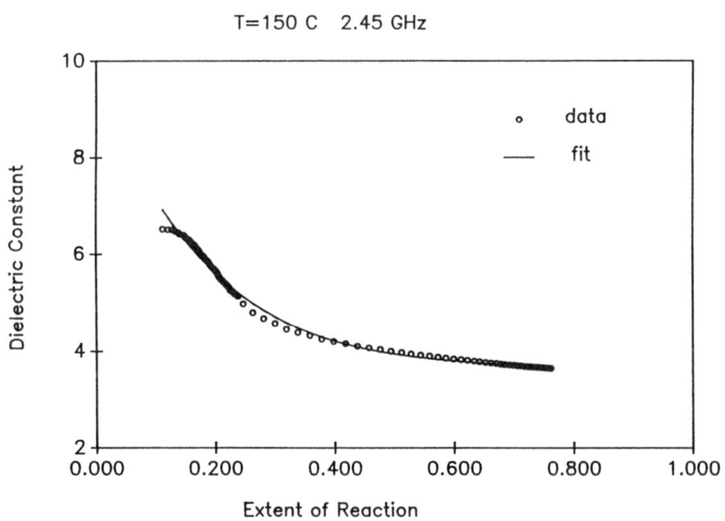

Figure 14. The dielectric constant data and their fit.

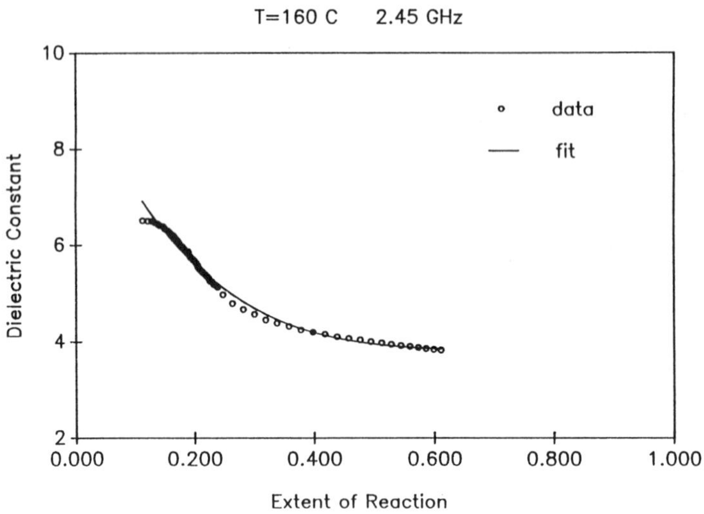

Figure 15. The dielectric constant data and their fit.

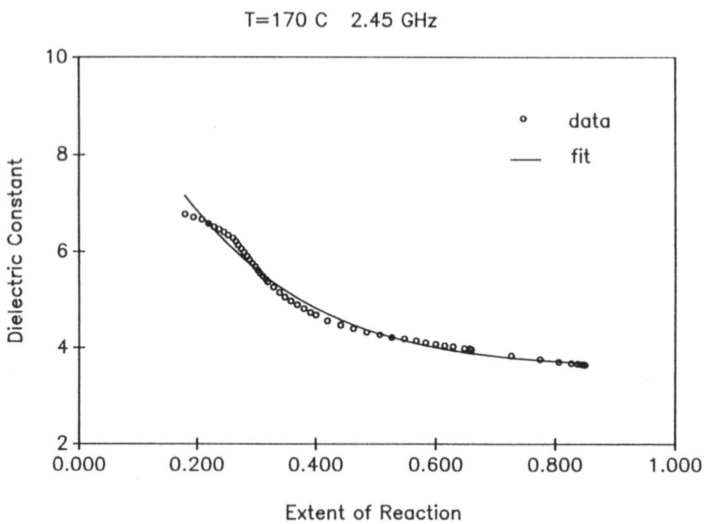

Figure 16. The dielectric constant data and their fit.

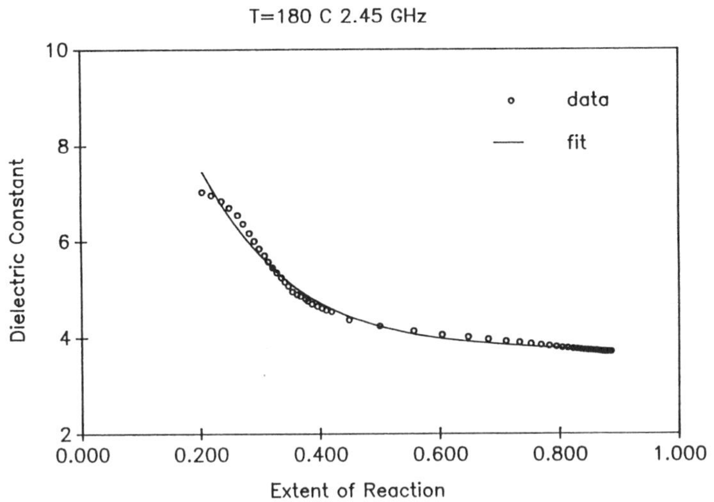

Figure 17. The dielectric constant data and their fit.

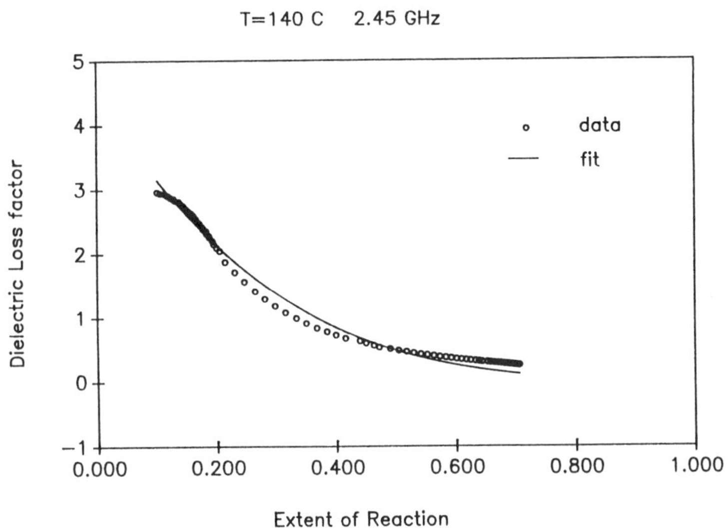

Figure 18. The dielectric loss factor data and their fit.

Acknowledgements

We greatfully acknowledge the support of Air Force/DARPA contact F33615-85-C-5153.

References

1. Norman F. Sheppard Jr., Advances in Polymer Science, Vol. 80, Springer-Verlag, Berlin, Heidelberg, pg. 1-47, (1986) and references cited therein.
2. A. J. Bur, Pllymer, 26, 963-977, (1985).
3. V. Frosini, E. Butta, J. Applied Polym. Sci., 11, 527-551 (1967).

4. W. G. Potter, "Epoxide Resins," Springer-Verlag, New York, (1970).
5. S. Matsuoka, X. Quan, H. E. Bair, and D. J. boyle Macromolecules, 22, 4093-4098 (1989).
6. M. Teffal and A. Gourdenne, Eur. Polym. J., 19, 543 (1983).
7. Q. LeVan and A. Gourdenne, Eur. Polym. J., 23 (10), 777 (1987).
8. J. Jow, M. Finzel, J. Asmussen, and M.C. Hawley, IEEE MIT-S Digest, V.1, 465, (1987).
9. J.C. Hedrick, D A. Lewis, T.C. Ward, and J.E. McGrath, Polym. Preprints, 29 (1), 363 (1988) and references cited therein.
10. J. Mijovic and J. Wijaya, Macromolecules, 23 (15), 3671-3674, (1990).
11. M.A. Rzepecka, J. Microwave Power, 8 (1), 3 (1973).
12. H. Altschuler, "Dielectric Constant", Handbook of Microvave measurements, Vol 2., M. Suchor, J. Fox, ed., Polytechnic Press, Brooklyn, New York, 1963.
13. CRC Handbook of Chemistry and Physics, CRC Press.
14. K.A. Buckinghan and W. Reddish, Proc. IEE, 114, 1810 (1967).
15. M.S. Heise and G.C. Martin, Macromolecules, 22, 99-104, (1989).
16. A.L. McClellan, "Tables of Experimental Dipole Moments," Vol 3., Rahara Enterprises, El Cerrito, CA, 1989.
17. A.K. Blythe, "Electrical Properties of Polymers," Cambridge University Press, Cambridge, (1979).
18. N.G. McCrum, B.E. Read and G. Williams, "Anelastic and Dielectric Effects in Polymeric Solids, "John Wiley and Sons, London, 1967.
19. G. Adams and J.H. Gibbs, J. Chem. Phys., 43 (1), 139-146 (1965).
20. Fitting routine adapted from P.R. Bevington's "Data Reduction and Error Analysis"

RECEIVED June 19, 1991

Degradation and Stabilization of Irradiated Polymers

POLYMERS ARE USED IN A GROWING NUMBER OF APPLICATIONS in which they receive exposure to ionizing radiation. Long-term degradation of the materials is of vital concern. Relevant applications include both post-exposure effects in materials subjected to radiation processing (such as cross-linking or sterilization) and instances in which materials receive low-level radiation exposure throughout their useful lifetime (as in a nuclear plant). In some applications, as with space vehicles, materials are exposed to radiation in the absence of O_2. Most application environments, however, involve exposure to air, either during irradiation or subsequent to radiation exposure. The effects of irradiation under oxidizing conditions can be very different from those occuring under an inert atmosphere and can involve very complicated effects such as a strong dependence on dose rate. Many studies on the radiation degradation of polymers during the last decade have thus focused on radiation–oxidation phenomena.

Important goals for technologies in which the radiation aging of polymers is problematic are the effective stabilization of materials and the reliable prediction of material lifetimes. Achieving these goals is predicated upon an understanding of the chemistry and physics of radiation-degradation processes. This section begins with an overview of radiation-degradation issues. The overview is followed by a comparative discussion of the early events of polymer radiation–chemical processes with the somewhat analogous (and much more extensively studied) photochemical processes, and with an emphasis on energy-transfer mechanisms. The discussion of polymer radiation–chemical processes are followed by chapters detailing the chemistry of radiation-induced free radicals, including radical-mediated oxidation reactions. Several chapters examine the role of oxygen-diffusion effects in determining radiation-induced property changes in polymers. Detailed studies of radiation oxidation using highly specialized analytical techniques, as well as investigations into radiation-induced mechanical property changes and discussions of methodologies for predicting long-term aging, are reported. Other chapters describe progress in stabilizing polymers in radiation environments and in the design of polymer-based scintillators that are resistant to radiation-induced discoloration.

Chapter 24

Chemistry of Radiation Degradation of Polymers

James H. O'Donnell

Polymer Materials and Radiation Group, Department of Chemistry, University of Queensland, Brisbane, Queensland, Australia 4072

High energy radiation produces ionization and excitation in polymer molecules. These energy-rich species undergo dissociation, abstraction and addition reactions in a sequence of reactions leading to chemical change. Scission and crosslinking of the polymer molecules, formation of small molecules and modification of the chemical structure of the polymer are responsible for the changes in material properties. The identification and measurement of these changes utilizing solution characterization methods, ESR spectroscopy and pulse radiolysis of radical and ionic intermediates, GC/MS analysis of volatile products and NMR and FTIR spectroscopy of chemical structure will be considered. The yields of the different chemical changes assist in understanding the mechanism of radiation degradation. Relationships between the chemical structure of the polymer and the radiation resistance are of fundamental importance for practical applications.

Knowledge of the chemistry of radiation degradation of polymers is becoming of increasing importance on account of the utilization of polymeric materials in a variety of radiation environments, the use of radiation for sterilization of medical equipment, and the development of radiation processes for modifying the properties of polymeric materials.

There have been investigations of the effects of radiation on polymeric materials since the construction of the first nuclear power plants in the 1950's, and this application has provided a major incentive for research in the field. There is now renewed interest in the radiation degradation of polymers with the need to establish the safe service lifetimes of polymer materials as the ages of these nuclear facilities increase.

Moreover, there are new applications of polymer materials in radiation environments which require greater knowledge of the anticipated rate of deterioration in their performance. Particle accelerators for high energy nuclear physics require polymeric materials to be used in radiation environments at extremely low temperatures, e.g. in super-conducting magnets. Satellites, space stations and space vehicles require light weight polymeric materials resistant to radiation under conditions of high vacuum and alternating high and low temperatures.

0097–6156/91/0475–0402$06.00/0

Types and Characteristics of Radiation

Radiation, in the present context, means radiation with sufficient energy to produce ionization (*1,2*). In practice, this is most likely to be γ-rays from radioactive ^{60}Co (the radioactive isotope of cobalt produced by neutron irradiation of ^{59}Co in a nuclear reactor) or from ^{137}Cs (the radioactive isotope of caesium recovered from the fission products of uranium in nuclear reactor fuel), or electrons produced as high energy beams in accelerators. There are heavier particles, including protons, alpha particles, and ions of higher elements in the periodic table produced by radioactive isotopes and, increasingly, by accelerators. To summarize:

Photon	Particle
X-rays	electrons
γ-rays	protons
	α-particles
	heavy ions

Absorption of radiation

Photons transfer energy to substrate molecules by three processes. Very high energy photons (>4 MeV) produce an electron and a positron (pair production) which subsequently undergo annihilation; this is not normally a significant process for irradiation of polymers. Most energy is deposited in the substrate by Compton scattering, whereby the ejection of a valence electron is accompanied by deflection of the incident photon by the electron cloud around the atom. At lower energies, the incident photon is completely absorbed by the substrate atom to produce ionization. Thus, the radiation chemistry of photon radiation (γ and X rays) is largely due to electrons.

The energy absorption in a polymer from irradiation by electrons occurs through the two processes of ionization and excitation. Both ions and excited species may be involved in subsequent chemical reactions or they may be neutralized or deactivated. The relative importance of chemical change resulting from ionic, radical and excited state intermediates is an important aspect of research into the radiation chemistry of polymers.

Measurement of Dose Absorbed

Any quantitative estimates of chemical changes produced by radiation require measurement of the energy absorbed in the substrate. Therefore, dosimetry is a vital part of radiation chemistry. A variety of techniques have been used to measure absorbed dose; in practice the choice depends on convenience. Some common dosimeters are:

Calorimetry	Ionization chamber
Faraday cup	Solid-state electronics
Photographic film	Dyes in plastic
Coloured ions in glass	Ferrous ion (Fricke)
Ceric ion	Alanine (ESR)
Polyethylene (H_2)	

Colour development in plastic film is very convenient in industrial facilities and especially for electron accelerators. Ferrous and ceric ion dosimeters are chemical systems which are widely used for γ-irradiation.

Radiation Units

The classical unit for the intensity of a radiation source was the curie (Ci) = 3.7 x 10^{10} disintegrations per second; the modern SI unit is the becquerel (Bq) = 1 disintegration/s. The absorbed dose was measured in electron volts (eV) per g (convenient for electron accelerators), and then in rad (=0.01J/kg), and the SI unit is the gray (Gy) = 1 joule/kg. Thus, the amount of chemical change can be related to the energy absorbed using mol per kg. The G value equal to the number of events per 100 eV (equivalent to 16 atto joules; 1 aJ = 10^{-18}J) of energy absorbed has been customarily used to measure radiation chemical yield, but μmol/J is now recommended (1 μmol/J = 10 G).

Penetration of Radiation

The depth profile of the absorbed energy is extremely important for electrons and X-rays in the mm range and for positive ions, when the penetration depth decreases rapidly into the μm range with increasing mass. A typical dose versus depth profile for electrons is shown in Figure 1. The surface effect, i.e. the variation in dose by about 60% between the surface and the maximum, and the decrease in dose beyond the maximum at about 50% penetration cause substantial inhomogeneity in dose through a thick sample (relative to the penetration range) and a serious over-estimation of the dose for thin samples unless the dosimeter has the same thickness as the sample.

Aspects of Radiation Chemistry

Direct and Indirect Action. High-energy radiation is absorbed in a polymer by interaction with the valence electrons of the atoms. It is non-specific spatially and with respect to different atoms and groups, except for their electron densities, in contrast to photochemistry, where the photons are absorbed by specific chromophores, so that there is high specificity of energy absorption within molecules. An important example is irradiation of polymers in solution. High-energy radiation is absorbed by both the polymer and the solvent producing ions, radicals and excited states. The chemical changes in the polymer will result from reactions of these species formed in the polymer (direct action) and in the solvent (indirect action). By contrast, the solvent is normally chosen to be non-absorbing in photochemistry.

Energy Transfer. Although the absorption of energy is spatially and molecularly random, chemical change may be quite non-random. One important process enabling this spatial selectivity is energy transfer, which may be intra- or inter-molecular. Energy transfer along polymer molecules and energy trapping is currently an important field of research in radiation chemistry of polymers. As a result of energy transfer, molecular components present in only small amounts may be the main sites of chemical change.

Hydrogen Transfer. Polymers contain hydrogen atoms as a major component of their molecular structures and C-H scission is an important radiation chemical reaction followed by abstraction reactions by the liberated H atoms. The intermediate radical and ionic species may undergo H atom transfer, a process driven by the excess energy in the system and by the formation of radicals of lower energy. H transfer can proceed in a sequence of steps, resulting in movement of radical and ionic sites over long distances, and this process may play an important role in crosslinking.

Primary and Secondary Products. Some molecular structures are particularly reactive with intermediate species in radiation chemistry. A very important example is the reaction of C=C double bonds with radicals (scavenging). The C=C bonds may be present in the original polymer, or formed during irradiation, e.g. by elimination of H_2. Thus, an unsaturated molecule containing C=C bonds may be a primary product of irradiation, and these C=C bonds may react with radical intermediates to form secondary products. The yields of both the primary and secondary products will be dose dependent as shown in Figure 2.

Temperature Rise from Energy Absorption. Absorption of radiation in a polymer is the first stage in a sequence of processes which leads to chemical change. However, only a portion of the energy is utilized in the chemistry and most is converted into molecular excitation or heat. Polymers have low coefficients of thermal transfer so that the absorption of energy can lead to significant temperature rises. This will be particularly important for electron irradiation at high dose rates.

Not only do the rates of chemical reactions increase with temperature (the normal Arrhenius behaviour), but if the glass transition (Tg) or melting (Tm) temperatures are exceeded, dramatic changes in material properties can occur. Small molecular products are always formed on irradiation of polymers. They are frequently trapped in glassy or crystalline polymers, but cause foaming above Tg or Tm.

Post-Irradiation Effects. Some chemical change occurs during irradiation of polymers and some occurs subsequent to irradiation. There are several causes of these post-irradiation effects. Radicals are trapped in irradiated polymers and react over a period of time, the rate depending on the reactivity of the radicals, the mobility of the matrix, and the diffusion of oxygen into the sample. Oxidative reactions are normally degradative, i.e. lead to scission, and cause deterioration in mechanical properties. Post-irradiation reactions of radicals can be observed by electron spin resonance (ESR) spectroscopy if the polymer is irradiated below Tg or Tm.

Irradiation in air leads to the formation of peroxides and these compounds have characteristic rate versus temperature relationships for decomposition, usually with significant rates in the range 50-150 ^{o}C. Therefore, polymers irradiated in air are likely to deteriorate slowly on standing at ambient temperature and more rapidly if heated.

Small molecule products formed during irradiation will have slow rates of diffusion in glassy and crytalline polymer, except for H_2. Consequently, hydrocarbons, carbon monoxide, carbon dioxide and monomers will be trapped in the polymer, causing internal stresses which may be very large. This internal stress results in the development of crazing and of cracks in the polymer with time. This phenomenon is readily demonstrated in glassy polymers such as poly(methyl methacrylate).

Radiation Environment. There are two important characteristics of the environment in which a polymer is subjected to irradiation: (1) chemical, and (2) thermal. The chemical environment is normally vacuum or gaseous. The gas may be nitrogen, helium, air, oxygen, chlorine, sulfur dioxide, etc. Chemically "inert" gases may not be inert to radiation, as energy transfer can occur to other components and result in excitation and chemical reaction. Oxygen readily forms peroxy species by addition to radicals and has a major enhancement effect on the radiation degradation of most polymers. The use of reactive gas environments has great potential for radiation chemical modification of polymers, especially for surface treatments, but has not been investigated widely.

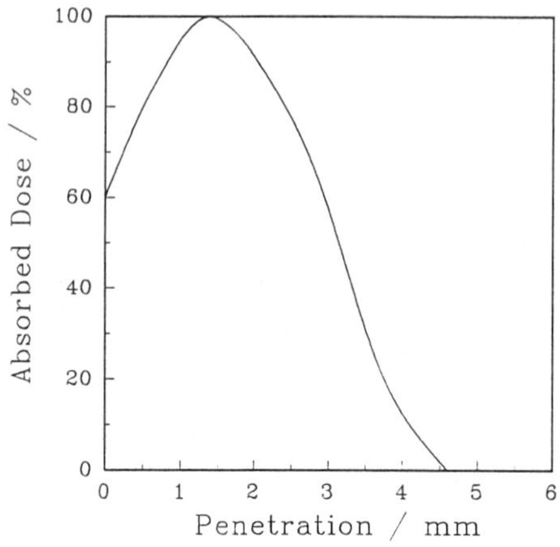

Figure 1. Depth profile for the dose absorbed in a substrate of density $1 g/cm^3$ for 1 MeV electrons.

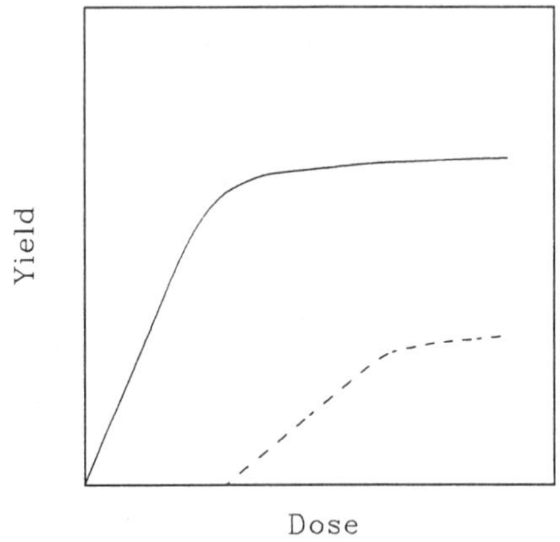

Figure 2. The dose dependence of primary (—) and secondary (- - -) products in radiolysis.

The Sequence of Events in Radiation Chemistry

The effects of radiation on polymers result from a sequence of events which can be divided into: (1) physical - energy absorption and transfer, (2) physico-chemical - ionization and excitation, (3) chemical - radical-molecule, ion-molecule, radical-radical and ion-ion reactions, (4) morphological structural changes, (5) material properties.

Effects of Radiation on Polymeric Materials

The properties of polymeric materials are affected by radiation as a result of the chemical changes in the polymer molecules. These changes include molecular weight and structure. Secondary effects are on crystallinity, and on molecular architecture in the amorphous phase. "Degradation" of polymers is frequently taken to mean reduction in molecular weight, or deterioration in some desirable property, but these are unnecessarily narrow definitions, as an increase in modulus due to crosslinking may be beneficial in one application, but the accompanying decrease in fracture strain may be a disadvantage in another application. Therefore, it is appropriate to consider degradation as any change in the molecular, morphological or material properties of a polymer resulting from irradiation.

Radiation-Induced Changes in Polymers

The molecular changes produced in polymers by radiation may be classified into:

1. scission and crosslinking of the polymer molecules, leading to decrease or increase in molecular weight and the possible development of an insoluble or gel (g) fraction of the polymer above the gel dose, Dg, which increases with dose up to a limit, which depends on the ratio of scission to crosslinking.

2. evolution of small molecular products, such as H_2, CO, CO_2, CH_4, depending on the composition of the polymer. Identification and measurement of these volatile products provides valuable information about the mechanism of the radiation chemical reactions. However, they can contaminate the environment of the polymer - a serious problem in space applications of polymeric materials. Also, for irradiations at temperatures below Tg of amorphous polymers, the normally volatile products are likely to be trapped in the polymer, and the internal stresses will cause post-irradiation deterioration of the polymer through crazing and cracking.

3. changes in the molecular composition and structure of the polymer molecules, including loss and formation of unsaturation. An example is coloration in poly(vinyl chloride), PVC, from transparent through yellow and orange to black after low radiation doses due to the formation of conjugated C=C unsaturation.

Measurement Techniques

Molecular Weight. Scission and crosslinking decrease or increase, respectively, the molecular weights of the polymer molecules. Therefore, measurements of the changes in molecular weight averages or distribution with dose can quantify these processes. Viscometric measurements in solution to give $[\eta]$ indicate whether crosslinking or scission predominates. If only scission occurs and the mol. wt. distribution is initially the most probable, then $[\eta]$ values will give Mv, provided that K and a for the Mark-Houwink equation are known. Mn may be derived from Mv if the breadth of the MWD is known, which then enables the scission yield, G(S), to be calculated.

However, [η] values can be very misleading if the initial mol. wt. distribution is different from the most probable. An initially broad distribution will narrow to the most-probable distribution with irradiation if there is only scission, and G(S) will be over-estimated from [η] values. Crosslinking reduces the ratio of hydrodynamic volume to mol. wt. compared to linear molecules and [η] values will then lead to underestimation of the mol. wt. Estimated corrections have been published.

Gel permeation chromatography, GPC, is used very widely to determine average molecular weights (and the mol. wt. distribution) on account of the experimental simplicity of the technique. Calibrations using monodisperse polystyrene standards must be corrected to the polymer under investigation, either by Q factors or the universal calibration, although characterized, polydisperse samples can be used. GPC suffers from the same hydrodynamic volume problem as viscometry when crosslinking occurs, but this problem is frequently ignored.

Measurements of Mn by osmometry or of Mw by light scattering provide absolute average molecular weights which enable reliable values of G(S) and G(X) to be derived for the polymer. However, these methods require careful experimental technique, and are limited by the solubility of the polymers in suitable solvents, especially for compatibility with the membrane in osmometry. The techniques are also limited by the mol. wt. range of the instruments.

Ultracentrifugation has been largely neglected as a technique for measuring the mol. wt. averages and distributions of irradiated polymers, perhaps because it is a technique used mainly by biochemists. Sedimentation velocity measurements can give the mol. wt. distribution and Mw. Sedimentation equilibrium enables Mw and Mz to be calculated. Greater use of this technique is warranted.

Soluble Fractions. Polymers in which crosslinking predominates over scission, as determined by $G(S)<4G(X)$, become incompletely soluble above the gel dose, Dg, and it is possible to derive values of G(S) and G(X) from measurements of the soluble, or insoluble (gel) fraction of the polymer as a function of dose. A variety of techniques are available for this measurement, including (1) conventional Soxhlet extraction of the polymer placed in a paper thimble with condensed solvent vapour, (2) immersion of the sample contained in a stainless steel mesh bag in boiling solvent, with thermal stabilizer added if necessary, (3) centrifugation of the polymer-solvent mixture to separate the soluble and gel fractions, (4) GPC of an aliquot of the liquid phase after extraction.

There are difficulties with all of these methods: powdered samples escape from the Soxhlet tube and the mesh bag, solvent is difficult to remove from thick samples during drying, soft gel from samples near the gel dose fractures readily, and GPC columns are readily clogged by microgel. Nevertheless, gel measurements are a widely used method for obtaining G(S) and G(X).

Calculation of G(S) and G(X) From Mol. Wt.

The relationship between Mn and G(S) and G(X) is given in equation 1 and applies to all initial mol. wt. distributions. Equation 2 is the relationship between Mw and G(S) and G(X) for an initial most-probable mol. wt. distribution ($A = 1.04 \times 10^{-10}$ and $B = 5.18 \times 10^{-11}$ for dose, D, in Gy). If scission only occurs, then either equation 1 or 2 alone can be used to derive G(S). When both scission and crosslinking occur, G(S) and G(X) can be obtained from the combination of equations 1 and 2, provided the initial distribution is the most probable. Typical plots of Mn and Mw according to equations 1 and 2 for polystyrene irradiated in vacuum are shown in Figure 3.

$$1/Mn(D) = 1/Mn(0) + A\,[G(S)-G(X)]\,D \qquad (1)$$

$$1/Mw(D) = 1/Mw(0) + B\,[G(S) - 4G(X)]\,D \qquad (2)$$

An equation for the relationship between Mw and G(S) and G(X) for any initial mol. wt. distribution is available, but it does not have an analytical solution. Various approximations have been made to give a linear relationship between 1/Mw and dose. Care must be taken in the use of experimental data for equation 2 because crosslinking causes the MWD to broaden and experimental measurements of Mw tend to underestimate the true value through neglect of the high mol. wt. tail of the distribution. There are difficulties in combining equations 1 and 2 to derive G(S) and G(X) if either G(S) or G(X) is small and uncertain or even negative values may be obtained.

Winzor et al. (*3*) have shown that the relationships between both Mw and Mz and dose can be simplified into a form, which enables G(S) and G(X) to be obtained from plots of $[M_i(0) / M_i(D) - 1] / D$ versus D, where i = w or z.

Calculation of G(S) and G(X) from MWD

There is more information in a complete MWD than in two average mol. wts. However, simple equations giving linear relationships with dose are not available for MWD. Treatments of the changes in MWD with irradiation are based on the integro-differential equations of Saito, either by analytical solutions, which require numerical integration, or by iterative computer simulation (*4*). It is possible to treat any initial MWD by these methods, but the procedures have been used infrequently.

Calculation of G(S) and G(X) from Soluble Fractions

The values of G(S) and G(X) for the irradiation of a polymer when $G(S) < 4G(X)$ are usually calculated from the Charlesby-Pinner (*5*) equation 3, which predicts a linear relationship between $s + s^{1/2}$ and 1/D. This equation is based on the assumptions that (1) the unirradiated polymer has a most probable MWD, (2) crosslinking occurs by a H-link mechanism, (3) crosslinking and scission occur with random spatial distributions in the polymer without any clustering. A typical Charlesby-Pinner plot, for polystyrene irradiated in vacuum, is shown in Figure 4.

The effect of an initial MWD which differs from the most probable MWD has been considered. Tabulations are available (*4*) for the relationship between the gel fraction and the density of crosslinks for different breadths of the MWD and ratios G(S)/G(X) for Schulz-Zimm and Wesslau distributions to derive G(S) and G(X).

$$s + s^{1/2} = G(S)/2G(X) + [4.82 \times 10^9] / [G(X)\,Mn(0)\,D] \quad (3)$$

If crosslinking occurs by a Y-linking mechanism, then the relationship between s and D is given by equation 4. Y-linking was proposed many years ago, but only recently has evidence been found from NMR spectra for Y-link structures in irradiated polymers. The role of H and Y links must now be considered an important question in the radiation chemistry of polymers; undoubtedly the molecular structure of the polymer will be important, but there

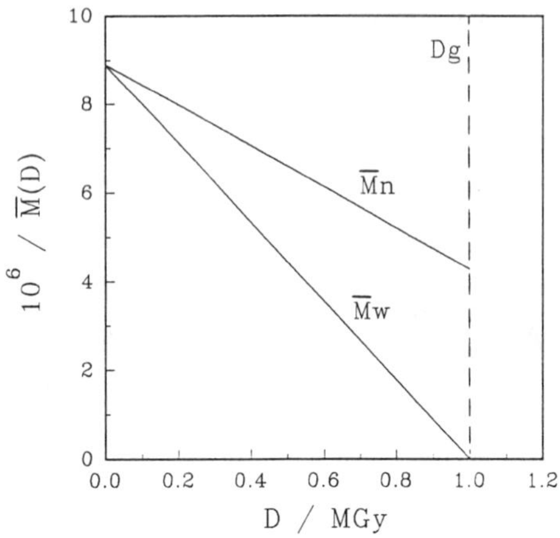

Figure 3. Mn and Mw plots according to equations 1 and 2 for polystyrene irradiated in vacuum at 30 °C which provide values for G(S) and G(X).

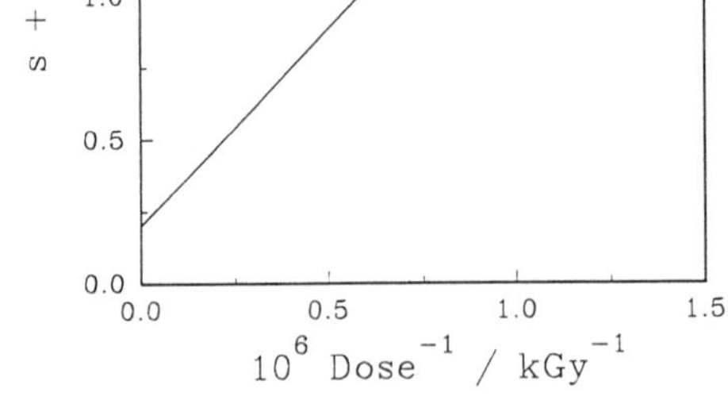

Figure 4. Charlesby-Pinner plot according to equation 4 of soluble fraction data for irradiation of polystyrene in vacuum at 30 °C.

may also be a dose dependence.

$$s + s^{1/2} = 2G(S)/G(X) + [1.93 \times 10^{7}] / [G(X)\,Mn(0)\,D] \quad (4)$$

NMR of Structural Changes

The methods used to determine crosslinking and scission in irradiated polymers, utilizing changes in mol. wt., soluble fraction, elastic modulus, swelling ratio, etc. are indirect, and depend on many assumptions. They give no information about the chemistry of scission and crosslinking. NMR spectroscopy provides the potential to obtain this information and with the increasing sensitivity and selectivity of NMR techniques, this potential is increasingly being realized. ^{1}H NMR and ^{13}C NMR are the most common spectra used. ^{19}F and ^{29}Si are excellent for polymers containing F and Si atoms. Solution spectra have the highest resolution, but solid-state NMR is continually improving. The solid-state ^{13}C NMR spectra of ethylene-propylene copolymer before and after γ-irradiation in Figure 5 shows resonances due to new methyl end groups from scission and resonances due to crosslinks.

DEPT and INEPT pulse sequence techniques enable ^{13}C sub-spectra to be obtained for different types of carbon atoms, e.g. $-CH_3$ and $-CH_2-$. 2-Dimensional NMR is increasingly used to identify different structures in molecules and, with increasing sensitivity, radiation-induced crosslinks can be observed. Disappearance of radiation-sensitive groups, such as carbonyl, and formation of new structures, e.g. C=C, can be determined quantitatively. NMR has greatly expanded the information previously obtained from infrared spectra, which has also been greatly improved by the advent of FTIR.

Small Molecular Products

The chemical reactions which occur during irradiation of polymers lead to the formation of a variety of small molecular products, including H_2, CH_4, alkanes, alkenes, CO, CO_2, SO_2, and monomer, depending on the composition of the polymer. Measurement of these products can provide valuable information on the radiation chemistry.

Gas chromatography is a direct method for observation of volatile molecules. GC-MS enables a wider range of products to be identified. HPLC is useful for measurement of molecular products of low volatility. There is an immense variety of possible reactions which can take place during irradiation, as every bond can potentially be cleaved. Consequently, there are very small amounts of many products formed, but they do not necessarily assist greatly in identifying the significant reactions.

Determination of Intermediate Species

ESR spectra. The chemical reactions which occur on irradiation of a polymer must involve ions and radicals, and give rise to the observed scission and crosslinking, small molecular products, and structural changes in the polymer. However, the reaction mechanism will be largely based on speculation without evidence for the identity and concentration of the intermediate species in the reactions.

ESR spectroscopy has been widely used for many years to investigate radical intermediates in irradiated polymers. These species are highly reactive and will mainly have extremely short lifetimes at ambient temperature. Cryogenic trapping and thermal annealing techniques are used to observe the ESR spectra of radicals and their reactions. At low temperatures, e.g. 77 K, broad singlet and

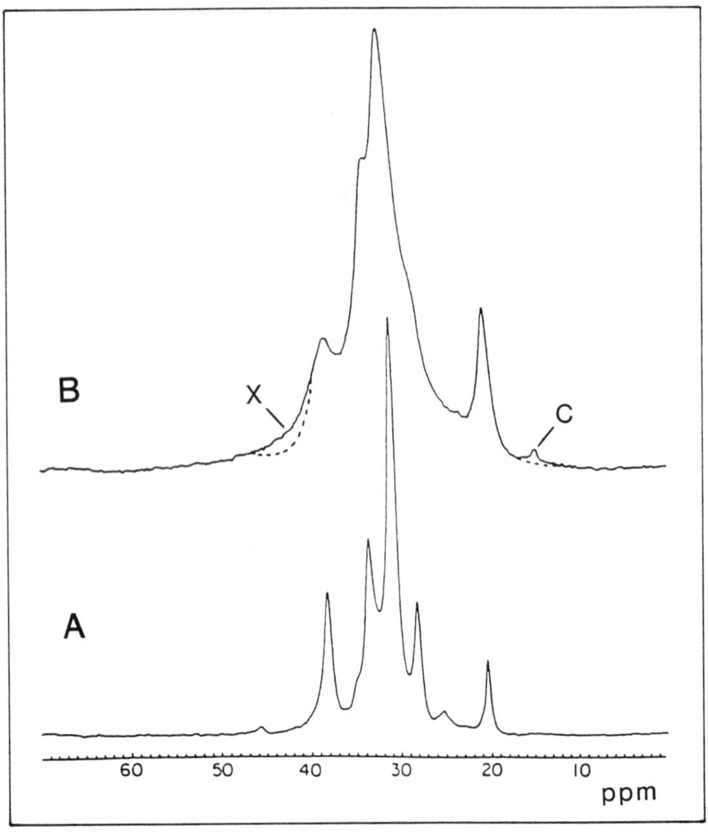

Figure 5. Solid-state ^{13}C NMR spectrum of ethylene-propylene copolymer before (A) and after (B) γ-irradiation in vacuum, showing new chain ends (C) and crosslinks (X).

doublet components are observed in the spectra and selective photobleaching can be used to show that they are due to radical-anions and radical-cations.

Conversion of one radical into another can sometimes be observed and attributed to an abstraction reaction to give a more stable radical. Different radicals disappear at particular temperatures, which may be correlated with vibrational and rotational motion of substituent groups, and especially with Tg and Tm.

Pulse Radiolysis: The short-lived ionic and radical intermediates in radiolysis can also be observed with very fast spectroscopic techniques, using pulses of radiation of μs, ns or even ps duration. The technique has been extensively used for aqueous solutions of small molecules, but applications to polymers are increasing. The information already obtained for high molecular weight paraffins and polyolefins has greatly increased knowledge of their radiation chemistry. This is a field of research that requires highly sophisticated instrumentation, but is essential for studying the early stages of radiolysis.

Molecular Structure and Radiation Sensitivity

The radiation chemistry of polymers is remarkably sensitive to their molecular structure, despite the non-selective absorption of energy. G values vary from 0.1 to 10 for low to high resistance to radiation. Higher G values indicate a kinetic chain reaction. Radiation sensitive groups include -COOH (decomposition to CO and CO_2), -SO_2- (two C-S scissions and evolution of SO_2), -Hal (scission of the C-Hal bond), C=C (scavenging of radicals, especially H).

It has been well established that scission of the backbone chain of polymer molecules is highly favoured by the presence of -C- structures (attributed to steric strain, low energy of monomer addition, and prevention of H migration), whereas -CH- and -CH_2- favour crosslinking.

Polymers containing aromatic (phenyl) rings have greater resistance to radiation than aliphatic structures and hence the high-performance, high-temperature polymers, such as the polysulfones, polyether ketones and polyimides, show high resistance to radiation. Aromatic rings provide both an intra- and inter-molecular protective effect. There is considerable interest in the relative resistance of different aromatic structures, e.g. phenyl and diphenyl, to radiation.

Acknowledgements

The author is grateful to his colleagues, Dr D.J.T. Hill and Dr P.J. Pomery for their contributions to all his activities; to the Australian Research Council and the Australian Institute of Nuclear Science and Engineering for supporting his research on radiation degradation of polymers and the Australian Nuclear Science and Technology Organization for providing irradiation facilities.

Literature Cited

1. O'Donnell, J.H.; Sangster, D.F. *Principles of Radiation Chemistry*; Edward Arnold: London, 1970.
2. Swallow, A.J. *Radiation Chemistry: An Introduction*; Longman: London, 1973.
3. O'Donnell, J.H.; Winzor, C.L.; Winzor, D.J. *Macromolecules* **1990**, 23, 167.
4. Saito, O. In *The Radiation Chemistry of Macromolecules*; Dole, M., Ed.; Academic Press: New York, 1973, Vol. 11, pp 223-261.
5. Charlesby, A. *Atomic Radiation and Polymers*; Pergamon Press: New York, 1960.

RECEIVED May 3, 1991

Chapter 25

Role of Excitation Transfer in the Photochemistry and Radiation Chemistry of Solid Polymers

Studies of Styrene and Vinylnaphthalene Copolymers with Vinyl Ketones

J. E. Guillet, S. A. M. Hesp[1], and Y. Takeuchi

Department of Chemistry, University of Toronto, Toronto, Ontario M5S 1A1, Canada

Radiation studies of styrene and vinyl naphthalene copolymers with vinyl ketones show that the presence of the naphthalene group imparts increased radiation stability to the polymers. Photochemical studies on the same polymers and model compounds lead to the conclusion that the naphthalene quenches singlet or triplet excitons migrating in the solid polymer. Mechanisms are proposed to explain this effect.

The radiation chemistry of polymers has been investigated extensively, largely as a result of interest in the ionizing radiation obtained from nuclear reactors and because of the extensive work done on nuclear weaponry and power systems. More recently it has become important in the design of materials for space vehicles and the manufacture of microelectronic and electro-optical devices. It is perhaps unfortunate that the early work in this field was done almost exclusively without the benefit of information about the photochemistry of the same macromolecules, and in fact this remains a major problem in the interpretation of the results of the radiation chemistry of polymers, even at the present time.

The field of radiation chemistry of polymers has been extensively reviewed (*1-4*) and in this chapter we will not attempt to summarize the field at its present stage but will simply review briefly a number of accounts of experimental investigations which attempt to draw correlations between radiation chemistry and the conventional photochemistry of macromolecules. In particular, we will restrict our attention to chemical changes caused by electromagnetic and electron-beam radiation which relate to the role of energy transfer in the solid state.

[1]Current address: Department of Chemistry, Queen's University, Kingston, Ontario K7L 3N6, Canada

0097–6156/91/0475–0414$06.00/0

Absorption of High Energy Radiation

As the wavelength of radiation is reduced below that of the far ultraviolet the energy of the incident photon increases drastically. Table I shows the wavelength ranges for various types of high energy radiation and the associated energy per quantum (*5*). As a result of the high energy of the x- or γ-ray photons, it is unlikely that the first step of interaction with matter will be with a valence or outer shell electron. In fact, the first step in the absorption of high energy radiation by matter is usually the formation of a high energy ion by ejection of an inner electron and the recombination of this ion with another electron gives rise to a stream of Compton electrons. These electrons cascade by collision with other molecules in the system until they reach lower energy levels and can be captured by the outer shell electrons. For example, a single γ-ray photon can give rise to many excitations resulting from the collision of Compton electrons with the matter surrounding the original absorption site. For this reason chemical events induced by high energy irradiation often follow tracks or spurs as the particle moves through the condensed phase.

Table I. Relation Between Wavelength and Photon Energy (*5*)

Wavelength λ (nm)	Type of radiation	Photon energy, ϵ_p eV	kcal	kJ
1250	Infrared	1	23	96
125	Ultraviolet	10	230	960
12.5		100	2300	9600
1.25	Soft X-rays	1000	2.3×10^4	9.6×10^4
0.125		10,000	2.3×10^5	9.6×10^5
0.0125	X-rays	10^5	2.3×10^6	9.6×10^6
0.001	γ-rays (^{60}Co)	1.2×10^6	2.9×10^7	12.1×10^7

The interaction of a Compton electron with a molecule AB may take one of two paths, either equation 1, the primary ionization, or some form of excitation, equation 2.

$$AB \rightsquigarrow AB^+ + e^- \quad \text{ionization} \tag{1}$$

$$AB \rightsquigarrow AB^* \quad \text{excitation} \tag{2}$$

The electrons ejected during the ionization will lose their excess kinetic energy by initiation of further ionization and excitation until the electrons finally reach thermal energy. At this point, neutralization of a cation or attachment can occur:

$$e^- + AB^+ \longrightarrow AB^* \tag{3}$$

or attachment to a neutral molecule

$$e^- + AB \longrightarrow AB^- \qquad (4)$$

Cationic species may also have sufficient excess energy to decompose further by one of two mechanisms:

$$AB^+ \longrightarrow A^+ + B \qquad (5)$$

$$AB^+ \longrightarrow A^+ + B\cdot + e^- \qquad (6)$$

Energy Transfer Processes (*4a,5*)

It has long been recognized that in certain cases of radiation-induced reactions, the amount of reaction occurring is much larger than expected (*5,6*). This phenomenon can be accounted for by the transfer of energy. which can occur by means of mass transfer, charge transfer, excitation transfer or a combination of these. The different mechanisms are not always distinguishable. However, in solid polymer matrices at low temperatures, mass transfer contributions are usually negligible.

Charge Transfer. Charge transfer can occur by the transport of electrons, positive charges or ions through the solid matrix. Electronic transfer occurs when electrons or positive charges are the only migrating entities. Ionic transfer involves a combination of mass and charge transfer, i.e., protons or larger charged entities may migrate through the polymer matrix.

The migration of electrons or positive charges can further be divided into localized and non-localized diffusion of charge. Positive or negative charges localized on particular groups on the polymer chain can be exchanged between like groups by resonant charge transfer without the loss of energy. The non-localized diffusion of charge can result in an annihilation of a positive and negative charge pair to result in a highly excited group in the polymer. This excitation energy can then migrate further through the polymer.

Excitation Transfer. Excitation energy transfer occurs when an excited molecular group in the polymer loses its excitation energy to an acceptor molecular group (equation 7). This transfer can

$$D^* + A \longrightarrow D + A^* \qquad (7)$$

occur either by the emission and reabsorption of a photon, called "trivial transfer", or by radiationless transfer. Depending on the distance between donor and acceptor and their respective energy levels and spin state, the energy exchange can be described by different models. For the exchange of electronic excitation energy, the mechanisms are exchange transfer, Förster transfer and excimer transfer.

Exchange Transfer. Exchange transfer occurs only over short distances since it requires an overlap of the orbitals of the groups involved. The greater the overlap, the more efficient is the transfer of electronic excitation. In the region of overlap, an excited electron can be associated with the donor as well as with the acceptor molecule.

According to Dexter (*7*), the rate of exchange energy transfer is given by

$$k_{D^* \to A} = (4\pi^2/h)Z^2 \int f_D(\nu) f_A(\nu) d\nu \tag{8}$$

where $Z^2 = k^2 \exp(-2R/L)$, k is a constant, L is the effective average Bohr radius, $f_A(\nu)$ is the acceptor absorption spectrum, $f_D(\nu)$ is the donor emission spectrum, and R is the donor-acceptor distance. From this equation it can be seen that the rate of transfer is exponentially dependent on the donor-acceptor distance.

Förster Transfer. Förster transfer (also referred to as resonance transfer) is due to the Coulombic interaction between two groups that are at a distance of several times their van der Walls radii. According to Förster (*8*), the rate of resonance transfer is given by

$$k_{D^* \to A} = \frac{8.8 \times 10^{-25} \kappa^2 \phi(D)}{n^4 R^6 \tau(D)} \int_0^\infty F_D(\nu) \epsilon(\nu) \frac{d\nu}{\nu^4} \tag{9}$$

where $k_{D^* \to A}$ is the rate constant for energy transfer (s^{-1}), R is the donor-acceptor distance, κ^2 is an orientation factor. For a random distribution of donors and acceptors $\kappa^2 = 2/3$, ϕ_D is the quantum yield for donor emission, n is the refractive index, $\tau(D)$ is the donor state lifetime, $F_D(\nu)$ is the normalized spectral distribution of donor emission, and $\epsilon(\nu)$ is the molar extinction coefficient of the acceptor. One can conclude that resonance or Förster transfer is less dependent on the donor-acceptor distance than exchange transfer.

In the case of Förster transfer, one usually defines a distance at which there is a 50% chance of effective tranfer between donor and acceptor. This distance, R_0, can be predicted from

$$R_0^6 = \frac{8.8 \times 10^{-25} \kappa^2 \phi(D)}{n^4} \int_0^\infty F_D(\nu) \epsilon(\nu) \frac{d\nu}{\nu^4} \tag{10}$$

Using equation 10, the R_0 value for naphthalene-to-naphthalene transfer, R_0, is about 7 Å, whereas for phenyl-to-phenyl it is about 3.5 Å.

Chemical Processes in Solid Polymers. The radiation chemistry of polymers will depend on the reactions of the ions, radicals or excited species created by the primary events. In polymers, not only does radiation induce chemical changes but extensive changes of physical properties may also occur as a result of crosslinking or chain scission. Concern with these two reactions of crosslinking and scission dominates the earliest studies of the radiation chemistry of polymeric materials (*9*). In general, the chemical structure of a polymer determines which of the reactions will predominate.

The efficiency of a radiochemical process is usually described in terms of the G value, which is the number of chemical or physical events per hundred electron volts of energy absorbed (9600 kJ). For a polymer with a normal molecular weight distribution, scission will predominate over crosslinking if the G value for scission, G_s, exceeds the G value for crosslinking, G_x, by a factor greater than two (*10*). When crosslinking predominates, the molecular weight rises until a critical dose, called the gel dose, r_g, is reached, above which an insoluble gel network of crosslinked polymer is formed in increasing amounts. As a general rule, vinyl polymers on which each atom of the main chain carries at least one hydrogen atom, will tend to crosslink, while, if the polymer contains a tetrasubstituted carbon atom, the polymer will usually degrade (*11*). Presumably, this is due to the higher probability of the β-scission process in tertiary hydrocarbon radicals:

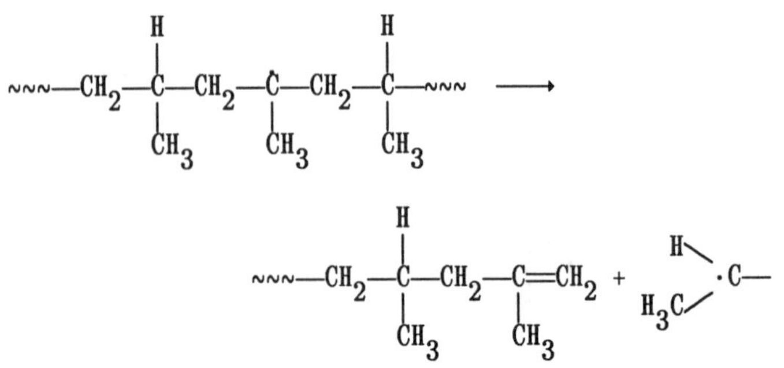

(11)

There are, however, many notable exceptions to this general rule.

From this definition of the G value, it is obvious that the relationship of this to a photochemical quantum efficiency, ϕ, is given by

$$G = \phi(100/\epsilon_p) \tag{12}$$

where ϵ_p is the energy in eV per photon of the light used. For light of 313 nm, for example, one photon has an energy of 3.96 eV.

The radiation dose absorbed by a polymer is usually expressed in terms of the Rep (Röntgen equivalent physical) or, more commonly, Rad. The amount of radiation absorbed per unit time can be calculated from the known energy flux of the source multiplied by the absorbance of the polymer for the particular radiation used, or can be determined directly with a suitable actinometer. Usually a good approximation for γ-radiation is that the absorbance is proportional to the density of the polymer. Experimental details on these procedures can be found in standard reference works on polymer radiation chemistry (*1-4,9*).

Comparison of UV and γ-Ray Photochemistry of Polymeric Ketones

Studies of Model Compounds. Only a few studies have been made of the correlations which can be drawn between UV photochemistry and radiation chemistry in polymer systems. Among the earliest of these are studies by Schultz (*12*) who showed that solid films of poly(methyl isopropenyl ketone) (PMIPK) degraded rapidly in

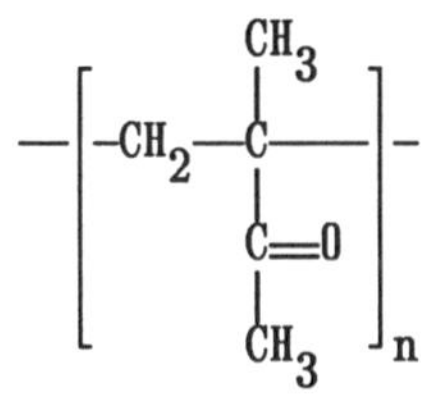

PMIPK

molecular weight when exposed to both γ-rays and UV photons. However, the G_s value for cobalt-60 γ-rays was 1.1, as compared to 4.5 for UV photolysis at 254 nm (ϕ = 0.22). The γ-ray G_s value increased to 2.4 in the presence of air. Slivinskas and Guillet studied a series of model compounds containing ketone functional groups. The first of these studies involved a series of symmetrical aliphatic ketones subjected to γ-radiolysis (^{60}Co) under vacuum, the products being analyzed by GLC. The results were then compared with the UV photochemistry of the same model molecules (*13*).

The absorption of high-energy radiation does not follow the same selection rules and spin conservation laws that apply to the formation of electronically excited states in UV absorption. With high-energy γ-ray photons, the probability of absorption is, in practice, directly proportional to the mass of the stopping atoms and consequently a rather random absorption of energy should take place in the first instance. Surprisingly, in spite of this, it appears from experimental results, that a significant portion of the energy is localized in the same excited states as are induced by the absorption of the UV photon. For example, Pitts and Osborne (*14*) observed that the major products arising from γ-irradiated aliphatic methyl ketones were those of the classical Norrish type I and II processes.

The primary events after interaction of γ-rays with an aliphatic ketone are usually considered to be

$$R-\overset{\overset{O}{\|}}{C}-R \rightsquigarrow R-\overset{\overset{O^+}{\|}}{C}-R + e^- \qquad (13)$$

$$\mathrm{R{-}\overset{\displaystyle O}{\overset{\|}{C}}{-}R} + \mathrm{R{-}\overset{\displaystyle O^{+}}{\overset{\|}{C}}{-}R} \xrightarrow{\mathrm{RH}} \mathrm{R{-}\overset{\displaystyle O}{\overset{\|}{C}}{-}\dot{R}H^{+}} + \mathrm{\cdot R'H{-}\overset{\displaystyle O}{\overset{\|}{C}}{-}R} \qquad (14)$$

$$\mathrm{R{-}\overset{\displaystyle O}{\overset{\|}{C}}{-}\dot{R}H^{+}} \longrightarrow \mathrm{R{-}\underset{\bullet}{\overset{\displaystyle OH^{+}}{\overset{\|}{C}}}{-}R} \qquad (15)$$

The free electrons released by the ionization will undergo additional collisions and will probably react to neutralize some of the ionic species present by

$$\mathrm{R{-}\overset{\displaystyle O}{\overset{\|}{C}}{-}R} + e^{-} \longrightarrow \mathrm{R{-}\overset{\displaystyle O^{-}}{\overset{\|}{C}}{-}R} \qquad (16)$$

$$\mathrm{R{-}\overset{\displaystyle O}{\overset{\|}{C}}{-}R^{+}} + e^{-} \longrightarrow \mathrm{R{-}\overset{\displaystyle O^{*}}{\overset{\|}{C}}{-}R} \qquad (17)$$

Additional excited molecules will be formed by reactions between cationic and anionic species present. Thus

$$\mathrm{R{-}\overset{\displaystyle O^{-}}{\overset{\|}{C}}{-}R} + \mathrm{R{-}\overset{\displaystyle O}{\overset{\|}{C}}{-}\dot{R}H^{+}} \longrightarrow \mathrm{R{-}\overset{\displaystyle O}{\overset{\|}{C}}{-}R} + \mathrm{R{-}\underset{\bullet}{\overset{\displaystyle OH}{\overset{\|}{C}}}{-}R} \qquad (18)$$

$$\mathrm{R{-}\overset{\displaystyle O^{-}}{\overset{\|}{C}}{-}R} + \mathrm{R{-}\overset{\displaystyle O}{\overset{\|}{C}}{-}R{+}} \longrightarrow \mathrm{2R{-}\overset{\displaystyle O^{*}}{\overset{\|}{C}}{-}R} \qquad (19)$$

Based on the experimental evidence, most of the excited ketone species will react by the four classical processes elucidated in the UV photochemistry of aliphatic and aromatic ketones (*15*):

Norrish type I

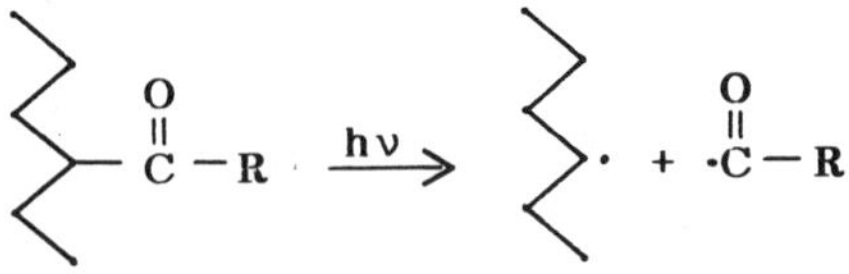

(20)

Norrish type II

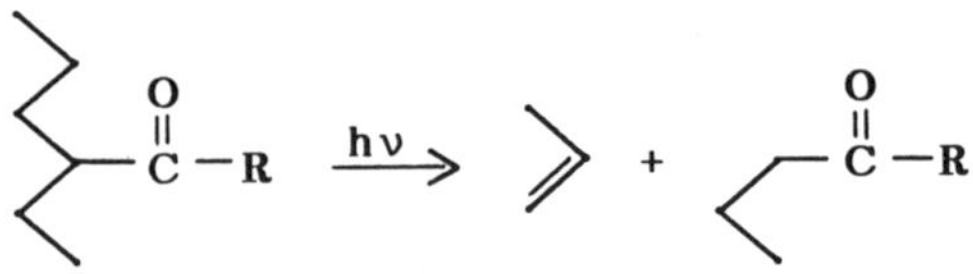

(21)

Hydrogen abstraction

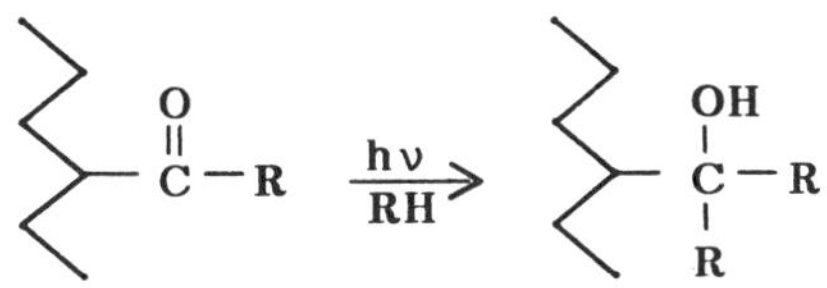

(22)

Cyclobutanol formation

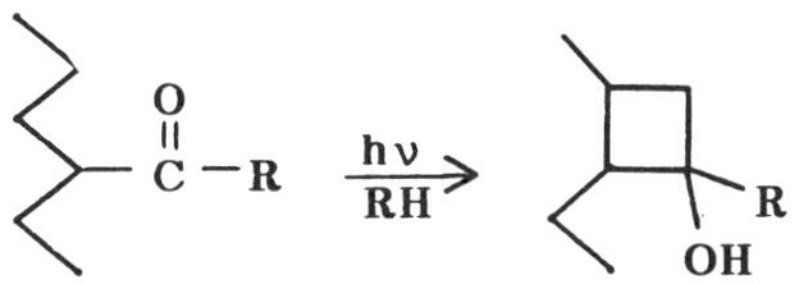

(23)

Although the primary excited species (e.g., ions and electrons) formed by radiation processes are seldom those created by selective absorption of UV photons, they have very short lifetimes (*16*) and the path of the reaction appears to be nearly the same. It seems likely that lower energy excited states such as those accessible by UV absorption are eventually populated by a series of energy transfer processes. In any case, the products produced from the excited ketone molecules are predominantly those expected from the Norrish type I and II processes (*13*,*14*).

Polymeric Ketones. In the case of vinyl ketone copolymers with polystyrene, reaction of the carbonyl group competes with that of the methylene groups in the hydrocarbon backbone and the phenyl groups in the side chains. Analysis of the products of the reaction in simple model aliphatic ketones can give an accurate estimate of the relative proportions of the reactions of the carbonyl and methylene groups. Table II shows that the carbonyl group is about four times as reactive towards γ-rays as would be predicted from its mass absorption coefficient. For example, although in 4-heptanone the carbonyl would be expected to absorb only 25% of the radiation, the Norrish type I and II processes accounted for 84% of the total reaction products observed. Similarly, the ratio of Norrish type I and II products to total reaction products remains constant at about 4 for the series from 4-heptanone to 7-tridecanone (*13*).

It was also noted, both in photochemical and radiochemical investigations, that crystalline ketones, such as 8-pentadecanone and 12-tricosanone, gave no measurable amounts of either type I, or type II or of normal disproportionation products (*13*). This is presumably because the molecular motion possible in a crystalline lattice is insufficient to allow escape of the radical products or the formation of the cyclic six-membered ring intermediate required for the Norrish type II process. When the ketones were diluted with hexane or other hydrocarbon solvents it was found that the yield of type I and II products was independent of concentration down to about 50% dilution, after which there was some decrease in the products (Figure 1). This shows that there was an effective mechanism for the

Table II. Radiolysis Yields of Symmetrical Aliphatic Ketones[a]

Ketone (phase)	G, products from ketone: Type I Alkane	Type II Methyl ketone	Type II Olefin	G(A)[b]	G(B)[c]	$G(CH_2)$	G(total products)	$\frac{G_I+G_{II}}{G_{total}}$	f[d]	$\left(\frac{G_I+G_{II}}{G_{total}}\right)\left(\frac{1}{f}\right)$
4-Heptanone (*l*)	1.42	0.49	0.49	0.15	0.32	0.47	2.87	0.84	0.25	3.4
5-Nonanone (*l*)	0.86	0.51	0.30	0.10	0.27	0.37	2.04	0.82	0.20	4.1
6-Undecanone (*l*)	0.56	0.45	0.24	0.17	0.40	0.57	1.82	0.69	0.16	4.2
7-Tridecanone (*l*)	0.36	0.28	0.17	0.07	0.58	0.65	1.44	0.55	0.14	3.9
8-Pentadecanone (*s*)	$<1\times10^{-4}$	$<1\times10^{-4}$	$<1\times10^{-4}$							
12-Tricosanone (*s*)	$<1\times10^{-4}$	$<1\times10^{-4}$	$<1\times10^{-4}$							

[a] Neat, under vacuum, at 35°C with ^{60}Co γ rays. [b] A = sum of all low molecular weight products.
[c] B = sum of all high molecular weight products. [d] f = fraction of radiation absorbed by CO.

transport of the excitation energy from the solvent to the ketone in this instance. The mechanism by which energy originally deposited in the methylene groups of the solvent is transported to the ketone carbonyl is not yet completely understood.

The most important conclusions can be drawn when the results for γ-radiolysis are compared with those of UV photolysis of the same ketones (*13*). Table III for example shows comparisons of the G-value for photolysis, calculated from equation 11, as compared to that from γ-radiolysis. In these experiments, estimates of the G values for the type I process were estimated from the yield of carbon monoxide in the photolysis and for the alkane in radiolysis, i.e.,

$$R-\overset{O}{\overset{\|}{C}}-R \xrightarrow{h\nu} CO + 2R\cdot \xrightarrow{RH} CO + 2RH \text{ (alkane)} \qquad (24)$$

thus

$$G_{CO} = G_{alkane}/2 \qquad (25)$$

The ratio of the G-value for the type I radical process to the type II intramolecular rearrangement is very much higher in the case of radiolysis than it is with UV photolysis. This was explained as being due to the fact that, in radiolysis, higher vibrational levels of the electronically excited states of the ketone are achieved, causing a significant increase in the quantum yield or G-values for radical formation. It was suggested that this excess energy of excitation allowed more facile escape of the radical pairs from the solvent cage. As expected from earlier UV studies (*17*) the quantum yields for both processes decrease with increasing length of the hydrocarbon chain.

Styrene-Vinyl Ketone Copolymers. Polystyrene (PS) is one of the most resistant polymers to radiochemical damage and for this reason

Table III. Comparison of G Values for Photolysis (313 nm) and Radiolysis of Symmetrical Aliphatic Ketones at 35°C[a]

Ketone	G (photolysis)			G (radiolysis)		
	$-CO_{(I)}$	Methyl $ketone_{(II)}$	Ratio G_I/G_{II}	$(Alkane/2)_{(II)}$	Methyl $ketone_{(II)}$	Ratio G_I/G_{II}
4-Heptanone	0.44	3.8	0.12	0.71	0.49	1.45
5-Nonanone	0.025	2.8	0.0090	0.43	0.51	0.88
6-Undecanone	0.020	2.4	0.0083	0.28	0.45	0.62
7-Tridecanone	0.017	2.0	0.0085	0.18	0.26	0.69
8-Pentadecanone	0.015	1.7	0.0088	0.065	0.092	0.71
12-Tricosanone	0.012	1.5	0.0080	0.040	0.081	0.69

[a]Ref. 13.

has often been used as a base for scintillators and other plastic devices exposed to high energy radiation.

Early studies on the crosslinking of polystyrene have been reviewed by Chapiro (*2*). In subsequent work, Slivinskas and Guillet (*18*) studied a series of copolymers of styrene with methyl vinyl ketone (MVK). Substantial differences were observed between UV photolysis and γ-radiation, particularly in that the polymers are very much less sensitive to γ-radiation than to UV photolysis. PS itself is known to undergo crosslinking very slowly when exposed to γ-radiation, but when progressively copolymerized with vinyl ketones, the amount of crosslinking decreases such that, when the amount of ketone exceeds about 2%, the polymer degrades rather than crosslinks. This is shown in Figure 2. In contrast to the results with polyethylene-carbon monoxide copolymers (*19*), no change in the yield of hydrogen as a function of carbonyl content occurred, an observation explained on the basis that the excitation of PS is localized in the phenyl rings and thus no large-scale migration of the energy is possible in this polymer by the exciton band mechanism which applies to polyethylene. It is interesting to note that the quantum yield for chain scission in PS-0.5% MVK copolymer, calculated from the G values, is 0.002, whereas for UV photolysis, the quantum yield would be 0.2, two orders of magnitude higher. It is obvious that, in the polystyrene copolymer the phenyl rings are protecting the carbonyl groups from excitation by the high energy radiation, whereas no such protective effect is observed when the carbonyl group is directly excited by UV light.

In recent work we have shown the importance of energy delocalization in the radiation chemistry of aromatic polymers (*5*,*19*,*20*). In dilute solution, this delocalization occurs along a single chain, and is often called "the antenna effect" (*5*). Efficient energy delocalization also occurs in solid aromatic polymers, but because of extensive interchain penetration it is difficult to distinguish between inter- and intramolecular energy migration. However, practical use of the effect has been made in the design of scintillators and other devices for detection of high-energy radiation (*6*). By including a reactive energy trap in the polymer, it is possible to increase the radiation yield of a radiochemical process. Alternatively, stable energy traps may retard chemical reactions which lead to the deterioration of physical properties and thus impart enhanced resistance to damaging radiation.

The present work relates to studies of the radiation chemistry of copolymers of styrene (ST) and vinyl naphthylene (VN) with and without methyl isopropenyl ketone (MIPK) as a comonomer. MIPK introduces the radiation-sensitive ketone moiety which reacts by the two classical Norrish reactions when raised to its first excited singlet or triplet state.

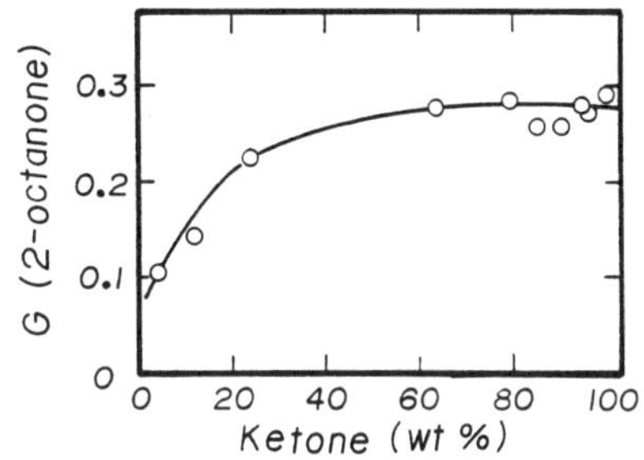

Figure 1. G values for 2-octanone (type II) product from radiolysis of solutions of 7-tridecanone in octane as a function of ketone concentration. (Reproduced with permission from reference 13. Copyright 1973 by John Wiley & Sons.)

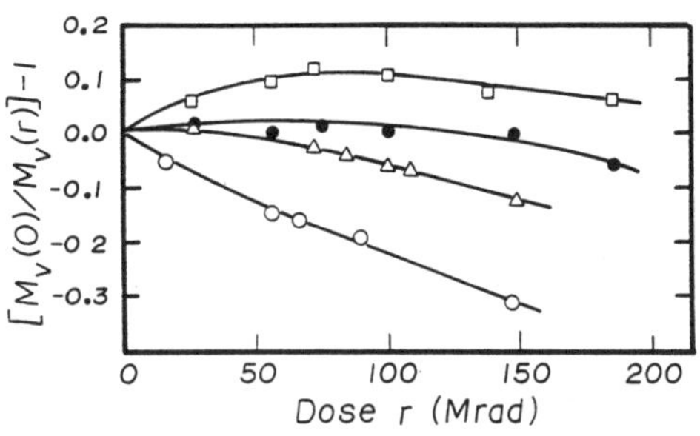

Figure 2. Changes in M_v versus radiation dose R for MVK–PS copolymers irradiated under vacuum in solid phase at 35 °C. (Reproduced with permission from reference 18. Copyright 1973 by John Wiley & Sons.)

Type I

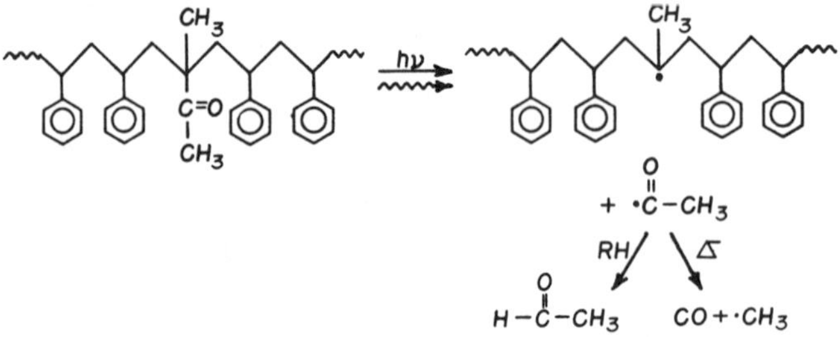

(26)

Type II

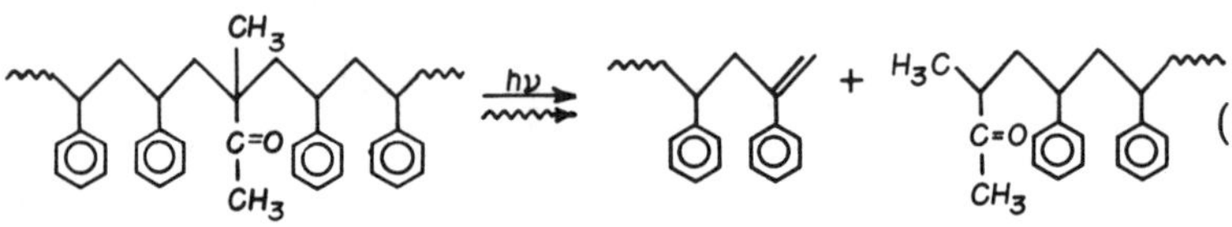

(27)

β-Scission

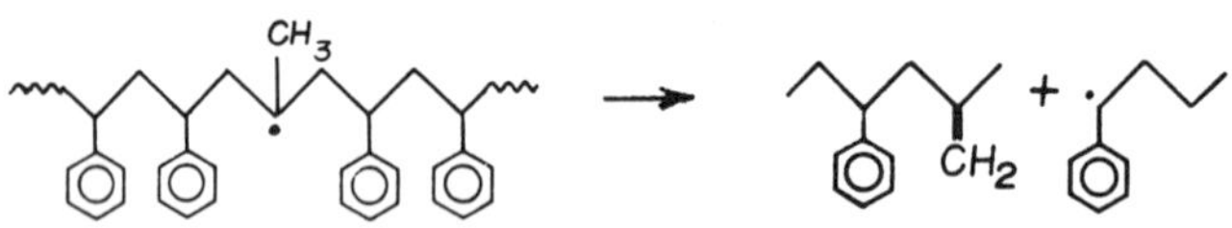

(28)

The type I reaction can be detected by the loss of carbonyl groups, since the acyl radical will either decarbonylate to give CO and $\cdot CH_3$ or abstract a proton from the surroundings to give acetaldehyde which is lost by volatilization. The type II process results in a decrease in polymer molecular weight, which can be monitored by viscometry and other techniques. A change in molecular weight also occurs by β-scission of the polymer radical produced in the type I process (equation 28) (*21*). In the solid phase, crosslinking can occur by recombination of any of the polymeric species generated in the type I or β-scission processes.

Theoretical studies by Kells and Guillet (*22*) indicate that precise estimation of the radiation yields (i.e., G_s and G_x) by conventional methods of molecular weight determination such as GPC or light scattering are extremely difficult unless the starting polymer is monodisperse. In this work, we rely primarily on FTIR measurements of the loss of ketone carbonyl to estimate the efficiency of the Norrish type I process in solid polymers containing ketone groups.

The extent of energy migration can be measured by the efficiency of the reaction of ketone traps excited by migrating excitons in solid polymers (*17*). This paper deals with the study of copolymerized stable traps such as the naphthalene group.

Experimental

Polymer Synthesis and Characterization. Copolymers of Methyl Isopropenyl Ketone, Vinyl Naphthalene, and Styrene. A number of copolymers of MIPK, VN and ST with varying ketone content were made by the procedures given below. Characterization data are summarized in Table IV. Molecular weights were determined by GPC using PS standards. Intrinsic viscosities ([η]) were determined by precision automatic viscometery (*23*).

MIPK monomer (EcoPlastics Ltd.) was fractionally distilled below room temperature before use. ST monomer was distilled at room temperature under vacuum. The copolymers were prepared using azobis-(isobutyronitrile) (AIBN) as a free radical initiator and dioxane as solvent.

AIBN (*ca.* 30 mg) and 10 g of monomer (in various ratios of ST to MIPK) were dissolved in 50 mL of dioxane. The glass reaction ampules were connected to a vacuum and the mixtures were degassed using three freeze-thaw cycles. After sufficient degassing, the ampules were sealed under vacuum and left at 65°C for 48 h.

The polymers were separated from their monomers and solvent by three precipitations from dioxane into methanol. The remaining product was dried in vacuum overnight and freeze-dried from a dioxane solution to remove any residual monomer or solvent molecule. Characterization of the polymers is summarized in Table IV.

Copolymers of Methyl Isopropenyl Ketone and β-Vinylnaphthalene. The MIPK-VN copolymer was made as follows. β-Vinylnaphthalene (2.6

Table IV. Characterization of ST, VN and MIPK Copolymers

Copolymer	[η]	$M_n \times 10^{-3}$	$M_w \times 10^{-3}$	M_w/M_n	% Conversion
PS-10.3% MIPK		37	68	1.84	~50
PS-17.8% MIPK		37	65	1.76	~50
PS-25.0% MIPK		23	48	2.09	~50
PS-37.4% MIPK		16	28	1.75	~50
PS-42.3% MIPK		8	17	2.12	~50
PS-57.7% MIPK					~50
PVN-12% MIPK	0.32				~50
PS-1.4% VN	0.32	35.2	54.2	1.54	43
PS-6.2% VN	0.31	42.5	65.1	1.53	40
PS-13% VN	0.31	36.8	56.9	1.55	45
PS-24% VN	0.28	41.3	65.8	1.59	43
PS-36% VN		36.2	56.9	1.57	47
PS-58% VN		34.8	53.1	1.52	51
P2VN	0.25	42.7	66.5	1.55	61
PS	0.33	36.2	56.1	1.54	44

g), 0.3 g of MIPK and 10 mg of AIBN initiator were dissolved in 15 mL of dioxane. After three freeze-thaw cycles, the ampule was sealed and left at 60°C for 24 h. The polymer was precipitated in approximately 1.5 L of methanol, using a standard technique. The conversion was 50%.

The MIPK content was determined using high-resolution ^{1}H-NMR. The ketone $-CH_3$ at 2.10 ppm and the naphthalene -H around 7 ppm gave a ketone content of approximately 12%. The polydispersity of the polymer sample, M_w/M_n, was measured by GPC.

Copolymers of β-Vinylnaphthalene and Styrene. β-Vinylnaphthalene was purified over alumina. Styrene was distilled under reduced pressure at room temperature. AIBN (60 mg) and 20 g of monomer (in various ratios of ST and VN) were dissolved in 100 mL of 1,4-dioxane. The ampules were degassed before polymerization of the monomers at 65°C. After 48 h, the polymers were separated from their monomers and solvent by three precipitations from benzene into methanol. The remaining product was completely dried under high vacuum for 72 h.

Results and Discussion

Poly(styrene-*co*-methyl isopropenyl ketone) Copolymers. The series of PS-MIPK copolymers was subjected to electron-beam irradiation, and the loss of carbonyl was measured by FTIR spectroscopy (*20*).

The UV irradiation of these films showed a quantum yield of $\phi_{-CO} = 0.24 \pm 0.02$ regardless of composition. By contrast, the electron-beam studies of the same copolymers showed strong evidence for energy migration and transfer to the carbonyl groups. The G values were plotted as a function of MIPK content (Figure 3). The plot deviates significantly from the linear relationship (dotted line) expected if the radiation energy is partitioned equally among absorbing chromophores. It seems clear that excitation transfer must occur from excited styrene units, particularly at lower ketone contents.

Poly(vinylnaphthalene-*co*-12% methyl isopropenyl ketone) (VN-MIPK). By replacing the phenyl group in the styrene with a naphthyl group, one introduces an additional photophysical pathway for the dissipation of excitation energy. A further effect of replacing a phenyl group with a naphthyl is the increased transfer length for singlet excitation energy in naphthalene. The Förster radius for naphthalene to naphthalene is about twice as great as for phenyl to phenyl.

The results for this polymer showed that no significant loss of ketone carbonyl occured, even at relatively high doses. This effect is very similar to that observed in photochemistry of ketone polymers. For example, Holden et al. (*24*) showed that copolymerization of 1- and 2-naphthyl methacrylate with vinyl ketones yields polymers with very low photochemical reactivity.

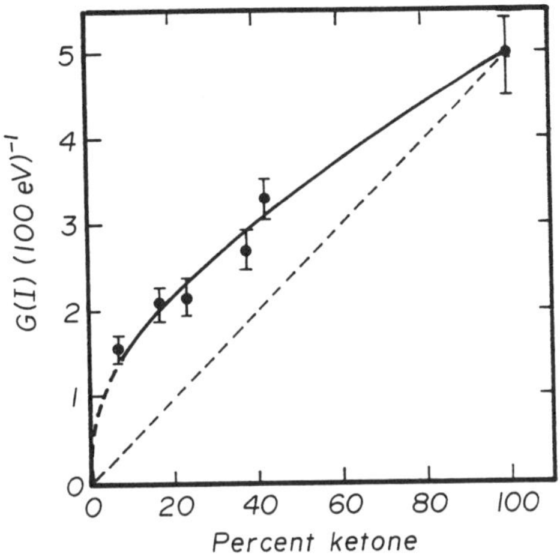

Figure 3. G(-CO) for PS–MIPK copolymers as a function of MIPK content. (Reproduced with permission from reference 16. Copyright 1990 by Harper & Row.)

In the photochemical experiment, this was explained with reference to the energy levels of the singlet and triplet excited states, shown in Figure 4. In this case, energy originally absorbed by the ketone group gives rise to an excited singlet state. Rapid intersystem crossing gives a ketone triplet state which is higher in energy than that of naphthalene. Quenching of the ketone triplet by naphthalene gives triplet naphthalene which disposes of the excitation energy by photophysical pathways, i.e., luminescence or intersystem crossing to the ground state, thus reducing the chemical yields from the ketone chromophore.

A similar mechanism could be invoked to explain the electron beam results for the VN-MIPK copolymer if we assume that the migrating excitons in the polymer film are primarily triplets. Since the triplet level of naphthalene is lower in energy than that of the ketone carbonyl, no excitation of the carbonyl would be expected and thus no photochemistry occurs. Also, direct excitation of the carbonyl by radiation would result in triplet species which would be expected to be quenched by nearby naphthalene groups.

The lower singlet and triplet energy levels of the phenyl rings in polystyrene are also included in Figure 4. Since the triplet level of the phenyl groups in polystyrene is well above that of the ketone group, triplet reactions will not be quenched and efficient photo and radiochemical reactions should be expected in ST-MIPK copolymers as has been observed experimentally.

γ-Radiation Studies. Since much of the chemistry induced by γ-radiation is due to electrons produced by Compton scattering, it seems likely that effects similar to those of direct e-beam studies may be observed. The first excited singlet and triplet states of naphtha-

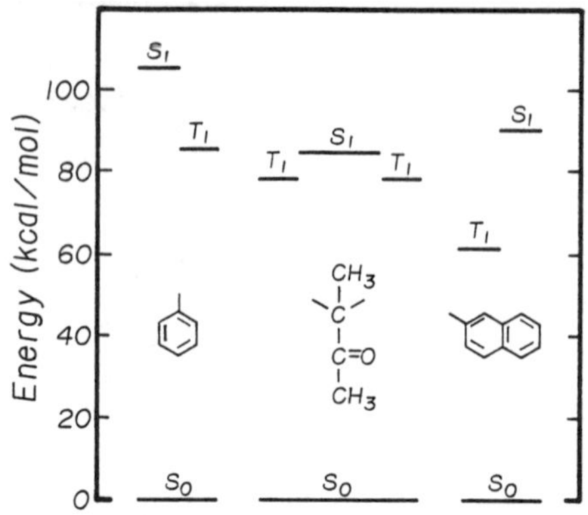

Figure 4. Energy level diagram for styrene and vinylnaphthalene copolymerized with methyl isopropenyl ketone. (Reproduced with permission from reference 16. Copyright 1990 by Harper & Row.)

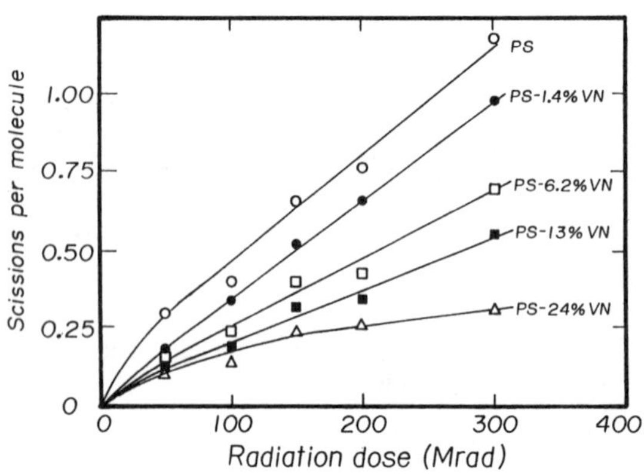

Figure 5. Rate of scission of γ-irradiated copolymers.

lene lie below those of the corresponding states of the phenyl ring in polystyrene. Since PVN is known to be more radiation stable than PS (*25*), one might expect to see an increase in stability in styrene copolymers with VN. Preliminary data on the rate of scission of a series of such copolymers exposed in air to γ-radiation is shown in Figure 5. There does indeed seem to be significant stabilization of the VN-ST copolymers compared to pure PS, although more work will be required to confirm the increased stability under very long term exposure to high-energy radiation in these and other copolymers.

These applications illustrate the usefulness of photochemical principles in explaining radiochemical processes in certain polymer systems.

Acknowledgments

We are grateful to the Natural Sciences and Engineering Research Council of Canada and to the Ontario Centre for Materials Research for their financial support of this research. J. E. G. thanks the Canada Council for support in the form of a Killiam Research Fellowship.

Literature Cited

1. Bovey, F. A. *The Effects of Ionizing Radiation on Natural and Synthetic High Polymers*; Polymer Reviews; Interscience, New York, 1958; vol. 1.
2. Chapiro, A. *Radiation Chemistry of Polymer Systems*, Interscience, New York, 1962, vol. 15.
3. Ausloos, P. *Fundamental Processes in Radiation Chemistry*, Interscience, New York, 1968, vols. 1 and 2.
4. *The Radiation Chemistry of Macromolecules*; Dole, M., Ed.; Academic Press, New York; (a) 1972, vol. 1; (b) 1973, vol. 2.
5. Guillet, J. E. *Polymer Photophysics and Photochemistry*; Cambridge University Press: Cambridge, 1985; Chapt. 14.
6. Birks, J. B. *The Theory and Practice of Scintillation Counting*; Pergamon Press: Oxford, 1964
7. Dexter, D. L. *J. Chem. Phys.* **1953**, *21*, 836.
8. Förster, Th. *Ann. Phys.* **1948**, *2*, 55.
9. Charlesby, A. *Atomic Radiation and Polymers*; Pergamon Press: Oxford, 1960.
10. MGinnis, V. D. *Encycl. Polym. Sci. Eng.* **1986**, *4*, 418.
11. Clough, R. *Encycl. Polym. Sci. Eng.* **1988**, *13*, 673.
12. Schultz, A. R. *Encycl. Polym. Sci. Techn.* **1966**, *4*, 398.
13. Slivinskas, J. A.; Guillet, J. E. *J. Polym. Sci., Polym. Chem. Ed.*, **1973**, *11*, 3043.
14. Pitts, J. N.; Osborne, A. D. *J. Am. Chem. Soc.* **1961**, *83*, 3011.
15. Guillet, *op. cit.*, Ch. 12.
16. Tabata, Y. In *Polymers for Microelectronics -- Science and Technology*; Tabata, Y.; Mita, I.; Nonogaki, S., Eds.; Kodansha Scientific Publishers: Tokyo, 1990.
17. Hartley, G.; Guillet, J. E. *Macromolecules* **1968**, 1, 413.
18. Slivinskas, J. A.; Guillet, J. E. *J. Polym. Sci., Polym. Chem. Ed.*, **1973**, *11*, 3057.
19. Slivinskas, J. A.; Guillet, J. E. *J. Polym. Sci., Polym. Chem. Ed.*, **1974**, *12*, 1469.
20. Hesp, S. A. M.; Guillet, J. E. In *Polymers for Microelectronics -- Science and Technology*; Tabata, Y.; Mita, I; Nonogaki, S., Eds.; Kodansha Scientific Publishers: Tokyo, 1990.
21. Guillet, J. E. In *Advances in Photochemistry*; Volman, D. H.; Hammond, G. S.; Gollnick, K., Eds.; Wiley, New York, **1988**, vol. 14.
22. Kells, D. I. C.; Guillet, J. E. *J. Polym. Sci., Part A-2*, **1969**, *7*, 1895.
23. Kilp, T.; Houvenaghel-Defoort, B.; Panning, W.; Guillet, J. E. *Rev. Sci. Instrum.* **1976**, *47*, 1496.
24. D. A. Holden, X.-X. Ren and J. E. Guillet, *Macromolecules*, **1984**, 17, 1500.
25. Gunder, O. A.; Koda, V. S., *Radiokhimiya*, **1969**, 11, 119.

RECEIVED June 20, 1991

Chapter 26

Stabilization of Polyolefins to Gamma Irradiation

Role of the Initial Radicals

D. J. Carlsson, S. Chmela[1], and J. Lacoste[2]

Division of Chemistry, National Research Council, Ottawa, Ontario K1A 0R9, Canada

The initial, atypical peroxyl radicals which start the free-radical chain oxidations have been identified in γ-irradiated polyolefins. These radicals were quantified in total by electron spin resonance and trapped by reaction with nitric oxide at -78°C to give nitrates which could be individually identified by infrared spectroscopy. Loss in physical properties was correlated with the accumulation of oxidation products formed during post-irradiation storage for polypropylene containing several stabilizer combinations. The best retention of impact strength corresponded to the minimum formation of oxidation products although none of the stabilizer combinations was able to completely suppress the oxidation processes.

The exposure of polyolefins to high energy radiations (γ, X-rays, electron beams, etc.) causes the formation of various free radicals (*1,2*). Some of these radicals combine with oxygen dissolved in the polymers to initiate chain reactions and form thermally unstable products. Degradation of polymers exposed to γ-rays results both from the direct radiation effects (C-C scission caused directly by radiation reactions) and from the much slower oxidative chain scissions (*3*). These continue after the end of irradiation, during the subsequent storage of the polymers. The oxidative process will involve peroxyl radicals derived directly from the radiation reactions (reaction 1) and also from the attack of these initial peroxyl radicals on the polymer (reactions 2 and 3). The peroxyl species in reaction 3 are "normal" in that their structure is determined by the selectivity of peroxyl radicals, as in any oxidation process (UV,

[1]Current address: Polymer Institute, Slovak Academy of Science, Bratislava, Czechoslovakia
[2]Current address: Université Blaise-Pascal, Clermont-Ferrand, France

0097–6156/91/0475–0432$06.00/0

thermal, mechano, or radiation initiated). The peroxyl $rO_2\cdot$ from reaction 1 is atypical in that its structure is determined by the cascade of post-irradiation events although its selectivity in reaction 2 will be identical to that of $PO_2\cdot$ from reaction 3 (*4*). Both types of hydroperoxide (rOOH and POOH) are believed to slowly breakdown at room temprature to initiate the post-irradiation degradation(*2*).

$$\text{Polyolefin} \xrightarrow{\gamma} r\cdot \xrightarrow{O_2} rO_2\cdot \quad (1$$

$$rO_2\cdot + \underset{(PH)}{\text{Polyolefin}} \longrightarrow rOOH + P\cdot \quad (2$$

$$P\cdot + O_2 \longrightarrow PO_2\cdot \quad (3a$$

$$PO_2\cdot + PH \longrightarrow POOH + P\cdot \xrightarrow{O_2} PO_2\cdot \quad (3b$$

$$rO_2\cdot, PO_2\cdot \longrightarrow \text{Some chain scission products} \quad (4$$

$$rOOH, POOH \xrightarrow{RT} \text{free radicals, rOH, POH} \quad (5$$

Prevention of radiation induced oxidation will depend on the reduction of radical formation and/or the scavenging of free-radical intermediates ($r\cdot$, $rO_2\cdot$, $P\cdot$, $PO_2\cdot$). An important necessity for medical equipment which is to be γ-sterilized is the complete absence of real or perceived contamination. This effectively eliminates all color-forming antioxidants such as phenols (yellow-brown products) or aromatic amines (red-brown products) from plastic medical ware (*5*). Hindered amines such as these based on 2,2,6,6-tetramethylpiperidine operate as antioxidants at ambient temperatures in light-stabilization packages. Furthermore they and their products are colorless or only very weakly absorbing. These aliphatic amines have been previously shown to function as stabilizers to post-γ-irradiation oxidation (*2,6*). Highly efficient stabilizer combinations might possibly suppress oxidation to the point where the atypical product(s) (rOOH) dominate. Post-irradiation oxidation is largely dependent upon initiation by the slow thermal decomposition of hydroperoxides (rOOH and/or POOH) (*2,6*). Hydroperoxide decomposition by an additive to give harmless products will also prevent this oxidative degradation.

γ-Initiation of polymer oxidation is superficially a very clean method for studying polymer degradation mechanisms. The possible dominance of atypical species ($r\cdot$, $rO_2\cdot$, rOOH) questions the universal nature of the results from this type of degradation and their application to other types such as those initiated by photo and thermal processes. In this study we have identified and quantified several of the key intermediates in γ-irradiated polyolefins, examined the effects of various stabilizers on the product yields and compared yields of products from efficiently stabilized systems with these of the initial, atypical products.

Experimental

Polyethylene (high density, Union Carbide, 25 μm) and isotactic polypropylene (Himont Profax, 25 μm) were acetone extracted to remove antioxidants. Atactic polypropylene films (Hercules experimental resin, zero crystallinity and helical content) were cast from toluene. Stabilized polypropylene films were prepared by melt compounding and hot pressing or melt extrusion. Samples were γ-irradiated in an AECL Gammacell 220, (dose rate ~1.0 Mrad/h). Products were measured by electron spin resonance (Varian E4) and Fourier transform infrared (IR, Perkin Elmer 1500) spectroscopies. Peroxyl radicals were preserved at low temperature then reacted with gaseous NO for 2-3 hours at -78°C to give the corresponding nitrates which could be identified by IR. Hydroperoxides were observed as their corresponding nitrates after prolonged (>100h) reaction with NO at -20°C (*7*). Ketonic species were estimated by IR after reaction of all-OH species in oxidized films with SF_4 (*7*). Impact properties of melt extruded plaques were measured by drop-weight impact, after exposure to the accelerated test conditions of 60°C storage in air.

Stabilizers studied included bis 4-(2,2,6,6-tetramethylpiperidyl) decandioate (HALS-I, Ciba Geigy), bis(4-methylphenyl)carbinol (benzhydrol-I, Milliken XA-100), 2-hydroxy-4-octoxybenzophenone (2HOϕCOϕ, American Cyanamid) and an alkane oil (oil-I, Witco 300) (*6*).

Results and Discussion

Structure of the Initial Peroxyl Radical. The initial peroxyl radicals ($rO_2\cdot$) were generated by irradation of polymers at -78°C in air. These radicals were then studied at ≤ -60°C when they neither propagate nor terminate in polyethylene and polypropylene under air or vacuum (*3*). The total peroxyl yield was measured by electron spin resonance (esr) together with the yield of any surviving macro alkyl radicals. However e.s.r. can give no information on the precise structure of the peroxyl radical in the solid state.

More information in the structure of the initial peroxyl radicals may be obtained if they can be trapped to give products which can be easily identified. At Dry Ice temperature (-78°C), peroxyl radicals were found to react smoothly and quantitatively with gaseous nitric oxide to give organo-nitrates (reaction 6). After the complete removal of all oxygen, e.s.r. showed that the peroxyl signal was lost within about 5 h. upon exposure to ~200 torr of NO at -70°C. We have previously shown that organo-nitrates have sharp, intense IR absorptions which can be used to

$$\left.\begin{matrix} rO_2\cdot \\ \\ PO_2\cdot \end{matrix}\right\} + NO \rightarrow \left\{\begin{matrix} rONO_2 \\ \\ PONO_2 \end{matrix}\right. \qquad (6$$

differentiate and quantify primary, secondary and tertiary nitrates (*7*). Sec.-nitrates formed from sec-hydroperoxides absorb at 1276 cm^{-1}, tert.-nitrates absorb at 1294 cm^{-1} whereas primary nitrates absorb at 1276 cm^{-1} and ~1641 cm^{-1} (Table I). The syn nitrate absorption at ~1630 cm^{-1} is very intense and sensitive to nitrate structure but unfortunately sensitive to the polarity of its immediate environment (Table I).

The IR spectra from i-PP and HDPE films after γ-irradiation at -78°C in air followed by NO exposure at -78°C are shown in Figure 1. From Figure 1, negligible amounts of carbonyl species (at ~1715 cm^{-1}) were detectable, consistent with the complete absence of peroxyl self reactions. For HDPE, the IR changes are consistent with the presence of sec. nitrate (from sec. peroxyl), whereas PP shows the presence of some tert. nitrate together with the dominant sec. nitrate. The twin peaks at 1631 and 1623 cm^{-1} in PP may imply the presence of peroxyl radicals in two different environments.

With the reasonable assumption that the radicals formed at ≤-60°C are similar to these produced at ambient temperatures, the NO reaction leads to the initial peroxyl yields shown in Table II. The peroxyl yields from polypropylene are markedly different from the primary: secondary: tertiary yields predicted from the selectivity of the oxidative propagation reactions (reaction 3) (primary: secondary: tertiary CH attack ~1:20:200 for liquid model hydrocarbons) (*4*). Product yields from

Table I Model Nitrate IR Absorption Maxima

Nitrate Structure	Absorption	Maximum	(cm^{-1})	Environment
Primary				
$CH_3(CH_2)_{10}$-ONO_2	1641	1278	850	Hexane[a]
CH_3-CH_2-CH-CH_2-CH_3 (with CH_2-ONO_2 on CH)	1635	1278	860	CH_3CN[a], Hexane[b]
Secondary				
$CH_2(CH_2)_4CH(-ONO_2)CH_2CH_3$; $CH_3(CH_2)_7CH(-ONO_2)CH_3$	1633	1276	867-860	Hexane[a]
$CH_3CH(CH_3)CH(-ONO_2)$-$CH(CH_3)CH_3$; $CH_3CH(CH_3)CH_2$-$CH(-ONO_2)$ (with CH_2-$CH(CH_3)CH_2$-$CH(CH_3)_2$ on CH)	1623	1276	867-860	Hexane[b], CH_3CN[a], or Tornac[c]
Tertiary				
CH_3-$C(CH_3)_2$-CH_2-$C(CH_3)_2$-ONO_2	1630	1294	862	Hexane[a]
CH_3-CH_2-$C(CH_3)$-CH_2-CH_3 (with ONO_2 on C)	1623	1305,1294	861	Hexane[b], CH_3CN[a]

a Dilute solution, ~5 x 10^{-3} mole/l

b Concentrated solution, ~2 mole/l

c Hydrogenated, random butadiene-acrylonitrile copolymer

atactic and isotactic polypropylene were identical within experimental error (Table II). Also shown in Table II are macroalkyl radical yields for PP and HDPE measured after irradiation under vacuum at -196°C together with the persistent alkyl level found even in the presence of O_2 for HDPE. The mixed peroxyl/alkyl radicals were quantified by recording spectra at low (0.5 mw) and high (10 mw) power levels. At 0.5 mw both radicals contribute fully to the signal, whereas at 10 mw only peroxyl radicals contribute fully and (in our equipment) macroalkyl radicals contribute at 40% of their true level.

The persistent macroalkyl signal in HDPE may come from radicals produced in the O_2- impermeable crystalline phase. For PP, the more open (lower density) crystal is apparently O_2- permeable.

The sec.-nitrate from HDPE, detected as the dominant species implies the formation of only radical I (reaction 7a). The absence of peroxyl II from backbone cleavage implies that this reaction is rare or recombination of the macroalkyl pair

Table II Products from the γ-irradiation of Polyolefins

	Products (10^3 mole/kg/Mrad)				
Polyolefin	[R·][a]	[RO_2·][b]	Primary $RONO_2$[c]	sec. $RONO_2$[c]	tert. $RONO_2$[c]
HDPE	2.8 ± 0.5	2.1 ± 0.5[d]	0	1.9 ± 0.2	0
i-PP	3.6 ± 0.5	3.5 ± 0.5	0.3 ± 0.1	1.7 ± 0.2	0.9 ± 0.2
a-PP	ND	3.2 ± 0.5	0.4 ± 0.1	1.6 ± 0.2	0.7 ± 0.2

a From esr, irradiation under vacuum at -196°C.
b From esr, irradiation under 0_2 or air at -78°C followed by ~15 h at -78°C under 0_2 or air. Identical yield when vacuum irradiated sample exposed to 0_2 or air for ~15 h at -78°C.
c From FTIR, after $R0_2$ exposure to NO at -78°C.
d Also with (0.6 ± 0.2) x 10^{-3} mole/kg of trapped macroalkyl radicals.

is rapid. A 1,6-migration reaction of the alkyl to give the radical precursor to III can occur, but is believed to be slow in comparison with O_2 interception (*8*). For PP a more complex range of possibilities exist (reaction 8). The tert. nitrate probably results from IV above. The dominant sec. nitrate may come largely from reaction 8c as methane yields (reaction 8a) are very low from PP (*9*).

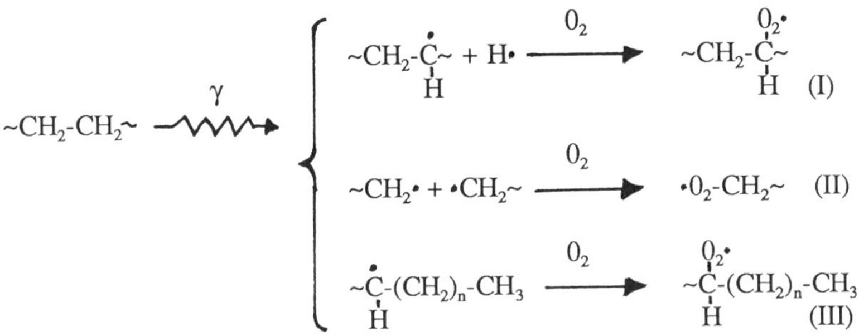

(7a

(7b

(7c

Oxidation Product Quantification. We have previously shown that the various (primary, secondary and tertiary) hydroperoxide and alcohol species can be identified in oxidized polyolefin by exposure to NO subsequent to the oxidation (*7*). All hydroperoxides quantitatively form the respective nitrates whereas alcohols form the corresponding nitrites (reactions 9 and 10). These again have the IR absorptions shown in Table I. We have recently found that it is essential to work at -20°C to accurately quantify tert-nitrates, which decompose quite rapidly at room temperature (*10*).

Products from the γ-initiated oxidative degradation of polypropylene at room temperature were measured for the additive free polymer and in the presence of several effective stabilizer systems. Stabilizer effectiveness was measured by retention of drop wight impact strength after an accelerated, post-irradiation test, involving storage at 60°C in air. Oxidation products were quantified by esr (total initial peroxyl radicals generated) and IR (ketones after SF_4 treatment and hydroperoxides after NO reaction) as described previously (*7*). The IR-identified products were measured after 120 hours at 21°C, subsequent to the γ-irradiation. This time was initially chosen because by then most of the radicals had decayed completely. The hydroperoxide mix from γ-irradiated polypropylene is markedly different from that expected from the initial peroxyl radicals. This is shown by the IR spectra in Figure 2, where the tert-nitrate (1294 cm^{-1}) dominates over the sec. nitrate (1276 cm^{-1}) after room temperature oxidation. From Figure 2 it is also apparent that the best stabilizer system markedly suppresses the formation of oxidation products. (It is important to note the differing absorbance scales used in Figure 2).

Oxidation products from the series of stabilized polypropylene films were measured using the NO and SF_4 treatments to improve sensitivity and selectivity. These data are collected in Table III. None of the stabilizer systems affected the initial peroxyl yields (within the errors of esr double integration). Primary hydroperoxide groups were not detected. None of the additives, except HALS-1, materially affected the product yields or impact retention when used alone, as compared with the additive free sample.

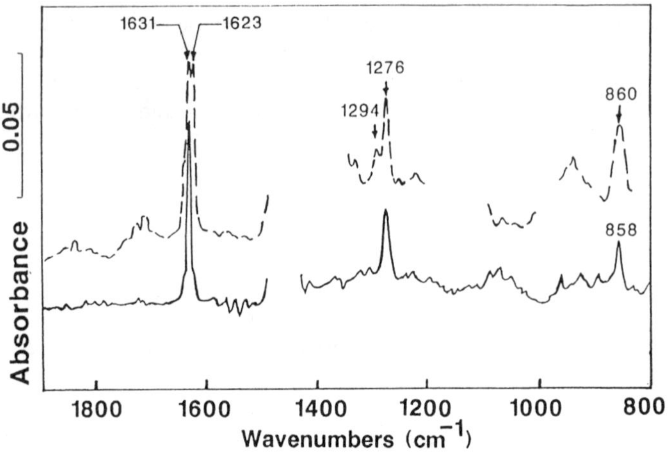

Figure 1 Nitrate products from peroxyl radicals by FTIR
Polyolefin film samples exposed to 6 Mrad γ-irradiation at -78°C in air, followed by 5h. NO exposure at -78°C after removal of all 0_2. Spectra obtained by subtracting the spectrum of the starting film from that of the treated film in all cases.
- - - i-PP samples: 4 x 25 μm films; — PE samples: 4 x 25 μm films.

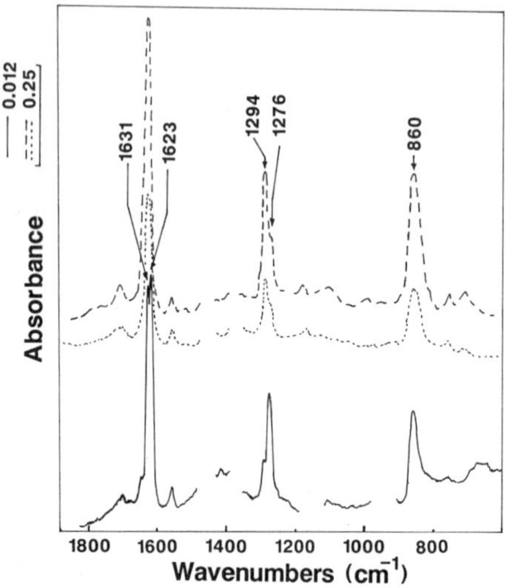

Figure 2 Comparison of FTIR spectra of nitrate products
Polypropylene (X4, 30μm films) spectra obtained by subtraction as in Figure 1.

— Unstabilized film, 2.5 Mrad at -78°C in air followed by 5h. NO at -78°C, 0_2 free.

- - - Unstabilized film, 2.5 Mrad at room temperature in air, followed by 120h. storage at room temperature in air, then 160h. NO at -20°C, 0_2 free.

.... Stabilized film (HALS-1 with Benzhydrol) treated as for previous samples.

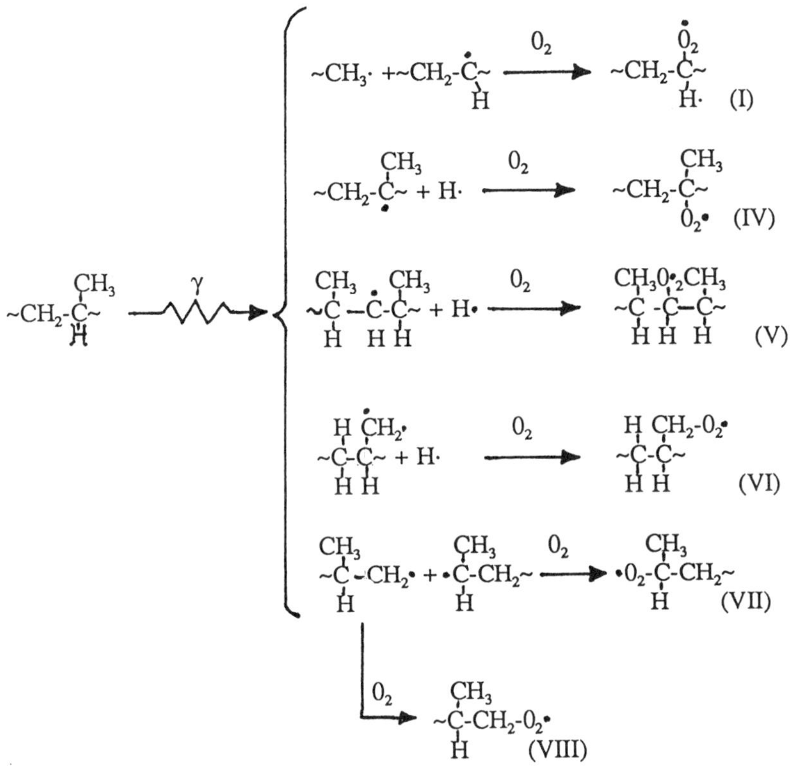

(8a

(8b

(8c

(8d

(8e

(8f

$$\mathrm{ROOH + NO \xrightarrow{-20^{\circ}C} RONO_2} \qquad (9$$

$$\mathrm{ROH + NO \xrightarrow{-20^{\circ}C} RONO} \qquad (10$$

None of the stabilizer systems affected the initial yields of peroxyl radicals produced at -78°C. It is proposed that this will also be the situation for irradiation at close to room temperature. Additive migration will be enhanced in the latter case but alkyl radical reactivity with 0_2 will also be higher. Because of the extremely rapid sequence of events leading to the first peroxyl radicals it is unlikely that any stabilizer can interfere with those events when present at only the 0.1 w% level.

From Table III it is apparent that stabilizer systems giving outstanding retention of impact in the 60°C accelerated test also markedly suppress the formation of both hydroperoxides and ketonic products. The latter originate predominantly from peroxyl self reaction, some of which cause backbone scission. Although consistent with enhanced impact retention, the suppression of oxidation products formation after 120 h and 21°C is proportionally much lower. We have recently found that this results from measuring the oxidation products at too early a stage in the oxidation (*11*). For much longer post-irradiation periods, the oxidation level of the unstabilized control increases dramatically whereas the good stabilizers hold the oxidation essentially at the low levels shown in Table III.

From the peroxyl radicals yields shown in Table II, for a 2.5 Mrad dose, a maximum of ~2.2 x 10^{-3} mole/kg of tert.-hydroperoxide and ~4.3 x 10^{-3} mole/kg of sec.-hydroperoxide could form from the initial, atypical radicals. From Table III these yields are only a small fraction of the hydroperoxide yields found for the best stabilizer system, which must be allowing some propagation before stabilizer interception of propagation species.

In conclusion, γ-irradiation of polyolefins gives rise to peroxyl radicals which are atypical of the subsequent chain oxidation process. The extensive chain oxidation leads to a mix of tert. and sec. hydroperoxides which approximates that expected from peroxyl reactivities. This mix overwhelms the low concentrations of products from the initial radicals even in well stabilized systems. Several stabilizer systems give high resistance to degradation during irradiation and storage and operate by suppressing the oxidative chain reactions.

Table III γ-Irradiation of Stabilized Polypropylene Films

Stabilizer System	Relative initial peroxyl Yield[a]	Oxidation Product (mole/kg)x10^3 [b] [tert-OOH]	[sec-OOH]	[>C=0]	Impact Retention (days)[c]
None	1.0	95	14	16	12
HALS-I[d]	1.1	52	10	9	200
HALS-I[d]+ Benzhydrol[d]	1.2	29	7	5	450
HALS-I[d]+ 2HOφCOφ [d]	1.0	43	8	7	>450
HALS-I[d]+ Oil-I[e]	1.1	49	11	8	320

a) 2.5 Mrad irradiation at -78°C
b) 2.5 Mrad irradiation at ambient followed by storage at 21°C for 120h.
c) On 40 mil polypropylene plaques, stored at 60°C after 3.0 Mrad irradiation at ambient. Time to cause 50% loss of initial impact strength.
d) At 0.1 wt%
e) at 3.5 wt%

Acknowledgments

The authors wish to thank Himont researchers, J.M. Cooke and R.F. Becker for help in sample preparation and in physical testing in the early stages of this work.

Literature Cited

1. Issued as NRCC #32518
2. Carlsson, D.J.; Dobbin, C.J.B.; Jensen, J.P.T. and Wiles, D.M. *Amer. Chem. Soc. Symp. Ser. 280*, 359 **1985**.
3. Carlsson, D.J.; Dobbin, C.J.B. and Wiles, D.M. *Macromolecules 18*, 2092, **1985**.
4. Howard, J.A. and Scaiano, J.C. in "Radical reaction rates in liquids: Oxy-, peroxy- and related radicals" ed. H. Fischer and K.H. Hellwege, Vol. *13* of *Landolt Bornstein*, New Series II, Springer Verlarg, Berlin **1987**, Part D.
5. Dunn, T.S. and Williams, J.L. *J. Ind. Irrad. Tech. 1*, 33, **1983**.
6. Becker, R.F.; Carlsson, D.J.; Cooke, J.M. and Chmela, S. *J. Polym. Deg. Stab. 22*, 313, **1988**
7. Carlsson, D.J.; Brousseau, R.; Zhang, C. and Wiles, D.M. *Amer. Chem. Soc. Symp. Ser. 364*, 376, **1988**.
8. Ranor, K.D.; Lusztyk, J. and Ingold, K.U. *J. Org. Chem. 53*, 5220 **1988**.
9. Hegazy, E.-S.A, Seguchi, T., Arakawa, K. and Machi, S. *J. Appl. Polym. Sci. 26*, 1361, **1981**.
10. Carlsson, D.J. and Lacoste, J. *J. Polym. Deg. Stab.* in press.
11. Carlsson, D.J.; Falicki, S.; Cooke, J.M. and Gosciniak, D.J. unpublished results.

RECEIVED March 4, 1991

Chapter 27

Irradiation Effects of Paraffins as Polymer Model Compounds

Tadao Seguchi

Takasaki Radiation Chemistry Research Establishment, Japan Atomic Energy Research Institute, Takasaki-shi, Gunma, 370–12 Japan

Irradiation effects of n-paraffins and squalane as the model compounds of polyetylene and ethylene-propylene copolymer, respectively, were investigated by products analysis. n-Paraffins of even and odd carbons as $C_{20}H_{42}$, $C_{21}H_{44}$, $C_{23}H_{48}$, and $C_{24}H_{50}$, and squalane($C_{30}H_{62}$) were irradiated by gamma rays under vacuum in crystalline, glassy, and liquid state. The evolved gases were analyzed by gas chromatograpy and the molecular weight changes were analyzed by liquid chromatograpy and mass spectroscopy. The G-values of crosslinking of n-paraffins were 1.2 in crystalline state and 1.7 in liquid state, and showed no difference between the odd and even carbons. The G-value of squalane was 1.7 in liquid state and 1.3 in glassy state at -77°C. The double bonds were much accumulated in the crosslinked products, especially in liquid state. The probability of chain scission was estimated negligible, except chain ends of the molecule.

Radiation effects of polymers have been widely studied in past four decades, and the most of the phenomena might be found out. Now, it is well recognized that the principal reactions to affect the properties of polymers are cross-linking, chain scission, and unsaturated bonds formation(1). However, the several important aspects such as the absolute G-values of their reactions, the reaction sites, and the effect of polymer morphorogy, seem to be remained unclear.

For the more understanding in the radiation effect of polymers, paraffins had been studied, and the radiation induced chemical yields such as free radicals, chain scission products, and isomers of dimer were determined(2,3,4,5). Nevertheless, in a great many of these works on paraffins and polyetylene, which were reviewed by Unger(6), the real aspects mentioned above seem to be still ambiguous even in paraffins. For understanding the complete range of radiation effects, it is necessary to combine the various data

0097–6156/91/0475–0442$06.00/0

determined from the same irradiated samples and reconstruct the reaction mechanisms. Owing to the application of new and improved analytical machines, the analysis of products induced by irradiation can be made easily and also more precisely. Hence, it is expected that the unknown aspects in the radiation effects of paraffins as polymer model compounds would be resolved.

In this article, the irradiation effects of normal paraffins from C_{20} to C_{24} and squalane as the model compound of polyethylene, and ethylene-propylene copolymer, respectively, were described by the products analysis. The same samples were investigated by the multiple analytical machines as gas chromatography(GC), liquid chromatography(LC), and mass spectrometry(MS). In order to observe the effect of morphology, the samples are irradiated in crystalline, liquid, and glassy states at the three different irradiation temperatures.

Experimental.

Samples: Normal paraffins of n-$C_{20}H_{42}$, n-$C_{21}H_{44}$, n-$C_{23}H_{48}$, and n-$C_{24}H_{50}$, and squalane($C_{30}H_{62}$) with branched-chain were used. The chemical stracture and properties are shown in table I. Each sample of 1 g was put in a glass ampoule, and sealed off after the evacuation.

Irradiation: Irradiation were carried out by Co-60 gamma rays at three different temperatures: 55°C(liquid state), 25°C (crystalline state except squalane), and -77°C(crystalline state for n-paraffins and glassy state for squalane). The dose rate at sample position in the glass ampoule was 2.78Gy/s(10kGy/h), which was determined using alanine dosimeter(7).

Gas analysis: After irradiation the ampoule containing the sample was connected to vacuum system which had a pressure gauge and gas reservoir of Gas Chromatography(GC). Initially, the total pressure and volume of evolved gases were measured, then the gases were analyzed by two GC machines(GC-1, GC-2). The column of GC-1 has molecular sieve 5A, which analyzes H_2, O_2, N_2, CH_4, and CO, and GC-2 with a column of Polapac-S analyzes CO_2, C_2(C_2H_2, C_2H_4, C_2H_6), C_3(C_3H_6, C_3H_8), and C_4 compounds. Helium was the carrier gas for both GC, and the temperature of both column was kept at 30°C. The details of GC system and the measurements were described in another paper(8).

Molecular weight analysis with liquid chromatography(LC): After the measurement of evolved gas, a part of the sample was dissolved into tetrahydroflan (THF) with a concentration of 1.00%. The LC machine has two detectors: refractive index(RI) and ultraviolet (UV) absorption. The column (JASCO Megapac Gel 201) of LC can separate the molecules of molecular weight(Mn) from 200 to 3,000 of the polystylene equivalent molecular weight. Temperature of the LC machine including column and injection-port was maintained at 25°C. The RI and UV (worked at 254 nm) detectors were operated parallel at the same time and the both spectra were recorded in a same chart. The details were described in another paper(9).

Molecular weight analysis by mass spectroscopy(MA): The mass spectrometer(Hitachi M-80B) can analyze the mass number(M/e) from 100 to about 4,000 without decomposition of the molecules during measurement by application of the ionization of field desorption technique. The details of this mass analysis was described in the paper(10).

Results and Discussion.

1. Gas analysis.

Total gas evolution was nearly proportional to dose up to about 4 MGy of the maximum dose for all samples at three different temperatures. The H_2 was the main products for n-paraffins and 99% of the total gas evolution, and the others were CH_4, C_2H_6, C_2H_4, and C_3H_8. The C_4-compounds, such as C_4H_{10} or C_4H_8, and the higher molecular weight compounds were less than the GC detection limit. For squalane, the H_2 is about 93% and CH_4, C_3H_8, and C_3H_6 were observed. The G-values of product gases for each sample are listed in Table II.

In a case of n-eicosane($C_{20}H_{42}$), the relations between the gas yield of H_2, CH_4, C_2H_4, C_2H_6, and C_3H_8 and dose are plotted in Fig.1(a,b,c,and d). The evolution of H_2 and CH_4 is proportional to dose and their yields are higher in liquid state at 55°C than in crystalline state at 25°C and -77°C. The C_2H_6 increases linearly with dose only in liquid state at 55°C, but it is negligible small at lower doses in crystalline state, whereas it increases at the higher doses above 1 MGy. The C_2H_4 and C_3H_8 are observed only in liquid state, but are not detected in crystalline state. For n-$C_{24}H_{50}$, the species of evolved gases are the same as those from n-$C_{20}H_{42}$, and their evolution behavior are also similar to Fig.1.

For n-$C_{21}H_{44}$ and n-$C_{23}H_{48}$, the species of evolution gases were the same with n-$C_{20}H_{42}$ and n-$C_{24}H_{50}$. The yields of CH_4 and C_2H_6 from n-$C_{21}H_{44}$ are plotted in Fig.2(a and b). The C_2H_6 increases linearly with dose in crystalline state, which is only different point from that of even carbon paraffins of n-$C_{20}H_{42}$ and n-$C_{24}H_{50}$.

For branched-chain paraffin of squalane ($C_{30}H_{62}$), the yield of CH_4 is about 5% of the total and C_3H_6 increases linearly with dose at any temperature, but C_2H_6 and C_2H_4 yields are less than those from n-paraffins. The yield of gases in liquid phase showed no difference between 25°C and 55°C, but the yields in glassy state at -77°C are less than those in liquid state.

The G-values of gas evolution listed in Table II were determined at zero dose extrapolation when the gas yield is not proportional to dose such as C_2H_6 in Fig.1.

Among the five paraffins, G(H_2) in the liquid state is almost independent on the molecular weight and molecular structure, but decreased to about 70% in crystalline state. The G(H_2) of squalane in glassy state at -77°C is higher than that of n-paraffins in crystalline state, so the glassy state might be near to liquid state rather than crystalline state in the gas evolution, if the effects of irradiation temperature were normalized.

The G(CH_4) tends to decrease with increase of molecular weight at any temperature, and the ratio of G(CH_4) / number of methyl group at chain ends per sample weight is nearly constant for n-paraffins, so CH_4 is the product of methyl group scission at the chain end. G(CH_4) of squalane is about 5 times that of n-paraffins in the liquid phase, and the number of methyl group per sample weight is about 3 times that of n-$C_{24}H_{50}$. Therefore, the probability of the chain scission at the chain side methyl group is larger than that of chain end methyl group.

The C_2H_6 from n-paraffins in the liquid state and C_3H_8 from squalane are the chain scission products at the second carbons from chain ends, then, their G-values are nearly proportional to the

Table I. Samples Used in Experiments

Sample	Symbol	Melt.temp. (°C)	Crystalline structure
n-Eicosane	n-$C_{20}H_{42}$	36.8	Triclinic
n-Heneicosane	n-$C_{21}H_{44}$	40.5	Orthorhombic
n-Tricosane	n-$C_{23}H_{48}$	47.6	Orthorhombic
n-Tetracosane	n-$C_{24}H_{50}$	51.0	Triclinic
Squalane	$C_{30}H_{62}$	-38	Glassy

NOTE: Reproduced with permission from reference 8. Copyright 1988 Pergamon.

Table II. G-values of gas evolution from n-paraffins and squalane

Sample	T*1	P*2	$G(H_2)$	$G(CH_4)$	$G(C_2H_6)$	$G(C_3H_8)$	$G(C_2H_4)$	$G(C_3H_6)$
n-$C_{20}H_{42}$	-77	C	2.14	0.022	0.000	0.00	0.00	0.00
	25	C	2.26	0.025	0.000	0.00	0.00	0.00
	55	L	3.32	0.034	0.057	0.040	0.01	0.00
n-$C_{21}H_{44}$	-77	C	2.16	0.021	0.024	0.00	0.00	0.00
	25	C	2.38	0.026	0.025	0.00	0.00	0.00
	55	L	3.22	0.033	0.056	0.045	0.02	0.00
n-$C_{23}H_{48}$	-77	C	2.25	0.020	0.016	0.00	0.00	0.00
	25	C	2.45	0.025	0.019	0.00	0.00	0.00
	55	L	3.28	0.034	0.053	0.040	0.015	0.00
n-$C_{24}H_{50}$	-77	C	2.16	0.017	0.000	0.00	0.00	0.00
	25	C	2.52	0.020	0.000	0.00	0.00	0.00
	55	L	3.22	0.031	0.045	0.034	0.01	0.00
$C_{30}H_{62}$ (Sq)	-77	G	2.79	0.125	0.003	0.035	0.01	0.008
	25	L	3.27	0.164	0.005	0.045	0.007	0.028
	55	L	3.26	0.165	0.005	0.045	0.007	0.029

T*1 Irradiation temperature(°C)
P*2 Phase of sample during irradiation, C:Crystal, L:Liquid, G:Glass

NOTE: Reproduced with permission from reference 8. Copyright 1988 Pergamon.

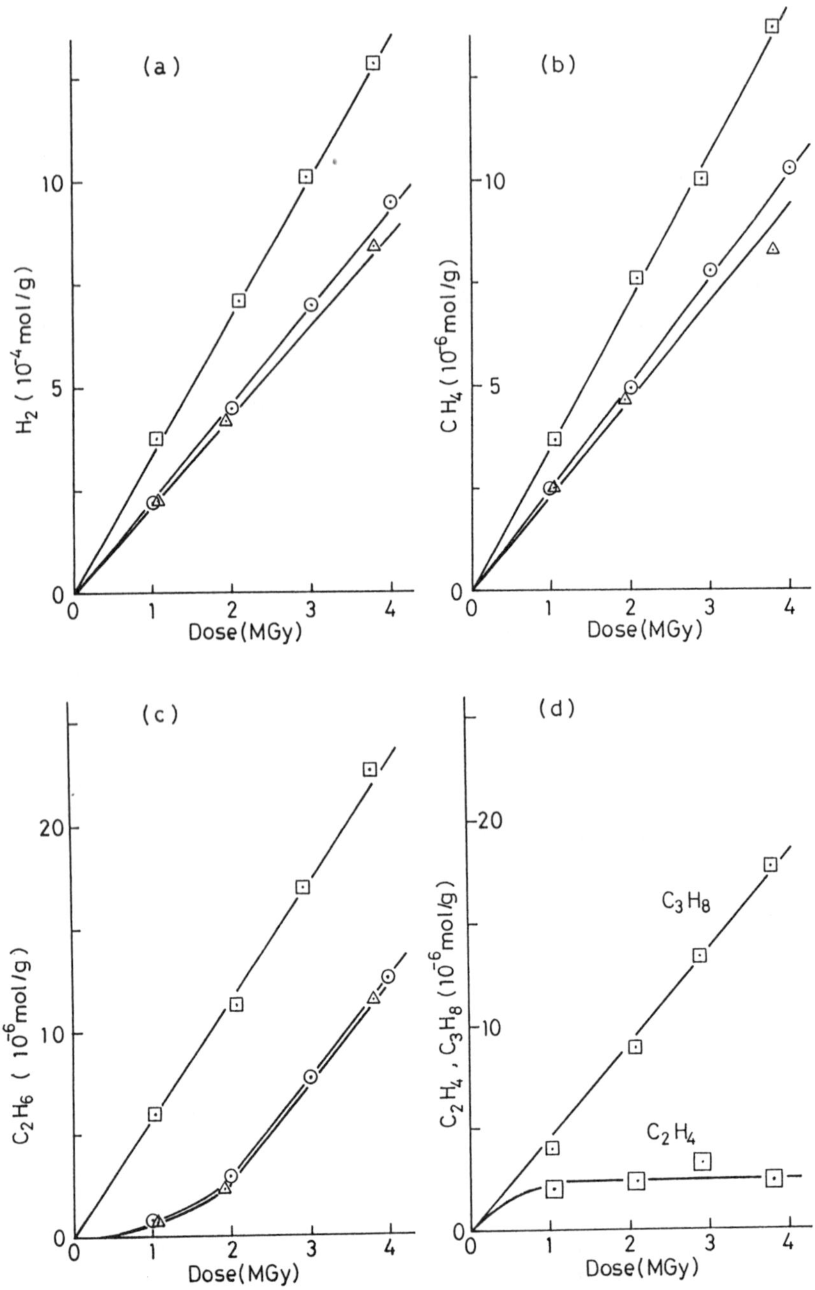

Fig.1 Yields of H_2(a), CH_4(b), C_2H_6(c), and C_2H_4, C_3H_8(d) vs. dose for n-$C_{20}H_{42}$ irradiated at 55°C(⊡), 25°C(⊙), and -77°C(△).

(Reproduced with permission from reference 8. Copyright 1988 Pergamon.)

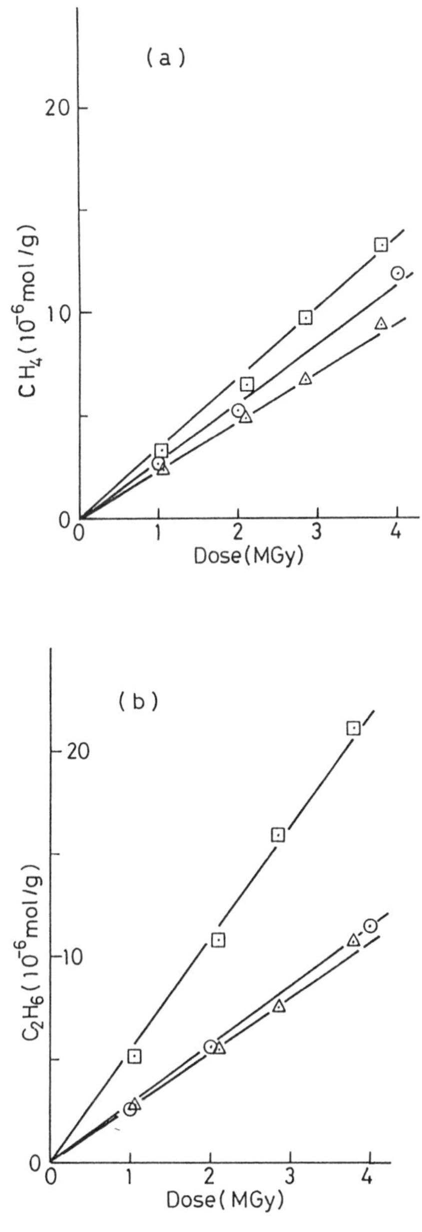

Fig.2 Yields of CH_4(a), and C_2H_6(b) vs. dose for n-$C_{21}H_{44}$ irradiated at 55°C(⊡), 25°C(⊙), and -77°C(△).

(Reproduced with permission from reference 8. Copyright 1988 Pergamon.)

number of chain ends per molecular weight. In the crystalline state of n-paraffins, $G(C_2H_6)$ is very different between even and odd carbons, where the even carbons belong to triclinic structure and odd carbons to orthorhombic structure. It is thought that the second carbons from chain ends are fixed in the triclinic crystal and have no mobility, like the inner carbons. However, by increasing the dose, the crystal surface would become defective as in a glassy state, because the chemical reactions such as crosslinking and dehydrogenation take place mainly at the crystalline surface. Then, the second carbons could gradually become mobile with increasing dose to proceed the chain scission. For orthorhombic crystals, the second carbons from the chain ends can mobile, like chain end methyl groups. The chain scission would take place only at the mobile carbons, and no chain scission at the fixed carbons in crystallite. The C_3H_8 was not produced in triclinic and orthorhombic crystalline states, because the third carbons from chain ends should be fixed in the crystallite and have no mobility. In liquid state, the chain scission at the third carbon proceeds, but the scission at the inner carbons more than third carbon should be decreased greatly, because the C_4 products were very low.

2. LC analysis

The LC spectra in RI detector of non-irradiated paraffins showed a single peak and the elution volume(EV) of n-paraffins is well separated according to the number of carbons. However, the spectrum in UV detector was observed only for the irradiated paraffins, then it should reflect the unsaturation products.

The LC spectra observed by RI and UV detectors for n-$C_{20}H_{42}$ irradiated up to 2 MGy in crystalline state at 25°C are shown in Fig.3(a). The UV absorption was clearly detected at the monomer peak in RI detector, and the intensity increased with dose. The several new peaks in the RI and UV spectra were detected and assigned to the crosslinked products; dimer, trimer, tetramer, and the more polymeric products as indicated in the figure. The UV spectrum reflect the unsaturated products in monomer and in crosslinked products.

The RI spectrum at dimer is composed of two peaks as shown in Fig.3(b-i) in enlarged spectrum, because it become more clear two peaks by decreasing the injection volume of sample as seen in Fig.3(b-ii). EV at the peaks of D(I) and D(II) is higher than EV at the peaks of n-$C_{40}H_{82}$, which means that the products are not the dimer crosslinked at chain end carbons. It is thought that the D(I) is the dimer crosslinked between the second carbons from chain end, and D(II) is the dimer crosslinked between the third carbons or second carbon and third carbon from the chain end. With increase of dose, the EV at peaks of crosslinked products in the spectrum was almost the same, and the intensity at polymeric products increased relatively. So, the polymeric products would be induced by the stepwise reactions of monomer with crosslinked products. At the monomer peak, there were no apparent change of EV, but decreases with dose.

When n-$C_{20}H_{42}$ is irradiated in the liquid state, the dimer in RI spectrum changes to a single peak which is nearly same with D(II) peak in Fig.3(b). The trimer, tetramer and pentamer have clear peaks in both of RI and UV spectrum, and their EV is shifted to lower molecular weight. Then, the crosslinking points at carbons in liquid state would be inner carbon from the chain ends than the case of crystalline state. The UV peak intensity increases more in the polymerized products, which means the unsaturation is much concentrated in the polymeric products.

The LC spectra of RI and UV from n-$C_{24}H_{50}$ irradiated in crystalline and in liquid state were similar to Fig.3, that is, (1) RI spectrum had two peaks at dimer in crystalline state and a single peak in liquid phase, and (2) UV spectrum had the higher intensity at the polymeric products, especially in the liquid state.

For odd carbons of n-$C_{21}H_{44}$ and n-$C_{23}H_{48}$ irradiated in the crystalline and liquid state, the LC spectra were almost the same with those from even carbons of n-$C_{20}H_{42}$ and n-$C_{24}H_{50}$ irradiated in both the crystalline and liquid states.

When squalane is irradiated in liquid state at 25°C, the LC spectra of RI and UV are assigned to monomer, dimer, trimer, and tetramer. The spectrum in glassy state irradiation was not so different from the liquid state, but the spectral intensity at polymeric products more than tetramer tends to increase in liquid state.

The yield of crosslinking products was determined from the RI spectral intensity, because the integrated peak intensity is proportional to the concentration in the solution of LC sample, especially in monomer peak. The spectrum was divided into 3 parts as shown in Fig.3(a): monomer(M), dimer(D), and polymer(P) which is the products more than trimer. As the sensitivity of RI detector depends on the molecular weight, the yield of crosslinked products was calculated by assuming that the sensitivity of dimer or trimer obeys the corresponding molecular weights of n-paraffins(9).

The yields of monomer(M), dimer(D), and polymer(P) were calculated for all samples irradiated under various conditions, and the summation of crosslinking products(D+P) and non-reacted monomer(M) was nearly constant for each sample, that is, (M+D+P) was a constant at any doses within experimental errors, so it may be reasonable that the total yield of (M+D+P) in irradiated sample is equal to unirradiated monomer yield(Mo). This result means that the yield of monomer consumption is equivalent to the yield of cross-linking products.

The calculated yields of D and P in RI spectrum are plotted against dose for n-$C_{20}H_{42}$ in Fig.4. The dimer yield increases with dose at initially, then levels off, especially in liquid state, and the polymeric products(P) increase greatly at higher dose. The yields of polymeric products increase via the formation of dimer and trimer as following reactions;

$$M^* + M^* \text{ ------ } M\text{-}M \text{ (D)} \quad (1)$$
$$M^* + D^* \text{ ------ } M\text{-}D \text{ (T)} \quad (2)$$
$$M^* + T^* \text{ ------ } M\text{-}T \text{ (Tetramer)} \quad (3)$$
$$\text{- - - - - -} \quad \text{------} \quad \text{- - - - -}$$
$$M^* + \text{Origomer}^* \text{ ------ Polymeric Products} \quad (4)$$

where, M^*, D^*, T^*, and Origomer* are the active species as free radicals produced by irradiation in monomer, dimer, trimer and origomer, respectively.

The UV spectral intensities of M, D, and P in irradiated paraffin

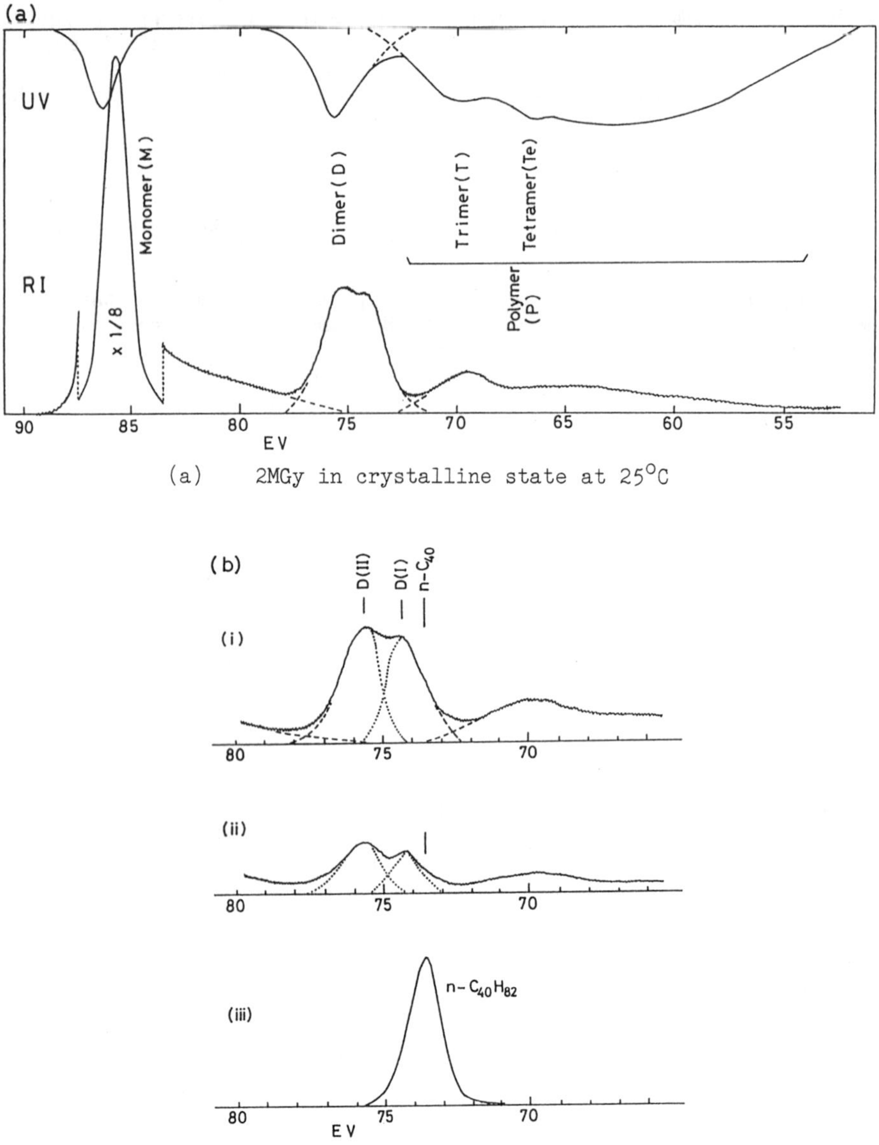

(b-i) Enlarged spectrum of dimer in RI detector,
(b-ii) Same spectrum at low concentration of (b-i),
(b-iii) Spectrum of $n\text{-}C_{40}H_{82}$

Fig.3 Liquid Chromatogram of irradiated $n\text{-}C_{20}H_{42}$ using RI and UV detectors.
(Reproduced with permission from reference 9. Copyright 1989 Pergamon.)

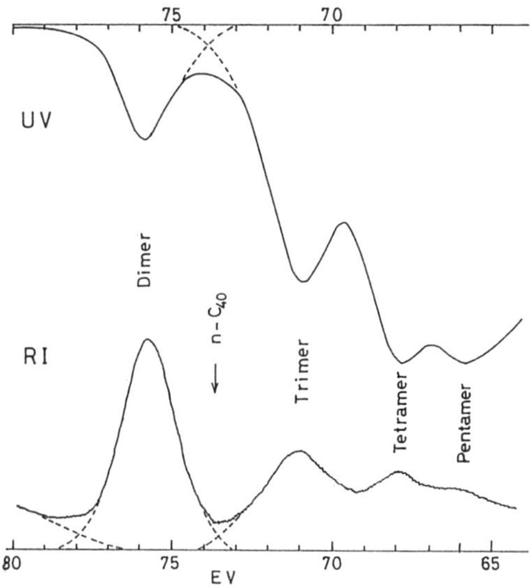

(c) 2MGy in liquid state at 55°C.

Figure 3. Continued.

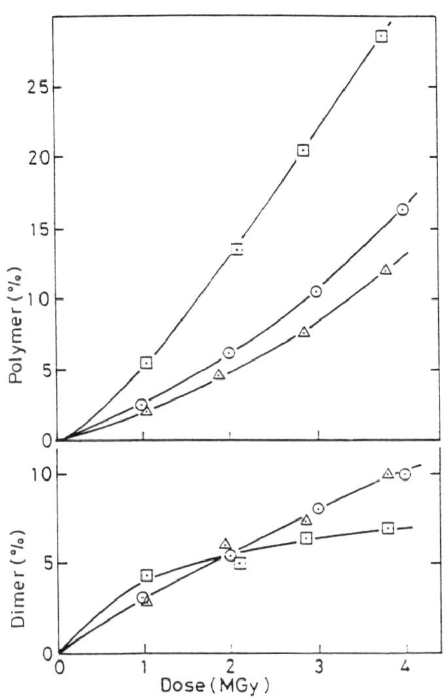

Fig.4 Yields of dimer and polymer vs. dose for $n-C_{20}H_{42}$ irradiated in crystalline state at 25°C(⊙) and -77°C(△), and in liquid state at 55°C(⊡).

(Reproduced with permission from reference 9. Copyright 1989 Pergamon.)

could be ascribed to unsaturation formed in the respective polymeric products. As the sensitivity of UV detector on molecular weight was not corrected, so the spectral intensity among M, D, and P might not show the exact concentration ratio between the unsaturated crosslinked products, but reflects the relative concentration. The unsaturation yield in monomer is detected well in crystalline state, but is very small in liquid state. With increasing dose, the relative spectral intensity at polymeric products increased greatly, which indicated that monomer and dimer with unsaturation would react preferentially with other molecules to convert into the more polymerized products including the unsaturation, especially in liquid state.

The G-value of crosslinking(G(x)) was determined from monomer consumption calculated in RI spectrum by the following assumptions; one crosslinking consumes two monomer molecules, the consumption rate per dose depends on the monomer concentration, and the consumption by reaction with crosslinked products is not accounted. The calculation methods were described in the paper(9). The G(x) of all samples at different irradiation temperature are listed in table III. The cross-linking probability is almost the same among normal paraffins from n-$C_{20}H_{42}$ to n-$C_{24}H_{50}$, and showed no difference between even and odd numbered carbons. The G(x) of squalane in liquid state is nearly equal to those of n-paraffins in liquid state, but G(x) in glassy state is larger than that in crystalline state of n-paraffins.

3. Mass analysis

The mass spectrum of original n-$C_{20}H_{42}$ showed the mass number corresponding to monomer and the isotopes of ^2H and ^{13}C. Figure 5 shows the spectrum of irradiated one up to 3.8 MGy in crystalline state at -77°C. The mass numbers of 282, 562, and 842 are corresponding to the saturated monomer, dimer, and trimer of n-$C_{20}H_{42}$, and the 283 and 284 are the monomer containing the isotopes of ^{13}C and/or ^2H. The 280 and 278 are the molecules detached of one and two hydrogen molecules from the monomer, respectively, which have one and two unsaturation. The 560, 558, and 556 are the dimer with one, two, and three unsaturation, respectively, and the others around dimer are the molecules with the isotopes of ^{13}C and ^2H. The 840, 838, 836, and 834 are the trimer components with unsaturation, and the others around trimer are the isotopes. The ions with the intermediate mass number between monomer and dimer or less than monomer were the background level.

When n-$C_{20}H_{42}$ was irradiated up to 4 MGy in crystalline state at 25°C, the spectrum was nearly same with Fig.5, but the intensity of dimer and trimer was rather higher. For n-$C_{24}H_{50}$ irradiated up to 4 MGy in crystalline state at 25°C and -77°C, the mass spectra were composed of ions of the monomer dimer and trimer including unsaturation and isotopes as same with the spectrum in n-$C_{20}H_{42}$. When n-$C_{20}H_{42}$ is irradiated up to 3.8 MGy in liquid state at 55°C, the ions from monomer, dimer, trimer, tetramer, and pentamer components are observed as shown in Fig.6. The ions of unsaturated molecules are much abundant in the polymeric molecules, and the ions of saturated molecules in pentamer are negligible. The spectrum from ions with intermediate numbers of carbon between monomer and dimer and less than monomer were very low intensity, and their total ion intensity was about few percent of dimer ion intensity.

Table III. G-values of crosslinking for n-paraffins and squalane

Sample	Irradiation temperature		
	-77°C	25°C	55°C
	Crystalline		Liquid
n-$C_{20}H_{42}$	1.01 ± 0.05	1.15 ± 0.05	1.66 ± 0.07
n-$C_{21}H_{44}$	1.02	1.15	1.72
n-$C_{23}H_{48}$	1.00	1.14	1.82
n-$C_{24}H_{50}$	1.01	1.15	1.70
Squalane	Glassy	Liquid	
$C_{30}H_{62}$	1.3 ± 0.1	1.6 ± 0.1	1.6 ± 0.1

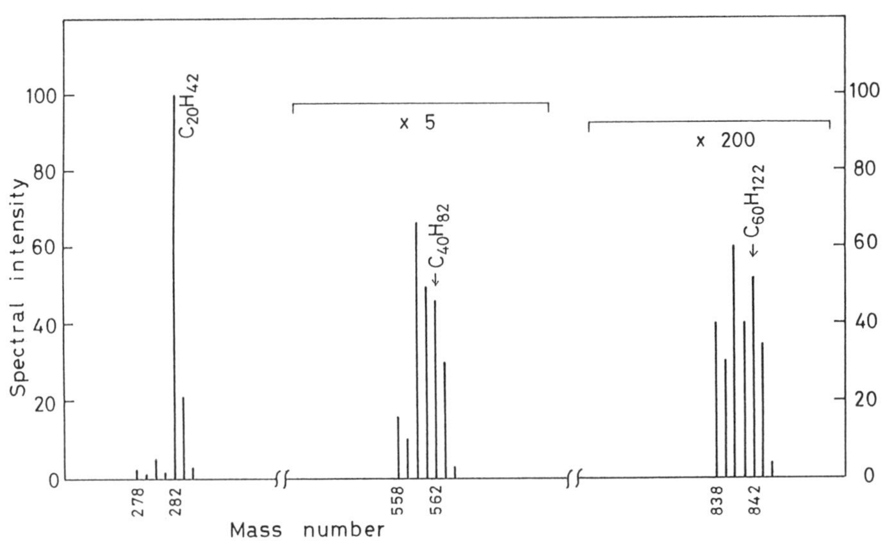

Fig.5 Mass spectrum of n-$C_{20}H_{42}$ irradiated to 3.8 MGy in crystalline state at -77°C.
(Reproduced with permission from reference 10. Copyright 1990 Pergamon.)

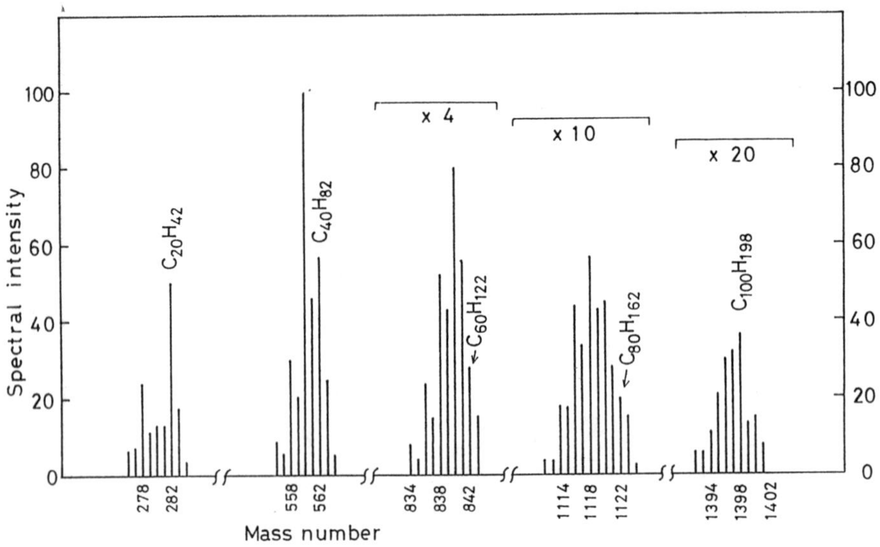

Fig.6 Mass spectrum for n-$C_{20}H_{42}$ irradiated to 3.8 MGy in liquid state at 55°C.

(Reproduced with permission from reference 10. Copyright 1990 Pergamon.)

In n-$C_{24}H_{50}$ irradiated up to 3.8 MGy at 55°C, the ions from monomer, dimer, trimer, and tetramer are observed, and the ions of unsaturated molecules are predominant in the polymeric products.

In the mass spectral analysis, the products induced by irradiation are the saturated and unsaturated parent (monomer) molecules and its crosslinked molecules as dimer, trimer, tetramer, and pentamer. The products with their intermediate molecular weight (carbon number) between monomer and dimer, trimer are very small in crystalline state, and their yields are few percent of crosslinked products in a case of liquid state irradiation. Therefore, it is supposed that the almost of molecules activated by irradiation should react with other activated molecules without decomposition such as the main chain scissions. A small amount of chain scission products observed at higher doses may be the chain scission product at the chain end or branched chain as crosslinked products. The counter products are methane, ethane, and propane observed in gas analysis, but their G-values were only few % of the G-value of crosslinking (see table II).

Conclusion.

The polymeric products as dimer, trimer, tetramer, and highly polymeric oligomers in crystalline, liquid, and glassy state were clearly observed by LC and MS, and the G-values of crosslinking can be determined for paraffins as the model compounds of polyethylene and ethylene-propylene copolymer. Although a small amount of the chain scission products could be detected by MS analysis, their yields were only few % of crosslinking yield in case of liquid state. In the gas evolution, the chain scission products of CH_4, C_2H_6, and C_3H_8 were detected, and their G(scission) were 0.02 - 0.04 for all n-paraffins even in liquid state, which are produced by scission at chain ends. In the case of n-$C_{24}H_{50}$ irradiated in liquid state, the counter products of C_{23}, C_{22}, C_{21} were expected to be 0.5% of monomer after 3.8 MGy irradiation from GC analysis, but the observed values by MS analysis were less than 0.1 % of monomer. Then, it is thought that the most of chain scission gases are produced by the decomposition at the branches of crosslinked molecules. G-values of crosslinking of paraffins are just a half of G-values of hydrogen yield for all samples at any conditions. It means that the 50 % of hydrogen are related to crosslinking formation and the rest of 50 % unsaturation formation. So, the crosslinking and unsaturation should be closely related each other. Infact, the unsaturations were much condensed in crosslinked products.

The crosslinking and chain scission behavior observed in n-paraffins and in squalane would be similar to those in polyethylene (PE) and in ethylene-propylene copolymer(EPR), respectively. The radiation induced reactions in crystalline and in amorphous parts of PE may be analogous to those in crystalline and liquid state of n-paraffins, and the reactions in amorphous EPR may be analogous to those in liquid and glassy squalane. When G-values of crosslinking of PE and EPR could be extrapolated from the values in table 2, the chain scission probability in PE and EPR would be negligibly small.

Generally, a concept that chain scission should occur in any polymer chains has been accepted, because radiation energy is too high to comparing with bond energy of polymers and the most of experiments indicate that chain scission probability does exist, as far as sol-fraction measurements are applied for an analysis using Charlesby-Pinner equation(11). However, there has been no clear evidence that main chain scission occurs in linear polyethylene by radiation as reported by Manderkern(12). Our studies using the model compounds of n-paraffins and of squalane are indicating that the main chain scission both in polyethylene and ethylene-propylene copolymer is negligible.

Acknowledgments.

This study was done for the joint research between JAERI and University of Tokyo. Author expresses his appreciation to the members of the research group (Dr.Tamura,N., Mr.Hayakawa,N., Dr.Arakawa,K., Prof.Tabata,Y., Dr.Katsumura,Y., and Dr.Hayashi,N.).

References.

(1) Radiation Chemistry of Macromolecules, Dole,M.,Editor, Academic Press, New York, 1972(Vol.1) and 1973(Vol.2)

(2) Radiation Chemistry of Hydrocarbons, Foldiak, G.,Editor, Studies in Physical and Theoretical Chemistry 14, Elsevier Science Publishing Company,1981

(3) Salovey,R., Falconer,W.E., J.Phys.Chem.1965, 69, p2345,1966,70, p3203

(4) Toriyama,K.,Iwasaki,M.,Fukaya,M.,J.Chem.Soc.,Chem.Commun. 1982, p1293

(5) Baudson,T., Tilquine,B., Radiat.Phys.Chem.,1984, 24, p589 (6) Ungar,G., J.Material Sci.,1981, 16, p2635

(7) Kojima,T., Tanaka,R., Morita,Y., Seguchi,T., Appl.Radiat. Isot., 1986, 37, p517

(8) Seguchi,T., Hayakawa,N., Tamura,N., Hayashi,N., Katsumura,Y., Tabata,Y., Radiat.Phys.Chem., 1988, 32, p753

(9) Seguchi,T., Hayakawa,N., Tamura,N., Hayashi,N., Katsumura,Y., Tabata,Y., Radiat.Phys.Chem.,1989, 33,p119

(10)Seguchi,T., Arakawa,K., Tamura,N., Katsumura,Y., Hayashi,N., Tabata,Y., Radiat.Phys.Chem.,1990, 36, p259

(11)Charlesby,A., Pinner,S.H., Proc.Royal.Soc.London, 1959,Ser.A,249, p363

(12)Manderkern,L., Radiation Chemistry of Macromolecules of Polymers, Editor Dole,M., in reference (1)

RECEIVED May 1, 1991

Chapter 28

Quantitative Confirmation of Simple Theoretical Models for Diffusion-Limited Oxidation

Kenneth T. Gillen and Roger L. Clough

Organic Materials Division (Org. 1811), Sandia National Laboratories, Albuquerque, NM 87185

Theoretical modelling of diffusion-limited oxidation leads to oxidation profile shapes governed by two parameters α and β plus a simple relationship relating these parameters to the oxygen permeability coefficient and the oxygen consumption rate underlying the oxidation conditions. To quantitatively test the predicted profile shapes and the governing theoretical relationship, a commercial EPDM rubber material of two thicknesses was radiation-aged (Co-60) in air. Experimental oxidation profiles were obtained by monitoring the density changes caused by oxidation. The experimental profile shapes and their dependence on sample thickness could be accurately fit with the theory, yielding values for α and β. Comparison of these results with the independently measured oxygen permeability coefficient and the oxygen consumption rate allowed a quantitative confirmation of the theoretical relationship.

Exposure of polymers to air during aging (radiation, thermal, UV) often results in inhomogeneously oxidized samples, a complication which impacts attempts both to understand the oxidation process and to extrapolate accelerated exposures to long-term conditions. The most important such complication involves diffusion-limited oxidation which can occur if the rate of oxygen consumption in a material is greater than the rate at which oxygen can be resupplied by diffusion processes from the surrounding air atmosphere. This scenario will usually lead to a heterogeneity in the oxidation across the material, with equilibrium oxidation (e.g., corresponding to air-saturated conditions) occurring at the sample surfaces and reduced or non-existent oxidation in the interior. The importance of this effect depends on material geometry coupled with the oxygen consumption rate and the oxygen permeability coefficient (*1,2*). Frequently, the percentage of the sample oxidized under shorter-term (e.g., hours to months) accelerated aging conditions is substantially lower than under longer-term application conditions (*2-8*). To extrapolate accelerated simulations to long-term, air-aging conditions, one must be able to monitor and quantitatively understand diffusion-limited oxidation effects.

0097–6156/91/0475–0457$06.00/0

Experimental techniques for monitoring diffusion-limited oxidation profiles have recently been reviewed (*2-11*). By comparing experimental and theoretical profiles, the theories can be verified and then confidently used to predict the importance of diffusion effects prior to the initiation of aging tests. This paper summarizes what we believe is the first rigorous quantitative experimental confirmation of diffusion-limited oxidation theories.

Experimental

Material. The commercial EPDM material used for these studies was supplied by Parker Seal Group and designated as E740-75. It was obtained as compression-molded sheets (15 cm by 15 cm) in several thicknesses. The formulation is described as a peroxide-cured terpolymer of ethylene and propylene in approximately equal amounts plus ~3% hexadiene; it contains ~35% by weight carbon black and a free-radical scavenger antioxidant.

Radiation Aging. Combined radiation-thermal exposures were carried out in an underwater cobalt-60 facility using water-tight aging cans (volume of ~1 liter), an arrangement that facilitated long-term exposures. By selecting positions relative to the cobalt-60 pencils, dose rates ranging from ~0.1 kGy/h to 7 kGy/h (10 krad/h to 700 krad/h) were available for the current experiments. Uncertainties in the dose rates are estimated to be ±10%. Can temperatures were kept at 70±0.5° C during the radiation exposures. A slow, steady flow (~30 cc/min) of either air or nitrogen was supplied to each can throughout the experiment. A detailed description of the aging facility has been published (*12*).

Density Profiling. Details of the density profiling technique have been published previously (*6*). The technique is based on the use of a density gradient column to obtain the density of successive thin slices across a sample. It depends on the fact that oxidation reactions often lead to substantial and easily measurable increases in polymer density. For the current material (nominal unaged density ~1.12 g/cc), gradient columns were made with a density range from ~1.10 to 1.15 g/cc using $Ca(NO_3)_2$-H_2O solutions. The experimental uncertainty for a single density measurement is estimated to be less than ±0.0004 g/cc.

Oxygen Consumption Measurements. In order to determine the amount of oxygen consumed by the EPDM sample as a function of the radiation dose, measured quantities of material were sealed with known amounts of oxygen in glass containers. The containers were then aged at 70° C in combination with a selected dose rate ranging from 0.12 kGy/h to 2.08 kGy/h. After exposure, a Tracor MT150g gas chromatograph, equipped with a molecular sieve column and a thermal conductivity detector, was used to determine the amount of oxygen remaining in the tube. The oxygen was quantified using primary standards. In general, the oxygen consumption rate will depend on the oxygen partial pressure surrounding the material. Since the analysis below will require an estimate of the consumption rate under air-aging conditions (~13.2 cmHg oxygen partial pressure in Albuquerque), and since the experimental procedure requires a measurable drop in oxygen pressure, attempts were made to select the ratio of sample weights to cell volumes such that the average of the initial and final oxygen pressures was reasonably close to the desired 13.2 cmHg value. For the seven reported consumption runs, the initial and final oxygen partial pressures in the cells averaged 16.1 cmHg and 6.4 cmHg, respectively.

Oxygen Permeability Coefficient Measurements. Unaged sheets of the EPDM material were sent to Mocon Modern Controls, Inc. of Minneapolis where oxygen permeability coefficients were measured at 23° C and 60° C. The instrument is based on the ASTM standard test method D-3985-81. To avoid detector saturation, the measurements were conducted with a test gas containing 0.5% oxygen.

Diffusion-Limited Oxidation Theories

In order to quantitatively model diffusion-limited oxidation profiles, one must combine expressions for oxygen consumption with diffusion equations. To derive oxygen consumption rates, detailed knowledge of the kinetics of oxidation must be available. This is a difficult requirement given the complexities of oxidation processes in polymers, especially those containing antioxidants (*13*). Most modelling attempts assume that the following scheme (*14,15*) represents a reasonable first order approximation for the oxidation of organic materials in various environments (thermal, mechanical, UV light, ionizing radiation, etc.).

INITIATION

$$\text{Polymer} \xrightarrow{R_i} R\bullet \quad (1)$$

PROPAGATION

$$R\bullet + O_2 \xrightarrow{k_2} RO_2\bullet \quad (2)$$

$$RO_2\bullet + RH \xrightarrow{k_3'} ROOH + R\bullet \quad (3)$$

BIMOLECULAR TERMINATION

$$R\bullet + R\bullet \xrightarrow{k_4} \text{PRODUCTS} \quad (4)$$

$$RO_2\bullet + R\bullet \xrightarrow{k_5} \text{PRODUCTS} \quad (5)$$

$$RO_2\bullet + RO_2\bullet \xrightarrow{k_6} \text{PRODUCTS} + O_2 \quad (6)$$

UNIMOLECULAR TERMINATION

$$RO_2\bullet \xrightarrow{k_7} \text{PRODUCTS} \quad (7)$$

$$R\bullet \xrightarrow{k_8} \text{PRODUCTS} \quad (8)$$

BRANCHING

$$ROOH \xrightarrow{k_9} RO\bullet + \bullet OH \quad (9)$$

Environmental differences occur mainly in the details of the free-radical producing initiation step. It is particularly easy to control the initiation rate in the gamma-initiated case, since R_i is typically independent of aging time and linear with the radiation dose rate. In addition, branching reactions, which may complicate the oxidation processes occurring at the high temperatures used for thermal oxidation studies, are often unimportant for lower-temperature, relatively short-term gamma-initiated oxidation. Any growth in importance of branching reactions as degradation proceeds will cause the oxidation rate to increase with time, further complicating attempts at quantitative modelling. For these reasons, gamma-initiated oxidation is the best choice for initial attempts at quantitatively modelling diffusion-limited oxidation effects. Complications, such as time-dependent oxidation rates, may then be addressed once confidence exists in models derived for the simpler situations.

For gamma-initiated oxidations involving moderate temperatures, we will therefore assume that the chemistry is adequately represented by reactions 1-8. The scheme involving the bimolecular termination steps (reactions 1-6), originally derived for the oxidation of organic liquids, has been invoked for many years for explaining the oxidation of polymers. Using a steady-state analysis and assuming long kinetic chain lengths (many propagation cycles compared to termination reactions) and $k_5^2 = 4k_4k_6$, the oxygen consumption rate is given by (*14*)

$$\frac{d[O_2]}{dt} = \frac{C_{1b}\,[O_2]}{1 + C_{2b}[O_2]} \qquad (10)$$

where the constants C_{1b} and C_{2b}, appropriate to the bimolecular theory, are given by

$$C_{1b} = \frac{k_2\,R_i^{0.5}}{(2k_4)^{0.5}} \quad \text{and} \quad C_{2b} = \frac{k_6^{0.5}k_2}{k_4^{0.5}k_3} \qquad (11)$$

and $k_3 = k_3'[RH]$.

For many polymeric materials, unimolecular termination reactions are often found to be dominant. For instance, in the presence of sufficient radical-scavenger antioxidant, the radical species ($RO_2\bullet$ and $R\bullet$) can terminate in pseudo first-order reactions, yielding an oxidation scheme consisting of reactions 1-3 plus 7 and 8. The rate of oxidation (oxygen consumption rate) is then given by (*2*)

$$\frac{d[O_2]}{dt} = \frac{C_{1u}\,[O_2]}{1 + C_{2u}[O_2]} \qquad (12)$$

This expression is identical in form to the bimolecular result given in equation 10, except that the constants C_{1u} and C_{2u} from the unimolecular analysis are given by

$$C_{1u} = \frac{k_2R_i}{k_8} \quad \text{and} \quad C_{2u} = \frac{k_2k_7}{k_8(k_3+k_7)} \qquad (13)$$

Although the bimolecular and unimolecular schemes yield oxygen consumption rates that have identical functional dependencies on the oxygen concentration, their dependencies on the initiation rate, R_i, are quite different (half- and first-order, respectively). Thus, when data at various initiation rates are available, one can easily distinguish between the two schemes.

By assuming that the oxygen consumption rate and the oxygen permeability coefficient are independent of time, Cunliffe and Davis (*1*) used the bimolecular kinetic expression for oxygen consumption and showed that it could be combined with diffusion expressions to derive theoretical oxidation profiles for slabs of material of thickness L. Assuming one dimensional diffusion (L much smaller than the other two slab dimensions), Fickian behavior and steady state conditions with the x-coordinate aligned along the thickness direction (*16*),

$$D \frac{d^2[O_2(x)]}{dx^2} = -R[O_2(x)] \tag{14}$$

where $[O_2(x)]$ refers to the concentration of oxygen at the point x, $R[O_2(x)]$ denotes the rate of removal of oxygen by reaction at x and D is the diffusion coefficient for oxygen in the material. Substituting the kinetic modelling results given in either equation 10 or equation 12 for $R[O_2(x)]$ yields

$$D \frac{d^2[O_2(x)]}{dx^2} = \frac{C_1[O_2(x)]}{1 + C_2[O_2(x)]} \tag{15}$$

At the surfaces of the sample (x = 0 and L), the oxygen concentration must be given by

$$[O_2(0)] = [O_2(L)] = [O_2]_s = Sp \tag{16}$$

where the surface (equilibrium) oxygen concentration is assumed to follow Henry's law and will therefore be given by the product of S and p, the solubility of oxygen in the polymer and the oxygen partial pressure surrounding the sample, respectively. Another boundary condition occurring at the center of the sample is

$$\frac{d[O_2(L/2)]}{dx} = 0 \tag{17}$$

Transforming the above with the following reduced variables

$$X = \frac{x}{L} \quad \text{and} \quad \theta = \frac{[O_2]}{[O_2]_s} \tag{18}$$

Cunliffe and Davis (*1*) showed that equation 15 could be written as

$$\frac{d^2\theta}{dX^2} = \frac{\alpha}{1+\beta\theta} \tag{19}$$

where

$$\alpha = C_1L^2/D \quad \text{and} \quad \beta = C_2Sp = C_2[O_2]_s \tag{20}$$

Although they applied this result to the bimolecular kinetic result (C_1 and C_2 are equal to C_{1b} and C_{2b}, respectively), it is clear that identical results hold for unimolecular kinetics if C_{1u} and C_{2u} are used. The boundary conditions now become $\theta = 1$ at $X = 0$ and $X = 1$ and $d\theta/dX = 0$ at $X = 0.5$. Numerical methods lead to a family of solutions for the relative (e.g., normalized to the surface value) oxygen concentration versus spatial position. If θ_i refers to the relative oxygen concentration at point i in the sample, combining either equation 10 or equation 12 with the definitions of α and β leads to the following expression for the oxidation rate at point i

$$R[\theta_i] = \frac{\alpha D\theta_i [O_2]_s}{L^2(1+\beta\theta_i)} \qquad (21)$$

Similarly, the oxidation rate at the surface (R_s), which is identical to the equilibrium oxygen consumption rate, is given by

$$R_s = \frac{\alpha D\ [O_2]_s}{L^2(1+\beta)} \qquad (22)$$

Thus the relative oxidation at point i is given by

$$R_i = \frac{R[\theta_i]}{R_s} = \frac{\theta_i(1+\beta)}{1+\beta\theta_i} \qquad (23)$$

We have used the above analysis to derive theoretical oxidation profiles (plots of R_i versus normalized cross-sectional position) and the corresponding profiles of the relative oxygen concentration (θ_i versus position) (*2*). Representative results for three values of β and various values of α are given in Figures 1-3, where P gives the percentage of the distance from one oxygen-exposed surface of the sheet to the opposite oxygen-exposed surface. For small values of β, the oxidation is proportional to the oxygen concentration, yielding "U-shaped" profiles. When β is large, oxidation is insensitive to oxygen concentration until the latter has dropped significantly, resulting in "step-shaped" profiles. For intermediate values of β, profile shapes of intermediate character result. By rearranging equation 22 and noting that

$$D[O_2]_s = pP_{ox} \qquad (24)$$

where P_{ox} is the oxygen permeability coefficient through the material, the following very useful theoretical relationship is obtained

$$R_s L^2/(pP_{ox}) = \alpha/(\beta+1) \qquad (25)$$

Quantitative Tests of the Theoretical Oxidation Profiles

If experimental oxidation profiles can be successfully fit to the above theories, values of α and β resulting from the fits can then be tested to see whether they are related to independently measured oxygen consumption and permeability coefficient values as predicted from equation 25. Validation of this relationship will allow it to be used as a simple means of predicting the potential importance of diffusion-limited oxidation prior to the initiation of any aging or degradation studies. Since both the bimolecular and unimolecular kinetic schemes yield the same profile shapes versus α and β plus the identical governing theoretical relationship (equation 25), the easiest approach to testing the theory is to conduct experiments in air at constant temperature plus constant initiation rate. In such instances, β will be a constant and α will depend only on the sample thickness. The shapes predicted by the theories and equation 25 can then be tested without the necessity of knowing the underlying mechanistic details.

We begin by noting that the theory was derived assuming a time-independent R_s and a time-independent P_{ox} during the aging. Therefore, a successful

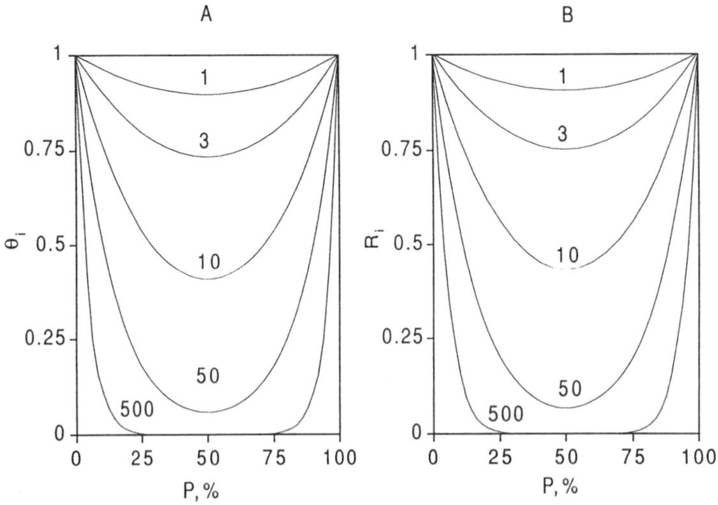

Figure 1. Theoretical profiles of Θ_i, the relative oxygen concentration (A) and R_i, the relative oxidation (B) for various values of α (indicated on figure) with $\beta = 0.1$. P represents the percentage of the distance from one oxygen-exposed surface of the material slab to the opposite oxygen-exposed surface.

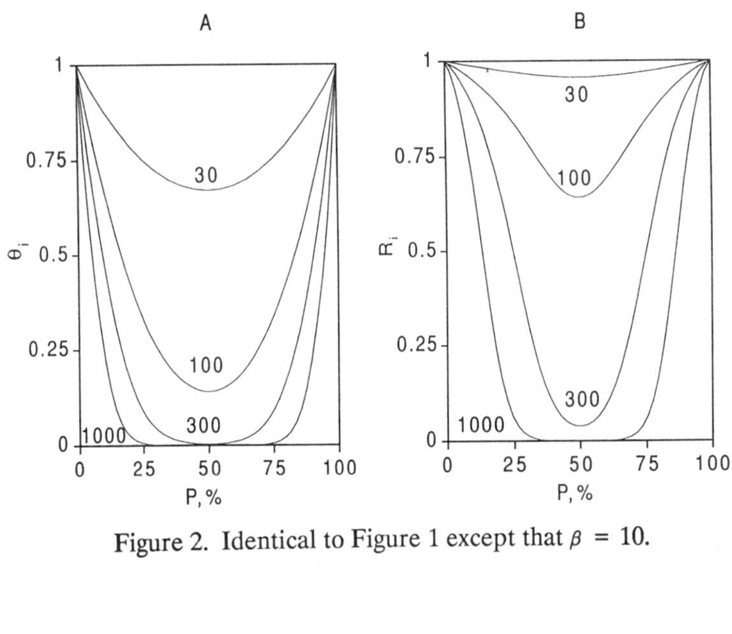

Figure 2. Identical to Figure 1 except that $\beta = 10$.

validation requires a system in which R_S and P_{OX} are known and are relatively constant with aging time. For a number of reasons, this assumption often breaks down for oven aging situations at high temperatures, especially for elastomeric materials and other materials above their glass transition temperature (*17*). For one thing, thermal aging can lead to significant increases in modulus values with corresponding decreases in oxygen permeability coefficients. In addition, oxygen consumption rates usually increase with aging time, sometimes significantly after a so-called induction period. The time-dependent increase of the consumption rate often correlates with important branching reactions, as denoted by reaction 9 above.

In contrast to many thermal aging situations, the time-independent assumptions can hold in high energy radiation environments. For example, in gamma-initiated oxidations of certain commercial elastomers, R_S was found to be relatively constant up to moderate total doses (*3,8*). In addition, based on an extensive study of several polyethylene, EPR, EPDM, chlorosulfonated polyethylene, chlorinated polyethylene and chloroprene rubber materials, it is apparent that P_{OX} values seldom change by more than 10-20% for doses of 0.5 MGy (*18*).

Oxygen Consumption and Permeability Coefficient Results. Table I summarizes the dose rates and doses used plus the amounts of oxygen consumed for each of the seven consumption experiments. Since equilibrium oxygen consumption values are required in order to test equation 25, the conditions used for the consumption experiments should guarantee homogeneous oxidation. Table I lists the thicknesses used for the seven experiments; in a later section of this paper, we will show that diffusion-limited oxidation effects are unimportant for these thicknesses.

Table I. Summary of Oxygen Consumption Experiments[a]

Dose Rate kGy/h	Dose MGy	ΔO_2 moles O_2/cc	$\phi(O_2)$ moles O_2/cc Gy	R_S ccSTP O_2/cc s	Thickness cm	L_c[b] cm
0.12	0.040	2.26×10^{-5}	5.69×10^{-10}	4.24×10^{-7}	0.3	1.3
1.22	0.203	1.09×10^{-4}	5.36×10^{-10}	4.07×10^{-6}	0.08	0.32
1.22	0.291	1.64×10^{-4}	5.64×10^{-10}	4.28×10^{-6}	0.08	0.32
1.22	0.41	2.36×10^{-4}	5.76×10^{-10}	4.37×10^{-6}	0.08	0.32
1.22	0.495	3.15×10^{-4}	6.36×10^{-10}	4.83×10^{-6}	0.08	0.31
2.08	0.659	4.64×10^{-4}	7.04×10^{-10}	9.11×10^{-6}	<0.05	0.24
2.08	1.11	1.33×10^{-3}	1.12×10^{-9}	1.45×10^{-5}	<0.05	0.16

[a]Temperature equals 70°C.
[b]from equation 30 using average values of p and R_S during experiment.

The oxygen consumption results from Table I are plotted in Figure 4. The approximately linear behavior of the 1.22 kGy/h results out to ~0.4 MGy coupled with a similar consumption rate per Gy for the low dose rate result at ten times lower dose rate offer preliminary evidence for the absence of dose-rate effects (e.g., the unimolecular model). We will, in fact, assume that the consumption rate is independent of dose rate (unimolecular kinetics), thereby

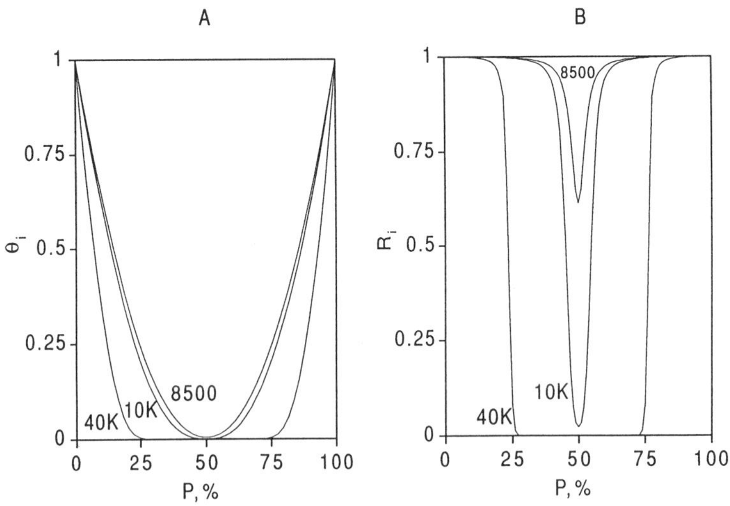

Figure 3. Identical to Figure 1 except that β = 1000.

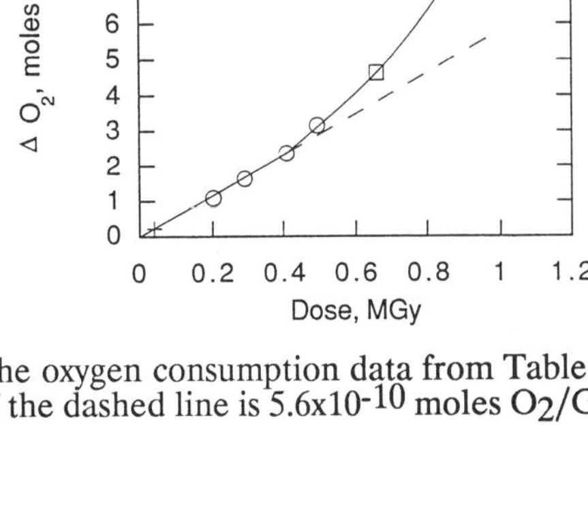

Figure 4. Plot of the oxygen consumption data from Table I versus radiation dose. The slope of the dashed line is 5.6×10^{-10} moles O_2/Gy/cc.

using the Table I results to obtain the oxygen consumption rate under the conditions of interest for testing equation 25. This preliminary evidence for the unimolecular model will be confirmed in a future publication (*19*) which will describe profile results as a function of initiation rate.

The results of Figure 4 indicate that the oxygen consumption rate can be considered approximately constant only for doses up to ~0.4 MGy. Since the theory was derived under the assumption of a constant oxygen consumption rate, we will therefore confine our attention to EPDM samples that were exposed to 0.32 MGy total dose. For these samples, the consumption rate R_s is given by the slope of the dashed line in Figure 4, e.g.,

$$R_s = (5.6 \pm 0.2)\times10^{-10} \text{ moles } O_2/\text{Gy/cc at } 70°\text{C}$$

We also need a value for the oxygen permeability coefficient, P_{ox} at 70° C in order to test the theoretical relationship given in equation 25. Experimentally, P_{ox} was found to be 1.2×10^{-9} ccSTP/cm/s/cmHg at 23° C and 5.1×10^{-9} ccSTP/cm/s/cmHg at 60° C (estimated uncertainties of ±10%). Assuming an Arrhenius temperature dependence (7.7 kcal/mol activation energy from the above data), the log of P_{ox} was plotted against the inverse absolute temperature and then extrapolated the short distance to 70° C, yielding

$$P_{ox} = (7.2 \pm 0.7)\times10^{-9} \text{ ccSTP/cm/s/cmHg at } 70°\text{C}$$

Although P_{ox} will in general decrease with radiation dose, results on similar EPDM materials (*18*) indicate that any drop will be minor (on the order of 5%) for the doses that will be used in testing the theory.

Thus, up to the limit of where the theory would be expected to hold for this material (~0.4 MGy), the left hand side of equation 25 gives

$$R_s L^2/(pP_{ox}) = 132\ L^2 I \qquad (26)$$

where L is the sample thickness (in cm) and I is the radiation dose rate (in Gy/s) used for the experiment.

Density Profile Fitting. EPDM samples of two thicknesses (0.302 cm and 0.18 cm) were radiation-aged in a 70° C air environment to a total dose of 0.32 MGy at a dose rate of 6.65 kGy/h. Since density changes are predominantly caused by oxidation (*6*), and since density changes were readily measured for these samples, the density profile technique (*2,6*) was chosen to experimentally monitor diffusion-limited oxidation. To correct for the small density changes which occur under inert aging conditions, a sample was aged in a nitrogen environment at 70° C to a total dose of 0.38 MGy. The density increase relative to the unaged material was equal to $\sim5\times10^{-4}$ g/cc; assuming linear behavior, this predicts that a density change of $\sim4\times10^{-4}$ g/cc occurs under inert environments after 0.32 MGy radiation dose. This small change was subtracted from the density changes found in the presence of oxygen in order to derive the results plotted as crosses in Figure 5. These crosses therefore represent the changes due to oxidation versus cross-sectional position on the sample; the vertical span of each cross represents the estimated experimental uncertainty while the horizontal span denotes the position and thickness of each slice. Figure 5 gives results for the two different thicknesses (0.302 cm and 0.18 cm) of sample aged under identical conditions (0.32 MGy dose at 6.65 kGy/h); as expected, diffusion effects are less important for the thinner sample.

In order to fit the experimental data with the theoretical results, we first note that the experimental conditions (isothermal in air, constant initiation rate, total dose in the linear region of oxygen consumption) imply that theoretical fits to the experimental results shown in Figure 5 will have a constant value of β. As seen from equation 20, values of α for the thick (0.302 cm) and thin (0.18 cm) samples (α_1 and α_2, respectively) must be related by the squares of the sample thicknesses. In other words,

$$\alpha_2 = (0.18/0.302)^2 \alpha_1 = 0.355 \alpha_1 \quad (27)$$

By comparing the general shapes of the experimental profiles in Figure 5 with the theoretical profiles given in Figures 1-3, it is clear that an intermediate value of β will be appropriate. The actual fitting procedure begins with the selection of a trial value of β; one then determines whether a set of α values (α_1, α_2), constrained by equation 27, can be found that collectively give good fits to both sets of experimental data. For the present data, the best fits occur for a β value around 6, with α_1 and α_2 given by 154.5 and 55, respectively (*20*). Calculated profiles using these values for the theoretical parameters are shown as the solid curves in Figure 5. The comparisons show that the theory does a superb job of characterizing the profile shapes and their dependence on L.

Although the best fits are achieved for a $\beta \sim 6$, reasonable fits also result for slightly smaller and slightly larger values of β. In terms of the overall shapes, fits become marginal as β approaches approximately 2 on the low side and 14 on the high side. Figure 6 shows the best fits for a β of 2 (α_1 and α_2 given by 66 and 23.4, respectively). Figure 7 shows the best results for a β of 14 ($\alpha_1 = 340$ and $\alpha_2 = 121$).

In addition to the three sets of theoretical fits shown in Figures 5-7, additional comparisons of theory and experiment were conducted at other selected β values between 2 and 14 in order to find both the best fit values of α_1 and α_2 for each β, and the range of α values which give reasonable fits for each β. The results from these analyses are shown in Figure 8, where for each value of β investigated, the corresponding best fit value of α_1 is plotted as a circle and the range of α_1 allowed is indicated by the error bars. Not surprisingly, the error bars allowed for α_1 are larger in regions where excellent fits are possible (e.g., around $\beta = 6$) and get much smaller at the lower ($\beta = 2$) and upper ($\beta = 14$) limits where marginal fits occur.

Confirmation and Use of Theoretical Expressions. The results clearly show that the two parameter theory for diffusion-limited oxidation does an excellent job of characterizing profile shapes and their dependence on sample thickness. The optimum fits occur when $\beta = 6$ and $\alpha_1 = 154.5$; thus for the sample of thickness L = 0.302 cm,

$$\frac{\alpha}{\beta+1} = 22.1 \pm 3 \quad (28)$$

We can now compare this result, which represents the right-hand-side of the governing theoretical relationship given in equation 25, with the left-hand-side. From equation 26 with L = 0.302 cm and I = 1.847 Gy/s

$$R_s L^2/(pP_{ox}) = 22.2 \pm 3 \quad (29)$$

This result, which was derived independently from oxygen consumption and permeability coefficient measurements, is in excellent agreement with the best fit result derived in equation 28.

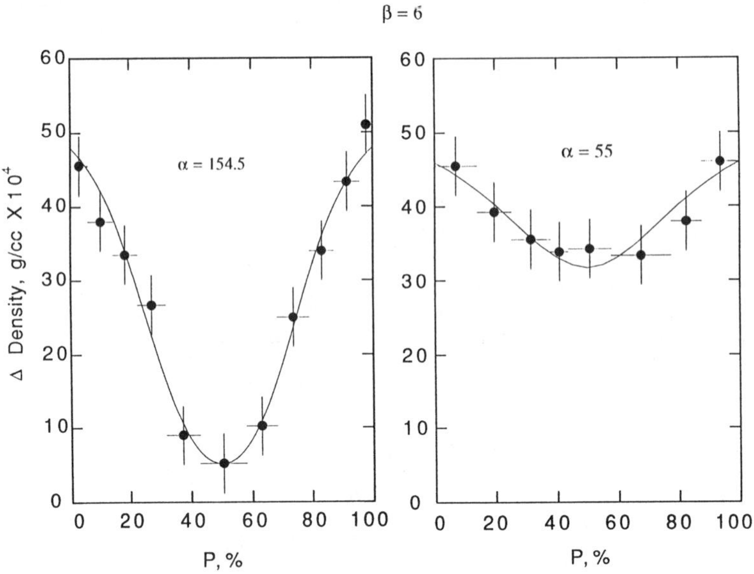

Figure 5. Experimental profiles of density changes (solid circles with crosses) caused by oxidation for 0.32 cm (left) and 0.18 cm (right) thick EPDM samples after aging in air to a dose of 0.32 MGy at 6.65 kGy/h plus 70° C. The curves through the data give the best theoretical fits for β = 6.

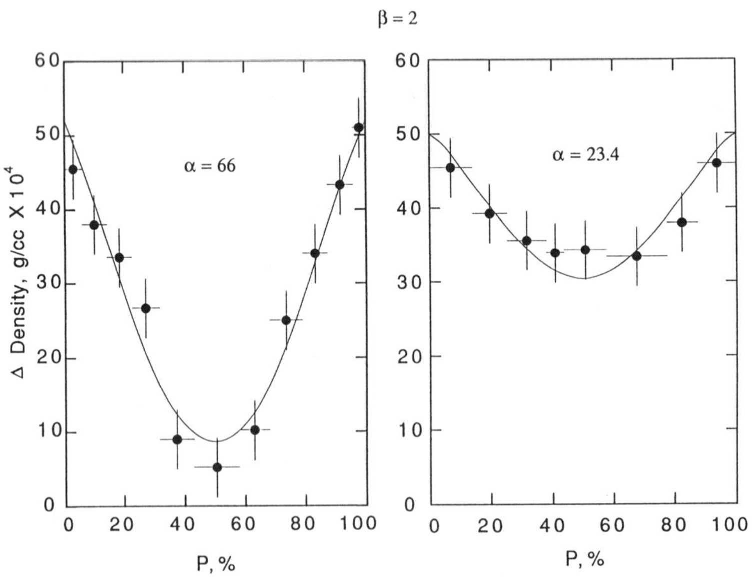

Figure 6. Same experimental data as in Figure 5 fit with theory using β = 2.

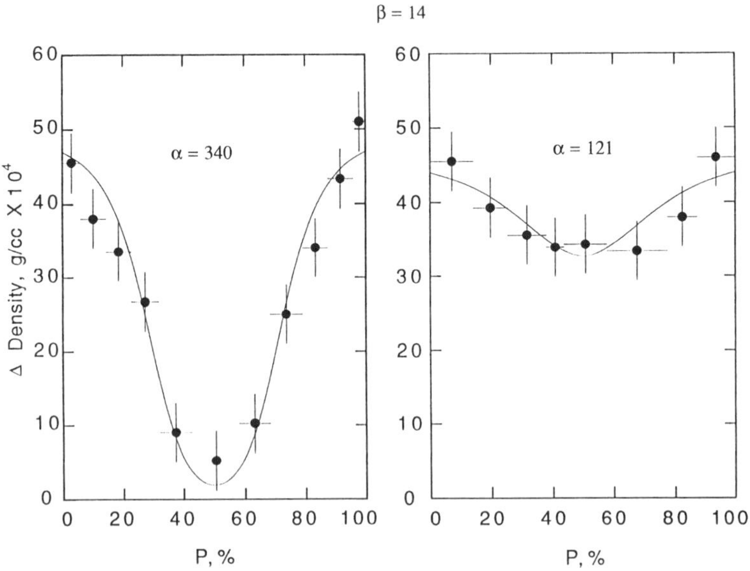

Figure 7. Same experimental data as in Figure 5 fit with theory using β = 14.

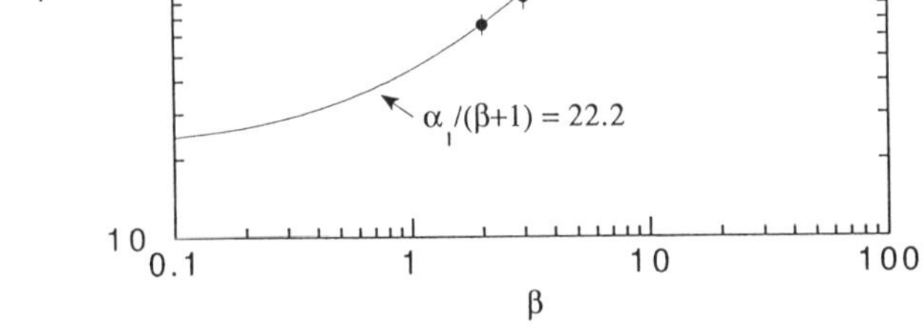

Figure 8. Circles plot α_1 values that give the best theoretical fits to the experimental data of Figure 5 for a number of β values in the range where reasonable fits occur (β ~ 2 to 14). Solid curve plots the function $\alpha_1/(\beta+1)$ = 22.2.

Since we earlier determined that a range of values of β can be used to fit the experimental results, it is of interest to note that this quantitative confirmation of the governing theoretical relationship is independent of the value of β selected. This can be easily seen in Figure 8 where a curved line is drawn corresponding to $\alpha_1/(\beta+1) = 22.2$. This curve agrees extremely well with the sets of α_1 and β determined earlier to gives the best fits to the data as a function of β. These observations instill even more confidence in the appropriateness of the theoretical modelling.

Now that we have confidence in the theory, it is of interest to combine the theoretical results together with the theoretical relationship in equation 25 in order to derive a general criterion for estimating when diffusion-limited oxidation effects are important. We define a critical sample thickness L_c as the limiting sample thickness if one wants to guarantee that the integrated oxidation across the sample will be at least 90% of the homogeneously oxidized case. From equation 25, L_c will be related to α_c (the value of α corresponding to 90% integrated oxidation) by

$$L_c = \left[\frac{\alpha_c\, p\, P_{ox}}{(\beta+1)\, R_s}\right]^{0.5} \tag{30}$$

Values of $\alpha_c/(\beta+1)$ are plotted versus β in Figure 9; they vary from ~1.3 to 9.7 as β varies from low to high values. Some researchers (*1,21*) have assumed a value of 8, reasonable for large values of β; others (*22*) have used a value of 1, more reasonable for smaller values of β. If one were trying to be conservative in estimating the conditions necessary to eliminate diffusion-limited oxidation anomalies, the lowest value ($\alpha_c/(\beta+1)$~1) would be appropriate. Since the scant evidence available to date seems to indicate that most commercial polymers in air have intermediate values of β (~3 and higher), a more pragmatic approach would be to choose $\alpha_c/(\beta+1)$~4. In any case, as more confidence is gained in the theoretical models, and enhanced knowledge of β values appropriate to air aging of various materials is derived, even more confident estimates of the importance of diffusion effects from equation 30 will be possible.

Earlier we noted that reliable values of R_s could only be obtained from the oxygen consumption experiments (Table I) if these experiments were conducted under conditions guaranteeing homogeneously oxidized samples. Since β in air for the current EPDM is equal to ~6, which corresponds to a value of $\alpha_c/(\beta+1)$~6 from Figure 9, we can now use equation 30 to estimate values of L_c for the experimental conditions used in the consumption experiments. The results are summarized in Table I. Comparisons of these L_c values with the sample thicknesses used for the various consumption experiments (adjacent column in Table) offer strong evidence that these experiments were conducted under conditions where homogeneous oxidation occurs. Similar calculations for the aging conditions appropriate to the samples used for density profiling (results plotted in Figures 5-7) yield L_c~0.16 cm. This result is consistent with the important diffusion effects found for the 0.32 cm thick sample and the moderate diffusion effects observed for the 0.18 cm thick sample.

Conclusions

Two simple theoretical models for diffusion-limited oxidation were successfully fit to experimental oxidation profiles for a commercial EPDM rubber formulation, yielding values for the two theoretical parameters, α and β,

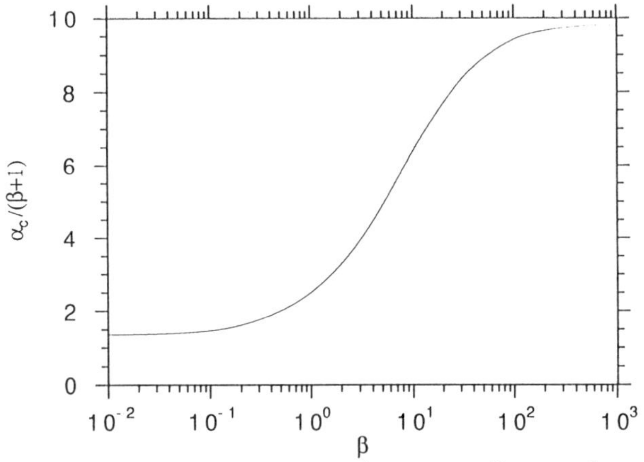

Figure 9. Plot of $\alpha_C/(\beta+1)$ versus β where α_C refers to the value of α corresponding to 90% integrated oxidation.

common to both models. Using independently measured values for the oxygen permeability coefficient and the oxygen consumption rate for this material, a theoretical expression relating these quantities to α and β (common to both models) was quantitatively verified. In order to choose between the two models and derive kinetic information on such things as the kinetic chain length and ratios of rate constants, work currently in progress is assessing the effects of both initiation rate and oxygen pressure dependence on the profile shapes (*19*).

Acknowledgments

This work was performed at Sandia National Laboratories supported by the U. S. Department of Energy under contract number DE-AC04-76DP00789. Able technical assistance was provided by P. D. Silva and G. M. Malone.

Literature Cited

1. Cunliffe, A. V.; Davis, A. *Polym. Degrad. Stab.* **1982,** *4,* 17.
2. Gillen, K. T.; Clough, R. L. In *Handbook of Polymer Science and Technology, Vol. 2: Performance Properties of Plastics and Elastomers*; Cheremisinoff, N. P., Ed.; Marcell Dekker, Inc: New York, New York, 1989, pp. 167-202.
3. Gillen, K. T.; Clough, R. L. In *The Effects of Ionising Radiations on Polymeric Materials*; Collyer, A. A.; Clegg, D. W., Eds.; Elsevier, in press.
4. Clough, R. L.; Gillen, K. T.; Quintana, C. A. *J. Polym. Sci., Polym. Chem. Ed.* **1985,** *23,* 359.
5. Clough, R. L.; Gillen, K. T. In *Polymer Stabilization and Degradation, ACS Symposium Series No. 280*; Klemchuk, P. P., Ed.; American Chemical Society: Washington, D. C., 1984, pp. 411-422.
6. Gillen, K. T.; Clough, R. L.; Dhooge, N. J. *Polymer* **1986,** *27,* 225.
7. Gillen, K. T.; Clough, R. L.; Quintana, C. A. *Polym. Degrad. and Stab.* **1987,** *17,* 31.
8. Gillen, K. T.; Clough, R. L. *Polym. Eng. and Sci.* **1989,** *29,* 29.
9. Jouan, X.; Gardette, J. L. *Polym. Commun.* **1987,** *28,* 329.

10. Papet, G.; Jirackova-Audonin, L.; Verdu, J. *Radiat. Phys. Chem.* **1987,** *29,* 65.
11. Papet, G.; Jirackova-Audonin, L.; Verdu, J. *Radiat. Phys. Chem.* **1989,** *33,* 329.
12. Gillen, K. T.; Clough, R. L.; Jones, L. H. *Sandia Labs Report* **1982,** NUREG/CR-2877, SAND81-2613.
13. Grassie, N.; Scott, G. *Polymer Degradation and Stabilisation*; Cambridge University Press: Cambridge, England, 1985.
14. Bolland, J. L. *Proc. Roy. Soc.* **1946,** *186,* 218.
15. Bateman, L. *Quarterly Rev. (London)* **1954,** *8,* 147.
16. Crank, J. *The Mathematics of Diffusion*; Clarendon Press: Oxford, England, 1975.
17. Clough, R. L.; Gillen, K. T. *Polym. Mater. Sci. and Eng.* **1988,** *58,* 209.
18. Seguchi, T; Yamamoto, Y. *Japan Atomic Energy Research Inst. Report* **1986,** JAERI 1299.
19. Gillen, K. T.; Clough, R. L. to be published.
20. Gillen, K. T.; Clough, R. L. *Polym. Preprints* **1990,** *31, No. 2,* 387.
21. Seguchi, T.; Hashimoto, S.; Arakawa, K.; Hayakawa, N.; Kawakami, W.; Kuriyama, I. *Radiat. Phys. Chem.* **1981,** *17,* 195.
22. Billingham, N. C.; Calvert, P. D. In *Developments in Polymer Stabilisation*; Scott, G., Ed.; Applied Science Publishers, Ltd.: 1980, Vol. 3.

RECEIVED February 20, 1991

Chapter 29

Change in Mechanical Properties of Low-Density Polyethylene during Radiochemical Aging

L. Audouin and J. Verdu

Ecole Nationale Supérieure d'Arts et Métiers 151, Boulevard de l'Hôpital, 75013 Paris, France

Tensile samples of LDPE were irradiated in air and in nitrogen at various dose rates. By carbonyl FTIR analyses on microtome sections, it was established that the depth of the oxidized layer is a decreasing function of the dose rate. The analysis of the change of tensile properties is complicated owing to the fact that chain scission and crosslinking (having opposite effect on ultimate tensile properties) predominate respectively in the superficial and in the core zone. A sharp decrease of the ultimate elongation seems to occur when the depth of the oxidized layer becomes higher than 170-180 μm. This behavior, which was already observed in the case of HDPE, could be explained in terms of fracture mechanics.

A great amount of research work has been devoted to the radiochemical ageing of polyethylene (PE) in the past half century. The dual character of ageing mechanisms is now well recognized. In anaerobic conditions, crosslinking and crystal destruction are the main causes of properties changes (1-6), whereas in the presence of oxygen, chain scission in the amorphous phase governs essentially the change of thermomechanical properties (7-15).

From a practical viewpoint, embrittlement in the presence of oxygen is the main problem. It is well know that the embrittlement dose (for instance dose to reach to 50 % of the initial ultimate elongation) is an increasing function of the dose rate δ. In other words, the "yield" of the embrittlement process is a decreasing function of δ. Although branching kinetics could be involved in this dependance (13,14), it is usually considered that it is mainly due to the oxidation kinetic control by oxygen diffusion. In this case, the sample degrades heterogeneously with the oxidation being restricted to a more or less extended superficial zone observable by various analytical methods (15-19).

Many analysis of diffusion controlled kinetics have been published in the past years (18-24). They are always based on the analysis of

0097-6156/91/0475-0473$06.00/0

oxygen concentration (C) changes in an elementary layer located at the distance x from the surface. All the proposed models start from the same type of differential equation based on the Fick's law :

$$\frac{\partial C}{\partial t} = D \frac{\partial^2 C}{\partial x^2} - r(C) \qquad (I)$$

where D is the oxygen diffusivity and r the local rate of oxygen consumption. The function linking r to the local oxygen concentration C and to the other internal or external variables can be in principle established from oxidation mechanisms. Especially important is the initiation step, whose rate r_i is in principle proportional to the dose rate. The study of various mechanistic schemes led Gillen and Clough (19) to distinguish the "bimolecular" and "unimolecular" cases where the whole oxidation rate is proportional to respectively the square root of dose rate or the dose rate.

It was shown that, in this case, the length of the plateau, i.e. practically the thickness of the oxidized layer (TOL) is proportional to $(D/r)^{1/2}$ (18).

When the oxidation rate is first order in O_2, the resolution of equation (I) would lead to an exponential profile whose characteristic depth would be proportional to $(D/k)^{1/2}$, k being the first order rate constant for oxygen consumption. Considering in a first approximation, that the zero order rate r or the first order rate constant k are proportional to the dose rate δ, gives :

$$\text{TOL} \quad \alpha \quad (D/\delta)^{1/2} \qquad (II)$$

This relationship is consistent with experimental results in the case of low density polyethylene (18).

When O_2 diffusion control operates, the classical structure-properties relationships cannot be used to predict the change of mechanical properties due to ageing. As a matter of fact, these relationships cannot be applied to heterogeneous samples. A treatment of this problem from fracture mechanics concepts was first proposed by Rolland et al (25) and then developed by Schoolenberg (26) in the case of polypropylene photooxidation. According to this theory, the oxidized layer can be considered as a notch capable of initiating the sample fracture beyond certain critical characteristics. As a consequence, there is a critical dose rate δ_c above which, for a given dose, the change of mechanical behavior is essentially governed by crosslinking in the core zone, and below which fracture initiation due to the embrittlement of the oxidized layer plays the key role. Evidence for the existence of a such critical dose rate was recently presented in the case of high density polyethylene (16).

The aim of this paper is to apply the same concepts to the study of the radiochemical ageing of low density polyethylene in order to try to establish semi-empirical relationships allowing the prediction of the TOL and of the embrittlement dose.

EXPERIMENTAL

MATERIALS Plates of 2.5 mm thickness were made by calendering from an industrial LdPE slightly stabilised against thermal oxydation (LACQTENE 1020 FN 24 supplied by ATOCHEM - FRANCE). Its melt index was 2.0, its melting point 110-120° C, its enthalpy of fusion : 206 $kJ.Kg^{-1}$ corresponding to a crystallinity degree of 39 %. Its density was 0.92. Dogbone samples of 110 mm length (calibrated section 25 x 25 x 2.5 mm) according to the French standard AFNOR NFT 51 034 were cut in the direction perpendicular to the processing axis.

CHARACTERIZATION IR spectra were taken with a Perkin Elmer FTIR 1710 apparatus. The results will be expressed in absorbance per thickness unit (cm^{-1}). Microtome sections of 20 µm were taken, in bulk samples, with a Rirchert-Jung microtome equipped with steel blades.
Tensile testing was made using an INSTRON model 4500 machine at a 50 $mm.min^{-1}$ tensile rate, at 21°C. The reported data correspond to average values for 5-10 samples.
Sol-gel and swelling experiments were performed on the core zone of the samples, after elimination of the oxidized layer (0.5 mm). The measurements were made in boiling xylene for 48 hours. $\overline{M_c}$ was determined from the Flory-Rhener relationship (27), using a value of 0.122 for the interaction coefficient.

EXPOSURE The samples were irradiated by Co^{60} gamma rays in air, at 32 ± 2°C, in a ventilated chamber.
Various dose rates ranging from 0.18 to 9.30 $Gy.s^{-1}$ (with a relative uncertainty of ± 10 %) were studied. Some samples were irradiated between two thick PE plates in order to avoid oxidation effects. Particular attention will be paid to the samples irradiated with a dose of 150 kGy whatever the dose rate.

RESULTS

1) Depth profile of carbonyl concentration Some examples of depth profiles of carbonyl concentration determined by IR analyses on microtome section are presented in Fig. 1. They display similar shape as previously found with another LdPE sample (18) with a "skin" zone in which the CO concentration decreases ; an almost horizontal plateau, and a "transition" zone in which the concentration tends to zero. The thickness of oxidized layer (TOL) is here defined as the thickness where the carbonyl absorbance reaches 50% of its horizontal plateau value.

For the highest dose rates, i.e. for the lowest TOL values, the plateau tends to disappear, and the carbonyl concentration decreases continuously from the surface to a depth typically lower than 120 µm beyond which it is practically zero.

As previously observed, TOL is practically independant of the dose.

TOL was plotted against $(1/\delta)^{1/2}$ in order to check the validity of rel. II (Fig. 2).

For dose rate $\delta < 4$ $Gy.s^{-1}$, the points are close to a straightline of equation.

$$TOL \approx 95 + \frac{85}{\sqrt{\delta}}$$

For dose rates above 4 $Gy.s^{-1}$, the experimental values of TOL are significantly lower than those extrapolated from the above relationship. It could be tentatively supposed that in this domain, TOL is practically proportional to the dose rate :

$$TOL \approx \frac{240}{\sqrt{\delta}}$$

It is not proved that LDPE differs qualitatively from HDPE. As a

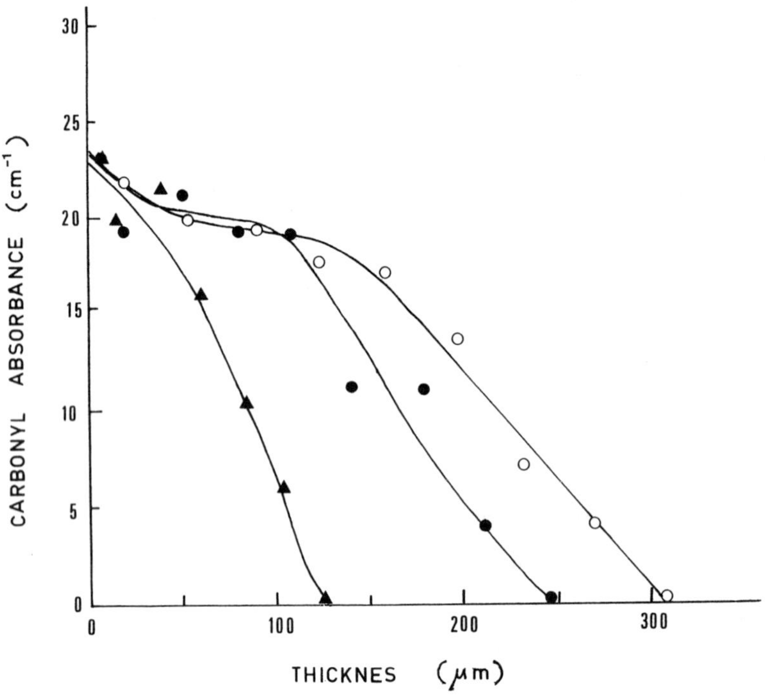

Fig. 1 Example of depth profiles of carbonyl concentraion for three distinct dose rates ▲ 9.31 Gy.s^{-1} ; ● 1.88 Gy s^{-1} ; ○ 0.64 Gy s^{-1}.

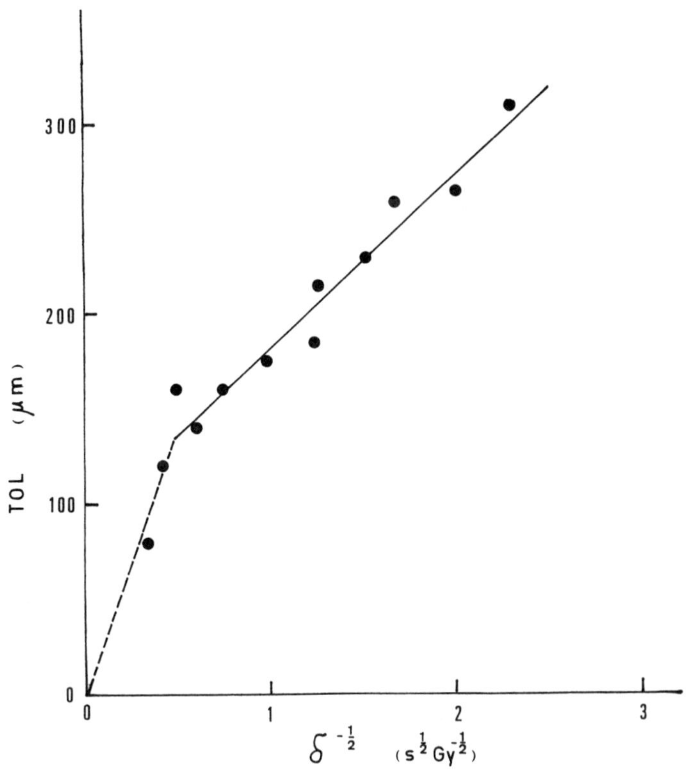

Fig. 2 Thickness of the oxidized layer against reciprocal of the square root of the dose rate.

mather of fact the investigated range of dose rates is considerably larger for LDPE than for HDPE.

2) Crosslinking process The gel fraction G, the reciprocal of the swelling ratio Q^{-1} and the average molar weight of the network segments M_c for a 150 kGy irradiation are reported in Table 1. It appears that, in the core zone, the yield of crosslinking is practically independant of the dose rate within experimental errors.
The change of gel fraction with dose is presented in Fig. 3. The gelation dose is about 50 kGy as previously found (16).

3) Change of mechanical properties The tensile properties were determined after irradiation (150 kGy). The results are reported in Table 1. As previously observed (16), there is practically no change in yield stress, whereas the rupture coordinates vary in the opposite way since σ_R increases and ϵ_R decreases for the dose under study.

In Fig. 4, ϵ_R was plotted against dose rate. These results can be summarized as follows :

i) As expected, the curve displays a transition near to 3.0 ± 0.3 $Gy.s^{-1}$. The sample ductility decreases considerably (ϵ_R is practically divided by 2), when the dose rate becomes lower than this critial value.

The samples irradiated in anaerobic conditions have however ultimate properties considerably higher than those irradiated in air, even at the highest dose rate. It must be noted that these samples have a significantly higher crosslink density (Table 1). Furthermore, crosslinking of the superficial layer plays probably an inhibiting role in crack initiation.

ii) The critical dose rate $\delta_c = 3\ Gy.s^{-1}$ corresponds to a critical TOL value as shown by Fig. 5 in which the ultimate elongation was plotted against the measured TOL. The critical TOL value is about 140 ± 20 µm.

DISCUSSION

Concerning the shape of the depth distribution of carbonyl concentration and the TOL values, the above results agree reasonably with previously reported ones (18).

a) The depth distributions of carbonyls approximate a plateau characteristic of kinetic order with respect to oxygen concentration approaching zero. The sharp increase in the superficial layer (≈ 50 µm), which was previously found (18), remains unexplained.

b) The TOL values are of the same order of magnitude : Here TOL ≈ 180 µm for a dose rate of 1 $Gy.s^{-1}$, against ≈ 130 µm in the preceding work (18). Indeed, a change of morphology or composition can affect the O_2 diffusivity or the oxidation rate, leading to a variation of TOL according to rel. II.

c) The TOL values are also of the same order of magnitude as for HDPE (16). In this latter case, the oxygen permeability P_{OX} was lower than for LDPE, but the oxidation rate was also lower. As a matter of fact for practically the same dose (150-160 kGy), A_{CO} ≈ 2 cm^{-1} in the superficial zone of HdPE (18) against ≈ 20 cm^{-1} (Fig. 1) in the same zone of LDPE. Thus, the ratio P_{OX}/r which controls TOL (eq. II) takes practically the same value for both polymers.

Table 1. LDPE sample characteristics after 150 kGy irradiation and unirradiated

N°	Dose rate $Gy.s^{-1}$	TOL (µm)	G		Q^{-1}		$\overline{M}_c$ $g.mol^{-1}$	σ_y (MPa)	σ_r (MPa)		ϵ_r (%)	
1	0.19	310	0.69		0.22		2800	10.9	9.7		300	
2	0.25	265	0.61		0.19		3200	10.7	9.5		336	
3	0.36	260	0.64		0.21		2900	11.2	11.0		430	
4	0.44	230	0.59		0.19		3500	10.1	10.0		374	
5	0.64	185	–		–		—	—	—		–	
6	0.64	215	0.62		0.19		3400	11.7	6.6		275	
7	1.03	175	–		–		—	—	—		–	
8	1.88	160	0.65		0.21		2800	11.7	7.3		330	
9	2.78	140	–	(0.77)	–	(0.35)	—	10.4	10.2	(18.2)	436	(1007)
10	4.17	160	0.63	(0.74)	0.21	(0.28)	2650	10.2	12.0	(18.7)	715	(1125)
11	5.55	120	0.65	(0.70)	0.22	(0.23)	2400	10.4	12.1	(20.3)	686	(1161)
12	9.31	80	0.67		0.21		2800	11.4	12.2		692	
13	0	0	0		–		uncrosslinked	11.0	9.5		750	

Source: Data from references 1–13.
Key: TOL: thickness of the oxidized layer; G: gel fraction; R^{-1}, reciprocal of the swelling ratio; $\overline{M}_c$: molar weight of the network segments; σ_r: ultimate stress; ϵ_r: ultimate elongation.
Note: The values between parentheses correspond to anaerobic exposures.

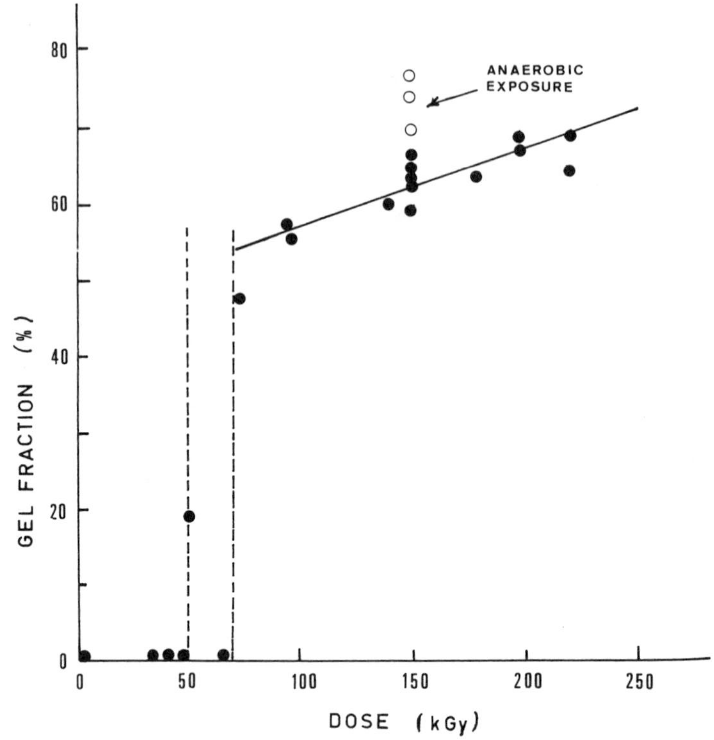

Fig. 3 Gel fraction of core samples against irradiation dose for various dose rates.

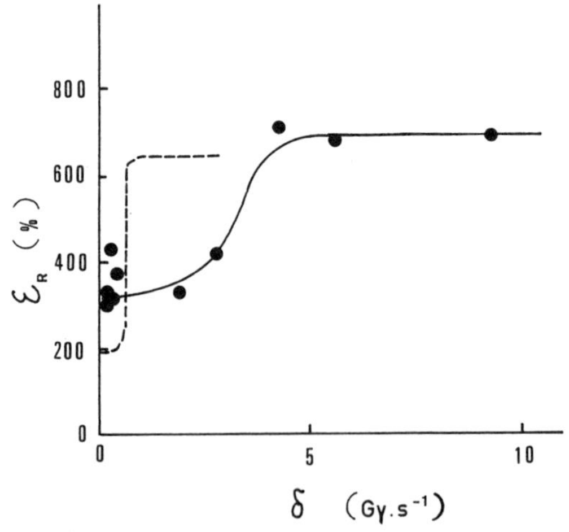

Fig. 4 Ultimate elongation against dose rate for an irradiation dose of 150 kGy. Dashed line : Results obtained on HDPE after ref. (15).

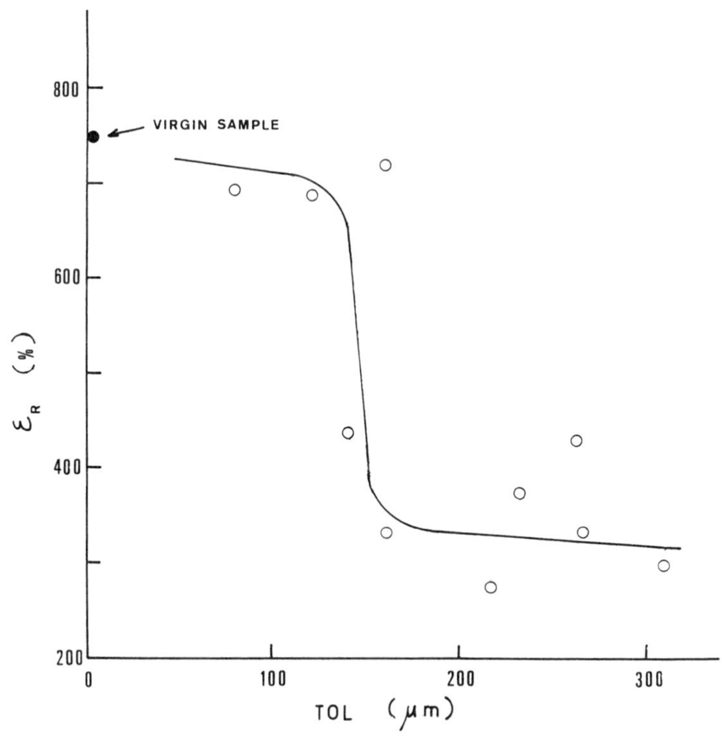

Fig. 5 Ultimate elongation against thickness of the oxidized layer for an irradiation dose of 150 kGy.

Concerning the crosslinking process in the core zone of the sample, it must be noted that :
d) It reaches an asymptotic value lower than for HdPE (gel content 0.64 ± 0.05 against 0.85 (1)).
e) Since the network characteristics in the core zone are practically independant of the dose rate (Table 1), samples irradiated for the same dose differ only by the characteristics of the oxidized layer. As shown in Fig. 1, the conversion of the oxidation reaction, i.e. the extent of degradation in this layer, seems to be almost independant of the dose rate. Thus, for a given dose, for instance 150 kGy, the samples irradiated at various dose rates differ essentially by their TOL.

Let us now consider the change of mechanical properties, essentially the ultimate elongation, on the basis of the fracture mechanics model (25-26).

According this model, a crack is initiated in the superficial (oxidized) layer and propagates at an increasing rate (in the case of tensile testing) through this layer. When it reaches to the interface between oxidized and non oxidized ("transition") zone, there are two possibilities :
i) The core zone is ductile for the crack propagation rate reached at the interface, and crack blunting and arrest occurs.
ii) The core zone is brittle and the crack propagates through the core zone until the complete fracture of the sample. This mechanism suggests that there are critical characteristics of the oxidized layer, corresponding to a critical crack propagation rate at the "skin-core" interface, i.e. to a ductile-brittle transition for the core zone. In principle at least three ageing dependant characteristics are involved in the change of mechanical properties upon irradiation.
f) Crosslinking in a thin superficial layer. Crack initiation occurs at the sample surface. As it is well known, for instance in the field of stress-cracking in tensio-active media, crosslinking inhibits crack initiation by increasing the concentration of tie molecules (28). This effect can explain at least partially the considerable difference existing between the ultimate properties of the samples irradiated respectively in air and in anaerobic conditions (Table 1). It is noteworthy that the ultimate elongation of the samples irradiated in air at high dose rates, is of the same order as for the virgin (uncrosslinked) sample (750 %).
g) Crosslinking in the core zone : Indeed, the ability of the core zone to arrest the crack propagation depends on its crosslink density. In the conditions under study, crosslinking seems to increase the polymer strength since the ultimate stress of samples irradiated at high dose rates ($\sigma_R \approx$ 12 MPa), is higher than for the virgin sample (9.5 MPa) whereas the rate of crack initiation is presumably the same.
h) Thickness of the oxidized layer. As expected, TOL plays a very important role. As shown in Fig. 5, the critical TOL, above which the "sensitizing" effect of the oxidized layer becomes important, is about 140 µm.

It appears thus that a combination of the theories of diffusion controlled kinetics and fracture mechanics offers an interesting basis for a tentative of lifetime prediction in the case under study.

Concerning the first step of this approach, i.e. the TOL calculation, a problem remains unresolved : the decrease of the carbonyl concentration in the superficial layer, until the horizontal plateau. A preliminary analysis showed that simple explanations involving for instance an ozone attack or an initial sample heterogeneity (21) are to be rejected (18). Whatever the cause of this phenomenon, the corresponding oxygen overconsumption in the superficial layer might be taken into account, especially for high dose rates where it occurs in the whole oxidized layer. In these cases, kinetics cannot be considered zero order and a significant departure form the model represented by rel. II can be expected. It is noteworthy that the slope change observed in the curve of TOL rs $\delta^{-1/2}$ (Fig. 2) occurs precisely at the TOL value below which the horizontal plateau of the carbonyl depth distribution disappears completely. Further research on this mechanisms of oxygen overconsumption would be needed to interprete this behavior.

A precise TOL modelling would in principle take into account the local changes of oxygen permeability induced by oxidative degradation and crosslinking. Previously reported results (29) show however that these changes can be considered negligible in the dose range under study.

For the second step of this approach, i.e. the determination of ultimate properties form the knowledge of TOL, it will be difficult to go beyond a qualitative or semi-quantitative level. This is because classical fracture mechanics concepts are not well adapted to the cases where the samples undergo very large plastic deformations before rupture as for PE.

It seems thus that in the near future, the best way for lifetime predictions will consist to determine empirically the critical TOL value.

ACKNOWLEDGEMENTS The samples were supplied by ATOCHEM, (CERDATO, Serquigny, France), which is gratefully acknowledged.

LITERATURE CITED

1. Charlesby A., Proc. Roy. Soc. **1952**, A 215, 187.
2. Lawton E.J., Bueche A.M. and Balwitt J.S., Nature **1953**, 172, 76.
3. Charlesby A., Proc. Roy. Soc. **1954** A 222, 60.
4. Dole M., Keeling C.D. and Rose D.G., J. Am. Chem. Soc. **1954**, 76, 4304.
5. Lawton, E.J., Semany P.D. and Balwitt J.S., J. Am. Chem. Soc. **1954**, 76, 3437.
6. Dole M. and Howard W.H., J. Phys. Chem. **1957**, 61, 137.
7. Chapiro A., J. Chim. Phys. **1955**, 52, 246.
8. Decker C., Mayo F.R. and Richardson M., J. Polym. Sci Polym. Chem. Ed. **1973**, 11, 2879.
9. Petruj J. and Marchal J., Radiat. Phys. Chem. **1980**, 16, 27.
10. Hori Y., Shimad S. and Kashiwahara H., Polymer **1977**, 18, 151.
11. Wilski H., Kolloïd ZZ Polym. **1973**, 251, 703.
12. Narkis M., Baiter I., Shloknik S., Siegmann A. and Eyerer P., J. Macromol. Sci. Phys. **1987**, 26, 37.

13. Clough R.L. and Gillen K.T., J. Polym. Sci. Poly. Chem. Ed. **1981**, 19, 2041.
14. Gillen T. and Clough R.L., J. Polym. Sci. Polym. Chem. Ed. **1985**, 23, 2683.
15. Papet G., Jirackova-Audouin L. and Verdu J., Radiat. Phys. Chem. **1987**, 29, 65.
16. Audouin-Jirackova L., Papet G. and Verdu J., Europ. Polym J. **1989**, 25, 181.
17. Skorida V.D., Belousova M.V., Maklakov A.K., Zgadzat O.E.H., Doroginitskij M.M., Potapova I.V., and Romanov R.S. Vysokomolek. Soyed. Ser. B **1984**, 26, 765.
18. Papet G., Jirackova-Audouin L. and Verdu J., Radiat. Phys. Chem. **1989**, 33, 329.
19. Gillen K.T. and Clough R.L., Techniques monitoring heterogeneous oxidation of polymers in handbook of polymer science and technology vol. 2 (N.P. Cheremisinoff. Ed.), **1989**, Marcel Dekker.
20. Cunliffe A.V. and Davis A. Polym. Degrad. Stab. **1982**, 4, 17
21. Furneaux G.C., Ledbury K.S. and Davis A., Polym. Degrad. Stab. **1981** 3, 431.
22. Fairgreve S.P. and Mc Callum J.R., Polym. Degrad. Stab. **1985**, 11, 251,
23. Puig J.R. Les techniques de l'Ingénieur - Génie nucléaire (Paris) **1981**, 1B, 3770.
24. Seguchi T., Arakawa K., Hayakawa N., Watanabe Y. and Kuriyama I., Radiat. Phys. Chem. **1982**, 19, 321.
25. Rolland L., Thomson R., Mostovoy S. and Broutman L. Int. Conf. Deformation, yield and fracture - **1982**, Cambridge.
26. Schoolenberg G.E., J. Mater. Sci. **1988**, 23, 1580.
27. Flory P.J. and Rehner J., J. Chem. Phys. **1943**, 11, 11.
28. Lustiger A. in "Failure of Plastics" Eds. W. Brostow and R.D. Corneliussen, **1986** Chap. 16, p. 305. SPE. Hanser. Pub. Munich.
29. Seguchi T. and Yamamoto Y., Japanese Atomic Energy Research Institute Publication JAERI, **1986**, 1299.

RECEIVED January 24, 1991

Chapter 30

Study of Radiation- and Thermal-Oxidized Regions in Ethylene–Propylene Rubber Using an X-ray Microanalyzer

Y. Morita, T. Yagi, and W. Kawakami

Takasaki Radiation Chemistry Research Establishment, Japan Atomic Energy Research Institute, Takasaki-shi, Gunma 370–12, Japan

An electron-probe X-ray microanalyzer was used to monitor the oxidation profile across the cross section of ethylene-propylene rubber sheets which were irradiated and/or thermal-aged in air. The measured oxidation profiles were compared with calculated ones based on both chemical kinetic scheme and oxygen diffusion process in the rubber sheet, and the parameters containing oxidation-rate constants in solid-state reactions were estimated.

The apparent activation energies of degradations of the rubber sheet, obtained from tensile elongation measurements, depend on whether oxidation in the sheet proceeds homogeneously or heterogeneously, which can be discriminated by monitoring the oxidation-profiles. It is concluded that oxidation-profile studies can be an extremely useful aid for interpreting accelerated aging studies of polymer materials.

In recent years, various polymer materials have been used as insulators of cables, O-rings and connectors, etc., in nuclear power plants and irradiation facilities containing accelerators. These polymer materials are subjected in most cases to the radiation environments: dose rates of ca. 10^{-2}~1Gy/h at 60~90°C for a long time, i.e., 40 years and up*(1)*. At these dose rates which are much lower than those in laboratories(10^{2}~10^{4}Gy/h), radiation degradation of polymers is mainly caused by radiation-oxidation reactions.

A significant subject in the study of radiation oxidation on solid-state polymers is heterogeneous oxidation, which is measured as profiling spatial variations in oxidation across a polymer sheet or a film, and caused by diffusion-limited oxidation in polymers*(2,3)*. This heterogeneous oxidation takes place in most polymer materials under irradiation in air, for example, polyethylene, epoxy resin, polyimide and poly

0097–6156/91/0475–0485$06.00/0

ether ether ketone. Recently, it has become possible to monitor the oxidation profiles quantitatively, using a number of experimental techniques: density profiling*(4)*, modulus profiling*(5)* and micro-FTIR spectroscopy*(6)*.

In this paper, we present an oxidation-profiling method based on the electron - probe X-ray microanalyzer(XMA), apply this method to profiling oxidized regions after radiation and thermal oxidation on the sheet's cross section of ethylene-propylene rubber, and discuss the measured profiles by comparing them with theoretical oxidation profiles. We also discuss the relationships between the oxidation profiles and both the tensile elongation results and the apparent activation energies of degradation of ethylene-propylene rubber.

Experimental

Materials. Ethylene-propylene rubber(EPDM; propylene content 26wt%, diene content 6.3wt%), purchased from Japan Synthetic Rubber Co. Ltd., was used as a base polymer. EPDM was mixed with 2Phr of dicumyl peroxide by a mixing roll at 120°C, and hot-pressed at 180°C for 10min to make sheets of 0.5~2mm thicknesses. The obtained sheets were cut into a chip of 10x10mm^2 for XMA measurements, and into a Dumbbell shape of JIS No.5 size for tensile tests.

Irradiation and Aging Procedures. Chip specimens for XMA measurements were irradiated by Co-60 gamma rays(JAERI, Takasaki) using dose rates of 0.1~10kGy/h at room temperature in air. Dumbbell specimens for tensile tests were also irradiated with a dose rate of 2Gy/h at various temperatures from 70°C to 140°C in air using a small gear oven.

Both specimens were aged under thermal-only conditions at temperatures of 70~125°C in air, as references for the simultaneous radiation-thermal exposures.

Electron-Probe X-ray Microanalysis. Figure 1 shows a specimen for XMA measurements. The specimens were prepared from the chips(10x10mm^2) mentioned above, by cutting into slices of ca. 5x2mm^2 size in parallel to the sheet's cross section not from the edge but from an inner part of the sheet. These small slice specimens were dipped into potassium hydroxide isopropanolic(KOH-isopropanolic) solutions of various concentration at room temperature and 70°C, washed with water, and then dried in air over a period of one night and completed by vacuum drying. To avoid charge built-up, Au(gold) was deposited on the cross-section's surface of the dried slice specimens.

The potassium distribution on the treated surface of the slice specimen, which corresponds to the oxidation profile on the cross section of the sheet, was measured with an electron-probe X-ray microanalyzer(XMA; Japan Electric Optical Laboratory Co. Ltd., JXA-733) using the following conditions: accelerating voltage of 25kV, probe current of 2×10^{-8}A and scanning speed of 0.2mm/min. To obtain a reproducibility of the oxidation profiles, XMA counts of potassium were taken with continuously adjusting focus on the observing surface using an adjunct optical microscope. In addition, some of these profiles were checked by scanning measurements of three times.

Tensile Testing. Dumbbell specimens were gripped by the jaws which were placed 2.5cm apart at the start of the test, and the cross-head speed was 100mm/min(Shimazu Autograph Model DSC-500). Five specimens were tested for each treated condition and the data were averaged to obtain the results used in the analysis.

Results and Discussion

XMA Measurement. Figure 2 shows the influence of treatment conditions of the irradiated slice specimen with the KOH-alcoholic solutions on XMA counts of potassium at the sheet surface. The XMA counts increase with increasing the dipping time in the solutions, and reach the saturated value after ca.16~24h in 0.05N KOH-isopropanolic solution. But, the XMA counts increase at slower rate for 0.05N KOH(aqueous) solution diluted by isopropanol(1:1) and 0.1N KOH-isopropanolic solution than those for 0.05N KOH-isopropanolic solution. This difference is explained as follows. The probe-electron beam(3μmϕ) is estimated to penetrate the sheet to the depth of 8~10 μm*(7)*. The XMA counts of potassium become larger when the surface part of the specimen is swollen by the solutions, and reacted with potassium in the depth direction. The swelling of the surface part is not sufficient for the treatment by 0.05N KOH solution having a poor affinity with EPDM, and 0.1N KOH-isopropanolic solution having high viscous property. In the following experiments, we treated the specimens for XMA measurement by dipping them into 0.05N KOH-isopropanolic solution for 16h at room temperature.

Figure 3 shows XMA count profiles of the EPDM specimens irradiated at 1kGy/h in air at room temperature. The abscissa indicates the depth in the sheet's cross section from the surface(from one side to the other). The XMA counts of potassium correspond to the amounts of irradiation oxidation in the rubber, as described below.

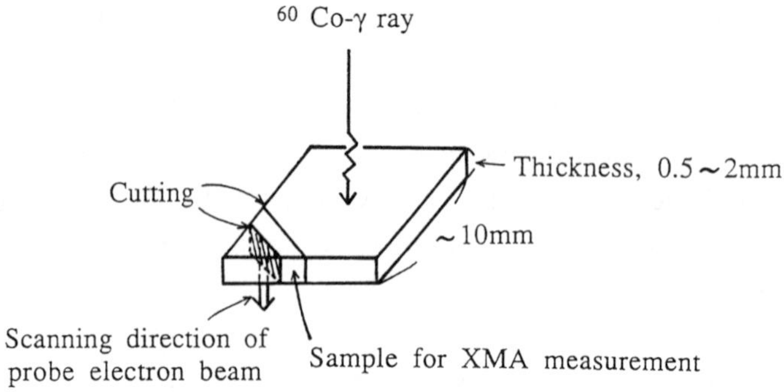

Figure 1. Specimen for XMA measurements.

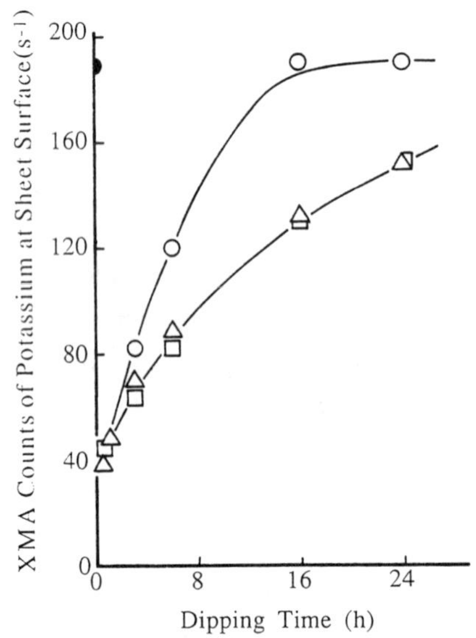

Figure 2. Effect of dipping time on XMA counts at sheet surface.
Specimen: EPDM sheet(2mm thickness) irradiated by 200kGy(1kGy/h) in air at room temperature(r.t.).
Dipping condition:

○; 0.05N KOH-$(CH_3)_2$CHOH solution at r.t.
□; 0.1N KOH-$(CH_3)_2$CHOH solution at r.t.
△; 0.1N KOH(aqueous):$(CH_3)_2$CHOH=1:1 solution at r.t.
●; 0.05N KOH-$(CH_3)_2$CHOH solution at 70°C for 5min.

Therefore, the shapes in this figure show profiles of the radiation-oxidized region on the cross section of the EPDM sheet.

The XMA count of an unirradiated EPDM sheet(Original) is nearly zero. In the irradiated EPDM specimens, there are two regions: the oxidized region in superficial part of the sheet, and the unoxidized region in central part(core part). The boundary between these regions is clear, and the thickness of the oxidized region does not change with total dose.

Figure 4 shows the relationship between the XMA counts at the surface of the EPDM sheet and total dose. The XMA counts of potassium increase linearly with increasing total dose. This means that the irradiation forms the oxidized groups such as a carboxyl group and a peroxide group, which are capable of reacting with potassium:

$$-CH_2\overset{O}{\overset{\|}{C}}OH,\ -CH_2COOH + KOCH(CH_3)_2 \longrightarrow -CH_2\underset{O}{\underset{\|}{C}}OK,\ -CH_2COOK + HOCH(CH_3)_2$$

The XMA counts of potassium is, therefore, considered as representing the degree of radiation oxidation in the EPDM sheet. The XMA count profile also represents the profile of the oxidized region in the sheet's cross section, i.e., the oxidation profile of the sheet.

Oxidation Profile of EPDM. Figure 5 shows the effects of dose rate and thickness of the sheet on the oxidation profile of EPDM specimens. The lower in the dose rate(Figure 5a), the wider the oxidation region. The whole cross section of the specimen is oxidized for a thickness of 1mm at a dose rate of 0.1kGy/h. With respect to the effect of thickness(Figure 5b), the unoxidized region of the core increases with increasing thickness of the sheet. Similar profiles have been obtained by density profiling*(4)*, modulus profiling*(5)* and micro-FTIR spectroscopy*(6)*. Compared with these methods, the XMA method can get continuous profiles of the oxidized region by a simple process.

This heterogeneous oxidation is known to be due to oxygen-diffusion-limited effects*(3)*, i.e., the rate of oxygen consumption in a polymer exceeds the rate at which oxygen from the surrounding atmosphere can be resupplied to the interior by diffusion processes. More information is available from Figure 5a. With decreasing dose rate, the XMA counts(oxidation degree) at the sheet surface where diffusion of

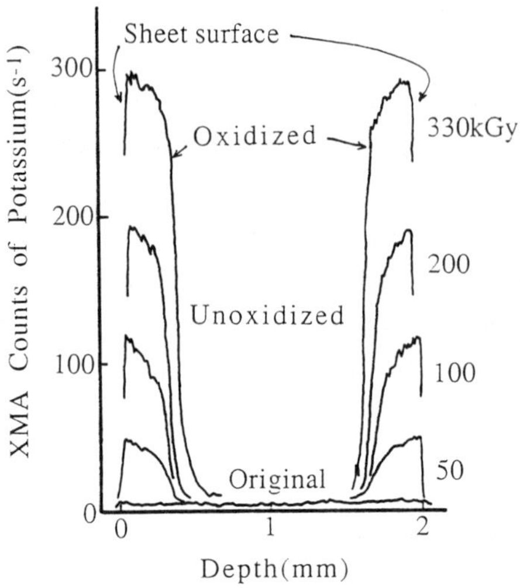

Figure 3. Effect of total dose on oxidation profile.
Specimen: EPDM sheet(2mm thickness) irradiated at 1kGy/h in air at r.t.

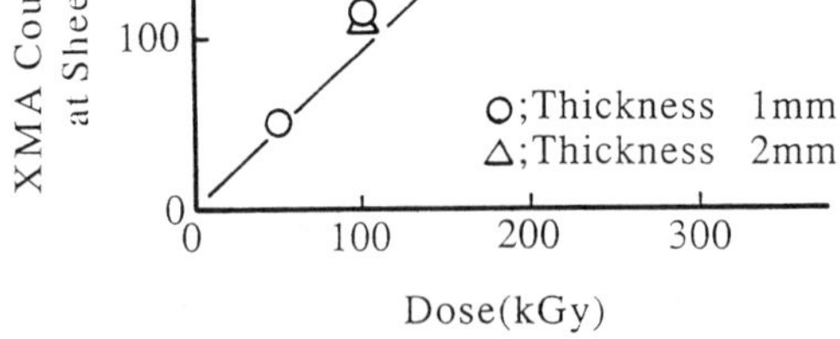

Figure 4. Relationship between total dose and XMA counts at sheet surface.
Specimen: EPDM sheet irradiated at 1kGy/h in air at r.t.

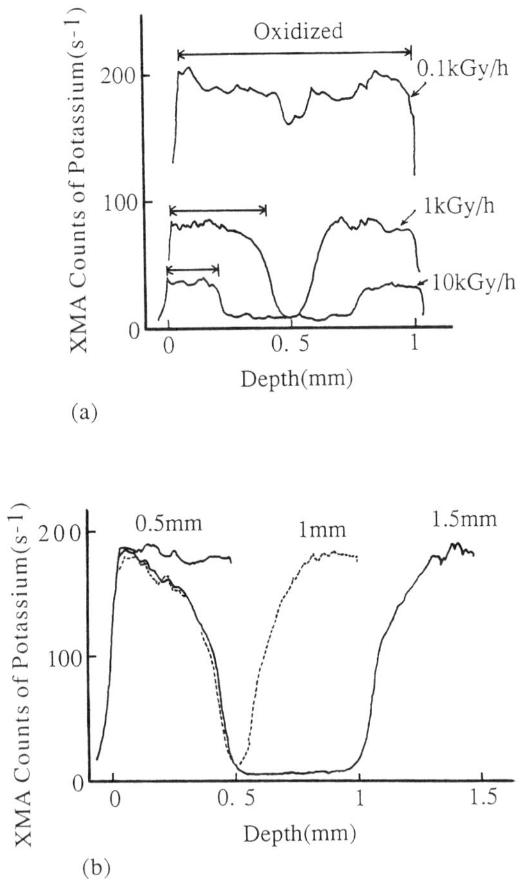

Figure 5. Effects of dose rate and thickness on oxidation profiles of EPDM sheet.
(a) Effect of dose rate; Irradiation: 74kGy in air at r.t.
(b) Effect of thickness; Irradiation: 200kGy(1kGy/h) in air at r.t.

oxygen has no effect, increase substantially for a constant total dose of 74kGy. This point will be discussed below.

To analyze the diffusion-limited oxidation, the relationship between oxidation profile shapes and the details of underlying chemical kinetics has been studied*(2,3)*. There are two type of termination kinetics in critical consensus reactions: bimolecular termination is appropriate for an unstabilized polymer, and unimolecular termination is also appropriate for a well-stabilized polymer. The clear difference between these two terminations is that the oxygen consumption rate depends on a square root of dose rate for the bimolecular termination, compared with a first order dependence on dose rate for the unimolecular termination. The EPDM specimens used in our experiments are unstabilized, and, from the dependence of the XMA counts at the sheet surfaces on dose rates(Figure 5a), the oxidation degree correlated to oxygen consumption rate increases nearly proportional to the square root of dose rate, as shown later. So, we adopt the kinetics containing the bimolecular termination to analyze the results of the EPDM experiments. These kinetics consist of the critical consensus reaction as follow:

Initiation:

$$\text{PH(polymer)} \xrightarrow{\phi I} \text{P}\cdot$$

Propagation:

$$\text{P}\cdot + O_2 \xrightarrow{k_2} \text{POO}\cdot$$

$$\text{POO}\cdot + \text{PH} \xrightarrow{k'_3} \text{POOH} + \text{P}\cdot$$

$$(k_3 = k'_3(\text{PH}))$$

Bimolecular termination:

$$2\text{P}\cdot \xrightarrow{k_4} \text{Products}$$

$$\text{P}\cdot + \text{POO}\cdot \xrightarrow{k_5} \text{Products}$$

$$2\text{POO}\cdot \xrightarrow{k_6} \text{Products}$$

Oxygen consumption rate(r) for this reaction scheme is obtained as equation 1 *(3)* under the assumption of steady state kinetics with respect to the active species: $d(P\cdot)/dt=d(POO\cdot)/dt=0$.

$$r = k_2[P\cdot]C = C_{1b}C/[1+C_{2b}C] \qquad (1)$$

where

$$C_{1b} = k_2(\phi I)^{0.5}/(2k_4)^{0.5} \qquad [h^{-1}]$$

$$C_{2b} = k_2 k_6^{0.5}/(k_3 k_4^{0.5}) \qquad [cm^3/mole]$$

C : Oxygen concentration [mole/cm^3(polymer)]
ϕ : Initiation rate factor(=1.04x10^{-4}Giρ) [mole/MGy/cm^3]
I : Dose rate [MGy/h]
Gi: G-value of initiation [1/100eV]
ρ : Density [g/cm^3]

Oxygen concentration in the sheet at steady state is represented by dimensionless form:

$$d^2\eta/dX^2 = \alpha\eta/(1+\beta\eta) \qquad (2)$$

where
η : Dimensionless oxygen concentration(C/C_0)
C_0: Oxygen solubility [mole/cm^3(polymer)]
X : Dimensionless depth(x/l)
l : Thickness of sheet [cm]
D : Diffusion coefficient of oxygen [cm^2/h]
α : Dimensionless oxygen consumption rate(l^2C_{1b}/D)
β : C_0C_{2b}

Finally, we obtained the dimensionless expression of amount of oxygen consumed in the sheet*(8)*:

$$W(X) = \int_0^\lambda \alpha\eta/(1+\beta\eta)d\lambda \qquad (3)$$

where
λ : Dimensionless time(tl^2/D)
t : Irradiation time [h]

By fitting equation 3 to the XMA profile data, we obtain α and β. Figure 6 shows the results of fitting to the XMA data of the EPDM specimens. The circles, the squares and the diamonds are the values read from the XMA profiles. The lines are calculated according to equation 3. For the equation of the bimolecular termination, we change the exponential parameter of dose rate to more general form:

$$\alpha = (l^2/D)(k_2/(2k_4)^{0.5})(\phi I_0)^{0.5}(I/I_0)^a \qquad (4)$$

The lines fit well to the XMA profile data in Figure 6. In Table I , we list the values of parameters containing rate constants, obtained by fitting the XMA data. To obtain 0.6 as "a" value near 0.5 indicates that the scheme of the bimolecular termination is a reasonable approximation for the unstabilized EPDM, since the bimolecular termination will have the square-root dependence on dose rate for the surface oxidation(Figure 6a).

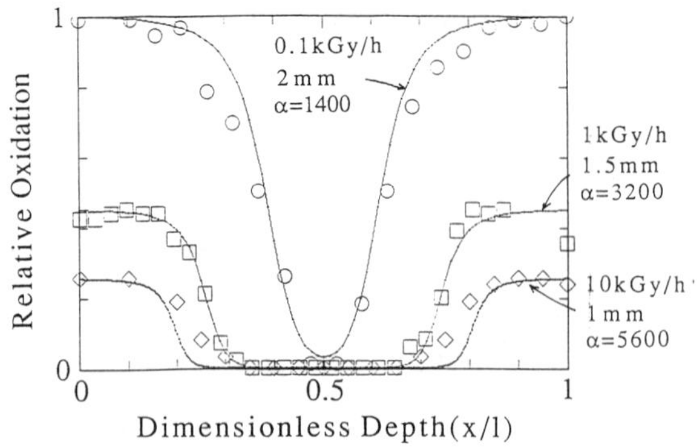

(a) Effect of Dose Rate

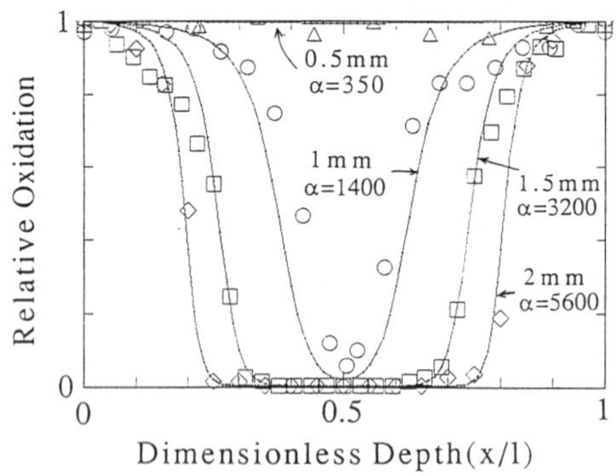

(b) Effect of Thickness

Figure 6. Fitting equation 3 to profile data of EPDM sheet.
○,□,◇; Data from XMA profiles(Figure 5).

Table I. Parameters of Equation 3 for EPDM

	Thickness (mm)	*Dose rste (KGy/h)*	*Dose (KGy)*	*a*	*α*	*β*
Effect	2	0.1	74	0.6	1400	100
of	1.5	1	"	"	3200	"
dose rate	1	10	"	"	5600	"
Effect	0.5	1	200	0.6	350	100
of	1	"	"	"	1400	"
thick-	1.5	"	"	"	3200	"
ness	2	"	"	"	5600	"

$\alpha = (L^2/D)(k_2/(2k_4)^{0.5})(\phi I_0)^{0.5}(I/I_0)^a$

$\beta = (k_2C_0/k_3)(k_6/k_4)^{0.5}$

where $I_0 = 0.1$kGy/h

However, as oxidation proceeds even when irradiated polymers are allowed to stand in room temperature, more experiments considered the effect of total elapsed time may be necessary to determine the termination scheme of this oxidation system.

Relationship between Oxidation Profile and Mechanical Property of EPDM. Figure 7 shows oxidation profiles of EPDM specimens after thermal-only aging at various temperatures. At the higher temperature of 115°C, the oxidation profile shows heterogeneous oxidation: a highly oxidized superficial layer and a slightly oxidized core zone. With decreasing aging temperatures, however, the oxidation proceeds in the core zone, as shown in Figure 7c where the vertical scale is double. In this case(70°C), the oxidation proceeds over the entire cross section, i.e., nearly homogeneous oxidation takes place.

Figure 8 shows the curves of ultimate tensile elongation versus aging time for the same EPDM specimens as used for the XMA measurements. The elongations degrade gradually in the regions of homogeneous oxidation(e.g., at 70°C). On the other hand, they drop steeply when heterogeneous oxidation becomes important(e.g., at 100°C and during the later stage of the curve at 70°C). The lines labeled $E=E_0/2$ and $E=100\%$, refer to the locations where elongation equals half of the initial and 100%, respectively.

Figure 9 shows the Arrhenius plot for the thermal aging data of the EPDM specimen. The logarithmic values of the times required to reach $E=E_0/2$ and $E=100\%$ are plotted versus T^{-1}. The plots have a break point at ca. 100°C on each curve. In the higher temperature regions in which heterogeneous oxidation takes place, the apparent activation energy is 128kJ/mole(30.5kcal/mole); this is known usually as this energy of degradation for EPDM*(9)*. In the lower temperature regions in which homogeneous oxidation mainly takes place, the apparent activation energy is 57kJ/mole(13.5kcal/mole) for the case of T_h(defined on Figure 9). In other word, at higher temperatures, the oxidation reaction in the superficial layer has a higher reaction rate and a large temperature dependence; this changes the surface of the sheet to brittle. Cracks generated in this superficial layer during tensile testing propagate quickly through the Dumbbell specimen. At lower temperatures, the oxidation reaction proceeds homogeneously in the whole of the sheet at a lower rate, and has a smaller temperature dependence. The life time(T_h) at 70°C is obtained to be $2.3x10^3$h, while, if one uses the extrapolated curve based on the higher temperature(over 100°C) experiments, it is estimated to be $1.9x10^4$h which is much longer than that obtained from the lower temperature data.

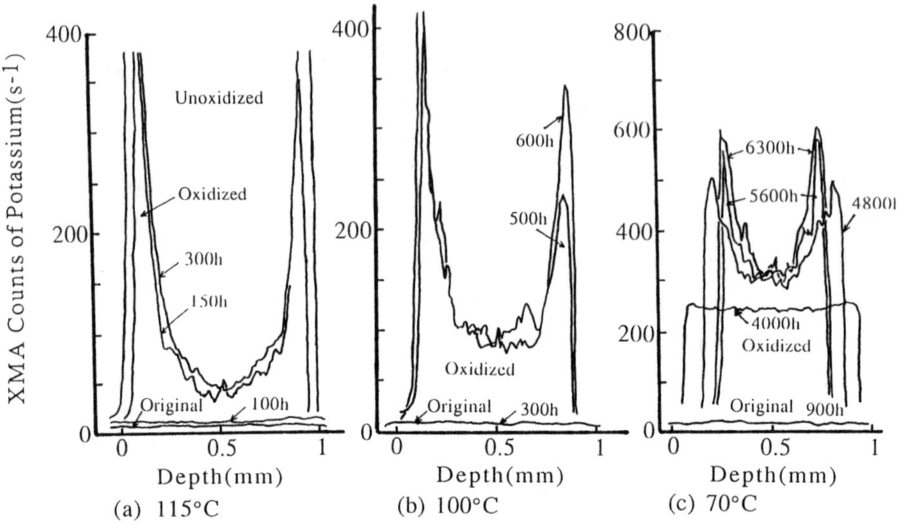

Figure 7. Oxidation profiles of thermal-aged(in air) EPDM sheet. Thickness of sheet: 1mm.

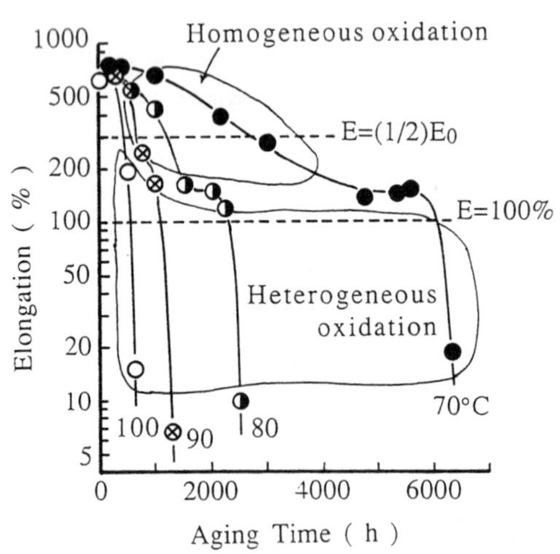

Figure 8. Ultimate tensile elongation versus aging time(in air). Thickness of sheet: 1mm. E_0: Initial elongation.

Figure 10 shows the Arrhenius plot of EPDM specimens obtained by simultaneous treatment of irradiation(2Gy/h) and heating. In this case, the break point of the Arrhenius plot shifts from ca.100°C for the thermal-only treatment to ca.120°C for the simultaneous treatment; however, the apparent activation energy at lower temperatures(under 120°C) is smaller than that at higher temperatures, similar to the thermal-only case.

The heterogeneous oxidation occurs when a consumption rate of oxygen (oxidation rate) is larger than a diffusion rate of oxygen in the EPDM sheets. By the adding radiation in the simultaneous treatment, the oxidation rate becomes larger, while the diffusion rate of oxygen has no effect in the irradiation of this low level. It is expected in this treatment, therefore, that the break point of the Arrhenius plot shifts to lower temperature, compared with that of thermal-only treatment. However, the shift of the break point in Figure 10 is opposite direction, i.e., the more higher temperature. We consider about this opposite shift of the break point that the radiation leads to decomposition of some species which accelerate heat-oxidation reactions in the superficial layer of the sheet.

Consequently, the apparent activation energies of degradation obtained by tensile elongations depend on whether oxidation proceeds in the homogeneous or the heterogeneous mode. And so, it is important to measure the oxidation profiles for accelerated tests of polymer specimens, at higher dose rates and temperatures, to screen for and understand heterogeneous oxidation.

Conclusion

The electron-probe X-ray microanalyzer(XMA) is used to measure the semi-quantitative oxidation profile across ethylene-propylene rubber(EPDM) sheets which are irradiated and/or thermally-aged in air. The XMA method can get continuous profiles of the oxidized region by a simple process. The measured oxidation profiles were compared with calculated ones based on both chemical kinetic scheme and oxygen diffusion process in the rubber sheet, and the parameters containing oxidation-rate constants in solid-state reactions were estimated.

For the irradiated and/or thermally-aged EPDM sheets, there are two different apparent activation energies of degradation; these are understood from oxidation profiles.

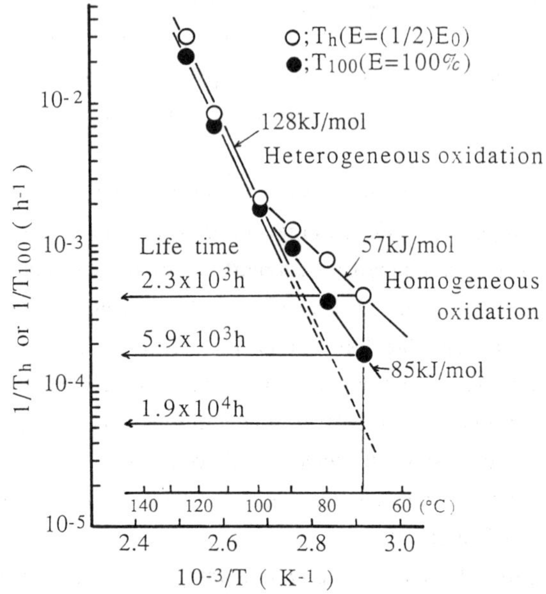

Figure 9. Arrhenius plots of thermal aging data from Figure 8.

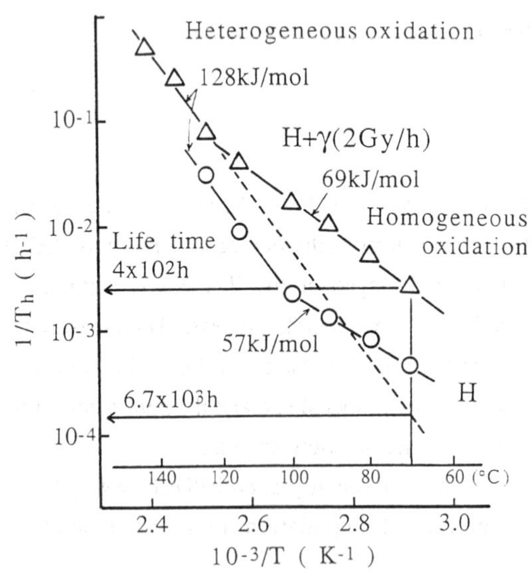

Figure 10. Arrhenius plots of simultaneous treatment for EPDM sheet. Thickness of sheet: 1mm. Treatment: Heat and 2Gy/h in air.

Acknowledgments

The authors thank Mr. Hiroshi Ito in this laboratory about the suggestions of measuring condition of XMA.

Literature Cited

1.Bonzon,L.L.;Wyant,F.J.;Bustard,L.D.;Gillen,K.T.*Status Report on Equipment Qualification Issues Research and Resolution*; NUREG/CR-4301, SAND85-1309; Sandia National Laboratory:USA,1986;pp2.1-1.

2.Cunliffe,A.V.;Davis,A.*Polym. Degrad. Stab.*,**1982**,*4*,pp17.

3.Gillen,K.T.;Clough,R.L.*Handbook of Polymer Science and Technology*; Marcel Dekker, Inc.:New York and Basel,USA,1989;pp168-175.

4.Gillen,K.T.;Clough,R.L.;Dhooge,N.J.*Polymer*,**1986**,*27*,pp225.

5.Gillen,K.T.;Clough,R.L.;Quintana,C.A.*Polym. Degrad. Stab.*,**1987**,*17*,pp31.

6.Jouan,X.;Adam,C.;Fromageot,D.;Gardette,J.;Lemaire,J.*Polym. Degrad. Stab.*, **1989**,*25*,pp247.

7.for example, Uchiyama,I.;Watanabe,A.;Kimoto,S.*X-ray Microanalyzer*; Nikkan Kogyo Shinbunsya Co. Ltd.:Tokyo,Japan,1973,pp131.

8.Morita,Y.;Yagi,T.;Kawakami,W.*International Symposium of Radiation Degradation of Polymers and the Radiation Resistant Materials*; Japan Atomic Energy Research Institute:Japan,1989;pp81-90.

9.Walter,G.*Rubber Chem. Tech.*,**1976**,*49*,pp809-811.

RECEIVED March 22, 1991

Chapter 31

Influence of Ionizing Radiation on Thermoset Plastics

H. Wilski[1]

Hoechst AG, 6230 Frankfurt 80, Germany

Thermoset plastics were irradiated under two very different conditions. One set of samples was irradiated with electrons up to 10 MGy under the exclusion of oxygen, using a high dose rate. The second set was irradiated with gamma-rays up to 1 MGy, in air, using a dose rate of only 13.8 Gy/h. The latter experiment lasted longer than 10 years. The flexural strengths of the inorganic filled thermosets (epoxy, melamine-formaldehyde, unsaturated polyester and phenol-formaldehyde) remained nearly unchanged under these conditions though an influence of the dose rate was clearly seen in both the latter cases. Organic filled thermosets were deteriorated, the influence of the dose rate being stronger. The deflection temperature was changed to the worse in all cases (independent of the dose rate) thus indicating chain scission.

Irradiation deteriorates all organic plastics if the dose is sufficiently high. For thermoplastics it is known that if the irradiation takes place in air the degradation is the more severe the lower the dose rate is. The interrelation between the dose rate and the degree of damage was thoroughly investigated for many thermoplastics in the past. Compilations of the papers published can be found elsewhere *(1 - 4)*. This dependence of the deterioration on the dose rate was not understood for a long time. Today it is known that the oxygen concentration in the interior of the samples is the reason. The dissolved oxygen reacts with the radiation induced radicals and builds peroxides. These peroxides are not stable; they decay slowly under chain scission. The longer the irradiation time is, the more complete is the break down of the peroxides and the damage to the material. Gillen and Clough were the first to explain this reaction *(5)*.

[1]Current address: Am Sportplatz, 27 D–6231, Stulzbach, Germany

0097–6156/91/0475–0500$06.00/0

The radiation resistance of thermoset plastics was also investigated several times. These experiments were always carried out with high dose rates. The thermoset plastics proved themselves to be very stable in comparison with most of the usual thermoplastics *(6 - 9)*. But the influence of the dose rate was neglected so far in nearly all of these experiments. Consequently it is not yet known whether the dose rate has any influence on the result if thermosets are irradiated in air. An answer to this question is important if thermoset plastics are to be used for a long period of time in a radiation environment. Only if the influence of the dose rate is known reliably, one can estimate the prospective lifetime of components made from these materials.

To learn more about the influence of the dose rate a number of thermoset plastics of different chemical compositions was irradiated with two very different dose rates. In this paper the influence of the irradiation on the flexural strength and the deflection temperature will be discussed.

Experimental

The thermoset plastics used in this work were made from resin moulding compounds which are defined in the German standard DIN 7708 (1968). Detailed descriptions of these plastics can be found in the literature *(10)*. The moulding compounds which we used were products of Hoechst AG, Frankfurt, Germany, with the trade name "Hostaset". Similar products are manufactured in many countries of the western world. The chemical composition of the resin moulding compounds used is given in the following Table I (all figures in percent; the rest mounting up to 100 % contains additives not shown here).

From these compounds bars of thermoset plastics were made by compression moulding. The bars had a rectangular cross-section, 120 mm x 15 mm x 10 mm. The short term irradiation under (nearly) complete exclusion of oxygen was done with a 3 MeV electron beam accelerator (irradiation from both sides). The long term irradiation experiments were performed with cobalt-60 sources with an average dose rate of only 13.8 Gy/h. These experiments lasted over a period of 10.51 years, ending with a dose of 1.27 MGy. The dose rate could not be held accurately constant during this long period of time. In fact it changed between 21 and 9.3 Gy/h. A few bars remained in the irradiation chamber after the end of the main part of the experiment. After 16.01 years of irradiation these bars had received a dose of 1.79 MGy.

The flexural strength was tested according to DIN 53 452 (ISO 178) with five bars at 23 °C, the deflection temperature was tested according to DIN 53 458 and 53 462 ("Martens method") with two bars only. In this test a bar is held at the top with a 240 mm long lever at a bending tension of 5 N/mm^2 and heated at 50 K/h. The temperature, at which the end of the lever has sunk by 6 mm, is called the "deflection temperature".

Table I. Chemical Composition of the Thermosets

Inorganic Filled Thermosets

Epoxy plastic (laboratory product) "EP"

28.2 EP-resin (made of epichlorohydrin and diphenylolpropane), 3.0 m-phenylene diamine, 1.3 m-toluylene diamine, 12.0 lime stone, 27.0 China clay, 26.0 glass fibers.

Unsaturated polyester "L 1405"

24.0 Unsaturated polyester resin (made of terephthalic acid, fumaric acid and butandiol-1.4), 2.0 diallyl phthalate, 46.5 lime stone, 10.0 China clay, 15.0 glass fibers.

Melamine-formaldehyde MF "Typ 158"

50.0 Melamine-formaldehyde resin, 46.0 asbestos fibers, 3.0 glass fibers.

Phenol-formaldehyde PF "Typ 15"

30.3 Novolak, 4.1 hexamethylene tetramine, 55.0 asbestos fibers, 6.3 cotton threads, 2.7 wood flour.

Organic Filled Thermosets

Melamine-formaldehyde MF "Typ 150"

39.0 Melamine-formaldehyde resin, 8.0 urea-formaldehyde resin, 15.0 dust from Typ 150, 33.2 wood flour, 3.1 chalk.

Phenol-formaldehyde PF "Typ 31"

38.6 Novolak, 5.5 hexamethylene tetramine, 47.2 wood flour, 3.6 chalk.

Results

The test specimens as well as all other thermoset plastics were moulded from compounds containing very coarse components, which are never ideally homogenized. A relatively large scatter of the properties is the result. To minimize this scatter, the standards for the thermosets prescribe relatively thick test specimens. These thick pieces are, of course, not very suitable for the investigation of the influence of oxygen during a long term experiment. But there was no other choice for the test specimens since otherwise the reproducibility of the measurements could by no means be guaranteed.

The flexural strength of the inorganic filled thermoset plastics as a function of the dose is shown in Figure 1. In this figure the points drawn in the ordinate and marked with "B" (Beginning) show the flexural strengths of the non-irradiated materials prior to the beginning of the experiments. The measured points "E" (End) stand for the flexural strengths of the non-irradiated materials measured after a storage time of more than 16 years. The two

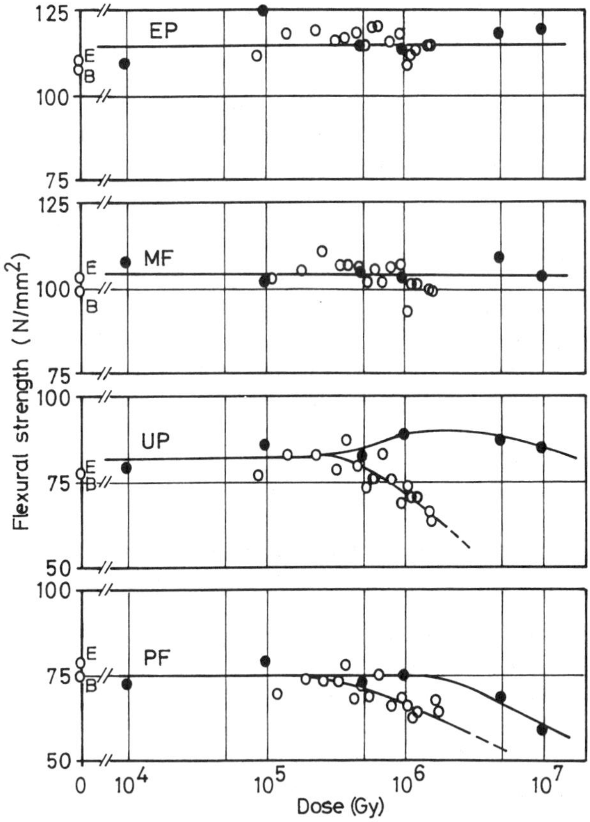

Figure 1. Flexural strength of thermoset plastics with inorganic filler as a function of dose. ● Irradiation with high dose rate under exclusion of air. ○ Irradiation with extremely low dose rate in air. EP = epoxy plastic, MF = melamine-formaldehyde Typ 158, UP = unsaturated polyester L 1405, PF = phenol-formaldehyde Typ 15.

corresponding values are always the same within the limits of the experimental error. An influence of the storage time (with no irradiation) therefore can be excluded for all thermosets investigated.

In case of the epoxy thermoset plastic there is also no influence of the irradiation with high dose rate up to a dose of 10 MGy. This was indeed expected from the many experiments published by Schönbacher et al. *(11)*. The long term irradiation as well has no influence on the flexural strength in this case. The same is true for melamine-formaldehyde plastic. The picture changes in case of the unsaturated polyester plastic. There is a small improvement of the flexural strength at high doses, but the material becomes deteriorated above 0.5 MGy if it is irradiated with the very low dose rate of only 13.8 Gy/h in air. The same seems to happen to the phenol-formaldehyde plastic. But in this case it has to be realized that the thermoset not only contains inorganic fillers but also 9 % of organic components. Unfortunately this thermoset plastic was not available with only inorganic filler. Therefore the observed changes to a certain extent are not the consequence of radiation damage of the phenolic plastic itself, but originate from the damage of the relatively small amount of the organic filler material. Evidence is brought for this in the Figures 2 and 3, which are shown on the same scale as Figure 1. The melamine-formaldehyde plastic and the phenol-formaldehyde plastic, both filled with considerable amounts of wood flour, are readily deteriorated by irradiation. In both cases an influence of the dose rate is also seen clearly (the fat symbols denote high dose rate, the empty circles denote low dose rate).

Because there was not almost seen a change in the flexural strength of the (inorganic filled) thermoset plastics, there is the question if there is no radiation induced change in the molecular structure of the plastic material itself. Since all the thermoset plastics are cross-linked with a high network density they are completely insoluble and consequently cannot be easily investigated with usual methods. But the heat deflection temperature may give some information about changes in the network density. As a matter of fact it is well known, that after-curing reactions, induced by annealing, can lead to a considerable improvement of the deflection temperature. For example, 15 hours annealing at 120 °C raises the deflection temperature for the phenol-formaldehyde plastic Typ 31 from 128 °C to 142 °C, in other words by 14 °C *(12)*. This means it has to be expected that radiation induced cross-linking will lead to an increase in the deflection temperature. Figure 4 shows the result of the deflection temperature measurements. (The data points of the MF Typ 158 are not shown since they coincide with those of PF Typ 15.) A remarkable decrease of the deflection temperature is seen for all thermosets. This only can so far be explained by chain scission. It is also seen that the decrease of the deflection temperature does *not* depend on the dose rate. This may be explained by the extremely small diffusion coefficient for oxygen found for all of these thermosets (Pauly, S. Polymer, in press). During the long term irradiation only a very thin outer layer may be penetrated by oxygen and thus radiation chemical oxidized. The flexural strength measured at room temperature in the glassy state of the polymers may

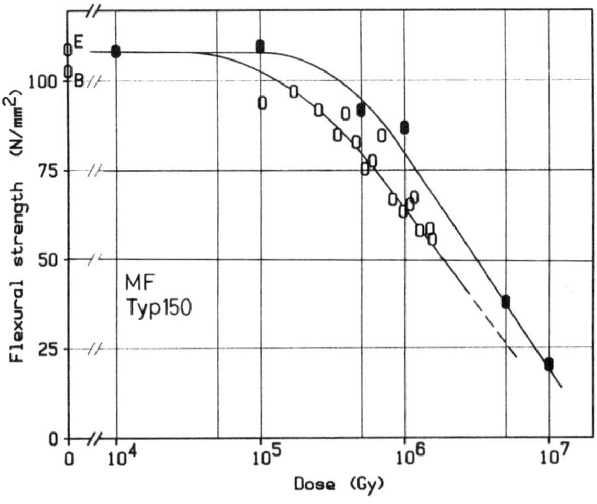

Figure 2. Flexural strength of melamine-formaldehyde plastic Typ 150 with organic filler (33.2 % wood flour) as a function of dose.

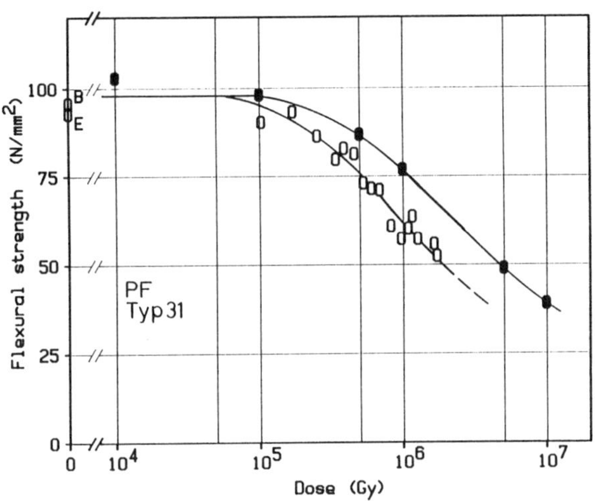

Figure 3. Flexural strength of phenol-formaldehyde plastic Typ 31 with organic filler (47.2 % wood flour) as a function of dose.

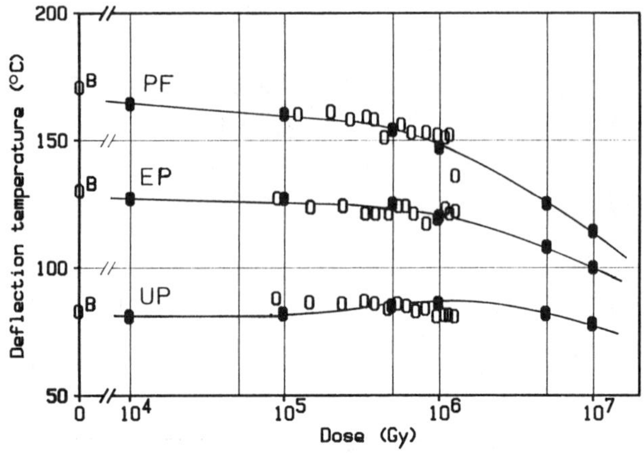

Figure 4. Deflection temperature of thermoset plastics with inorganic filler as a function of dose. Notation as in Figure 1.

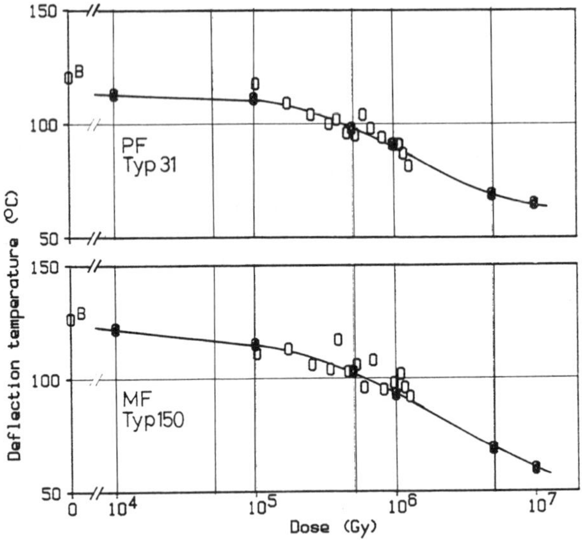

Figure 5. Deflection temperature of thermoset plastics with organic filler as a function of dose.

be affected by the brittle outer layer which was built in this way. On the other hand the deflection temperature lies well above the respective glass transition temperatures of the thermosets themselves and their oxidized outer layers. In this temperature region there exists no more brittleness and therefore no more any difference between high and low dose rate experiments. - A decrease of the heat distortion temperature of epoxy plastics was already observed by Colichman et al. *(13)*.

Figure 5 shows the heat deflection temperature of the organic filled thermosets MF Typ 150 and PF Typ 31. In these cases there was also no influence of the dose rate observed. The reason is the same as before. The decrease in deflection temperature of the (mostly) inorganic filled PF Typ 15 at 10 MGy is 56 °C, the decrease of the organic filled PF Typ 31 at the same dose is 55 °C. This is within the limits of error the same value, and that would mean the deflection temperature is really a property of the thermosets themselves and not of the filler material. This fact provides support for the assumption that the phenolic thermosets do not cross-link by irradiation but in fact degrade.

From a practical point of view it seems to be important that irradiation makes the thermosets more sensitive to the influence of heat - an aspect neglected so far.

Concluding Remarks

The experimental results described in this paper show that the dose rate is of importance in some cases, in others it is insignificant. But it shall be mentioned that other properties, e.g. the surface resistance and the tracking resistance are strongly influenced by the dose rate as a consequence of the surface chemistry of the thermosets. A detailed discussion of these results as well as a precise description of the experimental conditions belonging to this paper will be published in the near future (Gilfrich, H.P.; Rösinger, S.; Wilski, H.).

Acknowledgments

Thanks are due to Dipl.-Ing. R. Börst, Hoechst AG, for making the computer drawings.

Literature Cited

1 Clough, R.L.; Gillen, K.T.; Campan, J.-L.; Gaussens, G.; Schönbacher, H.; Seguchi, T.; Wilski, H.; Machi, S. *Nuclear Safety* **1984**, *25*, 238.
2 Wündrich, K. *Radiat. Phys. Chem.* **1985**, *24*, 503.
3 Wilski, H. *Radiat. Phys. Chem.* **1987**, *29*, 1.
4 Wündrich, K. in *Polymer Handbook*; Brandrup, J. and Immergut, E.H., Ed., John Wiley, New York, N.Y., **1989** VI/463.
5 Gillen, K.T.; Clough, R.L. *J. Polym. Sci. Polym. Chem. Ed.* **1985**, *23*, 2683.
6 Collins, C.G.; Calkins, V.P. *Radiation damage to elastomers, plastics, and*

organic liquids; APEX 261, Techn. Publications GE-ANPD, Evendale, Ohio **1956**.
7 Rauhut, K.; Rösinger, S.; Wilski, H. *Kunststoffe* **1980**, *70*, 89.
8 Schönbacher, H.; Stolarz-Izycka, A. CERN 79-08 **1979**.
9 Beynel, P.; Maier, P.; Schönbacher, H. CERN 82-10 **1982**.
10 *Kunststoff Handbuch Duroplaste*; Becker, G.W.; Braun, D.; Woebcken, W., Eds.; Hanser, München, Germany **1988**, Vol. 10.
11 Schönbacher, H.; Schreiber, B.; Stierli, R. *Kunststoffe* **1986**, *76*, 759.
12 Loc. cit. 10, p. 252.
13 Colichman, E.L.; Strong, J.D. *Mod. Plastics* **1957**, *35* (October), 180.

RECEIVED May 29, 1991

Chapter 32

Assessment of Radiation Damage to a Halogen-Free Cable Jacket

M. Tavlet[1], H. Schönbacher[1], R. Cameron[2], and C. G. Richardson[2]

[1]European Organization for Nuclear Research (TIS), 1211 Geneva 23, Switzerland
[2]BP Chemicals, 1217 Meyrin 2, Switzerland

The EVA-based compound (BP D 2983) supplied by BP Chemicals is used for many control cable sheaths installed in CERN's Large Electron-Positron Collider (LEP). The present paper reports the effects of ionizing radiation on the mechanical properties of this flame-retardant cable jacketing material. The samples were taken from different development plaques, as well as from prototype and production cables. In addition to mechanical tests (IEC Standard 544), special tests and analyses were carried out to investigate antioxidant and filler contents in extruded materials.

The results presented clearly show the difference between accelerated and long-term tests as well as between prototype and production cables. The most important influence of the filler on the properties of the material seems to be the poor homogeneity. The results confirm also that antioxidant does not play an important role during high dose rate irradiations, but is efficient during long-term irradiations.

In 1980, a safety instruction of the European Organization for Nuclear Research (CERN) implemented the use of halogen-free materials for cable-insulating and sheathing materials, for the reasons explained hereafter.

In addition to the usual electrical, mechanical, thermal, and environmental endurance properties of these materials, there were also the following requirements: i) they must be flame retardant; ii) in the event of fire, the resultant gases must be non-toxic and non-corrosive; iii) the smoke emittance must be low. As a result of these requirements, some commonly used materials had to be excluded : e.g. PVC, Hypalon, Neoprene, fluorocarbons and other halogenated and sulphur containing compounds.

It was further required that the materials be radiation resistant, and in the specification for the Large Electron- Positron Collider (LEP) an elongation-at-break that retains more than 50% of its initial value after an irradiation of $5x10^5$ Gy at high dose rate was stipulated. For the last 10 years, since the introduction of these halogen-free materials, more than 300 cable compounds have been tested at CERN (*1*).

The replacement of the halogen-containing materials was achieved in two steps: EPR/EPDM-based rubber compounds were used for medium- and high-voltage power cables, and polyolefin-based compounds were used for control and low-voltage cables.

0097–6156/91/0475–0509$06.00/0

The main supplier of control cables for LEP, was a Norwegian cable manufacturer, who used an EVA-based material (BPD 2983) supplied by BP Chemicals. This material was tested on plaques and was approved. Also, the sheath of a prototype black cable was tested and found to conform to CERN specifications. Samples were subsequently taken from production cables and tested. These results showed a drastic decrease in the radiation resistance of the sheathing material. The tests were repeated and the susceptibility to radiation was confirmed. In order to find the reasons for this, special tests and analyses were made to check the antioxidant and filler contents in extruded materials.

The present paper reports the studies of the effects of ionizing radiation on the mechanical properties of the EVA jacket material taken from different development plaques, as well as from prototype and production cables.

Experimental

Materials and Test Samples. The base material is an EVA copolymer, produced from a mixture of ethylene and vinyl acetate (28%). The polymeric chains are similiar to those of polyethylene, with acetate side groups [CH_3-C(O)-O-] and also random carbon side chains.

The material contains aluminium trihydrate (60%), which is used as a flame-retardant mineral filler. The flame-retardancy property is rarely affected by radiation although the bond between the filler and the polymer may be modified.

The antioxidant is of the unpolymerized phenolic type; the typical content is 0.2%. It is from two different suppliers with two trade names. The base molecule is shown here below. With its two substituted aromatic groups, it acts as a radical-scavenger and as an energy-scavenger, which neutralizes radicals and excited sites produced in the polymeric chain.

Antioxidant;

Santonox R,

or *TBM /60*

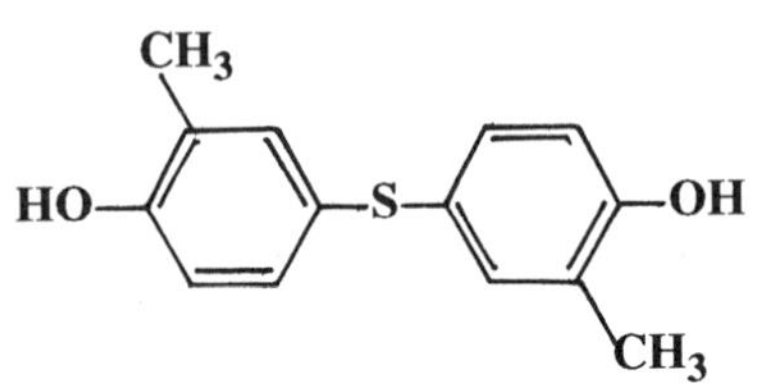

The material of the prototype cable as well as that of a further plaque contains a carbon black batch (2.5%) present both as colorant and as UV stabilizer. Its presence seems to have only a minor influence on the radiation resistance of the material.

The first moulded plaques of this material were received by CERN for testing in 1982 (each sample was given a number e.g. CERN 652).

After approval by CERN, the material was used by the Norwegian cable manufacturer to produce cable sheaths. The first prototype cable supplied for the LEP project was black (CERN 766). Later, tests were carried out on other moulded plaques supplied by BP (CERN 885) and on many production cables supplied by the cable factory. Table I presents most of the tested materials and their relevant characteristics.

In an attempt to explain the observed discrepancies in the test results, tensile samples were taken from different parts of one cable. After production, the cables are rolled on drums, and they can retain their curved shape for several weeks or months until installation. A permanent deformation may therefore take place. Samples were taken from the outside of the curved cable (CERN 930 O), from the side of the curve (CERN 930 S), and from the inside (CERN 930 I).

In order to check whether post-irradiation oxidative degradation takes place, two sets of 10 samples were taken from the same sheath (CERN 942), and were irradiated at the same time and under identical conditions. The first set was tested just after irradiation; the second set was tested three months later.

Three other materials (CERN 943, 944, and 945) were also taken from production cables. To improve the statistics, ten samples were cut for each dose and dose rate.

Finally, more plaques were prepared with different amounts of antioxidant (and with the same nominal content of mineral filler) (CERN 946 to 950). They were irradiated in the reactor and in the cobalt source. The influence of the antioxidant on the mechanical properties of the irradiated plaques was then evaluated.

Table I : List of some tested materials (Cable type gives the number of conductors. The irradiation conditions are explained hereafter)

CERN No	Test samples (type)	Date	Thickness (mm)	Irradiation conditions
652	BP original plaque	1982	2.00 ± 0.05	R L
885	BP plaque	1987	2.00 ± 0.05	R L
766	Prototype cable (black)	1985	1.50 ± 0.1	R L
860	Production cable NF-48	1986	1.85 ± 0.1	R L
894	Production cable NE-48	IX-1986	1.50 ± 0.1	R C
895	Production cable NE-2	I- 1986	1.40 ± 0.2	R C
909	Production cable NH-7	XI-1986	1.30 ± 0.25	R C
910	Production cable NF-12	IV-1987	1.45 ± 0.1	R C
911	Production cable NE-18	I- 1986	1.10 ± 0.15	R C
912	Production cable NE-18	IX-1986	1.40 ± 0.1	R C
930	Production cable (O,I,S)	1985	1.60 ± 0.05	C
942	Production cable NH-25	1986	1.50 ± 0.05	C L
943	Production cable NE-48	I- 1986	1.12 ± 0.07	R C C
944	Production cable NH-48	XI-1986	1.30 ± 0.1	R C C
945	Production cable NF-48	II- 1987	1.50 ± 0.1	R C C
946	Plaque without antioxidant	1987	1.5 ± 0.05	R C
947	Plaque with 0.1 % antioxidant	1987	1.5 ± 0.05	R C
948	Plaque with 0.2 % antioxidant	1987	1.5 ± 0.05	R C
949	Plaque with 0.3 % antioxidant	1987	1.5 ± 0.05	R C
950	Plaque 0.2% antiox.+ carbon	1987	1.5 ± 0.05	R C

Irradiation Conditions. Three irradiation sources were used:
i) The ASTRA reactor: 95% of the dose absorbed by the plastic materials comes from gamma-rays, with an energy between 1 and 2 MeV; the remaining 5% comes from neutrons. The dose rate in this reactor is between $7x10^4$ and $2.7x10^5$ Gy/h. The maximum temperature of the samples is 62°C; the irradiation medium is air. This temperature is low enough and the time of irradiation so short that we do not need to take the temperature-degradation effect into consideration.

ii) A ^{60}Co source with a dose rate between 3 and 4.5 kGy/h. This irradiation condition was considered as an alternative to the reactor. According to the IEC 544 Standards (2), it is at the lower end of the 'high dose-rate' range (> 1 Gy/s).
The samples are irradiated in air at ambiant temperature.
iii) A ^{60}Co source with a dose rate between 70 and 100 Gy/h. This source is used for the assessment of the long-term ageing effects, as recommended by IEC 544.

These three sources are marked respectively 'R', 'C', and 'L' in Table I. A fourth source, which is also a ^{60}Co source, with a dose rate of 11,500 Gy/h, was used only for the three materials CERN 943, 944, and 945.

In our routine programme, samples are irradiated at $2x10^5$, $5x10^5$ and $1x10^6$ Gy, plus occasionally $5x10^6$ Gy in the reactor. For the long-term irradiations, only the two lower doses are retained, plus a lower dose for some materials. The exact doses and dose rates are given in the individual tables of results.

Radiation Effects. The free radicals with unpaired spins, and the loosely bound trapped electrons, produce optical coloration of the irradiated polymers. In addition, the unsaturations resulting from hydrogen evolution or from other decomposition may yield conjugated groups with electron excitation levels in the visible spectral range. For the material under investigation, the modifications of the optical properties were not important, just a small darkening was observed after irradiation.

During irradiation, the electrons that are excited to the conduction band, and the hydrogen radicals, increase the conductivity of the irradiated polymer. This phenomenon is limited to a short period after irradiation. Permanent changes of the electrical resistivity of irradiated polymers can be produced by cracks, bubbles, and similar flaws that result from radiation damage. As permanent changes of the electrical properties result from the degradation of the mechanical properties, it is consistent to test the latter in order to estimate the degradation of a cable insulation material (2).

Gas evolution in the irradiated polymer may lead to bubble formation, which initiates cracks under stress. This occurs only when irradiating with high-LET particles at high dose rates.

In low-density polyethylene and in EVA, the two mechanisms of chain scission and cross-linking compete. In general, it seems that for low irradiation doses, the cross-linking occurs first, and that chain scission predominates for higher doses. Degradation is promoted during long-term irradiation in the presence of oxygen (*3*).

Test Methods. Following the recommendations of the International Electrotechnical Commission (2), the materials were submitted to tensile tests. These tests were carried out on an Instron 1026 machine. Elongation-at-break and tensile strength were registered. Dumb-bell samples (normalized S2 according to ISO/R 37/2) were cut from the plaques or from the extruded sheaths. The mechanical tests were carried out on both irradiated and non-irradiated samples. The hardness Shore D was also measured, using a Volpert apparatus.

Furthermore, irradiated and non-irradiated samples were submitted to ESR spectroscopy in a Varian E3 spectrometer. From previous studies (*4*) we find that in this kind of material the ESR signal comes mainly from the irradiated aluminium trihydrate. Therefore the height of the signal at a given dose (and a given sample mass) is assumed to be proportional to the filler content.

The antioxidant content in the samples was checked by the technique of UV absorption (at 280 nm) and also by solvent extraction.

Results

The detailed results are given in the form of graphs plus tables for materials 652, 766, 860, 930, and 942 (Figures 1 to 5). For the other materials, some results appear in the tables shown below.

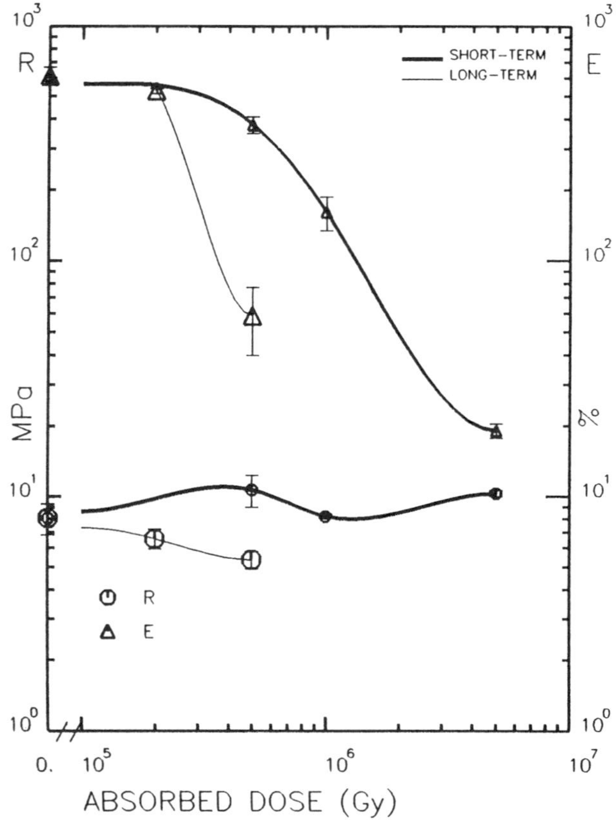

CERN MATERIAL No. : 652

MATERIAL : Thermoplastic EVA
TYPE : MCA 319 (D 2983 FR)
SUPPLIER : BP Chemicals
REMARKS : Plates 1982

TEST RESULTS :

Dose (MGy)	Dose rate (Gy/H)	Tensile properties: Strength R (MPa)	Tensile properties: Elongation E (%)	Hardness Shore D H (Degree)
0.000	–	8.0 ± 1.2	610.0 ± 55.6	42.0 ± 0.0
0.500	1.7×10^5	10.6 ± 1.7	377.5 ± 30.5	41.5 ± 0.0
1.000	1.7×10^5	8.2 ± 0.2	160.0 ± 26.2	43.0 ± 0.0
5.000	1.7×10^5	10.3 ± 0.3	19.0 ± 1.4	47.0 ± 0.0
0.000	–	8.0 ± 1.2	610.0 ± 55.6	42.0 ± 0.0
0.200	100.0	6.6 ± 0.6	526.5 ± 13.6	41.0 ± 0.0
0.500	100.0	5.3 ± 0.4	58.5 ± 18.7	35.5 ± 0.0

RADIATION INDEX (1.7×10^5 Gy/h) : 5.8
RADIATION INDEX (100.0 Gy/h) : 5.4

Figure 1: Degradation of mechanical properties of material 652 with absorbed dose.

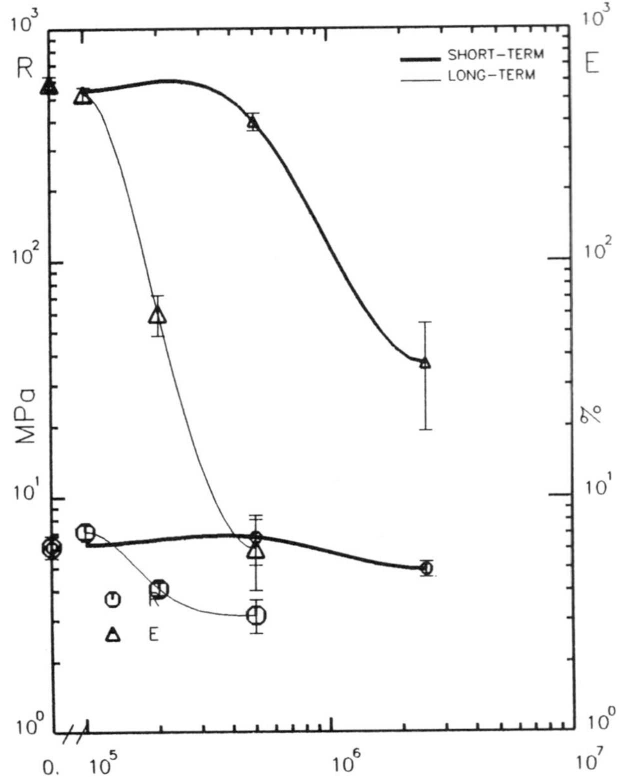

CERN MATERIAL No. 766

MATERIAL : Polyolefin thermoplastic
TYPE : BP D 2983 FR (black)
SUPPLIER : Norsk Kabelfabrik EB
REMARKS : Taken from LEP prototype cable

TEST RESULTS :

Dose (MGy)	Dose rate (Gy/H)	Tensile properties Strength R (MPa)	Elongation E (%)	Hardness Shore C H (Degree)
0.000	—	6.2 ± 0.7	583.3 ± 45.5	35.0 ± 0.0
0.500	1.7×10^5	6.7 ± 1.6	397.0 ± 35.1	46.0 ± 0.0
2.500	1.7×10^5	4.9 ± 0.4	36.7 ± 17.6	49.0 ± 0.0
0.000	—	6.2 ± 0.7	583.3 ± 45.5	25.0 ± 0.0
0.100	1000.0	7.1 ± 0.5	526.0 ± 36.1	28.0 ± 0.0
0.200	100.0	4.1 ± 0.2	60.0 ± 11.8	29.0 ± 0.0
0.500	100.0	3.1 ± 0.5	6.0 ± 2.0	31.0 ± 0.0

RADIATION INDEX (1.7×10^5 Gy/h) : 5.8
RADIATION INDEX (550.0 Gy/h) : 5.1

Figure 2: Degradation of mechanical properties of material 766 with absorbed dose.

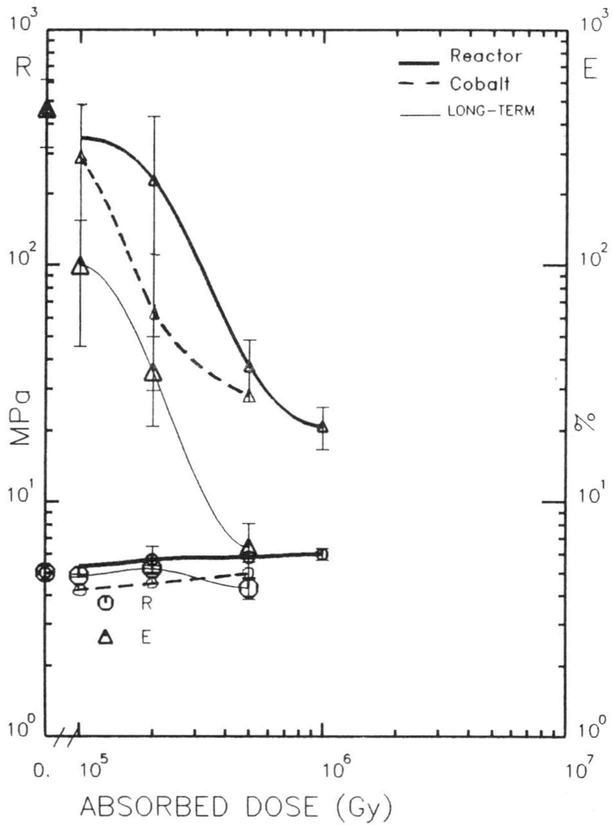

CERN MATERIAL No. : 860

MATERIAL : Polyolefin
TYPE : BP D 2983 FR (white)
SUPPLIER : Norsk Kabelfabrik EB
REMARKS : Taken from NF-48 LEP cable (C)

TEST RESULTS :

Dose (MGy)	Dose rate (Gy/H)	Tensile properties		Hardness Shore D
		Strength R (MPa)	Elongation E (%)	H (Degree)
0.000	–	5.0 ± 0.3	464.8 ± 147.8	25.0 ± 0.0
0.200	1.7×10^5	5.7 ± 0.8	229.3 ± 199.9	29.0 ± 0.0
0.500	1.7×10^5	5.8 ± 0.3	37.2 ± 10.9	30.0 ± 0.0
1.000	1.7×10^5	6.0 ± 0.3	20.8 ± 4.1	35.0 ± 0.0
0.000	–	5.0 ± 0.3	464.8 ± 147.8	25.0 ± 0.0
0.100	7110.0	4.2 ± 0.2	287.2 ± 194.8	25.0 ± 0.0
0.200	7110.0	4.5 ± 0.2	62.4 ± 46.4	27.0 ± 0.0
0.500	7110.0	4.9 ± 0.2	29.6 ± 3.6	32.0 ± 0.0
0.000	–	5.0 ± 0.3	464.8 ± 147.8	25.0 ± 0.0
0.100	1000.0	4.9 ± 0.1	99.2 ± 53.9	25.0 ± 0.0
0.200	108.0	5.2 ± 0.2	35.2 ± 14.5	27.0 ± 0.0
0.500	100.0	4.3 ± 0.5	6.4 ± 1.7	31.0 ± 0.0

RADIATION INDEX (1.7×10^5 Gy/h) < 5.3
RADIATION INDEX (7110.0 Gy/h) : 5.1
RADIATION INDEX (100.0 Gy/h) < 5.0

Figure 3: Degradation of mechanical properties of material 860 with absorbed dose.

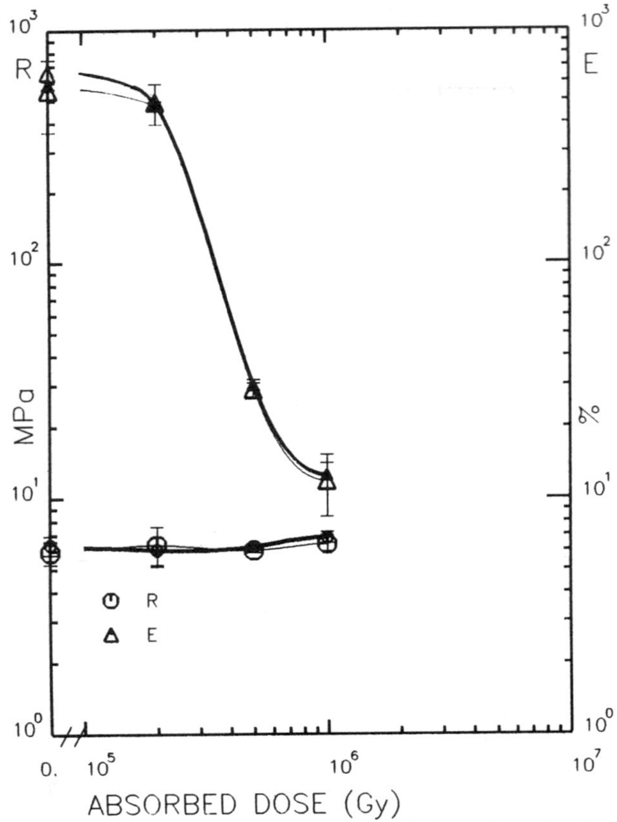

CERN MATERIAL No. 930

MATERIAL : POLYOLEFIN
TYPE : BP D 2983 FR
SUPPLIER : NORSK
REMARKS : Taken from NG 28 LEP cable

TEST RESULTS :

Dose (MGy)	Dose rate (Gy/H)	Tensile properties Strength R (MPa)	Elongation E (%)	Position of samples
0.000	–	6.3 ± 0.6	582.9 ± 22.3	Outside (930 O)
0.200	1120.0	6.0 ± 0.8	481.7 ± 14.1	
0.500	2220.0	6.2 ± 0.1	30.0 ± 1.6	
1.000	2710.0	6.8 ± 0.4	12.5 ± 1.6	
0.000	–	5.9 ± 0.7	553.9 ±189.9	Side (930 S)
0.200	1120.0	6.3 ± 1.2	489.3 ± 96.8	
0.500	2220.0	6.0 ± 0.1	28.6 ± 2.0	
1.000	2710.0	6.4 ± 0.4	11.8 ± 3.5	
0.000	–	6.2 ± 0.2	645.7 ± 14.6	Inside (930 I)
0.200	1120.0	6.3 ± 1.0	475.0 ±145.2	
0.500	2220.0	5.8 ± 0.2	27.1 ± 6.2	
1.000	2710.0	6.7 ± 0.3	14.2 ± 2.0	

Figure 4: Degradation of mechanical properties of material 930 with absorbed dose.

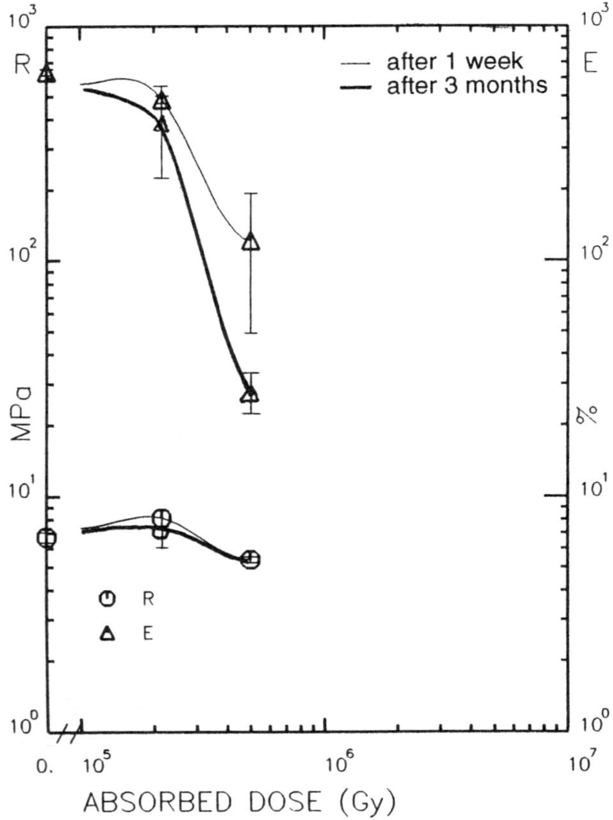

CERN MATERIAL No. : 942

MATERIAL : Polyolefin
TYPE : BP D 2983 FR
SUPPLIER : NORSK-Kabel
REMARKS : Taken from NH 25 LEP cable

TEST RESULTS :

Tested 1 week after irradiation

Dose	Dose rate	Tensile properties		Hardness
		Strength	Elongation	Shore D
(MGy)	(Gy/H)	R (MPa)	E (%)	H (Degree)
0.000	–	6.7 ± 0.3	636.2 ± 21.0	30.0 ± 0.0
0.216	3500.0	8.1 ± 0.6	485.0 ± 17.1	35.0 ± 0.0
0.500	3764.0	5.4 ± 0.1	120.6 ± 71.6	39.0 ± 0.0

Tested 3 months after irradiation

Dose	Dose rate	Tensile properties		Hardness
		Strength	Elongation	Shore D
(MGy)	(Gy/H)	R (MPa)	E (%)	H (Degree)
0.000	–	6.7 ± 0.3	636.2 ± 21.0	30.0 ± 0.0
0.216	3500.0	7.2 ± 1.5	392.1 ±176.5	33.0 ± 0.0
0.500	3764.0	5.2 ± 0.2	26.9 ± 5.3	35.0 ± 0.0

Figure 5: Degradation of mechanical properties of material 942 with absorbed dose.

The values of the mechanical properties are known to within a precision of 10% for the materials 652, 766, and 885, taken from moulded plaques and prototype cable (Figures 1 and 2). Within experimental error, the results of mechanical tests after short-term irradiation of material 766 taken from an extruded sheath agree with those of material 652 taken from a moulded plaque. The results for the long-term irradiation of the 652 were very satisfactory.

The radiation test results of material 860 taken from a production cable sheath showed large discrepancies in the individually measured values and a much greater sensitivity to radiation compared with previous samples (Figure 3).

The tests were repeated on two other production cable sheaths. The high radiation sensitivity was confirmed and the standard deviations were still higher; for example: Elongation = (449.3 ± 225.7) % for 894 before irradiation,
and Elongation = (58.4 ± 30.5) % for 895 at 1 MGy

The long-term irradiation of material 766 revealed a great sensitivity to low dose-rate irradiation (Figure 2).

The results of the samples taken from the cable on the drum (930 O, 930 I, and 930 S) showed that the initial stress influences only on the elongation measured before irradiation. The samples taken from the outside curvature of the cable show a lower elongation, whilst the samples taken from the side are less regular because of their lateral deformation. After irradiation, no significant differences were found in the results (Figure 4). We may therefore assume that the initial stress is of little importance on the radiation resistance.

On the other hand, a big difference was observed between the results of the samples taken from cable 942 tested just after irradiation and those tested three months later. This shows the importance of the oxidative degradation and of radio-induced chemical reactions (Figure 5).

Samples from the same cable (942) were also irradiated to 5×10^5 Gy at the low dose rates of 1000 Gy/h and 100 Gy/h. These irradiations ended in May and August 1989 respectively. The samples were tested together in October and the results were almost the same, within experimental errors, except that the tensile strength seems to be lower in the case of the 100 Gy/h irradiation.

Physicochemical analyses were carried out on samples (909 to 912) and compared with the results of the mechanical radiation tests. It appears that the antioxidant content and the mineral filler content can vary significantly.

Neither the antioxidant content nor the filler content could be related exactly to the remaining mechanical properties of the cable sheaths after irradiation (Table II). The influence of the antioxidant content is clearly seen in the comparison of the remaining elongation after irradiation of the materials 946 to 950 in the cobalt source (see the last column of the Table II and the Figure 6).

Discussion

Dose-Rate Effect. Conditions such as dose rate, irradiation medium, and ambiant temperature may be of great importance when irradiating polymers. In particular, our polyolefin-based material is very sensitive to dose-rate effect and therefore to long-term irradiation. When irradiated at a low dose rate in the presence of oxygen, the degradation of this material is much more significant.

Tables III and IV show respectively the relative elongation and the relative tensile strength after irradiation at 5×10^5 Gy at different dose rates. From these tables we see the sharp reduction of elongation and the reduction in tensile strength after irradiation at low dose rate, because of the oxidative degradation. On the other hand, short-term irradiations at this dose usually increased the tensile strength, because of

Table II : composition and elongation of some materials. (n.m = not measured; * indicates big discrepancies)

CERN No	Amounts (in % of nominal value) of antioxidant and filler		Relative elong. E/E_0 after 0.5 MGy Reactor or ^{60}Co	
860	50	128	0.08 *	0.06 *
894	n.m	93	0.12 *	0.08 *
895	n.m	66	0.44	0.04 *
909	61	94	0.64	0.09
910	130	125	0.45 *	0.12 *
911	47	74	0.37	0.04
912	68	99	0.51	0.09
946	0	100	0.28	0.04
947	50	100	0.29 *	0.08
948	100	100	0.38	0.21
949	150	100	0.50	0.44
950	100	100	0.26 *	0.26

Table III : relative elongation (compared to initial elongation) after irradiation at $5x10^5$ Gy

	E/E_0 after irradiation at 0.5 MGy			
CERN No	Reactor $2x10^5$Gy/h	Cobalt 60 10^4 Gy/h	Cobalt 60 4000 Gy/h	Long-term 100 Gy/h
652	0.62	-	-	0.10
885	0.59	-	-	0.02
766	0.68	-	-	0.01
860	0.08	-	0.06	0.01
943	0.16	0.15	0.11	-
945	0.12	0.09	0.07	-

Table IV : relative tensile strength (compared to initial value) after irradiation at $5x10^5$ Gy

	R/R_0 after irradiation at 0.5 MGy			
CERN No	Reactor $2x10^5$Gy/h	Cobalt 60 10^4 Gy/h	Cobalt 60 4000 Gy/h	Long-term 100 Gy/h
652	1.32	-	-	0.66
885	1.07	-	-	0.56
766	1.08	-	-	0.50
860	1.16	-	0.98	0.56
943	1.04	1.19	1.00	-
945	0.78	0.93	0.75	-

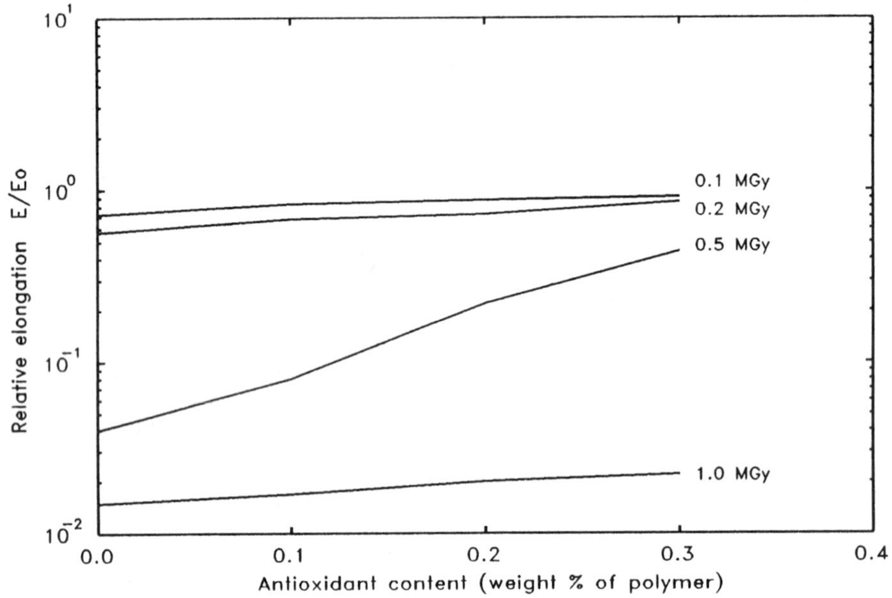

Figure 6: Relative elongation versus antioxidant content for different irradiation doses at 1 Gy/s, for materials 946 to 950.

the cross-linking. This confirms the results of Hanisch et al. (*3*) regarding LDPE. Futher information on this may be found in (*1*, *3*, and *5-7*).

If we consider that the degradation may be extrapolated to lower dose rates (*6*) and (*7*), then the degradation of the cables used in real accelerators where the dose rate is expected to be around 10 to 15 Gy/h, could be a cause for concern.

Thickness. As all our extruded samples were of comparable thicknesses (Table I), we may not see any influence of this parameter on the oxidation of the samples. However, the oxidation seems to be heterogeneous after irradiation in the reactor or in the cobalt source (see the results of samples 942, where degradation continued after irradiation). This is in agreement with the measurements of Gillen and Clough (*6*), who found that oxidation is heterogeneous for dose rates higher than 100 Gy/h.

Note that if the material is used as a cable sheath, then the oxygen comes only from the outside of the sheet, the inside face being more protected. Therefore, we may expect a lower influence of the oxidation.

Antioxidant Content. Figure 6 shows that the radiation resistance increases as a function of the antioxidant content. This is particularly obvious for the intermediate dose and intermediate dose rate ($5x10^5$ Gy at 1 Gy/s). This Figure is related to samples taken from moulded plaques.

Table II and Figure 7 show the values of the relative elongation (compared with the initial value), after irradiation at $5x10^5$ Gy, versus the antioxidant content (in percentage of nominal value). These results confirm that the antioxidant does not play an important role during high dose- rate irradiations in the reactor. It becomes effective during the 100 h irradiation in the cobalt source (this is also in agreement with results presented in (*3*)).

As the oxygen penetration in these materials seems to be important, perhaps we can hope that the antioxidant will remain active for longer irradiation times, and that this will be sufficent to prevent further oxidation and degradation of the material.

Filler Content. Table II and Figure 8 show the values of the relative elongation (compared with the initial value), after irradiation at $5x10^5$ Gy, versus the filler content. Our method of measuring the filler content is based on the intensity of the ESR signal after irradiation. Note that the intensities do not seem to be influenced by various antioxidant contents. The reliability of our method is proved by the fact that the materials 946 to 950 give comparable signal intensities after irradiation at the same dose.

The ESR spectrum shows that the mineral filler absorbs a part of the energy brought about by the radiation; it forms radicals. The proportions of the energy absorbed by the filler and by the polymer are not known. The flame-retardant property of the filler is not affected by irradiation.

The most important influence of the filler on the properties of the material seems to be the poor homogeneity, i.e. the materials with a high filler content presented discrepencies in mechanical tests [principally, materials 860, 894, and also 910, where E(0.2 MGy) = (293.7 $\pm$ 239.2) %].

Others Remarks. It should be mentioned that the ESR spectrum is not exactly the same after irradiation in the reactor or with the ^{60}Co source. We do not try to explain this observation here.

As the small samples used for the ESR spectroscopy or for the antioxidant measurement were not taken adjacent to the dumb-bell samples (for the mechanical tests, about 40 samples were cut from along the cable), we may suspect that the composition of the materials varies between all these positions, and may also vary from batch to batch.

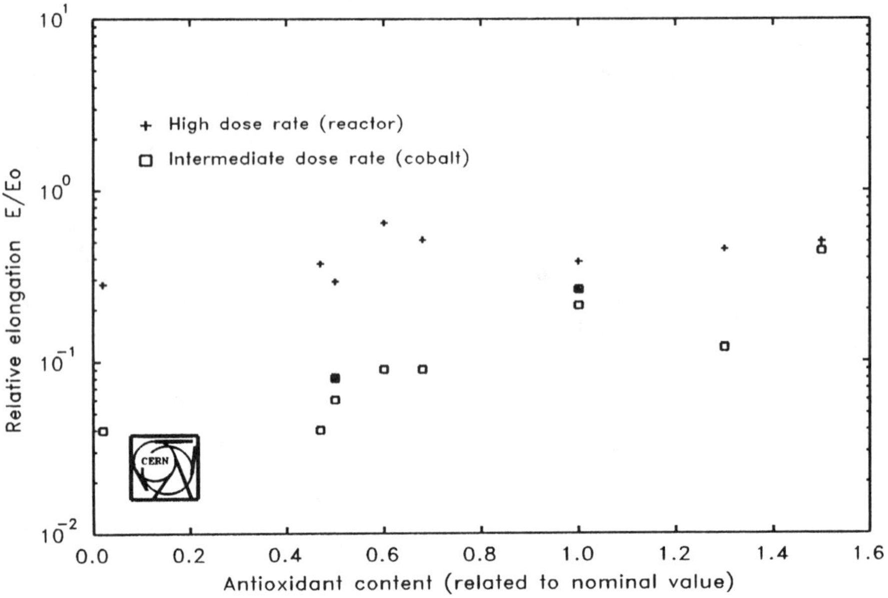

Figure 7: Relative elongation versus antioxidant content after irradiaiton at 0.5 MGy (two dose rates) for all materials.

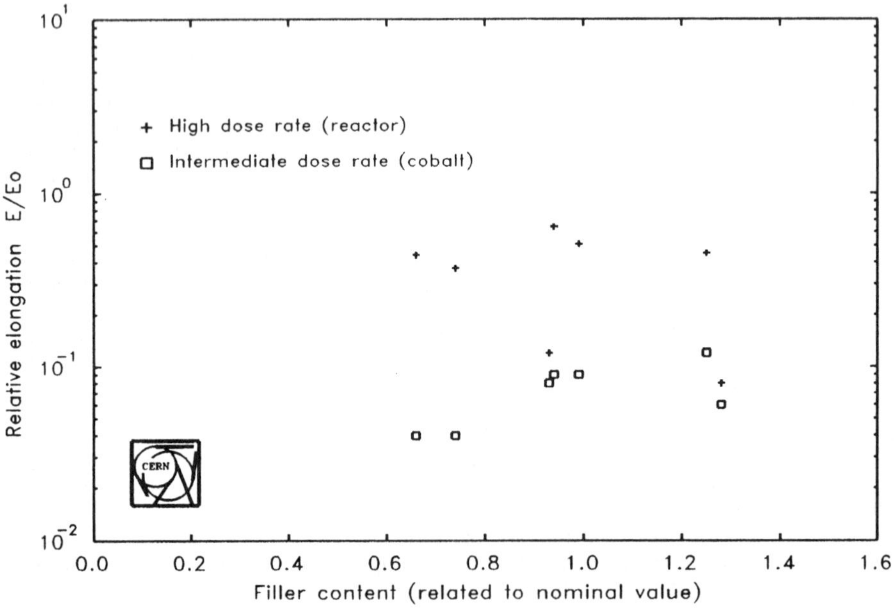

Figure 8: Relative elongation versus filler content after irradiaiton at 0.5 MGy (two dose rates) for some materials.

Conclusions

This study is an example of routine irradiation tests for a cable sheath. It extends over six years, from the time when the specification was drawn up until the cable was installed.

The results presented clearly show

i) the difference between accelerated and long-term tests, which was expected for this type of polyolefin material where oxygen degradation plays an important role;

ii) very big differences between prototype and production cable, which are more or less still unexplained, except with respect to the suspected inhomogeneity of the production cable material;

iii) the influence of the antioxidant and the filler contents on the radiation resistance, too much filler seems to affect the homogeneity of the material and hence its radiation resistance; the benefit of the antioxidant during long-term irradiation is clearly seen from plaques 946 to 949;

iv) the addition of 2.5% of carbon black may be effective as UV stabilizer but does not improve the radiation resistance.

A lot of experience has been gained in the past with PVC cable materials. The replacement of halogen-containing sheath materials by halogen-free ones, based on polyolefins, reduced the radiation resistance. Most of the cable-insulating materials used in the LEP tunnel are considerably more sensitive to radiation than are the PVC materials used in other CERN accelerators.

Note that the BP product BPD 2983 was one of the first compounds to be approved by CERN, in the early 1980s. Since this time, compound technology has progressed, and BPD 2983 has now to be considered obsolete and is replaced by superior products.

From the present study it can be assumed that the limit of usability for these cable sheathing materials in LEP will be around $2x10^5$ Gy. Moreover significant levels of ozone are produced in the LEP tunnel, and this will increase the degradation of the cables. To confirm this, further tests will be carried out on samples taken after service. More studies (*7*) are being undertaken regarding the extrapolation of radiation degradation to service dose rate.

Literature Cited

(*1*) Schönbacher, H.; Tavlet, M. *Compilation of radiation damage test data; Part I, 2nd edition; Halogen-free cable-insulating materials.* CERN 89-12 (1989)

(2) IEC Publication 544; *Guide for determining the effects of ionizing radiation on insulating materials.* (Four part, 1977 to 1985)

(*3*) Hanisch,H. et al. *Radiat.Phys.Chem.* 1987, 30-1, pp.1-9

(*4*) Coche,M. *Polymères irradiés testés par résonance de spin électronique.* 1988, CERN-TIS-RP/IR/88-05

(*5*) Maier,P.; Stolarz-Izycka,A. *Long-term radiation effects on commercial cable-insulating materials irradiated at CERN.* 1983, CERN 83-08

(*6*) Gillen,K.T.; Clough,R.L. *Polym.Degrad.Stability.* 1989, 24, pp.137-168

(*7*) Danielsson,H.; Schönbacher,H.; Tavlet,M. *Estimate of cable life times in the radiation environment of LEP 200.* CERN, in preparation.

RECEIVED January 14, 1991

Chapter 33

A Practical Model for Prediction of the Lifetime of Elastomeric Seals in Nuclear Environments

S. G. Burnay

Advanced Engineering Materials Department, AEA Industrial Technology, Harwell Laboratory, Didcot, Oxfordshire OX11 0RA, United Kingdom

A predictive model of radiation degradation in elastomeric materials has been developed, based on time-temperature-dose rate superposition. Examples of the application of the model to practical elastomers used in the nuclear industry are given. The model does not require a detailed knowledge of the degradation mechanisms.

Elastomeric seals are widely used in nuclear power stations and other nuclear plant; such seals can be exposed to elevated temperatures and low dose rate irradiation for times up to the lifetime of the plant. In assessing the long term behaviour of elastomeric seals, accelerated tests are necessary to enable estimates of lifetime to be made under the anticipated service conditions. It is well established that many polymeric materials exhibit dose rate effects or synergism between temperature and radiation; any seal lifetime prediction method used needs to take these effects into account.

The physical parameters which are most relevant to seal leakage under irradiation have been previously assessed (*1*); compression set has been found to be the most useful parameter for routine monitoring of seal performance.

This paper describes the development of a lifetime prediction model based on time-temperature-dose rate superposition which has proved to be useful in assessing the behaviour of elastomeric seals in nuclear environments. The new model is of practical use to design engineers in assessing maintenance schedules for replacement of such seals.

Lifetime Prediction Models

Empirical Model. An empirical model based on time-temperature-dose rate superposition was developed at Harwell for lifetime prediction of elastomeric seals (*2*). This early model attempted to separate the thermal and radiation components of the degradation by determining the shift factors under a range of conditions. The thermal shift factors were determined experimentally at a series of dose rates, all relative to the same reference temperature, producing a series of master curves at each dose rate. The radiation shift factors were then determined by shifting these master curves to a single master curve relative to zero dose rate.

The empirical model used this single master curve of compression set versus time combined with a set of empirical equations relating the thermal and radiation shift

0097–6156/91/0475–0524$06.00/0

factors to the temperature and dose rate (*2*). The master curve was determined experimentally by superposition of data and its time dependence was an arbitrary function. In this early work it was found that the thermal and radiation shift factors were not independent, the thermal shift factor having an apparent dose rate dependent activation energy. This early model contains 4 empirical parameters which must be determined experimentally for each elastomer; the model is described in more detail in reference (*2*).

The model parameters in this empirical model are specific to the elastomer formulation tested. Although this simple model has proved useful in practical applications, there are several disadvantages to the technique -

(i) the number of parameters needed to describe the model mean that a large amount of experimental data are required to fully characterise any one elastomer formulation;

(ii) because of the empirical nature of the model, the parameters do not have a physical significance and cannot be used to predict the behaviour of similar materials;

(iii) the arbitrary shape for the master curve means that interpolation of data can only be done graphically;

(iv) the model cannot fully explain the behaviour of some materials.

New Superposition Model. The empirical equations describing the thermal and dose rate shift factors can be simplified by using the approach of Rudd (*3*) who suggests the use of a single shift factor a(T,D) relative to the reference conditions to superpose data (Figure 1), rather than the two shift factors of the empirical model. This single shift factor is equivalent to $a_T.a_R$ in the earlier model. His approach makes use of the fact that superposition is a symmetry property of the damage parameter; this will determine the functional form required of the temperature and dose rate dependence of the shift factor a(T,D) for superposition to be possible.

Rudd describes two types of superposition that are possible, designated type 1 and type 2. For type 1 superposition, plots of a damage parameter (such as compression set) versus time will superpose with a shift factor a(T,D) whose functional form is not critical. For type 2 superposition, plots of dose to equivalent damage versus dose rate will superpose with a shift factor b(T) which is a function of temperature only. This is the type of superposition behaviour which has been observed and modelled by Gillen and Clough (*4,5*). The two types of superposition are independent, but if a system displays both types then the type 1 shift factor a(T,D) and the type 2 shift factor b(T) are related and must have the form,

$$a(T,D) = \frac{f(b(T).D)}{b(T)} \qquad (1)$$

where b(T) is some function of temperature and D is the dose rate.

Rudd (*4*) has suggested an alternative empirical relationship which will satisfy equation (1),

$$a(T,D) = \exp \frac{-E}{R}\left(\frac{1}{T_{ref}} - \frac{1}{T}\right)\left[1 + k.\exp\frac{Ex}{R}\left(\frac{1}{T_{ref}} - \frac{1}{T}\right).D^{x}\right] \qquad (2)$$

where E, k and x are empirical constants for the new model. Implied in this equation is that the type 2 shift factor, b(T), is given by

$$b(T) = \exp\frac{E}{R}\left(\frac{1}{T_{ref}} - \frac{1}{T}\right) \quad (3)$$

Equation (2) describes the new empirical model, which has fewer disposable parameters than the earlier model. The main advantage of this type of approach is that characterisation of an elastomer requires considerably less data than the earlier model since there are only three model parameters to be determined - E, k and x. The parameter E can be determined from degradation in the absence of irradiation, leaving only k and x to be determined from irradiation data. The functional form of the new model fits Rudd's criterion, equation (1), so that both types of superposition are valid.

Figure 2 shows how a(T,D) will vary with dose rate and temperature for the new model. At low dose rates the degradation is dominated by the thermal component; at high dose rates, temperature has little effect on the degradation. This decrease in temperature dependence with increasing dose rate, which was interpreted as a change in apparent activation energy in the earlier empirical model (*2,3*), is now seen to arise automatically from equation (2), without having to invoke a dose rate dependent activation energy.

For the majority of elastomers that have been studied, the parameter x has been found to be equal to one. For this case, equation (2) simplifies to a two parameter model,

$$a(T,D) = \exp\frac{-E}{R}\left(\frac{1}{T_{ref}} - \frac{1}{T}\right) + k.D \quad (4)$$

which can be interpreted in physical terms as the sum of a thermal component, described by the Arrhenius relationship with an activation energy E, and a dose rate component, with k representing a reaction constant for radiation degradation. This description of the shift factor also implies that degradation involves competing thermally induced and radiation induced initiation processes to form radicals which can then participate in similar subsequent reactions. The model does not require a detailed knowledge of the reaction mechanisms but assumes that a single mechanism dominates the degradation behaviour.

Dose to Equivalent Damage. An alternative way of approaching the model is to examine how the dose required to reach a particular level of the damage parameter will vary with temperature and dose rate, i.e. type 2 superposition. At a temperature T and dose rate D, the dose to equivalent damage, i.e. DED, is given by

$$DED = \frac{t_m.D}{a(T,D)} \quad (5)$$

where t_m is the time required to reach the damage level on the master curve at the reference condition of D = 0 and $T = T_{ref}$. The parameter t_m is only a function of the damage level selected and is independent of T and D.

Substituting for the shift factor a(T,D) and taking the special case of $T = T_{ref}$ this equation simplifies to

$$DED = \frac{D.t_m}{1 + k.\ D^x} \qquad (6)$$

At high enough dose rates, when $k.D^x >> 1$, DED will be proportional to D^{1-x}. This implies that for materials where x = 1, DED will be independent of dose rate at high dose rates. However, at low dose rates there will be a marked dose rate effect, irrespective of the value of x, since the material will ultimately be limited by thermal degradation. This is illustrated in Figure 3 which shows the DED plotted as a function of the dose rate; the curve is equation (6) using the same values of E, k and x as used in Figure 2. As the dose rate decreases, the curve tends assymptotically towards the straight line given by purely thermal degradation. In Figure 3 this is shown as a dashed line; for the example shown, thermal degradation dominates for dose rates less than about 0.01 Gy/s.

Since DED is being plotted as a function of dose rate, i.e. type 2 superposition, data obtained at temperatures other than T_{ref} can be superposed by using the type 2 shift factor b(T) given by equation (3). A plot of DED versus D/b(T) will then yield a single curve for data at different temperatures. This is precisely the form of the superposition model put forward by Gillen and Clough (*5,6*) where they used the activation energy for the breakdown of peroxides to shift their data on PVC.

Time-Dependence of Compression Set. Another limitation to the earlier 4-parameter empirical model (*2*), the arbitrary shape of the master curve, can also be improved. Examination of a large number of data from several different elastomers has shown that the time dependence of compression set is frequently a simple power law. For a temperature T and dose rate D, the observed compression set is given by

$$set = Q\,[a(T,D).t]^q \qquad (7)$$

where q < 1, typically of the order of 0.5 or less, and Q is constant for a particular formulation. This can be used in conjunction with the superposition model to calculate the predicted compression set.

Application to Practical Elastomeric Seals

Practical Examples. Some examples of how the superposition model has been applied to elastomeric seal materials will illustrate the range of degradation behaviour which can be modelled in practice. These examples are for materials in use in the UK nuclear industry, i.e. they are compounded formulations not model systems.

The type of behaviour which is observed will be determined by the relative values of the reaction constant k for radiation degradation and the activation energy E for thermal degradation. If k is sufficiently large for the radiation component in equation (2) to dominate in the temperature range of interest, then temperature is found to have little effect (Figure 4). However, if k is relatively small, temperature will have a much larger effect (Figure 5). Both of the materials illustrated in Figures 4 and 5 have been fitted to the simplified model, equation (4).

The first two examples were for a limited temperature range but the model can be used over a much wider temperature range provided that a single degradation mechanism is operating. Figure 6 illustrates a high temperature elastomer which has been modelled over a temperature range from 80°C to 250°C. For this particular

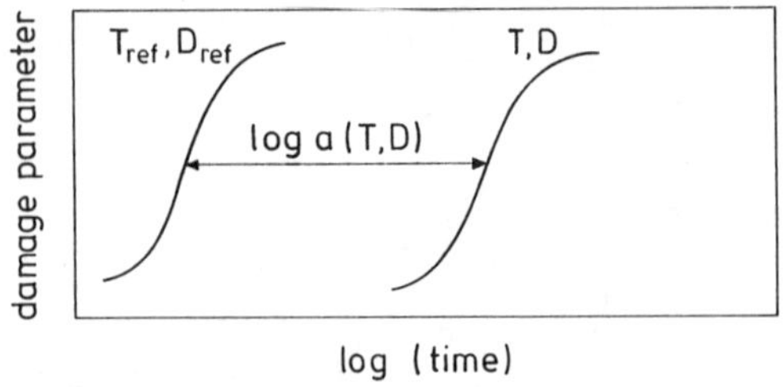

Figure 1. Superposition principle - schematic.

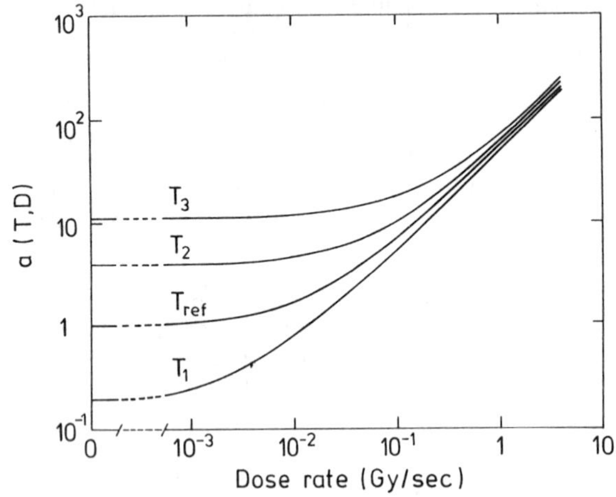

Figure 2. Calculated shift factors a(T,D) for compression set; temperatures $T_3 > T_2 > T_{ref} > T_1$.

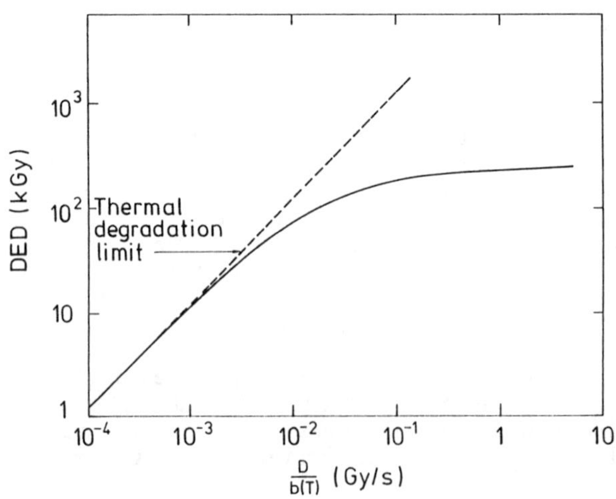

Figure 3. Dose to equivalent damage (DED) for compression set at T_{ref}.

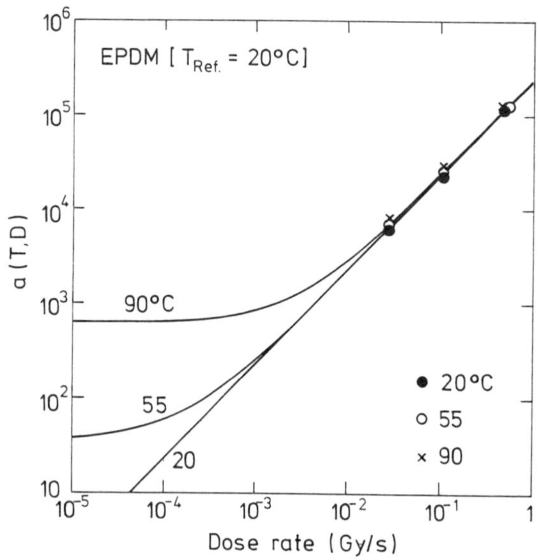

Figure 4. Shift factor a(T,D) as a function of dose rate and temperature - EPDM.

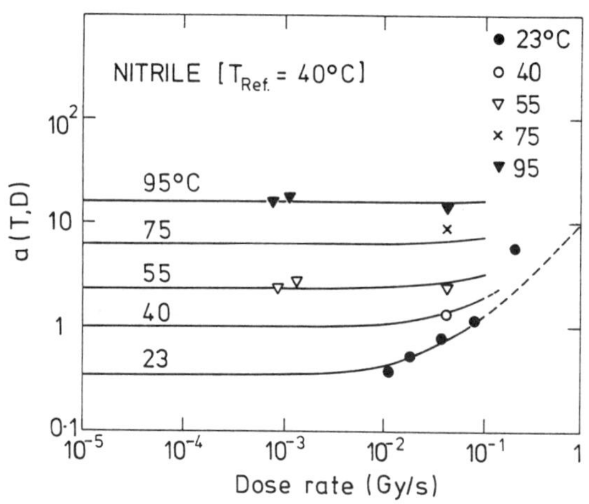

Figure 5. Shift factor a(T,D) as a function of dose rate and temperature - nitrile.

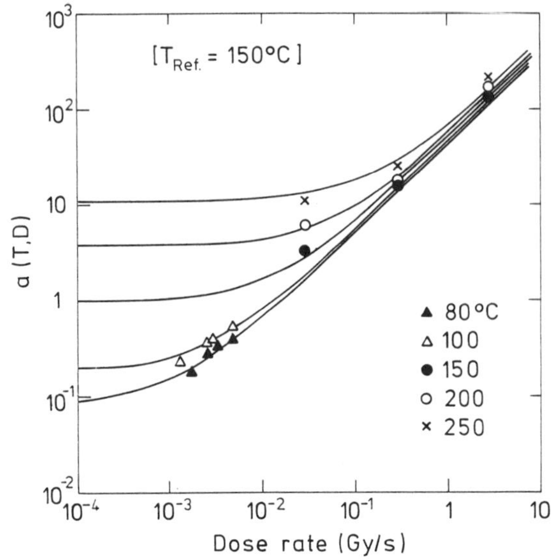

Figure 6. Shift factor a(T,D) as a function of dose rate and temperature - high temperature elastomer.

material, data are available at high temperatures but superposition is known to break down for T > 250°C as the degradation mechanisms change (*1*). The original raw data for this material (*1*) has been reassessed using a power law time dependence for compression set, rather than an arbitrary curve shape, to redefine the shift factor a(T,D) in the new model for each data set.

Plots of dose to equivalent damage (DED) versus dose rate can also be used to illustrate the model; the type 2 shift factor, equation (3), has been used to superpose data obtained at different temperatures. For the EPDM rubber, DED is virtually independent of dose rate down to 10^{-5} Gy/s at 20°C (Figure 7) whereas for the nitrile rubber DED is strongly dose rate dependent at 40°C below 1 Gy/s (Figure 8). In the high temperature elastomer shown in Figure 9, DED is still slightly dose rate dependent even at high dose rates; this is because the exponent x in equation (6) is 0.95 rather than the value of 1.0 found for the EPDM and nitrile.

Limitations to the Model. Although the superposition model does not require a detailed knowledge of the reaction schemes underlying the degradation, it does assume that a single mechanism is dominant. If this assumption breaks down, then the model will be invalid. An example of this is seen in the nitrile elastomer shown in Figure 5 where superposition breaks down at dose rates $\gtrsim$ 0.1 Gy/s.

A similar breakdown in superposition has previously been observed (*1*) in the high temperature elastomer illustrated in Figure 6. In this material there is a temperature limitation of 250°C for the validity of the superposition model.

In general, the model has proved to be of practical use where extrapolations are made towards lower temperatures and lower dose rates than those tested experimentally.

In combined environment radiation experiments such as these, diffusion limited oxidation is likely to play an important part (*5,6*), particularly at high dose rates. The breakdown of the model at dose rates > 0.1 Gy/s in the nitrile elastomer is likely to be due to such effects. Although measurements of oxidation profiles have not been carried out on the elastomeric materials tested so far, it is recognised that heterogeneous oxidation is likely to be significant at the highest dose rates

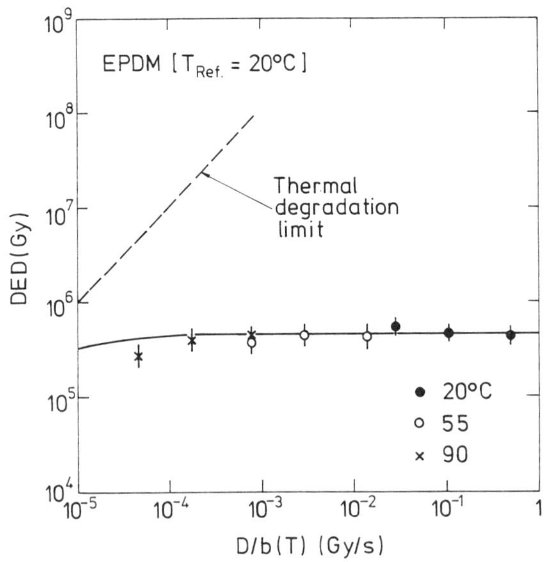

Figure 7. Dose to 50% compression set versus dose rate for EPDM at 20°C.

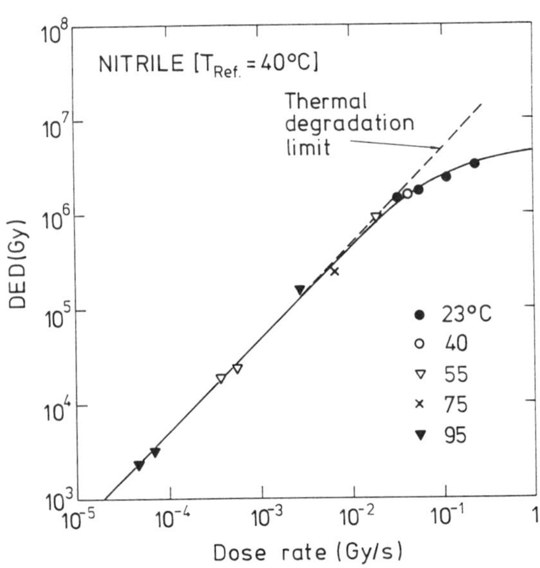

Figure 8. Dose to 50% compression set versus dose rate for nitrile at 40°C.

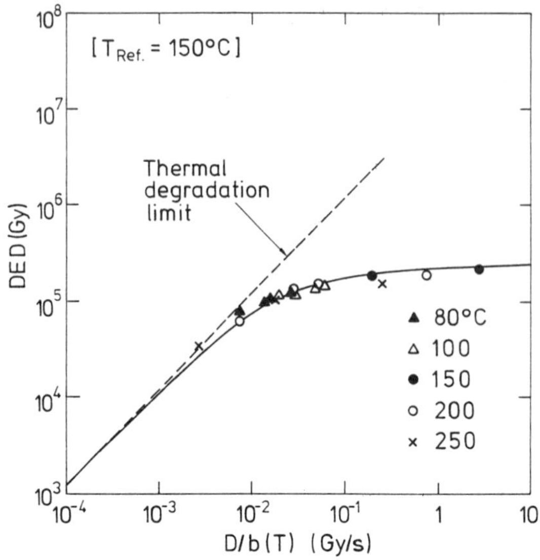

Figure 9. Dose to 80% compression set versus dose rate for high temperature elastomer at 150°C.

used. For this reason, best estimates of the model parameters have been made based on the lower dose rate data where possible.

Conclusions

A new empirical superposition model has been developed with three disposable parameters. For many elastomers, this improved model can be further simplified to a two parameter model which can be interpreted as the sume of a thermal component and a radiation component. This model can be applied to a number of elastomeric materials of interest to the nuclear industry.

In the new model, the temperature dependence which is observed under irradiation is determined by the relative values of the model parameters E and k. The dose rate dependence is determined by the parameter x, particularly at high dose rates.

The use of a simple power law to describe the time dependence of compression set has been found to be a useful adjunct to the new superposition model.

Acknowledgements

This work was carried out under the Underlying Research Programme of the United Kingdom Atomic Energy Authority. The author would like to thank Nuclear Electric and British Nuclear Fuels plc for permission to use data on their materials.

Literature Cited

(1) Burnay, S.G.; Hitchon, J.W. J. Nucl. Materials, **1985**, 131, p. 197.
(2) Burnay, S.G.; Hitchon, J.W. in *Influence of Radiation on Material Properties: 13th Inter. Symposium (Part II)*, ASTM STP 956, **1987**, eds.

F.A. Garner et al., American Society for Testing and Materials, Philadelphia, p. 609-614.
(3) Rudd, H.J. Time temperature dose rate superposition behaviour in irradiated polymers, AERE-R13746, **1990**, Harwell Laboratory, Didcot, Oxon, UK.
(4) Rudd, H.J. Harwell Laboratory, unpublished information.
(5) Gillen, K.T.; Clough, R.L. Polym. Degrad. and Stabil., 24, **1989**, p. 137.
(6) Gillen, K.T.; Clough, R.L. J. Polym. Sci., Polym. Chem. Ed., 23, **1985** p. 2683.

RECEIVED February 20, 1991

Chapter 34

Delayed Luminescence from Organic Compounds Induced by X-Radiation

Xingzhou Hu and G. David Mendenhall[1]

Department of Chemistry, Michigan Technological University, Houghton, MI 49931

We have observed a delayed emission, ascribed to charge-recombination luminescence, up to 300 min after X-irradiation of simple and polymeric hydrocarbons. The emission was concentrated at red wavelengths (500-600nm), and the initial intensities spanned four orders of magnitude. The decay of the emission could be represented adequately (r 0.98-1.00) in every case with a rate law At^{-n}, with $0.6<n<2.1$ over a given decade. Sudden changes in the temperature during the luminescence decay of irradiated poly(propylene) changed the emission intensity in a way corresponding to an activation energy of 4-5kcal which corresponds to a mean energy for the trapped charge-carriers, probably e^-. The yield of light from poly(ethylene) was estimated as $8x10^{-4}$ per X-ray photon ($G = 9x10^{-6}$). Branched-chain and hydroxy- or halo-substituted hydrocarbons gave about 10^2 less recombination luminescence than did solid, linear paraffins. Purification of tetracosane tended to increase the rate of decay of radiation-induced luminescence, and this behavior is consistent with impurities acting as traps for mobile charges. Luminescence from polymer samples containing tetramethyl-1,2-dioxetane or free-radical initiators occurred at distinctly shorter wavelengths, ca. 300-450 nm.

During the past dozen years or so, we have carried out a number of empirical[2] and fundamental[3,4] studies of the chemiluminescence (CL) from hydrocarbon oxidation. Much of our work has concerned poly(propylene) (PP). In a series of efforts we have (1) used CL to rank correctly eight formulated PP samples toward

[1]Address correspondence to this author.

0097-6156/91/0475-0534$06.00/0

their thermal oxidative stability at 150°C[5], (2) shown that the CL from extruded PP fiber probably arose from an oxidized surface layer[6], (3) demonstrated correlations, valid within a given formulation, between peroxide content of γ-irradiated PP and CL obtained by heating the fiber with an IR laser[7], and (4) shown qualitative correlations 7-30 days after irradiation between CL from formulated, γ-irradiated PP samples and impact strengths measured during the same time period.[8]

In an attempt to improve on the last-named study, we irradiated PP briefly with X-rays and examined the light emission immediately after the irradiation. However, the correlation of the rates of loss of physical properties (days after γ-irradiation) with the parameters describing the luminescence (minutes after X- or γ-irradiation) was not as good.[9] Further investigation, reported here, led us to conclude that the luminescence at room temperature immediately following exposure of PP and other organic materials to ionizing radiation arises largely from charge-recombination rather than from recombination of peroxy radicals.

Experimental

Poly(propylene) was in the form of 2mm sheets (Himont USA Profax 6801, >95% isotactic) containing 150 ppm of a phenolic processing stabilizer (Irganox 1076). The polymer was purified by extraction with dichloromethane for 26h followed by drying 26h at 0.01 torr, although this treatment made little difference in the decay curves. Peroxidic compounds were decomposed in some samples by heating them above the melting point at 0.01 torr.

Other compounds were reagent grade (Polysciences, Inc. or Aldrich) and were examined without further purification. The luminescence from X-irradiated tetracosane was also studied after purifying it by washing a solution in hexanes successively with two portions of sulfuric acid, followed by aqueous base, deionized water, and crystallization at -60°C. The product was examined a third time after distilling it under vacuum (<1 torr). The branched hydrocarbon f(1.2), or 5,5-bis(3',3'-dimethylbutyl)-2,2,8,8-tetramethylnonane, was synthesized as described.[10] 2-Propyl and *tert*-butyl hyponitrites (IPN and BHN) were synthesized according to published procedures.[11,12] The 3,3,4,4-tetramethyl-1,2-dioxetane was synthesized by Dr. Craig A. Ogle. A few milligrams of each solute were added directly (IPN) or were evaporated from pentane or hexane solutions as required onto the polymer.

Poly(ethylene), n-alkanes, and some of the other materials were melted under argon in a glass vial with I.D. 1.5cm, solidified with an ice bath, and the pellet was recovered by breaking the vial. Other materials (adamantane, f(1.2)) were irradiated as a thin layer of powder. Films of poly(N-vinylcarbazole) (PVK, Polysciences) were prepared by solvent-evaporation.[13] Each sample was placed in the well of a poly(methyl methacrylate) sample holder attached to a shielded Phillips Electronics Co. Model 52354/0 X-ray source with a tungsten target. This apparatus was designed for scattering measurements, so the X-ray beam struck the sample at an angle of about 70°. Comparable results were obtained with two other X-ray sources. Limited work was carried out with a ^{60}Co source (Gamma Cell 200, Canadian Atomic Energy, Inc.) at a dose rate of 3.06 krads/hr, but the weak radiation flux and smaller absorption coefficient made it difficult to measure decay curves with reasonable irradiation times except for poly(ethylene).

After irradiation, powdered or pelletized samples were transferred to a glass vial (O.D. 2.5cm) and placed in the well of the Luminometer for measurement of the luminescence. We did not correct the luminescence for self-absorption effects due to opacity, which varied considerably with the different samples. We tested a typical translucent PP plaque by measuring a luminescence decay with the irradiated face and then the opposite face toward the PM tube. The plot of log (emission) vs. log (time) showed small positive and negative deviations that became apparent when we drew a line drawn through all of the data, corresponding to the face-up or face-down condition. The deviations were of the same magnitude as those we observed when the sample was repositioned laterally in the well of the apparatus.

Direct measurements with a thermocouple revealed no temperature rise (<0.1°C) in a sample of poly(propylene) after several minutes of X-irradiation. The X-ray flux was measured with aqueous $FeSO_4$ dosimetry[14], and the dose to hydrocarbon samples was calculated from the sample thickness with correction for atomic composition of water vs the sample.[15]

For irradiation and light measurements under a single atmosphere, a sample container was constructed from a steel tube 27.5x47 mm OD with one end sealed with a flat. The sample was placed in the tube at the sealed end, and flushed 1h with Ar or O_2 from a gas inlet 10mm from the closed end. The open end of the tube was then placed against the X-ray port, and gas flushing was continued during and after irradiation. The tube was subsequently transferred open-end down into the well of the luminescence apparatus. Appropriate shielding from scattered X-rays was maintained with lead sheets and a steel pipe.

The X-irradiations of PP at 0°C and at -78°C were carried out with samples in the well of an Al rod hollowed out from both ends so that a flat metal partition separated the sample from ice or dry ice in the other side. A snugly-fitting metal collar pressed the sample against the partition. The temperature-jump experiments were carried out by switching from circulating water to or from ice in the chamber of a similar apparatus. Condensation of water on the exposed PP surface at -78°C was minor.

UV irradiations of PVK were carried out with a Spectronics Corp 22w low-pressure Hg lamp 6cm from the sample.

Light emission was measured with a Turner Designs, Inc. Model 20e Luminometer with circulating water through the sample holder at 25.0±0.2°C. Data were measured manually with a 5s-20s acquisition time, with longer times only in the very late stages of the decay, and the time coordinate was measured from the end of the irradiation to the midpoint of each data-collection period.

Absorption measurements were carried out with an HP Model 8451A diode-array spectrometer. Regression analyses of the data were performed with an HP15A hand calculator, and fits to the integrated equation were carried out on an IBM mainframe with SAS software.

Results

Decay curves. In Figure 1 a number of decay curves of light emitted from X-radiated hydrocarbons on a time scale of minutes are plotted on a log-log scale. The parameters describing these decay curves and others at short times are summarized in Table 1. Over extended time ranges the plots approached linearity, and the overall shapes are extremely similar to published curves of radiation-induced luminescence usually at sub-second time scales.[16] The term n in the power function At^{-n} was between 0.6 and 2.1 in all cases, but usually close to 1.0, and was very similar to the range observed by ourselves[13,17] and by others[18] for recombination luminescence stimulated in other ways.

The chemical structure had a strong effect on the intensity of the delayed luminescence, and in the same direction as observed for thermoluminescence of hydrocarbons,[19] so that straight-chain paraffins and polymers gave off much more light than branched ones. The morphology of the sample apparently has less effect, since (rubbery) PIB showed a luminescence intensity lower than that from (crystalline) tetracosane but higher than (semicrystalline) PP. The intensities are not easily compared because of differences in transparencies. Samples of the same compound prepared for observation in different ways usually showed somewhat different slopes in the radiation-stimulated decay curves.

The decay of luminescence from X-irradiated poly(styrene) proceeded as $t^{-0.695}$, in fortuitously good agreement with a $t^{-0.693}$ time-dependence measured from a different source of PS after stimulation with incandescent light.[17]

Charge-recombination luminescence has been extensively characterized in poly(N-vinylcarbazole)[13,20] and there is little ambiguity about the nature of the luminescence. After X-radiation (3.0 min, 452 rads/s) a film of this polymer showed luminescence with decay parameters (1-22 min) of A= 4.687, n = 0.788 (r^2 =

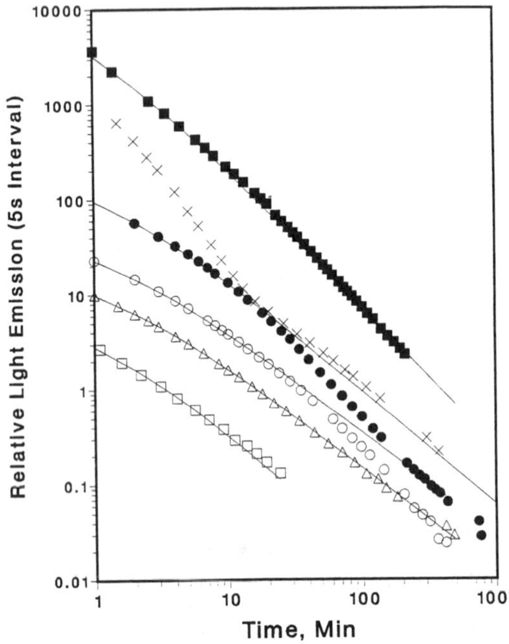

Figure 1. Characteristic decays of unfiltered light emission after X-radiation of (top to bottom) poly(ethylene), tetracosane, powdered poly(ethylene), poly(isobutylene), isotactic poly(propylene), and 5,5-bis(3,3-dimethylbutyl)-2,2,8,8-tetramethylnonane. The solid lines were calculated from eq. 1 and the following values (top to bottom) of n, Δt, and A': 1.464, 1.3, 4721.1; 1.195, 3.0, 78.65; 1.087, 3.0, 17.56; 7.624, 3.0, 1.093; 1.295, 3.0, 2.480.

1.000), compared with A= 1.563, n = 0.800 (r^2 = 0.998) for the decay stimulated in the *same* film by a 3min-exposure to incandescent light.

The effect of purification of tetracosane on its luminescence is shown in Table 2. There is a rough trend toward higher values of the exponential term n with increasing purity of the hydrocarbon.

The same table also shows that the measured decay parameters A and n depend on the time of irradiation. If the charge-recombination is a localized process represented by At^{-n}, then for irradiation times Δt the luminescence emission intensity I_t at any time t during the subsequent decay will be the sum of light emitted during times t to t + Δt from the end of a hypothetical, instantaneous irradiation. We have

$$I_t = \int_t^{t+\Delta t} A' t^{-n} dt = [A'/(1-n)][(t + \Delta t)^{1-n} - t^{1-n}] \qquad (1)$$

For large Δt,

$$\begin{aligned} I_t &\simeq [A'/(1-n)](\Delta t^{1-n} - t^{1-n}) \\ &= A'' - Bt^{1-n} \end{aligned} \qquad (2)$$

For a PE pellet the decay of the luminescence induced by a 1.3min exposure to X-rays approached a $t^{-1.558}$ time-dependence for long times (Figures 1 and 2). A second pellet was irradiated for 24.2h in a weak γ-source, and the data were analyzed by plotting I_t for short times after irradiation vs t^{1-n} (i.e. $t^{-0.558}$), which gave from linear regression:

$$A'' = -6.178 \text{ min}^{-1} \qquad B = 96.568 \text{ min}^{-1}$$

From the intercept A" and eq (1) we calculate A' = 41.5, and from the slope B we obtain A' = 37.8 in reasonable agreement. The log-log plot of the luminescence from the 24-hr irradiation was strongly curved, however. Since the temperature was not controlled during the γ-irradiation, and the γ-irradiated sample could not be removed quickly from the ^{60}Co source, the extent of agreement of the two data sets with eq (1) and (2) is rather satisfactory. The general trend toward lower slopes of the log-log plots with longer irradiation times can be explained in terms of the above relations.

The downward curvature of four of the plots in Figure 1 was duplicated fairly well when the curve was fitted to equation (1), although the fits to that equation were still not perfect. In two of the curves the luminescence falls off more rapidly than predicted. This result may reflect temperature drifts or spatial inhomogenieties of irradiation.

The shape of the log-log curve from irradiated, *purified* tetracosane (Figure 1, second from top) did not resemble the curves

Table 1. Fitted parameters for luminescence decay after X-irradiation [a]

Compound	Wt, g	Form	A, min^{-1}	n	t range, min
Eicosane	0.4416	disk,	5115	2.082	0.5-24
Tetracosane	0.3535	disk,	2031	1.819	1-21
Hexatriacontane	0.5091	disk,	3452	1.250	0.5-24
f(1.2)[b]	0.0520	powder	3.127	.977	1-24
Poly(ethylene)	0.9001	disk	5741	1.403	1-31
Poly(propylene)	0.5514	sheet	59.80	1.175	0.5-31
Poly(isobutene)	0.8639	sheet	20.48	0.908	1-32
Poly(α-methylstyrene)	0.6922	disk	6.238	0.783	0.5-26
Stearyl bromide	0.698	disk	22.67	1.05	0.6-24
Tetradecanol	0.4055	disk	9.428	0.803	0.6-25
Adamantane	0.0526	powder	36.87	0.975	2-38
Teflon[c]	0.9380	disk	33.79	1.029	1-22
Poly(N-vinylcarbazole)	0.0093	film	2.577	0.666	0.6-27

a. X-ray dose 3min. at 2.71×10^4 rads/min.
b. 5,5-Bis(3,3-dimethylbutyl)-2,2,8,8-tetramethylnonane.
c. Cleaned with $K_2Cr_2O_7/H_2SO_4$.

Table 2. Effect of purification on delayed emission from tetracosane [a]

Treatment	wt, g	X-ray time, min	A, min^{-1}	n	t range, min
None	0.3555	5	1955	1.298	1-29
	0.3538	5	1847	1.316	1-30
			avg.	(1.31±0.14)	
	0.3628	0.33	94.43	1.315	1-29
	0.3331	0.33	104.3	1.461	1-29
			avg.	(1.39±0.10)	
H_2SO_4-wash	0.2628	5	1448	1.774	1-31
	0.2467	0.33	173	1.492	0.5-26
Distilled	0.3573	5	1169	1.571	1.5-31
	0.3056	0.33	179.7	1.696	0.5-30

a. $r^2 > 0.99$ except for two entries with >0.98.

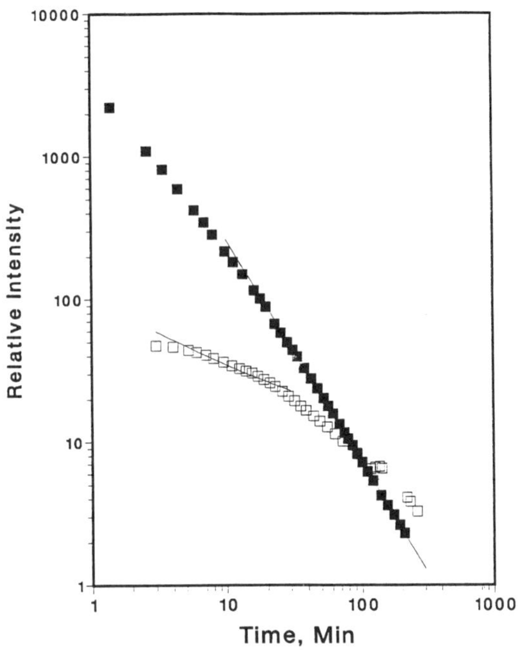

Figure 2. Luminescence decay curves of PE X-irradiated 3.0 min at 520 rads/min, showing the regression fit ($t^{-1.558}$) to the points at long times, and γ-irradiated 24h at 51 rads/min. The line through the curve was from the equation $I = At^{-0.558} + B$, with A and B constants, fitted to the data between 2 and 20 minutes.

from other samples of the same compound. A contribution from a luminescent higher-order process in the early stages, possibly from mobile vs. localized charged pairs, may be responsible.

Effect of Dose. As an approximation we have plotted (Figure 3) values of the parameter A for the fitted equation At^{-n} against $V^{1.5}$ for the X-ray source. A monotonic increase in emission with irradiation intensity is apparent.

Effect of repeated irradiation. In order to see if the accumulated oxidation products in PP affected the luminescence decay after X-irradiation, the same plaques of unstabilized PP were irradiated four different times within a 26 hr period. The values of the decay parameters (Table 3) show differences that are within the normal reproducibilities of the experiment. Similar results were found for two PP samples irradiated two weeks apart.

Effect of temperature. Poly(propylene) after X-irradiation at 0°C or at -78°C showed light-emission that also apparently decayed according to a power function, although there was some uncertainty about the control of the temperature. This observation is further evidence that reaction between two molecular species is not the source of the luminescence, since activation energies for bimolecular reactions should become very difficult at the lower temperatures because of lower diffusion rates. The relative intensity of the luminescence did drop by about a factor of 10 at 78°C. If the luminescent species are generated by extensive fragmentation of the parent molecules, their formation should be hindered by cooling.

A second set of experiments with poly(propylene) plaques revealed no differences in intensities within normal reproducibility (10%) following irradiation at 25°C or 40°C.

On the other hand, when the temperature was changed *during* the decay of the luminescence, the emission intensity showed a marked increase or decrease corresponding to an activation energy of 4-5 kcal (0.17-0.22eV, Figure 4). This value is in the range of estimates of electron trap energies in other polyolefins.[21]

Effect of atmosphere. Much of the work described here and previously[9] with CL from ionizing radiation had been carried out under the assumption that free-radical oxidation was responsible for the light emission. A strong ESR signal due to ROO· radicals is a well-known effect of ionizing radiation on polyolefins in the presence of oxygen,[22] and the signal decreases slowly on standing.[23] In one control experiment, we flushed the sample compartment attached to the X-ray unit containing a poly(propylene) disk with argon for 20 minutes before irradiation. Although this time would be insufficient to remove all oxygen from the sample, we were surprised that there was no significant effect on the subsequent luminescence decay in air (Table 4). When both irradiation and decay were carried out in the same atmosphere with samples equilibrated for several hours, the emission in argon was about 70%

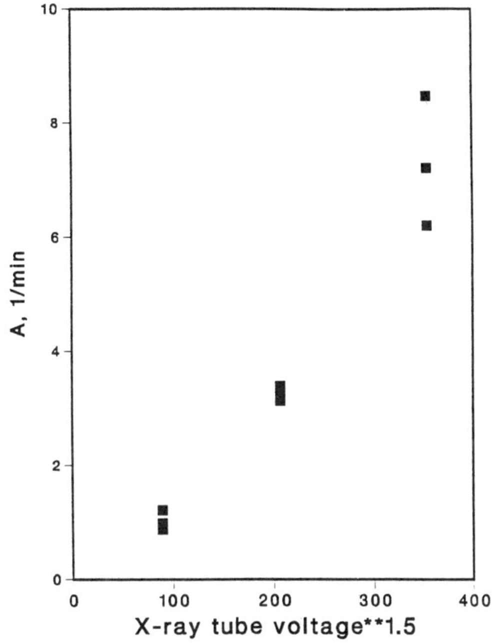

Figure 3. Plot of preexponential factor A (min^{-1}) for luminescence from poly(propylene) vs. X-ray intensity. The irradiation times were 5.0 min. in each case.

Table 3. Effect of repeated X-irradiations on delayed emission from poly(propylene)

Hours[a]	A, min^{-1}	n
Sample 1:		
0	7.406	0.934
1	9.738	1.009
2	8.918	0.924
26	10.163	0.978
Sample 2:		
0	8.316	0.981
1	11.191	1.023
2	9.383	0.943
26	9.983	0.984

a. Irradiated 5 min; total dose 8.13×10^4 rads. Decays analyzed as At^{-n} for t<60 min showed $r^2 > 0.99$ in all cases.

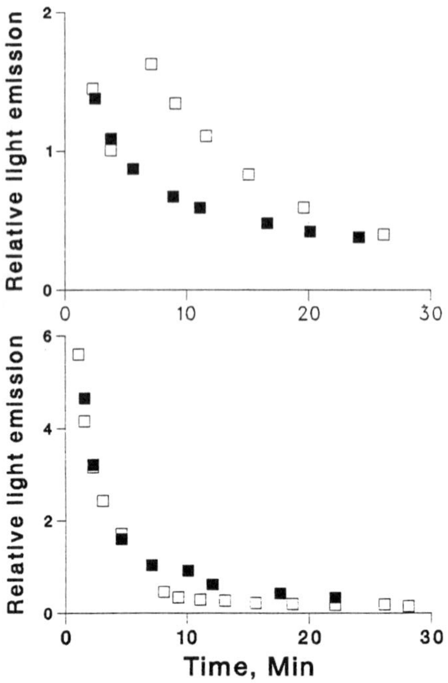

Figure 4. Luminescence decay of unstabilized PP plaques X-irradiated 3.0 min. Top: samples at 0°C, with one sample warmed to 30°C after 5.0 min. Bottom: samples at 30°C, with one sample cooled to 0°C after 5.0 min.

Table 4. Emission decay from X-irradiated poly(propylene) in different atmospheres [a]

Atmosphere		A, min^{-1}	n	t range, min
Air		9.097	1.045	0.5-25
		11.046	0.998	0.6-27
		8.752	1.012	1-28
	avg	9.6 ± 1.2	1.02 ± 0.02	
O_2		8.541	1.039	1-28
		9.111	0.996	1-28
		8.287	0.978	1-29
	avg	8.6 ± 0.4	1.00 ± 0.03	
Ar		8.433	0.880	0.5-28
		10.131	0.885	0.6-28
		7.446	0.926	0.6-29
	avg	8.7 ± 1.4	0.90 ± 0.03	

a. $r^2 > 0.996$ for all cases. X-radiation 5min at dose rate of 2.71×10^4 rads/min.

as large as it was in air. When argon replaced oxygen *during* the decay of luminescence from X-irradiated thin films of PE, there was little effect on the luminescence intensity. From these results we infer that a non-oxygen-dependent pathway was a major component of the observed luminescence.

In all of our X-ray studies the amount of radiation did not significantly change the nature of the sample as determined by other techniques. For instance, UV-visible spectra recorded before and after irradiation, or immediately after irradiation and a few hours later did not show any significant differences, although we would expect differences to appear on longer standing.

Effect of 9,10-dibromoanthracene and fluorene. The anthracene (DBA) is a well-known chemiluminescence enhancer but had an irregular effect on the short-term charge-recombination luminescence stimulated with X-rays (Table 5). At very high DBA loadings, we observed previously a 3-fold increase of luminescence compared with samples without DBA after long-term γ-irradiations.[7,24] The CL from poly(propylene) under other conditions, where reactions of peroxidic species predominate, was greatly enhanced by fluorescent additives.[7]

Fluorene was expected to intercept free-radicals to give the 9-fluorenyl radical. The reaction of this radical with oxygen, followed by subsequent termination reaction, were expected to give some excited singlet fluorenone, which emits light with a high quantum efficiency. However, the presence of increasing fluorene *lowered* the intensity of the emission after X-irradiation. Possibly the hydrocarbon acts as a trap for charge-carriers and depresses the generation of luminescent charge-pairs.

Emission wavelengths. Since the emission under study here was rather weak, changed rapidly in the early stages, and required large sample areas, we did not attempt to use a conventional spectrometer for spectral resolution. We instead measured the fraction of light from irradiated samples that passed through various color filters (Figure 5, Table 6). The measured fractions will differ from the true fractional transmittance because of the spectral response function of the detector, but plots of the fractions against each other from different sources with the same instrument reveal whether the emission wavelengths are similar or not. If the spectra are identical, the fractions plotted against each other will fall on a diagonal line, independent of the detailed spectral characteristics of the emitter, color filter, or detector response function.

From consideration of the data in Table 6 and the response curve of the PM tube, which falls rapidly at longer wavelengths, we infer that most of the light from X- and γ-irradiated samples lies between 500 and 600nm. This is strikingly different from emission centered around 400-500 nm that is characteristic of chemiluminescence from simple hydrocarbons and polyolefins.[25] Thermoluminescence spectra from saturated hydrocarbons commonly display a bimodal distribution with one peak around 500nm.[19a]

Table 5. Effect of DBA and Fluorene on Luminescence from X-irradiated Poly(propylene) [a]

Additive, wt%	A, min^{-1}	n
DBA:		
0.010	3.64 ± 0.03	1.08 ± 0.01
0.030	4.00 ± 0.30	1.06 ± 0.02
0.050	4.6 ± 0.3	1.04 ± 0.02
0.10	3.10 ± 0.05	1.02 ± 0.01
Fluorene:		
0.010	6.5 ± 0.3	1.08 ± 0.04
0.030	5.5 ± 0.1	1.06 ± 0.02
0.050	3.9 ± 0.1	1.04 ± 0.03
0.20	3.2 ± 0.1	0.99 ± 0.02

a. Average parameters fitted to $I = At^{-n}$ from 3 decays after X-ray exposure for 3.0 min. Control sample with no DBA or fluorene run separately and showed (average of 3 decays) $A = 7.0 \pm 1.1\ min^{-1}$, $n = 0.94 \pm 0.06$.

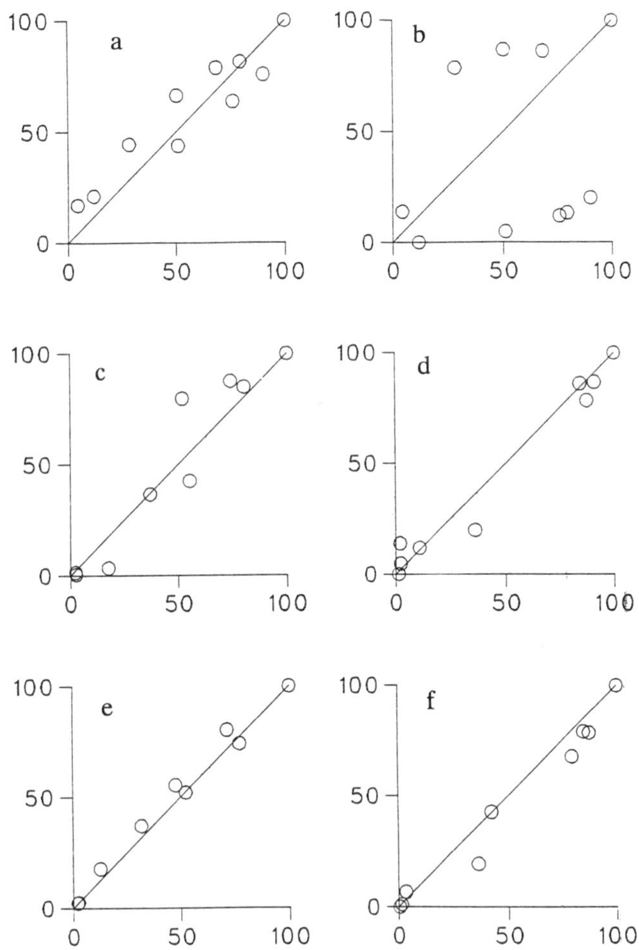

Figure 5. Plots of percentages of detected light from X- or γ-irradiated materials transmitted through various color filters. a. PP with tetramethyldioxetane vs PP with BHN(80°C), b. PP, BHN vs PP, X-ray c. PE, X-ray, vs tetracosane, X-ray, d. Hexatriacontane, X-ray, vs PP, X-ray, e. PE, γ-ray vs PE, X-ray, f. TC (distilled), X-ray vs TC (before distillation), X-ray.

Table 6. Transmittance of Luminescence through Color Filters [a]

Transparent Range of Filter (nm)	Transmitted part of emission (%)							
	PP with TMD (25°C)	PP with BHN (80°C)	PP, X-ray	TC, X-ray	PE, X-ray	HTC, X-ray	PE, -ray	TC[b] X-ray
>310	100	100	100	100	100	100	100	100
>380	78.7	68.5	86.2	85.0	80.2	84.5	71.5	79.1
>430	66.3	50.3	86.8	87.6	74.1	90.9	77.2	78.6
>480	44.4	28.4	78.6	79.7	52.0	87.5	52.4	67.8
340-560	75.7	90.0	20.0	42.5	55.3	36.2	47.7	42.7
340-530	63.7	76.0	11.9	36.6	37.1	11.0	31.9	19.3
360-520	81.4	79.4	13.3	-	-	-	-	-
310-470	43.8	50.9	4.8	3.4	17.7	2.2	12.7	6.7
>620	17.0	4.4	13.9	1.4	2.3	2.0	2.1	1.1
240-410	21.1	11.9	0	0.5	2.5	1.3	2.4	0

a. Data for Figure 5.

b. Distilled.

The predominance of longer wavelengths immediately allows us to explain the lack of effect of 9,10-dibromoanthracene on the emission, since the excited state energies would be only comparable or less than the energy levels of the fluorescer, and efficient energy transfer to DBA would not be expected.

Plots of the transmitted fractions of luminescence also showed that thermal sources of excited carbonyl species (a dioxetane, 2-propyl hyponitrite, or tert-butyl hyponitrite) were all similar (but not identical), and all of these were very different from the emission from X- or γ-irradiated compounds (Figure 5 and Table 6).

The emissions from straight and branched hydrocarbons after X-radiation were similar (Figure 5d), although there were some differences outside the error limits. Differences in emission wavelengths from irradiated tetracosane of different purities led us to conclude that more than one emitter is present in the system. Some of the emission does arise at shorter wavelengths and may be ascribed to a contribution from oxyluminescence.

The wavelength distributions from X- and γ-radiated poly(ethylene) were virtually identical (Figure 5e), although the dose rates and irradiation times were very different.

Functionalized and other hydrocarbons. Free electrons are expected to be scavenged by organic halides with formation of halide ion, and we thought that cationic species (H^+, RH_2^+, etc.) might be trapped by alcohols. We observed well-defined, albeit somewhat weak curves ascribed to recombination luminescence, however, when stearyl bromide ($C_{18}H_{37}Br$) and tetradecanol were X-irradiated (Table 1).

We thought that excited H-atoms might play a role in the emission, but poly(tetrafluoroethylene) was highly luminescent after exposure to X-rays, and the decay of the luminescence followed the characteristic power function. Another possibility is that extended, conjugated species[26] are the luminescent bodies, but adamantane, which lacks a straight-chain structure, gave relatively intense recombination luminescence. The energies of the X-ray interactions are of course sufficient to cleave C-C bonds, so that adamantane and poly(isobutene) are not precluded from forming conjugated structures by deep-seated molecular changes.

Estimate of G-value for poly(ethylene). An approximate, extrapolated value of 9×10^8 photons/s at time zero was made from the parameters of an alternate equation $I(t) = A/(Bt^n + 1)$ fitted to the decay curve of poly(ethylene) together with the estimated efficiency of our detecting system. From the dose rate (452 rads/s) and the weight of the sample we calculate that $G = 9.3 \times 10^{-6}$ photons/100eV from charge-recombination luminescence. This value compares very favorably with an estimate of 10^{-5} for thermoluminescence from γ-irradiated PE in the presence of air.[19a]

If we make an approximation that all of the X-rays are from the largest peaks in the tungsten spectrum at 1.48Å (L_a, 8.40keV) and 1.28Å (L_b, 9.67keV) with $I(L_a)/I(L_b)$ = 1.43, we calculate that the extrapolated value of I_o corresponds to about $8x10^{-4}$ photons per absorbed X-ray.

Discussion

The experiments described here represent an extension of reports of light emission immediately after exposure of hydrocarbons to ionizing radiation at low temperatures. Boustead and Charlesby[19b] ascribed the emission from γ-irradiated PE at 77°K to charge-recombination and mentioned that it did not decay according to a simple rate law. Markiewicz and Fleming[27] also observed light emission after X-irradiation of PE under vacuum at 77°K. The delayed luminescence is well-known in the literature on thermoluminescence under the term "pre-glow".[18] In addition, it should be noted that Woo et al. used X-rays to stimulate conventional fluorescence spectra of solutions of aromatic hydrocarbons.[28]

Kadir et al.[29] published plots of luminescence decay obtained at 40°C and 50°C from γ-irradiated poly(propylene). We replotted their data on a log-log scale and obtained acceptably linear plots. The slopes of the plots were between -0.54 and -1.0, which are within the limits observed in this study.

Theoretical treatments have been derived to describe the decay of charge or luminescence according to a t^{-n} law in terms of competitive first-order processes of matrix relaxation[19], escape of electrons from traps at different energies,[30] a distribution of distances for tunnelling of an electron,[31] and in terms of diffusive exploration of a particle on a fractal surface.[32] Our only evidence bearing on the legitimacy of these theories is (1) the temperature-jump experiment indicates that an activated process is involved, i.e. the luminescence is not from a temperature-independent tunnelling process, and (2) matrix relaxation processes are unlikely in the high-melting, crystalline solids that emitted light after X-irradiation, so that electron or proton-hopping are the more viable possibilities in these materials. We note also that Bassler et al[20] have simulated the characteristics of charge-recombination luminescence without resorting explicitly to fractal concepts.

Nature of the emitters. The source of the emission from hydrocarbons in the present work or in related TL phenomena is not clear. Pulsed UV irradiation at 380 nm of virgin or X-irradiated poly(propylene) samples gave a weak signal at 580nm with a width of only 4nm. This band may correspond to part of the emission after X-irradiation. Correspondingly, we could stimulate very weak, delayed emission from poly(propylene) by exposure to unfiltered light from a high-pressure xenon lamp. We were unable to see significant changes in the UV-visible absorption spectrum of PP before and after X-radiation when we used the diode-array spectrometer in the subtractive mode.

We suspect that the emitter from X-rayed organics are excited radicals formed by neutralization of the corresponding cation, which in simple hydrocarbons may be a conjugated species. Although alkane cation radicals are observed as intensely-colored species after exposure of hydrocarbons to ionizing radiation,[33] they are too unstable under ambient conditions to be the emitters in this study,[34] and would give cations and radicals by the sequence:

$$RH \xrightarrow{\text{rad.}} RH^{\cdot +} + e^-$$
$$RH + RH^{\cdot +} \longrightarrow R^{\cdot} + RH_2^{+}$$
$$RH_2^{+} \longrightarrow R^{+} + H_2$$

Moreover, recombination of e^- and a dication RH^{+2} would be required to generate an excited $RH^{\cdot +}$, which is highly improbable on the time scale of our observations. The reaction of e^- and R^+ to give $R^{\cdot}$ would be sufficiently exothermic to produce red light,[35] although other modes of excitation, such as fragmentation of a hot olefin from neutralization of an olefin cation radical, are possible. Since cations do not usually react with oxygen, we can also explain why samples under ambient conditions exposed to X-rays in an inert atmosphere show the same luminescence decay subsequently in oxygen or argon, since the oxygen-sensitive radical is not formed until after luminescence has been emitted. Fluorescence emission at visible wavelengths has in fact been observed from benzylic radicals[36] and from trifluoromethyl radical[37] (cf. PTFE).

Since the putative radical from the charge-recombination process in paraffins can react with oxygen to initiate autoxidation, the slow release and recombination of deeply-trapped electrons in irradiated polymers may contribute to their characteristic delayed loss of physical properties.[23b, 38] Since the self-reactions of peroxy radicals so produced also can lead to chemiluminescence emission, the process in turn may account for the minor, short-wavelength components of the light from X-irradiated materials in this study.

Conclusions

Our experiments show that charge-recombination luminescence from X-ray exposure is an extremely widespread and easily measured property of organic compounds under ambient conditions provided that a sufficiently large sample is examined. The kinetics of the decay of the emission are consistent with some current theories of charge transport in dielectrics.

The sensitivity of the X-ray stimulated luminescence decay curves to sample purity and composition, and their relative insensitivity to atmosphere and temperature, is behavior that is entirely similar to charge-recombination luminescence we described earlier from a limited set of polymer samples after stimulation with incandescent light.[17] The major difference is that in the present experiments the emission was observed from every single sample that we irradiated. Since X-rays in contrast to UV-visible photons, are absorbed by materials entirely at the atomic level[39], the presence of molecular chromophores or impurities is not necessary for charge-separation to occur.

It is apparent that either charge-recombination luminescence or free-radical luminescence may predominate in saturated organic materials exposed to ionizing radiation, depending on the time scale and conditions of examination. The different spectra from each source provide a natural way to separate their contributions and will help us relate the light emission to polymer stability.

Acknowledgment

This work was generously supported by Himont USA, Inc. We also thank Dr. Y-S. Chao and Dr. S.M. Fernandez of Scientific Research Associates, Inc., CT for carrying out the fluorescence measurement, and Drs. D.E. Mikkola and S.D. McDowell for assistance with the X-irradiations.

References

1. Visiting Scientist from the Academia Sinica, Beijing, PRC.
2. a. G.D. Mendenhall, Ang. Chem. Int. Ed. 1977, 16, 225-32. b. G.D. Mendenhall, Ang. Chem. Int. Ed. 1990, 29, 362-73.
3. Visible CL: a. E.M.Y. Quinga and G.D. Mendenhall, J. Amer. Chem. Soc. 1986, 108, 474-8. b. X.C. Sheng and G.D. Mendenhall, 198th American Chemical Society Meeting, Miami Beach, FL, September, 1989, Abstract Org-109. c. S.-H. Lee and G.D. Mendenhall, J. Amer. Chem. Soc. 1988, 110, 4318-23.
4. IR CL: a. Q. Niu and G.D. Mendenhall, J. Amer. Chem. Soc. 1990, 112, 1656-7. b. S.H. Lee, Q. Niu, X.C. Sheng, and G.D. Mendenhall, Photochem. Photobiol., 1989, 50, 251.
5. Proprietary work for Exxon Chemical Co, 1978; see reference 25.
6. K.R. Flaherty, W.M. Lee, and G.D. Mendenhall, J. Polym. Sci.: Polym. Lett. 1984, 22, 665-7
7. G. D. Mendenhall, H. Byun, and J. E. Cooke, in "Advances in Polyolefins", R. B. Seymour and T. Cheng, eds., Plenum Press, NY, 1987, pp. 405-434.
8. G.D. Mendenhall, H.K. Agarwal and J.M. Cooke, T.S. Dziemianowicz, in Polymer Stabilization and Degradation, ACS Symp. Ser. 280, 1985, pp 373-85.
9. X. Hu, G.D. Mendenhall, and R. Becker, Polym. Prepr. (Am. Chem. Soc., Div. Polym. Chem.), 1990, 31(2), 323-4.
10. G.D. Mendenhall, S.X. Liang, and E. H-T. Chen, J. Org. Chem. 1990, 55, 3697.
11. C.A. Ogle, S.W. Martin, M.P. Dziobak, M.W. Urban and G.D. Mendenhall, J. Org. Chem. 1983, 48, 3728.
12. G.D. Mendenhall, Tetrahedron Lett. 1983, 24, 451-2.
13. X. Guo and G.D. Mendenhall, Chem. Phys. Lett. 1988, 152, 146-50.
14. a. H. Fricke and S. Morse, Phil. Mag. 1929, 7, 129-141. b. Radiation Dosimetry, Vol II, F.H. Attix, W.C. Roesch, ed., Academic Press, N.Y., 1966, pp. 167-97.
15. Handbook of Spectroscopy, Vol. I, J.W. Robinson, ed., CRC Press, Inc., Cleveland, 1974, p. 181 and p. 237.
16. A.K. Jonscher and A. dePolignac, J. Phys.C.: Solid State Phys.,1984,17, 6493-6519.
17. G.D. Mendenhall and H.K. Agarwal, J. Appl. Poly. Sci. 1987, 33, 1259 and references therein.
18. S.W.S. McKeever, "Thermoluminescence of Solids", Cambridge University Press, New York, 1985.

19. a. A. Charlesby and R.H. Partridge, Proc. Roy. Soc. 1963, 170-87; 188-206. b. I. Boustead and A. Charlesby, Proc. Roy. Soc. London 1970, A316, 291-302.
20. a. B. Ries, G. Schonherr, H. Bassler and M. Silver, Phil. Mag. 1984, B49, 259. b. F. Stolzenburg, B. Ries and H. Bassler, Ber. Bunsenges. Physik. Chem. 1987, 91, 853.
21. G. Sawa, M. Ieda and K. Kitagawa, Elect. Lett. 1974, 10, 49-51.
22. P. J. Wunsche, Macromol. Sci.-Phys. 1984, B23, 65-84.
23. eg. a. M.G. Alonso-Amigo and S. Schlick, Macromolecules, 1987, 20, 795-801. b. R.F. Becker, D.J. Carlsson, J.M. Cooke and S. Chmela, Polym. Degrad. Stab. 1988, 22, 313-323.
24. G. D. Mendenhall, H. Byun, and J. E. Cooke, Polym. Prepr. (Am. Chem. Soc., Div. Polym. Chem.), 1985, 26, 376-7.
25. G.E. Ashby, J. Poly. Sci. 1961, 50, 99.
26. B.J. Kelsall and L. Andrews, J. Phys. Chem. 1984, 88, 2723-9.
27. a. A. Markiewicz and R.J. Fleming, J. Polym. Sci. B: Polym. Physics, 1986, 24, 1713-24. b. A. Markiewicz and R.J. Fleming, J. Phys. D. Appl. Phys. 1988, 21, 349-55.
28. C.S. Woo, A.P. D'Silva, V. A. Fassel, and G.J. Oestreich, Environ. Sci. Tech., 1978, 12, 173-4.
29. Z.A. Kadir, F. Yoshii, K. Makuuchi, and I. Ishigaki, Polymer 1989, 30, 1425-32. See also ref. 15.
30. J.T. Randall and M.H.F. Wilkins, Proc. Roy. Soc. (London) 1945, 184, 390-407.
31. a. Reference 14b, p. 100. b. W. Lefik, A. Plonka, and J. J. Kroh, Radioanal. Nucl. Chem. 1986, 101, 267-73.
32. a. P.W. Klymko and R. J. Kopelman, Chem. Phys. 1983, 87, 75. b. L.A. Dissado, Chem. Phys. Lett. 1986, 124, 206-210.
33. W.H. Hamill, J. Phys. Chem. 1978, 82, 2073-7.
34. a. K. Toriyama, N. Keichi and M.J. Iwasaki, Am. Chem. Soc. 1987, 109, 4496-4500. b. D.W. Werst and A.D. Trifunac, J. Phys. Chem. 1988, 92, 1093-1103. c. A.D. Trifunac, D.W. Werst and L.T. Percy, Radiat. Phys. Chem. 1989, 34, 547-552.
35. From $\Delta H_f(t\text{-}Bu^+)$ = 167 kcal/mole,[35b] $\Delta H_f(t\text{-}Bu^\cdot)$ = 8.4 kcal/mole,[35c]. and ϵ(PP) = 2 we calculate $e^- + t\text{-}Bu^+$ ---> $t\text{-}Bu^\cdot$ + 79kcal/mole corresponding to 362nm. b. H.M. Rosenstock, K. Draxl, B.W. Steiner, J.T. Herron, J. Phys Chem. Ref. Data, 6, Suppl 1, 1977. c. S.W. Benson, "Thermochemical Kinetics", 2nd ed., John Wiley and Sons, NY, 1976, p. 299.
36. B.B. Craig and M.F. Sonnenschein, J. Lumin. 1989, 43, 227.
37. R. Hermann, Radiat. Phys. Chem. 1989, 34, 369-74.
38. T.S. Dunn and J.L. Williams, J. Indus. Irrad. Tech., 1983, 1, 33.
39. The X-ray *absorption* spectra of polymers with C,H,O atoms only are very similar: M. Przybylski, M. Stamm, R. Zietz, J. Physique, 1987, 48, 1351-6

RECEIVED July 17, 1991

Chapter 35

Stability of Polypropylene to Gamma Irradiation

J. L. Williams

Becton Dickinson Research Center, 21 Davis Drive, Research Triangle Park, NC 27709

Irradiation of polypropylene in air leads to severe oxidative degradation, discoloration, and eventually embrittlement. The mechanisms involved have been examined and methods of stabilization have been explored in this study. Although this work has concentrated on radiation effects on polypropylene, the methods for stabilization are applicable to other polymers.

Ionizing radiation is rapidly becoming the preferred method for sterilizing medical products. Unlike gaseous sterilization, radiation eliminates chemical residuals and the need to quarantine large volumes of product. As a general rule, substantial degradation occurs during the radiation sterilization of a plastic product, and special grades of plastic must be employed to prevent excess degradation.

The need for materials that will withstand ionizing radiation during sterilization of medical products has increased dramatically due to regulatory problems associated with ethylene oxide residuals found after ethylene oxide sterilization. As a consequence, the need for better stabilization packages that can tolerate a sterilizing dose of 25 kGy has increased along with the requirement to understand the mechanism of why a plastic embrittles or discolors during irradiation. A guideline to selecting plastics for use in medical products is given in (*1*) and will not be repeated in this chapter. However, polypropylene which readily degrades during irradiation will be used as a mechanistic example along with experimental results to support the theory of polymer stabilization by different approaches.

In order to stabilize polypropylene to high energy radiation, solutions to the problems of post irradiation embrittlement, discoloration and thermal instability must be found. Radiation results in degradation of the polymer by direct energy transfer, subsequent autoxidative reaction kinetics and thermal degradation due to processing temperatures, gamma and electron local heating, and post irradiation high temperature storage. Control of radical reaction kinetics is dependent on the polymer morphology, in general, and specifically on the degree of crystallinity, molecular weight distribution and main chain mobility. Control of both autoxidation

0097–6156/91/0475–0554$06.00/0

and heat-initiated oxidation require the addition of free radical scavengers, peroxide decomposers, and other stabilizers to the polymer formulation. The necessary condition for use of resulting formulations is that the stabilizer provides the desired characteristics yet does not discolor the polymer. Since the radiation induced radicals and peroxides formed in solid polymers are long lived, the potential for chain cleavage remains for years following irradiation. The stabilizer system must therefore remain active for storage not only at room temperature but also at high temperatures which may be encountered in warehouses in hot climates.

Materials and Methods

The details of the experimental procedures employed for most data presented are given in the respective references. Where unpublished results appear, ESR data were recorded at 9.25 GHZ and 3303.5G using a Varian E-Line Century Series Electron Spin Resonance Spectrometer. Spin concentration measurements were calculated from the double integration of spectra obtained on a 200G sweep width at a power level of 2 mW, a modulation amplitude of 4.0G, a time constant of 0.5 second, and a sweep time of 2 min.

Mechanical property measurements on test bars or molded syringes were carried out on a conventional Instron (Model 1122). In the case of post irradiation time studies, the samples were allowed to age at room temperature. The samples were irradiated with ^{60}Co at a 10 kGy/hr dose rate in an AECL Gammabeam 150 at ambient conditions.

Discussion and Background

Semicrystalline polymers such as polypropylene are nonhomogeneous. To a first approximation, however, polypropylene becomes a two-phase system consisting of amorphous and crystalline domains including a discontinuity at the crystal interface. The degree of crystallinity ranges from 0.35 to 0.65 with spherulite diameters of 0.1 - 300 microns. The densities in the two phases are $\rho_a = 0.854$ g/cm^3 and $\rho_c = 0.946$ g/cm^3 (2). During irradiation, energy is deposited uniformly and radicals are formed throughout the polymer in both the amorphous and crystalline domains. Different chemistry can result from energy deposition in the two regions since oxygen, stabilizers, and specific active radical species are excluded from the crystalline phase.

As a result, the local chemical reaction rates will be different from the average which results in a distribution of rate constants and activation energies in the different phases and at the crystal interface. The reaction kinetics will take three forms: those for polymer radicals alone, oxidation reactions and stabilizer reactions. All three forms of radical reactions are tied to the kinetics of molecular motion in solid polymers. (*2,3*)

Finally, the energy deposited in the polymer system results in an equally complex initiation of chemical changes to the additives employed in the polymer. Often, a particular additive cannot be used due to the detrimental effects brought about by direct radiation modifications.

Color Formation. Pure polypropylene when irradiated in air from 10 to 100 kGy does not discolor. However, a characteristic yellow discoloration occurs in this dose range in most commercial grade polypropylenes. The formation of color is due primarily to the presence of phenolic additives which are included in the formulation to inhibit oxidation of alkyl and peroxy radicals formed during irradiation or exposure to high temperature.

Phenolic antioxidants are generally trisubstituted phenols which generate stable phenoxyl radicals upon oxidation. The phenoxyl radicals formed can dimerize, disproportionate or continue to react with other polymer radicals (*4*). Reactions with polymer peroxy radicals result in the formation of peroxycyclohexadienones and quinonoid compounds, such as quinone methide and spiroquinolide ether. These conjugated structures absorb in the visible region of the spectrum and discolor the polypropylene. For applications where color is the primary concern and the polymer does not see a significant temperature cycle, phenolics can either be removed or added at very low concentrations. Also, consideration should be given to additives which do not discolor but still behave as primary radical scavengers or peroxide decomposers, such as dilaurylthiodipropionate (DLTDP). Most translucent medical products such as syringes and bottles would fall into this category of non-discoloring products.

Radical Reactions. Since energy is deposited uniformly throughout the polymer, radicals are formed inside the crystalline lattice as well as in the amorphous regions. The form of the radicals, however, is different. The main radical species formed in polypropylene during irradiation in vacuum and air are listed in Table I (*5,6,7*). The respective ESR spectra are shown in Figure 1.

Table I

Radical Species Formed During Irradiation of Polypropylene

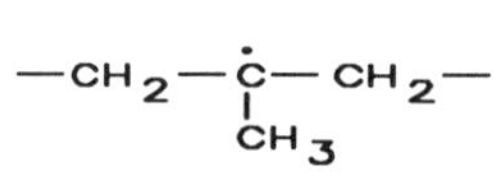

Alkyl

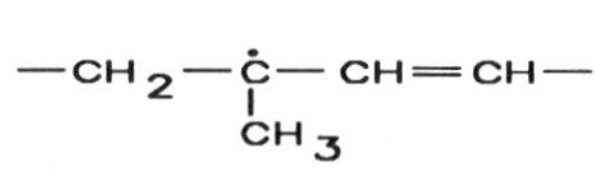

Allyl

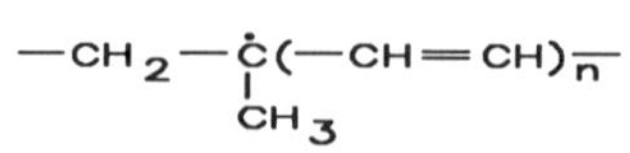

Polyenyl

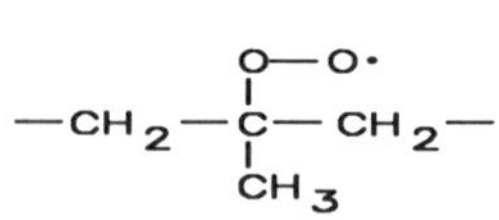

Peroxy

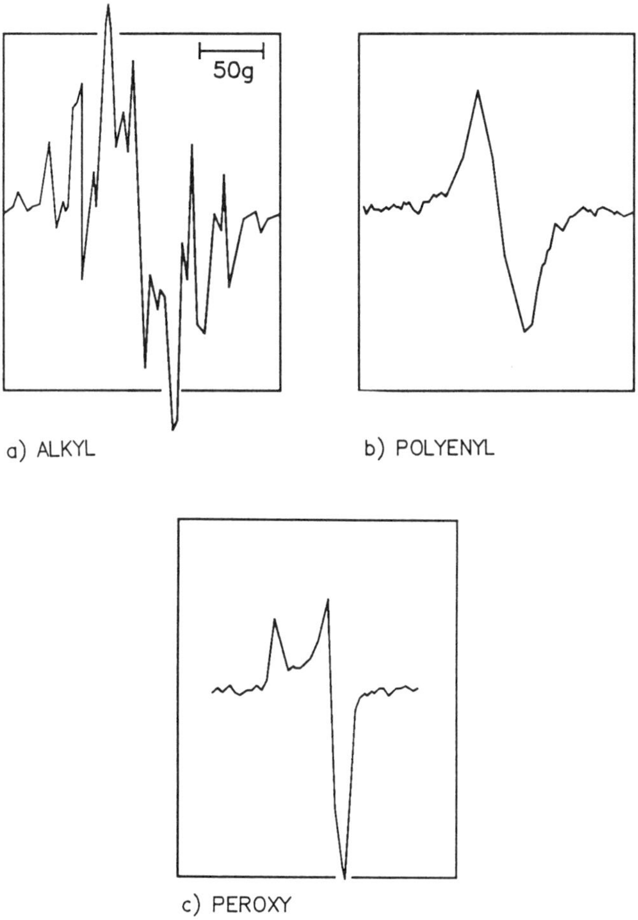

Figure 1. ESR spectra of polypropylene radical species.

The alkyl radical is the dominant species formed during irradiation in both the crystalline and amorphous domains. When the sample is maintained under vacuum, the radicals slowly convert to the allylic species and eventually to the more resonant stabilized polyenyl radical. These radicals must be located at the crystalline interface since radicals containing double bonds would be excluded from the crystal lattice structure resulting in a defect (*8*). This fact has been substantiated by the rapid oxidation of the polyenyl radical formed in vacuum when exposed to air as illustrated in Figure 2 (*6*). Since oxygen is restricted from the crystal, it follows that chain transfer mechanisms must dominate for the radicals to move to the crystal surface and react with oxygen. Otherwise more evidence of alkyl radicals should be present in the ESR spectrum which is not the case. The polyenyl radical would then have to be at the interface in order to react with oxygen forming the dominant peroxy radical observed at longer times by ESR. In a similar manner, alkyl radicals formed in the amorphous phase will also rapidly oxidize to peroxy radicals. These oxidative reactions are dependent on oxygen diffusion and are discussed in more detail in the references. (*2,3,9*)

A list of the various radical reactions occurring with polymer radicals, oxygen and stabilizer appears in Table II (*8,10-13*). Various aspects of these reactions and their extent determine the severity of the degradative reactions before an additive or termination reaction disrupts the kinetic chain and reduces the degradation of the polymer.

Table II

Radiation Induced Polymer/Stabilizer Radical Reactions

<u>Initiation</u>

$RH \rightarrow R\cdot$

<u>Polymer Radical Oxidative Reactions</u>

$R\cdot + O_2 \rightarrow RO_2\cdot$

$RO\cdot + RH \rightarrow ROH + R\cdot$

$RO_2\cdot + RH \rightarrow ROOH + R\cdot$

$ROOH \rightarrow RO\cdot + HO\cdot$

$HO\cdot + RH \rightarrow HOH + R\cdot$

<u>Radical Reactions During Irradiation</u>

$RO_2\cdot \rightarrow R\cdot + O_2$

$R'RO_2\cdot \rightarrow R-\overset{O}{\overset{\|}{C}}H + R'O\cdot$

$R'ROOH \rightarrow R'RO\cdot + HO\cdot \rightarrow RC=O + R'\cdot + HO\cdot$

$R'ROOR \rightarrow R'RO\cdot + RO\cdot \rightarrow RC=O + R'\cdot + RO\cdot$

$AOH \rightarrow AO\cdot$

$AOH \rightarrow R\cdot + AO\cdot$

<u>Termination Reactions</u>

$R\cdot + R \rightarrow R-R$

$R\cdot + RH \rightarrow RH + R\cdot$

$RO_2\cdot + RO_2\cdot \rightarrow ROOR + O$

$R\cdot + AOH \rightarrow RH + AO\cdot$

$RO\cdot + AOH \rightarrow ROH + AO\cdot$

$RO_2\cdot + AOH \rightarrow ROOH + AO\cdot$

$R\cdot + AO\cdot \rightarrow AOR$

$RO\cdot + AO\cdot \rightarrow AO-AO$

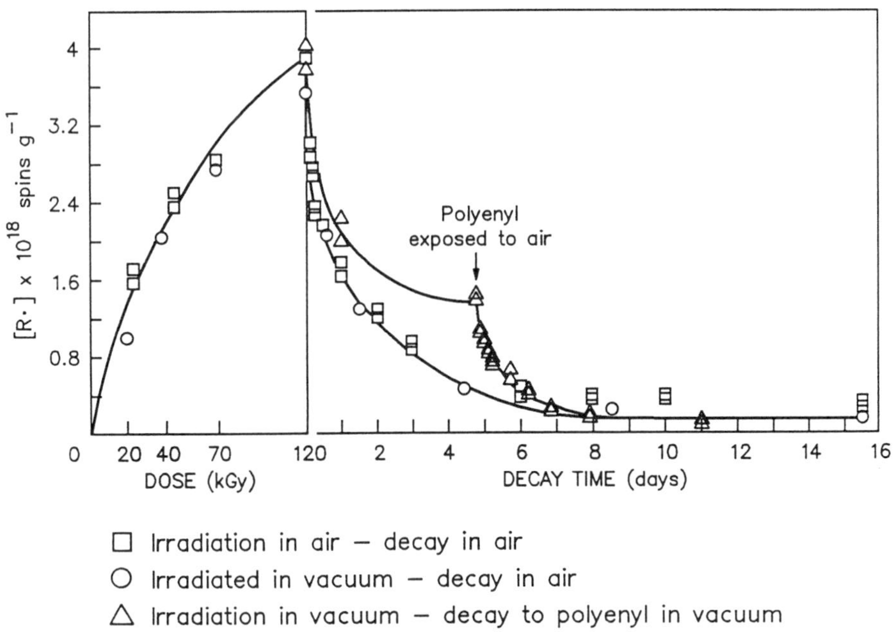

Figure 2. Radical buildup and decay of gamma-irradiated polypropylene.

Polymer Morphology. The preparation of molded parts from the melt will modify the morphology of polypropylene. Various aspects of the semicrystalline morphology are equally important to the control of polymer degradation. For instance, it has been shown in the literature that the degree of crystallinity increases in the first step of heat treatment with the rate of the increase dependent on the initial sample crystallinity (2). In addition, the oxidation rate of semicrystalline polymers depends on the degree of crystallinity (*2,14*). Adams (*7*) has shown that the molecular weight distribution varies with time exposure to temperature. Control of polymer morphology is of primary significance to direct radiation effects on polymers and subsequent degradation.

Primary radical termination may occur by chain segmental motion, hydrogen atom transfer or oxidative propagation. Consider the following equations from Table II:

$$R\bullet + R\bullet \rightarrow R\text{-}R \tag{1}$$

$$R\bullet + RH \rightarrow RH + R\bullet \tag{2}$$

These termination and transfer reactions are important to the stabilization of polymers. It is generally assumed that equation 2 dominates because physical chain motion is too slow a process for measurable rates of chain termination (*3*). From studies of gamma-irradiated polypropylene in vacuum, it has been shown that chain termination can be effectively enhanced by the addition of mobilizing additives to the polymer such as mineral oil. (*15,16*) As illustrated in Figure 3 for the same % crystallinity, the addition of mobilizing additive increases the chain termination constant by a factor of four. This result has been further substantiated by studying the effect of degree of crystallinity on the reaction rate of equation 1 as illustrated in Figure 4. These data illustrate that the low percent crystallinity samples yield a higher termination rate constant which is further enhanced in the mobilized samples (*17*).

During irradiation in air where oxygen is present, peroxy radicals are easily formed as follows:

$$R\bullet + O_2 \rightarrow RO_2\bullet \tag{3}$$

The reaction rate of equation 3 is much more rapid than for main chain radical termination given in equation 1 due to ease of oxygen diffusion to the radical site. However, the effect of enhancing main chain termination by addition of mobilizing additives is a significant and a measurable factor as shown in Figure 5. Although the mobilizer increases oxygen diffusion, it is less an effect than the increase in the main chain radical termination reaction, which is greatly aided by the additional segmental chain mobility. As a consequence, the addition of mobilizer confers additional radiation stability to the polymeric system depending mainly on the level of mobilizer added.

The peroxy radicals formed during radiation lead to the formation of peroxides as follows:

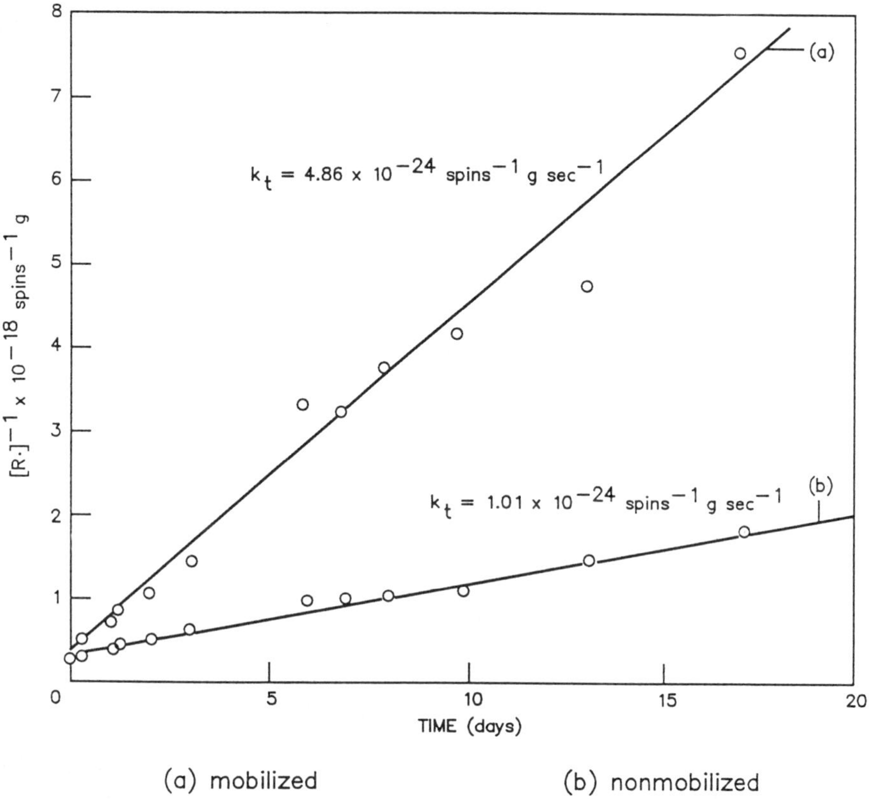

Figure 3. Second order termination rates in vacuum (48% crystalline homopolymer).

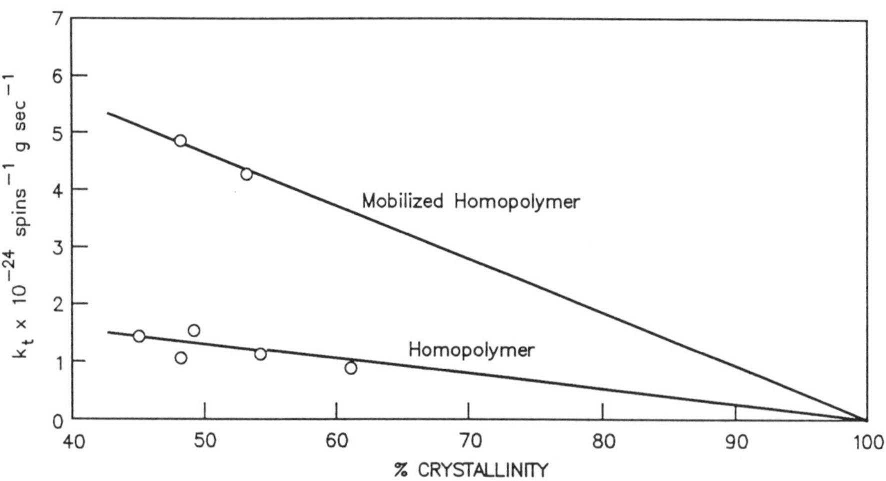

Figure 4. Dependence of second order termination rate on percent crystallinity of gamma irradiated polypropylene.

$$RO_2\bullet + RO_2\bullet \rightarrow ROOR + O_2 \quad (4)$$

$$RO_2\bullet + RH \rightarrow ROOH + R\bullet \quad (5)$$

The peroxides remain in the polymer for long periods of time and upon dissociation result in chain scissions. The effect is continued embrittlement as a function of long term aging as illustrated in Figure 6 (*17*). It follows that increased mobility alone will not totally stabilize the polymer. As a consequence, inhibitor or aromatic stabilizer molecules, e.g., AOH, should be available in the polymer to suppress primary and oxidized polymer radicals.

$$R\bullet + AOH \rightarrow RH + AO\bullet \quad (6)$$

$$RO\bullet + AOH \rightarrow ROH + AO\bullet \quad (7)$$

$$RO_2\bullet + AOH \rightarrow ROOH + AO\bullet \quad (8)$$

Primary radical scavengers and other stabilizers are a necessary addition to stable formulations.

Primary Methods of Polymer Stabilization

Mobilized Polymer. As discussed above, the addition of mobilizer increases main chain radical termination significantly reducing polymer chain autoxidation. The effect of the overall reduction in polymer embrittlement by the addition of the mobilizer is shown graphically in Figure 7. The angle at break of a syringe flange is plotted as a function of time up to 12 months storage. It is evident that the mobilized sample does not embrittle significantly even after 20 kGy and aging for nine months. On the other hand, the non mobilized sample starts to embrittle immediately. The increased stability of the mobilized sample is brought about mainly by the reduction in radical sites for main chain oxidative degradation.

Molecular Weight Distribution. A severely radiation embrittled polypropylene object will regain its original strength if remolded. The explanation for the renewed strength in this case resides in the fact that new tie molecules are formed between crystallites, upon remolding. An immediate conclusion is that the bulk strength of the polymer is dependent on the number of tie molecules between crystallites and not the average chain length. Any factor which increases the number of tie molecules should, in principle, increase the radiation stability of the semi-crystalline polymer.

In polypropylene with narrow molecular weight distribution, the chain lengths are more uniform and nearer to the mean value than polypropylene with broad molecular weight distributions. As a consequence, more load bearing chains or tie molecules are present in the amorphous phase between crystallites in the polypropylene with narrow molecular weight distributions than in broad molecular weight distribution, of the same melt flow. This fact results in better maintenance

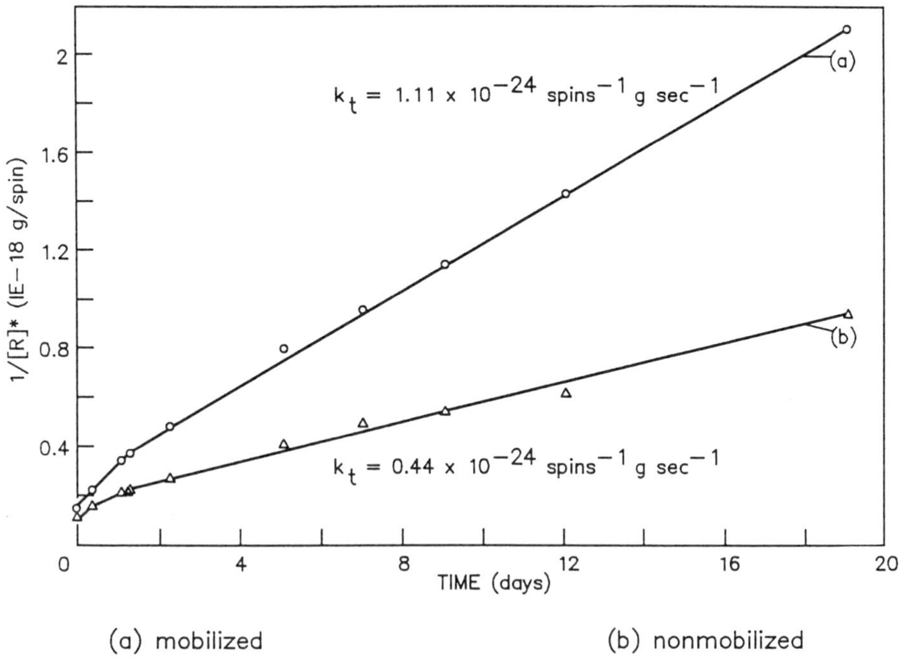

Figure 5. Second order termination rates in air (48% crystalline homopolymer).

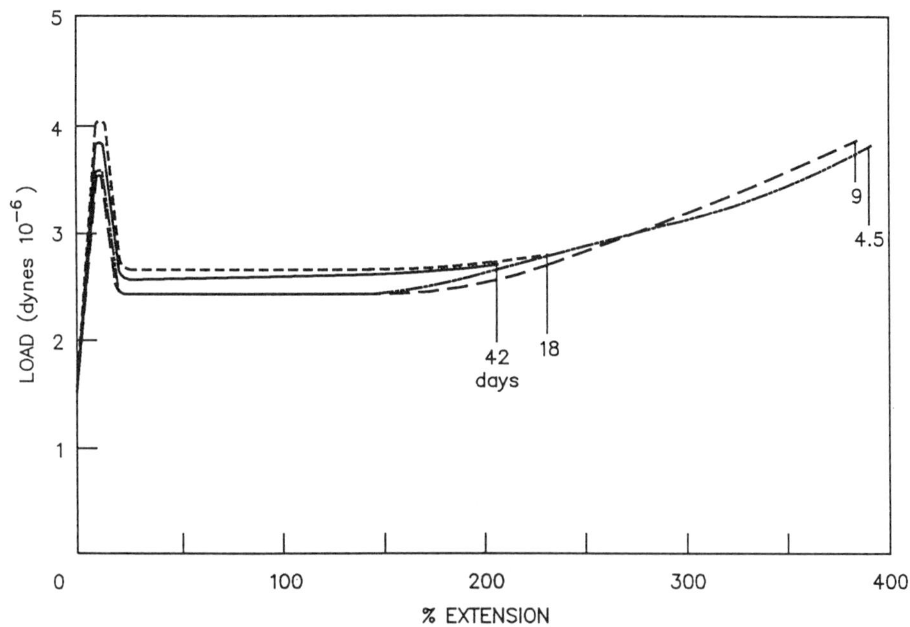

Figure 6. Embrittlement of irradiated polypropylene test bars as a function of aging time.

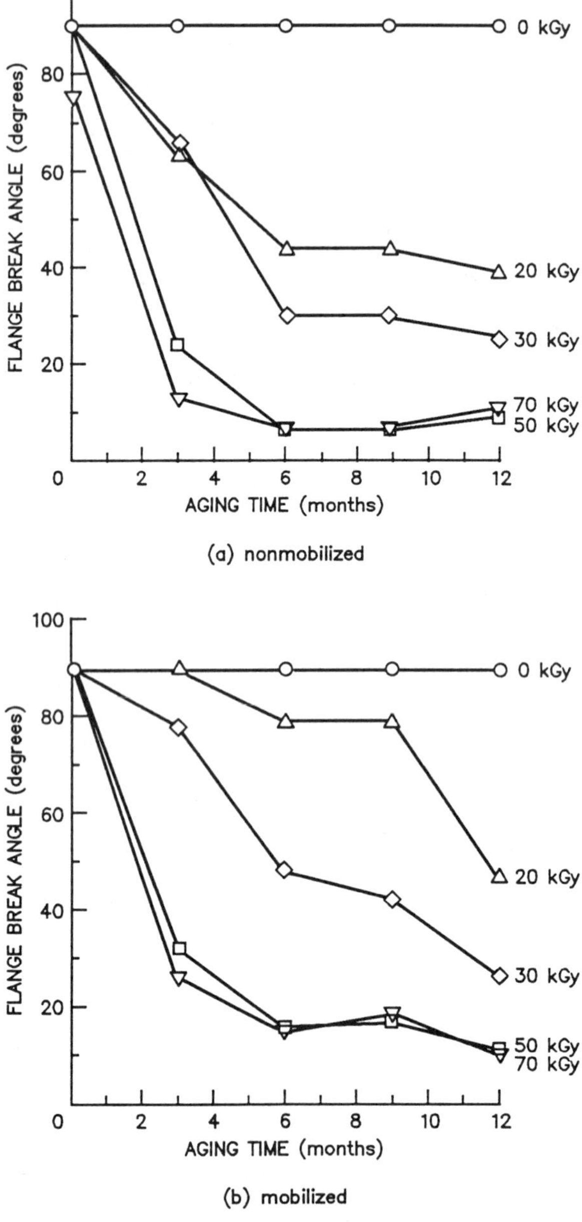

Figure 7. Improvement in radiation stability with mobilized homopolymer at ambient conditions.

of mechanical properties after irradiation, as illustrated in Figure 8, particularly when coupled with the beneficial effects of the mobilizing additive discussed above (*18*).

Antioxidants and Stabilizers. Polyolefins are sensitive to thermally induced oxidative degradation even without irradiation. The use of stabilizers becomes even more important to radiation stable formulations due to the molding, processing, and storage at high temperature. Stabilizers shown to be beneficial to both thermal and UV stability include phenolic antioxidants (*19-21*), hindered piperidine derivatives (*22-24*), sulfur (*25*), and phosphorous containing compounds (*26,27*).

Radical scavenging activity of various stabilizers was studied individually and in various combinations as a function of temperature (*28,29*). The combination of hindered phenol and piperidine yielded the highest activity and demonstrated the best thermal stability. The combination of DLTDP to these same compounds significantly reduces both the activity and thermal stability. These data were substantiated by aging studies of the mechanical properties of irradiated polypropylene (*28,30*).

The beneficial effect of adding a primary radical scavenger to molded syringe barrels is demonstrated in Figure 9. Narrow distribution formulations containing a primary radical scavenger are compared with and without the addition of a mobilizer. It is clear that the mobilizer enhances the radical scavenging activity and yields a polypropylene stable to irradiation effects, up to a dose of 70 kGy after 12 months aging.

In order to be effective, the stabilizer must be both soluble and mobile in the polymer. The semicrystalline morphology excludes the stabilizers from the crystalline phase, therefore concentrating them in the amorphous regions. Experimental data indicates that the limiting step for inhibited oxidation is the molecular diffusion of the stabilizer. The addition of the mobilizer facilitates the diffusion of these large stabilizer molecules and thus enhances the reaction rates and resulting stabilization of the polymer (*3*).

Conclusions and Future Trends

From the data discussed here, several methods have been identified that are effective for the stabilization of polypropylene. Maintaining the lowest practical degree of crystallinity is important. Homopolymers with narrow molecular weight distribution are preferable. The inclusion of mobilizing additives reduces polymer oxidative degradation by facilitating both main chain termination and additive diffusion rates. To achieve maximum autoxidative inhibition and thermal stability with minimal discoloration, a primary radical scavenger should be included in the formulation. The combination of these various factors in a given formulation yields a high probability of successful stabilization of polypropylene to radiation degradation.

In this context, polypropylene has been used as an example, but all the theory and mechanisms would apply to any organic material.

The economic advantages of radiation sterilization will fuel the growth of this general form of sterilization in the future for both gamma and e-beam. Gamma will continue to be the primary source for sterilization due to the outstanding penetration characteristics of this form of radiation compared to electron irradiation.

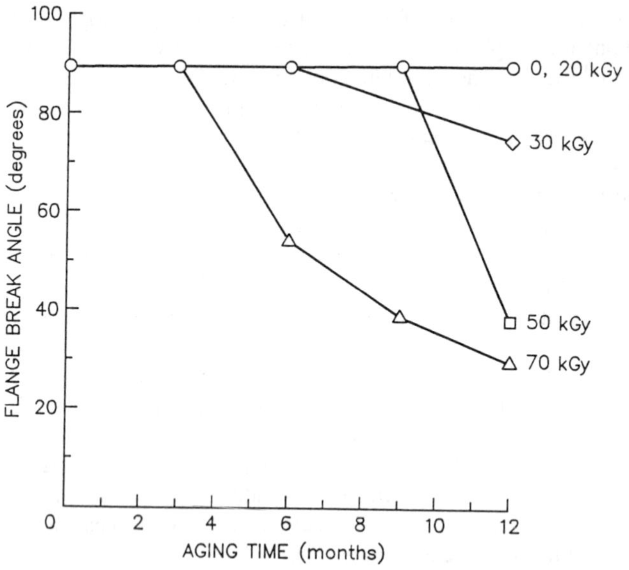

Figure 8. Narrow molecular weight distribution homopolymer with mobilizing additive at ambient conditions.

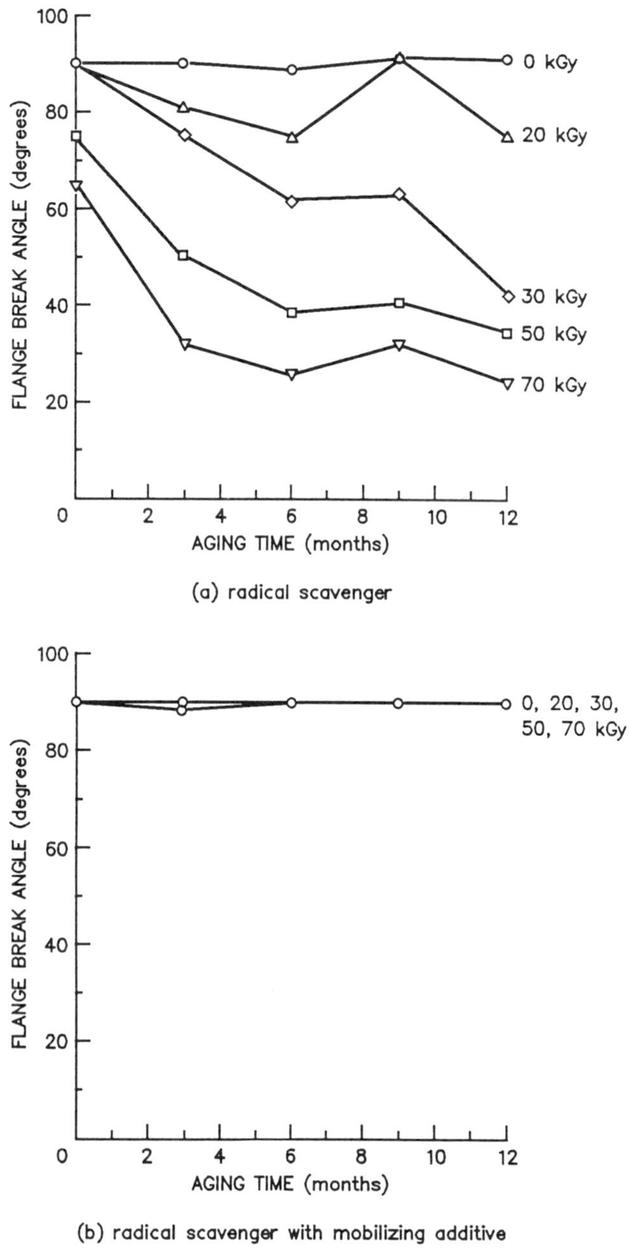

Figure 9. Enhanced radiation stability with addition of a radical scavenger to the polypropylene formulation at ambient conditions.

The future will require continued improvements in materials especially in the polyolefin class which readily degrade during irradiation if not properly stabilized. It is anticipated that better stabilizing systems could be developed.

Literature Cited

1. Skiens, W.E., Williams, J.L. In: *Biocompatible Polymers*; Szycher, Michael, Ed., Technomic Publishing Co.: Westport, CT, 1982.
2. Mucha, M., Kryszewski, M. *Colloid Polym. Sci.* ***1980,*** *258*, pp 743.
3. Buchachenko, A.L. *J. Polym. Sci.*, Polym. Symp. No. 57 **1976**, p 299.
4. Westfahl, J.C., Carman, C.J., Layer, R.W. *Rubber Chem. Technol.* **1972,** *45*, pp 402.
5. Dunn, T.S., Williams, J.L., Sugg, H.W. *Adv. Chem. Ser. 169* **1978,** pp 151.
6. Dunn, T.S., Epperson, B.J., Sugg, H.W., Stannett, V.T., Williams, J.L. *Radiat. Phys. Chem.* **1979,** *14*, pp 625.
7. Adams, J.H. *J. Polym. Sci.*, Part A-1 **1970,** *8*, pp 1077.
8. Kryszewski, M., Nadolski, B., Zujewicz, M. *J. Polym, Sci.,* Polym. Symp. No. 57 **1976,** pp 129.
9. Hori, Y., Shimada, S., Kashiwabara, H. *Polymers* **1977,** *18*, pp 151.
10. Chapiro, A. *Radiation Chemistry of Polymeric Systems*; Interscience Publishers: New York, NY, 1962.
11. Dole, M. *The Radiation Chemistry of Macromolecules*; Academic Press: New York, NY, 1973.
12. Ranby, B.; Rabek, J. *Photodegradation, Photo-oxidation and Photo-stabilization of Polymers*; John Wiley and Sons: New York, NY, 1975.
13. Pospisil, J. *Pure Appl. Chem.* **1973,** *36*, pp 207.
14. Billingham, N.; Prentice, P.; Walker, T. *J. Polym. Sci.,* Polym. Symp. No. 57, **1976,** pp 287.
15. Williams, J.L.; Dunn, T.S.; Stannett, V.T. *Radiat. Phys. Chem.* **1982,** *19*, pp 291.
16. Williams, J.L.; Dunn, T.S.; Stannett, V.T. *U.S. Patent 4 110 185* **1978.**
17. Dunn, T.S.; Williams, E.E.; Williams, J.L. *Radiat. Phys. Chem.* **1982,** *19*, pp 287.
18. Williams, J.L.; Dunn, T.S.; Stannett, V.T. *U.S. Patent 4 274 932* **1981**
19. Roginskii, V.A.; Dubinskii, V.A.; Shlyapnikova, I.A.; Miller, V.B. *Eur. Polym. J.* **1977,** *13*, pp 1043.
20. Rysavy, D. *Polymer* **1967,** *8*, pp 449.
21. de Jonge, C.R.H.I.; van der Maeden, F.P.B.; Biemond, M.E.F.; Huysmans, W.G.B.; Mijs, W.J. *J. Polym. Sci.,* Polym. Symp. No. 57, **1976,** pp 197.
22. Felder, B.; Schumacher, R.; Sitek, F. *ACS Symp. Ser. 151*, **1981,** pp 65.
23. Allen, N.S.; Fatinikun *Polym. Degrad. Stab.* **1980,** *3*, pp 243.
24. Carlsson, D.J.; Chan, K.H.; Wiles, D.M.; *ACS Symp. Ser. 151*, **1981,** pp 51.
25. Jirackova, L.; Pospisil, J. *Angew. Makromol. Chem.* **1979,** *82*, pp 197.
26. de Paolo, P.A.; Smith, H.P. *Adv. Chem. Ser. 85*, **1968,** pp 202.
27. Bentrude, W.G. *Tetrahedron Lett.* **1965,** *40*, pp 3543.
28. Dunn, T.S.; Williams, E.E.; Williams, J.L. *J. Polym. Sci.* **1982,** *20*, pp 1599.
29. Williams, J.L.; Williams, E.E.; Dunn, T.S. *J. Appl. Polym. Sci.* **1982,** *27*, pp 951.
30. Williams, J.L.; Williams, E.E.; Dunn, T.S.; *Radiat. Phys. Chem.* **1982,** *19*, pp 189.

RECEIVED June 28, 1991

Chapter 36

Designing Radiation-Hard Plastic Scintillators

Stan Majewski and Carl Zorn

Physics Division, Continuous Electron Beam Accelerator Facility, 12,000 Jefferson Avenue, Newport News, VA 23606

Plastic scintillators are a ubiquitous component of modern particle physics detectors. With the planned construction of a new generation of high intensity, high energy, colliding beam accelerators, it is known that unprecedented high levels of radiation will be present in the experimental halls. Some components of the detector (specifically those utilizing scintillator) will have to endure annual doses of 10^4 to 10^5 Gy or more. Standard scintillators will not survive such conditions. This is a survey of research started several years ago with the goal of developing plastic scintillators appropriate to such an environment.

Particle physics has the goal of trying to decode the interrelationship of matter and energy at its deepest level. The experimental tools used are essentially huge microprobes that analyze the decay products of immensely high energy processes. Their essential purpose is to (a) track the path of charged particles, (b) identify the particles, and (c) measure the momenta and energy of the particles. The energy of the particles is measured by calorimeters. An ideal calorimeter would be a material that could both absorb and detect the particle's presence. Crystalline scintillators, such as NaI(Tl) are one such device, but are too expensive to be made in the quantities needed, and have too slow a response time (in general). For a variety of reasons, the best compromise has been the *sampling calorimeter*. This device consists of alternate series of a high-Z material (lead is the most common) and a particle-sensitive medium. The former initiates *electromagnetic* or *hadronic* showers in the material. These are the result of the initial particle's interaction with the high-Z material and result in a forward spray of particles carrying the energy and momentum of the incident particle. This spray of particles also passes through the the sensitive medium, resulting in excitation of the material, and

0097–6156/91/0475–0569$06.00/0

subsequent detection of the de-excitation products. A variety of sampling media are possible, but the most popular is the plastic scintillator. It has the advantage of being cheap, fast in response (<5 ns), and is easy to handle. In this case, the particles in the shower excite the plastic resulting in some energy being passed to fluorescent components present in the plastic. The final output is photons, some fraction of which are detected by photo-detectors (usually photomultiplier tubes).

It is seen that the essential purpose of the calorimeter is to absorb an incident particle's energy, and convert it to a detectable quantity. This implies that calorimeters are literally irradiated by the products of the inital reaction region. In the case of the new generation of particle accelerators (SSC in the U.S.A., and LHC in Europe), reliable simulations have shown that such devices will experience annual operating doses that can easily exceed 10^4 Gy *(1,2)*. A variety of evidence indicates that standard plastic scintillators will not be able to perform well under such conditions. For this reason, a research program was initiated with the goal of improving the radiation resistance of plastic scintillators.

Basic Phenomenology

A brief review of the mechanism of scintillation in plastics is useful in formulating a radiation-resistant scintillator. The base plastics polystyrene (PS) and polyvinyltoluene (PVT) are weak fluorescent emitters in the UV (330 nm). By doping the plastic with a primary (e.g., PTP) whose absorption matches the emission band of the plastic (Förster overlap), the emission can be shifted to a longer wavelength (350 nm). At low concentrations of the primary (10^{-4} M), the energy transfer is radiative. By increasing the concentration (10^{-1}-10^{-2} M), a *non-radiative*, Coulombic, dipole-dipole process becomes dominant (Förster transfer). Since the resultant wavelength is too short (resulting in poor coupling to the photo-detector response function and large absorption by the plastic), a secondary fluor is added to shift the final emission to the blue (420 nm). The latter is kept at a small concentration in order to minimize self-absorption (while large enough to maximize light output). Hence the energy transfer between the primary and secondary is necessarily *radiative*.

From the physicist's point of view, the basic phenomenon noted upon severe irradiation of a scintillator (>10^4 Gy) is the loss in light output. The cause of this loss is rooted in two principal components: (i) the permanent destruction of fluorescent components, and (ii) the permanent change in the spectral transmission of the base plastic where the fluor(s) emit. The latter has a subtle aspect to it in that it can simulate the first. A study *(3)* of several common scintillators found that after irradiation, the resultant loss in light output could not be explained by either damage to the fluors (which was minimal) nor to direct absorption of the emission light by the base plastic (as the change in transmission of the scintillator was minimal). Rather, the base plastic was found to have formed new absorption centers deep in the UV corresponding to those wavelengths where the primary fluor radiatively transferred energy to the secondary. Another study *(4)* came to the same conclusion, but by studying radiation damage to liquid scintillators.

There are a variety of parameters of concern when irradiating scintillators, but the three key ones are (i) absorbed dose, (ii) dose rate, and (iii) the presence or absence of oxygen. The effect of the first is obvious, but the latter two are interrelated. Although not all the details have been worked out, there is enough data to indicate the general trend. For some regime of dose rates (>10 Gy/hr), it seems that the observed level of damage is independent of dose rate when irradiations are carried out in air *(5)* for polystyrene-based scintillators. But at very low dose rates, the damage level climbs to very high levels *(6)* (observed in acrylic-based scintillator). At least one reason for this seems to lie in the presence of oxygen. At the low dose rates, oxygen is able to participate fully in the permanently damaging reactions *(7)*. For a variety of reasons, there is a current revived interest in using scintillators in the form of optical fibers. These consist of a doped polystyrene core surrounded by a thin cladding of lower refractive index plastic, namely acrylic. Although it is believed that the polystyrene will probably not be affected additionally at lower dose rates, there is a need to check if the radiation sensitive acrylic cladding has any additional effects upon the optical performance of the fiber *(8)*.

At the higher dose rates, one is able to observe an annealing (or recovery) phenomena. That is, after a high rate irradiation, an observed level of damage is reached. If the material is exposed to oxygen, the damage level *decreases* over a period of time (depending upon temperature, plastic, dye(s), oxygen level, etc.) that is on the order of several weeks or months to a plateau of permanent residual damage. There is experimental evidence to indicate that over some range of dose rates ($10\text{-}10^4$ Gy/hr), the residual damage is independent of dose rate *(9)*. This may even be the case for very low dose rates (say <0.1 Gy/hr) *(9)* for polystyrene-based scintillators.

Improving Radiation Resistance

An examination of the basic mechanism of scintillation provides one method of improving the radiation resistance of the plastic scintillator. Scintillators function by transferring energy from the plastic to a primary fluor (non-radiatively). The primary generally has a long UV emission peak which is easily absorbed by the plastic base. By adding an appropriate secondary fluor, the primary emission is radiatively transferred to the secondary which emits in the blue (a region also matching the photo-detector response function). Since the residual damage in the irradiated plastic is in the form of additional absorption in the UV, this implies that the primary cause for loss in scintillation output is due to the absorption of the primary emission by the additionally absorptive plastic base. This assumes that damage to the fluor(s) is minimal. Several recent studies of radiation damage to fluors *(3,10)* have indicated that the commonly used fluors are highly resistant to radiation damage. Hence the secondary causes of scintillation output losses are damaged dopants and absorption of the secondary fluor's emission by the base plastic.

Once this idea of "hidden absorption" *(3)* by the base plastic is appreciated, the first order solution is clear: replace the two component fluor system with a single large Stokes shift primary with an emission wavelength at least as large as the original system. In fact, one would prefer an even longer range of wavelengths (>500 nm) to avoid the residual absorption seen between 400 and 500 nm. The ideal fluor would have the following characteristics: (a) radiation-hard, (b) high solubility, (c) large Stokes shift (>10,000 cm^{-1}), (d) high quantum efficiency, (e) excellent Förster overlap with the base plastic, (f) low self-absorption, (g) inherently large extinction coefficient, and (h) fast decay time (<5 ns). (It is assumed that all such properties are relative to the dopant being in the specific base matrices of polystyrene or polyvinyltoluene.)

Three such examples have been attempted:(i) a blue emitting (peak=425 nm) fluor PMP (1-phenyl-3-mesityl-2-pyrazoline), (ii) a green-emitting (peak=530 nm) fluor 3HF (3-hydroxyflavone) and (iii) another green-emitting fluor HBT (2-(2′-hydroxyphenyl)benzothiazole). PMP has almost all the required characteristics, except that a study revealed that PMP is radiation soft when compared to standard dopants *(11)*. Similarly, HBT revealed a sensitivity to irradiation *(12)*. The fluor 3HF has been the most successful dopant for creating a more radiation tolerant scintillator *(13,14)* despite some inherent drawbacks, namely a poor quantum efficiency (0.31±0.05), relatively high self-absorption, and mediocre Förster overlap with polystyrene *(12)*, and somewhat slow decay time (8 ns) *(J.M. Flournoy, EG&G Energy Measurements, Inc., private communication)*. One should point out that, in fact, the problem of self-absorption and poor Förster overlap for 3HF has resulted in use of the fluor, not as the primary, but as a secondary with PTP (para-terphenyl) as the primary *(15,16)*. In this way, the brightness of the resultant scintillator is maximized. Hence the improvements in radiation tolerance are due mostly to shifting the final emission beyond most of the additionally induced absorption in the base plastic. A true large Stokes shift system should display considerable improvements over this initial test case. Such systems (with fast decay times) are in development *(12,17-20)*.

Figure 1 displays the effect of shifting the emission wavelength to longer values for plastic scintillators in the form of 1 mm diameter optical fibers (polystyrene core and acrylic cladding). The "shadow" region is a portion of the fiber that was shielded from iradiation for the purpose of determining the intrinsic scintillation loss (see figure 3). The fibers were subjected to a typical accelerated test: a high dose rate irradiation (10-30 Mrad/hr) by a 3 MeV electron beam with the fibers in an inert atmosphere (argon), followed by 2 weeks of recovery in air. Note that attenuation changes are presented here—the measurements were normalized to the end of the fiber abutting the phototube faceplate (PMT). The phototube ends of the fiber were painted with a 1 inch long strip of black paint to absorb the cladding light which makes a significant contribution to the observed signal near the phototube. Furthermore, an appropriate optical filter was used to eliminate the additional contribution from the short wavelength absorption by the plastic base that is present over the first meter of fiber. Excitation of the fiber was accomplished by producing a collimated beam from a small X-ray generator which could be moved along the length of the fiber in a controlled fashion. The fiber in figure 1a is a proprietary fiber *(17)*, the fiber in figure 1b uses PTP as a primary and 3HF as a secondary, while the fiber in figure 1c is doped with rubrene in addition to PTP and 3HF.

Another approach is to search for alternative plastic bases that display greater radiation resistance than the standard bases (PS and PVT). Figure 2 displays two examples of such experimentation with the polymer base to form a more radiation tolerant *blue-emitting* plastic scintillating fiber. (The components of these fibers are proprietary *(17)*). Research is continuing on this approach.

A final word must be said about an ultimate application of this research: the

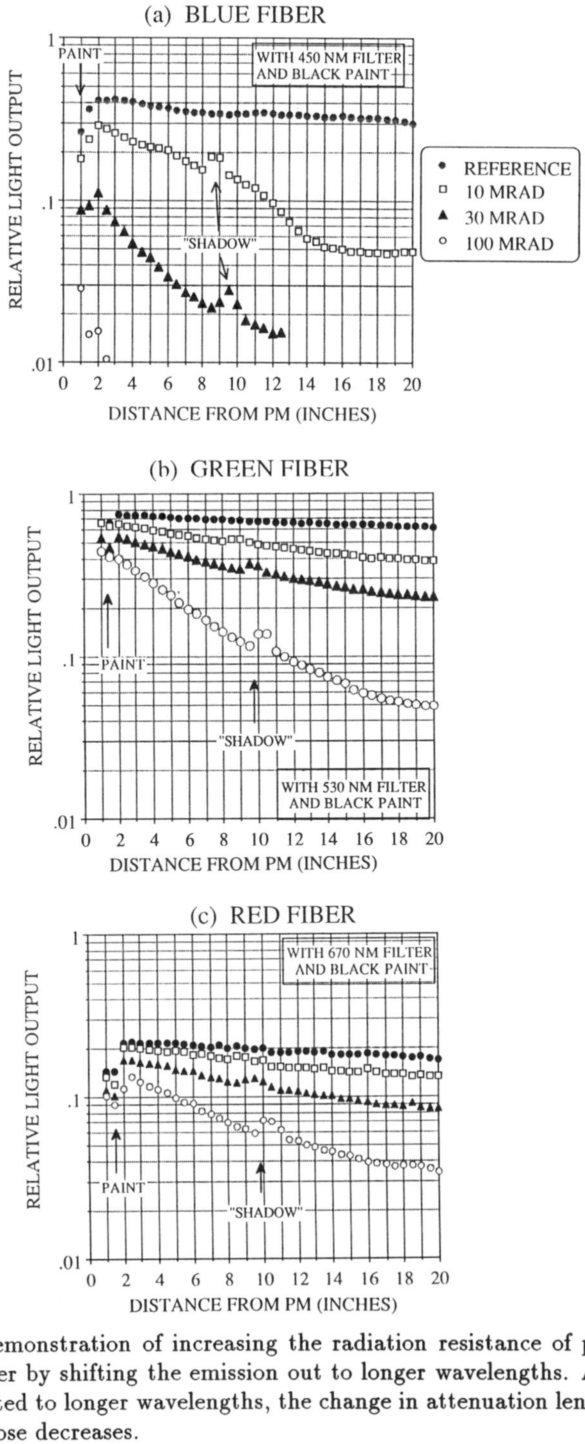

Figure 1 A demonstration of increasing the radiation resistance of plastic scintillating fiber by shifting the emission out to longer wavelengths. As the emission is shifted to longer wavelengths, the change in attenuation length as a function of dose decreases.

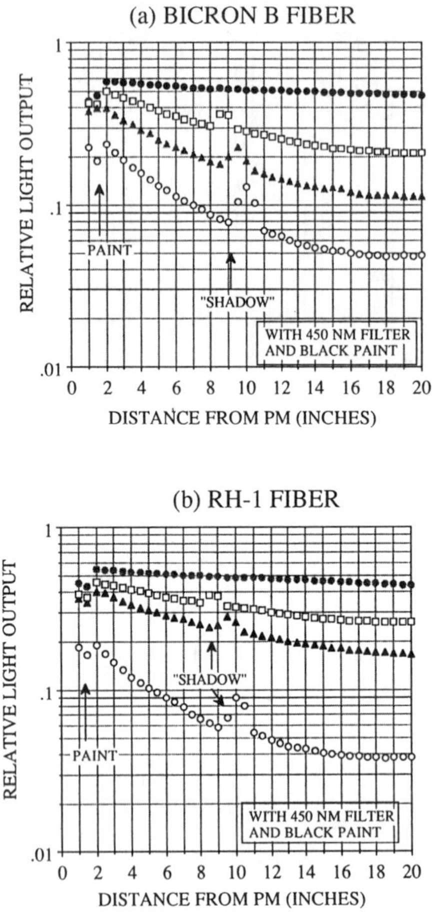

Figure 2 Attenuation curves for prototype radiation hard, *blue emitting* fibers. In this case, experimentation with the polymer base has produced greater radiation tolerance. The legend of figure 1 is applicable here.

use of plastic scintillating fibers in SSC detectors. These would be in the same form as optical fibers: a central (scintillating) core surrounded by a thin cladding of lower refractive index with an overall diameter of 0.5-1.0 mm. As the sampling element of electromagnetic calorimeters, the fibers face their greatest challenge in terms of radiation-resistance. Although most research has concentrated on the problem of improving the fibers' resistance to irradiation via the improvement of the fibers' attenuation characteristics *(15,16)*, recent simulation work *(21)* has indicated that due to the relatively concentrated longitudinal deposition of energy within the calorimeter, the intrinsic scintillation light output losses have the greatest effect upon the performance of the calorimeter. In contrast, with hadronic calorimeters, their fundamentally larger size and the large longitudinal spread of the energy deposition will make it essential to maximize the attenuation performance of the fiber.

The problem of intrinsic light output loss after irradiation translates into two requirements: (i) the use of intrinsically radiation-hard fluors, and (ii) increasing the concentration of the secondary fluor (if no sufficiently large Stokes shift primary is available) in order to reduce the amount of primary emission absorbed by the damaged plastic. Figure 3 displays the result of irradiating several fibers, each with radically different concentrations of the fluor 3HF. A short section of fiber (between 8 and 9 inches from the PMT) was protected by a lead shield during irradiation. This portion remains as "virgin" scintillating fiber. Hence the size of the "step-function" after irradiation indicates the relative loss in intrinsic scintillation efficiency. (Any differences in output among the fibers was normalized out at the start of the step.) As can be seen in figure 3, increasing the concentration of the secondary fluor can significantly decrease this loss in light output. This effect can also be interpreted as due to the saturation in light output with high concentrations of dopant. If one sits on the plateau in light output versus fluor concentration, then destruction of a given amount of fluor will not affect the overall light output as would happen at a small concentration of fluor. Of course, other effects will have to be weighed when using this method, namely, (a) if the destroyed fluor adds additional absorption centers, and (b) the effect of increased self-absorption on the attenuation character of the scintillator due to the higher concentration of dopant.

Summary

[1] Shifting the emission wavelength from the blue to the green via the use of large Stokes shift fluors (most notably 3HF) can significantly improve the radiation tolerance of plastic scintillators in terms of changes in attenuation after irradiation (and recovery).

[2] In addition, increasing the concentration of the fluor can decrease the loss in scintillation efficiency after irradiation. This will have to take into account any self-absorption increase due to the larger concentration.

[3] It may yet be possible to produce radiation-resistant, blue-emitting scintillators by synthesizing radiation-hard plastics that are also suitable as base materials for scintillators. By keeping the emission in the blue, then only standard (and cheap) photo-detectors need be utilized. This also circumvents the problem that the longer wavelength fluors have larger decay times (i.e., the scintillator is slower).

[4] Most tests of radiation tolerance have been based upon accelerated, high dose rate experiments. Although there is good evidence to suggest that for polystyrene-based scintillators, a correspondance can be made from high dose rates to realistic rates, it will be necessary to make low dose rate tests in order to verify this hypothesis, particularly with regard to scintillating optical fibers.

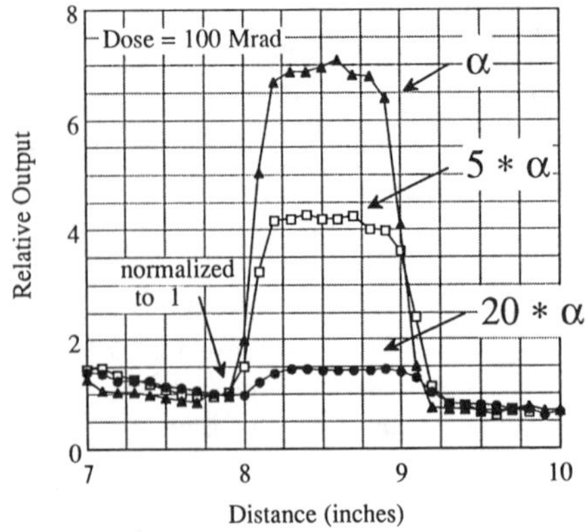

Figure 3 The effect of the concentration α (= 0.01% w/w) of the fluor 3HF (concentration of the primary PTP was fixed) upon the loss in intrinsic scintillation yield after irradiation (plus 2 weeks of recovery in air).

Literature Cited

[1] D.E. Groom, "Radiation Levels in SSC Detectors," *Proceedings of the Summer Study on High Energy Physics in the 1990s*, ed. S. Jensen, (World Scientific, Singapore 1989) pp. 711-716.

[2] A. Van Ginneken, "Estimated Radiation Levels in SSC Detectors," *Radiation Effects at the SSC*, ed. G.D. Gilchriese, SSC report SR-1035, (LBL, Berkeley, 1988) pp. 409-428.

[3] A.D. Bross and A. Pla-Dalmau, "Radiation Induced Hidden Absorption Effects in Polystyrene Based Plastic Scintillator," *these proceedings*.

[4] C. Zorn et al., *IEEE Trans. Nucl. Sc.* **NS-37** (1990).

[5] U. Hölm and K. Wick, *IEEE Trans. Nucl. Sc.* **NS-36** (1989) 579.

[6] Y. Sirois and R. Wigmans, *Nucl. Instr. Meth.* **A240** (1985) 262.

[7] R. Clough, "Radiation-Resistant Polymers," *Encyclopedia of Polymer Science and Engineering*, Vol.13, (John Wiley and Sons, 1988) pp. 667-708.

[8] C. Zorn et al.,"Results of a Preliminary Low Dose Rate Irradiation of Selected Plastic Scintillating Fibers," contribution to the Symposium on Detector Research and Development for the Superconducting Supercollider, Fort Worth, TX, October 15-18, 1990 (to be published).

[9] B. Bicken et al.,"Recovery and Permanent Radiation Damage of Plastic Scintillators at Different Dose Rates," contribution to the 1990 IEEE Nuclear Science Symposium, Arlington, VA, October 23-27 1990 (to be published).

[10] R. Clough et al.,"Radiation Degradation of Organic Scintillators," see reference [8].

[11] S. Majewski et al., *Nucl. Instr. Meth.* **A281** (1989) 497.

[12] A. Pla-Dalmau,"Design of Fluorescent Compounds for Scintillation Detection," doctoral dissertation, Northern Illinois University (1990).

[13] C. Zorn et al., *Nucl. Instr. Meth.* **A273** (1988) 108.

[14] C. Zorn et al., *Nucl. Instr. Meth.* **A276** (1989) 58.

[15] S. Majewski et al., "Comparative Study of the Radiation Resistance of Selected Plastic Scintillating Fibers," *Proceedings of the ECFA Study Week on Instrumentation Technology for High-Luminosity Hadron Colliders*, ed. E. Fernandez, G. Jarlskog, CERN report 89-10 (1989).
[16] K.F. Johnson et al., *IEEE Trans. Nucl. Sc.* **NS-37** (1990).
[17] Bicron Corporation, Newbury, Ohio.
[18] A.D. Bross et al.,"New Scintillation Materials for Fiber Tracking and Calorimetry," see reference [9].
[19] J. Kauffman, "Design of Radiation-Hard Fluors," *Proceedings of the Workshop on Scintillating Fiber Detector Development for the SSC*, Fermilab, Batavia, IL, Nov. 1988, pp. 675-740.
[20] J. Kauffman,"Progress on Design of Radiation-Hard Fluors," contribution to the Workshop on Radiation Hardness of Plastic Scintillator, Florida State University, Tallahassee, FL, March, 1990 (in press).
[21] H. Paar, "Status of Scintillating Fiber Calorimetry," see ref. [15], pp. 207-230.

RECEIVED January 29, 1991

Chapter 37

Radiation-Induced Hidden-Absorption Effects in Polystyrene-Based Plastic Scintillators

A. D. Bross and A. Pla-Dalmau

Fermi National Accelerator Laboratory, P.O. Box 500, Batavia, IL 60510

Radiation damage studies were performed on polystyrene based plastic scintillators. Three multi-component systems, each containing two organic fluorescent compounds in a polystyrene matrix, were analyzed. In these systems, p-terphenyl was used as the primary dopant with either dimethyl-POPOP, 3HF, or BBQ as the secondary dopant. In addition, undoped polystyrene samples were also irradiated. Absorbance, fluorescence, and beta excitation measurements were carried out on each sample before, immediately after, and two weeks after irradiation. Radiation-induced absorption in the polymer was detected in the 300–400 nm spectral region. However, no substantial damage was noted in the dopants for scintillator samples irradiated at 10 Mrad.

Plastic scintillators have been commonly used as particle detectors in nuclear and high energy physics experiments. They are based on a polymer matrix typically doped with two fluorescent compounds referred to as the primary dopant and the secondary dopant or wavelength shifter [1]. Plastic scintillators can be fabricated into many forms such as sheets, rods, and even optical fibers rendering them suitable for many detector geometries. In addition, they exhibit a short fluorescence decay time which allows them to be used in high count rate experiments. Currently, their main drawback is their susceptibililty to radiation damage [2].

This study focuses on the radiation-induced damage to the polymer base and its effect on the scintillation light. The irradiation results will show how damage to the polymer matrix plays a major role in the scintillation light losses for the systems studied. Damage to the dopants does not appear to be the major contributing effect in these systems.

0097–6156/91/0475–0578$06.00/0

Experimental Section

Styrene was purified utilizing a *tert*-butylcatechol removal column supplied by Aldrich Chemical Co. and was then vacuum distilled. Three multi-component samples were prepared, each consisting of a styrene solution doped with p-terphenyl (1.25% by weight) and a wavelength shifter (0.01% by weight). In this study, the wavelength shifters were 1,4-bis-2-(4-methyl-5-phenyl-oxazolyl)benzene (dimethyl-POPOP or DMPOPOP), 3-hydroxyflavone (3HF), and 7H-benzimidazo[2,1–a]benz[de]isoquinoline-7-one (BBQ). Samples containing a single dopant in a styrene solution were also made. The concentration of each dopant was the same as in the multi-component samples just mentioned, i.e., 1.25% for p-terphenyl, and 0.01% for each of the wavelength shifters. In addition, solutions of undoped styrene were also prepared. The solutions were placed in Pyrex test tubes, degassed by repeated freeze-pump-thaw cycles, and then polymerized in a silicone oil bath. The polymerization cycle was 24 h. at 110 °C, 48 h. at 125 °C, 12 h. at 140 °C, and then a ramp down at 10 °C/h. to 90 °C. The polymerized samples were machined into 1 cm thick disks and then polished.

The samples that were then to be irradiated were first placed in stainless steel cans. The cans were then evacuated for a period of a week in order to remove all moisture and dissolved gases from the samples. Finally, the cans were back-filled with dry nitrogen and sealed. The samples were irradiated at the Nuclear Reactor Laboratory of the University of Michigan using a ^{60}Co source. The doped samples were irradiated to an integrated dose of 10 Mrad and the pure polystyrene samples to doses of 1, 5, 10, 50, and 100 Mrad. In all exposures, the dose rate was approximately 1 Mrad/h.

Absorbance, fluorescence, and beta excitation measurements were performed before, immediately after, and two weeks after irradiation. This last measurement was recorded after the samples were annealed in an oxygen atmosphere. Absorption/transmittance and fluorescence spectra were recorded using a Hewlett-Packard 8451A spectrophotometer. For all absorption/transmittance measurements, unirradiated pure polystyrene was used as the reference. The light yield measurements used a ^{207}Bi beta source (1 MeV electrons) for excitation. The scintillation light was viewed by a Hamamatsu R669 photomultiplier tube which was coupled directly to a Lecroy Model 3001 qVt multi-channel analyzer.

Results and Discussion

Irradiation of pure polystyrene. Pure polystyrene samples were irradiated at five dose levels. Immediately after irradiation the transmittance spectra (Figure 1) show a significant increase in absorption in the 350–600 nm range. All spectra indicate the formation of an absorption band at 525 nm whose strength increases with increasing radiation levels. After these measurements, the samples were annealed in oxygen for two weeks. The transmittance spectra recorded after the annealing process (Figure 2) prove that the samples undergo a substantial

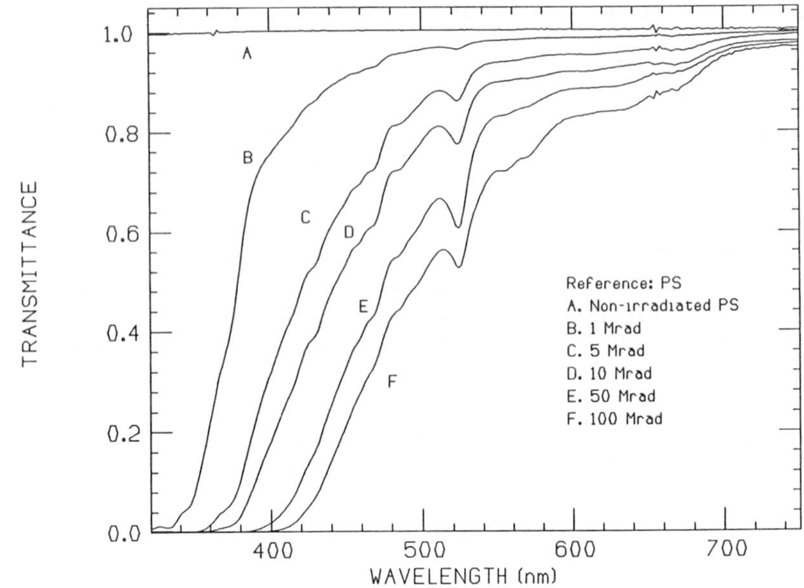

Figure 1. Transmittance data for pure polystyrene immediately after irradiation.

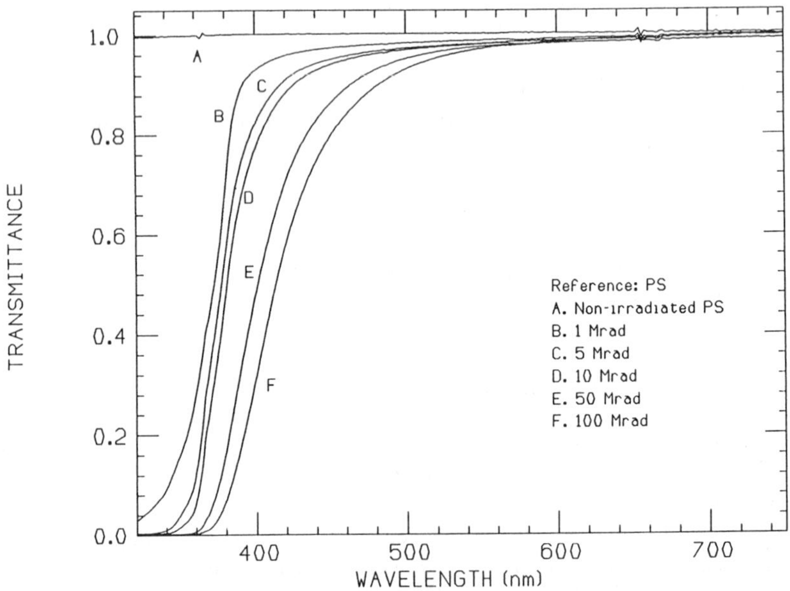

Figure 2. Transmittance data for irradiated pure polystyrene after the annealing process.

recovery, particularly for those irradiated at higher doses. However, considerable permanent absorption does remain in the polymer as a result of the radiation.

Another prominent feature is observed in the samples immediately after irradiation. When excited using 313 nm light, the samples fluoresce in the wavelength region between 500 and 600 nm (Figure 3). Before irradiation, polystyrene shows little fluorescence at wavelengths longer than 320 nm when excited with 313 nm light. Unirradiated polymer has a fluorescence peak at approximately 320 nm. After irradiation, the normal 320 nm fluorescence appears at longer wavelengths (380 nm), probably due to the increase in conjugation of the polymer. In addition, a new fluorescence band is detected at 550 nm. This fluorescence decreases as the absorption band at 525nm disappears. Exciting this species with 500 nm light yields a fluorescence spectrum (Figure 4) that agrees with the 550 nm peak seen in Figure 3. Irradiation favors the formation of several excited species. Among these new species, one is clearly detected. It absorbs at 525 nm, fluoresces at 550 nm, and is long-lived. This excited species is observed for several weeks if the sample is kept in a nitrogen atmosphere. If the sample is placed in an oxygen atmosphere, however, the diffusion of oxygen into the sample appears to quench this fluorescent species instantly. The oxygen induced quenching of this species can be monitored under UV light. We observed that the volume within the sample that continued to exhibit green fluorescence decreases with time when the sample remained in an oxygen environment.

Pulse radiolysis measurements on polystyrene films and in cyclohexane solution [3, 4] have assigned a similar absorption to singlet excimers of polystyrene. However, this species is short-lived. Other absorptions detected in this region have been assigned to charge transfer complexes with longer decay times.

Irradiation of doped polystyrene. The three doped polystyrene samples were irradiated to a dose of 10 Mrad. Figures 5-7 give transmittance data for these three scintillator samples before irradiation, immediately after irradiation, and after a two week oxygen anneal. For reference, the fluorescence distribution for the sample is also included with its transmittance data. As in the case of the pure polymer, we also see significant annealing in these samples. In addition, the induced absorption seen at 525 nm in the undoped polymer is also present in the doped samples.

Light yield measurements were performed on these samples using beta excitation as described in Section 2. The reduction in light output after annealing was 19±1.5% for the DMPOPOP sample, 22±1.5% for the 3HF sample, and 19±1.5% for the BBQ sample . However, looking at the transmittance data in Figures 5-7, we see that transmission losses cannot account for all of the light loss in any of the three samples. Transmission loss can only account for a 3±2% loss in the DMPOPOP sample, 1±2% loss in the 3HF sample, and a 6±2% loss in the BBQ sample.

One possible explanation for the observed light loss is that the dopants are affected by the radiation. In order to study this, we used front-surface UV excitation to directly stimulate fluorescence of the various dopants. Samples of polystyrene doped only with a single compound were prepared so that we could

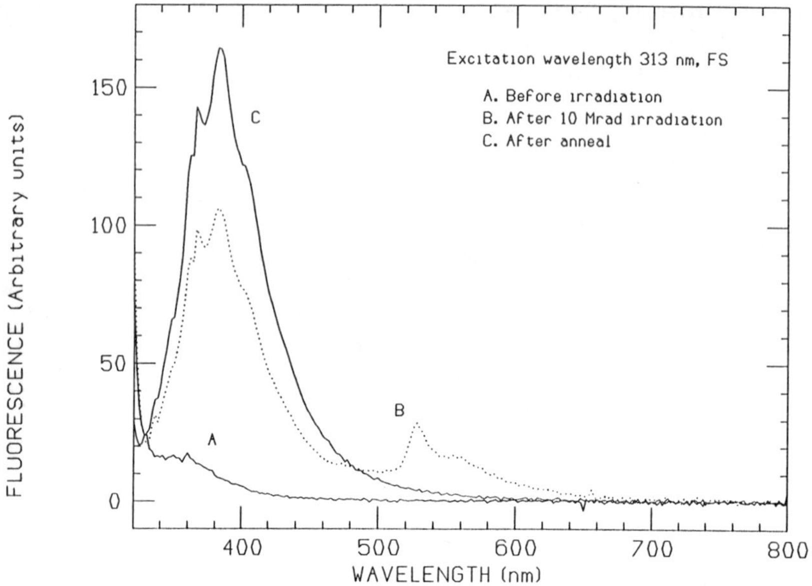

Figure 3. Fluorescence spectra of pure polystyrene: before (A), immediately after(B), and two weeks after 10 Mrad irradiation (C). Excitation wavelength 313 nm.

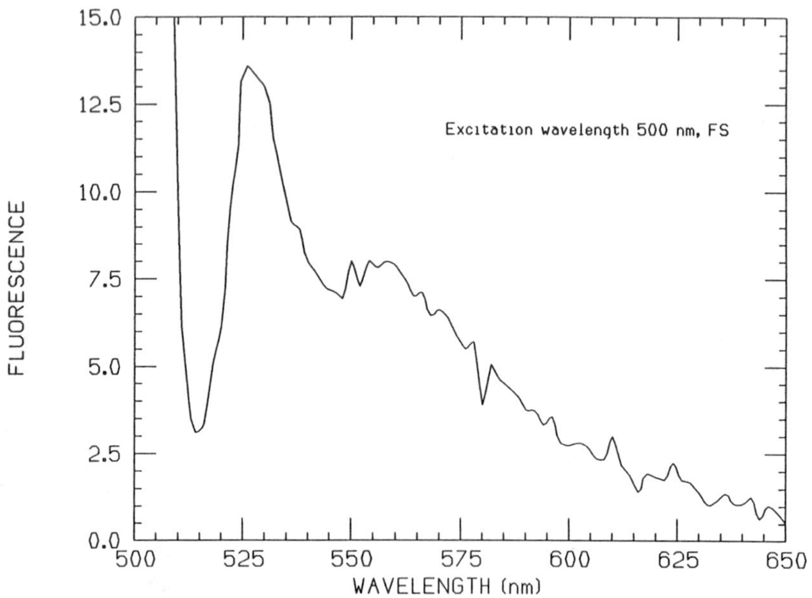

Figure 4. Fluorescence spectrum of irradiated pure polystyrene. Excitation wavelength 500 nm.

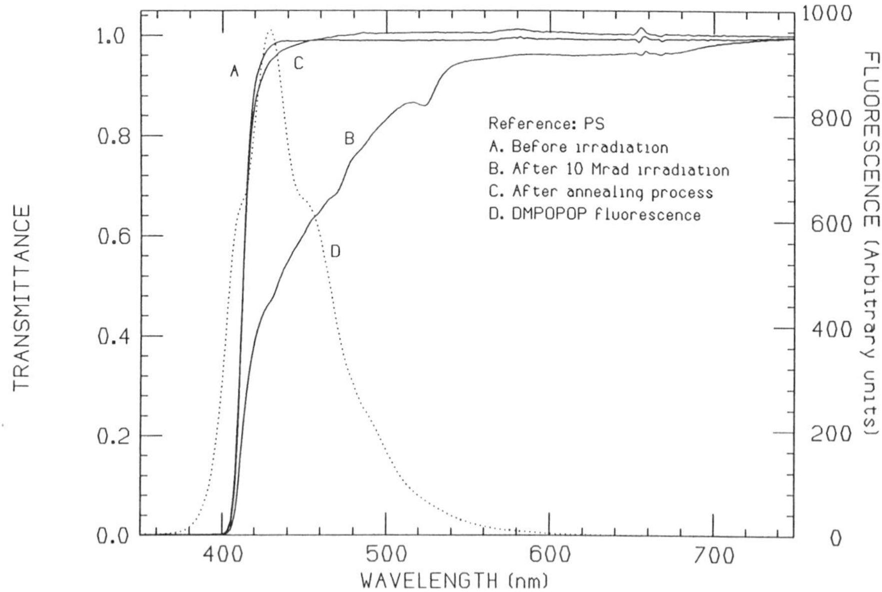

Figure 5. Transmittance data for the p-terphenyl plus DMPOPOP sample: before (A), immediately after (B), and two weeks after 10 Mrad irradiation (C). Dotted line indicates DMPOPOP fluorescence.

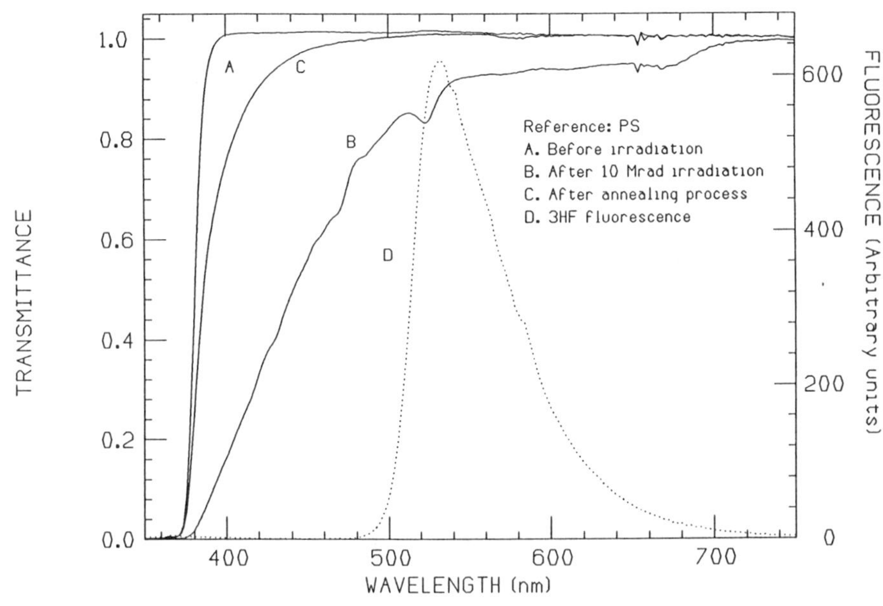

Figure 6. Transmittance data for the p-terphenyl plus 3HF sample: before (A), immediately after (B), and two weeks after 10 Mrad irradiation (C). Dotted line indicates 3HF fluorescence.

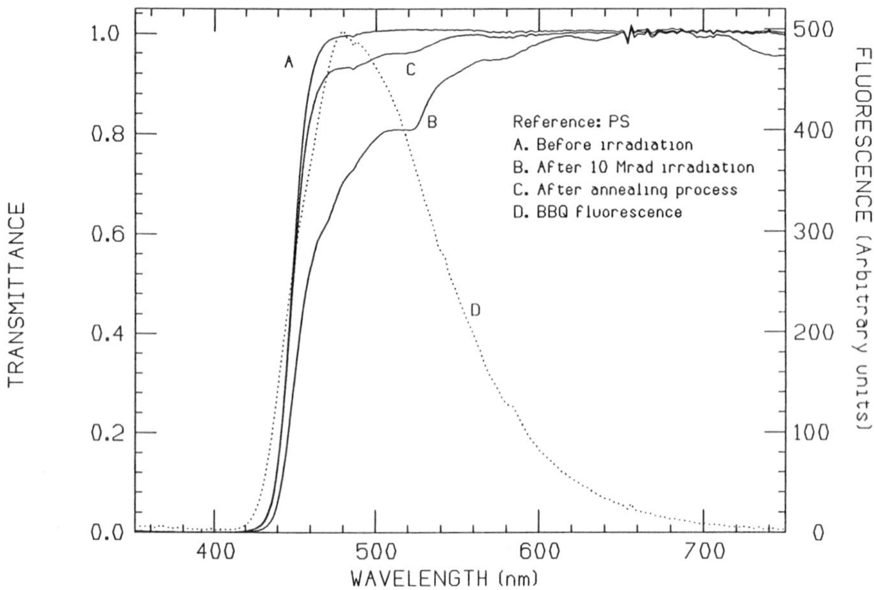

Figure 7. Transmittance data for the p-terphenyl plus BBQ sample: before (A), immediately after (B), and two weeks after 10 Mrad irradiation (C). Dotted line indicates BBQ fluorescence.

study the primary and the various secondaries independently. Figure 8 shows the fluorescence from p-terphenyl in a sample with 1.25% p-terphenyl in polystyrene. Its fluorescence distribution is centered at approximately 350 nm. The excitation wavelength was 313 nm. The data show the fluorescence distributions before irradiation and after irradiation (10 Mrad) and annealing of the sample. As can be seen, there is little change in the fluorescence. The integrated intensity changes by 2±2%. The p-terphenyl fluorescence area is integrated between 320–380 nm since after irradiation the polymer fluorescence is detected at 380 nm. The intensity of the excitation source was monitored by using a fluorescence standard and the curves in Figure 8 have been corrected for changes in the excitation source intensity. Figures 9-11 show the same data for samples prepared with 0.01% of the three secondaries, DMPOPOP, 3HF, and BBQ, in polystyrene. In all three figures we see a pronounced change in fluorescence intensity after the 10 Mrad irradiation and anneal. The drop in integrated fluorescence yield for DMPOPOP, 3HF, and BBQ is 42±2%, 51±2%, and 43±2% respectively. However, we also see evidence for polystyrene fluorescence. In the data for 3HF and BBQ, the 380 nm fluorescence seen in the irradiated pure polystyrene samples (Figure 3) is also clearly evident. In the DMPOPOP sample, the polystyrene fluorescence and the fluorescence from DMPOPOP overlap, so no clear second peak is seen. However, we do see increased fluorescence after irradiation in the band between 320 and 395 nm. Again, this would indicate a contribution from polystyrene fluorescence. The radiation induced absorption in polystyrene at 313 nm is competing with absorption by the secondary. Figure 12 shows absorption data for samples of unirradiated pure polystyrene (reference: air), irradiated polystyrene, and unirradiated polystyrene doped with DMPOPOP, 3HF, and BBQ. As can be seen from these data, polystyrene absorption in the region between 320 and 380 nm has increased significantly after irradiation and does compete favorably with absorption by the secondaries. This effect does not eliminate the possibility, however, that some of the drop in light yield for the three secondaries could still be due to radiation induced degradation of the dopant.

In order to explore this in more detail, additional fluorescence measurements were performed using excitation light of longer wavelength. In these cases, the wavelength of the excitation light is less strongly attenuated by the radiation induced absorption in the polystyrene. Figure 13 gives the fluorescence data for DMPOPOP doped polystyrene using $\lambda_{ex} = 380$ nm. The difference in integrated intensity is now only 10%. Figure 14 shows data for the 3HF sample with $\lambda_{ex} =$ 360 nm and Figure 15 shows data for the BBQ sample with $\lambda_{ex} = 450$ nm. The loss in integrated intensity is 22% and 4% for 3HF and BBQ respectively. For the 3HF case with $\lambda_{ex} = 360$ nm and to a lesser extent for the DMPOPOP case with $\lambda_{ex} = 380$ nm, polymer absorption is still significant (Figure 12.) The 22% and 10% drops in intensity can, therefore, only be interpreted as upper bounds on the light loss due to dopant degradation. The actual losses are likely much less since we see such dramatic changes in the loss ratios when 313 nm excitation light is used versus 360 nm or 380 nm light. Polymer absorption in the wavelength region of p-terphenyl fluorescence competes with the transfer between the primary

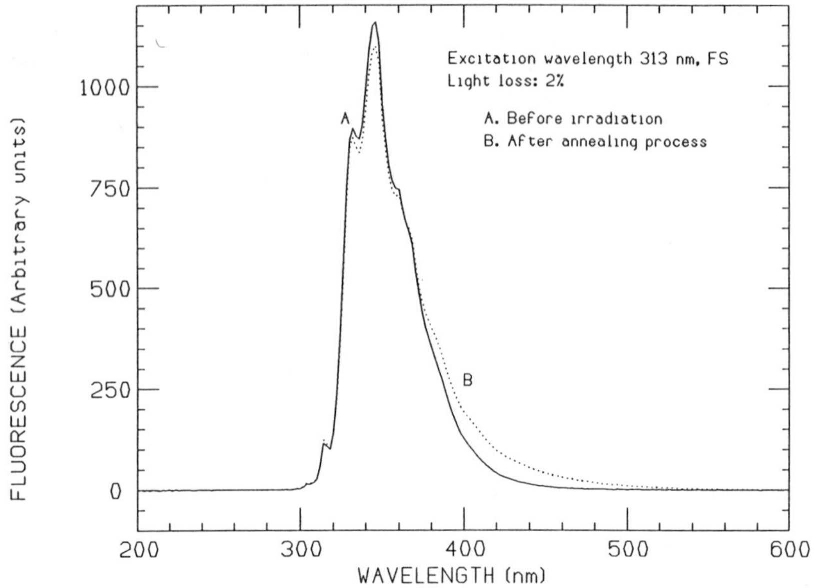

Figure 8. p-Terphenyl fluorescence in polystyrene: before 10 Mrad irradiation (A) and after the annealing process (B). Excitation wavelength 313 nm.

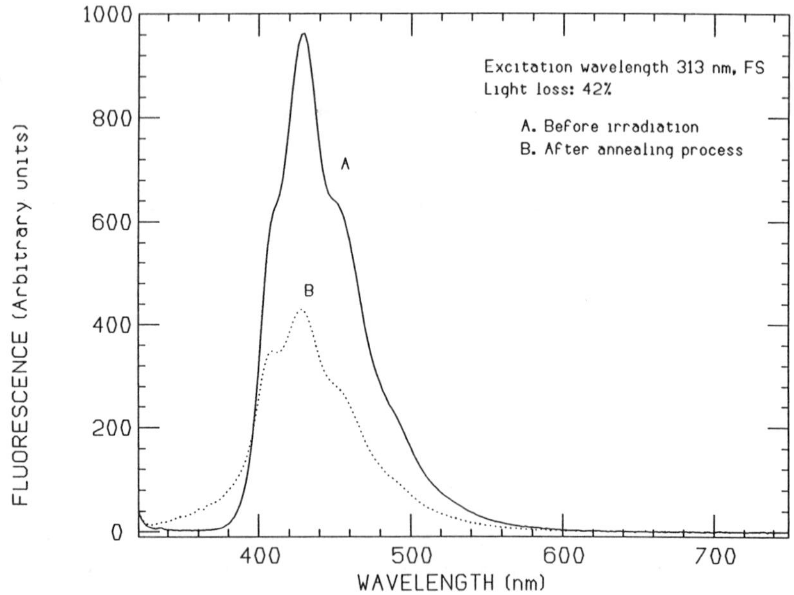

Figure 9. DMPOPOP fluorescence in polystyrene: before 10 Mrad irradiation (A) and after the annealing process (B). Excitation wavelength 313 nm.

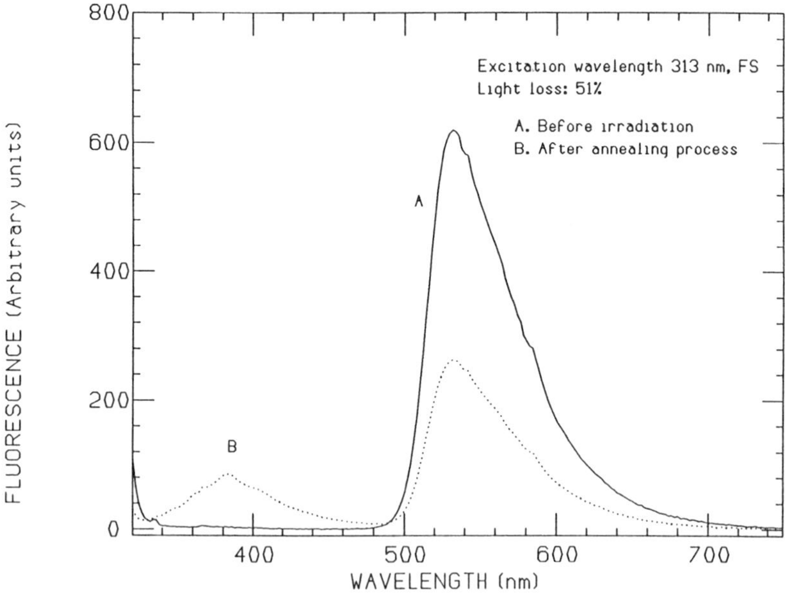

Figure 10. 3HF fluorescence in polystyrene: before 10 Mrad irradiation (A) and after the annealing process (B). Excitation wavelength 313 nm.

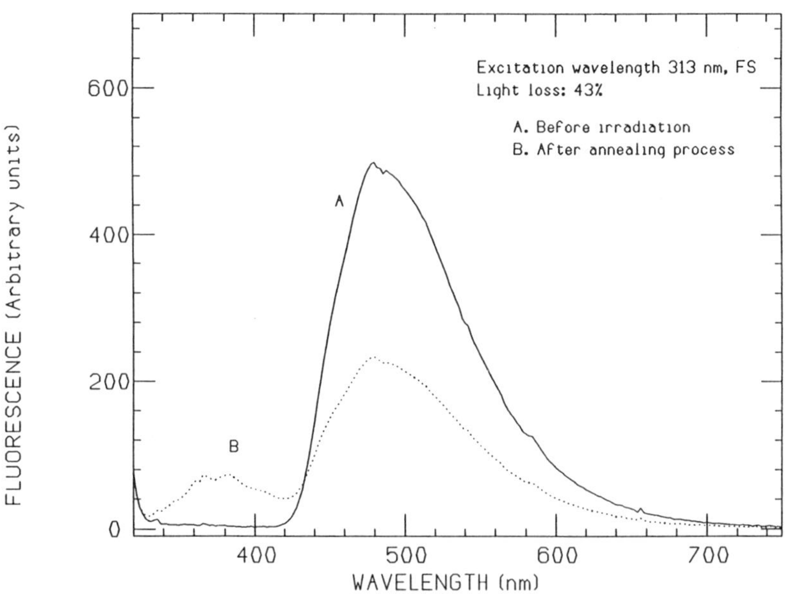

Figure 11. BBQ fluorescence in polystyrene: before 10 Mrad irradiation (A) and after the annealing process (B). Excitation wavelength 313 nm.

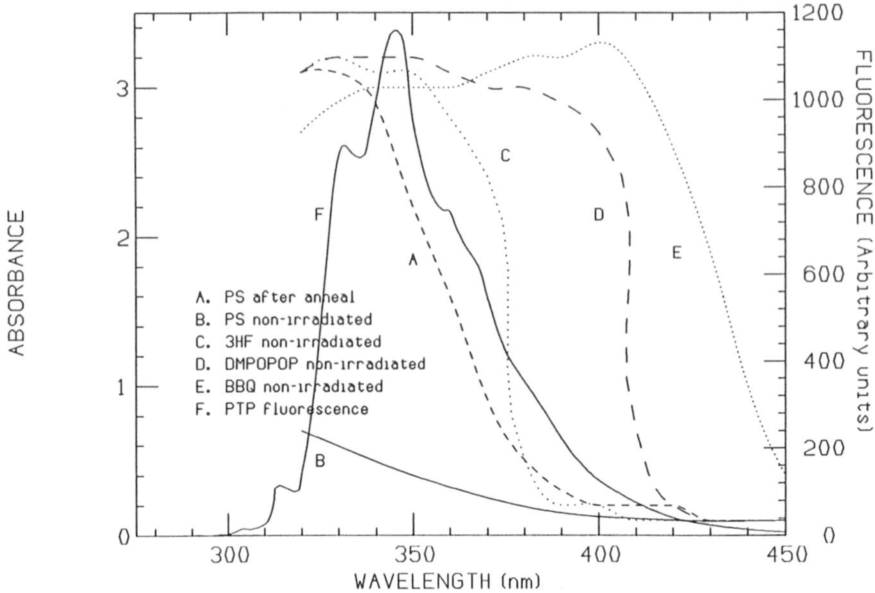

Figure 12. Absorption spectra of irradiated polystyrene (A) and non-irradiated samples of polystyrene (B), 3HF (C), DMPOPOP (D), and BBQ (E) in the region of p-terphenyl fluorescence (F).

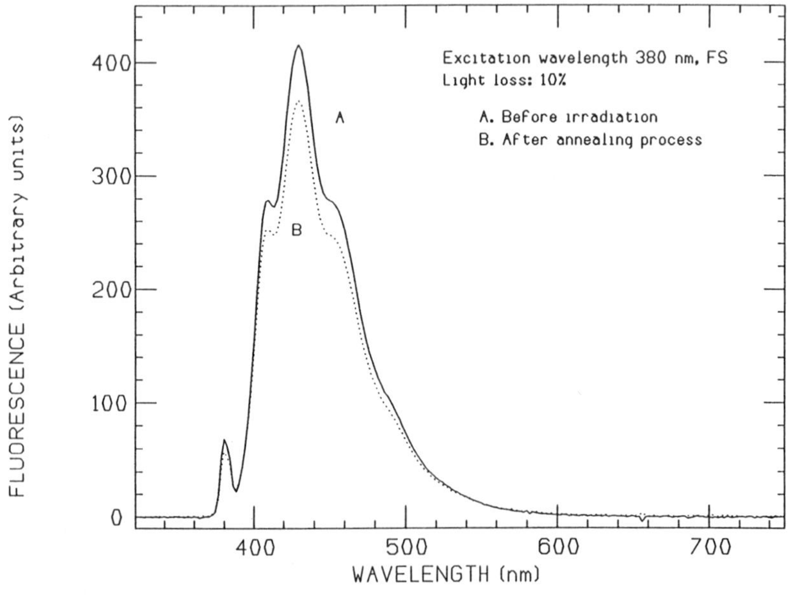

Figure 13. DMPOPOP fluorescence in polystyrene: before 10 Mrad irradiation (A) and after the annealing process (B). Excitation wavelength 380 nm.

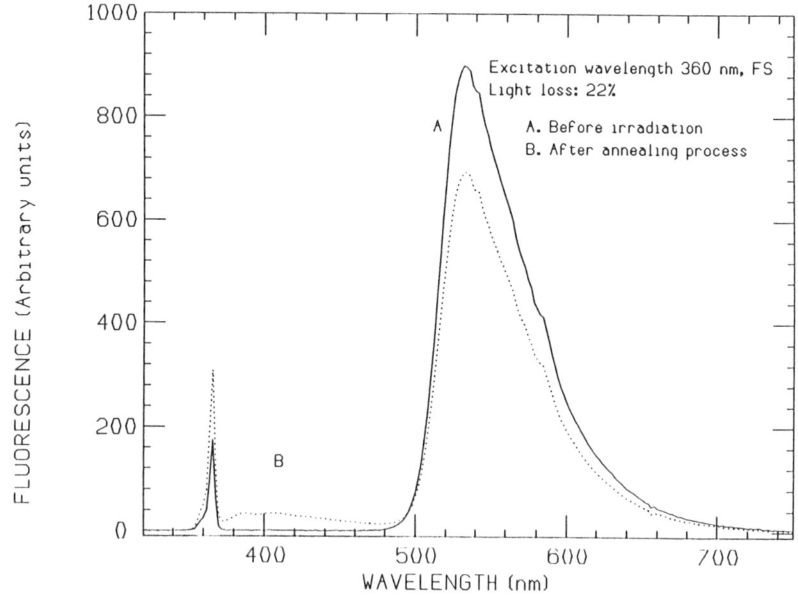

Figure 14. 3HF fluorescence in polystyrene: before 10 Mrad irradiation (A) and after the annealing process (B). Excitation wavelength 360 nm.

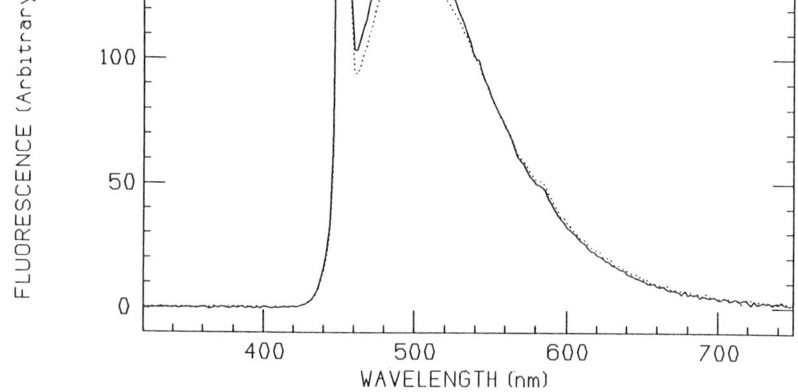

Figure 15. BBQ fluorescence in polystyrene: before 10 Mrad irradiation (A) and after the annealing process (B). Excitation wavelength 450 nm.

(p-terphenyl) and the secondary. One can estimate the size of this effect by comparing the fluorescence distributions (using $\lambda_{ex} = 350$ nm) before and after irradiation for a given secondary dopant. The loss factors are 20±2%, 26±2%, and 22±2% for DMPOPOP, 3HF, and BBQ respectively.

Conclusions

The light loss that was measured in the three test scintillators using beta excitation (Section 3) can be explained by the following mechanism. Upon excitation by ionizing radiation, energy is transferred in a plastic scintillator in a two step process. First, polystyrene excitation is transferred to p-terphenyl via Förster transfer. This is then followed by p-terphenyl fluorescence and then by reabsorption and emission by the secondary. Polymer absorption in the wavelength region of p-terphenyl fluorescence (330–380 nm) competes with the transfer between the p-terphenyl and the secondary. The radiation-induced polymer absorption, as represented by the losses measured for the fluorescence intensity of the three secondaries under 350 nm excitation, can completely account for the light loss observed in the three test scintillators under beta excitation.

References

[1] Birks, J. B. *The Theory and Practice of Scintillation Counting*; Pergamon Press, The McMillan Co.: New York, N. Y., 1964.

[2] Marini, G. *et. al.*, *Radiation Damage to Organic Scintillation Materials*, CERN pub. 85-08, 1985.

[3] Washio, M.; Tagawa, S.; Tabata, Y. *Radiat. Phys. Chem.* **1983**, *21*, 239.

[4] Tabata, Y.; Tagawa, S.; Washio, M. In *Materials for Microlithography*; Thompson, L. F.; Wilson, C. G.; Frechet, J. M. J., Eds., ACS Symposium Series; American Chemical Society: Washington, D. C., 1984, Vol. 266; pp. 151-163.

RECEIVED January 14, 1991

Chapter 38

Radiation-Degradation Studies of the Mechanical Properties of Polymer Matrix Composites Used in Fusion Magnets

S. Egusa[1], Tadao Seguchi[1], M. Hagiwara[1], H. Nakajima[2], S. Shimamoto[2], M. A. Kirk[3], and R. C. Birtcher[3]

[1]Takasaki Radiation Chemistry Research Establishment, Japan Atomic Energy Research Institute, Takasaki-shi, Gunma, 370–12 Japan
[2]Naka Fusion Research Establishment, Japan Atomic Energy Research Institute, Naka-machi, Naka-gun, Ibaraki, 311–02 Japan
[3]Materials Science and Technology Division, Argonne National Laboratory, Argonne, IL 60439

When polymer matrix composites are used in superconducting magnets for fusion reactors, one of the most serious concerns is their resistance to changes in mechanical and electrical properties upon simultaneous irradiation with neutrons and γ-rays at 4.2 K. This chapter mainly describes the mechanical properties of polymer matrix composites tested at 77 K, 4.2 K, and at room temperature after ^{60}Co γ-ray and neutron irradiations at room temperature and at 5 K. Based on the degradation behavior, the mechanisms underlying the irradiation effects on polymer matrix composites are discussed with respect to factors such as composite type, test temperature, radiation type, and irradiation temperature.

In the construction of superconducting magnets for Tokamak and Mirror type fusion reactors, large amounts of polymer matrix composites are used as mechanical supporters and as electrical and thermal insulators. The magnets will be subjected to substantial quantities of neutrons and γ-rays during the fusion-reactor operation, thus leading to significant degradation of the magnet component materials such as insulators, stabilizers, and superconductors (1-4). Probably the degradation is most serious for composite organic insulators, because organic materials are usually less radiation resistant than inorganic materials (5). Thus the operating lifetime of the fusion magnets may be virtually determined by the radiation resistance of the composite insulators.

From this point of view, several groups of workers have studied the irradiation effects on the mechanical and electrical properties of polymer matrix composites (6-11). The authors' group has also started such studies in 1983, mainly with regard to the composite mechanical properties (12-24). This chapter presents a review on

0097–6156/91/0475–0591$06.00/0

the progress made by the authors in understanding the irradiation effects at a fundamental level. This chapter mainly describes the mechanical properties of polymer matrix composites tested at 77 K, 4.2 K, and at room temperature after ^{60}Co γ-ray and neutron irradiations at room temperature and at 5 K. Based on the degradation behavior, the mechanisms underlying the irradiation effects on polymer matrix composites are discussed with respect to factors such as composite type, test temperature, radiation type, and irradiation temperature.

Experimental Section

Most composites studied here were specially prepared using the reinforcing fabrics shown in Table I, and matrix resins of epoxy and polyimide. The epoxy resin was tetraglycidyl diaminodiphenyl methane (TGDDM) cured with diamino diphenyl sulfone (DDS), or diglycidyl ether of bisphenol A (DGEBA) cured with diamino diphenyl methane (DDM). The polyimide resin was polyaminobismaleimide (Kerimid 601). The composites thus prepared are shown in Table II.

Commercially available E-glass fiber composites such as G-10CR, G-11CR (Spaulding Fibre Company, Inc.), and TIL-G1000 ('TORAY') were also studied. The matrix resin in G-10CR was a solid DGEBA cured

Table I. Plain-Woven Glass Fabrics Used as Reinforcing Filler

Fabric	Fiber Type[a]	Fiber Diameter (μm)	Number of Fibers in a Yarn	Number of Yarns per 25 mm	
				Warp	Weft
KS-1210	E-Glass	7	200	53	48
KS-1600	E-Glass	9	400	41	32
WTX-116E	T-Glass	7	200	60	58
WTA-18W	T-Glass	9	400	44	34

[a]E-glass composition (wt%): SiO_2(55.2), Al_2O_3(14.8), CaO(18.7), MgO(3.3), B_2O_3(7.3), Na_2O+K_2O(0.5), Fe_2O_3(0.3), F_2(0.3), TiO_2(0.1). T-glass composition (wt%): SiO_2(65), Al_2O_3(23), CaO(<0.01), MgO(11), B_2O_3(<0.01), Na_2O+K_2O(<0.1), Fe_2O_3(<0.1), Zr_2O_3(<1.0).

Table II. Polymer Matrix Composites Prepared for This Study

Composite	Reinforcing Fabric	Matrix Resin	Volume Fraction of Fibers (%)	Average Thickness (mm)
Glass/Epoxy I	KS-1210	TGDDM/DDS	63	1.81±0.07
Glass/Epoxy II	KS-1600	TGDDM/DDS	66	1.98±0.04
Glass/Epoxy III	WTX-116E	TGDDM/DDS	63	2.12±0.01
Glass/Epoxy IV	WTA-18W	TGDDM/DDS	58	1.99±0.02
Glass/Epoxy V	KS-1210	DGEBA/DDM	55	2.44±0.05
Glass/Polyimide I	KS-1210	Kerimid 601	62	2.05±0.06
Glass/Polyimide II	WTX-116E	Kerimid 601	50	1.99±0.03
Glass/Polyimide III	WTA-18W	Kerimid 601	60	1.96±0.01

with dicyanodiamide, while that in G-11CR was a liquid DGEBA cured with DDS. The matrix resin in TIL-G1000 was a polyimide. The 1.7-3.2 mm thick sheets of the composite samples were cut into rectangular specimens of 6.4 mm width and 70 mm length. The cutting was done such that the 70 mm axis was in a 0° or 45° orientation with respect to the warp and weft, thus obtaining 0° or 45° specimens.

^{60}Co γ-ray irradiations were carried out in air or in argon at room temperature with dose rates of 0.012-0.017 MGy/hr. For some specimens, 2 MeV electrons from a Cockcroft-Walton type accelerator were used in place of γ-rays. In this chapter, however, no distinction between γ-rays and electrons is drawn because the two types of radiation are practically equivalent so far as the composite mechanical properties are concerned (22,24). Neutron irradiations were carried out in the Intense Pulsed Neutron Source (IPNS) at Argonne National Laboratory (25). The in-air room-temperature irradiations were made in horizontal thimbles called "Rabbit" and "H2", while the 5 K irradiations in liquid helium were made in a vertical thimble called "VT2".

Three-point bend tests were conducted at 77 K, 4.2 K, and at room temperature. For specimens irradiated at 5 K with neutrons, the tests were performed after warmup to room temperature. The failure tests were made at a crosshead speed of 0.5 or 0.6 mm/min with a span length of 20 or 24 mm. The tests were usually repeated three times to obtain an average value and the standard deviation for each data point.

Failure Modes in Three-Point Bend Tests

Typical load-deflection curves for 0° and 45° specimens tested at 77 K and at room temperature are shown in Figure 1 for the glass/epoxy I composite before irradiation. For both test temperatures, the curves are almost linear up to the failure point for the 0° specimens, whereas for the 45° specimens the curves display a strong nonlinearity.

In general, a rectangular specimen loaded at three points can fail in a flexural mode at the specimen surface or in a shear mode at the neutral axis. A mixed failure mode can also occur when the flexural and shear failures take place simultaneously. Hanna and Steingiser reported that the failure mode can be determined from a load-deflection curve of the three-point bend test (26). When a flexural failure occurs, the specimen fails abruptly without an appreciable decrease in the slope of the load-deflection curve. When a shear failure occurs, on the other hand, the slope of the curve decreases gradually with increasing deflection, and approaches a zero value before the specimen fails completely.

According to this criterion, the 0° specimen tested at 77 K or at room temperature fails in a flexural mode, whereas the 45° specimen fails in a shear or shear/flexural mixed mode, as seen from Figure 1 for the glass/epoxy I composite. For the other kinds of composites also, the load-deflection curves revealed that the 0° specimen fails in a flexural mode, whereas the 45° specimen fails in a shear or mixed mode for both test temperatures. In the present work, for this reason, the ultimate flexural strength is calculated from $3P_f(\ell/h)/2bh$ for a 0° specimen, while for a 45° specimen the ultimate interlaminar shear strength is calculated from $3P_f\ell/4bh^2$, where P_f is the applied load at failure, ℓ is the span length, b is the specimen width, and h is the specimen depth.

Effect of the Degree of Matrix Resin Cure

Glass/epoxy composites (filler: KS-1210; matrix: TGDDM/DDS) having various degrees of cure of the matrix resin were prepared by changing the cure conditions (20). The ultimate flexural strength of a 0° specimen tested at 77 K or at room temperature is plotted in Figure 2 as a function of the glass transition temperature of the matrix resin, T_g. The T_g is closely related to the degree of cure of the matrix resin and its mechanical properties.

The composite strength at room temperature (Figure 2) scarcely depends on the T_g, although in the T_g range below 140℃ the strength has a tendency to decrease slightly. This result can be explained by a fiber-controlled failure mode such that the failure of a composite occurs when the composite strain is reached at the ultimate strain of fibers (22). In this failure mode, the composite strength is independent of the ultimate strain of the matrix as long as this strain is higher than the ultimate strain of the fibers. Thus the composite strength at room temperature scarcely depends on the T_g, because usually the matrix resin is more ductile than the fiber.

The composite strength at 77 K (Figure 2) is lower than that at room temperature in the T_g range below 140℃. The strength at 77 K, however, begins to increase with increasing T_g at about 140℃, and then levels off at T_g values of 180℃ or above, approaching a value about twice that at room temperature. These results can be explained by a matrix-controlled failure mode such that the failure of a composite occurs or at least begins when the composite strain is reached at the ultimate strain of the matrix (22). In this failure mode, the composite strength is virtually determined by the matrix ultimate strain instead of the fiber ultimate strain. Thus the composite strength at 77 K depends on the T_g, because the matrix resin becomes brittle on cooling. These considerations lead to a conclusion that when the TGDDM/DDS epoxy matrix composite is fabricated as a low-temperature material, the T_g value should be used for monitoring quality control. In the present work, for this reason, only the composites having a T_g above 210℃ were used as starting material for the study of irradiation effects.

The approximate two-fold increase in the composite strength on cooling to 77 K in the T_g range above 180℃ (Figure 2) is due, for the most part, to a cooling-induced increase in the bundle strength of E-glass fibers in the composite (27). In fact, such an increase in the composite strength on cooling occurs also for other kinds of E-glass fiber composites such as the glass/epoxy V, glass/polyimide I, G-10CR, and G-11CR composites (16,19,22). For carbon or aramid fiber composites, however, no increase in the composite strength occurs on cooling from room temperature to 77 K (22,24).

Degradation Behavior for γ-Ray Irradiation

The ultimate flexural strength of a 0° specimen tested at 4.2 K is plotted in Figure 3 as a function of the absorbed dose in the matrix

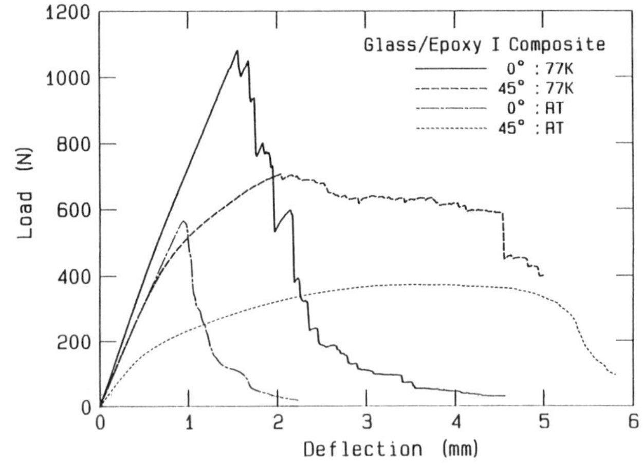

Figure 1. Load-deflection curves obtained in three-point bend tests at 77 K and at room temperature for the 0^0 and 45^0 specimens of the glass/epoxy I composite before irradiation.
(Reproduced with permission from reference 23. Copyright 1990 Chapman and Hall London.)

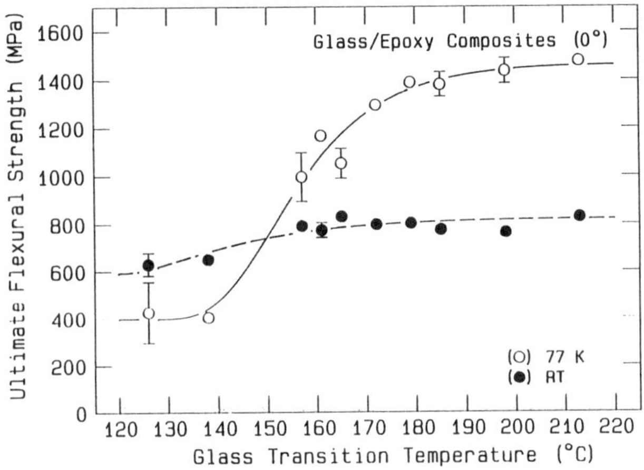

Figure 2. Plot of the ultimate flexural strength at 77 K and at room temperature versus the glass transition temperature of the matrix resin for the 0^0 specimens of the glass/epoxy composites prepared under the various cure conditions (filler: KS-1210; matrix: TGDDM/DDS).
(Reproduced with permission from reference 20. Copyright 1988 Chapman and Hall London.)

for the five kinds of E-glass fiber composites irradiated with ^{60}Co γ-rays at room temperature (15). If the radiation resistance of these composites is evaluated on the basis of the absorbed dose at which the composite strength decreases to one-half its initial value, the radiation resistance is found to increase in the order: G-10CR < G-11CR ≒ glass/epoxy V < glass/epoxy I < glass/polyimide I composites. This result strongly suggests that the radiation resistance of the matrix resin increases in the order: DGEBA type epoxy < TGDDM type epoxy < polyimide, because for all of these composites the reinforcing filler is E-glass fabric with a comparable volume fraction of fibers.

The ultimate flexural strengths of these composites tested at 77 K are essentially the same as those shown in Figure 3 both before and after irradiation (16). This result appears to be due mainly to the temperature dependence of the mechanical properties of the matrix resin. The stiffness of epoxy resins is, in fact, almost temperature-independent between 4.2 and 77 K, although between 300 and 77 K the stiffness increases by a factor of about 3 on cooling (28). The ultimate flexural strength tested at room temperature, on the other hand, differs from that tested at 4.2 or 77 K, as shown in Figure 4 for the glass/epoxy I and glass/polyimide I composites (22). Comparison of the 77 K and room-temperature data points for each composite shows that the initial strength at 77 K is about twice that at room temperature, and that a decrease in the strength by irradiation is appreciably greater at 77 K than at room temperature. It is concluded, therefore, that the dose dependence of the composite flexural strength depends not only on the matrix resin in the composite but also on the test temperature.

The ultimate interlaminar shear strength of a 45^{0} specimen tested at 77 K or at room temperature is plotted in Figure 5 as a function of the absorbed dose in the matrix for the glass/epoxy I and glass/polyimide I composites (23). Comparison of the 77 K and room-temperature data points for each composite shows that the initial strength at 77 K is about twice that at room temperature, and that a decrease in the strength by irradiation is appreciably greater at 77 K than at room temperature. It is particularly worth noting that the dose dependence at each temperature follows an identical pattern for the two composites. These findings indicate that the dose dependence of the composite shear strength is dependent only on the test temperature, and is much less dependent on the matrix resin in the composite.

Degradation Mechanism of Composite Strength

In general, the radiation-induced degradation of a polymer matrix composite could be due to radiation damage produced in the matrix, at the fiber/matrix interface, and in the fiber. Usually, such damage in the fiber can be neglected, because the radiation resistance of inorganic reinforcing fibers is much higher than that of organic matrix resins (5). Therefore, the degradation mechanism of the composite strength should be interpreted in terms of radiation damage in the matrix, at the fiber/matrix interface, or both.

Composite Flexural Strength. Figures 3 and 4 show that the radia-

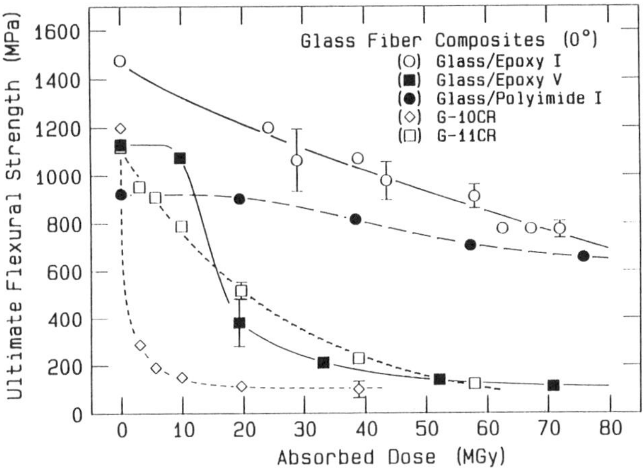

Figure 3. Plot of the ultimate flexural strength at 4.2 K versus the absorbed dose in the matrix for the 0° specimens of the glass/epoxy I, glass/epoxy V, glass/polyimide I, G-10CR, and G-11CR composites irradiated with ^{60}Co γ-rays.
(Reproduced with permission from reference 15. Copyright 1986 Elsevier.)

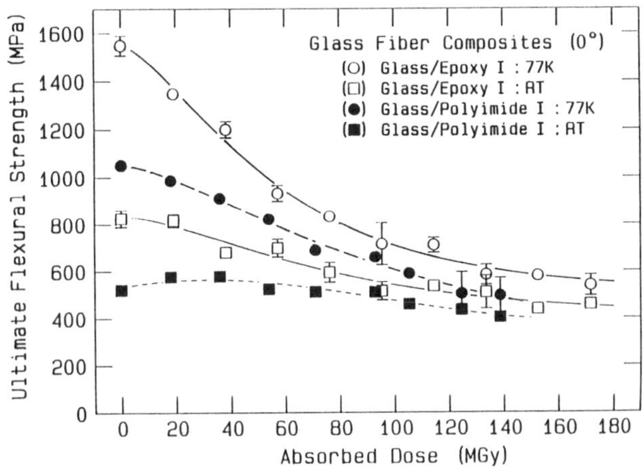

Figure 4. Plot of the ultimate flexural strength at 77 K and at room temperature versus the absorbed dose in the matrix for the 0° specimens of the glass/epoxy I and glass/polyimide I composites irradiated with ^{60}Co γ-rays or 2 MeV electrons.
(Reproduced with permission from reference 23. Copyright 1990 Chapman and Hall London.)

tion resistance in the composite flexural strength at each test temperature differs strikingly from one composite to another. This wide variation in the radiation resistance of a composite is ascribed primarily to a difference in the matrix resin and its radiation resistance. Thus it may be possible that the degradation of the composite flexural strength is dominated by the radiation damage in the matrix.

As one possibility of the dominant factor determining the degradation of the composite flexural strength, let us consider the ultimate strain of the matrix. Figure 6 shows a relationship between the composite ultimate strain and the matrix ultimate strain for the 0° specimens of the glass/epoxy I and glass/polyimide I composites tested at 77 K and at room temperature. The relationship was calculated by using the experimental data for these composites and for the pure TGDDM/DDS epoxy and polyimide resins irradiated up to the same doses as the composites (22). The composite ultimate strain is seen to increase with an increase in the matrix ultimate strain in a similar pattern for both of these composites at 77 K. At room temperature, on the other hand, the composite ultimate strain appears to increase at first, and then levels off at a matrix ultimate strain of about 4% or above.

These findings strongly suggest that the polymer matrix composites studied here have at least two different failure modes when tested in the warp direction of the reinforcing fabrics. In Figure 6, the composite ultimate strain which increases linearly with the matrix ultimate strain is most likely assigned to the matrix-controlled failure mode described in the previous section of this chapter. The composite ultimate strain which is almost independent of the matrix ultimate strain, on the other hand, is assigned to the fiber-controlled failure mode.

In order to discuss this point quantitatively, we have tried to derive an expression for the relationship between the composite ultimate strain and the matrix ultimate strain (22). The broken lines in Figure 7 show the matrix- and fiber-controlled failure modes. The solid curve, on the other hand, shows a complex failure mode such that the failure of the composite is caused by interactions between the matrix- and fiber-controlled failure modes. The solid curve indicates that the composite ultimate strain increases with an increase in the matrix ultimate strain at first, and then approaches gradually the fiber ultimate strain. This characteristic of the complex failure mode is, in fact, observed for the glass/epoxy I and glass/polyimide I composites, as seen in Figure 6. It is reasonably concluded, therefore, that the dose dependence of the composite flexural strength is primarily determined by a change in the matrix ultimate strain due to irradiation.

Composite Shear Strength. Figure 5 shows that the glass/epoxy I and glass/polyimide I composites are quite similar to each other in the dose dependence of the ultimate interlaminar shear strength at each test temperature. The identical dose dependence of the shear strength may reflect a similarity of the fiber/matrix interface between the two composites. The surface treatment of the glass fiber is, in fact, the same for both of these composites (23). These facts suggest that the degradation of the composite shear neglecting the presence of reinforcing fibers in the composite.

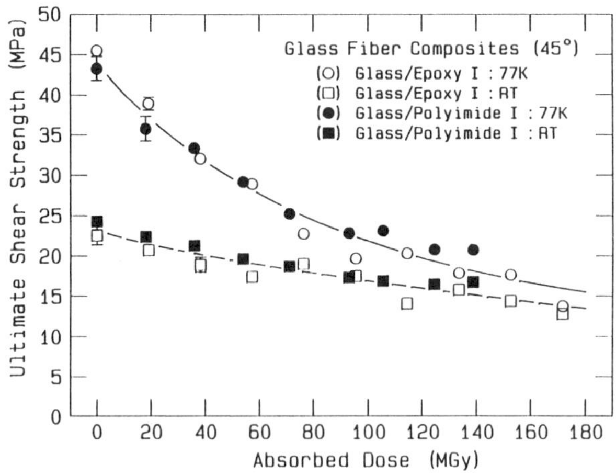

Figure 5. Plot of the ultimate interlaminar shear strength at 77 K and at room temperature versus the absorbed dose in the matrix for the 45° specimens of the glass/epoxy I and glass/polyimide I composites irradiated with ^{60}Co γ-rays or 2 MeV electrons. (Reproduced with permission from reference 23. Copyright 1990 Chapman and Hall London.)

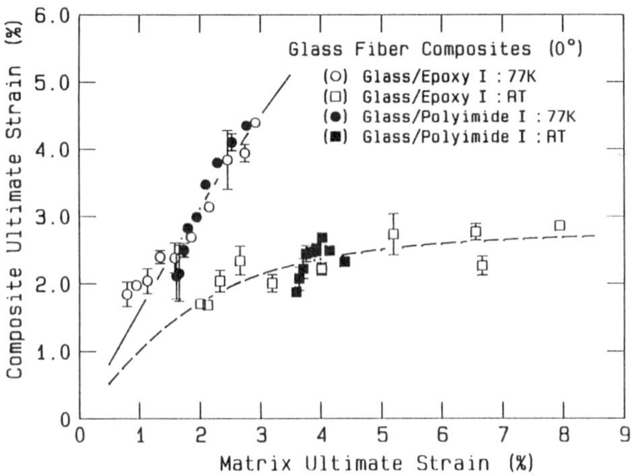

Figure 6. Relationship between the composite ultimate strain and the matrix ultimate strain for the 0° specimens of the glass/epoxy I and glass/polyimide I composites tested at 77 K and at room temperature. (Reproduced with permission from reference 22. Copyright 1988 Elsevier.)

strength is dominated by the radiation damage at the fiber/matrix interface rather than that in the matrix.

This idea is consistent with the scanning electron microscope observation reported by Takeda et al. for the glass/epoxy I composite (29). According to their observation made after interlaminar shear tests, the fracture surface of the irradiated composite displays separation or debonding between the fiber and the matrix, thus reflecting the radiation-induced decomposition of surface-treating compounds at the fiber/matrix interface. It will be reasonable to conclude, therefore, that the degradation of the composite shear strength is primarily determined by a change in the fiber/matrix bond strength due to irradiation.

Degradation Behavior for Neutron Irradiation

Absorbed-Dose Calculation. The neutron dose absorbed in a material can be calculated if the total neutron fluence and the conversion factor to absorbed dose are known. In the present work, the total neutron fluence was determined by measuring the radioactivity of iron foil put into the container of composite specimens (17). As to the conversion factor, Greenwood calculated this, assuming that the equilibrium or the spatial homogeneity of recoil particles is established in a material (30). For a polymer matrix composite, however, this assumption may fail depending on how large the specimen is. This is because the range of recoil particles can be comparable with the linear dimensions of matrix and fiber in the composite, depending on the energy of the neutrons. In such a case, some of the recoil particles generated in the matrix escape into the fiber, and vice versa. For the exact evaluation of neutron dose in a composite, therefore, the spatial distribution of absorbed dose in the composite must be calculated.

The calculation was performed by taking into account the range of recoil particles and by assuming a hexagonal array of equally spaced fibers shown in Figure 8. A typical example of the absorbed-dose distribution calculated for recoil protons is shown in Figure 9 for the glass/epoxy I composite in the IPNS thimble of VT2 (17). It is seen that a large number of recoil protons escape from the matrix into the surrounding fibers, thus considerably decreasing the amount of energy deposition in the matrix compared to an infinite matrix. In Figure 10, the absorbed-dose distribution calculated for α and ^{7}Li particles is shown for the glass/epoxy I composite (17). The α and ^{7}Li particles are produced by a ^{10}B(n, α)^{7}Li reaction when the reinforcing filler contains boron, as is the case with E-glass fibers (see Table I). Figure 10 shows that a large number of α and ^{7}Li particles escape from the fiber into the matrix, thus adding a significant amount of energy deposition to the matrix.

On the basis of these results, the conversion factor from total neutron fluence to absorbed dose for a matrix of a composite was calculated for composite specimens irradiated with IPNS neutrons (17). For composite specimens irradiated with ^{60}Co γ-rays, on the other hand, it was confirmed experimentally that the dose absorbed in a matrix of a composite is almost the same as that absorbed in an infinite matrix (17). Accordingly, the conversion factor from γ-ray exposure to absorbed dose was calculated in the usual way of

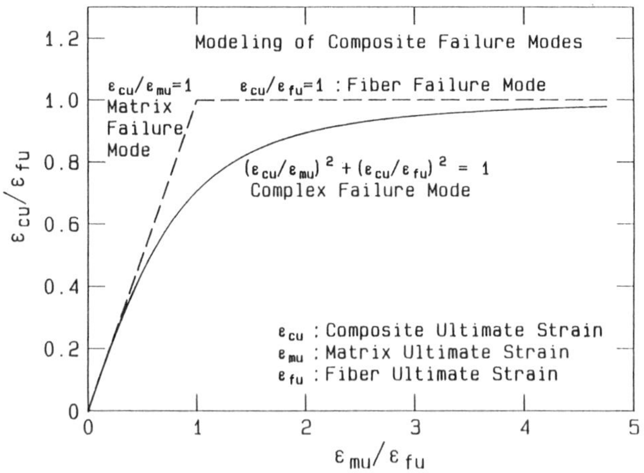

Figure 7. Modeling of composite failure modes based on a relationship between the composite ultimate strain and the matrix ultimate strain.
(Reproduced with permission from reference 22. Copyright 1988 Elsevier.)

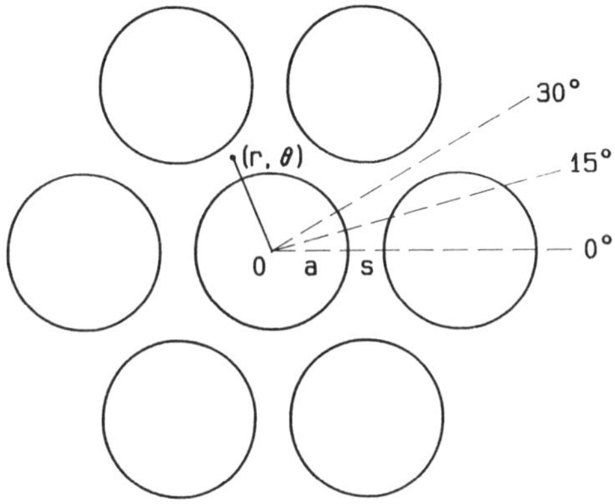

Figure 8. Hexagonal array of fibers used in calculating the spatial distribution of neutron absorbed dose in a composite.
(Reproduced with permission from reference 17. Copyright 1987 Elsevier.)

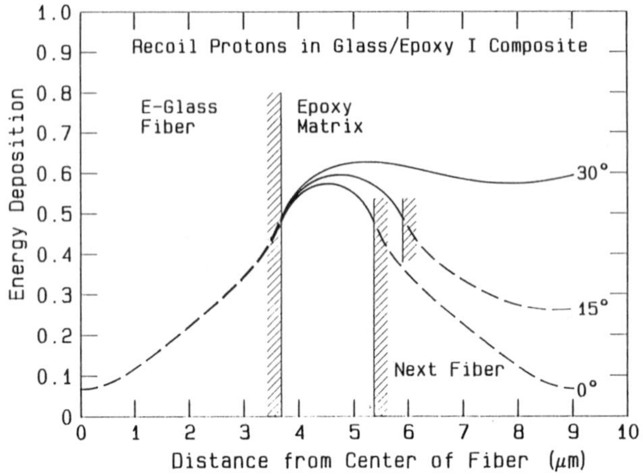

Figure 9. Spatial distribution of energy deposition due to recoil protons calculated by assuming a hexagonal array of fibers (see Figure 8). This example is for the glass/epoxy I composite in the IPNS thimble of VT2.

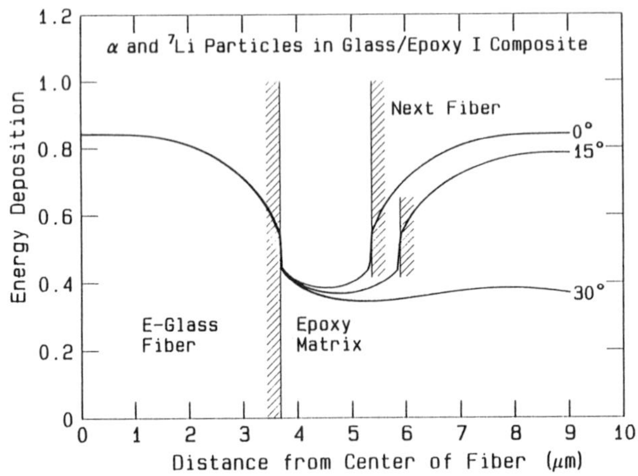

Figure 10. Spatial distribution of energy deposition due to α and ^{7}Li particles calculated for the glass/epoxy I composite by assuming a hexagonal array of fibers (see Figure 8).

Degradation of Flexural Strength. The ultimate flexural strength of a 0° specimen tested at 77 K after the IPNS neutron irradiation is plotted in Figures 11 and 12 for the glass/epoxy I and glass/polyimide I composites, respectively, as a function of the absorbed dose in the matrix due to the recoil particles generated in the matrix, D_m (18). For comparison, the composite flexural strength tested at 77 K after ^{60}Co γ-ray irradiation is also plotted as a function of the γ-ray dose absorbed in the matrix, D_{γ}. It is seen from Figure 11 or 12 that a decrease in the composite flexural strength caused by the same amount of energy deposition is greater for the Rabbit or VT2 neutrons than for γ-rays, thus indicating a higher radiation sensitivity of these composites towards neutrons than γ-rays.

From a simplistic standpoint, this result suggests that the radiation sensitivity of epoxy and polyimide resins is higher for recoil protons than for γ-rays. This is because although not only recoil protons but also heavier recoils of carbon, nitrogen, oxygen, etc. are generated in the matrix resin, the contribution of these recoils to the total neutron dose is most significant for recoil protons (74-79% for the epoxy matrix and 67-71% for the polyimide matrix). The higher sensitivity of these resins towards recoil protons than γ-rays is compatible with the finding by Burns and Jones that the decomposition efficiency of aromatic substances such as benzene and biphenyl increases with an increase in the linear energy transfer (LET) in the order: γ-rays < recoil protons (31). This LET effect is ascribed to a difference in the microscopic inhomogeneity of energy deposition between the two types of radiation (32).

Effect of Irradiation Temperature. Comparison of the VT2 and Rabbit data points in Figure 11 or 12 suggests that the dose dependence of the composite flexural strength follows an identical pattern regardless of the 5 K and room-temperature irradiations. This finding and the similarity of the neutron spectrum for the VT2 and Rabbit thimbles (25) strongly suggest that the irradiation temperatures of 5 K and room temperature have no significant influence on the degradation behavior of a polymer matrix composite.

It should be pointed out, however, that this suggestion may be true only when the composite specimen irradiated at 5 K is warmed up to room temperature before the mechanical test at low temperatures. At present, it is completely uncertain whether this is also the case when the mechanical test is performed at 4.2 or 77 K without warmup to room temperature.

Effect of Reinforcing Fabric Type

Comparison of the H2 data point with the Rabbit or VT2 data point in Figure 11 or 12 shows that a decrease in the composite strength by irradiation is apparently greater for the H2 neutrons for both the glass/epoxy I and glass/polyimide I composites. In Figures 11 and 12, the neutron data point is plotted as a function of the D_m dose only, neglecting the absorbed dose in the matrix due to the α and ^{7}Li particles produced by a $^{10}B(n, \alpha)^{7}Li$ reaction in the fiber. The

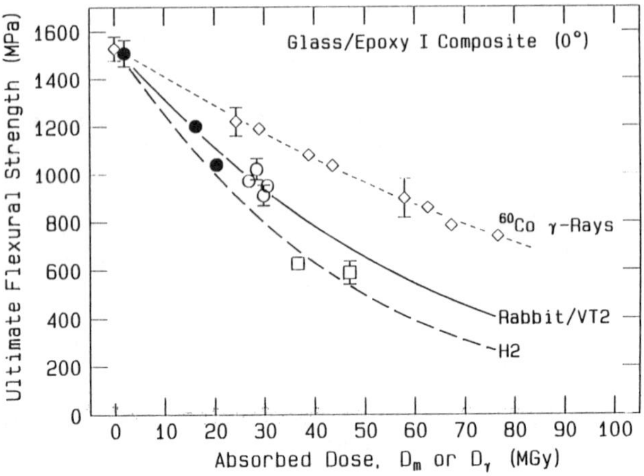

Figure 11. Plot of the ultimate flexural strength at 77 K versus the absorbed dose in the matrix for the 0° specimen of the glass/epoxy I composite irradiated with neutrons or ^{60}Co γ-rays. The open points (○,□) and the solid point (●) are the data for room-temperature and 5 K neutron irradiations, respectively. (Reproduced with permission from reference 18. Copyright 1987 Elsevier.)

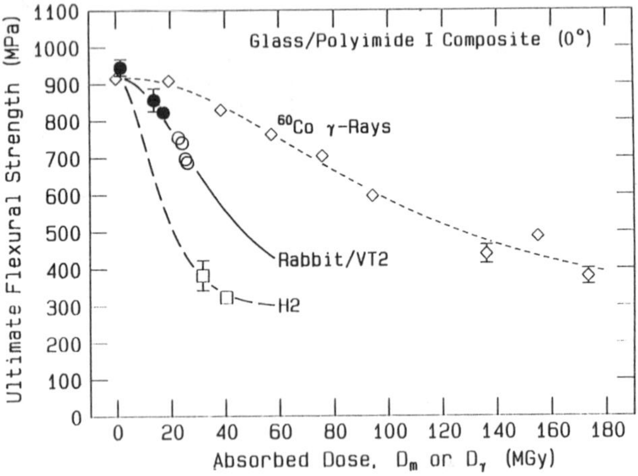

Figure 12. Plot of the ultimate flexural strength at 77 K versus the absorbed dose in the matrix for the 0° specimen of the glass/polyimide I composite irradiated with neutrons or ^{60}Co γ-rays. See the caption for Figure 11 for the symbols of data points. (Reproduced with permission from reference 18. Copyright 1987 Elsevier.)

enhanced decrease in the composite strength for the H2 neutrons is definitely ascribed to the additional radiation damage in the composite due to the α and ^{7}Li particles (6,8,13,17,18). This is because the ^{10}B reaction has a large cross section for low-energy neutrons, and the flux of such neutrons is much higher for the H2 neutrons than for the Rabbit or VT2 neutrons (25). As component materials to be used in fusion magnets, therefore, boron-free glass fiber composites are recommended over the boron-containing glass fiber composites.

From this point of view, the effects of the kind of glass fibers and the type of fabric weave on the composite degradation behavior were studied by using the reinforcing fabrics shown in Table I. The E-glass fabrics of KS-1210 and KS-1600 (Kanebo Ltd.) were so selected as to differ from each other in the number of fibers in a yarn and in the number of yarns per 25 mm in the warp and weft directions. The boron-free T-glass fabrics of WTX-116E and WTA-18W (Nittobo Co.) were so selected as to be the counterparts of KS-1210 and KS-1600, respectively, with regard to the type of fabric weave.

The ultimate flexural strength of a 0° specimen tested at 77 K after ^{60}Co γ-ray irradiation is plotted in Figure 13 as a function of the absorbed dose in the matrix for the glass/epoxy I–IV composites (24). It is seen that the initial strength of the KS-1210 or WTX-116E fabric composite is 28-40% higher than that of the KS-1600 or WTA-18W fabric composite, thus indicating that the initial strength is dependent on the type of fabric weave. It is also seen that the initial strength is less dependent on the kind of glass fibers. Following irradiation the strengths of these composites decrease monotonically with increasing absorbed dose. Roughly speaking, the dose dependence appears to follow a rather similar pattern for all of these composites, thus suggesting that the degradation behavior of the composite flexural strength depends neither on the type of fabric weave nor the kind of glass fibers. This result is consistent with a degradation mechanism such that the dose dependence of the composite flexural strength is primarily determined by a change in the matrix ultimate strain due to irradiation (22). In agreement with this mechanism, the composite degradation behavior is essentially independent of the type of reinforcing fabric for the glass/polyimide I–III composites also (24).

Effect of Irradiation Atmosphere

Composite insulators used in fusion superconducting magnets are subjected to radiation under an inert atmosphere of liquid helium. If the effect of irradiation atmosphere on the composite degradation behavior is significant, the results obtained from irradiation in air can not be used as design data for fusion magnets. As an approach to this problem, polymer matrix composites having various radiation resistance were irradiated with ^{60}Co γ-rays in air and in argon under comparable dose rates.

The ultimate flexural strength of a 0° specimen tested at 77 K is plotted in Figure 14 as a function of the absorbed dose in the matrix for the G-10CR, G-11CR, and TIL-G1000 composites (24). Comparison of the argon and air data points for each composite shows that the degradation behavior follows an identical pattern regard-

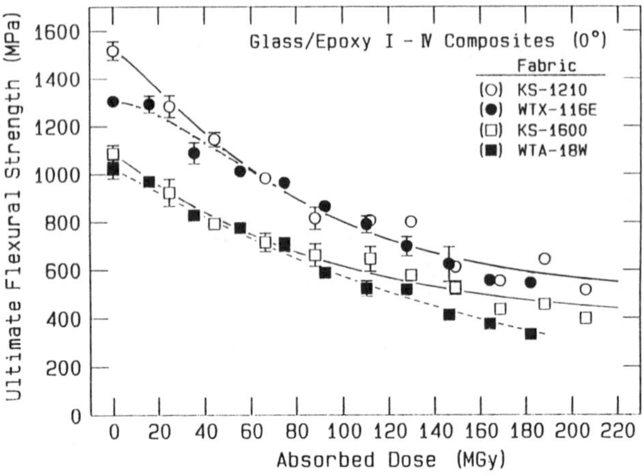

Figure 13. Plot of the ultimate flexural strength at 77 K versus the absorbed dose in the matrix for the 0° specimens of the glass/epoxy I–IV composites irradiated with ^{60}Co γ-rays.
(Reproduced with permission from *Cryogenics*, **1991**, *31*, 7–15. Copyright 1991 Butterworth.)

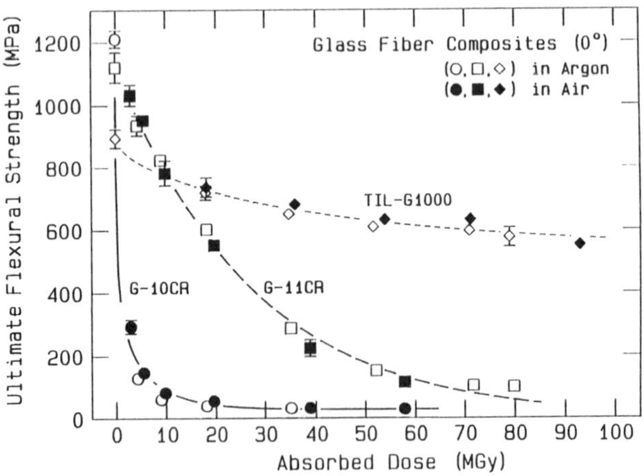

Figure 14. Plot of the ultimate flexural strength at 77 K versus the absorbed dose in the matrix for the 0° specimens of the G-10CR, G-11CR, and TIL-G1000 composites irradiated with ^{60}Co γ-rays in argon and in air.
(Reproduced with permission from *Cryogenics*, **1991**, *31*, 7–15. Copyright 1991 Butterworth.)

less of the irradiation atmosphere for all of these composites. This is also the case for other kinds of composites such as the glass/epoxy V and glass/polyimide I composites (24). It is reasonable to conclude, therefore, that the presence or absence of oxygen during irradiation has essentially no influence on the degradation behavior of a polymer matrix composite.

It should be pointed out, however, that irradiation in air versus argon must produce some differences in the radiation damage at the specimen surface by the presence and absence of oxidative degradation (33). The fact that the composite degradation behavior is still independent of the irradiation atmosphere (Figure 14) strongly suggests that the composite strength is fairly insensitive to small flaws which will be formed at the specimen surface by oxidative degradation during irradiation in air. Such insensitivity is most likely ascribed to the presence of reinforcing fibers in the composite. The presence of fibers is, in fact, known to interfere with the propagation of matrix cracking in the composite (34).

Conclusions

We have reviewed the mechanical properties of polymer matrix composites tested at 77 K, 4.2 K, and at room temperature after ^{60}Co γ-ray and neutron irradiations at room temperature and at 5 K. The most general characteristic of the composite degradation behavior is that the dose dependence of the composite flexural strength depends not only on the matrix resin in the composite but also on the test temperature. The complicated dose dependence can be explained systematically by a mechanism in which the degradation of the composite flexural strength is primarily determined by a change in the matrix ultimate strain due to irradiation. In agreement with this mechanism, the type of fabric weave and the kind of glass fibers have essentially no influence on the dose dependence of the composite flexural strength.

Another important characteristic of the composite degradation behavior is that the radiation sensitivity of the epoxy and polyimide matrix composites is higher for neutrons than ^{60}Co γ-rays, thus indicating that the decomposition efficiency of the matrix resin depends on the type of radiation. Thus it is concluded that the results of simulation irradiation with ^{60}Co γ-rays or accelerated electrons are unreliable as sources of design data for fusion superconducting magnets. As to the effect of irradiation temperature, the data points obtained so far suggest that there is no significant difference in the composite degradation behavior for 5 K and room-temperature irradiations.

Literature Cited

(1) Brown, B.S. J. Nucl. Mater. **1981**, 97, 1.
(2) Wiffen, F.W. J. Nucl. Mater. **1985**, 133 & 134, 32.
(3) Scott, J.L.; Clinard, F.W. Jr.; Wiffen, F.W. J. Nucl. Mater. **1985**, 133 & 134, 156.
(4) Kulcinski, G.L.; Dupouy, J.M.; Ishino, S. J. Nucl. Mater. **1986**, 141-143, 3.
(5) Hay, R.D.; Rapperport, E.J. Oak Ridge National Laboratory Report; ORNL/TM-2643 (1976).

(6) Coltman, R.R. Jr.; Klabunde, C.E. J. Nucl. Mater. **1983**, 113, 268.
(7) Takamura, S.; Kato, T. Advan. Cryogenic Engrg. Mater. **1984**, 30, 41.
(8) Tucker, D.S.; Fowler, J.D. Jr.; Clinard, F.W. Jr. Fusion Technol. **1985**, 8, 2696.
(9) Weber, H.W.; Kubasta, E.; Steiner, W.; Benz, H.; Nylund, K. J. Nucl. Mater. **1983**, 115, 11.
(10) Yamaoka, H.; Miyata, K. Advan. Cryogenic Engrg. Mater. **1986**, 32, 161.
(11) Nishijima, S.; Okada, T.; Miyata K.; Yamaoka, H. Advan. Cryogenic Engrg. Mater. **1988**, 34, 35.
(12) Egusa, S.; Kirk, M.A.; Birtcher, R.C.; Hagiwara, M.; Kawanishi, S. J. Nucl. Mater. **1983**, 119, 146.
(13) Egusa, S.; Kirk, M.A.; Birtcher, R.C. J. Nucl. Mater. **1984**, 126, 152.
(14) Egusa, S.; Kirk, M.A.; Birtcher, R.C.; Hagiwara, M. J. Nucl. Mater. **1985**, 127, 146.
(15) Egusa, S.; Nakajima, H.; Oshikiri, M.; Hagiwara, M.; Shimamoto, S. J. Nucl. Mater. **1986**, 137, 173.
(16) Egusa, S.; Hagiwara, M. Cryogenics **1986**, 26, 417.
(17) Egusa, S.; Kirk, M.A.; Birtcher, R.C. J. Nucl. Mater. **1987**, 148, 43.
(18) Egusa, S.; Kirk, M.A.; Birtcher, R.C. J. Nucl. Mater. **1987**, 148, 53.
(19) Egusa, S.; Udagawa, A.; Hashimoto, O.; Ono, T.; Yamamoto, Y.; Sonoda, K. J. Mater. Sci. Lett. **1988**, 7, 503.
(20) Egusa, S.; Seguchi, T.; Sugiuchi, K. J. Mater. Sci. Lett. **1988**, 7, 973.
(21) Egusa, S. J. Mech. Behavior of Mater. **1988**, 1, 1.
(22) Egusa, S. J. Mater. Sci. **1988**, 23, 2753.
(23) Egusa, S. J. Mater. Sci. **1990**, 25, 1863.
(24) Egusa, S. Cryogenics, in press.
(25) Birtcher, R.C.; Blewitt, T.H.; Kirk, M.A.; Scott, T.L.; Brown, B.S.; Greenwood, L.R. J. Nucl. Mater. **1982**, 108 & 109, 3.
(26) Hanna, G.L.; Steingiser, S. In Composite Materials: Testing and Design; American Society for Testing and Materials, ASTM STP 460; 1969, 182-191.
(27) Nishijima, S.; Nishiura, T.; Ikeda, T.; Okada, T.; Hagihara, T. Advan. Cryogenic Engrg. Mater. **1988**, 34, 75.
(28) Hartwig, G.; Knaak, S. Cryogenics **1984**, 24, 639.
(29) Takeda, N.; Kawanishi, S.; Udagawa, A.; Hagiwara, M. J. Mater. Sci. **1985**, 20, 3003.
(30) Greenwood, L.R. Proc. 4th ASTM-EURATOM Symp. Reactor Dosimetry; Gaithersburg, MD, 1982; 783-792.
(31) Burns, W.G.; Jones, J.D. Trans. Faraday Soc. **1964**, 60, 2022.
(32) Egusa, S.; Ishigure, K.; Tabata, Y. Macromolecules **1980**, 13, 171.
(33) Chapiro, A. Radiation Chemistry of Polymeric Systems; Interscience; New York, 1962; 360-361.
(34) Broutman, L.J. In Modern Composite Materials; Broutman, L.J.; Krock, R.H., Eds.; Addison-Wesley Publishing Company; Massachusetts, 1967; Chapter 13.

RECEIVED January 14, 1991

References and Indexes

References

1. Charlesby, A. *Atomic Radiation and Polymers*; Pergamon Press: Elmsford, NY, 1960.
2. Chapiro, A. *Radiation Chemistry of Polymeric Systems*; Wiley Interscience: New York, 1962.
3. *The Radiation Chemistry of Macromolecules*; Dole, M., Ed.; Academic Press: Orlando, FL, Vol. 1, 1972 and Vol. 2, 1973.
4. Makhlis, F. A. *Radiation Physics and Chemistry of Polymers*; John Wiley & Sons: New York, 1975.
5. Dolezel, B. *Die Bestandigkeit von Kunststoffen und Gummi*; Carl Hanser Verlag: Munich, 1978; Chapter 2.
6. Shalaby, S. W. "Radiative Degradation of Synthetic Polymers: Chemical, Physical and Environmental Considerations," *Macromol. Rev.* **1979,** *14*, 406.
7. Schnabel, W. *Polymer Degradation*; Carl Hanser Verlag: Munich, 1981; Chapter 5.
8. Hagiware, M.; Kagiya, T. "High Energy Degradation and Stabilization of Polymers," In *Degradation and Stabilization of Polymers*; Jellinek, H. H. G., Ed.; Elsevier: Amsterdam, 1983; Chapter 8.
9. *Proceedings of the Sixth Tihany Symposium on Radiation Chemistry*; Hedvig, P.; Nyikos, L.; Schiller, R., Eds.; Akademiai Kiado: Budapest, 1987; Part 4, Vol. 2.
10. Clough, R. L. "Radiation Resistant Polymers", In *Encyclopedia of Polymer Science and Engineering*; John Wiley & Sons: New York, 1988; Vol. 13, 2nd ed., p 667.
11. *The Effects of Radiation on High-Technology Polymers*; Reichmanis, E.; O'Donnell, J. H., Eds.; ACS Symposium Series 381; American Chemical Society: Washington, DC, 1989.

12. Clough, R. L.; Gillen, K. T. "γ-Radiation-Induced Oxidation, and Mechanisms of Its Inhibition," In *Oxidation Inhibition in Organic Materials*; Pospisil, J.; Klemchuk, P. P., Eds.; CRC Press: Boca Raton, FL, 1990; Vol. II, Chapter 4.

13. Collyer, A. A.; Clegg, D. W. *Irradiation Effects on Polymers*; Elsevier: Amsterdam, 1991.

Author Index

Affiliation Index

Subject Index

D

E

F

Q

R

S

Production: Margaret J. Brown
Indexing: Deborah H. Steiner
Acquisition: A. Maureen Rouhi
Cover design: Amy Hayes

Printed and bound by Maple Press, York, PA